# Manufacturing Processes for Engineering Materials

## THIRD EDITION

### Serope Kalpakjian

*Illinois Institute of Technology*

ADDISON-WESLEY PUBLISHING COMPANY

Menlo Park, California • Reading, Massachusetts • Harlow, England
Berkeley, California • Don Mills, Ontario • Sydney • Bonn •
Amsterdam • Tokyo • Mexico City

# WORLD STUDENT SERIES

*Sponsoring Editor:* Rob Merino
*Senior Production Editor:* Larry Olsen
*Manufacturing Supervisor:* Casi Kostecki
*Copy Editor:* Anna Reynolds Trabucco
*Production Assistant:* Emilie Bauer
*Text Designer:* Richard Kharibian
*Cover Designer:* Yvo Riezebos

ISBN 0-201-30411-2

1 2 3 4 5 6 7 8 9 10—MA— 00 99 98 97 96

To my grandson

**Brett Christopher**

# About the Author

Professor Serope Kalpakjian has been teaching and conducting research in manufacturing at the Illinois Institute of Technology since 1963. After graduating (with High Honors) from Robert College, Harvard University, and the Massachusetts Institute of Technology, he joined Cincinnati Milacron, Inc., where he was a research supervisor in charge of advanced metal-forming processes. He has published numerous papers in technical journals and is the author of several articles in the *Encyclopedia of Materials Science and Engineering*, the *McGraw-Hill Encyclopedia of Science and Technology*, and the *Marks' Standard Handbook for Mechanical Engineers*. He currently serves on the editorial boards of the *Journal of Manufacturing Systems*, the *Journal of Engineering Manufacture*, and the *Encyclopedia Americana*.

Professor Kalpakjian is also the author of the following books: *Mechanical Processing of Materials* (Van Nostrand, 1967); *Manufacturing Processes for Engineering Materials* (Addison-Wesley, 1984 and 1991, which received the M. Eugene Merchant Manufacturing Textbook Award in 1985); *Lubricants and Lubrication in Metalworking Operations* (with the late E.S. Nachtman; Marcel Dekker, 1985); and *Manufacturing Engineering and Technology* (third edition, Addison-Wesley, 1995, which received the M. Eugene Merchant Manufacturing Textbook Award in 1990); he has also edited *Tool and Die Failures: A Source Book* (ASM, 1982) and two conference proceedings. He is a Fellow of the American Society of Mechanical Engineers, ASM International, and the Society of Manufacturing Engineers, and is a full member of the International Institute for Production Engineering Research (CIRP). He is also a founding member and a past president of the North American Manufacturing Research Institution (NAMRI/SME).

The author has received a number of awards, including citations by the Forging Industry Educational and Research Foundation for Best Paper (1966) and the Society of Carbide and Tool Engineers (1977), the *Excellence in Teaching Award* from the Illinois Institute of Technology (1970), the *Centennial Medallion* from the American Society of Mechanical Engineers (1980), and the International *Education Award* from the Society of Manufacturing Engineers (1989).

# Preface

After many years of neglect, the subject of manufacturing engineering has finally acquired the academic status and significance that it so rightly deserves. Students, university administrators, and corporate executives, as well as the public at large, have come to the realization that without a sound manufacturing base, no nation can hope for economic survival in an increasingly competitive global marketplace.

Because it is a broad activity comprising many subjects, the proper method of teaching manufacturing processes in colleges and universities has been the subject of considerable discussion and evaluation. Beginning with the first edition of this book, the author firmly believed that, in addition to the applied and practical aspects, manufacturing processes should also be taught using a sound analytical basis, as are many courses in engineering curricula. Only through such approaches can the diverse topics and their complex interrelationships be truly understood and appreciated.

The favorable reception of the previous editions of *Manufacturing Processes for Engineering Materials* and the many valuable suggestions received from colleagues and reviewers has led the author to prepare this third edition. Although the book has basically the same format, balance, and introductory nature, it has certain additional features:

- More emphasis on analysis of the mechanics of manufacturing processes.
- Case studies have been added at the end of Chapters 5 through 12 to show the relevance of the topic to product manufacturing and analysis.
- The number of numerical examples in the text has been increased to help the student better understand the significance of the analytical approach to the subject matter.
- The number of questions and problems at the end of each chapter has been increased by about forty percent to better reflect the broad range of topics covered in the text and allow instructors to have a wider selection of questions and problems to assign as homework.
- Every attempt has been made to continue the presentation of the subject matter with a balanced coverage of the relevant fundamentals, analytical approaches, and applications to enable the student to properly assess the capabilities, limitations, and potentials of manufacturing processes and their competitive aspects.

This text has been written for students in mechanical, manufacturing, industrial, and metallurgical and materials engineering programs. We hope that students will appreciate that manufacturing engineering is a subject that is as exciting and challenging as any other engineering discipline.

# Acknowledgments

It is with great pleasure that I acknowledge the help of many people who assisted me in the preparation and publication of this third edition.

As always, many thanks to Laurie McGuire for her invaluable and constant help and very knowledgeable assistance in reviewing the second edition and in making numerous suggestions for this edition. I am grateful to Dan Joraanstad, Executive Editor, and Rob Merino, Acquisitions Editor, at Addition-Wesley for their constant and enthusiastic help. Larry Olsen, Senior Production Editor, has meticulously coordinated the production of this book.

Many thanks to my colleagues at various institutions for their help, their detailed reviews, comments, contributions, and many constructive suggestions for this edition:

R.J. Bowers (Edison Welding Institute)
J.G. Conley (Northwestern University)
R. Foley (Illinois Institute of Technology)
T. Hershberger and R. Redman (Biomet, Inc.)
K.M. Kulkarni (Advanced Metalworking Practices, Inc.)
Susan C. Mantell (University of Minnesota)
K.P. Rajurkar (University of Nebraska-Lincoln)
J.M. Samuel (University of Wisconsin-Madison)
S.R. Schmid (University of Notre Dame)
M.N. Tarabishy (Illinois Institute of Technology)
B.S. Thakkar (Lucent Technologies)
A.R. Thangaraj (Michigan Technological University)
R.H. Wagoner (Ohio State University)
N. Zabaras (Cornell University)

In addition, I appreciate the help of the many organizations who supplied me with numerous illustrations and photographs. I again express my deep gratitude to two eminent authorities in manufacturing who have had a great influence on my professional life, Milton C. Shaw and M. Eugene Merchant.

Finally, many thanks to my wife Jean for her help in the final preparations of this book.

Serope Kalpakjian
Chicago, Illinois

# Contents

**1  Introduction**                                                              1

1.1   What Is Manufacturing?   1
1.2   The Design Process and Concurrent Engineering   5
1.3   Design for Manufacture, Assembly, Disassembly, and Service   8
1.4   Selecting Materials   9
1.5   Selecting Manufacturing Processes   12
1.6   Computer-Integrated Manufacturing   15
1.7   Quality Assurance and Total Quality Management   19
1.8   Global Competitiveness and Manufacturing Costs   20
1.9   Lean Production and Agile Manufacturing   20
1.10  Environmentally Conscious Design and Manufacturing   21
1.11  Product Liability   22
1.12  Organization for Manufacture   23
Summary   23

**2  Fundamentals of the Mechanical Behavior of Materials**                      25

2.1   Introduction   25
2.2   Tension   26
2.3   Compression   45
2.4   Torsion   49
2.5   Bending (Flexure)   51
2.6   Hardness   52
2.7   Fatigue   58
2.8   Creep   59
2.9   Impact   60
2.10  Residual Stresses   61
2.11  Triaxial Stresses and Yield Criteria   65
2.12  Work of Deformation   76
Summary   80
Summary of Equations   81
Bibliography   83
Questions   83
Problems   85

**3**   Structure and Manufacturting Properties of Metals    89

   3.1   Introduction   89
   3.2   The Crystal Structure of Metals   90
   3.3   Deformation and Strength of Single Crystals   92
   3.4   Grains and Grain Boundaries   99
   3.5   Plastic Deformation of Polycrystalline Metals   102
   3.6   Recovery, Recrystallization, and Grain Growth   104
   3.7   Cold, Warm, and Hot Working   106
   3.8   Failure and Fracture   107
   3.9   Physical Properties   117
   3.10  Properties of Ferrous Alloys   122
   3.11  Properties of Nonferrous Metals and Alloys   128
   Summary   140
   Summary of Equations   141
   Bibliography   141
   Questions   142
   Problems   143


**4**   Surfaces, Dimensional Characteristics, Inspection, and Quality
       Assurance    145

   4.1   Introduction   145
   4.2   Surface Structure and Properties   146
   4.3   Surface Texture   148
   4.4   Tribology   151
   4.5   Surface Treatments, Coatings, and Cleaning   171
   4.6   Engineering Metrology and Instrumentation   178
   4.7   Dimensional Tolerances   186
   4.8   Testing and Inspection   188
   4.9   Quality Assurance   192
   Summary   204
   Summary of Equations   205
   Bibliography   206
   Questions   207
   Problems   209


**5**   Metal-Casting Processes and Equipment; Heat Treatment    211

   5.1   Introduction   211
   5.2   Solidification of Metals   212
   5.3   Cast Structures   221
   5.4   Fluid Flow and Heat Transfer   226
   5.5   Melting Practice and Furnaces   234

5.6    Casting Alloys   237
5.7    Ingot Casting and Continuous Casting   243
5.8    Casting Processes: Expendable Mold   246
5.9    Casting Processes: Permanent Mold   259
5.10   Processing of Castings; Heat Treatment   269
5.11   Casting Design   279
5.12   Economics of Casting   282
Case Study: Engine Crankcase Design and Casting Process Selection   284
Summary   286
Summary of Equations   287
Bibliography   288
Questions   288
Problems   290


**6   Bulk Deformation Processes**                                    **294**

6.1    Introduction   294
6.2    Forging   295
6.3    Rolling   324
6.4    Extrusion   342
6.5    Rod, Wire, and Tube Drawing   357
6.6    Swaging   369
6.7    Die Manufacturing Methods   369
6.8    Die Failures   371
Case Study: Development of Isothermally Forged Jet-Engine Blades   372
Summary   374
Summary of Equations   375
Bibliography   376
Questions   378
Problems   380


**7   Sheet-Metal Forming Processes**                                 **384**

7.1    Introduction   384
7.2    Sheet-Metal Characteristics   385
7.3    Shearing   389
7.4    Bending of Sheet and Plate   396
7.5    Stretch Forming   409
7.6    Bulging   411
7.7    Rubber Forming   412
7.8    Spinning   413
7.9    High-Energy-Rate Forming   418
7.10   Superplastic Forming   421
7.11   Various Forming Methods   422

7.12   Deep Drawing   424
7.13   Formability of Sheet Metals   436
7.14   Economics of Sheet-Metal Forming   445
Case Study: Forming of a Generalized Automotive Front Fender   449
Summary   445
Summary of Equations   450
Bibliography   450
Questions   450
Problems   452

**8   Material-Removal Processes:  Cutting**                                         **455**
8.1    Introduction   455
8.2    Mechanics of Chip Formation   456
8.3    Tool Wear   481
8.4    Surface Finish and Integrity   490
8.5    Machinability   493
8.6    Cutting-Tool Materials   496
8.7    Cutting Fluids   512
8.8    Cutting Processes and Machine Tools for Producing Round Shapes   516
8.9    Cutting Processes and Machine Tooles for Producing Various Shapes   535
8.10   Machining and Turning Centers   552
8.11   Vibration and Chatter   558
8.12   Machine-Tool Structures   560
8.13   Machining Economics   562
Case Study: Machining of Various Complex Parts   565
Summary   567
Summary of Equations   568
Bibliography   569
Questions   570
Problems   574

**9   Material Removal Processes:  Abrasive, Chemical, Electrical,
and High-Energy Beams**                                                             **576**
9.1    Introduction   576
9.2    Abrasives   577
9.3    Bonded Abrasives (Grinding Wheels)   578
9.4    Mechanics of Grinding   583
9.5    Grinding Wheel Wear   590
9.6    Grinding Operations and Machines   594
9.7    Ultrasonic Machining   602

9.8     Finishing Operations   604
9.9     Deburring   607
9.10    Chemical Machining   608
9.11    Electrochemical Machining   612
9.12    Electrochemical Grinding   615
9.13    Electrical-Discharge Machining   616
9.14    High-Energy-Beam Machining   621
9.15    Water-Jet Machining   625
9.16    Abrasive-Jet Machining   626
9.17    Process Economics   627
Case Study: Electrochemical Machining of a Biomedical Implant   628
Summary   630
Summary of Equations   631
Bibliography   632
Questions   633
Problems   634

## 10 Properties and Processing of Polymers and Reinforced Plastics   636

10.1    Introduction   636
10.2    The Structure of Polymers   638
10.3    Thermoplastics:  Behavior and Properties   646
10.4    Thermosets:  Behavior and Properties   653
10.5    Thermoplastics:  Applications   653
10.6    Thermosets:  Applications   655
10.7    High-Temperature Polymers, Electrically-Conducting Polymers, and Biodegradable Plastics   657
10.8    Elastomers (Rubbers):  Properties and Applications   658
10.9    Reinforced Plastics (Composite Materials)   659
10.10   Processing of Plastics   668
10.11   Processing of Reinforced Plastics   683
10.12   Design Considerations   689
10.13   Economics of Processing Plastics   690
10.14   Rapid Prototyping   692
Case Study: Manufacture of a Toothbrush   694
Summary   700
Summary of Equations   701
Bibliography   701
Questions   703
Problems   705

**11  Properties and Processing of Metal Powders, Ceramics, Glasses, and Composites**   **707**

11.1   Introduction   707
11.2   Production of Metal Powders   710
11.3   Compaction of Metal Powders   715
11.4   Sintering   723
11.5   Secondary and Finishing Operations   728
11.6   Design Considerations for Powder Metallurgy   729
11.7   Selective Laser Sintering (Rapid Prototyping)   731
11.8   Economics of Powder Metallurgy   733
11.9   Ceramics: Structure, Properties, and Applications   734
11.10  Processing of Ceramics   743
11.11  Glasses: Structure, Properties, and Applications   749
11.12  Processing of Glasses   753
11.13  Design Considerations for Ceramics and Glasses   757
11.14  Graphite and Diamond   757
11.15  Composites: Structure, Properties, and Applications   758
11.16  Processing Metal-Matrix and Ceramic-Matrix Composites   760
Case Study: Production of Automotive-Engine Main Bearing Caps by Powder Metallurgy   761
Summary   764
Summary of Equations   766
Bibliography   766
Questions   767
Problems   769

**12  Joining and Fastening Processes**   **770**

12.1   Introduction   770
12.2   Arc-Welding Processes: Consumable Electrode   772
12.3   Arc-Welding Processes: Nonconsumable Electrode   780
12.4   Laser-Beam and Electron-Beam Welding   782
12.5   The Welded Joint; Weld Design   785
12.6   Cold Welding   799
12.7   Ultrasonic Welding   800
12.8   Friction Welding   801
12.9   Resistance Welding   803
12.10  Explosion Welding   809
12.11  Diffusion Bonding (Welding)   810
12.12  Brazing and Soldering   812
12.13  Adhesive Bonding   818
12.14  Mechanical Fastening   823
12.15  Joining Plastics   827

Case Study: Weld Cracking   829
Summary   833
Bibliography   834
Questions   835
Problems   838

**13 Automation of Manufacturing Processes**                               **839**

13.1   Introduction   839
13.2   Automation   840
13.3   Numerical Control   845
13.4   Programming for Numerical Control   851
13.5   Adaptive Control   854
13.6   Material Handling and Movement   855
13.7   Industrial Robots   856
13.8   Sensor Technology   861
13.9   Flexible Fixturing   863
13.10  Assembly, Disassembly, and Service   864
Summary   867
Bibliography   868
Questions   869
Problems   869

**14 Computer-Integrated Manufacturing Systems**                           **871**

14.1   Introduction   871
14.2   Manufacturing Systems   872
14.3   Computer-Inetgrated Manufacturing   872
14.4   Computer-Aided Design and Engineering   874
14.5   Computer-Aided Manufacturing   877
14.6   Computer-Aided Process Planning   878
14.7   Computer Simulation of Manufacturing Processes and Systems   880
14.8   Group Technology   881
14.9   Cellular Manufacturing   887
14.10  Flexible Manufacturing Systems   889
14.11  Just-In-Time Production   891
14.12  Manufacturing Communications Networks   893
14.13  Artifical Intelligence   895
14.14  Factory of the Future   899
Summary   900
Bibliography   901
Questions   903
Problems   903

**15 Competitive Aspects of Manufacturing**      905

    15.1   Introduction  905
    15.2   Selection of Materials  906
    15.3   Product Design and Quantity of Materials  909
    15.4   Substitution of Materials  909
    15.5   Selection of Manufacturing Processes  911
    15.6   Process Capabilities  913
    15.7   Manufacturing Costs  916
    15.8   Value Engineering  921
    Summary  923
    Bibliography  924
    Questions  925
    Problems  925

Index      928

# 1

## Introduction

**1.1** What Is Manufacturing?

**1.2** The Design Process and Concurrent Engineering

**1.3** Design for Manufacture, Assembly, Disassembly, and Service

**1.4** Selecting Materials

**1.5** Selecting Manufacturing Processes

**1.6** Computer-Integrated Manufacturing

**1.7** Quality Assurance and Total Quality Management

**1.8** Global Competitiveness and Manufacturing Costs

**1.9** Lean Production and Agile Manufacturing

**1.10** Environmentally Conscious Design and Manufacturing

**1.11** Product Liability

**1.12** Organization for Manufacture

## 1.1 What Is Manufacturing?

As you read this Introduction, take a few moments to inspect the different objects around you: your watch, chair, stapler, pencil, calculator, telephone, and light fixtures. You will soon realize that all these objects have been transformed from various raw materials and assembled into the objects that you now see before you. Some objects, such as nails, bolts, and paper clips, are made of one material. However, most objects—aircraft engines (Fig. 1.1), ballpoint pens, toasters, bicycles, computers, and thousands more—are made of several parts from a variety of materials. A typical automobile, for example, consists of about 15,000 parts, a C–5A transport plane consists of more than 4 million parts, and a Boeing 747–400 has 6 million parts. All are made by various processes that we call manufacturing. **Manufacturing**, in its broadest sense, is the process of converting raw materials into products. It encompasses the design and manufacturing of goods using various production methods and techniques.

Manufacturing began about 5000 to 4000 B.C. with the production of various articles of wood, ceramic, stone, and metal (Table 1.1). The word *manufacturing* is derived from the Latin *manu factus*, meaning made by hand. The word *manufacture* first appeared in 1567, and the word *manufacturing*, in 1683. Modern manufacturing involves making products from raw materials by various processes, machinery, and operations, following a well-organized plan for each step.

1

**TABLE 1.1**

### HISTORICAL DEVELOPMENT OF MATERIALS AND MANUFACTURING PROCESSES

| | PERIOD | CASTING PROCESSES | FORMING PROCESSES |
|---|---|---|---|
| Egypt: ~ 3100 B.C. to ~ 300 B.C. / Greece: ~ 1100 B.C. to ~ 146 B.C. / Roman Empire: ~ 500 B.C. to A.D. 476 / Middle Ages: ~ 476 to 1492 / Renaissance: 14th to 16th centuries | Before 4000 B.C. | Gold, copper, meteoritic iron | Hammering |
| | 4000–3000 B.C. | Copper casting, stone and metal molds, lost wax process, silver, lead, tin, bronze | Stamping, jewelry |
| | 3000–2000 B.C. | Bronze casting and drawing; gold leaf | Wire by cutting sheet |
| | 2000–1000 B.C. | Wrought iron, brass | |
| | 1000–1 B.C. | Cast iron, cast steel | Stamping of coins |
| | A.D. 1-1000 | Zinc, steel | Armor, coining, forging, steel swords |
| | 1000–1500 | Blast furnace, type metals, casting of bells, pewter | Wire drawing, gold and silver smith work |
| | 1500–1600 | Cast iron cannon, tinplate | Water power for metalworking, rolling mill for coinage strips |
| Industrial Revolution: ~ 1750 to 1850 | 1600–1700 | Permanent mold castings, brass from copper and metallic zinc | Rolling (lead, gold, silver), shape rolling (lead) |
| | 1700–1800 | Malleable cast iron, crucible steel (iron bars and rods) | Extrusion (lead pipe), deep drawing, rolling |
| | 1800–1900 | Centrifugal casting, Bessemer process, electrolytic aluminum, nickel steels, babbitt, galvanized steel, powder metallurgy, tungsten steel, open-hearth steel | Steam hammer, steel rolling, seamless tube piercing, steel rail rolling, continuous rolling, electroplating |
| | 1900–1920 | | Tube rolling, hot extrusion |
| WWI | 1920–1940 | Die casting | Tungsten wire from powder |
| WWII | 1940–1950 | Lost wax for engineering parts | Extrusion (steel), swaging, powder metals for engineering parts |
| | 1950–1960 | Ceramic mold, nodular iron, semiconductors, continuous casting | Cold extrusion (steel), explosive forming, thermomechanical treatment |
| Space Age | 1960–1970 | Squeeze casting, single crystal turbine blades | Hydrostatic extrusion; electroforming |
| | 1970–1990s | Compacted graphite, vacuum casting, organically bonded sand, automation of molding and pouring, rapid solidification, metal-matrix composites, semi-solid metalworking, rheocasting | Precision forging, isothermal forging, superplastic forming, dies made by computer-aided design and manufacturing, rapid prototyping, net-shape forming |

*(continued)*

**TABLE 1.1 (CONTINUED)**

| JOINING PROCESSES | TOOLS, MACHINING, AND MANUFACTURING SYSTEMS | NONMETALLIC MATERIALS AND COMPOSITES |
|---|---|---|
| | Tools of stone, flint, wood, bone, ivory, composite tools | Earthenware, glazing, natural fibers |
| Soldering (Cu-Au, Cu-Pb, Pb-Sn) | Corundum | |
| Riveting, brazing | Hoe making, hammered axes, tools for ironmaking and carpentry | Glass beads, potter's wheel, glass vessels |
| Forge welding of iron and steel, gluing | Improved chisels, saws, files, woodworking lathes | Glass pressing and blowing |
| | Etching of armor | Venetian glass |
| | Sandpaper, windmill-driven saw | Crystal glass |
| | Hand lathe (wood) | Cast plate glass, flint glass |
| | Boring, turning, screw-cutting lathe, drill press | Porcelain |
| | Shaping, milling, copying lathe for gunstocks; turret lathe; universal milling machine; vitrified grinding wheel | Window glass from slit cylinder, light bulb, vulcanization, rubber processing, polyester, styrene, celluloid, rubber extrusion, molding |
| Oxyacetylene; arc, electrical resistance, and Thermit welding | Geared lathe, automatic screw machine, hobbing, high-speed steel tools, aluminum oxide and silicon carbide (synthetic) | Automatic bottle making, Bakelite, borosilicate glass |
| Coated electrodes | Tungsten carbide, mass production, transfer machines | Development of plastics, casting, molding, PVC, cellulose acetate, polyethylene, glass fibers |
| Submerged arc welding | | Acrylics, synthetic, rubber, epoxies, photosensitive glass |
| Gas metal-arc, gas tungsten-arc, and electroslag welding; explosive welding | Electrical and chemical machining, automatic control | ABS, silicones, fluorocarbons, polyurethane, float glass, tempered glass, glass ceramics |
| Plasma arc and electron beam, adhesive bonding | Titanium carbide, synthetic diamond, numerical control | Acetals, polycarbonate, cold forming of plastics, reinforced plastics, filament winding |
| Laser beam, diffusion bonding (also combined with super-plastic forming) | Cubic boron nitride, coated tools, diamond turning, ultraprecision machining, computer-integrated manufacturing, industrial robots, flexible manufacturing systems, untended factory | Adhesives, composite materials, optical fibers, structural ceramics, ceramic components for automotive and aerospace engines, ceramic-matrix composites |

Source: After J. A. Schey, C. S. Smith, R. F. Tylecote, T. K. Derry, T. I. Williams, and S. Kalpakjian.

Ti alloy
fan

Ti or Al alloy
low-pressure
compressor

Ti or Ni alloy
high-pressure
compressor

Ni alloy
combustion
chamber

Ni alloy
high-pressure
turbine

Ni alloy
low-pressure
turbine

Turbine
blades
Ni alloy

Turbine
exhaust
case
Ni alloy

Inlet
case
Al alloy

Accessory
section
Al alloy
or Fe alloy

**FIGURE 1.1**   Cross-section of a jet engine (PW2037) showing various components and the alloys used in making them. *Source:* Courtesy of United Aircraft Pratt & Whitney.

Manufacturing comprises approximately 20 to 30 percent of the value of all goods and services produced in industrialized nations. Generally, the higher the level of manufacturing in a country, the higher the standard of living of its people. Manufactured products are also used to make other products. Examples are large presses to form sheet metal for car bodies, metalworking machinery used to make parts for other products, and sewing machines for making clothing.

Because a manufactured item has undergone a number of changes in which raw material has become a useful product, it has **added value**, defined as monetary worth in terms of price. For example, clay has a certain value when mined. When the clay is used to make a ceramic dinner plate, cutting tool, or electrical insulator, value is added to the clay. Similarly, a wire coat hanger or a nail has added value over and above the cost of a piece of wire.

Manufacturing may produce **discrete products**, meaning individual parts or part pieces, or **continuous products.** Nails, gears, steel balls, beverage cans, and engine blocks are examples of discrete products. Wire, metal or plastic sheet, hose, and pipe are continuous products that may be cut into individual pieces and thus become discrete products.

Manufacturing is generally a complex activity involving people who have a broad range of disciplines and skills together with a wide variety of machinery, equipment, and tools with various levels of automation, including computers, robots, and material-handling equipment. Manufacturing activities must be responsive to several demands and trends:

- A product must fully meet **design requirements** and **specifications.**
- A product must be manufactured by the most **economical** methods in order to minimize costs.
- **Quality** must be built into the product at each stage, from design to assembly, rather than relying on quality testing after the product is made.
- In a highly competitive environment, production methods must be sufficiently **flexible** to respond to changing market demands, types of products, production rates, production quantities, and on-time delivery to the customer.
- New developments in **materials, production methods,** and **computer integration** of both technological and managerial activities in a manufacturing organization must constantly be evaluated with a view to their timely and economic implementation.
- Manufacturing activities must be viewed as a large **system,** each part of which is interrelated. Such systems can be modeled in order to study the effect of various factors, such as changes in market demands, product design, materials, costs, and production methods, on product quality and cost.
- The manufacturer must work with the customer to get timely feedback for continuous **product improvement.**
- The manufacturing organization must constantly strive for higher **productivity,** defined as the optimum use of all its resources: materials, machines, energy, capital, labor, and technology. Output per employee per hour in all phases must be maximized.

## 1.2  The Design Process and Concurrent Engineering

The design process for a product first requires a clear understanding of the functions and the performance expected of that product (Fig. 1.2). The product may be new, or it may be a newer model of an existing product. We all have observed, for example, how the designs and styles of radios, toasters, watches, automobiles, and washing machines have changed. The market for a product and its anticipated uses must be defined clearly, with the assistance of sales personnel, market analysts, and others in the organization. Product design is a critical activity because it has been estimated that 70 to 80 percent of the cost of product development and manufacture is determined at the initial design stages.

The design process begins with the development of an original product concept. An innovative approach to design is highly desirable—and even essential—for the product to be successful in the marketplace. Innovative approaches can also lead to major savings in material and production costs. The design engineer or product designer must be knowledgeable about materials, design, and manufacturing, as well as the overall economics of the operation.

Traditionally, design and manufacturing activities have taken place sequentially rather than concurrently or simultaneously. Manufacturing engineers were given the detailed drawings and specifications of the product and were asked to produce it. They often encountered difficulties because the design or product engineers did not anticipate the production problems that could occur. This situation has been improved greatly by the use of *simultaneous engineering* or **concurrent engineering** (CE).

**FIGURE 1.2**   Chart showing various steps involved in designing and manufacturing a product. Depending on the complexity of the product and the type of materials used, the time span between the original concept and marketing a product may range from few months to many years. *Concurrent engineering* combines these stages to reduce the time span and improve efficiency and productivity.

Concurrent engineering is a systematic approach integrating the design and manufacture of products with the view of optimizing all elements involved in the life cycle of the product. **Life cycle** means that all aspects of a product, such as design, development, production, distribution, use, and its ultimate disposal and recycling, are considered simultaneously. The basic goals of CE are to minimize product design and engineering changes and the time and costs involved in taking the product from design concept to production and introduction of the product into the marketplace. An extension of CE called **direct engineering (DE)** has recently been proposed. DE utilizes a database representing the engineering logic used in the design of each part of a product. If a design modification is made on a part, DE will determine the manufacturing consequences of that change.

Although the concept of concurrent engineering appears to be logical and efficient, its implementation can take considerable time and effort when those using it either do not work as a team or fail to appreciate its real benefits. It is apparent that for CE to succeed it must (1) have the full support of upper management, (2) have a multifunctional and interactive work team, including support groups, and (3) utilize all available technologies.

For both large and small companies, product design often involves preparing analytical and physical models of the product as an aid to analyzing factors such as forces, stresses, deflections, and optimal part shape. The necessity for such models depends on product complexity. Today, constructing and studying analytical models is simplified through the use of **computer-aided design, engineering, and manufacturing** techniques. On the basis of these models, the product designer selects and specifies the final shape and dimensions of the product, its dimensional accuracy and surface finish, and the materials to be used. The selection of materials is often made with the advice and cooperation of materials engineers, unless the design engineer is sufficiently experienced and qualified in this area. An important design consideration is how a particular component is to be assembled into the final product. Take apart a ballpoint pen or a toaster, or lift the hood of your car and observe how hundreds of components are put together in a limited space.

A powerful and effective tool, particularly for complex production systems, is **computer simulation** in evaluating the performance of the product and planning the manufacturing system to produce it. Several computer simulation languages using animated graphics and with various capabilities have been developed. Computer simulation also helps in early detection of design flaws, identifying possible problems in a particular production system, and optimizing manufacturing lines for minimum product cost.

The next step in the production process is to make and test a *prototype,* that is, an original working model of the product. An important new development is **rapid prototyping,** which relies on CAD/CAM and various manufacturing techniques (typically using polymers or metal powders) to produce prototypes in the form of a solid physical model of a part rapidly and at low cost. For example, prototyping new automotive components by traditional methods of shaping, forming, machining, etc. costs hundreds of million of dollars a year; some components may take a year to make. Rapid prototyping can cut these costs as well as development times significantly. These techniques are being advanced further to such an extent that they can now be used for low-volume economical production of parts.

Tests on prototypes must be designed to simulate as closely as possible the conditions under which the product is to be used. These include environmental conditions such as temperature and humidity, as well as the effects of vibration and repeated use and misuse of the product. Computer-aided engineering techniques are now capable of comprehensively and rapidly performing such simulations. During this stage, modifications in the original design, materials selected, or production methods may be necessary. After this phase has been completed, appropriate process plans, manufacturing methods (Table 1.2), equipment, and tooling are selected with the cooperation of manufacturing engineers, process planners, and all other involved in production.

**TABLE 1.2**

**SHAPES AND SOME COMMON METHODS OF PRODUCTION**

| SHAPE OR FEATURE | PRODUCTION METHOD |
|---|---|
| Flat surfaces | Rolling, planing, broaching, milling, shaping, grinding |
| Parts with cavities | End milling, electrical-discharge machining, electrochemical machining, ultrasonic machining, cast-in cavity |
| Parts with sharp features | Permanent-mold casting, machining, grinding, fabricating |
| Thin hollow shapes | Slush casting, electroforming, fabricating |
| Tubular shapes | Extrusion, drawing, roll forming, spinning, centrifugal casting |
| Tubular parts | Rubber forming, expanding with hydraulic pressure, explosive forming, spinning |
| Curvature on thin sheets | Stretch forming, peen forming, fabricating, assembly |
| Openings in thin sheets | Blanking, chemical blanking, photochemical blanking |
| Cross-sections | Drawing, extruding, shaving, turning, centerless grinding |
| Square edges | Fine blanking, machining, shaving, belt grinding |
| Small holes | Laser, electrical-discharge machining, electrochemical machining |
| Surface textures | Knurling, wire brushing, grinding, belt grinding, shot blasting, etching, deposition |
| Detailed surface features | Coining, investment casting, permanent-mold casting, machining |
| Threaded parts | Thread cutting, thread rolling, thread grinding, chasing |
| Very large parts | Casting, forging, fabricating, assembly |
| Very small parts | Investment casting, machining, etching, powder metallurgy |

## 1.3   Design for Manufacture, Assembly, Disassembly, and Service

As we have seen, design and manufacturing must be closely interrelated. They should never be viewed as separate disciplines or activities. Each part or component of a product must be designed so that it not only meets design requirements and specifications, but also can be manufactured economically and with relative ease. This approach improves productivity and allows a manufacturer to remain competitive.

This broad view has now become recognized as the area of **design for manufacture** (DFM). It is a comprehensive approach to production of goods and integrates the design process with materials, manufacturing methods, process planning, assembly, testing, and quality assurance. Effectively implementing design for manufacture requires that designers acquire a fundamental understanding of the characteristics, capabilities, and limitations of materials, manufacturing processes, and related operations, machinery, and equipment. This knowledge includes characteristics such as variability in machine performance, dimensional accuracy and surface finish of the workpiece, processing time, and the effect of processing method on part quality.

Designers and product engineers must be able to assess the impact of design modifications on manufacturing process selection, tools and dies, assembly, inspection, and product cost. Establishing quantitative relationships is essential in order to optimize the design for ease of manufacturing and assembly at minimum product cost (also called *producibility*). Computer-aided design, engineering, manufacturing, and process planning techniques, using powerful computer programs, have become indispensable to those conducting such analysis.

**FIGURE 1.3** Redesign of parts to facilitate automated assembly. *Source:* Reprinted from G. Boothroyd and P. Dewhurst, *Product Design for Assembly*, 1989 by courtesy of Marcel Dekker, Inc.

New developments include *expert systems*, which have optimization capabilities, thus expediting the traditional iterative process in design optimization.

After individual parts have been manufactured, they are assembled into a product. **Assembly** is an important phase of the overall manufacturing operation and requires consideration of the ease, speed, and cost of putting parts together (Fig. 1.3). Also, many products must be designed so that **disassembly** is possible with relative ease and less time consumption, enabling the products to be taken apart for maintenance, servicing, or recycling of their components. Because assembly operations can contribute significantly to product cost, **design for assembly (DFA)** as well as **design for disassembly** are now recognized as important aspects of manufacturing. Typically, a product that is easy to assemble is also easy to disassemble. The latest trend now includes **design for service**, ensuring that individual parts in a product are easier to reach and service.

Methodologies and computer software have been developed for DFA utilizing 3-D conceptual designs and solid models. In this way, subassembly and assembly times and costs are minimized while maintaining product integrity and performance; the system also improves the product's ease of disassembly. The trend now is to combine design for manufacture and design for assembly into the more comprehensive **design for manufacture and assembly (DFMA)**, which recognizes the inherent and important interrelationships between design and manufacturing.

## 1.4    Selecting Materials

An ever-increasing variety of materials is now available, each having its own characteristics, applications, advantages, and limitations (Fig. 1.4). The following are the general types of materials used in manufacturing today, either individually or in combination.

- **Ferrous metals:** carbon, alloy, stainless, and tool and die steels
- **Nonferrous metals and alloys:** aluminum, magnesium, copper, nickel, titanium, superalloys, refractory metals, beryllium, zirconium, low-melting alloys, and precious metals
- **Plastics:** thermoplastics, thermosets, and elastomers
- **Ceramics:** glass ceramics, glasses, graphite, and diamond

**FIGURE 1.4**   An outline of engineering materials.

- **Composite materials:** reinforced plastics, metal-matrix and ceramic-matrix composites, and honeycomb structures. These are also known as *engineered materials*.

As new materials are developed, important trends occur in their selection. Aerospace structures as well as products such as sporting goods (Fig. 1.5) have been at the forefront of new material usage. Because of the vested interest of the producers of different types of metallic and nonmetallic materials, there are constantly shifting trends in the usage of these materials, driven principally by economics. For example, by demonstrating steel's technical and economic feasibility, steel producers are countering the increased use of plastics in cars and aluminum in beverage cans. Likewise, aluminum producers are countering the use of various materials in cars. Several other such competitive applications can be cited, all of which ultimately benefit the consumer by reducing costs while maintaining product quality.

## Properties of materials

When selecting materials for products, we first consider their *mechanical properties*: strength, toughness, ductility, hardness, elasticity, fatigue, and creep. The strength-to-weight and stiffness-to-weight ratios of material are also important, particularly for aerospace and automotive applications. Aluminum, titanium, and reinforced plastics, for example, have higher strength-to-weight ratios than steels and cast irons. The mechanical properties specified for a product and its components should, of course, be appropriate for the conditions under which the product is expected to function.

We then consider the *physical properties* of density, specific heat, thermal expansion and conductivity, melting point, and electrical and magnetic properties. *Chemical properties* also play a significant role in hostile as well as normal environments. Oxidation, corrosion, general degradation of properties, toxicity, and flammability of materials are among the important factors to be considered. In some commercial airline disasters, for example, many deaths have been caused by toxic fumes from burning nonmetallic materials in the aircraft cabin.

*Manufacturing properties* of materials determine whether they can be cast, formed, shaped, machined, welded, or heat-treated with relative ease. Another important considera-

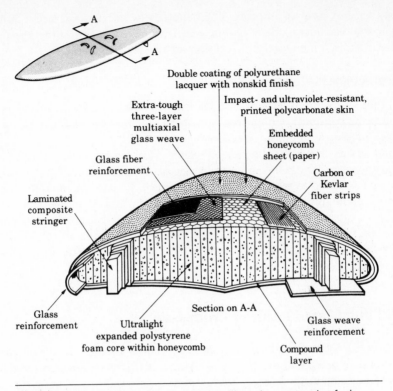

Double coating of polyurethane
lacquer with nonskid finish

Impact- and ultraviolet-resistant,
printed polycarbonate skin

Extra-tough
three-layer
multiaxial
glass weave

Embedded
honeycomb
sheet (paper)

Glass fiber
reinforcement

Carbon or
Kevlar
fiber strips

Laminated
composite
stringer

Glass
reinforcement

Ultralight
expanded polystyrene
foam core within honeycomb

Section on A-A

Glass weave
reinforcement

Compound
layer

**FIGURE 1.5** Cross-section of composite sailboard, an example of advanced materials construction. *Source:* Ken Easterling, *Tomorrow's Materials*, 2d ed., p. 133. Institute of Metals, 1990.

tion is that the methods used to process materials to the desired shapes may adversely affect the product's final properties, service life, and cost.

## Cost and availability

Cost and availability of raw and processed materials and manufactured components are major concerns in manufacturing. The economic aspects of material selection are as important as the technological considerations of properties and characteristics of materials.

If raw or processed materials or manufactured components are not commercially available in the desired shapes, dimensions, and quantities, substitutes or additional processing will be required, which can contribute significantly to product cost. For example, if we need a round bar of a certain diameter and it is not available in standard form, then we have to purchase a larger rod and reduce its diameter by some means, such as machining on a lathe, drawing though a die, or grinding.

Reliability of supply, as well as demand, affects cost. Most countries import numerous raw materials that are essential for production. The United States, for example, imports the majority of such raw materials as natural rubber, diamond, cobalt, titanium, chromium, aluminum, and nickel from other countries. The broad political implications of such reliance on other countries are self-evident.

Various costs are involved in processing materials by different methods. Some methods require expensive machinery, others require extensive labor, and still others require personnel with special skills, a high level of education, or specialized training.

### Appearance, service life, and recycling

The appearance of materials after they have been manufactured into products influences their appeal to the consumer. Color, feel, and surface texture are characteristics that we all consider when making a purchasing decision.

Time- and service-dependent phenomena such as wear, fatigue, creep, and dimensional stability are also important. These phenomena can significantly affect a product's performance and, if not controlled, can lead to total failure of the product. Similarly, compatibility of materials used in a product is important. Friction and wear, corrosion, and other phenomena can shorten a product's life or cause it to fail. An example is galvanic action between mating parts made of dissimilar metals, which causes them to corrode. Also, making materials-selection choices that facilitate recycling is increasingly a fundamental part of manufacturing design (see Section 1.10).

## 1.5   Selecting Manufacturing Processes

Many processes are used to produce parts and shapes (see Table 1.2). As you can see, there is usually more than one method of manufacturing a part from a given material. The broad categories of processing methods for materials are:

- **Casting:** expendable mold and permanent mold.
- **Forming and shaping:** rolling, forging, extrusion, drawing, sheet forming, powder metallurgy, and molding.
- **Machining:** turning, boring, drilling, milling, planing, shaping, broaching, grinding, ultrasonic machining; chemical, electrical, and electrochemical machining; and high-energy beam machining.
- **Joining:** welding, brazing, soldering, diffusion bonding, adhesive bonding, and mechanical joining.
- **Finishing operations:** honing, lapping, polishing, burnishing, deburring, surface treating, coating, and plating.

Selection of a particular manufacturing process depends not only on the component or part shape to be produced but also on many other factors. The type of material and its properties are basic considerations. Brittle and hard materials, for example, cannot be shaped easily, whereas they can be cast or machined by several methods. The manufacturing process usually alters the properties of materials. Metals that are formed at room temperature, for example, become stronger, harder, and less ductile than they were before processing.

Figure 1.6 shows two steel mounting brackets, one made by casting and the other by stamping of sheet metals. Note that there are some differences in the designs, although the parts are basically alike. Each of these two manufacturing processes has its own advantages, limitations, production rates, and cost.

Manufacturing engineers are constantly being challenged to find new solutions to manufacturing problems and cost reduction. For a long time, for example, sheet metal parts were

**FIGURE 1.6** Schematic illustrations of a steel mounting bracket that can be manufactured by two different processes.

(a) Casting      (b) Stamping

cut and made by traditional tools, punches, and dies. Although they are still widely used, these operations are now being replaced by laser cutting techniques. With advances in computer controls, we can automatically control the path of the laser, thus producing a wide variety of shapes accurately, repeatedly, and economically without the use of expensive tooling.

### Dimensional and surface finish considerations

Size, thickness, and shape complexity of the part have a major bearing on the process selected to produce it. Flat parts with thin cross-sections, for example, cannot be cast properly. Complex parts cannot be formed easily and economically, whereas they may be cast or fabricated from individual pieces or by powder metallurgy techniques.

Tolerances and surface finish obtained in hot-working operations cannot be as fine as those obtained in cold-working operations. Dimensional changes, warpage, and surface oxidation occur during processing at elevated temperatures. Some casting processes produce a better surface finish than others because of the different types of mold materials used.

The size and shape of manufactured products vary widely. For example, the main landing gear for the twin-engine, 400-passenger Boeing 777 jetliner is 4.3 m (14 ft) high, with three axles and six wheels, made by forging and machining processes. At the other extreme is the manufacturing of microscopic gears as small as 100 $\mu$m (0.004 in.) in diameter (Fig. 1.7). These gears power microrobots to repair human cells, microknives in surgery, and camera shutters for precise photography. The gears are made by a special electroplating and x-ray etching technique of metal plates coated with a polymer film. The center hole in these gears is so small that a human hair cannot pass through it. Such small-scale operations are called **nanotechnology** and **nanofabrication** (*nano* means one billionth).

**Ultraprecision manufacturing** techniques and machinery are now being developed and are coming into more common use. For machining mirrorlike surfaces, for example, the cutting tool is a very sharp diamond tip, and the equipment has very high stiffness and must be operated in a room where the temperature is controlled within less than 1°C. Highly sophisticated techniques such as molecular-beam epitaxy and scanning-tunneling engineering will be implemented to obtain accuracies on the order of the atomic lattice.

### Operational and manufacturing cost considerations

The design and cost of tooling, the lead time required to begin production, and the effect of workpiece materials on tool and die life are major considerations. The cost of tooling can be substantial, depending on tool size, design, and expected life. For example, a set of steel dies for stamping sheet-metal fenders for automobiles may cost about $2 million. For parts made

Human hair

**FIGURE 1.7**  Microscopic gear with a diameter on the order of 100 $\mu$m, made by a special etching process. *Source:* H. Guckel, Wisconsin Center for Applied Microelectronics, University of Wisconsin-Madison.

from expensive materials, the lower the scrap rate, the more economical the production process will be. Because machining generates chips, it may not be more economical than forming operations, all other factors being equal.

Availability of machines and equipment and operating experience within the manufacturing facility are also important cost factors. If some parts are not available, they may have to be manufactured by outside firms. Automakers, for example, purchase many parts from outside vendors or have them made by outside firms according to their specifications. The number of parts or products required and the desired production rate (pieces per hour) help determine the processes to be used and the economics of production. Beverage cans or transistors, for example, are consumed in numbers and at rates much higher than telescopes and propellers for ships.

The operation of machinery has significant environmental and safety implications. Depending on the type of operation and the machinery involved, some processes adversely affect the environment. Unless properly controlled, such processes can cause air, water, and noise pollution. The safe use of machinery is another important consideration, requiring precautions to eliminate hazards in the work place.

## Net-shape manufacturing

Since not all manufacturing operations produce finished parts or products, additional processes may be necessary. For example, a forged part may not have the desired dimensions or surface finish; thus additional operations such as machining or grinding may be necessary. Likewise, it may be difficult, impossible, or economically undesirable to produce a part with holes in it using just one manufacturing process, necessitating additional processes such as drilling. Also, the holes produced by a particular manufacturing process may not have the proper roundness, dimensional accuracy, or surface finish, creating a need for additional operations such as honing.

These additional processes can contribute significantly to the cost of a product. Consequently, there has been an important trend, called **net-shape** or **near-net-shape manufacturing,** in which the part is made as close to the final desired dimensions, tolerances, and specifications as possible. Typical examples of such manufacturing methods are near-net-shape forging and casting of parts, stamped sheet-metal parts, components made by powder metallurgy techniques, and injection molding of plastics.

## 1.6  Computer-Integrated Manufacturing

The major goals of *automation* in manufacturing facilities are to integrate various operations to improve productivity, increase product quality and uniformity, minimize cycle times, and reduce labor costs. Since the 1940s, automation has accelerated because of rapid advances in control systems for machines and in computer technology (Table 1.3).

**TABLE 1.3**

**DEVELOPMENTS IN THE HISTORY OF AUTOMATION OF MANUFACTURING PROCESSES**

| DATE | DEVELOPMENT |
|------|-------------|
| 1500–1600 | Water power for metalworking; rolling mills for coinage strips |
| 1600–1700 | Hand lathe for wood; mechanical calculator |
| 1700–1800 | Boring, turning, and screw-cutting lathe, drill press |
| 1800–1900 | Copying lathe, turret lathe, universal milling machine; advanced mechanical calculators |
| 1808 | Sheet-metal cards with punched holes for automatic control of weaving patterns in looms |
| 1863 | Automatic piano player (Pianola) |
| 1900–1920 | Geared lathe; automatic screw machine; automatic bottlemaking machine |
| 1920 | First use of the word *robot* |
| 1920–1940 | Transfer machines; mass production |
| 1940 | First electronic computing machine |
| 1943 | First digital electronic computer |
| 1945 | First use of the word *automation* |
| 1948 | Invention of the transistor |
| 1952 | First prototype numerical-control machine tool |
| 1954 | Development of the symbolic language APT(Automatically Programmed Tool); adaptive control |
| 1957 | Commercially available NC machine tools |
| 1959 | Integrated circuits; first use of the term *group technology* |
| 1960s | Industrial robots |
| 1965 | Large-scale integrated circuits |
| 1968 | Programmable logic controllers |
| 1970 | First integrated manufacturing system; spot welding of automobile bodies with robots |
| 1970s | Microprocessors; minicomputer-controlled robot; flexible manufacturing systems; group technology |
| 1980s | Artificial intelligence; intelligent robots; smart sensors; untended manufacturing cells |
| 1990s | Integrated manufacturing systems; intelligent and sensor-based machines; telecommunications and global manufacturing networks; fuzzy logic devices; artificial neural networks |

**FIGURE 1.8**   Machining a mold cavity for making plastic blender jars. Left: Three-dimensional wire-frame model of the mold, as designed and viewed on the monitor. Right: Machining the die cavity using a computer numerical control milling machine. *Source:* Model Threads Inc., and *Manufacturing Engineering,* March 1991, Society of Manufacturing Engineers.

Few developments in the history of manufacturing have had a more significant impact than computers. Beginning with computer graphics, the use of computers has been extended to computer-aided design and manufacturing and, ultimately, to **computer-integrated manufacturing systems.** The use of computers covers a very broad range of applications, including control and optimization of manufacturing processes, material handling, assembly, and automated inspection and testing of products.

## Machine control systems

**Numerical control (NC)** of machines is a method of controlling the movements of machine components by direct insertion of coded instructions in the form of numerical data (Fig. 1.8). Numerical control was first implemented in the early 1950s and was a major advance in automation of machines. With advances in computers, numerical control has expanded into **computer numerical control (CNC).**

In **adaptive control (AC),** process parameters are adjusted automatically to optimize production rate and product quality and minimize cost. Parameters such as forces, temperatures, surface finish, and dimensions of the part are monitored constantly. If they move outside the acceptable range, the system adjusts the process variables until the parameters again fall within the acceptable range.

Major advances have been made in **automated handling of materials** in various stages of completion such as the handling from storage to machines, from machine to machine, and at the points of inspection, inventory, and shipment. Introduced in the early 1960s, **industrial robots** are replacing humans in operations that are repetitive, boring, and dangerous, thus reducing the possibility of human error, decreasing variability in product quality, and improving productivity. Robots with sensory perception capabilities are being developed, with movements that are beginning to simulate those of humans. **Automated** and **robotics assembly** systems are replacing costly assembly by operators. Products are being designed or redesigned so that they can be assembled more easily by machine.

## Computer technology

Computers allow us to integrate virtually all phases of manufacturing operations, which consist of various technical as well as managerial activities (see Chapters 13 and 14). With sophisticated software and hardware, manufacturers are now able to minimize manufacturing costs, improve product quality, reduce product development time, and maintain a competitive edge in the domestic and global marketplaces.

**Computer-integrated manufacturing (CIM)** is particularly effective, with benefits including:

- Responsiveness to rapid changes in market demand, product modification, and shorter product life cycles.
- High-quality products at low cost.
- Better use of materials, machinery, and personnel, and reduced inventory.
- Better control of production and management of the total manufacturing operation.

**Computer-aided design (CAD)** allows the designer to conceptualize objects more easily without having to make costly illustrations, models, or prototypes. These systems are now capable of rapidly and completely analyzing designs, from a simple bracket to complex structures such as aircraft wings. Using **computer-aided engineering (CAE)**, the performance of structures subjected to static or fluctuating loads and temperatures can now be simulated, analyzed, and tested efficiently, accurately, and more quickly than ever on the computer. Designs can be optimized and modifications can be made directly and easily at any time. The information developed can be stored, retrieved, displayed, printed, and transferred anywhere in the organization.

**Computer-aided manufacturing (CAM)** involves all phases of manufacturing by utilizing and processing further the large amount of information on materials and processes collected and stored in the organization's database. Computers now assist manufacturing engineers and others in organizing tasks such as programming numerical control of machines; programming robots for material handling; designing tools, dies, and fixtures; and maintaining quality control.

**Computer-aided process planning (CAPP)** is capable of improving productivity in a plant by optimizing process plans, reducing planning costs, and improving the consistency of product quality and reliability. Functions such as cost estimating and work standards (time required to perform a certain operation) can also be incorporated into the system.

The high level of automation and flexibility achieved with numerically controlled machines has led to the development of **group technology (GT)** and **cellular manufacturing (CM)**. The concept of group technology is that parts can be grouped and produced by classifying them according to similarities in design and manufacturing processes. In this way, part designs and process plans can be standardized, and families of parts can be produced efficiently and economically.

**Flexible manufacturing systems (FMS)** integrate manufacturing cells into a large unit containing industrial robots and **automated guided vehicles (AGVs)** serving several machines, all interfaced with a central computer (Fig. 1.9). Flexible manufacturing systems have the highest level of efficiency, sophistication, and productivity in manufacturing. They are capable of producing parts randomly and quickly changing manufacturing sequences on different parts. Thus they can meet rapid changes in market demand for various types of products.

An important concept in manufacturing is **just-in-time production (JIT)**, in which supplies are delivered just in time to be used, parts are produced just in time to be made into

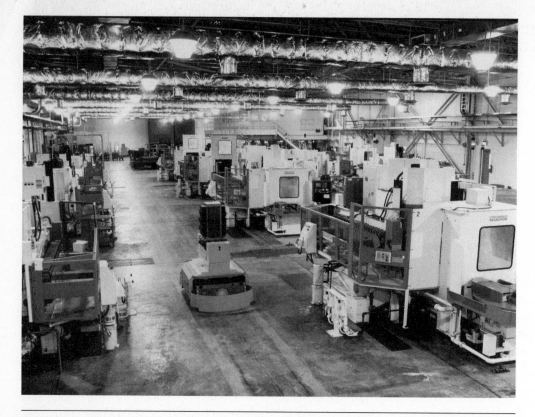

**FIGURE 1.9**   General view of a flexible manufacturing system, showing several machines (machining centers) and an automated guided vehicle (AGV) moving along the aisle. *Source:* Courtesy of Cincinnati Milacon, Inc.

subassemblies and assemblies, and products are finished just in time to be sold. In this way, inventory carrying costs are low, part defects are detected right away, productivity is increased, and high-quality products are made at low cost.

**Artificial intelligence** (**AI**) involves the use of machines and computers to replace human intelligence. Computer-controlled systems are becoming capable of learning from experience and making decisions that optimize operations and minimize costs. **Expert systems,** which are basically intelligent computer programs, are being developed rapidly with capabilities to perform tasks and solve difficult real-life problems as well as human experts would. With advances in **artificial neural networks** (**ANN**), which are being designed to simulate the thought processes of the human brain, these systems have the capability of modeling and simulating production facilities, monitoring and controlling manufacturing processes, diagnosing problems in machine performance, conducting financial planning, and managing a company's manufacturing strategy.

We can now envisage the *factory of the future* in which production takes place with little or no direct human intervention. The human role will be confined to supervision, maintenance, and upgrading of machines, computers, and software. The impact of such advances on the work force remains to be fully assessed.

# 1.7   **Quality Assurance and Total Quality Management**

Product quality has always been one of the most important aspects of manufacturing, as it directly influences the marketability of a product and customer satisfaction. Traditionally, quality assurance has been obtained by inspecting parts after they have been manufactured. Parts are inspected to ensure that they conform to a detailed set of specifications and standards such as dimensions, surface finish, and mechanical and physical properties.

However, quality cannot be inspected into a product after it is made. The practice of inspecting products after they are made is being replaced rapidly by the broader view that *quality must be built into a product*—from the design stage through all subsequent stages of manufacture and assembly. Because products are usually made by using several manufacturing processes that can have significant variations in their performance even within a short period of time, the control of processes is a critical factor in product quality. Thus, *we control processes and not products.*

Producing defective products can be very costly to the manufacturer, creating difficulties in assembly operations, necessitating repairs in the field, and resulting in customer dissatisfaction. Contrary to general public opinion, low-quality products do not necessarily cost less than high-quality products.

Although it can be described in various ways, **product integrity** is a term that can be used to define the degree to which a product (1) is suitable for its intended purpose, (2) fills a real market need, (3) functions reliably during its life expectancy, and (4) can be maintained with relative ease. Product integrity has also been defined as the total product experience of the customer, or as the totality of qualities needed to conceive, produce, and market the product successfully.

**Total quality management** (**TQM**) and quality assurance have become the responsibility of everyone involved in designing and manufacturing a product. Awareness of the technological and economic importance of built-in product quality has been heightened by pioneers in quality control, primarily Deming, Taguchi, and Juran. They pointed out the importance of management's commitment to product quality, pride of workmanship at all levels of production, and the use of powerful techniques such as **statistical process control** (**SPC**) and *control charts* for on-line monitoring of part production and rapid identification of sources of quality problems. Ultimately, the major goal is to prevent defects from occurring rather than detect defective products. We can now produce computer chips in which only a few parts out of a million may be defective. The importance of this trend in the reliability of products and customer satisfaction is self-evident.

Important developments in quality assurance include the implementation of **design of experiment,** a technique in which the factors involved in a manufacturing process and their interactions are studied simultaneously. Thus, for example, variables affecting dimensional accuracy or surface finish in a machining operation can be readily identified, allowing appropriate actions to be taken. The use of computers has greatly enhanced our capability to utilize such techniques.

The strong trend toward global manufacturing and competitiveness has created the need for international conformity and consensus regarding the establishment of quality control methods. This has resulted in the ISO 9000 series on Quality Management and Quality Assurance Standards. A company's registration for this standard, which is a quality process certification and not a product certification, means the company conforms to consistent practices as specified by its own quality system. ISO 9000 has permanently influenced the manner in which companies conduct business in world trade and is rapidly becoming the world standard for quality.

## 1.8   Global Competitiveness and Manufacturing Costs

Economics always has been a major factor in manufacturing and has become even more so as international competition (**global competitiveness**) for high-quality products and low prices becomes entrenched in worldwide markets. The markets have become multinational and dynamic, product lines have become extensive and technically complex, and demands by customers for quality have become commonplace (**world-class manufacturing**). To respond to these needs while keeping costs down is a constant challenge to manufacturing companies and is crucial to their very survival.

The cost of product is often the overriding consideration in its marketability and general customer satisfaction. Consequently, all costs involved in manufacturing a high-quality product must be thoroughly analyzed. Typically, manufacturing costs represent about 40 percent of a product's selling price.

The concepts of design for manufacture and assembly and concurrent engineering described in Sections 1.2 and 1.3 also include design principles for economic production:

- The design should be as simple as possible to manufacture, assemble, disassemble, and recycle.
- Materials should be chosen for their appropriate manufacturing characteristics.
- Dimensional accuracy and surface finish specified should be as broad as permissible.
- Because they can add significantly to cost, secondary and finishing processes should be avoided or minimized.

The total cost of manufacturing a product consists of costs of materials, tooling, and labor, as well as fixed and capital costs. Several factors are involved in each cost category. Manufacturing costs can be minimized by analyzing the product design to determine whether part size and shape are optimal and the materials selected are the least costly, while still possessing the desired properties and characteristics. The possibility of substituting materials is an important consideration in minimizing costs.

## 1.9   Lean Production and Agile Manufacturing

The trends we have described thus far have led to the concept of **lean production** or *lean manufacturing*. Although not a novel concept, it basically involves a major assessment of each of a company's activities: the efficiency and effectiveness of its various operations, the necessity of retaining some of its operations and managers, the efficiency of the machinery and equipment used in the operation while maintaining and improving quality, the number of personnel involved in a particular operation, and a thorough analysis to reduce the cost of each activity, including both productive and nonproductive labor.

This concept requires a fundamental change in corporate culture as well as understanding, cooperation, and teamwork between management and the work force. It does not necessarily mean cutting back resources, but aims at continuously improving efficiency and profitability of the company by removing all kinds of waste from the operations and dealing with problems right away, including environmental issues.

**Agile manufacturing** is a term that has been coined to indicate the use of the principles of lean production on a broader scale. The principle behind agile manufacturing is ensuring flexibility (agility) in the manufacturing enterprise so it can quickly respond to changes in product variety and demand and customer needs. This is to be achieved through people, equipment, computer hardware and software, reduced changeover time, and sophisticated communications systems. As an example, it has been predicted that the automotive industry will be able to configure and build a car in three days and that, eventually, the traditional assembly line will be replaced by a system in which a nearly custom-made car will be produced by connecting modules.

These approaches require that manufacturers **benchmark** their operations; this means understanding the competitive position of other manufacturers with respect to their own and setting realistic goals for the future. Thus, benchmarking is a reference from which various measurements can be made and compared.

## 1.10  Environmentally Conscious Design and Manufacturing

In the United States alone, nine million passenger cars and 285 million tires are discarded each year, about 100 million tires being reused in various ways. More than five billion kilograms of plastic products are discarded each year. Every three months, U.S. industries and consumers discard enough aluminum to rebuild the country's commercial air fleet. In Germany, 800,000 metric tons of old TV sets, radios, and computer equipment are discarded each year.

How do we recycle, treat, or dispose of this very large amount of waste? In manufacturing operations, we may cite numerous examples and raise a number of similar questions. Metalworking fluids such as lubricants and coolants are often used in machining, grinding, and forming operations. Various fluids and solvents are used in cleaning manufactured products. Some of these fluids pollute the air and waters during their use. Likewise, many by-products from manufacturing plants have for years been discarded: sand with additives used in metal-casting processes; water, oil, and other fluids from heat-treating facilities and plating operations; slag from foundries and welding operations; and a wide variety of metallic and nonmetallic scrap produced in operations such as sheet forming, casting, and molding.

The present and future adverse effects of these activities, their damage to our environment and to the earth's ecosystem, and ultimately their effect on the quality of human life are by now well recognized by the public as well as local and federal governments. Consider the effects of water and air pollution, acid rain, ozone depletion, the greenhouse effect, hazardous wastes, landfill seepage, and global warming.

In response to these major concerns, a wide range of laws and regulations have been promulgated by local, state, and federal governments as well as professional organizations in the United States and other industrial countries. These regulations are generally stringent, and their implementation can have a major impact on the economic operation and financial health of industrial organizations. Regardless, and notwithstanding arguments by some that environmental damage from these activities is overestimated, manufacturing engineers and the management of companies have great responsibility to plan and implement environmentally safe manufacturing operations.

Much can be gained by careful analysis of products, their design, materials used, and the manufacturing processes and practices employed in making these products. Certain guidelines can be followed in this regard, as outlined below.

- Reducing waste of materials at their source by refinements in product design and reducing the amount of materials used
- Conducting research and development of environmentally safe products and manufacturing technologies
- Reducing the use of hazardous materials in products and processes
- Ensuring proper handling and disposal of all waste
- Making improvements in recycling, waste treatment, and reuse of materials

Major developments have taken place regarding these important matters, and the term **environmentally conscious design and manufacturing** has now become common in the industry, indicating broad awareness of the problem. A major emphasis is on **design for the environment (DFE)** or **green design** (green meaning environmentally safe and friendly). This approach anticipates the possible negative environmental impact of materials, products, and processes so that they can be considered at the earliest stages of design and production. The main objectives now are preventing pollution at the source, recycling, and reuse rather than disposal. These goals have led to the concept of **design for recycling (DFR)**.

## 1.11  Product Liability

We are all familiar with the consequence of using a product that has malfunctioned, causing bodily injury, or even death, and financial loss to a person or organization. This important topic is commonly referred to as **product liability.** Because of the related technical and legal aspects, in which laws can vary from state to state and country to country, this is a complex subject and can have a major economic impact on the parties involved.

Designing and manufacturing safe products is an important and integral part of a manufacturer's responsibilities. All those involved with product design, manufacture, and marketing must be fully cognizant of the consequences of product failure, including failures occurring during possible misuse of the product. There has been a proliferation of product liability claims, of varying degrees of merit, made against manufacturers of consumer and industrial equipment, particularly in the United States.

Numerous examples of products that could involve liability can be cited: (1) a grinding wheel that shatters and blinds a worker, (2) a cable that snaps, allowing a platform to drop, (3) brakes that become inoperative due to the failure of one component, (4) machines with no guards or inappropriate guards, and (5) electric or pneumatic tools without proper warnings.

Furthermore, *human-factors engineering* and *ergonomic considerations* (human-machine interactions) are important aspects of the design and manufacture of safe products. Examples of suboptimal products include (1) an uncomfortable or unstable workbench or chair, (2) a mechanism that is difficult to operate manually, causing back injury, and (3) a poorly designed keyboard, causing pain to the user's hands and arms after repetitive use.

# 1.12 Organization for Manufacture

The various manufacturing activities and functions that we have described must be organized and managed efficiently and effectively in order to maximize productivity and minimize costs while maintaining high quality standards. Because of the complex interactions among the various factors involved in manufacturing—materials, machines, people, information, power, and capital—the proper coordination and administration of diverse functions and responsibilities are essential.

*Manufacturing engineers* traditionally have had several major responsibilities:

- Plan the manufacture of a product and the processes to be utilized. This function requires a thorough knowledge of the product, its expected performance, and specifications.

- Identify the machines, equipment, tooling, and personnel needed to carry out the plan. This function requires evaluation of the capabilities of machines, tools, and workers so that proper functions and responsibilities can be assigned.

- Interact with design and materials engineers to optimize productivity and minimize production costs.

- Cooperate with *industrial engineers* when planning plant floor activities such as plant layout, machine arrangement, material-handling equipment, time-and-motion study, production methods analysis, production planning and scheduling, and maintenance. Some of these activities are carried out under the name plant engineering and some are interchangeably performed by both manufacturing and industrial engineers.

Manufacturing engineers, in cooperation with industrial engineers, also are responsible for evaluating new technologies, their applications, and how they can be implemented. In view of the rapidly growing amount of technical information, this task in itself can present a major challenge. Gaining a broad perspective of computer capabilities, applications, and integration in all phases of manufacturing operations is important. This knowledge is particularly crucial for long-range production-facility planning because of the constantly changing global market demand and product mix.

## SUMMARY

- Manufacturing is the process of converting raw materials into products. (Section 1.1)

- Whereas design tasks were once seen as distinct from and prerequisite to manufacturing tasks, design is increasingly seen as an integral part of manufacturing, as evidenced by trends in concurrent engineering, design for manufacture, design for assembly, disassembly, and service, and environmentally conscious design and manufacturing. (Sections 1.2, 1.3, 1.10)

- A key task for manufacturing engineers is to select an optimal manufacturing method among multiple alternatives, given product design goals, process capabilities, and cost considerations. In addition, manufacturing engineers may participate in materials selection decisions and planning plant floor activities. (Sections 1.4, 1.5, 1.8, 1.12)

- Computer-integrated manufacturing techniques utilize computers to automate a wide range of design, analysis, manufacturing, and quality-control tasks. (Section 1.6)

- Ensuring product quality is, increasingly, a concurrent engineering process rather than a last step in the manufacture of a product. Total quality management and statistical process control techniques have increased our ability to build quality into a product at every step of the design and manufacturing process. (Section 1.7)

- Lean production and agile manufacturing are approaches that focus on the efficiency and flexibility of the entire organization in order to help manufacturers respond to competitive and economic pressures. (Section 1.9)

- Designing and manufacturing safe and environmentally conscious products is an important and integral part of a manufacturer's responsibilities. (Section 1.11)

# 2

# Fundamentals of the Mechanical Behavior of Materials

**2.1** Introduction
**2.2** Tension
**2.3** Compression
**2.4** Torsion
**2.5** Bending (Flexure)
**2.6** Hardness
**2.7** Fatigue
**2.8** Creep
**2.9** Impact
**2.10** Residual Stresses
**2.11** Triaxial Stresses and Yield Criteria
**2.12** Work of Deformation

## 2.1  Introduction

The manufacturing methods and techniques by which materials can be shaped into useful products were outlined in Chapter 1. In manufacturing, one of the most important groups of processes is **plastic deformation**, namely, shaping materials by applying forces in various ways (also known as *deformation processing*). It includes bulk deformation (forging, rolling, extrusion, and rod and wire drawing) and sheet-forming processes (bending, drawing, spinning, and general pressworking).

This chapter deals with the fundamental aspects of the mechanical behavior of materials during deformation. Individual topics are deformation modes, stresses, forces, effects of rate of deformation and temperature, hardness, residual stresses, and yield criteria.

In stretching a piece of metal to make an object such as an automobile fender or a length of wire, the material is subjected to tension. A solid cylindrical piece of metal is forged in the making of a turbine disk, thus subjecting the material to compression. Sheet metal undergoes shearing stresses when, for instance, a hole is punched through its cross-section. A piece of plastic tubing is expanded by internal pressure to make a bottle, thus subjecting the material to tension in various directions.

In all these processes, the material is subjected to one or more of the basic modes of deformation shown in Fig. 2.1, namely, tension, compression, and shear. The degree of deformation to which the material is subjected is defined as **strain**. For tension or compression, the **engineering strain**, or **nominal strain**, is defined as

$$e = \frac{\ell - \ell_o}{\ell_o}. \tag{2.1}$$

**25**

**FIGURE 2.1** Types of strain. (a) Tensile. (b) Compressive. (c) Shear. All deformation processes in manufacturing involve strains of these types. Tensile strains are involved in stretching sheet metal to make car bodies, compressive strains in forging metals to make turbine disks, and shear strains in making holes by punching.

Note that in tension the strain is positive and in compression it is negative. The **shear strain** in (c) is defined as

$$\gamma = \frac{a}{b}. \tag{2.2}$$

In order to change the geometry of the elements or bodies in Fig. 2.1, *forces* must be applied to them as shown by the arrows. The determination of these forces as a function of strain is very important in manufacturing processes. We have to know the forces in order to design the proper equipment, select the tool and die materials for proper strength, and determine whether a specific metalworking operation can be accomplished on certain equipment. Thus, the relation between a force and the deformation it produces is an essential parameter in manufacturing.

## 2.2  Tension

Because of its relative simplicity, the tension test is the most common test for determining the *strength-deformation characteristics* of materials. It involves the preparation of a test specimen (according to ASTM specifications) and testing it under tension on any of a variety of available testing machines.

The specimen has an original length $\ell_o$ and an original cross-sectional area $A_o$ (Fig. 2.2). Although most specimens are solid and round, flat sheet or tubular specimens are also tested under tension. The original length is the distance between **gage marks** on the specimen and is generally 50 mm (2 in.). Longer lengths may be used for larger specimens, such as structural members.

Typical results from a tension test are shown in Fig. 2.2. The **engineering stress,** or **nominal stress**, is defined as the ratio of the applied load to the original area,

$$\sigma = \frac{P}{A_o}, \tag{2.3}$$

**FIGURE 2.2** (a) Original and final shape of a standard tensile-test specimen. (b) Outline of a tensile-test sequence showing different stages in the elongation of the specimen.

and the engineering strain is

$$e = \frac{\ell - \ell_o}{\ell_o}.$$

When the load is first applied, the specimen elongates proportionately to load up to the **proportional limit**. This is the range of **linear elastic behavior**. The material will continue to deform elastically, although not strictly linearly, up to the **yield point**, $Y$. If the load is removed before the yield point is reached, the specimen will return to its original length. The **modulus of elasticity**, or **Young's modulus**, $E$, is defined as

$$E = \frac{\sigma}{e}. \tag{2.4}$$

This linear relationship between stress and strain is known as **Hooke's law**, the more generalized forms of which are given in Section 2.11.

The elongation of the specimen is accompanied by a contraction of its lateral dimensions. The *absolute* value of the ratio of the lateral strain to longitudinal strain is known as **Poisson's ratio**, $v$. Typical values for $E$ and $v$ for various materials are given in Table 2.1.

The area under the stress–strain curve up to the yield point $Y$ of a material is known as the **modulus of resilience**:

$$\text{Modulus of resilience} = \frac{Ye_o}{2} = \frac{Y^2}{2E}. \tag{2.5}$$

**TABLE 2.1**

**MECHANICAL PROPERTIES OF VARIOUS MATERIALS AT ROOM TEMPERATURE**

| METALS (WROUGHT) | $E$ (GPa) | $Y$(MPa) | UTS (MPa) | ELONGATION in 50 mm (%) | POISSON'S RATIO $\nu$ |
|---|---|---|---|---|---|
| Aluminum and its alloys | 69–79 | 35–550 | 90–600 | 45–4 | 0.31–0.34 |
| Copper and its alloys | 105–150 | 76–1100 | 140–1310 | 65–3 | 0.33–0.35 |
| Lead and its alloys | 14 | 14 | 20–55 | 50–9 | 0.43 |
| Magnesium and its alloys | 41–45 | 130–305 | 240–380 | 21–5 | 0.29–0.35 |
| Molybdenum and its alloys | 330–360 | 80–2070 | 90–2340 | 40–30 | 0.32 |
| Nickel and its alloys | 180–214 | 105–1200 | 345–1450 | 60–5 | 0.31 |
| Steels | 190–200 | 205–1725 | 415–1750 | 65–2 | 0.28–0.33 |
| Titanium and its alloys | 80–130 | 344–1380 | 415–1450 | 25–7 | 0.31–0.34 |
| Tungsten and its alloys | 350–400 | 550–690 | 620–760 | 0 | 0.27 |

| NONMETALLIC MATERIALS | | | | | |
|---|---|---|---|---|---|
| Ceramics | 70–1000 | – | 140–2600 | 0 | 0.2 |
| Diamond | 820–1050 | – | – | – | – |
| Glass and porcelain | 70–80 | – | 140 | 0 | 0.24 |
| Rubbers | 0.01–0.1 | – | – | – | 0.5 |
| Thermoplastics | 1.4–3.4 | – | 7–80 | 1000–5 | 0.32–0.40 |
| Thermoplastics, reinforced | 2–50 | – | 20–120 | 10–1 | – |
| Thermosets | 3.5–17 | – | 35–170 | 0 | 0.34 |
| Boron fibers | 380 | – | 3500 | 0 | – |
| Carbon fibers | 275–415 | – | 2000–3000 | 0 | – |
| Glass fibers | 73–85 | – | 3500–4600 | 0 | – |
| Kevlar fibers | 62–117 | – | 2800 | 0 | – |

Note: In the upper table the lowest values for $E$, $Y$, and UTS and the highest values for elongation are for pure metals. Multiply gigapascals (GPa) by 145,000 to obtain pounds per square in. (psi), megapascals (MPa) by 145 to obtain psi. For example, 100 GPa = 14,500 ksi, and 100 MPa = 14,500 psi.

This area has the units of **energy per unit volume** and indicates the **specific energy** that the material can store elastically. Typical values for modulus of resilience are 3 in. · lb/in.$^3$ $(2.1 \times 10^4 \, \text{N·m/m}^3)$ for annealed copper, $28(1.9 \times 10^5)$ for annealed medium-carbon steel, and $385(2.7 \times 10^6)$ for spring steel.

With increasing load, the specimen begins to yield; that is, it begins to undergo **plastic (permanent) deformation** and the relationship between stress and strain is no longer linear. The rate of change in the slope of the stress–strain curve beyond the yield point is rather small for most materials, so the determination of $Y$ may be difficult. The usual practice is to define the yield stress as the point on the curve that is *offset* by a strain of (usually) 0.2%, or 0.002 (Fig. 2.2b). Other offset strains may also be used and should be specified in reporting the yield stress.

It is important to note that yielding does not necessarily mean failure. In the design of structures and load-bearing members yielding is not acceptable since it leads to permanent deformation. However, yielding is essential in metalworking processes, such as bending,

**FIGURE 2.3**  Schematic illustration of loading and unloading of a tensile-test specimen. Note that during unloading the curve follows a path parallel to the original elastic slope.

forging, rolling, and sheet-metal forming, where materials have to be subjected to permanent deformation.

As the specimen continues to elongate under increasing load beyond $Y$, its cross-sectional area decreases permanently and uniformly through its gage length. If the specimen is unloaded from a stress level higher than $Y$, the curve follows a straight line downward and parallel to the original elastic slope (Fig. 2.3). As the load, hence the engineering stress, is further increased, it eventually reaches a maximum and then begins to decrease. The maximum stress is known as the **tensile strength** or **ultimate tensile strength** (UTS) of the material (Table 2.1). Ultimate tensile strength is a practical measure of the overall strength of a material.

When the specimen is loaded beyond its UTS, it begins to *neck* (Fig. 2.2) and the elongation is no longer uniform. That is, the change in the cross-sectional area of the specimen is no longer uniform along the length of the specimen but is concentrated locally in a "neck" formed in the specimen (known as **necking**, or *necking down*). As the test progresses, the engineering stress drops further and the specimen finally fractures in the necked region. The final stress level (marked by an X in Fig. 2.2b) is known as **breaking** or **fracture stress**.

### 2.2.1  Ductility

The strain at fracture is a measure of **ductility**, that is, how large a strain the material withstands before failure. Note from Fig. 2.2b that until the UTS is reached, elongation is *uniform*. The strain up to the UTS is known as **uniform strain**. The elongation at fracture is known as the **total elongation**. It is measured between the original gage marks after the two pieces of the broken specimen are placed together.

Two quantities commonly used to define ductility in a tension test are elongation and reduction of area. **Elongation** is defined as

$$\text{Elongation} = \frac{\ell_f - \ell_o}{\ell_o} \times 100 \qquad (2.6)$$

and is based on the total elongation (Table 2.1).

Necking is a *local* phenomenon. If we put a series of gage marks at different points on the specimen, pull, and break it under tension, and then calculate the percent elongation for each pair of gage marks, we find that with decreasing gage length the percent elongation increases (Fig. 2.4). The closest pair of gage marks undergo the largest elongation because they are closest to the necked region. However, the curves do not approach zero elongation with increasing gage length. The reason is that the specimens have all undergone a finite uniform and permanent elongation before fracture. It is thus important to report gage length in conjunction with elongation data, whereas other tensile properties are generally independent of gage length.

A second measure of ductility is **reduction of area,** defined as

$$\text{Reduction of area} = \frac{A_o - A_f}{A_o} \times 100. \qquad (2.7)$$

Thus, a material that necks down to a point at fracture, such as a glass rod at elevated temperature, has 100% reduction of area.

The elongation and reduction of area are generally related to each other for many common engineering metals and alloys. Elongation ranges approximately between 10 and 60%, and values between 20 and 90% are typical for reduction of area for most materials. *Thermoplastics* (Chapter 10) and *superplastic* materials (Section 2.2.7) exhibit much

**FIGURE 2.4** Total elongation in a tensile test as a function of original gage length for various metals. Because necking is a local phenomenon, elongation decreases with gage length. Standard gage length is usually 2 in. (50 mm), although shorter ones can be used if larger specimens are not available.

higher ductility. Brittle materials, by definition, have little or no ductility. Examples are glass at room temperature and a piece of chalk.

## 2.2.2 True stress and true strain

Because stress is defined as the ratio of force to area, **true stress** is defined as

$$\sigma = \frac{P}{A},$$ (2.8)

where $A$ is the actual (hence true) or instantaneous area supporting the load.

Likewise, the complete tension test may be regarded as a series of incremental tension tests where, for each succeeding increment, the original specimen is a little longer than the previous one. We can now define **true strain** (or *natural* or *logarithmic strain*) $\varepsilon$ as

$$\varepsilon = \int_{\ell_o}^{\ell} \frac{d\ell}{\ell} = \ln\left(\frac{\ell}{\ell_o}\right).$$ (2.9)

Note that, for small values of engineering strain, $e = \varepsilon$ since $\ln(1 + e) = \varepsilon$. For larger strains, however, the values diverge rapidly, as shown in Table 2.2.

The volume of a metal specimen in the plastic region of the test remains constant (Section 2.5). Hence the true strain within the uniform elongation range can be expressed as

$$\varepsilon = \ln\left(\frac{\ell}{\ell_o}\right) = \ln\left(\frac{A_o}{A}\right) = \ln\left(\frac{D_o}{D}\right)^2 = 2 \ln\left(\frac{D_o}{D}\right).$$ (2.10)

Once necking begins, the true strain at any point in the specimen can be calculated from the change in cross-sectional area at that point. Thus, by definition, the largest strain is at the narrowest region of the neck.

We have shown that, at small strains, the engineering and true strains are very close and so either one can be used in calculations. However, for the large strains encountered in metalworking, the true strain should be used. This is the true measure of the strain and can be illustrated by the following two examples.

Assume that a tension specimen is elongated to twice its original length. This deformation is equivalent to compressing a specimen to one-half its original height. Using the subscripts $t$ and $c$ for tension and compression, respectively, we find that $\varepsilon_t = 0.69$ and $\varepsilon_c = -0.69$, whereas $e_t = 1$ and $e_c = -0.5$. Thus, true strain is a correct measure of strain. For the second example, assume that a specimen 10 mm in height is compressed to a final thickness of zero. Thus $\varepsilon_c = -\infty$, whereas $e_c = -1$. Thus, we have deformed the specimen infinitely, which is exactly what the value of the true strain indicates. From these examples, true strains obviously are consistent with the actual physical phenomenon and engineering strains are not.

**TABLE 2.2**

**COMPARISON OF ENGINEERING AND TRUE STRAINS IN TENSION**

| $e$ | 0.01 | 0.05 | 0.1 | 0.2 | 0.5 | 1 | 2 | 5 | 10 |
|---|---|---|---|---|---|---|---|---|---|
| $\varepsilon$ | 0.01 | 0.049 | 0.095 | 0.18 | 0.4 | 0.69 | 1.1 | 1.8 | 2.4 |

---

**Example 2.1:   True strains and engineering strains**

A strip of metal is originally 1.5 m long. It is stretched in three steps: first to a length of 2.0 mm, then to 2.5 m, and finally to 3.0 m. Show that the total true strain is the sum of the true strains in each step—that is, the strains are additive. Show that, using engineering strains, the strain for each step cannot be added to obtain the total strain.

SOLUTION.   We first calculate the true strains for each step.

$$\text{Step 1: } \varepsilon_1 = \ln(2.0/1.5) = 0.288$$
$$\text{Step 2: } \varepsilon_2 = \ln(2.5/2.0) = 0.223$$
$$\text{Step 3: } \varepsilon_3 = \ln(3.0/2.5) = 0.182$$

Thus, the total true strain is

$$\varepsilon_{\text{total}} = 0.288 + 0.223 + 0.182 = 0.693.$$

The total true strain may also be calculated from the initial and final dimensions as

$$\varepsilon_{\text{total}} = \ln(3.0/1.5) = 0.693,$$

thus giving the same answer.

Engineering strains are calculated likewise. Thus,

$$\text{Step 1: } e_1 = (2.0 - 1.5)/1.5 = 0.333$$
$$\text{Step 2: } e_2 = (2.5 - 2.0)/2.0 = 0.250$$
$$\text{Step 3: } e_3 = (3.0 - 2.5)/2.5 = 0.200$$

The total engineering strain is $0.333 + 0.250 + 0.200 = 0.783$. However, when calculated from the initial and final dimensions, the total engineering strain is

$$e_{\text{total}} = (3.0 - 1.5)/1.5 = 1,$$

which is higher than the sum of the individual strains. Thus, engineering strains are not additive, whereas true strains are.

---

### 2.2.3   True stress–true strain curves

The relation between engineering and true values for stress and strain can now be used to construct true stress–true strain curves from a curve such as that in Fig. 2.2. The procedure for this construction is given in Example 2.2.

A typical true stress–true strain curve is shown in Fig. 2.5a. For convenience, such a curve can be approximated by the equation

$$\sigma = K\varepsilon^n. \tag{2.11}$$

Note that Eq. (2.11) indicates neither the elastic region nor the yield point $Y$ of the material. These quantities, however, are readily available from the engineering stress-strain curve. (Since the strains at the yield point are very small, the difference between the true and engineering yield stress is negligible for metals. The reason is that the difference in the cross-sectional areas $A_o$ and $A$ at yielding is very small.)

We can rewrite Eq. (2.11) as

$$\log \sigma = \log K + n \log \varepsilon.$$

Thus, if we now plot the true stress-true strain curve on a log-log scale, we obtain Fig. 2.5b. The slope $n$ is known as the **strain-hardening exponent,** and $K$ is known as the **strength coefficient.** Note that $K$ is the true stress at a true strain of unity.

Values of $K$ and $n$ for a variety of engineering materials are given in Table 2.3. The true stress-true strain curves for some materials are given in Fig. 2.4. Some differences between Table 2.3 and these curves may exist because of different sources of data and test conditions.

In Fig. 2.5a, $Y_f$ is the **flow stress,** defined as the true stress required to continue plastic deformation at a particular true strain, $\varepsilon_1$. Thus, for strain-hardening materials, the flow stress increases with increasing strain. Note from Fig. 2.5c that the elastic strains are much smaller than plastic strains. Consequently, and although both effects exist, we will ignore elastic

**TABLE 2.3**

**TYPICAL VALUES FOR $K$ AND $n$ IN EQ. (2.11) AT ROOM TEMPERATURE**

| MATERIAL | $K$ (MPa) | $n$ |
|---|---|---|
| Aluminum, 1100-O | 180 | 0.20 |
| 2024-T4 | 690 | 0.16 |
| 5052-O | 210 | 0.13 |
| 6061-O | 205 | 0.20 |
| 6061-T6 | 410 | 0.05 |
| 7075-O | 400 | 0.17 |
| Brass, 60–39–1 Pb, annealed | 800 | 0.33 |
| 70–30, annealed | 895 | 0.49 |
| 85–15, cold-rolled | 580 | 0.34 |
| Bronze (phosphor), annealed | 720 | 0.46 |
| Cobalt-base alloy, heat-treated | 2070 | 0.50 |
| Copper, annealed | 315 | 0.54 |
| Molybdenum, annealed | 725 | 0.13 |
| Steel, low-carbon annealed | 530 | 0.26 |
| 1045 hot-rolled | 965 | 0.14 |
| 1112 annealed | 760 | 0.19 |
| 1112 cold-rolled | 760 | 0.08 |
| 4135 annealed | 1015 | 0.17 |
| 4135 cold-rolled | 1100 | 0.14 |
| 4340 annealed | 640 | 0.15 |
| 17–4 P-H annealed | 1200 | 0.05 |
| 52100 annealed | 1450 | 0.07 |
| 302 stainless, annealed | 1300 | 0.30 |
| 304 stainless, annealed | 1275 | 0.45 |
| 410 stainless, annealed | 960 | 0.10 |

Note: 100 MPa = 14,500 psi.

**FIGURE 2.5**   (a) True stress–true strain curve in tension. Note that, unlike in an engineering stress–strain curve, the slope is always positive and that the slope decreases with increasing strain. Although in the elastic range stress and strain are proportional, the total curve can be approximated by the power expression shown. On this curve, $Y$ is the yield stress and $Y_f$ is the flow stress. (b) True stress–true strain curve plotted on a log-log scale. (c) True stress–true strain curve in tension for 1100-O aluminum plotted on a log–log scale. Note the large difference in the slopes in the *elastic* and *plastic* ranges. *Source:* After R. M. Caddell and R. Sowerby.

strains in our calculations for forming processes throughout the rest of this text. The plastic strain will be the total strain.

The area under the true stress–true strain curve is known as **toughness** and can be expressed as

$$\text{Toughness} = \int_{o}^{\varepsilon_f} \sigma d\varepsilon, \tag{2.12}$$

where $\varepsilon_f$ is the true strain at fracture. Toughness is defined as the energy per unit volume (specific energy) that has been dissipated up to the point of fracture. It is important to remember that this specific energy pertains only to that volume of material at the narrowest region of the neck. Any volume of material away from the neck has undergone less strain and hence has dissipated less energy.

As used here, *toughness* is different from the concept of fracture toughness as treated in textbooks on fracture mechanics. Fracture mechanics (the study of the initiation and propa-

**FIGURE 2.6** True stress–true strain curves in tension at room temperature for various metals. The point of intersection of each curve at the ordinate is the yield stress $Y$; thus, the elastic portions of the curves are not indicated. When the $K$ and $n$ values are determined from these curves, they may not agree with those given in Table 2.3 because of the different sources from which they were collected.

gation of cracks in a solid medium) is beyond the scope of this text. It is of limited relevance to metalworking processes, except in die design and die life.

---

**Example 2.2: Construction of true stress–true strain curve**

The following data are taken from a stainless steel tension test specimen:

$A_o = 0.056 \text{ in}^2$

$A_f = 0.016 \text{ in}^2$

$\ell_o = 2 \text{ in.}$

Elongation in 2 in. = 49%.

| Load, $P$ (lb.) | Extension, $\Delta\ell$ (in.) |
| --- | --- |
| 1600 | 0 |
| 2500 | 0.02 |
| 3000 | 0.08 |
| 3600 | 0.20 |
| 4200 | 0.40 |
| 4500 | 0.60 |
| 4600 (max.) | 0.86 |
| 3300 (fracture) | 0.98 |

Draw the true stress–true strain curve for the material.

SOLUTION.    In order to determine the true stress and true strain, we use the following relationships:

$$\text{True stress } \sigma = \frac{P}{A}$$

$$\text{True strain } \varepsilon = \ln\left(\frac{\ell}{\ell_o}\right) \text{ up to necking}$$

$$\varepsilon_f = \ln\left(\frac{A_o}{A_f}\right) \text{ at fracture}$$

$$\ell = \ell_o + \Delta\ell \text{ up to necking.}$$

Assuming that the volume of the specimen remains constant, we have

$$A_o\ell_o = A\ell.$$

Thus,

$$A = \frac{A_o\ell_o}{\ell} = \frac{(0.056)(2)}{\ell} = \frac{0.112}{\ell} \text{ in}^2.$$

Using these relationships, we obtain the following data:

| $\Delta\ell$ (in.) | $\ell$ (in.) | $\varepsilon$ | $A$ (in²) | True stress (psi) |
|---|---|---|---|---|
| 0 | 2.00 | 0 | 0.056 | 28,600 |
| 0.02 | 2.02 | 0.01 | 0.0555 | 45,000 |
| 0.08 | 2.08 | 0.039 | 0.054 | 55,800 |
| 0.20 | 2.20 | 0.095 | 0.051 | 70,800 |
| 0.40 | 2.40 | 0.182 | 0.047 | 90,000 |
| 0.60 | 2.60 | 0.262 | 0.043 | 104,500 |
| 0.86 | 2.86 | 0.357 | 0.039 | 117,300 |
| 0.98 | 2.98 | 1.253 | 0.016 | 206,000 |

The true stress and true strain are plotted (solid line) in Fig. 2.7. The point at necking is connected to the point at fracture by a straight line because there are no data on the instantaneous areas after necking begins. The correction on this curve (broken line) is explained in Section 2.11.2.

## 2.2.4    Instability in simple tension

As we noted previously, the onset of necking in a tension test corresponds to the ultimate tensile strength (UTS) of the material. The slope of the load-elongation curve at this point is zero (or $dP = 0$), and it is here that *instability* begins. That is, the specimen begins to neck and cannot support the load because the neck is becoming smaller in cross-sectional area.

Using the relationships

$$\varepsilon = \ln\left(\frac{A_o}{A}\right), \quad A = A_o e^{-\varepsilon}, \text{ and } P = \sigma A = \sigma A_o e^{-\varepsilon},$$

(a)

(b)

FIGURE 2.7   (a) True stress–true strain curve developed from the data given. Note that this curve has a positive slope, indicating that the material is becoming stronger as it is strained. (b) True stress–true strain curve plotted on log–log paper and based on the corrected curve. The steeper the slope $n$, the higher will be the capacity of the material to become stronger and harder as it is worked.

we determine $\dfrac{dP}{d\varepsilon}$. Thus,

$$\frac{dP}{d\varepsilon} = A_o\left(\frac{d\sigma}{d\varepsilon}e^{-\varepsilon} - \sigma e^{-\varepsilon}\right).$$

Because $dP = 0$ at the UTS where necking begins, we set this expression equal to zero and obtain

$$\frac{d\sigma}{d\varepsilon} = \sigma.$$

However,

$$\sigma = K\varepsilon^n,$$

and consequently,

$$nK\varepsilon^{n-1} = K\varepsilon^n$$

$$\varepsilon = n. \qquad\qquad (2.13)$$

Thus, the true strain at the onset of necking (termination of uniform elongation) is numerically equal to the strain-hardening exponent, $n$. Hence, the higher the value of $n$, the greater will be the strain to which a piece of material can be stretched before necking begins. Table 2.3 shows that metals such as annealed copper, brass, and stainless steel can be stretched uniformly to a greater extent than the other materials listed. These observations are covered in greater detail in Chapters 7 and 10 because of their relevance to sheet-forming processes. (For biaxial stretching instability, see Section 2.11.8.)

Instability in a tension test can be viewed as a phenomenon in which two competing processes are taking place simultaneously. As the load on the specimen is increased, its cross-sectional area becomes smaller, which is more pronounced in the region where necking begins. With increasing strain, however, the material is becoming stronger due to strain hardening. Because the load on the specimen is the product of area and strength, instability sets in when the rate of decrease in area is greater than the rate of increase in strength. This condition is also known as **geometric softening**.

---

**Example 2.3:**   Calculation of ultimate tensile strength

A material has a true stress–true strain curve given by

$$\sigma = 100{,}000\varepsilon^{0.5} \text{ psi.}$$

Calculate the true ultimate tensile strength and the engineering UTS of this material.

SOLUTION.   Since the necking strain corresponds to the maximum load and the necking strain for this material is

$$\varepsilon = n = 0.5,$$

we have as the true ultimate tensile strength

$$\sigma = Kn^n$$

$$= 100{,}000(0.5)^{0.5} = 70{,}710 \text{ psi.}$$

The true area at the onset of necking is obtained from

$$\ln\left(\frac{A_o}{A_{\text{neck}}}\right) = n = 0.5.$$

Thus,

$$A_{\text{neck}} = A_o e^{-0.5}$$

and the maximum load $P$ is

$$P = \sigma A = \sigma A_o e^{-0.5},$$

where $\sigma$ is the true ultimate tensile strength. Hence

$$P = (70{,}710)(0.606)(A_o) = 42{,}850 A_o \text{ lb.}$$

Since UTS $= P/A_o$,

$$\text{UTS} = 42{,}850 \text{ psi.}$$

---

### 2.2.5   Types of stress–strain curves

Every material has a differently shaped stress–strain curve. Its shape depends on its composition and many other factors to be treated in detail later. Some of the major types of curves are shown in Fig. 2.8 and have the following characteristics:

**FIGURE 2.8**  Schematic illustration of various types of idealized stress–strain curves. (a) Perfectly elastic. (b) Rigid, perfectly plastic. (c) Elastic, perfectly plastic. (d) Rigid, linearly strain hardening. (e) Elastic, linearly strain hardening. The broken lines and arrows indicate unloading and reloading during the test. Most engineering metals exhibit a behavior similar to curve (e).

1.  A perfectly elastic material behaves like a spring with stiffness $E$. The behavior of brittle materials, such as glass, ceramics, and some cast irons, may be represented by such a curve (Fig. 2.8a). There is a limit to the stress the material can sustain, after which it breaks. Permanent deformation, if any, is negligible.

2.  A rigid, perfectly plastic material has, by definition, an infinite value of $E$. Once the stress reaches the yield stress $Y$, it continues to undergo deformation at the same stress level. When the load is released, the material has undergone permanent deformation, with no elastic recovery (Fig. 2.8b).

3.  An elastic, perfectly plastic material is a combination of the first two: It has a finite elastic modulus and its undergoes elastic recovery when the load is released (Fig. 2.8c).

4.  A rigid, linearly strain-hardening material requires an increasing stress level to undergo further strain. Thus, its *flow stress* (magnitude of the stress required to maintain plastic deformation at a given strain) increases with increasing strain. It has no elastic recovery upon unloading (Fig. 2.8d). See Fig. 2.5a for flow stress.

5.  An elastic, linearly strain-hardening curve (Fig. 2.8e) is an approximation of the behavior of most engineering materials, with the modification that the plastic portion of the curve has a decreasing slope with increasing strain (Fig. 2.5a).

Note that some of these curves can be expressed by Eq. (2.11) by changing the value of $n$ (Fig. 2.9) or by other equations of a similar nature.

### 2.2.6  Effects of temperature

In this and subsequent sections, the various factors that have an influence on the shape of stress-strain curves will be discussed. The first factor is temperature. Although somewhat difficult to generalize, increasing temperature usually increases ductility and toughness and lowers the modulus of elasticity, yield stress, and ultimate tensile strength. These effects are shown in Fig. 2.10.

Temperature also affects the strain-hardening exponent $n$ of most metals, in that $n$ decreases with increasing temperature. Depending on the type of material and its composition and level of impurities, elevated temperatures can have other significant effects, as detailed in Chapter 3. However, the influence of temperature is best discussed in conjunction with strain rate for the reasons explained in the next section.

**FIGURE 2.9**  The effect of strain-hardening exponent *n* on the shape of true stress–true strain curves. When n = 1, the material is elastic, and when n = 0, it is rigid and perfectly plastic.

**FIGURE 2.10**  Typical effects of temperature on engineering stress–strain curves. Temperature affects the modulus of elasticity, yield stress, ultimate tensile strength, and toughness of materials.

### 2.2.7  Effects of strain rate

Depending on the particular manufacturing operation and equipment, a piece of material may be formed at low or at high speeds. In performing a tension test, the specimen can be strained at different rates to simulate the actual deformation process.

Whereas the **deformation rate** may be defined as the speed (in meters per second or feet per minute, for instance) at which a tension test is being carried out (e.g., the rate at which a rubber band is stretched), the **strain rate** is a function of the geometry of the specimen.

The **engineering strain rate** $\dot{e}$ is defined as

$$\dot{e} = \frac{de}{dt} = \frac{d\left(\dfrac{\ell - \ell_o}{\ell_o}\right)}{dt} = \frac{1}{\ell_o} \cdot \frac{d\ell}{dt} = \frac{v}{\ell_o}, \tag{2.14}$$

and the **true strain rate** $\dot{\varepsilon}$ as

$$\dot{\varepsilon} = \frac{d\varepsilon}{dt} = \frac{d\left[\ln\left(\dfrac{\ell}{\ell_o}\right)\right]}{dt} = \frac{1}{\ell} \cdot \frac{d\ell}{dt} = \frac{v}{\ell}, \tag{2.15}$$

where $v$ is the speed of deformation, e.g., the speed of the jaws of the testing machine in which the specimen is clamped.

Although the deformation rate $v$ and engineering strain rate $\dot{e}$ are proportional, the true strain rate $\dot{\varepsilon}$ is not proportional to $v$. Thus, in a tension test with $v$ constant, the true strain rate decreases as the specimen becomes longer. In order to maintain a constant $\dot{\varepsilon}$, the speed must be increased accordingly. However, for small changes in length of the specimen during a test, this difference is not significant.

Typical deformation speeds employed in various metalworking processes, and the strain rates involved, are shown in Table 2.4. Note that there are considerable differences in the

**TABLE 2.4**

**TYPICAL RANGES OF STRAIN, DEFORMATION SPEED, AND STRAIN RATES IN METALWORKING PROCESSES**

| PROCESS | TRUE STRAIN | DEFORMATION SPEED (m/s) | STRAIN RATE (s$^{-1}$) |
|---|---|---|---|
| Cold working | | | |
|   Forging, rolling | 0.1–0.5 | 0.1–100 | 1–10$^3$ |
|   Wire and tube drawing | 0.05–0.5 | 0.1–100 | 1–10$^4$ |
| Explosive forming | 0.05–0.2 | 10–100 | 10–10$^5$ |
| Hot working and warm working | | | |
|   Forging, rolling | 0.1–0.5 | 0.1–30 | 1–10$^3$ |
|   Extrusion | 2–5 | 0.1–1 | 10$^{-1}$–10$^2$ |
| Machining | 1–10 | 0.1–100 | 10$^3$–10$^6$ |
| Sheet-metal forming | 0.1–0.5 | 0.05–2 | 1–10$^2$ |
| Superplastic forming | 0.2–3 | 10$^{-4}$–10$^{-2}$ | 10$^{-4}$–10$^{-2}$ |

magnitudes. Because of this wide range, strain rates are generally quoted in orders of magnitudes, such as $10^2$ per second, $10^4$ per second, etc.

The typical effects of temperature and strain rate on the strength of metals are shown in Fig. 2.11. It clearly indicates that increasing strain rate increases strength and that the sensitivity of strength to the strain rate increases temperature, as shown in Fig. 2.12. Note, however, that this effect is relatively small at room temperature.

Figure 2.11 shows that the same strength can be obtained either at low temperature, low strain rate or at high temperature, high strain rate. These relationships are important in estimating the resistance of materials to deformation when processing them at various strain rates and temperatures.

**FIGURE 2.11**  The effect of strain rate on the ultimate tensile strength of aluminum. Note that as temperature increases, the slope increases. Thus, tensile strength becomes more and more sensitive to strain rate as temperature increases. *Source:* After J. H. Hollomon.

**FIGURE 2.12** Dependence of the strain–rate sensitivity exponent $m$ on the homologous temperature $T/T_m$ for various materials. $T$ is the testing temperature and $T_m$ is the melting point of the metal, both on the absolute scale. The transition in the slopes of the curve occurs at about the recrystallization temperature of the metals. *Source:* After F. W. Boulger.

The effect of strain rate on strength also depends on the particular level of strain: It increases with strain. The strain rate also affects the strain-hardening exponent $n$, because it decreases with increasing strain rate. The effect of strain rate on the strength of materials is generally expressed by

$$\sigma = C\dot{\varepsilon}^m, \tag{2.16}$$

where $C$ is the **strength coefficient**, similar to $K$ in Eq. (2.11), and $m$ is the **strain-rate sensitivity exponent** of the material.

A general range of values for $m$ is: up to 0.05 for cold working, 0.05 to 0.4 for hot working, and 0.3 to 0.85 for superplastic materials. For a *Newtonian* fluid, where the shear stress increases linearly with the rate of shear, the value of $m$ is 1. Some specific values for $C$ and $m$ are given in Table 2.5.

The term **superplastic** refers to the capability of some materials to undergo large uniform elongation prior to failure. The elongation may be on the order of a few hundred percent to as much as 2000 percent. An extreme example is hot glass. Among other materials exhibiting superplastic behavior are polymers at elevated temperatures, very fine-grain alloys of zinc-aluminum, and titanium alloys.

The magnitude of $m$ has a significant effect on necking in a tension test. Experimental observations show that, with high $m$ values, the material stretches to a greater length before it fails. This is an indication that necking is delayed with increasing $m$. When necking is about to begin, the region's strength with respect to the rest of the specimen increases due to strain hardening. However, the strain rate in the neck region is also higher than in the rest of the specimen, because the material is elongating faster there. Since the material in the necked region is becoming stronger as it is strained at a higher rate, this region exhibits

**TABLE 2.5**

**APPROXIMATE RANGE OF VALUES FOR *C* AND *m* IN EQ. (2.16) FOR VARIOUS ANNEALED METALS AT TRUE STRAINS RANGING FROM 0.2 TO 1.0**

| MATERIAL | TEMPERATURE, °C | C psi $\times$ 10³ | C MPa | m |
|---|---|---|---|---|
| Aluminum | 200–500 | 12–2 | 82–14 | 0.07–0.23 |
| Aluminum alloys | 200–500 | 45–5 | 310–35 | 0–0.20 |
| Copper | 300–900 | 35–3 | 240–20 | 0.06–0.17 |
| Copper alloys (brasses) | 200–800 | 60–2 | 415–14 | 0.02–0.3 |
| Lead | 100–300 | 1.6–0.3 | 11–2 | 0.1–0.2 |
| Magnesium | 200–400 | 20–2 | 140–14 | 0.07–0.43 |
| Steel | | | | |
|   Low-carbon | 900–1200 | 24–7 | 165–48 | 0.08–0.22 |
|   Medium-carbon | 900–1200 | 23–7 | 160–48 | 0.07–0.24 |
|   Stainless | 600–1200 | 60–5 | 415–35 | 0.02–0.4 |
| Titanium | 200–1000 | 135–2 | 930–14 | 0.04–0.3 |
| Titanium alloys | 200–1000 | 130–5 | 900–35 | 0.02–0.3 |
| Ti–6Al–4V* | 815–930 | 9.5–1.6 | 65–11 | 0.50–0.80 |
| Zirconium | 200–1000 | 120–4 | 830–27 | 0.04–0.4 |

*At a strain rate of $2 \times 10^{-4}$/s.
Note: As temperature increases, *C* decreases and *m* increases. As strain increases, *C* increases and *m* may increase or decrease, or it may become negative within certain ranges of temperature and strain.
Source: After T. Altan and F. W. Boulger, *Jr. Eng. Ind.* 95:1009–1019, 1973.

a greater resistance to necking. This increased resistance to necking thus depends on the magnitude of *m*.

As the test progresses, necking becomes more diffuse and the specimen becomes longer before fracture. Thus, total elongation increases with increasing *m* value (Fig. 2.13). As expected, the elongation after necking (postuniform elongation) also increases with increasing *m*. Moreover, the value of *m* decreases with metals of increasing strength.

The effect of strain rate on ductility is difficult to generalize. Because the formability of materials depends largely on their ductility, it is important to recognize the effect of temperature and strain rate on ductility. Generally, higher strain rates have an adverse effect on the ductility of materials. (The increase in ductility due to the strain-rate sensitivity of materials has been exploited in superplastic forming of metals, as described in Section 7.10.)

### 2.2.8 Effects of hydrostatic pressure

Note that the tests described thus far have been carried out at ambient pressure. Tests have been performed under hydrostatic conditions, with pressures ranging up to $10^3$ MPa ($10^5$ psi). Three important observations have been made concerning the effects of high hydrostatic pressure on the behavior of materials: (1) it substantially increases the strain at fracture (Fig. 2.14); (2) it has little or no effect on the shape of the true stress-true strain curve, but only extends it; and (3) it has no effect on the strain or the maximum load at which necking begins.

**FIGURE 2.13**   (a) The effect of strain–rate sensitivity exponent $m$ on the total elongation for various metals. Note that elongation at high values of $m$ approaches 1000%. *Source:* After D. Lee and W. A. Backofen. (b) The effect of strain–rate sensitivity exponent $m$ on the post uniform (after necking) elongation for various metals. *Source:* After A. K. Ghosh.

The increase in ductility due to hydrostatic pressure has also been observed in other tests, such as compression and torsion tests. This increase in ductility has been observed not only with ductile metals, but also with brittle metals and nonmetallic materials. Materials such as cast iron, marble, and various rocks acquire some ductility (or increase in ductility) and thus deform plastically when subjected to hydrostatic pressure. The level of pressure required to enhance ductility depends on the material. Experiments have shown that, generally, the mechanical properties of metals are not altered after being subjected to hydrostatic pressure.

### 2.2.9   Effects of radiation

In view of the nuclear applications of many metals and alloys, studies have been conducted on the effects of radiation on material properties. Typical changes in the mechanical properties of steels and other metals exposed to high-energy radiation are increased yield stress, increased tensile strength and hardness, and decreased ductility and toughness. The magnitudes of these changes depend on the material and its condition, temperature, and level of radiation.

**FIGURE 2.14** The effect of hydrostatic pressure on true strain at fracture in tension for various metals. Even cast iron becomes ductile under high pressure. *Source:* After H. Ll. D. Pugh and D. Green, *Proc. Inst. Mech. Eng.,* vol. 179, Part 1, No. 12, 1964–1965.

## 2.3 Compression

Many operations in metalworking, such as forging, rolling, and extrusion, are performed with the workpieces under compressive loads. The compression test, in which the specimen is subjected to a compressive load as shown in Fig. 2.1b, can give useful information for these processes, such as stresses required and the behavior of the material under compression. However, the deformation shown in Fig. 2.1b is ideal. The compression test is usually carried out by compressing a solid cylindrical specimen between two flat platens; the friction between the specimen and the dies is an important factor. Friction causes **barreling** (Fig. 2.15). In other words, friction prevents the top and bottom surfaces from expanding freely.

This situation makes it difficult to obtain relevant data and to construct properly the compressive stress–strain curve because (1) the cross-sectional area of the specimen changes along its height; and (2) friction dissipates energy, and this energy is supplied through an increased compressive force. Thus, it is difficult to obtain results that are truly indicative of the properties of the material. With effective lubrication or other means it is possible, of course, to minimize friction—and hence barreling—to obtain a reasonably constant cross-sectional area during this test.

The engineering strain rate $\dot{e}$ in compression is given by

$$\dot{e} = -\frac{v}{h_o},$$

(2.17)

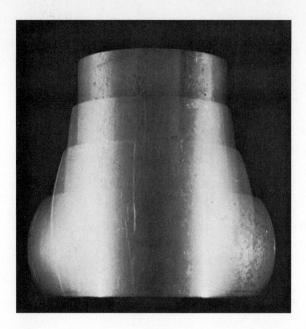

**FIGURE 2.15**  Barreling in compressing a round solid cylindrical specimen (7075-O aluminum) between flat dies. Barreling is caused by friction at the die-specimen interfaces, which retards the free flow of the material. See also Figs. 6.1 and 6.2. *Source:* K. M. Kulkarni and S. Kalpakjian.

where $v$ is the speed of the die and $h_o$ is the original height of the specimen. The true strain rate $\dot{\varepsilon}$ is given by

$$\dot{\varepsilon} = -\frac{v}{h},\tag{2.18}$$

where $h$ is the instantaneous height of the specimen. If $v$ is constant, the true strain rate increases as the test progresses. In order to conduct this test at a constant true strain rate, a *cam plastometer* has been designed that, through a cam action, reduces $v$ proportionately as $h$ decreases during the test.

The compression test can also be used to determine the ductility of a metal by observing the cracks that form on the barreled cylindrical surfaces of the specimen. Incidentally, hydrostatic pressure has a beneficial effect in delaying the formation of these cracks. With a sufficiently ductile material and effective lubrication, compression tests can be carried out uniformly to large strains. This result is unlike that of the tension test, where, even for very ductile materials, necking sets in after relatively little elongation.

### 2.3.1  Plane-strain compression test

Another test is the **plane-strain compression test** (Fig. 2.16), which simulates processes such as rolling. In this test, the die and workpiece geometries are such that the width of the specimen does not undergo any significant change during compression; that is, the material under

**FIGURE 2.16** Schematic illustration of the plane–strain compression test. The dimensional relationships shown should be satisfied for this test to be useful and reproducible. This test gives the yield stress of the material in plane strain, $Y'$. *Source:* After A. Nadai and H. Ford.

**FIGURE 2.17** True stress–true strain curve in tension and compression for aluminum. For ductile metals, the curves for tension and compression are identical. *Source:* After A. H. Cottrell, *The Mechanical Properties of Matter,* New York: Wiley, 1964, p. 289.

the dies is in the condition of *plane strain* (see Section 2.11.3). The yield stress of a material in plane strain $Y'$ is

$$Y' = \frac{2}{\sqrt{3}} Y = 1.15 \, Y, \tag{2.19}$$

according to the distortion-energy yield criterion (see Section 2.11.2).

As the geometric relationships in Fig. 2.16 show, the test parameters must be chosen properly to make the results meaningful. Furthermore, caution should be exercised in test procedures, such as preparing the die surfaces and aligning the dies, lubricating the surfaces, and accurately measuring the load.

When the results of tension and compression tests on the same material are compared, for *ductile* metals the true stress-true strain curves for both tests coincide (Fig. 2.17). However, this is not true for brittle materials, particularly in regard to ductility.

### 2.3.2 Bauschinger effect

Sometimes in deformation processing of materials a workpiece is first subjected to tension and then to compression, or vice versa. Examples are bending and unbending, roller leveling (see Section 6.3.3), and reverse drawing (see Section 7.12.2). When a metal with a tensile yield stress $Y$ is subjected to tension into the plastic range and then the load is released and applied in compression, the yield stress in compression is lower than that in tension (Fig. 2.18).

This phenomenon is known as the **Bauschinger effect** and is exhibited in varying degrees by all metals and alloys. This effect is also observed when the loading path is reversed,

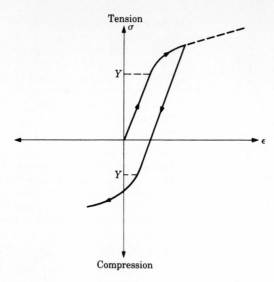

**FIGURE 2.18** Schematic illustration of the Bauschinger effect. Arrows show loading and unloading paths. Note the decrease in the yield stress in compression after the specimen has been subjected to tension. The same result is obtained if compression is applied first, followed by tension, whereby the yield stress in tension decreases.

i.e., compression followed by tension. Because of the lowered yield stress in the reverse direction of load application, this phenomenon is also called **strain softening** or **work softening**.

### 2.3.3   The disk test

For brittle materials such as ceramics and glasses, a **disk test** has been developed in which the disk is subjected to compression between two hardened flat platens (Fig. 2.19). When loaded as shown, tensile stresses develop perpendicular to the vertical centerline along the disk, fracture begins, and the disk splits vertically in half.

The tensile stress $\sigma$ in the disk is uniform along the centerline and can be calculated from the formula

$$\sigma = \frac{2P}{\pi dt},  \tag{2.20}$$

where $P$ is the load at fracture, $d$ is the diameter of the disk, and $t$ is its thickness. In order to avoid premature failure at the contact points, thin strips of soft metal are placed between the disk and the platens. These also protect the platens from being damaged during the test.

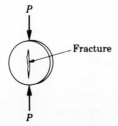

**FIGURE 2.19**  Disk test on a brittle material, showing the direction of loading and the fracture path. This test is useful for brittle materials, such as ceramics and carbides.

## 2.4  Torsion

Another method of determining material properties is the torsion test. In order to obtain an approximately uniform stress and strain distribution along the cross-section, this test is generally carried out on a tubular specimen (Fig. 2.20). The *shear stress* $\tau$ can be determined from the equation.

$$\tau = \frac{T}{2\pi r^2 t}, \tag{2.21}$$

where $T$ is the torque, $r$ is the mean radius, and $t$ is the thickness of the reduced section of the tube. The *shear strain* $\gamma$ is determined from the equation

$$\gamma = \frac{r\phi}{\ell}, \tag{2.22}$$

where $\ell$ in the length of the reduced section and $\phi$ is the angle of twist, in radians. With the shear stress and shear strain thus obtained from torsion tests, we can construct the shear stress-shear strain curve of the material (see Section 2.11.7).

In the elastic range, the ratio of the shear stress to shear strain is known as the **shear modulus** or the **modulus of rigidity**, $G$:

$$G = \frac{\tau}{\gamma}. \tag{2.23}$$

The shear modulus and the modulus of elasticity $E$ are related by the formula

$$G = \frac{E}{2(1 + v)}, \tag{2.24}$$

which is based on a comparison of **simple shear** and **pure shear** strains (Fig. 2.21). Note from this figure that the difference between the two strains is that simple shear is equivalent to pure shear plus a rotation of $\theta/2$ degrees. (See also Section 2.11.7 for the state of stress shown in the figure for pure shear.)

**FIGURE 2.20**  A typical torsion-test specimen. It is mounted between the two heads of a machine and is twisted. Note the shear deformation of an element in the reduced section.

**FIGURE 2.21**   Comparison of (a) simple shear and (b) pure shear.

In Example 2.4, we show that a thin-walled tube does not neck in torsion, unlike tension-test specimens as described in Section 2.2.4. Thus, we do not have to be concerned with changes in the cross-sectional area of the specimen in torsion testing. The shear stress–shear strain curves obtained from torsion tests increase monotonically just as they do in true stress–true strain curves.

Torsion tests are performed on solid round bars at elevated temperatures in order to estimate the forgeability of metals (see Section 6.2). The greater the number of twists prior to failure, the greater the forgeability of the material. Torsion tests have also been conducted on round bars that are compressed axially (Fig. 2.22). The maximum shear strain at fracture is measured as a function of the compressive stress. The shear strain at fracture increases substantially as the compressive stress increases. This observation again indicates the beneficial

**FIGURE 2.22**   The effect of axial compressive stress on the shear strain at fracture in torsion for various steels. Note that the effect on ductility is similar to that of hydrostatic pressure, shown in Fig. 2.14. *Source:* Based on data in P. W. Bridgman, *Large Plastic Flow and Fracture,* New York: McGraw-Hill, 1952.

effect of a compressive environment on the ductility of materials. Other experiments show that, with normal tensile stresses, the curves in Fig. 2.22 follow a trend downward and to the left, signifying reduced ductility.

The effect of compressive stresses in increasing the maximum shear strain at fracture has also been observed in metal cutting (see Section 8.2). The normal compressive stress has no effect on the magnitude of shear stresses required to cause yielding or to continue the deformation, just as hydrostatic pressure has no effect on the general shape of the stress-strain curve.

---

**Example 2.4:   Instability in torsion of a thin-walled tube**

Show that necking cannot take place in the torsion of a thin-walled tube made of a material whose true stress–true strain curve is given by $\sigma = K\varepsilon^n$.

SOLUTION.    According to Eq. (2.21), the expression for torque $T$ is

$$T = 2\pi r^2 \, t\tau$$

where for this cause the *shear stress* $\tau$ can be related to the *normal stress* $\sigma$ using Eq. (2.56) for the distortion-energy criterion as $\tau = \sigma/\sqrt{3}$ (see Section 2.11.7 for details). The criterion for instability in torsion would be

$$dT/d\varepsilon = 0.$$

Because $r$ and $t$ are constant, we have

$$dT/d\varepsilon = (2/\sqrt{3})\pi \, r^2 \, \tau \, d\sigma/d\varepsilon.$$

For a material represented by $\sigma = K\varepsilon^n$, we have

$$d\sigma/d\varepsilon = nK\varepsilon^{n-1}.$$

Therefore,

$$dT/d\varepsilon = (2/\sqrt{3}) \, \pi r^2 \, \tau n K\varepsilon^{n-1}.$$

Since none of the quantities in this expression are zero, $dT/d\varepsilon$ cannot be zero. Therefore, a tube in torsion does not undergo instability.

---

## 2.5   Bending (Flexure)

Preparing specimens from brittle materials, such as ceramics and carbides, is difficult because of the problems involved in shaping and machining them to proper dimensions. Furthermore, because of their sensitivity to surface defects and notches, clamping brittle test specimens for testing is difficult. Improper alignment of the test specimen may result in nonuniform stress distribution along the cross-section of the specimen.

A commonly used test method for brittle materials is the **bend** or **flexure test**, usually involving a specimen with a rectangular cross-section and supported at its ends (Fig. 2.23). The load is applied vertically—either at one or two points; hence these tests are referred to as three-point or four-point bending, respectively. The stresses in these specimens are tensile at

**FIGURE 2.23**  Two bend-test methods for brittle materials: (a) three-point bending; (b) four-point bending. The shaded areas on the beams represent the bending-moment diagrams, described in texts on the mechanics of solids. Note the region of constant maximum bending moment in (b), whereas the maximum bending moment occurs only at the center of the specimen in (a).

their lower surfaces and compressive at their upper surfaces. We calculate these stresses using simple beam equations described in texts on the mechanics of solids. The stress at fracture in bending is known as the **modulus of rupture**, or **transverse rupture strength**, and is obtained from the formula

$$\sigma = \frac{Mc}{I},\qquad(2.25)$$

where $M$ is the bending moment, $c$ is one-half of the specimen depth, and $I$ is the moment of inertia of the cross-section.

Note that there is a basic difference between the two loading conditions in Fig. 2.23. In 3-point bending, the maximum stress is at the center of the beam, whereas in the 4-point test, the maximum stress is constant between the two loading points. The stress is the same in both cases when all other parameters are maintained. However, there is a greater probability for defects and imperfections to be present in the volume of material between the loading points in the 4-point test than in the much smaller volume under the single load in the 3-point test. This means that the 4-point test is likely to result in a lower modulus of rupture than the 3-point test, which has been verified by experiments. Furthermore, the results of the 4-point test show less scatter than the 3-point test. The reason is that the strength of a small volume of brittle material (such as under the single load in the 3-point test) is subject to greater variations than is the strength of the weakest point in a larger volume (such as in the 4-point test).

## 2.6  Hardness

One of the most common tests for assessing the mechanical properties of materials is the hardness test. **Hardness** of a material is generally defined as its resistance to permanent indentation. Less commonly, hardness may also be defined as resistance to scratching or to wear. Various techniques have been developed to measure the hardness of materials using

| Test | Indenter | Shape of indentation Side view / Top view | | Load | Hardness number |
|------|----------|---------|---|------|------|
| Brinell | 10-mm steel or tungsten carbide ball | | | 500 kg 1500 kg 3000 kg | |
| Vickers | Diamond pyramid | 136° | | 1–120 kg | |
| Knoop | Diamond pyramid | $L/b = 7.11$ $b/t = 4.00$ | | 25 g–5 kg | |
| Rockwell | | | | kg | |
| A C D | Diamond cone | 120° | $t = $ mm | 60 150 100 | HRA HRC HRD $= 100 - 500t$ |
| B F G | $\frac{1}{16}$-in. diameter steel ball | | $t = $ mm | 100 60 150 | HRB HRF HRG $= 130 - 500t$ |
| E | $\frac{1}{8}$-in. diameter steel ball | | | 100 | HRE |

**FIGURE 2.24** General characteristics of hardness testing methods. The Knoop test is known as a microhardness test because of the light load and small impressions. *Source:* After H. W. Hayden, W. G. Moffatt, and V. Wulff.

different indenter materials and geometries. Because resistance to indentation depends on the shape of the indenter and the load applied, hardness is not a fundamental property. Among the most common standardized hardness tests are the Brinell, Rockwell, Vickers, Knoop (Fig. 2.24), and Scleroscope tests.

### 2.6.1 Brinell test

In the Brinell test, a steel or tungsten carbide ball 10 mm in diameter is pressed against a surface with a load of 500, 1500, or 3000 kg. The Brinell hardness number (HB) is defined as the ratio of the load $P$ to the curved area of indentation, or

$$ \text{HB} = \frac{2P}{(\pi D)(D - \sqrt{D^2 - d^2})} \, \text{kg/mm}^2, \tag{2.26} $$

where $D$ is the diameter of the ball and $d$ is the diameter of the impression in millimeters.

Depending on the condition of the material, different types of impressions are obtained on the surface after a Brinell hardness test has been performed. Annealed materials generally have a rounded profile, whereas cold-worked (strain-hardened) materials have a sharp profile

**FIGURE 2.25**    Indentation geometry for Brinell hardness testing: (a) annealed metal; (b) work-hardened metal. Note the difference in metal flow at the periphery of the impressions.

(Fig. 2.25). The correct method of measuring the indentation diameter $d$ for both cases is shown.

Because the indenter (with a finite elastic modulus) also undergoes elastic deformation under the applied load $P$, hardness measurements may not be as accurate as expected. One method of minimizing this effect is to use tungsten carbide balls, which, because of their high modulus of elasticity, deform less than steel balls. Also, since harder workpiece materials produce very small impressions, a 1500-kg or 3000-kg load is recommended in order to obtain impressions that are sufficiently large for accurate measurement. Tungsten carbide balls are generally recommended for Brinell hardness numbers higher than 500. In reporting the test results for these high hardnesses the type of ball used should be cited.

Because the impressions made by the same indenter at different loads are not geometrically similar, the Brinell hardness number depends on the load used. Thus the load employed should also be cited with the test results. The Brinell test is generally suitable for materials of low to medium hardness.

*Brinelling* is a term used to describe permanent indentations on a surface between contacting bodies—for example, a component with a hemispherical protrusion or a ball bearing resting on a flat surface. Under fluctuating loads or vibrations, such as during transportation or vibrating foundations, a permanent indentation may be produced on the flat surface by dynamic loading.

### 2.6.2    Rockwell test

In the Rockwell test, the depth of penetration is measured. The indenter is pressed on the surface, first with a minor load and then a major one. The difference in the depth of penetration is a measure of the hardness. There are several Rockwell hardness test scales that use different loads, indenter materials, and indenter geometries. Some of the more common hardness scales and the indenters used are listed in Fig. 2.24. The Rockwell hardness number, which is read directly from a dial on the testing machine, is expressed as follow: If the hardness number is 55 using the C scale, then it is written as 55 HRC. *Rockwell superficial hardness* tests have also been developed using lighter loads and the same type of indenters.

### 2.6.3    Vickers test

The *Vickers* test, also known as the *diamond pyramid hardness* test, uses a pyramid-shaped diamond indenter (see Fig. 2.24) with loads ranging from 1 to 120 kg. The Vickers hardness number (HV) is given by the formula

$$HV = \frac{1.854P}{L^2}.$$ 
(2.27)

The impressions are typically less than 0.5 mm on the diagonal. The Vickers test gives essentially the same hardness number regardless of the load. It is suitable for testing materials with a wide range of hardness, including very hard steels.

### 2.6.4 Knoop test

The Knoop test uses a diamond indenter in the shape of an elongated pyramid (see Fig. 2.24) with loads ranging generally from 25 g to 5 kg. The Knoop hardness number (HK) is given by the formula

$$HK = \frac{14.2P}{L^2}.$$ 
(2.28)

The Knoop test is a **microhardness** test because of the light loads utilized; hence it is suitable for very small or thin specimens and for brittle materials, such as gemstones, carbides, and glass. This test is also used in measuring the hardness of individual grains in a metal. The size of the indentation is generally in the range of 0.01 to 0.10 mm; thus, surface preparation is very important. Because the hardness number obtained depends on the applied load, test results should always cite the load applied.

### 2.6.5 Scleroscope

The Scleroscope is an instrument in which a diamond-tipped indenter (hammer), enclosed in a glass tube, is dropped on the specimen from a certain height. The hardness is determined by the rebound of the indenter; the higher the rebound, the harder the specimen. Indentation is slight. Because the instrument is portable, it is useful for measuring the hardness of large objects.

### 2.6.6 Mohs test

The Mohs test is based on the capability of one material to scratch another. Mohs hardness is on a scale of 10, with 1 for talc and 10 for diamond (the hardest substance known). Thus, a material with a higher Mohs hardness can scratch materials with a lower hardness. The Mohs scale is used generally by mineralogists and geologists. However, some of the materials tested are of interest to manufacturing engineers, as discussed in Chapter 8, 9, and 11. Although the Mohs scale is qualitative, good correlation is obtained with Knoop hardness. Soft metals have a Mohs hardness of 2 to 3, hardened steels about 6, and aluminum oxide (used in grinding wheels) a 9.

### 2.6.7 Durometer

The hardness of rubbers, plastics, and similar soft and elastic materials is generally measured with an instrument called a **durometer**. This is an empirical test in which an indenter is pressed against the surface, with a constant load applied rapidly. The depth of penetration is measured after one second, and the hardness is inversely related to the penetration. There are two different scales for this test. Type A has a blunt indenter, a load of 1 kg, and is used for softer materials. Type D has a sharper indenter, a load of 5 kg, and is used for harder materials. The hardness numbers in these tests range from 0 to 100.

### 2.6.8   Hardness and strength

Since hardness is the resistance to permanent indentation, testing for it is equivalent to performing a compression test on a small portion of a material's surface. Thus we would expect some correlation between hardness and yield stress $Y$ in the form of

$$\text{Hardness} = cY, \qquad (2.29)$$

where $c$ is a proportionality constant.

Theoretical studies, based on plane strain slip-line analysis with a smooth flat punch indenting the surface of a semi-infinite body, show that for a perfectly plastic material of yield stress $Y$, the value of $c$ is about 3. This value is in reasonably good agreement with experimental data (Fig. 2.26). Note that cold-worked materials (which are close to being perfectly plastic in their behavior) show better agreement than annealed ones. The higher $c$ value for the annealed materials is explained by the fact that, due to strain hardening, the average yield stress they exhibit during indentation is higher than their initial yield stress.

The reason that hardness, as a compression test, gives higher values than the uniaxial yield stress $Y$ of the material can be seen in the following analysis. If we assume that the volume under the indenter is a column of material, it would exhibit a uniaxial compressive yield

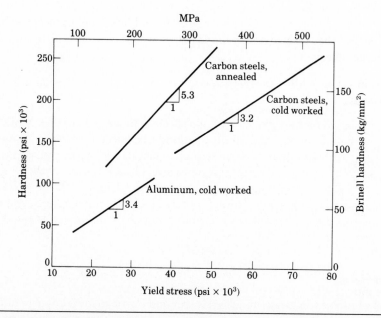

**FIGURE 2.26**   Relation between Brinell hardness and yield stress for aluminum and steels. For comparison, the Brinell hardness (which is always measured in kg/mm²) is converted to psi units on the left scale.

**FIGURE 2.27** Bulk deformation in mild steel under a spherical indenter. Note that the depth of the deformed zone is about one order of magnitude larger than the depth of indentation. For a hardness test to be valid, the material should be allowed to fully develop this zone. This is why thinner specimens require smaller impressions. *Source:* Courtesy of M. C. Shaw and C. T. Yang.

stress $Y$. However, the volume being deformed under the indenter is, in reality, surrounded by a rigid mass (Fig. 2.27). The surrounding mass prevents this volume of material from deforming freely. In fact, this volume is under *triaxial compression*. As we show in Section 2.11 on yield criteria, this material requires a normal compressive yield stress that is higher than the uniaxial yield stress of the material.

More practically, a relationship also has been observed between the ultimate tensile strength (UTS) and Brinell hardness number (HB) for steels:

$$UTS = 500(HB), \tag{2.30}$$

where UTS is in psi and HB in $kg/mm^2$ as measured with a load of 3000 kg. In SI units, the relationship is given by

$$UTS = 3.5(HB), \tag{2.31}$$

where UTS is in MPa.

**Hot hardness** tests can also be carried out using conventional testers with certain modifications, such as surrounding the specimen and indenter with a small electric furnace. The hot hardness of materials is important in applications where the materials are subjected to elevated temperatures, such as in cutting tools in machining and dies for metalworking.

**Example 2.5:** Calculation of modulus of resilience

A piece of steel is highly deformed at room temperature. Its hardness is found to be 300 HB. Estimate the modulus of resilience for this material in $lb/in^2$.

SOLUTION.   Since the steel has been subjected to large strains at room temperature, we may assume that its stress-strain curve has flattened considerably, thus approaching the shape of a perfectly plastic curve. According to Eq. (2.29) and using a value of $c = 3$, we obtain

$$Y = \frac{300}{3} = 100 \text{ kg/mm}^2 = 142,250 \text{ psi.}$$

The modulus of resilience is defined as in Eq. (2.5),

$$\text{Modulus of resilience} = \frac{Y^2}{2E}.$$

For steel, $E = 30 \times 10^6$ psi. Hence,

$$\text{Modulus of resilience} = \frac{(142,250)^2}{2 \times 30 \times 10^6} = 337 \text{ in.} \cdot \text{lb/in}^3.$$

## 2.7  Fatigue

Various structures and components in manufacturing operations, such as tools, dies, gears, cams, shafts, and springs, are subjected to rapidly fluctuating (cyclic or periodic) loads, as well as static loads. Cyclic stresses may be caused by fluctuating mechanical loads, such as gear teeth, or by thermal stresses, such as a cool die coming into repeated contact with hot workpieces. Under these conditions, the part fails at a stress level below which failure would occur under static loading. This phenomenon is known as **fatigue failure**, and is responsible for the majority of failures in mechanical components.

Fatigue test methods involve testing specimens under various states of stress, usually in a combination of tension and compression, or torsion. The test is carried out at various stress amplitudes ($S$), and the number of cycles ($N$) to cause total failure of the specimen or part is recorded. Stress amplitude is the maximum stress, in tension and compression, to which the specimen is subjected.

A typical plot, known as S-N *curves*, is shown in Fig. 2.28. These curves are based on complete reversal of the stress—that is, maximum tension, maximum compression, maxi-

**FIGURE 2.28**  Typical S-N curves for two metals. Note that, unlike steel, aluminum does not have an endurance limit.

**FIGURE 2.29** Ratio of endurance limit to tensile strength for various metals, as a function of tensile strength. Fatigue strength of metal can thus be estimated from its tensile strength.

mum tension, and so on, such as that obtained by bending an eraser or piece of wire alternately in one direction, then the other. The test can also be performed on a rotating shaft with a constant downward load. The maximum stress to which the material can be subjected without fatigue failure, regardless of the number of cycles, is known as the **endurance limit** or **fatigue limit**.

The endurance limit for metals is related to their ultimate tensile strength (Fig. 2.29). For steels, the endurance limit is about one-half their tensile strength. Although many metals, especially steels, have a definite endurance limit, aluminum alloys do not have one, and the *S-N* curve continues its downward trend. For metals exhibiting such behavior (most fcc metals), the fatigue strength is specified at a certain number of cycles, such as $10^7$. In this way, the useful service life of the component can be specified.

## 2.8 Creep

**Creep** is the permanent elongation of a component under a static load maintained for a period of time. It is a phenomenon of metals and certain nonmetallic materials, such as thermoplastics and rubbers, and can occur at any temperature. Lead, for example, creeps under a constant tensile load at room temperature. The thickness of window glass in old houses has been found to be greater at the bottom than at the top of windows, the glass having undergone creep by its own weight over many years.

For metals and their alloys, creep of any significance occurs at elevated temperatures, beginning at about 200°C (400°F) for aluminum alloys, and up to about 1500°C (2800°F) for refractory alloys. The mechanism of creep at elevated temperature in metals is generally

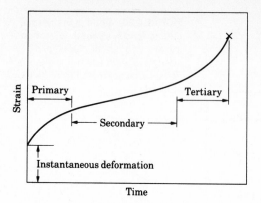

**FIGURE 2.30**  Schematic illustration of a typical creep curve. The linear segment of the curve (constant slope) is useful in designing components for a specific creep life.

attributed to grain-boundary sliding. Creep is especially important in high-temperature applications, such as gas-turbine blades and similar components in jet engines and rocket motors. High-pressure steam lines and nuclear-fuel elements also are subject to creep. Creep deformation also can occur in tools and dies that are subjected to high stresses at elevated temperatures during hot-working operations such as forging and extrusion.

A creep test typically consists of subjecting a specimen to a constant tensile load, hence constant engineering stress, at a certain temperature and measuring the change in length over a period of time. A typical creep curve (Fig. 2.30) usually consists of primary, secondary, and tertiary stages. This specimen eventually fails by necking and fracture, as in a tension test, which is called **rupture** or **creep rupture**. As expected, the creep rate increases with temperature and the applied load.

Design against creep usually involves a knowledge of the secondary (linear) range and its slope, since the creep rate can be determined reliably when the curve has a constant slope. Generally, resistance to creep increases with the melting temperature of a material, which serves as a general guideline for design purposes. Stainless steels, superalloys, and refractory metals and alloys are commonly used in applications where creep resistance is required.

**Stress relaxation** is closely related to creep. In stress relaxation, the stresses resulting from external loading of a structural component decrease in magnitude over a period of time, even though the dimensions of the component remain constant. Examples are rivets, bolts, guy wires, and similar parts under tension, compression, or bending. This phenomenon is particularly common and important in thermoplastics (see Section 10.3).

## 2.9  Impact

In many manufacturing operations, as well as during the service life of components, materials are subjected to **impact** (or *dynamic*) **loading**, as in high-speed metal-working operations such as drop forging. Tests have been developed to determine the impact toughness of metallic and nonmetallic materials under impact loading. A typical impact test consists of placing a notched specimen in an impact tester and breaking it with a swinging pendulum. In the Charpy test the specimen is supported at both ends (Fig. 2.31a), whereas in the Izod test it is

**FIGURE 2.31** Impact test specimens: (a) Charpy; (b) Izod.

supported at one end like a cantilever beam (Fig. 2.31b). From the amount of swing of the pendulum, the energy dissipated in breaking the specimen is obtained. This energy is the **impact toughness** of the material.

Impact tests are particularly useful in determining the ductile-brittle transition temperature of materials. Materials that have impact resistance are generally those that have high strength and high ductility, hence high toughness. Sensitivity to surface defects (**notch sensitivity**) is important, as it lowers impact toughness.

## 2.10 Residual Stresses

In this section we show that inhomogeneous deformation leads to **residual stresses**—stresses that remain within a part after it has been deformed and all external forces have been removed. A typical example of inhomogeneous deformation is the bending of a beam (Fig. 2.32). The bending moment first produces a linear elastic stress distribution. As the moment is increased, the outer fibers begin to yield and, for a typical strain-hardening material, the stress distribution shown in Fig. 2.32b is eventually obtained. After the part is bent (permanently, since it has undergone plastic deformation), the moment is removed by unloading. This unloading is equivalent to applying an equal and opposite moment to the beam.

As previously shown in Fig. 2.3, all recovery is elastic. Thus, in Fig. 2.32c, the moments of the areas *oab* and *oac* about the neutral axis must be equal. (For purposes of this treatment, we assume that the neutral axis does not shift.)

The difference between the two stress distributions produces the residual stress pattern within the beam. Note that there are compressive residual stresses in layers *ad* and *oe*, and tensile residual stresses in layers *do* and *ef*. With no external forces, the residual stress system in the beam must be in statis equilibrium. Although this example involves stresses in one direction only, in most situations in deformation processing the residual stresses are three dimensional.

In this example, the equilibrium of residual stresses may be distributed by altering the geometry of the beam, such as by removing a layer of material. The beam will then acquire a

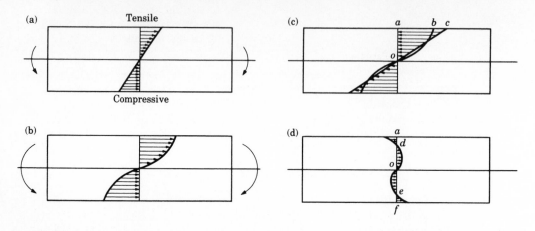

**FIGURE 2.32**   Residual stresses developed in bending a beam made of an elastic, strain-hardening material. Note that unloading is equivalent to applying an equal and opposite moment to the part, as shown in (b). Because of nonuniform deformation, most parts made by plastic deformation processes contain residual stresses. Note that the forces and moments due to residual stresses must be internally balanced.

new radius of curvature in order to ensure the balance of internal forces. Another example of this effect is the drilling of round holes on surfaces with residual stresses. It may be found that, as a result of removing this material, the equilibrium of the residual stresses is disturbed and the hole becomes oval.

Such disturbances of residual stresses lead to *warping*, some simple examples of which are shown in Fig. 2.33. The equilibrium of residual stresses may also be disturbed by *relaxation* of these stresses over a period of time, which results in instability of the dimensions and shape of the component. These dimensional changes can be an important consideration for precision machinery and measuring equipment.

Residual stresses are also caused by *phase changes* in metals during or after processing because of density differences between phases (such as between ferrite and martensite in steels). Phase changes cause microscopic volumetric changes and result in residual stresses. This phenomenon is important in warm and hot working and in heat treatment following cold working (such as deformation at room temperature).

Residual stresses can also be caused by *temperature gradients* within a body, as during the cooling cycle of a casting, when applying brakes to a railroad wheel, or in a grinding operation. The expansion and contraction due to temperature gradients are analogous to nonuniform plastic deformation.

### 2.10.1   Effects of residual stresses

Tensile residual stresses on the surface of a part are generally considered undesirable, because they lower the fatigue life and fracture strength of the part. A surface with tensile residual stresses can sustain lower additional tensile stresses (due to external loading) than can a surface that is free from residual stresses. This is particularly true for relatively brittle materials, where fracture takes place with little or no plastic deformation.

**FIGURE 2.33** Distortion of parts with residual stresses after cutting or slitting: (a) rolled sheet or plate; (b) drawn rod; (c) thin-walled tubing. Because of the presence of residual stresses on the surfaces of parts, a round drill may produce an oval-shaped hole because of relaxation of stresses when a portion is removed.

Tensile residual stresses in manufactured products can also lead to **stress cracking** or **stress-corrosion cracking** over a period of time. Conversely, compressive residual stresses on a surface are generally desirable. In fact, in order to increase the fatigue life of components, compressive residual stresses are imparted on surfaces by techniques such as **shot peening** and **surface rolling** (see Section 4.5.1).

### 2.10.2 Reduction of residual stresses

Residual stresses may be reduced or eliminated either by **stress-relief annealing** or by further *plastic deformation*. Given sufficient time, residual stresses may also be diminished at room temperature by *relaxation*. The time required can be greatly reduced by the temperature of the component. Relaxation of residual stresses by stress-relief annealing is generally accompanied by warpage of the part. Hence a "machining allowance" is commonly provided to compensate for dimensional changes during stress relieving.

The mechanism of reduction or elimination of residual stresses by plastic deformation is as follows: Assume that a piece of metal has the residual stresses shown in Fig. 2.34, namely, tensile on the outside and compressive on the inside. The part with these stresses, which are in the elastic range, is in equilibrium. Also assume that the material is elastic and perfectly plastic, as shown in Fig. 2.34d.

The levels of residual stress are shown on the stress-strain diagram, both being below the yield stress $Y$. If a uniformly distributed tension is applied to this specimen, points $\sigma_c$ and $\sigma_t$ in the diagram move up on the stress–strain curve, as shown by the arrows. The maximum level that these stresses can reach is the tensile yield stress $Y$. With sufficiently high loading, the stress distribution becomes uniform throughout the part, as shown in Fig. 2.34c. If the load is then removed, the stresses recover elastically and the part has no residual stresses. Note that very little stretching is required to relieve these residual stresses. The reason is that the elastic portions of the stress–strain curves for metals are very steep; hence the elastic stresses can be raised to the yield stress with very little strain.

The technique for reducing or relieving residual stresses by plastic deformation, such as by stretching as described, requires sufficient straining to establish a uniformly distributed stress in the part. A material such as the elastic, linearly strain-hardening type, Fig. 2.8e, therefore can never reach this condition since the compressive stress $\sigma_{c'}$ will always lag behind $\sigma_{t'}$. If the slope of the stress-strain curve in the plastic region is small, the difference between $\sigma_{c'}$ and $\sigma_{t'}$ will be rather small and little residual stress will be left in the part after unloading.

**FIGURE 2.34**   Elimination of residual stresses by stretching. Residual stresses can be also reduced or eliminated by thermal treatments, such as stress relieving or annealing.

---

**Example 2.6:   Elimination of residual stresses by tension**

Refer to Fig. 2.34 and assume that $\sigma_t = 140$ MPa and $\sigma_c = -140$ MPa. The material is aluminum and the length of the specimen is 0.25 m. Calculate the length to which this specimen should be stretched so that, when unloaded, it will be free from residual stresses. Assume that the yield stress of the material is 150 MPa.

SOLUTION.   Stretching should be to the extent that $\sigma_c$ reaches the yield stress in tension, $Y$. Thus, the total strain should be equal to the sum of the strain required to bring the compressive residual stress to zero and the strain required to bring it to the tensile yield stress. Hence,

$$\varepsilon_{\text{total}} = \frac{\sigma_c}{E} + \frac{Y}{E}. \tag{2.32}$$

For aluminum, let $E = 70$ GPa. Thus,

$$\varepsilon_{\text{total}} = \frac{140}{70 \times 10^3} + \frac{150}{70 \times 10^3} = 0.00414.$$

Hence, the stretched length should be

$$\ln\left(\frac{\ell_f}{0.25}\right) = 0.00414 \quad \text{or} \quad \ell_f = 0.2510 \text{ m.}$$

As the strains are very small, we may use engineering strains in these calculations. Thus,

$$\frac{\ell_f - 0.25}{0.25} = 0.00414 \quad \text{or} \quad \ell_f = 0.2510 \text{ m.}$$

---

## 2.11 Triaxial Stresses and Yield Criteria

In most manufacturing operations involving deformation processing, the material, unlike that in a simple tension or compression test specimen, is generally subjected to **triaxial stresses**. For example, in the expansion of a thin-walled spherical shell under internal pressure, an element in the shell is subjected to equal biaxial tensile stresses (Fig. 2.35a). In drawing a rod or wire through a conical die, an element in the deformation zone is subjected to tension in its length direction and to compression on its conical surface (Fig. 2.35b). An element in the flange in deep drawing of sheet metal is subjected to a tensile radial stress and compressive stresses on its surface and in the circumferential direction (Fig. 2.35c). As we show in subsequent chapters, many similar examples can be given in which the material is subjected to various normal and shear stresses during processing.

The three-dimensional stresses shown in Fig. 2.35 cause distortion of the elements. In the *elastic* range, the strains are represented by the *generalized Hooke's law* equations:

$$\varepsilon_1 = \frac{1}{E}[\sigma_1 - v(\sigma_2 + \sigma_3)] \tag{2.33a}$$

$$\varepsilon_2 = \frac{1}{E}[\sigma_2 - v(\sigma_1 + \sigma_3)] \tag{2.33b}$$

$$\varepsilon_3 = \frac{1}{E}[\sigma_3 - v(\sigma_1 + \sigma_2)]. \tag{2.33c}$$

Thus, for simple tension where $\sigma_2 = \sigma_3 = 0$, we have

$$\varepsilon_1 = \sigma_1/E,$$

(a)

(b)

(c)

**FIGURE 2.35** The state of stress in various metalworking operations. (a) Expansion of a thin-walled spherical shell under internal pressure. (b) Drawing of round rod or wire through a conical die to reduce its diameter. (c) Deep drawing of sheet metal with a punch and die to make a cup.

and

$$\varepsilon_2 = \varepsilon_3 = -v\sigma_1/E.$$

The negative sign indicates a contraction of the element in the 2 and 3 directions, respectively.

In a simple tension or compression test, when the applied stress reaches the uniaxial yield stress $Y$, the material will deform *plastically*. However, if the material is subjected to a more complex state of stress, there are relationships between these stresses that will predict yielding. These relationships are known as **yield criteria**, the most common ones being the maximum-shear-stress criterion and the distortion-energy criterion.

### 2.11.1  Maximum-shear-stress criterion

This criterion, also known as the **Tresca criterion**, states that yielding occurs when the *maximum shear stress* within an element is equal to or exceeds a critical value. As we show in Section 3.3, this critical value of the shear stress is a material property called **shear yield stress, $k$**. Thus, for yielding,

$$\tau_{max} \geq k. \tag{2.34}$$

The most convenient way of determining the stresses on an element is by the use of **Mohr's circles for stresses**. Some typical examples are shown in Fig. 2.36. Note that in these examples the normal stresses are **principal stresses**; in other words, *they act on planes on which there are no shear stresses*. Principal stresses and their directions can be determined easily from Mohr's circles or from stress-transformation equations, as described in various textbooks on the mechanics of solids.

If the maximum shear stress, as determined from Fig. 2.36 or from appropriate equations, is equal to or exceeds $k$, then yielding will occur. This condition can be best visualized

**FIGURE 2.36**  Mohr's circles for various states of stress: (a) uniaxial tension; (b) uniaxial compression; (c) biaxial tension (plane stress); (d) triaxial tension. These states of stress are encountered in most metalworking processes.

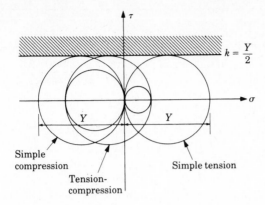

**FIGURE 2.37** Three sets of Mohr's circles representing stresses that are large enough to cause yielding. Note that there is an infinite number of states of stress that produce circles of the same maximum diameter, hence yielding. For any state of stress, the radius of the largest circle must be equal to the shear yield stress $k$ to cause yielding. The height of the ceiling depends on the particular material and its condition, temperature, and strain rate.

by the construction in Fig. 2.37. In order to cause yielding, the largest circle must touch the "ceiling" represented by the shear yield stress $k$. Note the many combinations of stresses (or states of stress) that can give the same maximum shear stress.

From the simple tension test, we note that

$$k = \frac{Y}{2}. \tag{2.35}$$

If, for some reason, we are unable to increase the stresses on the element in order to cause yielding, the solution is simply to lower the ceiling by raising the temperature of the material. This solution is the basis and one major reason for hot working of materials.

We assumed that the material is *continuous, homogeneous,* and *isotropic* (it has the same properties in all directions). Also, we assumed that tensile stresses are positive, compressive stresses negative, and the yield stress in tension and in compression are equal (see Bauschinger effect in Section 2.3.2). All are important assumptions.

We can now write the maximum-shear-stress criterion as

$$\sigma_{\max} - \sigma_{\min} = Y. \tag{2.36}$$

This means that the maximum and minimum normal stresses produce the largest circle and hence the *largest* shear stress. Consequently, *the intermediate stress has no effect on yielding.* It should be emphasized that the left-hand side of Eq. (2.36) represents the *applied stresses* and that the right-hand side is a *material property.*

### 2.11.2 Distortion-energy criterion

The **distortion-energy criterion**, also called the **von Mises criterion**, states that yielding occurs when the relationship between the principal applied stresses and uniaxial yield stress $Y$ of the material is

$$(\sigma_1 - \sigma_2)^2 + (\sigma_2 - \sigma_3)^2 + (\sigma_3 - \sigma_1)^2 = 2Y^2. \tag{2.37}$$

Note that, unlike the maximum-shear-stress criterion, the intermediate principal stress is included in this equation. Here again, the left-hand side of the equation represents the applied stresses and the right-hand side a material property.

**Example 2.7:   Yielding of thin-walled shell**

A thin-walled spherical shell is under internal pressure $p$. The shell is 20 in. in diameter and 0.1 in. thick. It is made of a perfectly plastic material with a yield stress of 20,000 psi. Calculate the pressure required to cause yielding of the shell according to both yield criteria discussed above.

SOLUTION.   For this shell under internal pressure, the membrane stresses are given by

$$\sigma_1 = \sigma_2 = \frac{pr}{2t}, \tag{2.38}$$

where $r = 10$ in. and $t = 0.1$ in. The stress in the thickness direction, $\sigma_3$, is negligible because of the high $r/t$ ratio of the shell. Thus, according to the maximum-shear-stress criterion,

$$\sigma_{max} - \sigma_{min} = Y$$

or

$$\sigma_1 - 0 = Y$$

and

$$\sigma_2 - 0 = Y.$$

Hence $\sigma_1 = \sigma_2 = 20,000$ psi.
   The pressure required is then

$$p = \frac{2tY}{r} = \frac{(2)(0.1)(20,000)}{10} = 400 \text{ psi.}$$

According to the distortion-energy criterion,

$$(\sigma_1 - \sigma_2)^2 + (\sigma_2 - \sigma_3)^2 + (\sigma_3 - \sigma_1)^2 = 2Y^2$$

or

$$0 \quad + \quad \sigma_2^2 \quad + \quad \sigma_1^2 \quad = 2Y^2$$

Hence $\sigma_1 = \sigma_2 = Y$.
   Thus, the answer is the same, or

$$p = 400 \text{ psi.}$$

**Example 2.8:   Correction factor for true stress–true strain curves**

Explain why a correction factor has to be applied in the construction of a true stress-true strain curve, shown in Fig. 2.38 from tensile-test data.

SOLUTION.   The stress distribution at the neck of a specimen, where there is a triaxial state of stress, is shown in Fig. 2.38. The reason for this state is that each element in the region has a different cross-sectional area; the smaller the area, the greater is the tensile stress on the element. Hence element 1 will contract more than element 2, and so on. However, element 1 is restrained from contracting freely by element 2, and element 2 is restrained by element 3, and so on. This restraint causes radial and circumferential tensile stresses in the necked region. This situation results in an axial tensile stress distribution as shown in Fig. 2.38.

The true uniaxial stress in tension is $\sigma$, whereas the calculated value of true stress at fracture is the average stress. Hence a correction has to be made. A mathematical analysis by P. W. Bridgman gives the ratio of true to average stress as

$$\frac{\sigma}{\sigma_{av}} = \frac{1}{(1 + 2R/a)[\ln(1 + a/2R)]}, \tag{2.39}$$

where $R$ is the radius of curvature of the neck and $a$ is the radius of the specimen at the neck. Since $R$ is not easy to measure during a test, an empirical relation has been established between $a/R$ and the true strain at the neck. The corrected true stress–true strain curve is shown in Fig. 2.7a, in Example 2.2.

**FIGURE 2.38** Stress distribution in the necked region of a tension-test specimen.

### 2.11.3 Plane stress and plane strain

Plane stress and plane strain are important in the application of yield criteria. **Plane stress** is the state of stress in which one or two of the pairs of faces on an elemental cube are free from stress. An example is the torsion of a thin-walled tube. There are no stresses normal to the inside or outside surface of the tube; hence the state of the stress of the tube is one of plane stress. Other examples are shown in Figs. 2.35a and 2.39 a, b, and c.

The state of stress where one of the pairs of faces on an element undergoes zero strain is known as **plane strain** (Fig. 2.39c and d). Another example is the **plane-strain compression** test shown in Fig. 2.16. There, by proper choice of specimen dimensions, the width of the specimen is kept essentially constant. (Note that an element does not have to be physically constrained on the pair of faces for plane-strain conditions to exist.) A third example is the torsion of a thin-walled tube in which the wall thickness remains constant (see Section 2.11.7).

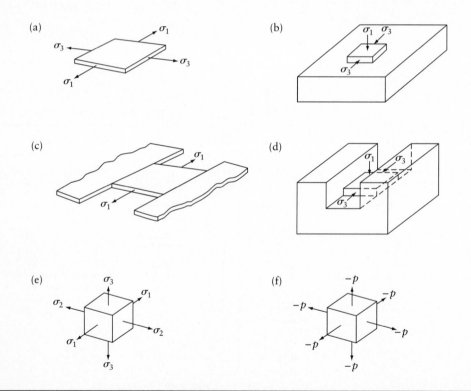

**FIGURE 2.39** Examples of states of stress: (a) plane stress in sheet stretching; there are no stresses acting on the surfaces of the sheet. (b) plane stress in compression; there are no stresses acting on the sides of the specimen being compressed. (c) plane strain in tension; the width of the sheet remains constant while being stretched. (d) plane strain in compression (see also Fig. 2.16); the width of the specimen remains constant due to the restraint by the groove. (e) Triaxial tensile stresses acting on an element. (f) Hydrostatic compression of an element. Note also that an element on the cylindrical portion of a thin-walled tube in torsion is in the condition of both plane stress and plane strain (see also Section 2.11.7).

A review of the two yield criteria previously outlined indicates that the *plane-stress* condition (where $\sigma_2 = 0$) can be represented by the diagram in Fig. 2.40. The maximum-shear-stress criterion gives an envelope of straight lines. In the first quadrant, where $\sigma_1 > 0$, $\sigma_3 > 0$, and $\sigma_2$ is always zero for plane stress, Eq. (2.36) reduces to $\sigma_{max} = Y$. Hence the maximum value that either $\sigma_1$ or $\sigma_3$ can acquire is $Y$; thus the straight lines in the diagram. In the third quadrant the same situation exists because $\sigma_1$ and $\sigma_3$ are both compressive. In the second and fourth quadrants, $\sigma_2$ (which is zero for the plane-stress condition) is the intermediate stress. Thus for the second quadrant, Eq. (2.36) reduces to

$$\sigma_3 - \sigma_1 = Y, \tag{2.40}$$

and for the fourth quadrant, reduces to

$$\sigma_1 - \sigma_3 = Y. \tag{2.41}$$

Equations (2.40) and (2.41) represent the 45° lines in Fig. 2.40.

The distortion-energy criterion for plane stress reduces to

$$\sigma_1^2 + \sigma_3^2 - \sigma_1\sigma_3 = Y^2 \tag{2.42}$$

and is shown graphically in the figure. Whenever a point (with its coordinates representing the two principal stresses) falls on these boundaries, the element will yield.

The three-dimensional *elastic* stress–strain relationships were shown in Eq. (2.33). When the stresses are sufficiently high to cause *plastic* deformation, the stress–strain relationships are obtained from **flow rules** (*Lévy-Mises equations*), described in detail in texts on plasticity. These relationships are in strain increments in the form of (see section 2.11.6)

$$d\varepsilon_1 = \frac{d\bar{\varepsilon}}{\bar{\sigma}}\left[\sigma_1 - \tfrac{1}{2}(\sigma_2 + \sigma_3)\right] \tag{2.43a}$$

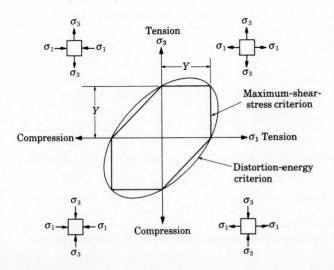

**FIGURE 2.40** Plane-stress diagrams for maximum-shear-stress and distortion-energy criteria. Note that $\sigma_2 = 0$.

$$d\varepsilon_2 = \frac{d\bar{\varepsilon}}{\bar{\sigma}}\left[\sigma_2 - \tfrac{1}{2}(\sigma_1 + \sigma_3)\right] \tag{2.43b}$$

$$d\varepsilon_3 = \frac{d\bar{\varepsilon}}{\bar{\sigma}}\left[\sigma_3 - \tfrac{1}{2}(\sigma_1 + \sigma_2)\right] \tag{2.43c}$$

Note that these expressions are similar to those given by the elastic stress–strain relationships of the generalized Hooke's law.

For the plane-strain condition shown in Fig. 2.39c and d, we have $\varepsilon_2 = 0$. Therefore,

$$\sigma_2 = \frac{\sigma_1 + \sigma_3}{2}. \tag{2.44}$$

Note that $\sigma_2$ is now an intermediate stress.

For the *plane-strain* compression of Fig. 2.39, the distortion-energy criterion (which includes the intermediate stress) reduces to

$$\sigma_1 - \sigma_3 = \frac{2}{\sqrt{3}}\, Y = 1.15Y = Y'. \tag{2.45}$$

Note that, whereas for the maximum-shear-stress criterion $k = Y/2$, for the distortion-energy criterion for the plane-strain condition we have $k = Y/\sqrt{3}$.

### 2.11.4  Experimental verification of yield criteria

The yield criteria described have been tested experimentally. A suitable specimen commonly used is a thin-walled tube under internal pressure and/or torsion. Under such loading, it is possible to generate different states of plane stress. Various experiments, with a variety of ductile materials, have shown that the distortion-energy criterion agrees better with the experimental data than does the maximum-shear-stress criterion.

These findings suggest use of the distortion-energy criterion for the analysis of metal-working processes, which generally make use of ductile materials. The simpler maximum-shear-stress criterion, however, can also be used, particularly by designers. The difference between the two criteria is negligible for most practical applications.

### 2.11.5  Volume strain

By summing the three equations of the generalized Hooke's law (Eq. 2.33), we obtain

$$\varepsilon_1 + \varepsilon_2 + \varepsilon_3 = \frac{1 - 2v}{E}(\sigma_1 + \sigma_2 + \sigma_3), \tag{2.46}$$

where the left-hand side of the equation can be shown to be the **volume strain** or **dilatation**, $\Delta$. Thus,

$$\Delta = \frac{\text{volume change}}{\text{original volume}} = \frac{1 - 2v}{E}(\sigma_1 + \sigma_2 + \sigma_3). \tag{2.47}$$

In the plastic range, where $v = 0.5$, the volume change is zero. Hence in plastic working,

$$\varepsilon_1 + \varepsilon_2 + \varepsilon_3 = 0, \tag{2.48}$$

which is a convenient means of determining a third strain if two strains are known.

The **bulk modulus** is defined as

$$\text{Bulk modulus} = \frac{\sigma_m}{\Delta} = \frac{E}{3(1 - 2\nu)}, \tag{2.49}$$

where $\sigma_m$ is the *mean stress*, defined as

$$\sigma_m = \tfrac{1}{3}(\sigma_1 + \sigma_2 + \sigma_3). \tag{2.50}$$

From Eq. (2.47), in the elastic range, where $0 < \nu < 0.5$, the volume of a tension-test specimen increases and that of a compression-test specimen decreases.

### 2.11.6  Effective stress and effective strain

A convenient means of expressing the state of stress on an element is the **effective** (*equivalent* or *representative*) **stress** $\bar{\sigma}$ and **effective strain** $\bar{\varepsilon}$. For the maximum-shear-stress criterion,

$$\bar{\sigma} = \sigma_1 - \sigma_3, \tag{2.51}$$

and for the distortion-energy criterion,

$$\bar{\sigma} = \frac{1}{\sqrt{2}}\left[(\sigma_1 - \sigma_2)^2 + (\sigma_2 - \sigma_3)^2 + (\sigma_3 - \sigma_1)^2\right]^{1/2}. \tag{2.52}$$

The factor $1/\sqrt{2}$ is chosen so that, for simple tension, the effective stress is equal to the uniaxial yield stress $Y$.

The strains are likewise related to the effective strain. For the maximum-shear-stress criterion,

$$\bar{\varepsilon} = \tfrac{2}{3}(\varepsilon_1 - \varepsilon_3), \tag{2.53}$$

and for the distortion-energy criterion,

$$\bar{\varepsilon} = \frac{\sqrt{2}}{3}\left[(\varepsilon_1 - \varepsilon_2)^2 + (\varepsilon_2 - \varepsilon_3)^2 + (\varepsilon_3 - \varepsilon_1)^2\right]^{1/2}. \tag{2.54}$$

Again, the factors $2/3$ and $\sqrt{2}/3$ are chosen so that for simple tension, the effective strain is equal to the uniaxial tensile strain. It is apparent that stress-strain curves may also be called effective stress–effective strain curves.

### 2.11.7  Comparison of normal stress–strain and shear stress–strain

Stress–strain curves in tension and torsion for the same material are, of course, comparable. Also, it is possible to construct one curve from the other since the material is the same, the procedure for which is outlined.

The following observations are made with regard to Fig. 2.41, which shows the tension and torsional state of stress:

- In the tension test, the uniaxial stress $\sigma_1$ is also the effective stress and the principal stress.
- In the torsion test, the principal stresses occur on planes whose normals are at $45°$ to the longitudinal axis; the principal stresses $\sigma_1$ and $\sigma_3$ are equal in magnitude but opposite in sign.

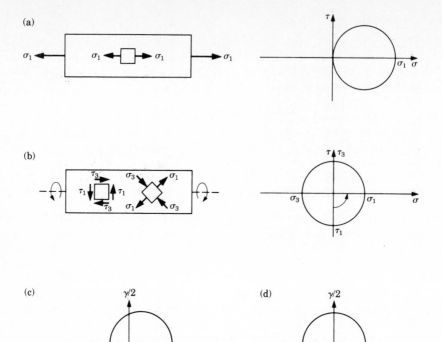

**FIGURE 2.41** Mohr's circle diagrams for stress and strain in uniaxial tension, (a) and (c), and torsion, (b) and (d). Guy wires for antennas, or the spokes of a bicycle, are subjected to uniaxial tension (because they are thin and long), whereas the drive shaft of an automobile is subjected to the stresses shown in (b).

• The magnitude of the principal stress in torsion is the same as the maximum shear stress.

We now have the following relationships:

$$\sigma_1 = -\sigma_3, \quad \sigma_2 = 0, \quad \text{and} \quad \sigma_1 = \tau_1.$$

Substituting these stresses into Eqs. (2.51) and (2.52) for effective stress, we obtain the following relationships: For the maximum-shear-stress criterion,

$$\bar{\sigma} = \sigma_1 - \sigma_3 = \sigma_1 + \sigma_1 = 2\sigma_1 = 2\tau_1, \qquad (2.55)$$

and for the distortion-energy criterion,

$$\bar{\sigma} = \frac{1}{\sqrt{2}}[(\sigma_1 - 0)^2 + (0 + \sigma_1)^2 + (-\sigma_1 - \sigma_1)^2]^{1/2} = \sqrt{3}\sigma_1 = \sqrt{3}\tau_1. \quad (2.56)$$

With regard to strains, we can construct the **Mohr's circle for strains**, as shown in Fig. 2.41c and d. We now make the following observations:

- In the tension test, $\varepsilon_2 = \varepsilon_3 = -\dfrac{\varepsilon_1}{2}$.

- In the torsion test, $\varepsilon_1 = -\varepsilon_3 = \dfrac{\gamma}{2}$.

- The strain in the thickness direction of the tube is zero, i.e., $\varepsilon_2 = 0$

Observation 3 is true because the thinning caused by the principal tensile stress is countered by the thickening under the principal compressive stress of the same magnitude. Hence $\varepsilon_2 = 0$. Since $\sigma_2$ is also zero, a thin-walled tube under torsion is both a plane stress and a plane-strain situation.

Substituting these strains into Eqs. (2.53) and (2.54) for effective strain, we obtain the following relationships: For the maximum-shear-stress criterion,

$$\bar{\varepsilon} = \tfrac{2}{3}(\varepsilon_1 - \varepsilon_3) = \tfrac{2}{3}(\varepsilon_1 + \varepsilon_1) = \tfrac{4}{3}\varepsilon_1 = \tfrac{2}{3}\gamma, \tag{2.57}$$

and for the distortion-energy criterion,

$$\bar{\varepsilon} = \frac{\sqrt{2}}{3}\left[(\varepsilon_1 - 0)^2 + (0 + \varepsilon_1)^2 + (\varepsilon_1 - \varepsilon_1)^2\right]^{1/2} = \frac{2}{\sqrt{3}}\varepsilon_1 = \frac{1}{\sqrt{3}}\gamma. \tag{2.58}$$

This set of equations provides a means by which tensile-test data can be converted to torsion-test data, and vice versa.

### 2.11.8   Instability in biaxial tension

In Section 2.2.4 we showed that in simple tension and for a material whose behavior can be represented by $\sigma = K\varepsilon^n$, instability occurs at a true strain of $\varepsilon = n$. Using the equations obtained from flow rules, it can be shown that for various conditions instability occurs at the strains shown in Table 2.6 (see Section 2.11.6).

**TABLE 2.6**

**PRINCIPAL STRAINS AT INSTABILITY FOR VARIOUS CONFIGURATIONS**

|  | $\varepsilon_1$ | $\varepsilon_2$ | $\varepsilon_3 = \ln\frac{t}{t_o}$ | $\bar{\varepsilon}$ |
|---|---|---|---|---|
| Simple tension | $n$ | $-\dfrac{n}{2}$ | $-\dfrac{n}{2}$ | $n$ |
| Balanced biaxial tension | $n$ | $n$ | $-2n$ | $2n$ |
| Spherical shell under internal pressure | $\dfrac{n}{3}$ | $\dfrac{n}{3}$ | $\dfrac{2n}{3}$ | $\dfrac{2n}{3}$ |
|  | Hoop strain | Axial strain |  |  |
| Open-end cylinder under internal pressure | $\dfrac{2n}{3}$ | $-\dfrac{n}{3}$ | $-\dfrac{n}{3}$ | $\dfrac{2n}{3}$ |
| Closed-end cylinder under internal pressure | $\dfrac{n}{2}$ | $0$ | $-\dfrac{n}{2}$ | $\dfrac{n}{\sqrt{3}}$ |

## 2.12 Work of Deformation

In this section we discuss the work required for plastic deformation of materials. Work is defined as the product of collinear force and distance, so a quantity equivalent to work per unit volume is the product of stress and strain. Because the relation between stress and strain in the plastic range depends on the particular stress-strain curve, this work is best calculated by referring to Fig. 2.42.

Note that the area under the true stress-true strain curve for any strain $\varepsilon_1$ is the **energy per unit volume u (specific energy)** of the material deformed. It is expressed as

$$u = \int_0^{\varepsilon_1} \sigma \, d\varepsilon. \tag{2.59}$$

As seen in Section 2.2.3, true stress-true strain curves can be represented by the expression

$$\sigma = K\varepsilon^n.$$

Hence Eq. (2.59) can be written as

$$u = K \int_0^{\varepsilon_1} \varepsilon^n d\varepsilon$$

or

$$u = \frac{K\varepsilon_1^{n+1}}{n+1} = \overline{Y}\varepsilon_1, \tag{2.60}$$

where $\overline{Y}$ is the **average flow stress** of the material.

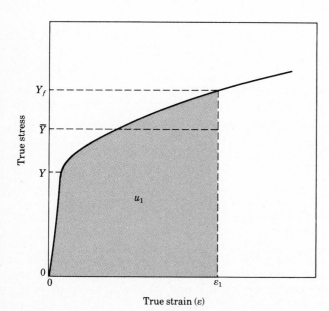

**FIGURE 2.42** Schematic illustration of true stress-true strain curve showing yield stress $Y$, average flow stress $\overline{Y}$, specific energy $u_1$ and flow stress $Y_f$.

This energy represents the work dissipated in uniaxial deformation. For triaxial states of stress, a general expression is given by

$$du = \sigma_1 d\varepsilon_1 + \sigma_2 d\varepsilon_2 + \sigma_3 d\varepsilon_3.$$

For an application of this equation, see Example 2.9. For a more general condition, the **effective stress** and **effective strain** can be used. The energy per unit volume is then

$$u = \int_0^{\bar{\varepsilon}} \bar{\sigma} \, d\bar{\varepsilon}. \tag{2.61}$$

To obtain the work expended, we multiply $u$ by the volume of the material deformed. Thus,

$$\text{Work} = (u)(\text{volume}) \tag{2.62}$$

The energy represented by Eq. (2.62) is the *minimum* energy or the *ideal* energy required for uniform (homogeneous) deformation. The energy required for actual deformation involves two additional factors. One is the energy required to overcome **friction** at the die-workpiece interfaces. The other is the **redundant work** of deformation, which is described as follows:

In Fig. 2.43a, a block of material is being deformed into shape by forging, extrusion, or drawing through a die, as described in Chapter 6. As shown in sketch 2.43b, this deformation is uniform, or homogeneous. In reality, however, the material more often than not deforms as in 2.43c from the effects of friction and die geometry. The difference between b and c is that c has undergone additional shearing along horizontal planes.

This shearing requires expenditure of energy, because additional plastic work has to be done in subjecting the various layers to shear strains. This is known as *redundant work*; the word redundant reflects the fact that this work does not contribute to the shape change of the material. (Note that b and c have the same overall shape and dimensions.)

The total specific energy required can now be written as

$$u_{\text{total}} = u_{\text{ideal}} + u_{\text{friction}} + u_{\text{redundant}}. \tag{2.63}$$

The efficiency of a process is defined as

$$\eta = \frac{u_{\text{ideal}}}{u_{\text{total}}}. \tag{2.64}$$

The magnitude of this efficiency varies widely, depending on the particular process, frictional conditions, die geometry, and other process parameters. Typical values are estimated to be 30–60% for extrusion and 75–95% for rolling.

(a)   (b)   (c)

**FIGURE 2.43** Deformation of grid patterns in a workpiece: (a) original pattern; (b) after ideal deformation; (c) after inhomogeneous deformation, requiring redundant work of deformation. Note that (c) is basically (b) with additional shearing, especially at the outer layers. Thus (c) requires greater work of deformation than (b).

**Example 2.9:   Expansion of a thin-walled spherical shell**

A thin-walled spherical shell made of a perfectly plastic material of yield stress $Y$, original radius $r_o$, and thickness $t_o$ is being expanded by internal pressure. Calculate the work done in expanding this shell to a radius of $r_f$. If the diameter expands at a constant rate, what changes take place in the power consumed as the radius increases?

SOLUTION.   The membrane stresses are given by

$$\sigma_1 = \sigma_2 = Y$$

(from Example 2.7), where $r$ and $t$ are instantaneous dimensions. The true strains in the membrane are given by

$$\varepsilon_1 = \varepsilon_2 = \ln\left(\frac{2\pi r_f}{2\pi r_o}\right) = \ln\left(\frac{r_f}{r_o}\right).$$

Because an element in this shell is subjected to equal biaxial stretching, the specific energy is

$$u = \int_0^{\varepsilon_1} \sigma_1\, d\varepsilon_1 + \int_0^{\varepsilon_2} \sigma_2\, d\varepsilon_2 = 2\sigma_1\varepsilon_1 = 2Y \ln\left(\frac{r_f}{r_o}\right).$$

Since the volume of the shell material is $4\pi r_o^2 t_o$, the work done is

$$W = (u)(\text{volume}) = 8\pi Y r_o^2 t_o \ln\left(\frac{r_f}{r_o}\right).$$

The specific energy can also be calculated from the effective stresses and strains. Thus, according to the distortion-energy criterion,

$$\bar{\sigma} = \frac{1}{\sqrt{2}}[(0)^2 + (\sigma_2)^2 + (-\sigma_1)^2]^{1/2} = \sigma_1 = \sigma_2$$

and

$$\bar{\varepsilon} = \frac{\sqrt{2}}{3}[(0)^2 + (\varepsilon_2 + 2\varepsilon_2)^2 + (-2\varepsilon_2 - \varepsilon_2)^2]^{1/2} = 2\varepsilon_2 = 2\varepsilon_1.$$

(The thickness strain $\varepsilon_3 = -2\varepsilon_2 = -2\varepsilon_1$ because of volume constancy in plastic deformation, where $\varepsilon_1 + \varepsilon_2 + \varepsilon_3 = 0$.) Hence

$$u = \int_0^{\bar{\varepsilon}} \bar{\sigma}\, d\bar{\varepsilon} = \int_0^{2\varepsilon_1} \sigma_1\, d\varepsilon_1 = 2\sigma_1\varepsilon_1.$$

Thus, the answer is the same.

Power is defined as work per unit time; thus

$$\text{Power} = \frac{dW}{dt}.$$

Since all other factors in the expression are constant, the expression for work can be written as being proportional to strain

$$W \propto \ln\left(\frac{r}{r_o}\right) \propto (\ln r - \ln r_o).$$

Hence

$$\text{Power} \propto \frac{1}{r}\frac{dr}{dt}.$$

Because the shell is expanding at a constant rate, $dr/dt =$ constant. Hence the power is related to the instantaneous radius $r$ by

$$\text{Power} \propto \frac{1}{r}.$$

### 2.12.1 Work and heat

Almost all the mechanical work of deformation in plastic working is converted into **heat**. This conversion is not 100% because a small portion of this energy is stored within the deformed material as elastic energy. This is known as **stored energy** (see Section 3.6). Stored energy is generally 5% to 10% of the total energy input. However, it may be as high as 30% in some alloys.

In a simple frictionless process—assuming that work is completely converted into heat—the temperature rise is given by

$$\Delta T = \frac{u_{\text{total}}}{\rho c}, \tag{2.65}$$

where $u_{\text{total}}$ is the specific energy from Eq. (2.63), $\rho$ is the density, and $c$ is the specific heat of the material. Higher temperatures are associated with large areas under the stress-strain curve and smaller values of specific heat.

The theoretical temperature rise for a true strain of 1 (such as 27-mm-high specimen compressed down to 10 mm) has been calculated to be as follows: aluminum, 75°C (165°F), copper, 140°C (285°F); low-carbon steel, 280°C (535°F); and titanium, 570°C (1060°F).

The temperature rise given by Eq. (2.65) is for an ideal situation, where there is no heat loss. In actual operations, heat is lost to the environment, to tools and dies, and to any lubricants or coolants used. If the process is performed very rapidly, these losses are relatively small.

Under extreme conditions, an adiabatic process is approached, with very high temperature rise, leading to **incipient melting**. This rise in temperature can be calculated if the stress-strain curve used is at the appropriate strain rate level. However, if the process is carried out slowly, the actual temperature rise will be a small portion of the calculated value. Properties such as specific heat and thermal conductivity also depend on temperature and they should be taken into account in the calculations.

---

**Example 2.10:   Temperature rise in compression**

A cylindrical specimen 1 in. in diameter and 1 in. high is being compressed by dropping a weight of 100 lb on it from a certain height. The material has the following properties: $K = 15{,}000$ psi, $n = 0.5$, density $= 0.1$ lb/in$^3$, and specific heat $= 0.3$ Btu/lb $\cdot$ °F. Assuming no heat loss and no friction, calculate the final height of the specimen.

SOLUTION.   The expression for heat is given by

$$\text{Heat} = (c_p)(\rho)(\text{Volume})(\Delta T).$$

Thus,

$$\text{Heat} = (0.3)(0.10)\big[(\pi)(1^2)/4\big](100)(778),$$

where 778 ft-lb = 1 BTU. Thus,

$$\text{Heat} = 1830 \text{ ft} \cdot \text{lb} = 22{,}000 \text{ in.} \cdot \text{lb}.$$

Also,

$$\text{Heat} = \text{Work} = (\text{Volume})(u) = (\pi/4)(15{,}000)(\varepsilon^{1.5})/1.5$$

and

$$\varepsilon^{1.5} = (22{,}000)(1.5)(4)/(\pi)(15{,}000) = 2.8;$$

hence,

$$\varepsilon = 1.99.$$

Using absolute values, we have

$$\ln(h_o/h_f) = \ln(1/h_f) = 1.99.$$

Thus, $h_f = 0.137$ in.

## SUMMARY

- Many manufacturing processes involve shaping materials by plastic deformation. Consequently, mechanical properties such as strength, elasticity, ductility, hardness, and the energy required for plastic deformation are important factors. The behavior of materials, in turn, depends on the particular material and its condition, as well as other variables, particularly temperature, strain rate, and the state of stress. (Section 2.1)

- Mechanical properties measured by tension tests include modulus of elasticity ($E$), yield stress ($Y$), ultimate tensile strength (UTS), and Poisson's ratio ($v$). Ductility, as measured by elongation and reduction of area, can also be determined by tension tests. True stress–true strain curves are important in determining such mechanical properties as strength coefficient ($K$), strain-hardening exponent ($n$), strain-rate sensitivity exponent ($m$), and toughness. Tension tests model such manufacturing processes as sheet forming and wire drawing. (Section 2.2)

- Compression tests model manufacturing processes such as forging, rolling, and extrusion. Properties measured by compression tests are subject to inaccuracy due to the presence of friction and barreling. (Section 2.3)

- Torsion tests are conducted on tubular specimens that are subjected to twisting forces. These tests model such manufacturing processes as shearing, cutting, and various machining processes. (Section 2.4)

- Bend or flexure tests are commonly used for brittle materials. The stress at fracture in bending is known as the modulus of rupture or transverse rupture strength. The forces applied in bending tests model those incurred in such manufacturing processes as forming of sheet and plate, as well as testing tool and die materials. (Section 2.5)

- A variety of hardness tests exist to test the resistance of a material to permanent indentation. Hardness is related to strength and wear resistance, but is itself not a fundamental property of a material. (Section 2.6)

- Fatigue tests model manufacturing processes whereby components are subjected to rapidly fluctuating loads. The desired measurement is the endurance limit or fatigue limit; that is, the maximum stress to which a material can be subjected without fatigue failure, regardless of the number of cycles. (Section 2.7)

- Creep is the permanent elongation of a component under a static load maintained for a period of time. A creep test typically consists of subjecting a specimen to a constant tensile load at a certain temperature and measuring its change in length over a period of time. The specimen eventually fails by necking and fracturing, as in a tension test. (Section 2.8)

- Impact tests model both manufacturing operations and service conditions in which materials are subjected to impact (or dynamic) loading, such as drop forging or the properties of tool and die materials. Impact tests determine the energy required to break the specimen; this energy is the impact toughness of the material. Impact tests are also useful in determining the ductile–brittle transition temperature of materials. (Section 2.9)

- Residual stresses are those which remain in a part after it has been deformed and all external forces have been removed; they are generally undesirable. The nature and extent of residual stresses in a part depend on the manner in which it has been plastically deformed. Residual stresses may be reduced or eliminated by stress-relief annealing, by further plastic deformation, or by relaxation. (Section 2.10)

- In metalworking operations, the workpiece material is generally subjected to three-dimensional stresses through dies and various tooling. Yield criteria have been developed to establish relationships between the uniaxial yield stress of the material and the stresses applied; these are maximum-shear-stress (Tresca) and distortion-energy (von Mises) criteria. Various relationships have been established between stresses and the resulting strains in plastic deformation. (Section 2.11)

- Because it requires energy to deform materials, the work of deformation per unit volume of material ($u$) is an important factor and is comprised of ideal, frictional, and redundant work components. In addition to supplying information on force and energy requirements, the work of deformation also indicates the amount of heat developed and hence the temperature rise in the workpiece. (Section 2.12)

## SUMMARY OF EQUATIONS

Engineering strain, $e = \dfrac{\ell - \ell_o}{\ell_o}$

Engineering strain rate, $\dot{e} = \dfrac{v}{\ell_o}$

Engineering stress, $\sigma = \dfrac{P}{A_o}$

True strain, $\varepsilon = \ln\left(\dfrac{\ell}{\ell_o}\right)$

True strain rate, $\dot{\varepsilon} = \dfrac{v}{\ell}$

True stress, $\sigma = \dfrac{P}{A}$

Modulus of elasticity, $E = \dfrac{\sigma}{e}$

Shear modulus, $G = \dfrac{E}{2(1 + v)}$

Modulus of resilience $= \dfrac{Y^2}{2E}$

Elongation $= \dfrac{\ell_f - \ell_o}{\ell_o} \times 100$

Reduction of area $= \dfrac{A_o - A_f}{A_o} \times 100$

Shear strain in torsion, $\gamma = \dfrac{r\phi}{\ell}$

Hooke's law, $\varepsilon_1 = \dfrac{1}{E}[\sigma_1 - v(\sigma_2 + \sigma_3)]$; etc.

Effective strain (Tresca), $\bar{\varepsilon} = \dfrac{2}{3}(\varepsilon_1 - \varepsilon_2)$

Effective strain (von Mises) $\bar{\varepsilon} = \sqrt{\dfrac{2}{3}}\left[(\varepsilon_1 - \varepsilon_2)^2 + (\varepsilon_2 - \varepsilon_3)^2 + (\varepsilon_3 - \varepsilon_1)^2\right]^{1/2}$

Effective stress (Tresca) $\bar{\sigma} = \sigma_1 - \sigma_3$

Effective stress (von Mises) $\bar{\sigma} = \dfrac{1}{\sqrt{2}}\left[(\sigma_1 - \sigma_2)^2 + (\sigma_2 - \sigma_3)^2 + (\sigma_3 - \sigma_1)^2\right]^{1/2}$

True stress–true strain relationship (power law), $\sigma = K\varepsilon^n$

True stress–true strain rate relationship, $\sigma = C\dot{\varepsilon}^m$

Flow rules, $d\varepsilon_1 = \dfrac{d\bar{\varepsilon}}{\bar{\sigma}}\left[\sigma_1 - \dfrac{1}{2}(\sigma_2 + \sigma_3)\right]$; etc.

Maximum-shear-stress criterion (Tresca), $\sigma_{\max} - \sigma_{\min} = Y$

Distortion-energy criterion (von Mises), $(\sigma_1 - \sigma_2)^2 + (\sigma_2 - \sigma_3)^2 + (\sigma_3 - \sigma_1)^2 = 2Y^2$

Shear yield stress, $k = Y/2$ for Tresca and $k = Y/\sqrt{3}$ for von Mises (plane strain)

Volume strain (dilatation), $\Delta = \dfrac{1 - 2v}{E}(\sigma_1 + \sigma_2 + \sigma_3)$

Bulk modulus $= E/3(1 - 2v)$

# BIBLIOGRAPHY

*ASM Handbook*, Vol. 8: *Mechanical Testing*. Materials Park, Ohio: ASM International, 1985.

Boyer, H.E. (ed.), *Atlas of Stress–Strain Curves*. Materials Park, Ohio: ASM International, 1986.

Budinski, K.G., *Engineering Materials: Properties and Selection*, 5th ed. Englewood Cliffs, New Jersey: Prentice-Hall, 1996.

Courtney, T.H., *Mechanical Behavior of Materials*. New York: McGraw-Hill, 1990.

Dieter, G.E., *Mechanical Metallurgy*, 3d ed. New York: McGraw-Hill, 1986.

Dowling, N.E., *Mechanical Behavior of Materials: Engineering Methods for Deformation, Fracture, and Fatigue*. Englewood Cliffs, New Jersey: Prentice-Hall, 1993.

Han, P. (ed.), *Tensile Testing*. Materials Park, Ohio: ASM International, 1992.

Hsu, T.H., *Stress and Strain Data Handbook*. Houston, Texas: Gulf Publishing Co., 1986.

Pohlandt, K., *Material Testing for the Metal Forming Industry*. New York: Springer, 1989.

# QUESTIONS

2.1 Can you calculate the percent elongation of materials based on the information given in Fig. 2.6 only? Explain.

2.2 Describe your observations regarding the contents of Table 2.1.

2.3 Referring to Fig. 2.4, explain if it is possible for the curves to reach zero percent elongation as the gage length is increased further.

2.4 Explain why the difference between engineering strain and true strain becomes larger as strain increases. Is this true for both tensile and compressive strains? Explain.

2.5 Using the same scale for stress, we note that the tensile true stress–true strain curve is higher than the engineering stress–strain curve. Explain whether this condition also holds for a compression test.

2.6 Which of the two tests, tension or compression, requires the higher capacity testing machine? Explain.

2.7 Explain how the modulus of resilience of a material changes, if at all, as it is strain-hardened: (1) for an elastic, perfectly plastic material, and (2) elastic, linearly strain-hardening material.

2.8 Take an element under the anvils in the deformation zone in the plane-strain compression test shown in Fig. 2.16. Is it subjected to simple shear or pure shear during deformation? Explain. (*Hint*: View this element from the sides, in the width direction, and compare with Fig. 2.21.)

2.9 If you pull and break a tensile-test specimen rapidly, where would the temperature be the highest? Explain why.

2.10 In the above question, comment on the temperature distribution if the specimen is pulled very slowly.

2.11 In a tension test, the area under the true stress–true strain curve is the work done per unit volume (specific work). We also know that the area under the load-elongation curve represents the work done on the specimen. If you divide this work by the volume of the specimen between the gage marks, you will determine the work done per unit volume (assuming that all deformation is confined between the gage marks). Will this specific work be the same as the area under the true stress–true strain curve? Explain. Will your answer be the same for any value of strain? Explain.

2.12 In the note at the bottom of Table 2.5, it is stated that as temperature increases, $C$ decreases and $m$ increases. Explain why.

2.13 You are given the $K$ and $n$ values of two different materials, respectively. Is this information sufficient to determine which material is the tougher? If not, what additional information do you need, and why?

2.14 Modify the curves in Fig. 2.8 to indicate the effects of temperature. Explain your changes.

2.15 With a specific example, show why deformation rate, say in m/s, and true strain rate are not the same.

2.16 Referring to Fig. 2.3, explain how the magnitude of elastic recovery will change as an elastic, linearly strain-hardening material is strained.

2.17 It has been stated that the higher the value of $m$, the more diffuse the neck is; and likewise, the lower the $m$ value, the more localized the neck is. Explain the reason for this behavior.

2.18 Explain why materials with high $m$ values, such as hot glass and silly putty, when stretched slowly undergo large elongations before failure. Consider events taking place in the necked region of the specimen.

2.19 You are running 4-point bending tests on a number of identical specimens of the same length and cross-section but with increasing distance between the upper points of loading (see Fig. 2.23b). What changes, if any, would you expect in the test results? Explain.

2.20 Would Eq. (2.10) hold true in the elastic range? Explain.

2.21 Why have different types of hardness tests developed? How would you measure the hardness of a very large object?

2.22 Which hardness tests and scales would you use for very thin strips of material, such as aluminum foil? Why?

2.23 Would the answer to Example 2.3 become double if the strength coefficient of the material is doubled? Explain.

2.24 List and explain the factors that you would consider in selecting an appropriate hardness test and scale for a particular application.

2.25 Describe the difference between creep and stress relaxation, giving two examples for each as they relate to engineering applications.

2.26 Referring to the impact tests shown in Fig. 2.31, explain how different the results would be if the specimens were impacted from the opposite directions.

2.27 If you remove the layer of material $ad$ from the part shown in Fig. 2.32, such as by machining or grinding, which way will the specimen curve? (*Hint*: Assume that the part in (d) is composed of four horizontal springs held at the ends. Thus, from the top down we have compression, tension, compression, and tension springs.)

2.28 Is it possible to completely remove residual stresses in a piece of material by the technique described in Fig. 2.34 if the material is elastic, linearly strain hardening? Explain.

2.29 Referring to Fig. 2.34, would it be possible to eliminate residual stresses by compression? Assume that the piece of material will not buckle under the uniaxial compressive force.

2.30 List and explain the desirable mechanical properties for (1) elevator cable, (2) paper clip, (3) leaf spring, (4) bracket for a bookshelf, (5) piano wire, (6) coat hanger, (7) gas-turbine blade, and (8) staple.

2.31 Using the generalized Hooke's law Eq. (2.33), show whether or not a thin-walled tube undergoes any thickness change when subjected to torsion in the elastic range.

2.32 Make a sketch showing the nature and distribution of residual stresses in Fig. 2.33 a and b before they were cut. Assume that the split parts are free from any stresses. (*Hint*: Force these parts back to the shape they were in before they were cut.)

2.33 It is possible to calculate the work of plastic deformation by measuring the temperature rise in a workpiece, assuming that there is no heat loss and the temperature distribution is uniform throughout. If the specific heat of the material decreases with increasing temperature, will the work of deformation calculated using the specific heat at room temperature be higher or lower than the actual work done? Explain.

2.34 Explain whether or not the volume of a homogeneous, isotropic specimen changes when subjected to a state of (a) uniaxial compressive stress and (b) uniaxial tensile stress, all in the elastic range.

2.35 How would you explain the general behavior shown in Fig. 2.11 to someone over the telephone?

2.36 We know that it is relatively easy to subject a specimen to hydrostatic compression, such as using a chamber filled with a liquid. Devise a means whereby the specimen (say, in the shape of a cube or a round disk) can be subjected to hydrostatic tension, or one approaching this state of stress. (Note that a thin-walled, internally pressurized spherical shell is not a correct answer because it is only subjected to a state of plane stress.)

2.37 Referring to Fig. 2.20, make sketches of the state of stress for an element in the reduced section of the tube when it is subjected to (1) torsion only, (2) torsion while the tube is internally pressurized, and (3) torsion while the tube is exter-

nally pressurized. Assume that the tube is closed-end.

2.38 How would you conduct the test shown in the insert to Fig. 2.22? Make a sketch of the setup you would use.

2.39 A penny-shaped piece of soft metal of a certain diameter is brazed to the ends of two flat round steel rods of the same diameter as the piece. The assembly is then subjected to uniaxial tension. What is the state of stress to which the soft metal is subjected? Explain.

2.40 A circular disk of soft metal is being compressed between two flat, hardened circular steel punches having the same diameter as the disk. Assume that the disk material is perfectly plastic and there is no friction or any temperature effects. Explain the change, if any, in the magnitude of the punch force as the disk is being compressed plastically to, say, a fraction of its original thickness.

## PROBLEMS

2.41 A paper clip is made of wire 1.5 mm in diameter. If the original material from which the wire is made is a rod 20 mm in diameter, calculate the longitudinal and diametrical engineering and true strains that the wire has undergone during processing.

2.42 A material has the following properties: UTS = 60,000 psi and $n = 0.4$. Calculate its strength coefficient $K$.

2.43 Based on the information given in Fig. 2.6, calculate the ultimate tensile strength of annealed copper.

2.44 Identify two materials in Fig. 2.6 that have the lowest and highest uniform elongations, respectively. Calculate these quantities as percentages of the original gage lengths.

2.45 Calculate the ultimate tensile strength (engineering) of a material whose strength coefficient is 700 MPa and a tensile-test specimen which necks at a true strain of 0.35.

2.46 A cable is made of three parallel strands of different materials, all behaving ac-

cording to the equation $\sigma = K\varepsilon^n$. Their properties and cross-sections are:

Material A: $K$ = 600 MPa,
$\qquad n = 0.5, A_o = 5$ mm$^2$.
Material B: $K$ = 300 MPa,
$\qquad n = 0.5, A_o = 4$ mm$^2$.
Material C: $K$ = 400 MPa,
$\qquad n = 0.5, A_o = 3$ mm$^2$.

(a) Calculate the maximum tensile force that this cable can withstand prior to necking.

(b) Explain how you would arrive at an answer if the $n$ values of the three strands were different from each other.

2.47 Using only Fig. 2.6, calculate the maximum load in tension testing of a 304 stainless steel specimen with an original diameter of 0.5 in.

2.48 Using the data given in Table 2.1, calculate the values of the shear modulus $G$ for the metals listed in the table.

2.49 Derive an expression for the toughness of a material represented by the equation

$$\sigma = K(\varepsilon + 0.2)^n,$$

whose fracture strain is denoted as $\varepsilon_f$.

2.50    A cylindrical specimen made of a brittle material 1 in. high and 1 in. in diameter is subjected to a compressive force along its axis. It is found that fracture takes place at an angle of 45° under a load of 30,000 lb. Calculate the shear stress and the normal stress, respectively, acting on the fracture surface.

2.51    What is the modulus of resilience of a highly cold-worked piece of steel with a hardness of 400 HB? For a piece of highly cold-worked copper with a hardness of 150 HB?

2.52    Calculate the work done in frictionless compression of a solid cylinder 50 mm high and 20 mm in diameter to a reduction in height of 50 percent for the following materials: (1) 1100-0 aluminum, (2) annealed copper, (3) annealed 4340 steel, and (4) 70-30 brass, as received.

2.53    In Example 2.2, calculate the toughness of the material by the area under the stress–strain curve. Then compare your answer by first determining the $K$ and $n$ values for this material and then obtaining the toughness by integration.

2.54    A material has a strength coefficient $K = 100,000$ psi and $n = 0.2$. Assuming that a tensile-test specimen made from this material begins to neck at a true strain of 0.2, show that the ultimate tensile strength of this material is 59,340 psi.

2.55    A tensile-test specimen is made of a material represented by the equation $\sigma = K(\varepsilon + n)^n$. (1) Determine the true strain at which necking will begin. (2) Show that it is possible for an engineering material to exhibit this behavior.

2.56    Take two solid cylindrical specimens of equal diameter but different heights. You compress both specimens (frictionless) by the same percent reduction, say 50 percent. Prove that the final diameters will be the same.

2.57    A horizontal rigid bar $c$-$c$ is subjecting specimen $a$ to tension and specimen $b$ to frictionless compression such that the bar remains horizontal (see figure below). The force $F$ is located at a dis-

tance ratio of 2:1. Both specimens have an original cross-sectional area of 1 in², and the original lengths are $a = 10$ in. and $b = 5$ in. The material for specimen $a$ has a true stress–true strain curve of $\sigma = 100,000\varepsilon^{0.5}$ psi. Plot the true stress–true strain curve that the material for specimen $b$ should have for the bar to remain horizontal.

2.58    Inspect the curve that you obtained in Problem 2.57. Does a typical strain-hardening material behave in that manner? Explain.

2.59    In a disk test performed on a specimen 25 mm in diameter and 5 mm thick, it fractures at a stress of 400 MPa. What was the load on it at fracture?

2.60    In sketch (a) in Fig. 2.34, let the tensile and compressive residual stresses both be 20,000 psi, and the modulus of elasticity of the material be $30 \times 10^6$ psi, with a modulus of resilience of 30 in.-lb/in³. If the original length in (a) is 30 in., what should be the stretched length in (b) so that, when unloaded, the strip will be free of residual stresses?

2.61    Show that you can take a bent bar made of an elastic, perfectly plastic material and straighten it by stretching it into the plastic range. (*Hint*: Observe the events shown in Fig. 2.34.)

2.62    A bar 0.5 m long is bent and then stress-relieved. The radius of curvature to the neutral axis is 0.25 m. The bar is 20 mm thick and is made of an elastic, perfectly plastic material with $Y = 400$ MPa and $E = 200$ GPa. Calculate the length to which this bar should be stretched so that, after unloading, it will become and remain straight.

2.63 Derive the three generalized Hooke's law equations given in Eq. (2.33).

2.64 A thin cylindrical sleeve of thickness $t$, length $L$, and inside radius $r$ is placed on a rigid shaft with no clearance (see figure below). The sleeve is then stretched elastically by tension $\sigma_a$ in the axial direction. Let the coefficient of friction at the sleeve–shaft interface be denoted as $\mu$. Derive an expression for the force necessary to strip the sleeve from the shaft. Assume that $\sigma_a$ does not affect the stripping force.

2.65 Assume that a material with a uniaxial yield stress $Y$ yields under a stress state of principal stresses $\sigma_1, \sigma_2, \sigma_3$, where $\sigma_1 > \sigma_2 > \sigma_3$. Show that the superposition of a hydrostatic stress $p$ on this system (such as placing the specimen in a chamber pressurized with a liquid) does not affect yielding. In other words, the material will still yield according to yield criteria.

2.66 Give two different and specific examples in which the maximum-shear-stress and the distortion-energy criteria give the same answer.

2.67 A cylindrical specimen 1 in. in diameter and 1 in. high is being compressed by dropping a weight of 100 lb. from a certain height. After deformation, it is found that the temperature rise in the specimen is $200°F$. Assuming no heat loss and no friction, calculate the final height of the specimen using the following data for the material: $K = 30,000$ psi, $n = 0.5$, density $= 0.1$ lb/in$^3$, and specific heat $= 0.3$ BTU/lb$°F$.

2.68 A thin-walled spherical shell with a yield stress $Y$ is subjected to an internal pressure $p$. With appropriate equations, show whether or not the pressure re-

quired to yield this shell depends on the particular yield criterion used.

2.69 Show that according to the distortion-energy criterion, the yield stress in plane strain is $1.15Y$ where $Y$ is the uniaxial yield stress of the material.

2.70 What would be the answer to Problem 2.69 if the maximum shear stress criterion is used?

2.71 A closed-end, thin-walled cylinder of original length $l$, thickness $t$, and internal radius $r$ is subjected to an internal pressure $p$. Using the generalized Hooke's law equations, show the change, if any, that occurs in the length of this cylinder when it is pressurized.

2.72 A round, thin-walled tube is subjected to tension in the elastic range. Show that both the thickness and diameter decrease as tension increases.

2.73 In Problem 2.72, assume that the tube is subjected to compression. Explain what changes, if any, take place in the thickness and diameter of the tube. Assume that the tube does not buckle.

2.74 Take a long cylindrical balloon and with a thin felt pen mark a small square on it. What will be the shape of this square after you blow up the balloon? (1) a larger square, (2) a rectangle with its long axis in the circumferential directions, (3) a rectangle with its long axis in the longitudinal direction, or (4) an ellipse? Perform this experiment and, based on your observations, explain it using appropriate equations. Assume that the balloon material is perfectly elastic and isotropic, and that this situation represents a thin-walled closed-end cylinder under internal pressure.

2.75 Take a cubic piece of metal with length of the sides of $l_o$ and deform it plastically to the shape of a rectangular parallelepiped of dimensions $l_1$, $l_2$, and $l_3$. Assuming that the material is rigid, perfectly plastic, show that volume constancy requires that the following expression be satisfied:

$$\varepsilon_1 + \varepsilon_2 + \varepsilon_3 = 0$$

2.76 What is the diameter of an originally 20 mm diameter solid steel ball when subjected to a hydrostatic pressure of 10 GPa?

2.77   Determine the effective stress and effective strain in plane-strain compression according to the distortion-energy criterion.

2.78   It can be shown that a thin-walled tube under longitudinal tension (in the elastic range) decreases in diameter. It is, of course, possible to internally pressurize the tube so that its diameter remains constant while being stretched. Assuming that the tube is open end, derive an expression for the pressure required to keep the diameter constant as the tube is stretched.

2.79   What is the magnitude of compressive stress for the plane-strain compression test shown in Fig. 2.16 according to the maximum-shear-stress criterion?

2.80   (a) Calculate the work done in expanding a thin-walled spherical shell 1 mm thick, from a diameter of 100 mm to 120 mm and made of a material where $\sigma = 200 + 50\varepsilon^{0.5}$ MPa. (b) Does your answer depend on the particular yield criterion used? Explain.

2.81   A cylindrical slug 1 in. in diameter and 1 in. high is placed at the center of a 2 in. diameter cavity in a rigid die (see figure below). The slug is surrounded by a compressible matrix the pressure of which is given by the relation

$$p_m = 50{,}000\Delta V / V_{om} \text{ psi},$$

where $m$ denotes matrix and $V_{om}$ is the original volume of the compressible matrix. Both the slug and the matrix are being compressed by a piston and without any friction. The initial pressure of the matrix is zero, and the slug material has the true stress–true strain curve of $\sigma = 20{,}000\varepsilon^{0.5}$. Obtain an expression for the force $F$ versus piston travel $d$ up to $d = 0.5$ in.

Compressible matrix

2.82   A specimen in the shape of a cube 10 mm on each side is being compressed without friction in a die cavity, as shown in Fig. 2.39d, where the width of the groove is 15 mm. Assume that the linearly strain-hardening material has the true stress–true strain curve given by

$$\sigma = 100 + 20\varepsilon \text{ MPa}.$$

Calculate the compressive force required when the height of the specimen is 3 mm, according to both yield criteria.

2.83   Assume that you are asked to give a quiz to students on the contents of this chapter. Prepare five quantitative problems and supply the answers.

# 3

# Structure and Manufacturing Properties of Metals

**3.1** Introduction

**3.2** The Crystal Structure of Metals

**3.3** Deformation and Strength of Single Crystals

**3.4** Grains and Grain Boundaries

**3.5** Plastic Deformation of Polycrystalline Metals

**3.6** Recovery, Recrystallization, and Grain Growth

**3.7** Cold, Warm, and Hot Working

**3.8** Failure and Fracture

**3.9** Physical Properties

**3.10** Properties of Ferrous Alloys

**3.11** Properties of Nonferrous Metals and Alloys

## 3.1 Introduction

The structure of metals greatly influences their behavior and properties. A knowledge of structures guides us in controlling and predicting the behavior and performance of metals in various manufacturing processes. Understanding the structure of metals also allows us to predict and evaluate their *properties*. This helps us make appropriate selections for specific applications under particular force, temperature, and environmental conditions. For example, you will come to appreciate the reasons for developing single-crystal turbine blades (Fig. 3.1) for use in jet engines; these blades have better properties than those made conventionally by casting or forging.

In this chapter, we shall first review the crystal structure of metals and the role of various imperfections as they affect plastic deformation and work-hardening characteristics. The importance of grain size, grain boundaries, and environment will also be described. The failure and fracture of metals, both ductile and brittle, will then be reviewed, together with factors that influence fracture, such as state of stress, temperature, strain rate, and external and internal defects in metals.

The rest of the chapter is devoted to a discussion of physical and other properties of ferrous and nonferrous metals and alloys relevant to their use as engineering materials. Some property data are also given as a guide in design considerations and material selection.

(a)                    (b)                    (c)

**FIGURE 3.1**   Turbine blades for jet engines, manufactured by three different methods: (a) conventionally cast; (b) directionally solidified, with columnar grains, as can be seen from the vertical streaks; and (c) single crystal. Although more expensive, single-crystal blades have properties at high temperatures that are superior to other blades. *Source:* Courtesy of United Technologies Pratt and Whitney.

## 3.2   The Crystal Structure of Metals

When metals solidify from a molten state, the atoms arrange themselves into various orderly configuration, called **crystals**. This arrangement of the atoms in the crystal is called **crystalline structure**. A crystalline structure is like metal scaffolding in front of a building under construction, with uniformly repetitive horizontal and vertical metal pipes and braces.

The smallest group of atoms showing the characteristic **lattice structure** of a particular metal is known as a **unit cell**. It is the building block of a crystal, and a single crystal can have many unit cells. Think of each brick in a wall as a unit cell. The wall has a crystalline structure; that is, it consists of an orderly arrangement of bricks. Thus the wall is like a single crystal, consisting of many unit cells.

The three basic patterns of atomic arrangement found in most metals are (a) **body-centered cubic (bcc)**; (b) **face-centered cubic (fcc)**; and (c) **hexagonal close-packed (hcp)** (Fig.3.2).

The order of magnitude of the distance between the atoms in these crystal structures is 0.1 nm ($10^{-8}$ in.). The way in which these atoms are arranged can have a significant influence on the properties of a particular metal. We can modify these arrangements by adding atoms of some other metal or metals. This is known as **alloying** and often improves the properties of the metal (see Chapter 5).

Of the three structures illustrated, the fcc and hcp crystals have the most densely packed configurations. In the hcp structure, the top and bottom planes are called *basal planes*. The reason that different crystal structures form is a matter of minimizing the energy required to form these structures. Thus tungsten forms a bcc structure because it requires less energy than other structures. Similarly, aluminum forms an fcc structure. However, at different

**FIGURE 3.2a** The body-centered cubic (bcc) crystal structure: (a) hard-ball model; (b) unit cell; and (c) single crystal with many unit cells. *Source:* W.G. Moffatt, et al., *The Structure and Properties of Materials,* Vol. 1, John Wiley & Sons, 1976.

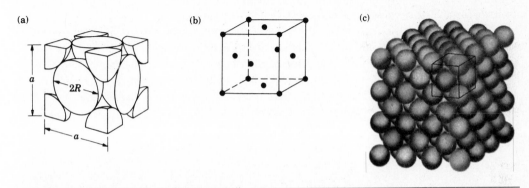

**FIGURE 3.2b** The face-centered cubic (fcc) crystal structure: (a) hard-ball model; (b) unit cell; and (c) single crystal with many unit cells. *Source:* W.G. Moffatt, et al., *The Structure and Properties of Materials,* Vol. 1, John Wiley & Sons, 1976.

**FIGURE 3.2c** The hexagonal close-packed (hcp) crystal structure: (a) unit cell; and (b) single crystal with many unit cells. *Source:* W.G. Moffatt, et al., *The Structure and Properties of Materials,* Vol. 1, John Wiley & Sons, 1976.

temperatures the same metal may form different structures because of the lower energy required at that temperature. For example, iron forms a bcc structure below 912°C (1674°F), and above 1394°C (2541°F), but an fcc structure between 912°C and 1394°C. As we describe in Section 5.10, these structural changes are an important aspect of heat treatment of metals and alloys.

## 3.3 Deformation and Strength of Single Crystals

When crystals are subjected to an external force, they first undergo **elastic deformation**; that is, they return to their original shape when the force is removed. An analogy to this type of behavior is a helical spring that stretches when loaded and returns to its original shape when the load is removed. However, if the force on the crystal structure is increased sufficiently, the crystal undergoes **plastic deformation** (or *permanent* deformation); that is, it does not return to its original shape when the force is removed.

There are two basic mechanisms by which plastic deformation may take place in crystal structures. One is the slipping of one plane of atoms over an adjacent plane—the **slip plane**—under a shearing force (Fig. 3.3). The deformation of a single-crystal specimen by slip is shown schematically in Fig. 3.4. This situation is much like sliding playing cards against each other.

Just as it takes a certain amount of force to slide playing cards against each other, so a crystal requires a certain amount of shear stress (**critical shear stress**) to undergo permanent

**FIGURE 3.3** Permanent deformation, also called plastic deformation, of a single crystal subjected to a shear stress: (a) structure before deformation; and (b) deformation by slip. The *b/a* ratio influences the magnitude of shear stress required to cause slip.

**FIGURE 3.4** (a) Permanent deformation of a single crystal under a tensile load. Note that the slip planes tend to align themselves in the direction of pulling. This behavior can be simulated using a deck of cards with a rubber band around them. (b) Twinning in tension.

**FIGURE 3.5**  Variation of shear stress in moving a plane of atoms over another plane.

deformation. Thus there must be a shear stress of sufficient magnitude within a crystal for plastic deformation to take place.

The maximum **theoretical shear stress** to cause permanent deformation in a perfect crystal is obtained as follows: When there is no stress, the atoms in the crystal are in equilibrium (Fig. 3.5). Under shear stress, the upper row of atoms is moved to the right, the position of the atom being denoted as $x$. Thus when $x = 0$ or $x = b$ the shear stress is zero. Each atom of the upper row is attracted to the nearest atom of the lower row, resulting in nonequilibrium at positions 2 and 4, where the stresses are maximum but opposite in sign. At position 3 the shear stress is again zero, since this is a symmetric position. In Fig. 3.5 we assumed that, as a first approximation, the shear stress varies sinusoidally. Hence the shear stress at a displacement $x$ is

$$\tau = \tau_{max} \sin \frac{2\pi x}{b}, \tag{3.1}$$

which, for small values of $x/b$, can be written as

$$\tau = \tau_{max} \frac{2\pi x}{b}.$$

From Hooke's law, we have

$$\tau = G\gamma = G\left(\frac{x}{a}\right).$$

Hence

$$\tau_{max} = \frac{G}{2\pi} \cdot \frac{b}{a}. \tag{3.2}$$

If we assume that $b$ is approximately equal to $a$, we obtain

$$\tau_{max} = \frac{G}{2\pi}. \tag{3.3}$$

More refined calculations have set the value of the maximum theoretical shear strength as between $G/10$ and $G/30$.

The shear stress required to cause slip in single crystals is directly proportional to the ratio $b/a$ in Fig. 3.5. We can therefore state that slip in a crystal takes place along planes of maximum atomic density, or that slip takes place in closely packed planes and in closely packed directions.

Because the $b/a$ ratio is different for different directions within the crystal, a single crystal has different properties when tested in different directions. Thus a crystal in anisotropic. A common example of **anisotropy** is woven cloth, which stretches differently when we pull it in different directions, or plywood, which is much stronger in the planar direction than along its thickness direction (it splits easily).

The second mechanism of plastic deformation is **twinning,** in which a portion of the crystal forms structurally a mirror image of itself across the plane of twinning (Fig.3.4b). Twins form abruptly and are the cause of the creaking sound (tin cry) when a tin or zinc rod is bent at room temperature. Twinning usually occurs in hcp and bcc metals by plastic deformation and in fcc metals by annealing (see Section 5.10.4).

### 3.3.1 Slip systems

The combination of a slip plane and its direction of slip is known as a **slip system.** In general, metals with five or more slip systems are ductile, whereas those with less than five are not ductile.

1. In body-centered cubic crystals there are 48 possible slip systems. Thus the probability is high that an externally applied shear stress will operate on one of these systems and cause slip. However, because of the relatively high $b/a$ ratio, the required shear stress is high. Metals with bcc structures, such as titanium, molybdenum, and tungsten, have good strength and moderate ductility.

2. In face-centered cubic crystals there are 12 slip systems. The probability of slip is moderate, and the required shear stress is low. These metals, such as aluminum, copper, gold, and silver, have moderate strength and good ductility.

3. The hexagonal close-packed crystal has three slip systems, thus having a low probability of slip. However, more systems become active at elevated temperatures. Metals with hcp structures, such as beryllium, magnesium, and zinc, are generally brittle at room temperature.

Note in Fig. 3.4a that the portions of the single crystal that have slipped have rotated from their original angular position toward the direction of the tensile force. Note also that slip has taken place along certain planes only. With the use of electron microscopy, it has been shown that what appears to be a single slip plane is actually a **slip band,** consisting of a number of slip planes (Fig. 3.6).

### 3.3.2 Ideal tensile strength of metals

In addition to obtaining the maximum theoretical shear strength, we can also calculate the theoretical or ideal **tensile** strength of materials. In the bar shown in Fig. 3.7, the interatomic distance is $a$ when no stress is applied. In order to increase this distance we have to apply a

Grain boundaries

Single crystal (grain)

Approximately 1000 atomic diameters

Slip lines approximately 100 atomic diameters

Slip band

~10,000 atomic diameters

Shear stress

**FIGURE 3.6** Schematic illustration of slip lines and slip bands in a single crystal subjected to a shear stress. A slip band consists of a number of slip planes. The crystal at the center of the upper part is an individual grain surrounded by other grains.

force to overcome the cohesive force between the atoms, the cohesive force being zero in the unstrained equilibrium condition. When the tensile stress reaches $\sigma_{max}$, the atomic bonds between two neighboring atomic planes break. This $\sigma_{max}$ is known as the **ideal tensile strength.**

We may approximate the curve for cohesive force by a sine curve of wavelength $\lambda$ to obtain

$$\sigma = \sigma_{max} \sin\frac{2\pi x}{\lambda}, \tag{3.4}$$

where $x$ is the displacement from $a$. For small values of $x$, we write

$$\sigma = \sigma_{max} \frac{2\pi x}{\lambda}.$$

Also, for small values of $x$ we have, from Hooke's law,

$$\sigma = \frac{E \cdot x}{a}$$

and hence

$$\sigma_{max} = \frac{E \cdot \lambda}{2\pi a}. \tag{3.5}$$

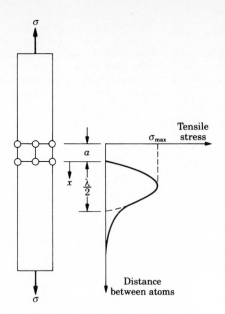

The work done per unit area of the tensile specimen is the area under the cohesive force curve. Thus,

$$\text{Work} = \int_0^{\lambda/2} \sigma_{max} \sin \frac{2\pi x}{\lambda} dx = \frac{\sigma_{max} \cdot \lambda}{\pi}. \tag{3.6}$$

The work consumed may be assumed to go into generating *two new surfaces* on the specimen where separation takes place. This involves surface energy. The atoms within the bulk of a solid or liquid have bonds to their neighbors in all directions; hence they are subjected to equal forces in all directions. The atoms at or near the surface, however, have about one-half the number of bonds. To bring an atom from within the bulk to the surface, work is required to overcome the unbalanced attractive forces. Thus the energy of an atom on the surface is higher than an atom in the bulk. The energy required to create a new surface is known as **surface energy** $\gamma$.

The surface energy involved in pulling a tensile specimen apart is

$$\text{Surface energy} = 2\gamma. \tag{3.7}$$

Combining Eqs. (3.5)–(3.7), we obtain

$$\sigma_{max} = \sqrt{E\gamma/a}. \tag{3.8}$$

For most solids, when we substitute appropriate values into this equation, we have roughly

$$\sigma_{max} \simeq E/10. \tag{3.9}$$

### 3.3.3  Imperfections

The actual strength of metals is approximately one to two orders of magnitude lower than the strength levels obtained from theoretical calculations. This discrepancy has been explained in terms of **imperfections** in the crystal structure. Unlike the idealized models we have

**FIGURE 3.8** Defects in a single-crystal lattice. Self-interstitial, vacancy, interstitial, and substitutional. *Source:* After Moffatt et al., *The Structure and Properties of Materials*, Vol. 1, John Wiley & Sons, 1976.

described, actual metal crystals contain a large number of defects and imperfections, which we may categorize as:

1. Point defects, such as a **vacancy** (missing atom), an **interstitial atom** (extra atom in the lattice), or an **impurity atom** (foreign atom) that has replaced the atom of the pure metal (Fig. 3.8).

2. *Linear* or *one-dimensional* defects, called **dislocations** (Fig. 3.9).

3. *Planar* or *two-dimensional* imperfections, such as **grain boundaries** or **phase boundaries.**

4. *Volume* or *bulk imperfections,* such as **voids, inclusions** (nonmetallic elements such as oxides, sulfides, and silicates), other **phases,** or **cracks.**

A slip plane containing a dislocation requires lower shear stress to cause slip than a plane in a perfect lattice (Fig. 3.10). One analogy used to describe the movement of an edge dislocation is the earthworm, which moves forward through a hump that starts at the tail and moves toward the head. A second is moving a large carpet by forming a hump at one end and

**FIGURE 3.9** (a) Edge dislocation, a linear defect at the edge of an extra plane of atoms. (After Guy and Hren, *Elements of Physical Metallurgy,* 1974.) (b) Screw dislocation (From: L. Van Vlack, *Materials for Engineering,* 4th edition, (Figure 4.44 from page 119), © 1980, Addison Wesley Publishing Company Inc. Reprinted by permission of Addison Wesley Longman, Inc.)

**FIGURE 3.10**   Movement of an edge dislocation across the crystal lattice under a shear stress. Dislocations help explain why the actual strength of metals is much lower than that predicted by theory.

moving it to the other end. The force required to move a carpet in this way is much less than that required to slide the carpet along the floor. Screw dislocations are so named because the atomic planes form a spiral ramp.

The *density of dislocations* (total length of dislocation line per unit volume, $mm/mm^3 = mm^{-2}$) increases with increasing plastic deformation by as much as $10^6$ at room temperature. The dislocation density for some conditions are:

Very pure single crystals, 0 to $10^3 mm^{-2}$
Annealed single crystals, $10^5$ to $10^6$
Annealed polycrystals, $10^7$ to $10^8$
Highly cold-worked metals, $10^{11}$ to $10^{12}$

The mechanical properties of metals, such as yield and fracture strength, as well as electrical conductivity, are affected by lattice defects. These are known as **structure-sensitive properties.** On the other hand, melting point, specific heat, coefficient of thermal expansion, and elastic constants are not sensitive to these defects, and are known as **structure-insensitive properties.**

### 3.3.4   Strain hardening (work hardening)

Although the presence of a dislocation lowers the shear stress required to cause slip, dislocations can (1) become entangled and interfere with each other and (2) be impeded by barriers, such as grain boundaries and impurities and inclusions in the material. Entanglement and impediments increase the shear stress required for slip. This entanglement is like moving two humps at different angles across a carpet with the two humps interfering with each other's movement, making it more difficult to move the carpet.

The increase in the shear stress required, and hence the increase in the overall strength of the metal, is known as **strain hardening** or **work hardening** (see Section 2.2.3). The greater the deformation, the more the entanglements, which increase the metal's strength. Work hardening is used extensively in strengthening metals in metalworking processes at ambient temperature. Typical examples are strengthening wire by reducing its cross-section by drawing it through a die, producing the head on a bolt by forging it, and producing sheet metal for automobile bodies and aircraft fuselages by rolling.

As shown by Eq. (2.11), the degree of strain hardening is indicated by the strain-hardening exponent $n$. Among the three crystal structures, hcp has the lowest $n$ value, followed by bcc and fcc, which has the highest.

We may now summarize the contents of this section as follows:

1. Plastic deformation in metals occurs by slip of atomic planes over one another.

2. In actual metals, the shear stress required to produce slip is one to two orders of magnitude lower than that required for perfect crystals; this is due to various imperfections in their crystal structure.

3. The mechanism of slip in actual metals involves dislocation motion.

4. Work hardening involves interactions between dislocations and other dislocations, dislocations and grain boundaries, dislocations and phase boundaries, and dislocations and solute atoms (minor components in a crystalline structure; see Section 5.2).

## 3.4 Grains and Grain Boundaries

Metals commonly used for manufacturing various products are composed of many individual, randomly oriented crystals (**grains**). We are thus dealing with metal structures that are not single crystals but **polycrystals,** that is, many crystals.

When a mass of molten metal begins to solidify, crystals begin to form independently of each other, with random orientations and at various locations within the liquid mass (Fig. 3.11). Each of these crystals grows into a crystalline structure or grain. The number and size of the grains developed in a unit volume of the metal depends on the rate at which **nucleation** (initial stage of formation of crystals) takes place. The number of different sites in which individual crystals begin to form (seven in Fig. 3.11a) and the rate at which these crystals grow are important in the size of grains developed. Generally, rapid cooling produces smaller grains, whereas slow cooling produces larger grains.

If the crystal nucleation rate is high, the number of grains in a unit volume of metal will be greater, and consequently grain size will be small. Conversely, if the rate of growth of the

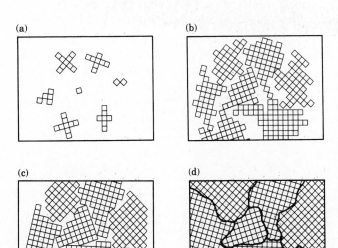

**FIGURE 3.11** Schematic illustration of the various stages during solidification of molten metal. Each small square represents a unit cell. (a) Nucleation of crystals at random sites in the molten metal. Note that the crystallographic orientation of each site is different. (b) and (c) Growth of crystals as solidification continues. (d) Solidified metal, showing individual grains and grain boundaries. Note the different angles at which neighboring grains meet each other. *Source:* W. Rosenhain.

**FIGURE 3.12**  Variation of tensile stress across a plane of polycrys- talline metal specimen subjected to tension. Note that the strength exhib- ited by each grain depends on its ori- entation.

crystals is high, compared to their nucleation rate, there will be fewer—but larger—grains per unit volume.

Note in Fig. 3.11 how the grains eventually interfere with and impinge upon one another. The surfaces that separate these individual grains are called **grain boundaries.** Each grain consists of either a single crystal, for pure metals, or a polycrystalline aggregate, for alloys. Note that the crystallographic orientation changes abruptly from one grain to the next across the grain boundaries. Recall from Section 3.3 that the behavior of a single crystal or single grain is anisotropic. The ideal behavior of a piece of polycrystalline metal (Fig. 3.12) is **isotropic** because the grains have random crystallographic orientations. Thus its properties do not vary with the direction of testing. In practice, however, this situation rarely exists.

### 3.4.1  Grain size

**Grain size** significantly influences the mechanical properties of metals. Large grain size is generally associated with low strength, hardness, and ductility. Large grains, particularly in sheet metals, also result in a rough surface appearance after being stretched. The yield strength $Y$ is the most sensi- tive property and is related to grain size by the empirical formula (*Hall-Petch equation*):

$$Y = Y_i + kd^{-1/2}, \tag{3.10}$$

where $Y_i$ is a basic yield stress that can be regarded as the stress opposing the motion of dis- locations, $k$ is a constant indicating the extent to which dislocations are piled up at barriers, such as grain boundaries, and $d$ is the grain diameter. Equation (3.10) is valid below the recrystallization temperature of the material.

We usually measure grain size by counting the number of grains in a given area or the number of grains that intersect a given length of a line randomly drawn on an enlarged photograph of the grains taken under a microscope on a polished and etched specimen. Grain size may also be determined by comparing it to a standard chart. The ASTM (The American Society for Testing and Materials) grain size number, $n$, is related to the number of grains, $N$, per square inch at a magnification of $100\times$ (equal to $0.0645$ mm$^2$ of actual area) by

$$N = 2^{n-1}. \tag{3.11}$$

Because grains are generally extremely small, many grains occupy a unit volume of metal (Table 3.1). Grain sizes between 5 and 8 are generally considered fine grains. A grain size of 7 is generally acceptable for sheet metals used to make car bodies, appliances, and kitchen utensils. Grains can be large enough to be visible to the naked eye, such as those of zinc on the surface of galvanized sheet steels.

### 3.4.2 Influence of grain boundaries

Grain boundaries have an important influence on the strength and ductility of metals. Because they interfere with the movement of dislocations, grain boundaries also influence strain hardening. These effects depend on temperature, rate of deformation, and the type and amount of impurities present along the grain boundaries. Grain boundaries are more reactive than the grains themselves. This is because the atoms along the grain boundaries are packed less efficiently and are more disordered and hence have a higher energy than the atoms in the orderly arrangement within the grains.

**TABLE 3.1**

**GRAIN SIZES**

| ASTM NO. | GRAINS/mm$^2$ | GRAINS/mm$^3$ |
|:---:|:---:|:---:|
| −3 | 1 | 0.7 |
| −2 | 2 | 2 |
| −1 | 4 | 5.6 |
| 0 | 8 | 16 |
| 1 | 16 | 45 |
| 2 | 32 | 128 |
| 3 | 64 | 360 |
| 4 | 128 | 1,020 |
| 5 | 256 | 2,900 |
| 6 | 512 | 8,200 |
| 7 | 1,024 | 23,000 |
| 8 | 2,048 | 65,000 |
| 9 | 4,096 | 185,000 |
| 10 | 8,200 | 520,000 |
| 11 | 16,400 | 1,500,000 |
| 12 | 32,800 | 4,200,000 |

At elevated temperatures and in materials whose properties depend on the rate of deformation, plastic deformation also takes place by means of **grain boundary sliding.** The **creep mechanism** (elongation under stress over a period of time, usually at elevated temperatures) results from grain boundary sliding.

### Grain boundary embrittlement

When brought into close atomic contact with certain low-melting-point metals, a normally ductile and strong metal can crack under very low stresses. Examples are aluminum wetted with a mercury-zinc amalgam or liquid gallium and copper at elevated temperature wetted with lead or bismuth (Fig. 3.13). These elements weaken the grain boundaries of the metal by **embrittlement.** We use the term **liquid-metal embrittlement** to describe such phenomena because the embrittling element is in a liquid state. However, embrittlement can also occur at temperatures well below the melting point of the embrittling element. This phenomenon is known as **solid-metal embrittlement.**

**Hot shortness** is caused by local melting of a constituent or an impurity in the grain boundary at a temperature below the melting point of the metal itself. When such a metal is subjected to plastic deformation at elevated temperatures (hot working), the piece of metal crumbles and disintegrates along the grain boundaries. Examples are antimony in copper and leaded steels and brass. To avoid hot shortness, the metal is usually worked at a lower temperature to prevent softening and melting along the grain boundaries. Another form of embrittlement is **temper embrittlement** in alloy steels, which is caused by segregation (movement) of impurities to the grain boundaries.

## 3.5   Plastic Deformation of Polycrystalline Metals

If a piece of polycrystalline metal with uniform equiaxed grains (having equal dimensions in all directions, as shown in the model in Fig. 3.14) is subjected to plastic deformation at room temperature (*cold working*), the grains become deformed and elongated. The deformation process may be carried out either by compressing the metal, as is done in forging, or by sub-

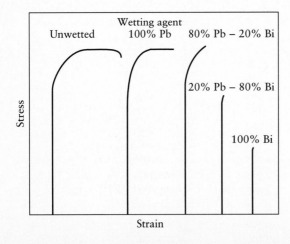

**FIGURE 3.13** Embrittlement of copper by lead and bismuth at 350°C (660°F). Embrittlement has important effects on the strength, ductility, and toughness of materials. *Source:* After W. Rostoker, et al., *Embrittlement by Liquid Metals.* Reinhold, 1960, p. 22.

jecting it to tension, as is done in stretching sheet metal. The deformation within each grain takes place by the mechanisms we described in Section 3.3 for a single crystal.

During plastic deformation, the grain boundaries remain intact and mass continuity is maintained. The deformed metal exhibits greater strength because of the entanglement of dislocations with grain boundaries. The increase in strength depends on the amount of deformation (strain) to which the metal is subjected; the greater the deformation, the stronger the metal becomes. Furthermore, the increase in strength is greater for metals with smaller grains because they have a larger grain-boundary surface area per unit volume of metal.

### 3.5.1 Anisotropy (texture)

Figure 3.14 shows that, as a result of plastic deformation, the grains have elongated in one direction and contracted in the other. Consequently, this piece of metal has become **anisotropic,** and its properties in the vertical direction are different from those in the horizontal direction.

Many products develop anisotropy of mechanical properties after they have been processed by metalworking techniques. The degree of anisotropy depends on how uniformly the metal is deformed. Note from the direction of the crack in Fig. 3.15, for example, that the ductility of the cold-rolled sheet in the vertical (transverse) direction is lower than in its longitudinal direction.

Anisotropy influences both mechanical and physical properties of metals. For example, sheet steel for electrical transformers is rolled in such a way that the resulting deformation imparts anisotropic magnetic properties to the sheet, thus reducing magnetic-hysteresis losses and improving the efficiency of transformers (see also "Amorphous alloys," Section 3.11.8). There are two general types of anisotropy in metals: **preferred orientation** and **mechanical fibering.**

**Preferred orientation.** Also called **crystallographic anisotropy,** preferred orientation can be best described by reference to Fig. 3.4. When a metal crystal is subjected to tension, the sliding blocks rotate toward the direction of pulling. Thus slip planes and slip bands tend to align themselves with the direction of deformation. Similarly, for a polycrystalline aggregate with grains in various orientations (Fig. 3.12), all slip directions tend to align themselves with the direction of pulling. Conversely, under compression, the slip planes tend to align themselves in a direction perpendicular to the direction of compression.

(a)    (b)

**FIGURE 3.14** Plastic deformation of idealized (equiaxed) grains in a specimen subjected to compression, such as is done in rolling or forging of metals: (a) before deformation; and (b) after deformation. Note the alignment of grain boundaries along a horizontal direction.

**FIGURE 3.15**   (a) Illustration of a crack in sheet metal subjected to bulging, such as by pushing a steel ball against the sheet. Note the orientation of the crack with respect to the rolling direction of the sheet. This material is anisotropic. (b) Aluminum sheet with a crack (vertical dark line at the center) developed in a bulge test. *Source:* J.S. Kallend.

**Mechanical fibering.**   Mechanical fibering results from the alignment of impurities, inclusions (stringers), and voids in the metal during deformation. Note that if the spherical grains in Fig. 3.14 were coated with impurities, these impurities would align themselves generally in a horizontal direction after deformation. Since impurities weaken the grain boundaries, this piece of metal would be weak and less ductile when tested in the vertical direction. An analogy would be plywood, which is strong in tension along its planar direction, but peels off easily when tested with tension in its thickness direction.

## 3.6   Recovery, Recrystallization, and Grain Growth

We have shown that plastic deformation at room temperature results in the deformation of grains and grain boundaries, a general increase in strength, and a decrease in ductility, as well as causing anisotropic behavior. These effects can be reversed and the properties of the metal brought back to their original levels by heating it in a specific temperature range for a period of time. The temperature range and amount of time depend on the material and several other factors. Three events take place consecutively during the heating process:

1.  **Recovery.** During recovery, which occurs at a certain temperature range below the **recrystallization temperature** of the metal, the stresses in the highly deformed regions are relieved and the number of mobile dislocations is reduced. Subgrain boundaries begin to form—called **polygonization**—with no appreciable change in mechanical properties, such as hardness and strength but some increase in ductility (Fig.3.16).

2.  **Recrystallization.** The process in which, at a certain temperature range, new equiaxed and strain-free grains are formed, replacing the older grains, is called *recrystallization*. The temperature for recrystallization ranges approximately between $0.3T_m$ and $0.5T_m$, where $T_m$ is the melting point of the metal on the absolute scale. The recrystallization temperature is generally defined as the temperature at which complete recrystallization occurs within approximately one hour. Recrystallization decreases the density of dislocations and lowers the strength but raises the ductility of the metal (Fig. 3.16). Metals

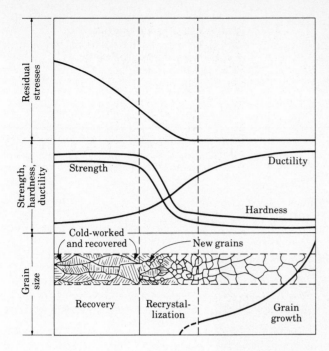

**FIGURE 3.16** Schematic illustration of the effects of recovery, recrystallization, and grain growth on mechanical properties and shape and size of grains. Note the formation of small new grains during recrystallization. *Source:* G. Sachs.

such as lead, tin, cadmium, and zinc recrystallize at about room temperature. Thus when these metals are *cold worked*, they do not work-harden.

Recrystallization depends on the degree of prior cold work (work hardening); the higher the amount of cold work, the lower the temperature required for recrystallization to occur. The reason is that as the amount of cold work increase, the number of dislocation and the amount of energy stored in dislocations (**stored energy**) also increase. This energy supplies the work required for recrystallization. Recrystallization is a function of time because it involves diffusion, that is, movement and exchange of atoms across grain boundaries.

The effects on recrystallization of temperature, time, and reduction in the thickness or height of the workpiece by cold working are as follows (Fig. 3.17):

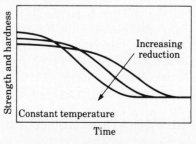

**FIGURE 3.17** Variation of strength and hardness with recrystallization temperature, time, and prior cold work. Note that the more a metal is cold worked, the less time it takes to recrystallize. This is because of the higher stored energy from cold working due to increased dislocation density.

**FIGURE 3.18** The effect of prior cold work on the recrystallized grain size of alpha brass. Below a critical elongation (strain), typically 5%, no recrystallization occurs.

**FIGURE 3.19** Surface roughness on the cylindrical surface of an aluminum specimen subjected to compression. *Source:* A. Mulc and S. Kalpakjian.

- For a constant amount of deformation by cold working, the time required for recrystallization decreases with increasing temperature.

- The more prior cold work done, the lower is the temperature required for recrystallization.

- The higher the amount of deformation, the smaller the grain size becomes during recrystallization (Fig. 3.18). This is a common method of converting a coarse-grained structure to one of fine grain, with improved properties.

- Anisotropy due to preferred orientation usually persists after recrystallization. To restore isotropy, a temperature higher than that required for recrystallization may be necessary.

3. **Grain Growth.** If we continue to raise the temperature of the metal, the grains begin to grow and their size may eventually exceed the original grain size. This phenomenon is known as *grain growth*, and it has a slightly adverse effect on mechanical properties (Fig. 3.16). However, large grains produce a rough surface appearance on sheet metals—called **orange peel**—when they are stretched to form a part or when a piece of metal is subjected to compression (Fig. 3.19), such as in forging operations.

# 3.7  Cold, Warm, and Hot Working

When plastic deformation is carried out at room temperature, it is called **cold working**, and when carried out at above the recrystallization temperature, it is called **hot working**. As the

**TABLE 3.2**

**HOMOLOGOUS TEMPERATURE RANGES FOR VARIOUS PROCESS**

| PROCESS | $T/T_m$ |
| --- | --- |
| Cold working | < 0.3 |
| Warm working | 0.3 to 0.5 |
| Hot working | > 0.6 |

name implies, **warm working** is carried out at higher than room temperature but lower than hot-working temperature. Thus warm working is a compromise between cold and hot working. The temperature ranges for these three categories of plastic deformation are given in Table 3.2 in terms of a ratio, where $T$ is the working temperature and $T_m$ is the melting point of the metal, both on the absolute scale. Although it is a dimensionless quantity, this ratio is known as the **homologous temperature**.

There are important technological differences in products that are processed by cold, warm, or hot working. For example, compared to cold-worked products, hot-worked products generally have less dimensional accuracy because of uneven thermal expansion and contraction during processing, and rougher surface appearance and finish because of the oxide layer that usually develops during heating. Other important manufacturing characteristics, such as formability, machinability, and weldability, are also affected by cold, warm, and hot working to various degrees.

## 3.8 Failure and Fracture

**Failure** is one of the most important aspects of a material's behavior because it directly influences the selection of a material for a certain application, the methods of manufacturing, and the service life of the component. Because of the many factors involved, failure and fracture of materials is a complex area of study. In this section, we consider only those aspects of failure that are of particular significance to selecting and processing materials. There are two general types of failure: (1) fracture and separation of the material, either through internal or external cracking; and (2) buckling (Fig. 3.20). Fracture is further divided into two general categories: ductile and brittle (Fig. 3.21).

(a)  (b)  (c)  (d)

Cracks

**FIGURE 3.20** Schematic illustration of types of failures in materials: (a) necking and fracture of ductile materials; (b) buckling of ductile materials under a compressive load; (c) fracture of brittle materials in compression; (d) cracking on the barreled surface of ductile materials in compression.

(a)    (b)    (c)    (d)

**FIGURE 3.21** Schematic illustration of the types of fracture in tension: (a) brittle fracture in polycrystalline metals; (b) shear fracture in ductile single crystals; (c) ductile cup-and-cone fracture in polycrystalline metals; (d) complete ductile fracture in polycrystalline metals, with 100 percent reduction of area.

Although failure of materials is generally regarded as undesirable, certain products are designed in such a way that failure is essential for their function. Typical examples are food and beverage containers with tabs or entire tops, which are removed by tearing the sheet metal along a prescribed path, and screw caps for bottles.

### 3.8.1  Ductile fracture

**Ductile fracture** is characterized by plastic deformation, which precedes failure of the part. In a tension test, highly ductile materials such as gold and lead may neck down to a point and then fail (Fig. 3.21d). Most metals and alloys, however, neck down to a finite area and then fail. Ductile fracture generally takes place along planes on which the *shear stress is a maximum*. In torsion, for example, a ductile metal fractures along a plane perpendicular to the axis of twist, that is, the plane on which the shear stress is a maximum. Fracture in shear is a result of extensive slip along slip planes within the grains.

Upon close examination of the surface of a ductile fracture (Fig. 3.22), we see a *fibrous* pattern with *dimples*, as if a number of very small tension tests have been carried out over the fracture surface (Fig. 3.23). Failure is initiated with the formation of tiny voids, usually around small inclusions or preexisting *voids*, which then *grow* and *coalesce*, developing cracks that grow in size and lead to fracture.

In a tension-test specimen, fracture begins at the center of the necked region from the growth and coalescence of cavities (Fig. 3.24). The central region becomes one large crack

**FIGURE 3.22** Surface of ductile fracture in low-carbon steel showing dimples. Fracture is usually initiated at impurities, inclusions, or preexisting voids in the metal. *Source:* K.-H. Habig and D. Klaffke. Photo by BAM Berlin/Germany.

**FIGURE 3.23**   A schematic illustration of the mechanism of ductile fracture. Note the formation of elongated dimples on the fracture surface. *Source:* After H. C. Rogers.

and propagates to the periphery of this necked region. Because of its appearance, the fracture surface of a tension-test specimen is call a **cup-and-cone fracture** (Fig. 3.21c).

### Effects of inclusions

Because they are nucleation sites for voids, **inclusions** have an important influence on ductile fracture and thus on the formability of materials (Fig. 3.25). Inclusions may consist of impurities of various kinds and second-phase particles, such as oxides, carbides, and sulfides. The extent of their influence depends on factors such as their shape, hardness, distribution, and volume fraction. The greater the volume fraction of inclusions, the lower will be the ductility of the material. Voids and porosity developed during processing, such as from casting, reduce the ductility of a material.

Two factors affect void formation. One is the strength of the bond at the interface of an inclusion and the matrix. If the bond is strong, there is less tendency for void formation during plastic deformation. The second factor is the hardness of the inclusion. If it is soft, such as manganese sulfide, it will conform to the overall change in shape of the specimen or work-

**FIGURE 3.24**   Sequence of events in necking and fracture of a tensile-test specimen: (a) early stage of necking; (b) small voids begin to form within the necked region; (c) voids coalesce, producing an internal crack; (d) rest of cross-section begins to fail at the periphery by shearing; (e) final fracture surfaces, known as cup- (top fracture surface) and-cone (bottom surface) fracture.

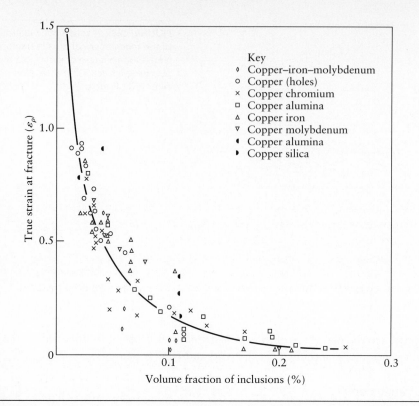

**FIGURE 3.25**   The effect of volume fraction of various second-phase particles on the true strain at fracture in a tensile test for copper. *Source:* After B.I. Edelson and W.M. Baldwin.

piece during plastic deformation. If it is hard, such as a carbide or oxide, it could lead to void formation (Fig. 3.26). Hard inclusions may also break up into smaller particles during deformation because of their brittle nature. The alignment of inclusions during plastic deformation leads to **mechanical fibering**. Subsequent processing of such a material must therefore involve considerations of the proper direction of working for maximum ductility and strength.

**FIGURE 3.26**   Schematic illustration of the deformation of soft and hard inclusions and their effect on void formation in plastic deformation. Note that hard inclusions, because they do not comply with the overall deformation of the ductile matrix, can cause voids.

**FIGURE 3.27** Schematic illustration of transition temperature. Note the narrow temperature range across which the behavior of the metal undergoes a major transition.

**Transition temperature.** Many metals undergo a sharp change in ductility and toughness across a narrow temperature range, called the **transition temperature** (Fig. 3.27). This phenomenon occurs in body-centered cubic and some hexagonal close-packed metals; it is rarely exhibited by face-centered cubic metals. The transition temperature depends on factors such as composition, microstructure, grain size, surface finish and shape of the specimen, and rate of deformation. High rates, abrupt changes in shape, and surface notches raise the transition temperature.

## Strain aging

*Strain aging* is a phenomenon in which carbon atoms in steels segregate to dislocations, thereby pinning them and thus increasing the resistance to dislocation movement. The result is increased strength and reduced ductility. The effects of strain aging on the shape of the stress-strain curve for low-carbon steel at room temperature are shown in Fig. 3.28.

**FIGURE 3.28** Strain aging and its effect on the shape of the true stress-true strain curve for 0.03% C rimmed steel at 60°C (140°F). *Source:* A.S. Keh and W.C. Leslie.

Curve *abc* shows the original curve, with upper and lower yield points typical of these steels. If the tension test is stopped at point *e* and the specimen is unloaded and tested again, curve *dec* is obtained. Note that the upper and lower yield points have disappeared. However, if 4 hours pass before the specimen is strained again, curve *dfg* is obtained. If 126 hours pass, curve *dhi* is obtained. Instead of taking place over several days at room temperature, this same phenomenon can occur in a few hours at a higher temperature, and is called **accelerated strain aging**.

An example of accelerated strain aging in steels is **blue brittleness**, so named because it occurs in the blue-heat range where the steel develops a bluish oxide film. This phenomenon causes a marked decrease in ductility and toughness and an increase in strength of plain carbon and some alloy steels.

### 3.8.2  Brittle fracture

Brittle fracture occurs with little or no gross plastic deformation preceding the separation of the material into two or more pieces. In tension, fracture takes place along a crystallographic plane, called a **cleavage plane**, on which the normal tensile stress is a maximum. Face-centered cubic metals usually do not fail in brittle fracture, whereas body-centered cubic and some hexagonal close-packed metals fail by cleavage. In general, low temperature and high rates of deformation promote brittle fracture.

In a polycrystalline metal under tension, the fracture surface has a bright granular appearance because of the changes in the direction of the cleavage planes as the crack propagates from one grain to another. An example of the surface of brittle fracture is shown in Fig. 3.29. Brittle fracture of a specimen in compression is more complex and may follow a path that is theoretically 45° to the applied force direction.

Examples of fracture along a cleavage plane are the splitting of rock salt or peeling of layers of mica. Tensile stresses normal to the cleavage plane, caused by pulling, initiate and control the propagation of fracture. Another example is the behavior of brittle materials, such as chalk, gray cast iron, and concrete. In tension, they fail in the manner shown

**FIGURE 3.29**  Typical surface of steel that has failed in a brittle manner. The fracture path is transgranular (through the grains). Compare with the ductile fracture surface in Fig. 3.22. Magnification: 200×. *Source:* Courtesy of B. J. Schulze and S. L. Meiley and Packer Engineering Associates, Inc.

in Fig. 3.21a. In torsion, they fail along a plane at 45° to the axis of twist, that is, along a plane on which the tensile stress is a maximum.

## Defects

An important factor in fracture is the presence of defects such as scratches, flaws, and external or internal cracks. Under tension, the tip of the crack is subjected to high tensile stresses, which propagate the crack rapidly because the material has little capacity to dissipate energy. The tensile strength of a specimen, with a crack perpendicular to the direction of pulling, is related to the length of the crack as follows:

$$\sigma \, \alpha \, \frac{1}{\sqrt{\text{crack length}}}. \tag{3.12}$$

The presence of defects is essential in explaining why brittle materials are so weak in tension compared to their strength in compression. The difference is on the order of 10 for rocks and similar materials, about 5 for glass, and about 3 for gray cast iron. Under tensile stresses, cracks propagate rapidly, causing what is known as catastrophic failure.

With polycrystalline metals, the fracture paths most commonly observed are **transgranular** (transcrystalline or intragranular), meaning that the crack propagates through the grain. **Intergranular** fracture, where the crack propagates along the grain boundaries (Fig. 3.30), generally occurs when the grain boundaries are soft, contain a brittle phase, or have been weakened by liquid- or solid-metal embrittlement.

The maximum crack velocity in a brittle material is about 38% of the elastic wave-propagation (or acoustic) velocity of the material. (This velocity is given by the formula $\sqrt{E/\rho}$, where $E$ is the elastic modulus and $\rho$ is the mass density.) Thus, for steel, the maximum crack velocity is 2000 m/s (6600 ft/s).

As Fig. 3.31 shows, cracks may be subjected to stresses in different directions. Mode I is tensile stress applied perpendicular to the crack. Modes II and III are shear stresses applied in two different directions. Tearing paper, cutting sheet metal with shears, or opening flip-top cans are examples of Mode III fracture.

**FIGURE 3.30**  Intergranular fracture, at two different magnifications. Grains and grain boundaries are clearly visible in this micrograph. The fracture path is along the grain boundaries. Magnification: left, 100×; right, 500×. *Source:* Courtesy of B. J. Schulze and S. L. Meiley and Packer Engineering Associates, Inc.

**FIGURE 3.31** Three modes of fracture. Mode I has been studied extensively because it is the most commonly observed situation in engineering structures and components. Mode II is rare. Mode III is the tearing process, similar to opening a flip-top can, tearing a piece of paper, or cutting materials with a pair of scissors.

**Example 3.1:** Brittle fracture

A 5-lb block is dropped from a height of 12 in. on a 1-in. cube-shaped specimen. The specimen is made of a perfectly brittle material with a surface energy of 0.01 in. · lb/in². Estimate the number of pieces into which the specimen will break.

SOLUTION.    Assume that each broken piece is a cube of dimension $x$ on each side and that all broken pieces are of equal size. Thus, let

$$x = \text{lateral dimension of each broken piece;}$$

$$n = \text{number of broken pieces.}$$

Then,

$$\text{surface area of each broken piece} = 6x^2,$$

$$\text{volume of each broken piece} = x^3, \text{ and}$$

$$\text{total volume of broken pieces} = 1 \text{ in}^3.$$

Hence

$$n = \frac{1}{x^3}.$$

The potential energy of the falling block is dissipated as surface energy, so

$$(5)(12) = (n)(6x^2)(0.01) = \frac{0.06}{x}.$$

Thus

$$x = 0.001 \text{ in.}$$

and

$$n = 10^9 \text{ pieces.}$$

Note that, because it represents a very small percentage of the energy involved, the original surface of the specimen (6 in$^2$) has been ignored in these calculations. Although $n$ appears to be a large number (because of the data in this problem), a piece of chalk, for example, indeed breaks into a large number of small pieces when stepped upon.

## Fatigue fracture

Fatigue fracture is basically of a brittle nature. Minute external or internal cracks develop at flaws or defects in the material, which then propagate and eventually lead to total failure of the part. The fracture surface in fatigue is generally characterized by the term **beach marks**, because of its appearance (Fig. 3.32). With large magnification, such as higher than 1000×, a series of **striations** can be seen on fracture surfaces, each beach mark consisting of a number of striations.

Fatigue life is greatly influenced by the method of preparation of the surfaces of the part or specimen (Fig. 3.33). The fatigue strength of manufactured products can be improved generally by the following methods:

1. Inducing compressive residual stresses on surfaces, such as by shot peening or roller burnishing
2. Surface (case) hardening by various means
3. Providing fine surface finish, thus reducing the effects of notches and other surface imperfections
4. Selecting appropriate materials and ensuring that they are free from significant amounts of inclusions, voids, and impurities

Conversely, the following factors and processes can reduce fatigue strength: decarburization, surface pits due to corrosion that act as stress raisers, hydrogen embrittlement, galvanizing, and electroplating.

**FIGURE 3.32** Typical fatigue fracture surface on metals, showing beach marks. Most components in machines and engines fail by fatigue and not by excessive static loading. Magnification: left, 500×; right, 1000×. *Source:* Courtesy of B. J. Schulze and S. L. Meiley and Packer Engineering Associates, Inc.

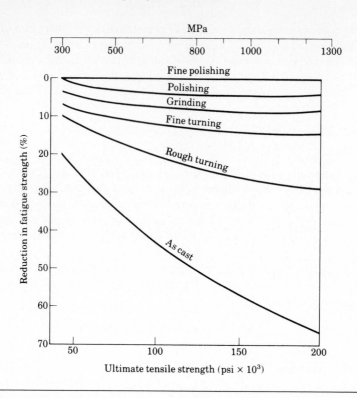

**FIGURE 3.33** Reduction in fatigue strength of cast steels subjected to various surface-finishing operations. Note that the reduction is greater as surface roughness and strength of the steel increase. *Source:* M.R. Mitchell.

**Stress-corrosion cracking.** An otherwise ductile metal can fail in a brittle manner by *stress-corrosion cracking* (*stress cracking or season cracking*). Parts free from defects after forming may, either over a period of time or soon after being made into a product, develop cracks. Crack propagation may be intergranular or transgranular.

The susceptibility of metals to stress-corrosion cracking depends mainly on the material, the presence and magnitude of tensile residual stresses, and the environment. Brass and austenitic stainless steels are among metals that are highly susceptible to stress cracking. The environment could be corrosive media such as salt water or other chemicals. The usual procedure to avoid stress-corrosion cracking is to stress relieve that part just after it is formed. Full annealing may also be done, but this treatment reduces the strength of cold-worked parts.

**Hydrogen embrittlement.** The presence of hydrogen can reduce ductility and cause severe embrittlement in many metals, alloys, and nonmetallic materials and cause premature failure. This phenomenon is known as **hydrogen embrittlement** and is especially severe in high-strength steels. Possible sources of hydrogen are during melting of the metal, during pickling (removing of surface oxides by chemical or electrochemical reaction), through electrolysis in electroplating, from water vapor in the atmosphere, or from moist electrodes and fluxes used during welding.

### 3.8.3 Size effect

Dependence of the properties of a material on its size is knows as **size effect**. Note from the foregoing discussions and from Eq. (3.12) that as the size decrease, defects, cracks, imperfections, and the like are less likely to occur. Hence the strength and ductility of a specimen increase with decreasing size. Strength is related to diameter, but the length of the specimen is also important. The greater the length, the greater the probability of defects. As an analogy, a long chain is likely to be weaker than a shorter one, because the probability of one of its links being weak increases with the number of links (length of chain). Size effect is also demonstrated by **whiskers**, which, because of their small size, are either free from imperfections or do not contain the types of imperfections that affect their strength. The actual strength in such cases approaches the theoretical strength.

## 3.9 Physical Properties

In addition to their mechanical properties, the *physical properties* of materials should also be considered in selecting and processing materials. Properties of particular interest are density, melting point, specific heat, thermal conductivity and expansion, electrical and magnetic properties, and resistance to oxidation and corrosion.

### 3.9.1 Density

*Density* of a material is its weight per unit volume. If expressed relative to the density of water, it is known as **specific gravity** and thus has no units. The range of densities for a variety of materials at room temperature is given in Table 3.3. The density of materials depends on atomic weight, atomic radius, and the packing of the atoms. Alloying elements generally have a minor effect on the density of metals, the effect depending on the density of the alloying elements.

The most significant role that density plays is in the **specific strength** (strength-to-weight ratio) and **specific stiffness** (stiffness-to-weight ratio) of materials and structures. Saving weight is particularly important for aircraft and aerospace structures, automotive bodies and components, and other products for which energy consumption and power limitations are major concerns. Substitution of materials for weight saving and economy is a major factor in advanced equipment and machinery as well as in consumer products such as automobiles.

Density is an important factor in the selection of materials for high-speed equipment, such as the use of magnesium in printing and textile machinery, many components of which usually operate at very high speeds. To obtain exposure time of 1/4000 s in cameras without sacrificing accuracy, the shutters of some high-quality 35-mm cameras are made of titanium. The resulting light weight of the components in these high-speed operations reduces inertial

**TABLE 3.3**

**PHYSICAL PROPERTIES OF SELECTED MATERIALS AT ROOM TEMPERATURE**

| METAL | DENSITY (kg/m³) | MELTING POINT (°C) | SPECIFIC HEAT (J/kg K) | THERMAL CONDUCTIVITY (W/m K) |
|---|---|---|---|---|
| Aluminum | 2700 | 660 | 900 | 222 |
| Aluminum alloys | 2630–2820 | 476–654 | 880–920 | 121–239 |
| Beryllium | 1854 | 1278 | 1884 | 146 |
| Columbium (niobium) | 8580 | 2468 | 272 | 52 |
| Copper | 8970 | 1082 | 385 | 393 |
| Copper alloys | 7470–8940 | 885–1260 | 337–435 | 29–234 |
| Iron | 7860 | 1537 | 460 | 74 |
| Steels | 6920–9130 | 1371–1532 | 448–502 | 15–52 |
| Lead | 11350 | 327 | 130 | 35 |
| Lead alloys | 8850–11350 | 182–326 | 126–188 | 24–46 |
| Magnesium | 1745 | 650 | 1025 | 154 |
| Magnesium alloys | 1770–1780 | 610–621 | 1046 | 75–138 |
| Molybdenum alloys | 10210 | 2610 | 276 | 142 |
| Nickel | 8910 | 1453 | 440 | 92 |
| Nickel alloys | 7750–8850 | 1110–1454 | 381–544 | 12–63 |
| Tantalum alloys | 16600 | 2996 | 142 | 54 |
| Titanium | 4510 | 1668 | 519 | 17 |
| Titanium alloys | 4430–4700 | 1549–1649 | 502–544 | 8–12 |
| Tungsten | 19290 | 3410 | 138 | 166 |

NONMETALLIC

| | | | | |
|---|---|---|---|---|
| Ceramics | 2300–5500 | — | 750–950 | 10–17 |
| Glasses | 2400–2700 | 580–1540 | 500–850 | 0.6–1.7 |
| Graphite | 1900–2200 | — | 840 | 5–10 |
| Plastics | 900–2000 | 110–330 | 1000–2000 | 0.1–0.4 |
| Wood | 400–700 | — | 2400–2800 | 0.1–0.4 |

forces that otherwise could lead to vibrations, inaccuracies, and even part failure over time. Because of their low density, ceramics are being considered for use in components of automated machine tools in high-speed operations. However, there are applications where weight is desirable. Examples are counterweights for various mechanisms (using lead and steel) and components for self-winding watches (using high-density materials such as tungsten).

### 3.9.2 Melting point

The **melting point** of a material depends on the energy required to separate its atoms. As Table 3.3 shows, the melting point of an alloy can have a wide range, unlike pure metals, which have a definite melting point. The melting points of alloys depend on their particular composition.

The melting point of a material has a number of indirect effects on manufacturing operations. The choice of a material for high-temperature applications is the most obvious consideration, such as in jet engines and furnaces, or where high frictional heat is generated, such as with high sliding speeds of machine components. Sliding down a rope or rubbing your hands together fast are simple demonstrations of how high temperatures can become with increasing sliding speed.

Since the recrystallization temperature of a metal is related to its melting point, operations such as annealing, heat treating, and hot working require a knowledge of the melting points of the materials involved. These considerations, in turn, influence the selection of tool and die materials in manufacturing operations. Another major influence of the melting point is in the selection of the equipment and melting practice in casting operations. The higher the melting point of the material, the more difficult the operation becomes. Also, the selection of die materials, as in die casting, depends on the melting point of the workpiece material. In the electrical-discharge machining process, the melting points of metals are related to the rate of material removal and tool wear.

### 3.9.3 Specific heat

**Specific heat** is energy required to raise the temperature of a unit mass of material by one degree. Alloying elements have a relatively minor effect on the specific heat of metals.

The temperature rise in a workpiece, resulting from forming or machining operations, is a function of the work done and the specific heat of the workpiece material. Thus the lower the specific heat, the higher the temperature will rise in the material. If excessive, temperature can have detrimental effects on product quality by adversely affecting surface finish and dimensional accuracy, can cause excessive tool and die wear, and can result in adverse metallurgical changes in the material.

### 3.9.4 Thermal conductivity

Whereas specific heat indicates the temperature rise in the material as a result of energy input, **thermal conductivity** indicates the ease with which heat flows within and through the material. Metallically bonded materials (metals) generally have high thermal conductivity, whereas ionically or covalently bonded materials (ceramics and plastics) have poor conductivity. Because of the large difference in their thermal conductivities, alloying elements can have a significant effect on the thermal conductivity of alloys, as we can see in Table 3.3 by comparing the metals with their alloys.

When heat is generated by plastic deformation or friction, the heat should be conducted away at a high enough rate to prevent a severe rise in temperature. The main difficulty in machining titanium, for example, is caused by its very low thermal conductivity. Low thermal conductivity can also result in high thermal gradients and thus cause inhomogeneous deformation in metalworking processes.

### 3.9.5 Thermal expansion

**Thermal expansion** of materials can have several significant effects. Generally, the coefficient of thermal expansion is inversely proportional to the melting point of the material. Alloying elements have a relatively minor effect on the thermal expansion of metals.

Shrink fits utilize thermal expansion. Other examples where relative expansion or contraction is important are electronic and computer components, glass-to-metal seals, struts on jet engines, and moving parts in machinery that require certain clearances for proper functioning. Improper selection of materials and assembly can cause **thermal stresses**, cracking, warping, or loosening of components in the structure during their service life.

Thermal conductivity, in conjunction with thermal expansion, plays the most significant role in causing thermal stresses, both in manufactured components and in tools and dies. This is particularly important in a forging operation, for example, when hot workpieces are placed over relatively cool dies, making the die surfaces undergo thermal cycling. To reduce thermal stresses, a combination of high thermal conductivity and low thermal expansion is desirable.

Thermal stresses can lead to cracks in ceramic parts and in tools and dies made of relatively brittle materials. **Thermal fatigue** results from thermal cycling and causes a number of surface cracks. **Thermal shock** is the term generally used to describe development of cracks after a single thermal cycle. Thermal stresses may be caused both by temperature gradients and by **anisotropy of thermal expansion**, which we generally observe in hexagonal close-packed metals and in ceramics.

Because of their resistance to high temperature and wear, ceramics have become attractive materials for components of heat engines. One of the most attractive ceramics is partially stabilized zirconia, which has thermal expansion characteristics close to those of cast iron. Thus, components such as cylinder liners do not undergo significant dimensional changes with respect to the engine block as engine temperature changes.

The influence of temperature on dimensions and modulus of elasticity can be a significant problem in precision instruments and equipment. A spring, for example, will have a lower stiffness as its temperature increases because of reduced elastic modulus. Similarly, a tuning fork or a pendulum will have different frequencies at different temperatures. Dimensional changes resulting from thermal expansion can be significant in measurements for quality control and in equipment such as measuring tapes for geodetic applications.

To alleviate some of the problems of thermal expansion, a family of iron-nickel alloys that have very low thermal-expansion coefficients has been developed, called **low-expansion alloys**. A typical composition is 64 percent iron–36 percent nickel. This alloy is called Invar, because the length of a component made of this alloy is almost invariable with respect to temperature, although over a narrow range. Consequently, these alloys also have good thermal-fatigue resistance.

### 3.9.6  Electrical and magnetic properties

Electrical conductivity and dielectric properties of materials are of great importance, not only in electrical equipment and machinery but also in manufacturing processes such as magnetic-pulse forming of sheet metals, and electrical-discharge machining and electrochemical grinding of hard and brittle materials.

The **electrical conductivity** of a material can be defined as a measure of the ease with which a material conducts electric current. Materials with high conductivity, such as metals, are generally referred to as **conductors**. The units of electrical conductivity are mho/m or mho/ft, where mho is the inverse of ohm, the unit for electrical resistance. **Electrical resistivity** is the inverse of conductivity. Materials with high resistivity are referred to as **dielectrics** or **insulators**. The influence of the type of atomic bonding on the electrical conductivity of materials is the same as that for thermal conductivity. Alloying elements have a major effect on the electrical conductivity of metals: the higher the conductivity of the alloying element, the higher the conductivity of the alloy. Dielectric strength of materials is the resistivity to direct electric current and is defined as the voltage required per unit distance for electrical breakdown. The unit are V/m or V/ft.

**Superconductivity** is the phenomenon of almost zero electrical resistivity that occurs in some metals and alloys below a critical temperature. The highest temperature at which superconductivity has to date been exhibited is with an alloy of lanthanum, strontium, copper, and oxygen, although other material compostions are also being studied. The highest temperature at which superconductivity has been exhibited to date is at about $-123°C$ ($-190°F$). Developments in superconductivity indicate that the efficiency of electrical components such as large high-power magnets, high-voltage power lines, and various other electronic and computer components can be markedly improved.

The electrical properties of materials such as single-crystal silicon, germanium, and gallium arsenide make them extremely sensitive to the presence and type of minute impurities as well as to temperature. Thus, by controlling the concentration and type of impurities (**dopants**), such as phosphorus and boron in silicon, electrical conductivity can be controlled. This property is utilized in **semiconductor** (solid state) devices used extensively in miniaturized electronic circuitry. They are very compact, efficient, and relatively inexpensive, consume little power, and require no warmup time for operation.

**Ferromagnetism** is the large and permanent magnetization resulting from an exchange interaction between neighboring atoms (such as iron, nickel, and cobalt) that aligns their magnetic moments in parallel. It has important applications in electric motors, generators, transformers, and microwave devices. **Ferrimagnetism** is permanent and large magnetization exhibited by some ceramic materials, such as cubic ferrites.

The **piezoelectric effect** (*piezo* from Greek, meaning to press) is exhibited by some materials, such as certain ceramics and quartz crystals, in which there is a reversible interaction between an elastic strain and an electric field. This property is utilized in making transducers, which are devices that convert the strain from an external force to electrical energy. Typical applications are force or pressure transducers, strain gages, sonar detectors, and microphones.

The phenomenon of expansion or contraction of a material when subjected to a magnetic field is call **magnetostriction**. Some materials, such as pure nickel and some iron-nickel alloys, exhibit this behavior. Magnetostriction is the principle behind ultrasonic machining equipment (see Section 9.7).

### 3.9.7 Corrosion resistance

We apply the word **corrosion** to deterioration of metals and ceramics, while calling similar phenomena in plastics **degradation**. Corrosion resistance is an important aspect of material selection for applications in the chemical, food, and petroleum industries, as well as in manufacturing operations. In addition to various possible chemical reactions from the elements and compounds present, environmental oxidation and corrosion of components and structures is a major concern, particularly at elevated temperatures and in automobiles and other transportation vehicles.

Resistance to corrosion depends on the particular environment, as well as the composition of the material. Corrosive media may be chemicals (acids, alkali, and salts), the environment (oxygen, pollution, and acid rain), and water (fresh or salt). Nonferrous metals, stainless steels, and nonmetallic materials generally have high corrosion resistance. Steels and cast irons generally have poor resistance and must be protected by various coatings and surface treatments.

Corrosion can occur over an entire surface, or it can be localized, such as **pitting**. It can occur along grain boundaries of metals as **intergranular corrosion** and at the interface of bolted or riveted joints as **crevice corrosion**. Two dissimilar metals may form a galvanic cell—that is, two electrodes in an electrolyte in a corrosive environment (including moisture)—and cause **galvanic corrosion**. Two-phase alloys are more susceptible to galvanic corrosion, because of the two different metals involved, than are single-phase alloys or pure metals. Thus, heat treatment can have an influence on corrosion resistance.

Moreover, corrosion may act in indirect ways. **Stress-corrosion cracking** is an example of the effect of a corrosive environment on the integrity of a product that, as manufactured, had residual stresses in it. Likewise, metals that are cold worked are likely to contain residual stresses and hence are more susceptible to corrosion, compared to hot-worked or annealed metals.

Tool and die materials also can be susceptible to chemical attack by lubricants and coolants. The chemical reaction alters their surface finish and adversely influences the metalworking operation. An example is carbine tools and dies having cobalt as a binder, in which the cobalt is attacked by elements in the cutting fluid (**selective leaching**). Thus, compatibility of the tool, die, and workpiece materials and the metalworking fluid under actual operating conditions is an important consideration in the selection of materials.

Chemical reactions should not be regarded as having adverse effects only. Nontraditional machining processes such as chemical machining and electrochemical machining are indeed based on controlled reactions. These processes remove material by chemical reactions, similar to the etchings of metallurgical specimens.

The usefulness of some level of **oxidation** is exhibited by the corrosion resistance of aluminum, titanium, and stainless steel. Aluminum develops a thin (a few atomic layers), strong, and adherent hard oxide film $(Al_2O_3)$ that protects the surface from further environmental corrosion. Titanium develops a film of titanium oxide $(TiO_2)$. A similar phenomenon occurs in stainless steels, which, because of the chromium present in the alloy, develop a similar protective film on their surfaces. These processes are known as **passivation**. When the protective film is scratched, thus exposing the metal underneath, a new oxide film forms in due time.

## 3.10  Properties of Ferrous Alloys

By virtue of their wide range of mechanical, physical, and chemical properties, **ferrous alloys** are among the most useful of all metals. Ferrous metals and alloys contain iron as their base metal; these metals are carbon and alloy steels, stainless steels, tool and die steels, cast irons, and cast steels. Ferrous alloys are produced as sheet steel for automobiles, appliances, and containers; as plates for ships, boilers, and bridges; as structural members, such as I-beams, bar products for leaf springs, gears, axles, crankshafts, and railroad rails; as stock for tools and dies; as music wire; and as fasteners such as bolts, rivets, and nuts.

**TABLE 3.4**

**ROOM-TEMPERATURE MECHANICAL PROPERTIES AND TYPICAL APPLICATIONS OF ANNEALED STAINLESS STEELS**

| AISI (UNS) | ULTIMATE TENSILE STRENGTH (MPa) | YIELD STRENGTH (MPa) | ELONGATION (%) | CHARACTERISTICS AND TYPICAL APPLICATIONS |
|---|---|---|---|---|
| 303 (S30300) | 550-620 | 240-260 | 50-53 | Screw-machine products, shafts, valves, bolts, bushings, and nuts; aircraft fittings; bolts; nuts; rivets; screws; studs |
| 304 (S30400) | 565-620 | 240-290 | 55-60 | Chemical and food processing equipment, brewing equipment, cryogenic vessels, gutters, downspouts, and flashings |
| 316 (S31600) | 550-590 | 210-290 | 55-60 | High corrosion resistance and high creep strength. Chemical and pulp handling equipment, photographic equipment, brandy vats, fertilizer parts, ketchup cooking kettles, and yeast tubs |
| 410 (S41000) | 480-520 | 240-310 | 25-35 | Machine parts, pump shafts, bolts, bushings, coal chutes, cutlery, finishing, tackle, hardware, jet engine parts, mining machinery, rifle barrels, screws, and valves |
| 416 (S41600) | 480-520 | 275 | 20-30 | Aircraft fittings, bolts, nuts, fire extinguisher inserts, rivets, and screws |

portation equipment, furnace and heat-exchanger parts, and components for severe chemical environments.

The **ferritic** (400 series) **steels** have a high chromium content: up to 27 percent. They are magnetic and have good corrosion resistance, but have lower ductility than austenitic stainless steels. Ferritic stainless steels are hardened by cold working and are not heat treatable. They are generally used for nonstructural applications, such as kitchen equipment and automotive trim.

Most of the **martensitic** (400 and 500 series) **steels** do not contain nickel and are hardenable by heat treatment. Their chromium content may be as high as 18 percent. These steels are magnetic and have high strength, hardness and fatigue resistance, and good ductility, but moderate corrosion resistance. Martensitic stainless steels are used for cutlery, surgical tools, instruments, valves, and springs.

The **precipitation hardening** (PH) **steels** contain chromium and nickel, along with copper, aluminum, titanium, or molybdenum. They have good corrosion resistance, ductility, and high strength at elevated temperatures. Their main application is in aircraft and aerospace structural components.

The **duplex structure steels** have a mixture of austenite and ferrite. They have good strength and higher resistance to corrosion in most environments and to stress-corrosion cracking than do the 300 series of austenitic steels. Typical applications are in water-treatment plants and heat-exchanger components.

### 3.10.3 Tool and die steels

Tool and die steels are specially alloyed steels (Tables 3.5 and 3.6) designed for high strength, impact toughness, and wear resistance at room and elevated temperatures. They are commonly used in forming and machining of metals.

**High-speed steels.**  High speed steels (HSS) are the most highly alloyed tool and die steels and maintain their hardness and strength at elevated operating temperatures. First developed in the early 1900s, there are two basic types of high-speed steels: the molybdenum type (M series) and the tungsten type (T series).

The M-series steels contain up to about 10 percent molybdenum, with chromium, vanadium, tungsten, and cobalt as other alloying elements. The T-series steels contain 12–18 percent tungsten, with chromium, vanadium, and cobalt as other alloying elements. The M-series steels generally have higher abrasion resistance than the T-series steels, have less distortion in heat treatment, and are less expensive. The M series constitutes about 95 percent of all high-speed steels produced in the United States. High-speed steel tools can be coated with titanium nitride and titanium carbide for better wear resistance.

**Hot-work, cold-work, and shock-resisting die steels.**  Hot-work steels (H series) are designed for use at elevated temperatures and have high toughness and high resistance to wear and cracking. The alloying elements are generally tungsten, molybdenum, chromium, and vanadium. **Cold-work steels** (A, D, and O series) are used for cold-working operations. They generally have high resistance to wear and cracking. These steels are available as oil-hardening or air-hardening types. **Shock-resisting steels** (S series) are designed for impact toughness and are used in applications such as header dies, punches, and chisels. Other properties of these steels depend on the particular composition. Various tool and die materials for a variety of manufacturing applications are presented in Table 3.7.

**TABLE 3.5**

**BASIC TYPES OF TOOL AND DIE STEELS**

| TYPE | AISI |
|------|------|
| High speed | M (molybdenum base) |
|  | T (tungsten base) |
| Hot work | H1 to H19 (chromium base) |
|  | H20 to H39 (tungsten base) |
|  | H40 to H59 (molybdenum base) |
| Cold work | D (high carbon, high chromium) |
|  | A (medium alloy, air hardening) |
|  | O (oil hardening) |
| Shock resisting | S |
| Mold steels | P1 to P19 (low carbon) |
|  | P20 to P39 (others) |
| Special purpose | L (low alloy) |
|  | F (carbon-tungsten) |
| Water hardening | W |

**TABLE 3.6**

**PROCESSING AND SERVICE CHARACTERISTICS OF COMMON TOOL AND DIE STEELS**

| AISI DESIGNATION | RESISTANCE TO DECARBURIZATION | RESISTANCE TO CRACKING | APPROXIMATE HARDNESS (HRC) | MACHINABILITY | TOUGHNESS | RESISTANCE TO SOFTENING | RESISTANCE TO WEAR |
|---|---|---|---|---|---|---|---|
| M2 | Medium | Medium | 60-65 | Medium | Low | Very high | Very high |
| T1 | High | High | 60-65 | Medium | Low | Very high | Very high |
| T5 | Low | Medium | 60-65 | Medium | Low | Highest | Very high |
| H11, 12, 13 | Medium | Highest | 38-55 | Medium to high | Very high | High | Medium |
| A2 | Medium | Highest | 57-62 | Medium | Medium | High | High |
| A9 | Medium | Highest | 35-56 | Medium | High | High | Medium to high |
| D2 | Medium | Highest | 54-61 | Low | Low | High | High to very high |
| D3 | Medium | High | 54-61 | Low | Low | High | Very high |
| H21 | Medium | High | 36-54 | Medium | High | High | Medium to high |
| H26 | Medium | High | 43-58 | Medium | Medium | Very high | High |
| P20 | High | High | 28-37 | Medium to high | High | Low | Low to medium |
| P21 | High | Highest | 30-40 | Medium | Medium | Medium | Medium |
| W1, W2 | Highest | Medium | 50-64 | Highest | High | Low | Low to medium |

Source: Adapted from *Tools Steels*. American Iron and Steel Institute, 1978.

**TABLE 3.7**

**TYPICAL TOOL AND DIE MATERIALS**

| PROCESS | MATERIAL |
| --- | --- |
| Die casting | H13, P20 |
| Powder metallurgy | |
|    Punches | A2, S7, D2, D3, M2 |
|    Dies | WC, D2, M2 |
| Molds for plastics and rubber | S1, O1, A2, D2, 6F5, 6F6, P6, P20, P21, H13 |
| Hot forging | 6F2, 6G, H11, H12 |
| Hot extrusion | H11, H12, H13 |
| Cold heading | W1, W2, M1, M2, D2, WC |
| Cold extrusion | |
|    Punches | A2, D2, M2, M4 |
|    Dies | O1, W1, A2, D2 |
| Coining | 52100, W1, O1, A2, D2, D3, D4, H11, H12, H13 |
| Drawing | |
|    Wire | WC, diamond |
|    Shapes | WC, D2, M2 |
|    Bar and tubing | WC, W1, D2 |
| Rolls | |
|    Rolling | Cast iron, cast steel, forged steel, WC |
|    Thread rolling | A2, D2, M2 |
|    Shear spinning | A2, D2, D3 |
| Sheet metals | |
|    Shearing | |
|      Cold | D2, A2, A9, S2, S5, S7 |
|      Hot | H11, H12, H13 |
|    Pressworking | Zinc alloys, 4140 steel, cast iron, epoxy composites, A2, D2, O1 |
|    Deep drawing | W1, O1, cast iron, A2, D2 |
| Machining | Carbides, high-speed steels, ceramics, diamond |

Notes: Tool and die materials are usually hardened to 55 to 65 HRC for cold working, and 30 to 55 for hot working.
Tool and die steels contain one or more of the following major alloying elements: chromium, molybdenum, tungsten, and vanadium.
For further details see the bibliography at the end of this chapter.

## 3.11   Properties of Nonferrous Metals and Alloys

**Nonferrous metals and alloys** cover a wide range of materials, from the more common metals, such as aluminum, copper, and magnesium, to high-strength, high-temperature alloys, such as tungsten, tantalum, and molybdenum. Although more expensive than ferrous metals, nonferrous metals and alloys have important applications because of their numerous mechanical, physical, and chemical properties.

A turbofan jet engine for the Boeing 757 aircraft typically contains the following nonferrous metals and alloys: 38 percent titanium, 37 percent nickel, 12 percent chromium, 6 percent cobalt, 5 percent aluminum, 1 percent niobium (columbium), and 0.02 percent tantalum. Without these materials, a jet engine (Fig. 3.34) could not be designed, manufactured, and operated at the energy and efficiency levels required.

Typical examples of the applications of nonferrous metals and alloys are aluminum for cooking utensils and aircraft bodies, copper wire for electricity and copper tubing for water in residences, zinc for carburetors, titanium for jet-engine turbine blades and prosthetic devices (such as artificial limbs), and tantalum for rocket engines. In this section, we discuss the general properties, production methods, and important engineering applications of nonferrous metals and alloys. We describe the manufacturing properties of these materials, such as formability, machinability, and weldability, in various chapters throughout this text.

### 3.11.1 Aluminum

The important factors in selecting **aluminum** (Al) and its alloys are their high strength-to-weight ratio, resistance to corrosion by many chemicals, high thermal and electrical conductivity, nontoxicity, reflectivity, appearance, and ease of formability and machinability; they are also nonmagnetic.

Principal uses of aluminum and its alloys, in decreasing order of consumption, are containers and packaging (aluminum cans and foil), buildings and other types of construction,

**FIGURE 3.34** Cross-section of a jet engine (PW2037) showing various components and the alloys used in making them. *Source:* Courtesy of United Aircraft Pratt & Whitney.

**TABLE 3.8**

**PROPERTIES OF SELECTED ALUMINUM ALLOYS AT ROOM TEMPERATURE**

| ALLOY (UNS) | TEMPER | ULTIMATE TENSILE STRENGTH (MPa) | YIELD STRENGTH (MPa) | ELONGATION IN 50 mm (%) |
|---|---|---|---|---|
| 1100 (A91100) | O | 90 | 35 | 35–45 |
| 1100 | H14 | 125 | 120 | 9–20 |
| 2024 (A92024 | O | 190 | 75 | 20–22 |
| 2024 | T4 | 470 | 325 | 19–20 |
| 3003 (A93003 | O | 110 | 40 | 30–40 |
| 3003 | H14 | 150 | 145 | 8–16 |
| 5052 (A95052) | O | 190 | 90 | 25–30 |
| 5052 | H34 | 260 | 215 | 10–14 |
| 6061 (A96061) | O | 125 | 55 | 25–30 |
| 6061 | T6 | 310 | 275 | 12–17 |
| 7075 (A97075) | O | 230 | 105 | 16–17 |
| 7075 | T6 | 570 | 500 | 11 |

**TABLE 3.9**

**MANUFACTURING PROPERTIES AND TYPICAL APPLICATIONS OF WROUGHT ALUMINUM ALLOYS**

| ALLOY | CHARACTERISTICS* | | | TYPICAL APPLICATIONS |
|---|---|---|---|---|
| | CORROSION RESISTANCE | MACHINABILITY | WELDABILITY | |
| 1100 | A | D-C | A | Sheet-metal work, spun hollow ware, tin stock |
| 2024 | C | C-B | C-B | Truck wheels, screw machine products, aircraft structures |
| 3003 | A | D-C | A | Cooking utensils, chemical equipment, pressure vessels, sheet metal work, builders' hardware, storage tanks |
| 5052 | A | D-C | A | Sheetmetal work, hydraulic tubes, and appliances; bus, truck, and marine uses |
| 6061 | B | D-C | A | Heavy-duty structures where corrosion resistance is needed, truck and marine structures, railroad cars, furniture, pipelines, bridge railings, hydraulic tubing |
| 7075 | C | B-D | D | Aircraft and other structures, keys, hydraulic fittings |

*A, excellent; D, poor

transportation (aircraft and aerospace applications, buses, automobiles, railroad cars, and marine craft), electrical (nonmagnetic and economical electrical conductors), consumer durables (appliances, cooking utensils, and furniture), and portable tools (Tables 3.8 and 3.9). Nearly all high-voltage transmission wiring is made of aluminum. Of their structural (load-bearing) components, 82 percent of a Boeing 747 aircraft and 79 percent of a Boeing 757 aircraft are made of aluminum.

As in all cases of material selection, because each aluminum alloy has its particular properties, their selection can be critical from manufacturing as well as economic considerations. For example, a typical aluminum beverage can consists of the following alloys, all sheet being in the H19 condition, which is the highest cold-worked state: 3004 or 3104 for the can body, 5182 for the lid, and 5042 for the tab.

Aluminum alloys are available as mill products—that is, wrought product made into various shapes by rolling, extrusion, drawing, and forging. Aluminum ingots are available for casting, as are powder metals for powder metallurgy applications. There are two types of wrought alloys of aluminum: (1) alloys that can be hardened by cold working (designated by the letter H) and are not heat treatable, and (2) alloys that are hardenable by heat treatment (designated by the letter T). Techniques have been developed whereby most aluminum alloys can be machined, formed, and welded with relative ease.

### 3.11.2  Magnesium

**Magnesium** (Mg) is the lightest engineering metal available and also has good vibration-damping characteristics. Its alloys are used in structural and nonstructural applications where weight is of primary importance. Magnesium is also an alloying element in various nonferrous metals.

Typical uses of magnesium alloys are for aircraft and missile components, material-handling equipment, portable power tools (such as drills and sanders), ladders, luggage, bicycles, sporting goods, and general lightweight components. These alloys are available either as castings or wrought products, such as extruded bars and shapes, forgings, and rolled plate and sheet. Magnesium alloys are also used in printing and textile machinery to minimize inertial forces in high-speed components.

Because it is not sufficiently strong in its pure form, magnesium is alloyed with various elements (Table 3.10) to impart certain specific properties, particularly a high strength-to-weight ratio. A variety of magnesium alloys have good casting, forming, and machining characteristics. Because they oxidize rapidly—that is, they are *pyrophoric*—a fire hazard exists and precautions must be taken when machining, grinding, or sand casting magnesium alloys. However, products made of magnesium and its alloys are not a fire hazard.

### 3.11.3  Copper

First produced in about 4000 B.C., **copper** (Cu) and its alloys have properties somewhat similar to those of aluminum alloys. In addition, they are among the best conductors of electricity and heat and have good corrosion resistance. Because of these properties, copper and its alloys are among the most important metals. They can be processed easily by various forming, machining, casting, and joining techniques.

**TABLE 3.10**

**PROPERTIES AND TYPICAL FORMS OF WROUGHT MAGNESIUM ALLOYS**

| ALLOY | COMPOSITION (%) | | | | CONDI-TION | ULTIMATE TENSILE STRENGTH (MPa) | YIELD STRENGTH (MPa) | ELONGA-TION IN 50 mm (%) | TYPICAL FORMS |
|-------|-----|-----|-----|-----|--------|------|------|-----|---------------|
|       | AL | ZN | MN | ZR | | | | | |
| AZ31B | 3.0 | 1.0 | 0.2 | | F | 260 | 200 | 15 | Extrusions |
|       |     |     |     | | H24 | 290 | 220 | 15 | Sheet and plates |
| AZ80A | 8.5 | 0.5 | 0.2 | | T5 | 380 | 275 | 7 | Extrusions and forgings |
| HK31A |     |     | 3Th | 0.7 | H24 | 255 | 200 | 8 | Sheet and plates |
| ZK60A |     | 5.7 |     | 0.55 | T5 | 365 | 300 | 11 | Extrusions and forgings |

Copper alloys often are attractive for applications where combined properties, such as electrical, mechanical, corrosion resistance, nonmagnetic, thermal conductivity, and wear resistance, are required. Applications include electrical and electronic components, springs, cartridges for small arms, plumbing, heat exchangers, and marine hardware, as well as consumer goods, such as cooking utensils, jewelry, and other decorative objects.

Copper alloys can acquire a wide variety of properties by the addition of alloying elements and by heat treatment to improve their manufacturing characteristics. The most common copper alloys are brasses and bronzes. **Brass**, which is an alloy of copper and zinc, is one of the earliest alloys developed and has numerous applications, including decorative objects (Table 3.11). **Bronze** is an alloy of copper and tin (Table 3.12). There are also other

**TABLE 3.11**

**PROPERTIES AND TYPICAL APPLICATIONS OF WROUGHT COPPER AND BRASSES**

| TYPE AND UNS NUMBER | NOMINAL COMPOSI-TION (%) | ULTIMATE TENSILE STRENGTH (MPa) | YIELD STRENGTH (MPa) | ELONG-ATION IN 50 mm (%) | TYPICAL APPLICATIONS |
|---------------------|---------|---------|---------|---------|----------------------|
| Electrolytic tough pitch copper (C11000) | 99.90 Cu, 0.04 O | 220–450 | 70–365 | 55–4 | Downspouts, gutters, roofing, gaskets, auto radiators, busbars, nails, printing rolls, rivets |
| Red brass, 85% (C23000) | 85.0 Cu, 15.0 Zn | 270–725 | 70–435 | 55–3 | Weather-stripping, conduit, sockets, fasteners, fire extinguishers, condenser and heat exchanger tubing |
| Cartridge brass, 70% (C26000) | 70.0 Cu, 30.0 Zn | 300–900 | 75–450 | 66–3 | Radiator cores and tanks, flashlight shells, lamp fixtures, fasteners, locks, hinges, ammunition components, plumbing accessories |
| Free-cutting brass (C36000) | 61.5 Cu, 3.0 Pb, 35.5 Zn | 340–470 | 125–310 | 53–18 | Gears, pinions, automatic high-speed screw machine parts |
| Naval brass (C46400 to C46700) | 60.0 Cu, 39.25 Zn, 0.75 Sn | 380–610 | 170–455 | 50–17 | Aircraft turnbuckle barrels, balls, bolts marine hardware, propeller shafts, rivets, valve stems, condenser plates |

**TABLE 3.12**

**PROPERTIES AND TYPICAL APPLICATIONS OF WROUGHT BRONZES**

| TYPE AND UNS NUMBER | NOMINAL COMPOSITION (%) | ULTIMATE TENSILE STRENGTH (MPa) | YIELD STRENGTH (MPa) | ELONGATION IN 50 mm (%) | TYPICAL APPLICATIONS |
|---|---|---|---|---|---|
| Architectural bronze (C38500) | 57.0 Cu, 3.0 Pb, 40.0 Zn | 415 (As extruded) | 140 | 30 | Architectural extrusions, store fronts, thresholds, trim, butts, hinges |
| Phosphor bronze, 5% A (C51000) | 95.0 Cu, 5.0 Sn, trace P | 325–960 | 130–550 | 64–2 | Bellows, clutch disks, cotter pins, diaphragms, fasteners, wire brushes, chemical hardware, textile machinery |
| Free-cutting phosphor bronze (C54400) | 88.0 Cu, 4.0 Pb, 4.0 Zn, 4.0 Sn | 300–520 | 130–435 | 50–15 | Bearings, bushings, gears, pinions, shafts, thrust washers, valve parts |
| Low silicon bronze, B (C65100) | 98.5 Cu, 1.5 Si | 275–655 | 100–475 | 55–11 | Hydraulic pressure lines, bolts, marine hardware, electrical conduits, heat exchanger tubing |
| Nickel silver, 65–10 (C74500) | 65.0 Cu, 25.0 Zn, 10.0 Ni | 340–900 | 125–525 | 50–1 | Rivets, screws, slide fasteners, hollow ware, nameplates |

bronzes, such as aluminum bronze, an alloy of copper and aluminum, and tin bronzes. Beryllium-copper, or **beryllium bronze**, and **phosphor bronze** have good strength and hardness for applications such as springs and bearings. Other major copper alloys are copper nickels and nickel silvers.

### 3.11.4 Nickel

**Nickel** (Ni), a silver-white metal discovered in 1751, is a major alloying element that imparts strength, toughness, and corrosion resistance. It is used extensively in stainless steels and nickel-base alloys, also called **superalloys**. These alloys are used for high-temperature applications, such as jet-engine components, rockets, and nuclear power plants, as well as in food-handling and chemical-processing equipment, coins, and marine applications. Because nickel is magnetic, nickel alloys are also used in electromagnetic applications, such as solenoids. The principal use of nickel as a metal is in electroplating for appearance and for improving corrosion and wear resistance.

Nickel alloys have high strength and corrosion resistance at elevated temperatures. Alloying elements in nickel are chromium, cobalt, and molybdenum. The behavior of nickel alloys in machining, forming, casting, and welding can be modified by various other alloying elements.

A variety of nickel alloys having a range of strength at different temperatures have been developed (Table 3.13). *Monel* is a nickel–copper alloy and *Inconel* is a nickel-chromium alloy. A nickel–molybdenum–chromium alloy (*Hastelloy*) has good corrosion resistance and high strength at elevated temperatures. *Nichrome* an alloy of nickel, chromium, and iron, has high oxidation and electrical resistance, and is used for electrical heating elements. *Invar*, an alloy of iron and nickel, has relatively low sensitivity to temperature (see Section 3.9.5).

**TABLE 3.13**

**PROPERTIES AND TYPICAL APPLICATIONS OF NICKEL ALLOYS (ALL ARE TRADE NAMES)**

| ALLOY (CONDITION) | PRINCIPAL ALLOYING ELEMENTS (%) | ULTIMATE TENSILE STRENGTH (MPa) | YIELD STRENGTH (MPa) | ELONGATION IN 50 mm (%) | TYPICAL APPLICATIONS |
|---|---|---|---|---|---|
| Nickel 200 (annealed) | None | 380–550 | 100–275 | 60–40 | Chemical and food processing industry, aerospace equipment, electronic parts |
| Duranickel 301 (age-hardened) | 4.4 Al, 0.6 Ti | 1300 | 900 | 28 | Springs, plastics extrusion equipment, molds for glass, diaphragms |
| Monel R-405 (hot-rolled) | 30 Cu | 525 | 230 | 35 | Screw-machine products, water meter parts |
| Monel K-500 (age hardened) | 29 Cu, 3 Al | 1050 | 750 | 30 | Pump shafts, valve stems, springs |
| Inconel 600 (annealed) | 15 Cr, 8 Fe | 640 | 210 | 48 | Gas turbine parts, heat-treating equipment electronic parts, nuclear reactors |
| Hastelloy C-4 (solution-treated and quenched) | 16 Cr, 15 Mo | 785 | 400 | 54 | High temperature stability, resistance to stress-corrosion cracking |

### 3.11.5 Superalloys

Superalloys are important in high-temperature applications, hence they are also known as **heat-resistant,** or **high-temperature, alloys.** Major applications of superalloys are in jet engines and gas turbines, with other applications in reciprocating engines, rocket engines, tools and dies for hot working of metals, and in the nuclear, chemical, and petrochemical industries.

These alloys are referred to as **iron-base, cobalt-base,** or **nickel-base superalloys.** They contain nickel, chromium, cobalt, and molybdenum as major alloying elements. Other alloying elements are aluminum, tungsten, and titanium. Superalloys are generally identified by trade names or by special numbering systems and are available in a variety of shapes. Most superalloys have a maximum service temperature of about 1000°C (1800°F) for structural applications. The temperatures can be as high as 1200°C (2200°F) for nonload-bearing components. Superalloys generally have good resistance to corrosion, mechanical and thermal fatigue, mechanical and thermal shock, creep, and erosion at elevated temperatures.

Iron-base superalloys generally contain 32–67 percent iron, 15–22 percent chromium, and 9–38 percent nickel. Common alloys in this group are the *Incoloy* series. Cobalt-base superalloys generally contain 35–65 percent cobalt, 19–30 percent chromium, and up to 35 percent nickel. Cobalt (Co) is a white-colored metal that resembles nickel. These superalloys are not as strong as nickel-base superalloys, but they retain their strength at higher temperatures. Nickel-base superalloys are the most common of the superalloys and are available in a wide variety of compositions (Table 3.14). The range of nickel is from 38 to 76 percent. They also contain up to 27 percent chromium and 20 percent cobalt. Common alloys in this group are the *Hastelloy, Inconel, Nimonic, René, Udimet, Astroloy,* and *Waspaloy* series.

**TABLE 3.14**

**PROPERTIES AND TYPICAL APPLICATIONS OF SELECTED NICKEL-BASE SUPERALLOYS AT 870°C (1600°F) (ALL ARE TRADE NAMES)**

| ALLOY | CONDITION | ULTIMATE TENSILE STRENGTH (MPa) | YIELD STRENGTH (MPa) | ELONGATION IN 50 mm (%) | TYPICAL APPLICATIONS |
|---|---|---|---|---|---|
| Astroloy | Wrought | 770 | 690 | 25 | Forgings for high temperature |
| Hastelloy X | Wrought | 255 | 180 | 50 | Jet engine sheet parts |
| IN-100 | Cast | 885 | 695 | 6 | Jet engine blades and wheels |
| IN-102 | Wrought | 215 | 200 | 110 | Superheater and jet engine parts |
| Inconel 625 | Wrought | 285 | 275 | 125 | Aircraft engines and structures, chemical processing equipment |
| Inconel 718 | Wrought | 340 | 330 | 88 | Jet engine and rocket parts |
| MAR-M 200 | Cast | 840 | 760 | 4 | Jet engine blades |
| MAR-M 432 | Cast | 730 | 605 | 8 | Integrally cast turbine wheels |
| René 41 | Wrought | 620 | 550 | 19 | Jet engine parts |
| Udimet 700 | Wrought | 690 | 635 | 27 | Jet engine parts |
| Waspaloy | Wrought | 525 | 515 | 35 | Jet engine parts |

### 3.11.6 Titanium

**Titanium** (Ti) was discovered in 1791 but was not commercially produced until the 1950s. Although expensive, the high strength-to-weight ratio of titanium and its corrosion resistance at room and elevated temperatures make it attractive for applications such as aircraft, jet-engine, racing-car, chemical, petrochemical, and marine components, submarine hulls, and biomaterials such as prosthetic devices (Table 3.15). Titanium alloys have been developed for service at 550°C (1000°F) for long periods of time and up to 750°C (1400°F) for shorter periods.

Unalloyed titanium, known as commercially pure titanium, has excellent corrosion resistance for applications where strength considerations are secondary. Aluminum, vanadium, molybdenum, manganese, and other alloying elements are added to titanium alloys to impart properties such as improved workability, strength, and hardenability.

The properties and manufacturing characteristics of titanium alloys are extremely sensitive to small variations in both alloying and residual elements. Thus, control of composition and processing are important, including prevention of surface contamination by hydrogen, oxygen, or nitrogen during processing. These elements cause embrittlement of titanium, resulting in reduced toughness and ductility.

The body-centered cubic structure of titanium (beta-titanium, above 880°C, 1600°F) is ductile, whereas its hexagonal close-packed structure (alpha-titanium) is somewhat brittle and is very sensitive to stress corrosion. A variety of other structures (alpha, near alpha, alpha–beta, and beta) can be obtained by alloying and heat treating, such that the properties can be optimized for specific applications. New developments include the so-called **titanium aluminide intermetallics**, TiAl and $Ti_3Al$. They have higher stiffness and lower density and can withstand temperatures higher than conventional titanium alloys.

**TABLE 3.15**

## PROPERTIES AND TYPICAL APPLICATIONS OF WROUGHT TITANIUM ALLOYS

| NOMINAL COMPOSITION (%) | UNS | CONDITION | ROOM TEMPERATURE | | | | VARIOUS TEMPERATURES | | | | | TYPICAL APPLICATIONS |
|---|---|---|---|---|---|---|---|---|---|---|---|---|
| | | | ULTIMATE TENSILE STRENGTH (MPa) | YIELD STRENGTH (MPa) | ELONGATION (%) | REDUCTION OF AREA (%) | TEMP (°C) | ULTIMATE TENSILE STRENGTH (MPa) | YIELD STRENGTH (MPa) | ELONGATION (%) | REDUCTION OF AREA (%) | |
| 99.5 Ti | R50250 | Annealed | 330 | 240 | 30 | 55 | 300 | 150 | 95 | 32 | 80 | Airframes; chemical, desalination, and marine parts; plate type heat exchangers |
| 5 Al, 2.5 Sn | R54520 | Annealed | 860 | 810 | 16 | 40 | 300 | 565 | 450 | 18 | 45 | Aircraft engine compressor blades and ducting; steam turbine blades |
| 6 Al, 4 V | R56400 | Annealed | 1000 | 925 | 14 | 30 | 300 | 725 | 650 | 14 | 35 | Rocket motor cases blades and disks for aircraft turbines and compressors; structural forgings and fasteners |
| | | | | | | | 425 | 670 | 570 | 18 | 40 | |
| | | | | | | | 550 | 530 | 430 | 35 | 50 | |
| | | Solution + age | 1175 | 1100 | 10 | 20 | 300 | 980 | 900 | 10 | 28 | |
| | | | | | | | | | | 12 | 35 | |
| | | | | | | | | | | 22 | 45 | |
| 13 V, 11 Cr, 3 Al, | R58010 | Solution + age | 1275 | 1210 | 8 | — | 425 | 1100 | 830 | 12 | — | High strength fasteners; aerospace components; honey comb panels |

### 3.11.7 Refractory metals

**Refractory metals** are molybdenum, columbium, tungsten, and tantalum. They are called *refractory* because of their high melting point. Although refractory metal elements were discovered about 200 years ago—and have been used as important alloying elements in steels and superalloys—their use as engineering metals and alloys did not begin until about the 1940s.

More than most other metals and alloys, these metals maintain their strength at elevated temperatures. Thus they are of great importance and use in rocket engines, gas turbines, and various other aerospace application, in the electronics, nuclear power, and chemical industries, and as tool and die materials. The temperature range for some of these applications is on the order of 1100°C to 2200°C (2000°F to 4000°F), where strength and oxidation are of major concern.

**Molybdenum.** *Molybdenum* (Mo), a silvery white metal, was discovered in the 18th century. It has a high melting point, a high modulus of elasticity, good resistance to thermal shock, and good electrical and thermal conductivity. Typical applications of molybdenum are in solid-propellent rockets, jet engines, honeycomb structures, electronic components, heating elements, and dies for die casting.

Principal alloying elements for molybdenum are titanium and zirconium. Molybdenum is used in greater amounts than any other refractory metal. It is an important alloying element in casting and wrought alloys, such as steels and heat-resistant alloys, and imparts strength, toughness, and corrosion resistance. A major disadvantage of molybdenum alloys is their low resistance to oxidation at temperatures above 500°C (950°F), thus necessitating the use of protective coatings.

**Niobium (columbium).** *Niobium* (Nb) or *columbium* (after the mineral columbite) possesses good ductility and formability and has greater oxidation resistance than other refractory metals. With various alloying elements, niobium alloys can be produced with moderate strength and good fabrication characteristics. These alloys are used in rockets and missiles, and nuclear, chemical, and superconductor applications. Niobium, first identified in 1801, is also an alloying element in various alloys and superalloys.

**Tungsten.** *Tungsten* (W, from wolframite) was first identified in 1781 and is the most plentiful of all refractory metals. Tungsten has the highest melting point of any metal (3410°C, 6170°F), and thus it is characterized by high strength at elevated temperatures. On the other hand, it has high density, brittleness at low temperatures, and poor resistance to oxidation.

Tungsten and its alloys are used for applications involving temperatures above 1650°C (3000°F), such as nozzle throat liners in missiles and in the hottest parts of jet and rocket engines, circuit breakers, welding electrodes, and spark-plug electrodes. The filament wire in incandescent light bulbs is made of pure tungsten, using powder metallurgy and wire-drawing techniques. Because of its high density, tungsten is used in balancing weights and counterbalances in mechanical systems, including self-winding watches. Tungsten is an important element in tool and die steels, imparting strength and hardness at elevated temperatures. Tungsten carbide, with cobalt as a binder for the carbide particles, is one of the most important tool and die materials.

**Tantalum.** Identified in 1802, *tantalum* (Ta) is characterized by a high melting point (3000°C, 5425°F), good ductility, and resistance to corrosion. However, it has high density and poor resistance to chemicals at temperatures above 150°C (300°F). Tantalum is also used as an alloying element.

Tantalum is used extensively in electrolytic capacitors and various components in the electrical, electronic, and chemical industries, as well as for thermal applications, such as in furnaces and acid-resistant heat exchangers. A variety of tantalum-base alloys is available in many forms for use in missiles and aircraft.

### 3.11.8   Various other nonferrous metals

**Beryllium.** Steel gray in color, *beryllium* (Be) has a high strength-to-weight ratio. Unalloyed beryllium is used in nuclear and x-ray applications because of its low neutron absorption characteristics, and in rocket nozzles, space and missile structures, aircraft disc brakes, and precision instruments and mirrors. Beryllium is also an alloying element, and its alloys of copper and nickel are used in applications such as spring (beryllium-copper), electrical contacts, and nonsparking tools for use in explosive environments, such as mines and metal powder production. Beryllium and its oxide are toxic and should not be inhaled.

**Zirconium.** *Zirconium* (Zr) is silvery in appearance, has good strength and ductility at elevated temperatures, and has good corrosion resistance because of an adherent oxide film. The element is used in electronic components and nuclear power reactor applications because of its low neutron absorption.

#### Low-melting metals

The major metals in this category are lead, zinc, and tin. **Lead** (Pb, after plumbium, the root for the word plumber) has properties of high density, resistance to corrosion (by virtue of the stable lead oxide layer that forms and protects the surface), softness, low strength, ductility, and good workability. Alloying with various elements, such as antimony and tin, enhances lead's properties, making it suitable for piping, collapsible tubing, bearing alloys, cable sheathing, roofing, and lead-acid storage batteries.

Lead is also used for damping sound and vibrations, radiation shielding against x-rays, in printing (type metals), ammunition, and weights and in the chemical and paint industries. The oldest lead artifacts were made about 3000 B.C. Lead pipes made by the Romans and installed in the Roman baths in Bath, England, two millennia ago are still in use. Lead is also an alloying element in solders, steels, and copper alloys and promotes corrosion resistance and machinability. Because of its toxicity, environmental contamination by lead is a significant concern.

Industrially, **zinc** (Zn, a bluish-white color), is the fourth most utilized metal, after iron, aluminum, and copper. Although known for many centuries, zinc was not studied and developed until the 18th century. It has two major uses: one is for galvanizing iron, steel sheet and wire, and the other is as an alloy base for casting. In **galvanizing,** zinc serves as the anode and protects the steel (cathode) from corrosive attack should the coating be scratched or punctured. Zinc is also used as an alloying element. Brass, for example, is an alloy of copper and zinc.

Major alloying elements in zinc are aluminum, copper, and magnesium. They impart strength and provide dimensional control during casting of the metal. Zinc-base alloys are used extensively in die casting for making products such as carburetors, fuel pumps, and grills for automobiles, components for household appliances (such as vacuum cleaners, washing machines, and kitchen equipment), various other machine parts, and photoengraving plates. Another use for zinc is in superplastic alloys, which have good formability characteristics by virtue of their capacity to undergo large deformation without failure. Very fine-grained 78% Zn–22% Al sheet is a common example of a superplastic zinc alloy that can be formed by methods used for forming plastics or metals.

Although used in small amounts, **tin** (Sn, after stannum) is an important metal. The most extensive use of tin, a silvery white, lustrous metal, is as a protective coating on steel sheet (*tin plate*), which is used in making containers–tin cans–for food and various other products. Inside a sealed can, the steel is cathodic and is protected by the tin (anode), so that the steel does not corrode. The low shear strength of the tin coatings on steel sheet also improves its performance in deep drawing and general pressworking operations

Unalloyed tin is used in applications such as lining material for water distillation plants, and as a molten layer of metal over which plate glass is made. Tin-base alloys (also called **white metals**) generally contain copper, antimony, and lead. The alloying elements impart hardness, strength, and corrosion resistance. Because of their low friction coefficients, which result from low shear strength and low adhesion, tin alloys are used as journal-bearing materials. These alloys are known as **babbitts** and contain tin, copper, and antimony. **Pewter** is an alloy of tin, copper, and antimony. It was developed in the 15th century and is used for tableware, hollow ware, and decorative artifacts. Organ pipes are made of tin alloys. Tin is an alloying element for type metals, dental alloys, and bronze (copper-tin alloy), titanium, and zirconium alloys. Tin–lead alloys are common **soldering** materials, with a wide range of compositions and melting points (see Section 12.12).

## Precious metals

Gold, silver, and platinum are the most important *precious* (that is, costly) metals and are also called **noble metals. Gold** (Au, after aurum) is soft and ductile, and has good corrosion resistance at any temperature. Typical applications include jewelry, coinage, reflectors, gold leaf for decorative purposes, dental work, and electroplating, as well as important applications as electric contact and terminals.

**Silver** (Ag, after argentum) is a ductile metal and has the highest electrical and thermal conductivity of any metal. It does, however, develop an oxide film that affects its surface properties and appearance. Typical applications for silver include tableware, jewelry, coinage, electroplating, and photographic film, as well as electrical contacts, solders, bearings, and food and chemical equipment. Sterling silver is an alloy of silver and 7.5 percent copper.

**Platinum** (Pt) is a soft, ductile, grayish-white metal that has good corrosion resistance, even at elevated temperatures. Platinum alloys are used as electrical contacts, spark-plug electrodes, catalysts for automobile pollution-control devices, filaments, nozzles, dies for extruding glass fibers, and thermocouples, and in the electrochemical industry. Other applications include jewelry and dental work.

**Shape-memory alloys.** Shape-memory alloys, after being plastically deformed at room temperature into various shapes, return to their original shapes upon heating. For example, a piece of straight wire made of this material can be wound into a helical spring. When heated with a match, the spring uncoils and returns to the original straight shape. A typical shape-memory alloy is 55% Ni–45% Ti. Other alloys under development are copper–aluminum–nickel, copper–zinc–aluminum, iron–manganese–silicon, and nickel–titanium. One potential application for these alloys is in structures, such as an antenna, for space travel. They would be folded at room temperature into a small volume to save space and for ease of transportation. After reaching their destination, the structures would be heated by some suitable means and return to their original shapes.

**Amorphous alloys.** A class of metal alloys that, unlike ordinary metals, does not have a long-range crystalline structure is called **amorphous alloys** (see also Section 5.9.8). These alloys have no grain boundaries, and the atoms are randomly and tightly packed. Because their structure resembles that of glasses, these alloys are also called **metallic glasses.** Amorphous alloys typically contain iron, nickel, and chromium, alloyed with carbon, phosphorus, boron, aluminum, and silicon. They are available in the form of wire, ribbon, strip, and powder.

These alloys exhibit excellent corrosion resistance, good ductility, high strength, and very low loss from magnetic hysteresis. The latter property is utilized in making magnetic steel cores for transformers, generators, motors, lamp ballasts, magnetic amplifiers, and linear accelerators, with greatly improved efficiency.

The amorphous structure was first obtained in the late 1960s by extremely rapid cooling of the molten alloy. One method is called *splat cooling*, in which the alloy is propelled at very high speed against a rotating metal surface. Since the rate of cooling is on the order of $10^6$ to $10^8$ K/s, the molten alloy does not have sufficient time to crystallize. If an amorphous alloy is raised in temperature and then cooled, it develops a crystalline structure.

## SUMMARY

- The manufacturing properties of metals and alloys depend largely on their mechanical and physical properties. These, in turn are governed mainly by the crystal structure of metals, grain boundaries, grain size, texture, and imperfections. (Sections 3.1, 3.2)

- Dislocations are responsible for the lower shear stress required to cause slip, and thus plastic deformation. However, when dislocations become entangled with one another, or are impeded by barriers such as grain boundaries, impurities, and inclusions, they increase the shear stress required to cause slip, and thus increase the strength of the material. This phenomenon is called work hardening or strain hardening. (Section 3.3)

- Grain size and grain boundaries affect the mechanical properties of metals, including strength, ductility, and hardness. Grain boundaries enhance strain hardening by interfering with the movement of dislocations. (Section 3.4)

- Plastic deformation at room temperature (cold working) of polycrystalline materials results in higher strength of the material as dislocations become entangled. It also generally causes anisotropy, whereby the mechanical properties are different in different directions. (Section 3.5)

- The effects of cold working can be reversed by heating the metal in a specific temperature range for a specific period of time, through the consecutive processes of recovery, recrystallization, and grain growth. (Section 3.6)

- Metals and alloys can be worked at room, warm, or high temperatures. Their overall behavior, force requirements, and workability depend largely on whether the working temperature is below or above the recrystallization temperature. (Section 3.7)

- Failure and fracture is an important aspect of a material's behavior when subjected to deformation in metalworking operations. Two types of fracture are ductile and brittle. Ductile fracture is characterized by plastic deformation preceding fracture and requires considerable energy. Because it is not preceded by plastic deformation, brittle fracture can be catastrophic, requiring much less energy than in ductile fracture. Impurities and inclusions play a major role in the fracture of metals and alloys. (Section 3.8)

- Physical and chemical properties of metals and alloys can significantly affect design considerations, service requirements, and compatibility with other materials, including tools and dies. (Section 3.9)

- A variety of metals and alloys is available with a wide range of properties, such as strength, toughness, hardness, ductility, and resistance to high temperatures and oxidation. They can be classified generally as ferrous alloys, nonferrous metals and alloys, refractory metals and their alloys, and superalloys. (Sections 3.10, 3.11)

## SUMMARY OF EQUATIONS

Theoretical shear strength of metals, $\tau_{\max} = G/2\pi$

Theoretical tensile strength of metals, $\sigma_{\max} = \sqrt{E\gamma/a} \simeq E/10$

Hall-Petch equation, $Y = Y_i + kd^{-1/2}$

ASTM grain size number, $N = 2^{n-1}$

Tensile strength vs. crack length, $\sigma \propto 1/\sqrt{\text{crack length}}$

## BIBLIOGRAPHY

### Materials science and engineering

Ashby, M.F., and D.R.H. Jones, Engineering Materials, Vol. 1, *An Introduction to Their Properties and Applications*, 1980; Vol. 2, *An Introduction to Microstructures, Processing and Design*, 1986; Vol. 3, *Materials Failure Analysis: Case Studies and Design Implications*, 1993. Oxford: Pergamon.

Askeland, D.R., *The Science and Engineering of Materials*, 3rd ed. Boston: PWS–KENT, 1994.

Callister, W.D., Jr., *Materials Science and Engineering*, 3rd ed. New York: Wiley, 1994.

Dieter, G.E., *Mechanical Metallurgy*, 3d ed. New York: McGraw-Hill, 1986.

Flinn, R.A., and P.K. Trojan, *Engineering Materials and Their Applications*, 4th ed. Boston: Houghton Mifflin, 1990.

Hertzberg, R.W., *Deformation and Fracture Mechanics of Engineering Materials*, 4th ed. New York: Wiley, 1996.

Shackelford, J.F., *Introduction to Materials Science for Engineers*, 4th ed. New York: Macmillan, 1996.

Smith, W.F., *Principles of Materials Science and Engineering*, 2d ed. New York: McGraw-Hill, 1990.

Van Vlack, L.H., *Elements of Materials Science and Engineering*, 6th ed. Reading, Mass.: Addison-Wesley, 1989.

## Material properties

*Aerospace Structural Metals Handbook*, 5 vols. Columbus, Ohio: Metals and Ceramics Information Center, Battelle, 1987.

*ASM Specialty Handbook*, various volumes. Materials Park, Ohio: ASM International.

Boyer, H.E. (ed.), *Selection of Materials for Component Design: Source Book*. Metals Park, Ohio: American Society for Metals, 1986.

Budinski, K.G., *Engineering Materials: Properties and Selection*, 5th ed. Englewood Cliffs, New Jersey: Prentice-Hall, 1996.

*Material Selector,* annual publication of *Materials Engineering Magazine*, Penton/IPC, Cleveland, Ohio.

Pollock, D.D., *Physical Properties of Materials for Engineers*, 2d ed. Boca Raton, FL: CRC Press, 1993.

Roberts, G.A., and R.A. Cary, *Tool Steels*, 4th ed. Metals Park, Ohio: American Society for Metals, 1980.

*Tool and Manufacturing Engineers Handbook*, 4th ed. Vol. 3: *Materials, Finishing and Coating*. Dearborn, Mich.: Society of Manufacturing Engineers, 1985.

## QUESTIONS

3.1 What is the difference between a unit cell and a single crystal?

3.2 Explain why we should study the crystal structure of metals.

3.3 What effects does recrystallization have on the properties of metals?

3.4 What is the significance of a slip system?

3.5 Explain what is meant by structure-sensitive and structure-insensitive properties of metals.

3.6 What is the relationship between nucleation rate and the number of grains per unit volume of a metal?

3.7 Explain the difference between recovery and recrystallization.

3.8 Is it possible for two piece of the same metal to have different recrystallization temperatures? Explain.

3.9 Is it possible for recrystallization to take place in some regions of a workpiece before other regions in the same workpiece? Explain

3.10 Describe why different crystal structures exhibit different strengths and ductilities.

3.11 Explain the difference between preferred orientation and mechanical fibering.

3.12 Draw some analogies to mechanical fibering, e.g., layers of thin dough sprinkled with flour.

3.13 A cold-worked piece of metal has been recrystallized. When tested, it is found to be anisotropic. Explain the probable reason.

3.14 Does recrystallization eliminate mechanical fibering in a workpiece? Explain

3.15 Explain why we may have to be concerned with orange peel effect on metal surfaces.

3.16 Do you think it might be important to know whether a raw material for a manufacturing process has anisotropic properties? What about anisotropy in the finished product? Explain, with appropriate examples.

3.17 How can you tell the difference between two parts made of the same metal, one shaped by cold working and the other by hot working? Explain the differences you might observe.

3.18 Explain why the strength of a polycrystalline metal at room temperature decreases as its grain size increases.

3.19 Describe the technique by which you would reduce the orange peel effect.

3.20 What is the significance of metals such as lead and tin having recrystallization temperatures at about room temperature?

3.21 You are given a deck of playing cards held together with one or several rubber bands of the same size. Which of the material-behavior phenomena described in this chapter could you demonstrate with these items?

3.22 Using the information in Chapters 2 and 3, describe the conditions that induce

3.23 A part fractures while being subjected to plastic deformation at an elevated temperature. List the factors that could have played a role in the fracture.

brittle fracture in an otherwise ductile piece of metal.

3.24 Make a list of metals that would be suitable for: (1) a paper clip, (2) a buckle for a safety belt, (3) a nail, (4) a battery cable, and 5) a gas turbine blade.

3.25 Explain the advantages and limitations of cold, warm, and hot working of metals, respectively.

3.26 Describe the significance of structures and machine components made of two materials with different coefficients of thermal expansion.

3.27 Explain why parts may crack when suddenly subjected to extremes of temperature.

3.28 From you own experience and observations, list three applications each for the following metals and their alloys: (1) steel, (2) aluminum, (3) copper, (4) lead, and (5) gold.

3.29 Name products that would not have been developed to their advanced stages, as we find them today, if alloys with high strength and corrosion and creep resistance at elevated temperatures had not been developed.

3.30 Inspect several metal products and components and make an educated guess as to what materials they are made from. Give reasons for your guess. If you list two or more possibilities, explain your reasoning.

3.31 List three engineering applications, each where the following physical properties would be desirable: (1) high density, (2) low melting point, and (3) high thermal conductivity.

3.32 Two physical properties that have a major influence on the cracking of workpieces or dies during thermal cycling are thermal conductivity and thermal expansion. Explain why.

3.33 What might be the adverse effects of using a material with low specific heat in manufacturing?

## PROBLEMS

3.34 The unit cells shown in Fig. 3.2 can be represented by hard-ball or hard-sphere models, similar to tennis balls arranged in various configurations in a box. In such an arrangement, *atomic packing factor* (APF) is defined as the ratio of the volume of atoms to the volume of the unit cell. Show that the packing factor for the bcc structure is 0.68 and for the fcc structure 0.74.

3.35 Show that the lattice constant $a$ in Fig. 3.2b is related to the atomic radius by the following expression:

$$a = 2\sqrt{2}R,$$

where $R$ is the radius of the atom as depicted by the hard-ball model.

3.36 Show that for the fcc unit cell, the radius $r$ of the largest hole is given by the expression

$$r = 0.414R.$$

Determine the size of the largest hole for the iron atom in the fcc structure.

3.37 Calculate the theoretical (1) shear strength and (2) tensile strength for aluminum, plain-carbon steel, and tungsten. Estimate the ratios of their theoretical/actual strengths.

3.38 A technician determines that the grain size of a certain etched specimen is 6. Upon further checking, it is found that the magnification used was 150, instead of 100, which is required by ASTM standards. What is the correct grain size?

3.39 Estimate the number of grains in a regular paper clip if the ASTM grain size is 5.

3.40 Plot the data in Table 3.3 in terms of grains/mm$^2$ vs. grains/mm$^3$ and state your observations.

3.41 Determine the maximum crack velocity in the three materials listed in Problem 3.37 (see Section 3.8.2, and Tables 2.1 and 3.3).

3.42  Calculate the maximum theoretical shear stress $\tau_{max}$, assuming that in Fig. 3.5 the shear stress $\tau$ varies linearly with $x$.

3.43  The natural frequency $f$ of a cantilever beam is given by the expression

$$f = 0.56 \sqrt{EIg/wL^4},$$

where $E$ is the modulus of elasticity, $I$ the moment of inertia, $g$ the gravitational constant, $w$ the weight of the beam per unit length, and $L$ the length of the beam. How does the natural frequency of the beam change, if at all, as its temperature is increased?

3.44  A strip of metal is reduced in thickness by cold working from 25 mm to 15 mm. A similar strip is reduced from 25 mm to 10 mm. Which one of these strips will recrystallize at a lower temperature? Why?

3.45  Assume that you are asked to give a quiz to students on the contents of this chapter. Prepare three quantitative problems and three qualitative questions, and supply the answers.

# 4

## Surfaces, Dimensional Characteristics, Inspection, and Quality Assurance

**4.1** Introduction
**4.2** Surface Structure and Properties
**4.3** Surface Texture
**4.4** Tribology
**4.5** Surface Treatments, Coatings, and Cleaning
**4.6** Engineering Metrology and Instrumentation
**4.7** Dimensional Tolerances
**4.8** Testing and Inspection
**4.9** Quality Assurance

## 4.1 Introduction

Because of the various mechanical, physical, thermal, and chemical effects induced by its processing history, the **surface** of a manufactured part generally has properties and behavior that are considerably different from those of its bulk. Although the **bulk** material generally determines the component's overall mechanical properties, the component's surface directly influences several important properties and characteristics of the manufactured part:

- Friction and wear properties of the part during subsequent processing when it comes directly into contact with tools and dies and when it is placed in service

- Effectiveness of lubricants during the manufacturing processes, as well as throughout the part's service life

- Appearance and geometric features of the part and their role in subsequent operations such as painting, coating, welding, soldering, and adhesive bonding, as well as the resistance of the part to corrosion

- Initiation of cracks because of surface defects, such as roughness, scratches, seams, and heat-affected zones, which could lead to weakening and premature failure of the part by fatigue or other fracture mechanisms

- Thermal and electrical conductivity of contacting bodies. For example, rough surfaces have higher thermal and electrical resistance than smooth surfaces

Friction, wear, and lubrication, now called **tribology**, are surface phenomena. Friction influences forces, power requirements, and surface quality of parts. Wear alters the surface geometry of tools and dies, which in turn adversely affects the quality of manufactured products and the economics of production. Lubrication, with few exceptions, is an integral aspect of all manufacturing operations, as well as in the proper functioning of machinery and equipment. We describe those aspects of lubrication relevant to manufacturing operations, such as the types of wear encountered, how to reduce wear, and the proper selection and application of lubricants and coolants.

Another important aspect in surface technology is the treatment of surfaces to modify their properties and characteristics. We describe several mechanical, thermal, electrical, and chemical methods that can be used to modify surfaces of parts to improve frictional behavior, effectiveness of lubricants, resistance to wear and corrosion, and surface finish and appearance.

Measurement of the relevant dimensions and features of parts is an integral aspect of interchangeable manufacture—the basic concept of standardization and mass production. We describe the principles involved and the various instruments used in measurement. Another important aspect is testing and inspection of manufactured products, using both destructive and nondestructive testing methods. One of the most critical aspects of manufacturing is product quality. We discuss the technologic and economic importance of building quality into a product, rather than inspecting the product after it is made. Finally, we describe modern trends in quality engineering using statistical techniques and control charts.

## 4.2   Surface Structure and Properties

Upon close examination of the surface of a piece of metal you will find that it generally consists of several layers (Fig. 4.1). Starting at the interior of the metal and moving outward to the surface is the bulk metal. Also known as the metal **substrate,** its structure depends on the composition and processing history of the metal. Above this bulk metal is a layer that usually has been plastically deformed and work hardened during the manufacturing process.

The depth and properties of the work-hardened layer—the **surface structure**—depend on factors such as the processing method used and frictional sliding to which the surface was subjected. Sharp tools and selection of proper process parameters produce surfaces with little or no disturbance. For example, if the surface is produced by machining with a dull tool or under poor cutting conditions—or is ground with a dull grinding wheel—this layer will be relatively thick. Also, nonuniform surface deformation or severe temperature gradients during manufacturing operations usually cause residual stresses in this work-hardened layer.

Contaminant
Adsorbed gas
Oxide
Work-hardened layer
Metal substrate

**FIGURE 4.1**   Schematic illustration of the cross-section of the surface structure of metals. The thickness of the individual layers depends on processing conditions and the environment. *Source:* After E. Rabinowicz.

Unless the metal is processed and kept in an inert (oxygen-free) environment—or it is a noble metal, such as gold or platinum—an **oxide layer** usually lies on top of the work-hardened layer. Some examples are:

1. *Iron* has an oxide structure with FeO adjacent to the bulk metal, followed by a layer of $Fe_3O_4$ and then a layer of $Fe_2O_3$, which is exposed to the environment.

2. *Aluminum* has a dense amorphous (without crystalline structure) layer of $Al_2O_3$ with a thick, porous hydrated aluminum-oxide layer over it.

3. *Copper* has a bright shiny surface when freshly scratched or machined. Soon after, however, it develops a $Cu_2O$ layer, which is then covered with a layer of CuO. This gives copper its somewhat dull color, such as we see in kitchen utensils.

4. *Stainless steels* are "stainless" because they develop a protective layer of chromium oxide, CrO (passivation).

Under normal environmental conditions, surface oxide layers are generally covered with adsorbed layers of gas and moisture. Finally, the outermost surface of the metal may be covered with contaminants, such as dirt, dust, grease, lubricant residues, cleaning-compound residues, and pollutants from the environment.

Thus, surfaces generally have properties that are very different from those of the substrate. The oxide on a metal's surface, for example, is generally much harder than the base metal. Hence oxides tend to be brittle and abrasive. This surface characteristic has several important effects on friction, wear, and lubrication in materials processing and coatings on products. The factors involved in the surface structure of metals are also relevant to a great extent to the surface structure of plastics and ceramics. The surface texture of these materials depends, as with metals, on the method of production. Environmental conditions also influence the surface characteristics of these materials.

### 4.2.1 Surface integrity

**Surface integrity** describes not only the topological (geometric) aspects of surfaces, but also their mechanical and metallurgical properties and characteristics. Surface integrity is an important consideration in manufacturing operations because it influences the properties of the product, such as its fatigue strength and resistance to corrosion, and its service life.

Several **defects** caused by and produced during component manufacturing can be responsible for lack of surface integrity. These defects are usually caused by a combination of factors, such as defects in the original material, the method by which the surface is produced, and lack of proper control of process parameters that can result in excessive stresses and temperatures. The following are definitions (listed in alphabetical order) of the major surface defects found in practice.

- *Cracks* are external or internal separations with sharp outlines. Cracks requiring a magnification of $10\times$ or higher to be seen by the naked eye are called microcracks.

- *Craters* are shallow depressions.

- *Folds* are the same as seams (see Seams on p. 148).

- *Heat-affected zone* is the portion of a metal subjected to thermal cycling without melting.

- *Inclusions* are small, nonmetallic elements or compounds in the metal.

- *Intergranular attack* is the weakening of grain boundaries by liquid-metal embrittlement and corrosion.

- *Laps* are the same as seams.

- *Metallurgical transformation* involves microstructural changes caused by temperature and pressure cycling. Included are phase transformations, recrystallization, alloy depletion, decarburization, and molten and recast, resolidified, or redeposited material, as in electrical-discharge machining.

- *Pits* are shallow surface depressions, usually the result of chemical or physical attack.

- *Plastic deformation* is a severe surface deformation caused by high stresses due to friction, tool and die geometry, worn tools, and method of processing.

- *Residual stresses* are surface stresses (tension or compression) caused by nonuniform deformation and temperature distribution.

- *Seams* are surface defects resulting from overlapping of the material during processing.

- *Splatter* consists of small resolidified molten metal particles deposited on a surface, as from welding.

## 4.3  Surface Texture

Regardless of the method of production, all surfaces have their own characteristics, which we refer to as **surface texture,** roughness, and finish. The description of surface texture as a geometrical property is complex. However, certain guidelines have been established for identifying surface texture in terms of well-defined and measurable quantities (Fig. 4.2).

**Flaws,** or **defects,** are random irregularities, such as scratches, cracks, holes, depressions, seams, tears, and inclusions.

**FIGURE 4.2**  Standard terminology and symbols to describe surface finish. The quantities are given in $\mu$in.

**Lay,** or **directionality,** is the direction of the predominant surface pattern and is usually visible to the naked eye.

**Roughness** consists of closely spaced, irregular deviations on a scale smaller than that for waviness. Roughness may be superimposed on waviness. Roughness is expressed in terms of its height, its width, and the distance on the surface along which it is measured.

**Waviness** is a recurrent deviation from a flat surface, much like waves on the surface of water. It is measured and described in terms of the space between adjacent crests of the waves (*waviness width*) and height between the crests and valleys of the waves (*waviness height*). Waviness may be caused by deflections of tools, dies, and the workpiece, warping from forces or temperature, uneven lubrication, and vibration or any periodic mechanical or thermal variations in the system during the manufacturing process.

### 4.3.1 Surface roughness

Surface roughness is generally described by two methods: arithmetic mean value ($R_a$); and root-mean-square average ($R_q$). The **arithmetic mean value ($R_a$;** formerly identified as AA for *arithmetic average* or CLA for *center-line average*) is based on the schematic illustration of a rough surface shown in Fig. 4.3. The arithmetic mean value $R_a$ is defined as

$$R_a = \frac{a + b + c + d + \cdots}{n},\tag{4.1}$$

where all ordinates, $a, b, c, \ldots$, are absolute values.

The **root-mean-square average ($R_q$;** formerly identified as RMS) is defined as

$$R_q = \sqrt{\frac{a^2 + b^2 + c^2 + d^2 + \cdots}{n}}\tag{4.2}$$

The datum line AB in Fig. 4.3 is located so that the sum of the areas above the line is equal to the sum of the areas below the line. The units generally used for surface roughness are $\mu$m (micrometer, or micron) or $\mu$in. (microinch), where 1 $\mu$m = 40 $\mu$in. and 1 $\mu$in. = 0.025 $\mu$m.

Additionally, we may also use the **maximum roughness height ($R_t$)** as a measure of roughness. It is defined as the height from the deepest trough to the highest peak. It indicates the amount of material that has to be removed to obtain a smooth surface by polishing or other means.

Because of its simplicity, the arithmetic mean value $R_a$ was adopted internationally in the mid-1950s and is widely used in engineeringing practice. Equations (4.1) and (4.2) show that there is a relationship between $R_q$ and $R_a$. For a surface roughness in the shape of a sine curve, $R_q$ is larger than $R_a$ by a factor of 1.11. This factor is 1.1 for most machining processes by cutting, 1.2 for grinding, and 1.4 for lapping and honing.

**FIGURE 4.3** Coordinates used for surface-roughness measurement, using Eqs. (4.1) and (4.2).

In general, we cannot adequately describe a surface by its $R_a$ or $R_q$ value alone, since these values are averages. Two surfaces may have the same roughness value but their actual topography may be quite different. A few deep troughs, for example, affect the roughness values insignificantly. However, differences in the surface profile can be significant in terms of fatigue, friction, and the wear characteristics of a manufactured product.

**Symbols for surface roughness.**   Acceptable limits for surface roughness are specified on technical drawings by the symbols shown around the check mark in the lower portion of Fig. 4.2, and their values are placed to the left of the check mark. Symbols used to describe a surface only specify the roughness, waviness, and lay; they do not include flaws. Whenever important, a special note is included in technical drawings to describe the method to be used to inspect for surface flaws.

## Measuring surface roughness

Various commercially available instruments, called **surface profilometers,** are used to measure and record surface roughness. The most commonly used instruments feature a diamond stylus traveling along a straight line over the surface (Fig. 4.4). The distance that the stylus travels, which can be varied, is called the **cutoff** (see Fig. 4.2).

To highlight the roughness, profilometer traces are recorded on an exaggerated vertical scale (a few orders of magnitude greater than the horizontal scale; Fig. 4.4), called **gain** on the recording instrument. Thus the recorded profile is significantly distorted, and the surface

**FIGURE 4.4**   (a) Measuring surface roughness with a stylus. The rider supports the stylus and guards against damage. (b) Path of stylus in surface roughness measurements (broken line) compared to actual roughness profile. Note that the profile of the stylus path is smoother than the actual surface profile. *Source:* D. H. Buckley. Typical surface profiles produced by (c) turning and (d) finish grinding processes. Note the difference between the vertical and horizontal scales. *Source:* D. B. Dallas (ed.), *Tool and Manufacturing Engineers Handbook,* 3d ed. Copyright © 1976, McGraw-Hill Publishing Company. Used with permission.

appears to be much rougher than it actually is. The recording instrument compensates for any surface waviness and indicates only roughness. A record of the surface profile is made by mechanical and electronic means.

We can observe surface roughness directly through an optical or scanning electron microscope. Stereoscopic photographs are particularly useful for three-dimensional views of surfaces, and we can use them to measure surface roughness.

### Surface roughness in practice

Surface roughness design requirements for typical engineering applications can vary by as much as two orders of magnitude for different parts. The reasons and considerations for this wide range include:

1. *Precision required on mating surfaces,* such as seals, fittings, gaskets, tools, and dies. For example, ball bearings and gages require very smooth surfaces, whereas surfaces for gaskets and brake drums can be quite rough.
2. *Frictional considerations,* that is, the effect of roughness on friction, wear, and lubrication
3. *Fatigue and notch sensitivity,* because the rougher the surface is, the shorter the fatigue life will be
4. *Electrical and thermal contact resistance,* because the rougher the surface is, the higher the resistance will be
5. *Corrosion resistance,* because the rougher the surface is, the greater will be the possibility of entrapped corrosive media
6. *Subsequent processing,* such as painting and coating, in which a certain roughness can result in better bonding
7. *Appearance*
8. *Cost considerations,* because the finer the finish is, the higher the cost will be

These factors should be carefully considered before a decision is made as to the recommendation about surface roughness for a certain product. As in all manufacturing processes, the cost involved in the selection should also be a major consideration.

## 4.4 Tribology

**Tribology** is the science and technology of interacting surfaces, thus involving friction, wear, and lubrication. All can significantly influence the technology and economics of manufacturing processes and operations.

### 4.4.1 Friction

**Friction** is defined as the resistance to relative sliding between two bodies in contact under a normal load. It plays an important role in all metalworking and manufacturing processes because of the relative motion and forces always present between tools, dies, and workpieces. Friction is an energy-dissipating process that results in the generation of heat. The subsequent rise in temperature can have a major effect on the overall operation. Furthermore, because it impedes free movement at interfaces, friction can significantly affect the flow and deformation of the material in many metalworking processes.

However, friction should not always be regarded as undesirable. The presence of friction is necessary for the success or optimization of many operations, the best example of which is the rolling of metals to make sheet and plate. Without friction it would be impossible to roll metals, just as it would be impossible to drive a car on the road. Various theories have been proposed to explain the phenomenon of friction. An early Coulomb model states that friction results from the mechanical interlocking of rough surfaces; this interlocking will require some force to slide the two bodies against each other.

For any model of friction to be valid, it must explain the frictional behavior of two bodies under different loads, speed of relative sliding, temperature, surface conditions, environment, etc., as observed in practice. Consequently, many models have been proposed with varying degrees of success. The most commonly used theory of friction is based on adhesion, because it agrees reasonably well with experimental observations.

### Adhesion theory of friction

The **adhesion theory of friction** is based on the observation that two clean, dry (unlubricated) surfaces, regardless of how smooth they are, contact each other at only a fraction of their apparent area of contact (Fig. 4.5). The static load at the interface is thus supported by the minute asperities in contact with each other. The sum total of the contacting areas is known as the **real area of contact**, $A_r$. For most engineering surfaces, the average angle that the asperity hill makes with a horizontal line is found to be between 5° and 15°. (Surface roughness illustrations are highly exaggerated because of the scales involved.)

Under light loads and with a large real area of contact, the normal stress at the asperity contact is *elastic*. As the load increases, the stresses increase and eventually *plastic* deformation takes place at the **junctions**. Furthermore, with increasing load, the asperities in contact increase in area; also, new junctions are formed with other asperities on the two surfaces in contact with each other. Because of the random height of the asperities, some junctions are in elastic and others in plastic contact.

The intimate contact of asperities creates an **adhesive bond**. The nature of this bond is complex, involving interactions on an atomic scale, and may include mutual solubility and diffusion. The strength of the bond depends on the physical and mechanical properties of the materials in contact, temperature, and the nature and thickness of any oxide film or other contaminants present on the surfaces.

**FIGURE 4.5**   (a) Schematic illustration of the interface of two contacting surfaces, showing the real areas of contact. (b) Sketch illustrating the proportion of the apparent area to real area of contact. The ratio of the areas can be as high as four to five orders of magnitude.

In metalworking processes, the load at the interface is generally high so that the normal stress on the asperity reaches the yield stress. Plastic deformation of the asperity and adhesion then take place at the junctions. In other words, the asperities form **microwelds.** The cleaner the interface, the stronger are the adhesive bonds. To pull the two surfaces apart requires a certain force, much like pulling apart a welded joint. The coefficient of adhesion is defined as

$$\text{Coefficient of adhesion} = \frac{\text{Tensile force}}{\text{Compressive force applied}}. \tag{4.3}$$

Experimental observations indicate that this coefficient can be quite high with softer metals and that, with few exceptions, it requires a certain finite force to pull the surfaces apart. The more ductile the materials, the greater is the plastic deformation and the stronger is the adhesion of the junctions.

## Coefficient of friction

Sliding between the two bodies with such an interface under a normal load $N$ is possible only by the application of a tangential force $F$. According to the adhesion theory, this is the force required to *shear* the junctions (friction force). The **coefficient of friction** $\mu$ at the interface is defined as

$$\mu = \frac{F}{N} = \frac{\tau A_r}{\sigma A_r} = \frac{\tau}{\sigma}, \tag{4.4}$$

where $\tau$ is the shear strength of the junction and $\sigma$ is the normal stress, which, for a plastically deformed asperity, is the yield stress of the asperity. Thus the numerator in Eq. (4.4) is a *surface* property, whereas the denominator is a *bulk* property. Because an asperity is surrounded by a large mass of material, the normal stress on an asperity is equivalent to the hardness of the material (see Section 2.6). Thus, the coefficient of friction $\mu$ can now be defined as

$$\mu = \frac{\tau}{\text{hardness}}. \tag{4.5}$$

Two additional phenomena take place in asperity interactions under plastic stresses: For a strain-hardening material, the peak of the asperity is stronger than the bulk material because of plastic deformation at the junction. Thus, under ideal conditions with clean surfaces, the breaking of the bonds under a tensile load will likely follow a path below or above the geometric interface.

Second, a tangential motion at the junction under load will cause *junction growth;* that is, the contact area increases. The reason is that, according to the yield criteria described in Section 2.11, the effective yield stress of the material decreases when subjected to a shear stress. Hence the junction has to grow in area to support the normal load. In other words, the two surfaces tend to approach each other under a tangential force.

If the normal load is increased further, the real area of contact $A_r$ in this model ideally continues to increase, especially with soft and ductile metals. In the absence of contaminants or fluids trapped at the interface, the real area of contact eventually reaches the apparent area of contact. This is the maximum contact area that can be obtained. Because the shear strength at the interface is constant, the friction force (shearing force) reaches a maximum and stabilizes (Fig. 4.6).

This condition, known as **sticking,** creates difficulty in defining friction. Sticking in a sliding interface does not necessarily mean complete adhesion at the interface, as in welding.

**FIGURE 4.6** · Schematic illustration of the relation between friction force (*F*) and normal force (*N*). Note that as the real area of contact approaches the apparent area, the friction force reaches a maximum and stabilizes. Most machine components operate in the first region. The second and third regions are encountered in metalworking operations because of the high contact pressures involved between sliding surfaces, i.e., die and workpiece.

Rather, it means that the frictional stress at the surface has reached the shear yield stress *k* of the material. However, when two clean surfaces are pressed together with sufficiently high forces, cold pressure welding can indeed take place.

As the normal load *N* is increased further, the friction force *F* remains constant and hence, by definition, the coefficient of friction decreases. This situation, which is an anomaly, indicates that there might be a different and more realistic means of expressing the frictional condition at an interface.

A more recent trend is to define a **friction factor** or **shear factor** *m* as

$$m = \frac{\tau_i}{k},\tag{4.6}$$

where $\tau_i$ is the shear strength of the interface and *k* is the *shear yield stress* of the softer material in a sliding pair. [The quantity *k* is equal to *Y*/2 according to Eq. (2.35) and to $Y/\sqrt{3}$ according to the distortion-energy criterion, where *Y* is the uniaxial yield stress of the material.]

The value of $\tau_i$ is difficult to measure because we have to know the real area of contact; average values will not be meaningful. However, note that in this definition when $m = 0$, there is no friction and when $m = 1$ complete sticking takes place at the interface. The magnitude of *m* is independent of the normal force or stress. The reason is that the shear yield stress of a thin layer of material is unaffected by the magnitude of the normal stress.

If the upper body in the model shown in Fig. 4.5 is harder than the lower one, or if its surface contains protruding hard particles, then as it slides over the softer body it can scratch and produce grooves on the lower surface. This is known as **ploughing** (or *plowing*) and is an important aspect in frictional behavior; in fact, it can be a dominant mechanism for situations where adhesion is not strong.

Ploughing may involve two different mechanisms. One is the generation of a *groove*, whereby the surface is deformed plastically. The other is the formation of a groove because the cutting action of the upper body generates a *chip* or *sliver*. Either of these processes involves work supplied by a force that manifests itself as friction force. The ploughing force can contribute significantly to friction and to the measured coefficient of friction at the interface.

The nature and strength of the interface is the most significant factor in friction. A strong interface requires a high friction force for relative sliding and one with little strength requires a low friction force. Equation (4.5) indicates that the coefficient of friction can be reduced not only by decreasing the numerator, but also by increasing the denominator.

This observation suggests that thin films of low shear strength over a substrate with high hardness is the ideal method for reducing friction. In fact, this is exactly what is achieved with a surface lubricated either with a fluid or solid lubricant.

With no low-shear-strength layer or contaminant present, and under ideal clean conditions with a rigid, perfectly plastic material pair, the coefficient of friction at an asperity can be calculated theoretically from Eq. (4.5) as

$$\mu = \frac{\tau}{\text{hardness}} = \frac{Y/\sqrt{3}}{3Y} \simeq 0.2. \tag{4.7}$$

If the number of asperities increases, such as with increasing load, the combined shear strength of the junctions may exceed the bulk shear strength of the softer metal. In this case, plastic flow takes place in a sublayer (sticking). The value of the normal stress in Eq. (4.4) then becomes that of the uniaxial compressive yield stress of the softer metal. The coefficient of friction is then

$$\mu = \frac{\tau}{\sigma} = \frac{Y/\sqrt{3}}{Y} = 0.577. \tag{4.8}$$

Equation (4.7) can be modified to include a term for ploughing. This term can be estimated roughly to be on the order of 0.1, which should be added algebraically to the value obtained from Eq. (4.7).

The sensitivity of friction to different factors can be illustrated by the following examples. The coefficient of friction of steel sliding on lead is 1.0; steel on copper is 0.9. However, for steel sliding on copper coated with a thin layer of lead, it is 0.2. The coefficient of friction of pure nickel sliding on nickel in a hydrogen or nitrogen atmosphere is 5, in air or oxygen it is 3, and in the presence of water vapor it is 1.6.

Coefficients of friction in sliding contact, as measured experimentally, vary from as low as 0.02 to 100, or even higher. This range is not surprising in view of the many variables involved in the friction process. In metalworking operations that use various lubricants, the range for $\mu$ is much narrower, as shown in Table 4.1.

**TABLE 4.1**

**COEFFICIENT OF FRICTION IN METALWORKING PROCESSES**

| | COEFFICIENT OF FRICTION ($\mu$) | |
|---|---|---|
| PROCESS | COLD | HOT |
| Rolling | 0.05–0.1 | 0.2–0.7 |
| Forging | 0.05–0.1 | 0.1–0.2 |
| Drawing | 0.03–0.1 | — |
| Sheet-metal forming | 0.05–0.1 | 0.1–0.2 |
| Machining | 0.5–2 | — |

The effects of load, temperature, speed, and environment on the coefficient of friction are difficult to generalize, as each situation has to be studied individually.

### Temperature rise caused by friction

Almost all the energy dissipated in overcoming friction is converted into heat. A small portion remains in the bodies as stored energy, and some of it goes into surface energy during generation of new surfaces and wear particles.

Frictional heat raises the interface temperature. The magnitude of this temperature rise and its distribution depends not only on the friction force, but also on speed, surface roughness, and the physical properties of the materials, such as thermal conductivity and specific heat. The interface temperature increases with increasing friction, speed, and with low thermal conductivity and specific heat of the materials. The interface temperature can be high enough to soften and melt the surface. (The temperature cannot, of course, exceed the melting point of the material.) A number of analytical expressions have been obtained to calculate the temperature rise in sliding.

### Reducing friction

Friction can be reduced by selecting materials that have low adhesion, such as carbides and ceramics, and by using surface films and coatings. Lubricants, such as oils, or solid films, such as graphite, interpose an adherent film between tool, die, and workpiece. The film minimizes adhesion and interactions of one surface to the other, thus reducing friction.

Friction can also be reduced significantly by subjecting the die–workpiece interface to *ultrasonic vibrations*, generally at 20 kHz. These vibrations momentarily separate die and workpiece, thus allowing the lubricant to flow more freely into the interface. An additional factor is the high-frequency variation of the relative velocity between die and workpiece, thus reducing the friction at the interface.

### Measuring friction

The coefficient of friction is usually determined experimentally, either during actual manufacturing processes or in simulated tests using small-scale specimens of various shapes. The techniques used generally involve measurement of either forces or dimensional changes in the specimen.

One test that has gained wide acceptance, particularly for bulk deformation processes such as forging, is the **ring compression test.** A flat ring is compressed plastically between two flat platens (Fig. 4.7). As its height is reduced, the ring expands radially outward. If friction at the interfaces is zero, both the inner and outer diameters of the ring expand as if it were a solid disk. With increasing friction, the inner diameter becomes smaller.

For a particular reduction in height, there is a critical friction value at which the internal diameter increases (from the original) if $\mu$ is low and decreases if $\mu$ is high. By measuring the change in the specimen's internal diameter, and using the curves shown in Fig. 4.8, which are obtained through theoretical analyses, we can determine the coefficient of friction.

Each ring geometry has its own specific set of curves. The most common geometry has outer diameter to inner diameter to height proportions of the specimen of 6:3:2. The actual

(a)

Good lubrication

(b)

Poor lubrication

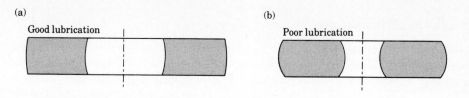

**FIGURE 4.7** Effect of lubrication on barreling in ring compression test: (a) with good lubrication, both the inner and outer diameters increase as the specimen is compressed; and (b) with poor or no lubrication, friction is high and the inner diameter decreases. The direction of barreling depends on the relative motion of the cylindrical surfaces with respect to the flat dies.

size of the specimen usually is not relevant in these tests. Thus, once you know the percentage reductions in internal diameter and height, you can determine $\mu$ from the appropriate chart.

**FIGURE 4.8** Charts to determine friction in ring compression tests: (a) coefficient of friction $\mu$; (b) friction factor $m$. Friction is determined from these charts from the percent reduction in height and by measuring the percent change in the internal diameter of the specimen after compression.

**Example 4.1:**    Determining coefficient of friction

In a ring-compression test, a specimen 10 mm in height with outside diameter (OD) = 30 mm and inside diameter (ID) = 15 mm is reduced in thickness by 50%. Determine the coefficient of friction $\mu$ and the friction factor $m$ if the OD after deformation is 38 mm.

SOLUTION.    We first have to determine the new ID. We do so from volume constancy, as follows:

$$\text{Volume} = \frac{\pi}{4}(30^2 - 15^2)10 = \frac{\pi}{4}(38^2 - ID^2)5.$$

From this, the new ID = 9.7 mm. Thus, the change in internal diameter is

$$\text{Change in ID} = \frac{15 - 9.7}{15} \times 100 = 35\% \text{ (decrease)}.$$

For a 50% reduction in height and a reduction in internal diameter of 35%, the following values are obtained from Fig. 4.8:

$$\mu = 0.21 \text{ and } m = 0.72.$$

---

**Friction in plastics and ceramics**

Although their strength is low compared to metals, plastics generally possess low frictional characteristics. This property makes polymers attractive for bearings, gears, seals, prosthetic joints, and general friction-reducing applications. In fact, polymers are sometimes called self-lubricating. The factors involved in friction and metal wear are also generally applicable to polymers. In sliding, the ploughing component of friction in thermoplastics and elastomers is a significant factor because of their viscoelastic behavior (exhibiting both viscous and elastic behavior) and subsequent hysteresis loss.

An important factor in plastics applications is the effect of temperature rise at the sliding interfaces caused by friction. Thermoplastics lose their strength and become soft as temperature increases. Their low thermal conductivity and low melting points are significant in terms of heat generation by friction. If the temperature rise is not controlled, sliding surfaces can undergo deformation and thermal degradation. In view of the importance of ceramics, their frictional behavior is now being studied intensively. Initial investigations indicate that the origin of friction in ceramics is similar to that in metals. Thus, adhesion and ploughing at interfaces contribute to the friction force in ceramics.

**4.4.2    Wear**

**Wear** is defined as the progressive loss or removal of material from a surface. Wear has important technologic and economic significance because it changes the shape of the workpiece, tool, and die interfaces. Hence it affects the process and size and quality of the parts produced. The magnitude of the wear problem is evident in the countless parts and components that continually have to be replaced or repaired.

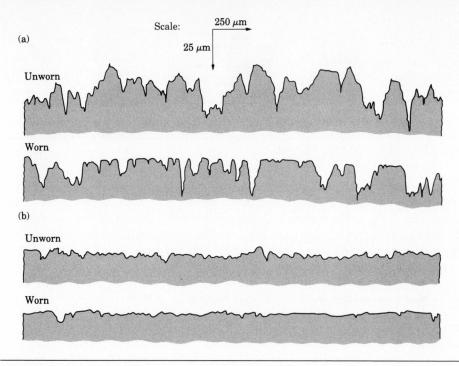

**FIGURE 4.9** Changes in originally (a) wire-brushed and (b) ground-surface profiles after wear. *Source:* E. Wild and K. J. Mack.

Examples of wear in manufacturing processes are dull drills that have to be reground, worn cutting tools that have to be indexed or resharpened, and forming tools and dies that have to be repaired or replaced. Important components in some metalworking machinery are *wear plates,* which are subjected to high loads. These plates, also known as *wear parts* because they are expected to wear, can be replaced easily.

Although wear generally alters the surface topography and may result in severe surface damage, it also has a beneficial effect: It can reduce surface roughness by removing the peaks from asperities (Fig. 4.9). Thus, under controlled conditions, wear may be regarded as a kind of smoothening or polishing process. The **running-in** period for various machines and engines produces this type of wear.

Wear is usually classified as adhesive, abrasive, corrosive, fatigue, erosion, fretting, and impact. We describe below the major types of wear relevant to manufacturing operations.

### Adhesive wear

If a tangential force is applied to the model shown in Fig. 4.5, shearing can take place either at the original interface or along a path below or above it (Fig. 4.10), causing **adhesive wear.** The fracture path depends on whether or not the strength of the adhesive bond of the asperities is greater than the cohesive strength of either of the two sliding bodies.

Because of factors such as strain hardening at the asperity contact, diffusion, and mutual solid solubility, the adhesive bonds are often stronger than the base metals. Thus, during

**FIGURE 4.10**   Schematic illustration of (a) two asperities contacting, (b) adhesion between two asperities, and (c) the formation of a wear particle.

sliding, fracture at the asperity usually follows a path in the weaker or softer component. A wear fragment is then generated. Although this fragment is attached to the harder component (upper member in Fig. 4.10), it eventually becomes detached during further rubbing at the interface and develops into a loose wear particle. This process is known as adhesive wear or sliding wear. In more severe cases, such as high loads and strongly bonded asperities, adhesive wear is described as scuffing, smearing, tearing, galling, or seizure. Oxide layers on surfaces greatly influence adhesive wear. They can act as a protective film, resulting in what is known as **mild wear,** consisting of small wear particles.

Based on the probability that a junction between two sliding surfaces would lead to the formation of a wear particle, an expression for adhesive wear (due to J. F. Archard) has been developed as

$$V = k \frac{LW}{3p},$$   (4.9)

where $V$ is the volume of material removed by wear from the surface, $k$ is the wear coefficient (dimensionless), $L$ is the length of travel, $W$ is the normal load, and $p$ is the indentation hardness of the softer body. In some references, the factor 3 is deleted from the denominator and incorporated in the wear coefficient $k$.

Typical values of $k$ for a combination of materials sliding in air are given in Table 4.2. The wear coefficient for the same pair of materials can vary by a factor of 3, depending on whether wear is measured as a loose particle or as a transferred particle (to the other body). Loose particles have a lower $k$ value. Likewise, when mutual solubility is a significant parameter in adhesion, similar metal pairs have higher $k$ values than dissimilar pairs.

Thus far we have based treatment of adhesive wear on the assumption that the surface layers of the two contacting bodies are clean and free from contaminants. Under these conditions, the adhesive wear rate can be very high, which is known as **severe wear.** As described in Section 4.2, however, metal surfaces are almost always covered with oxide layers, with thicknesses generally ranging between 20 and 100 Å. Although this thickness may at first be regarded as insignificant, the oxide layer has a profound effect on wear behavior.

Oxide layers are usually hard and brittle. When such a surface is subjected to rubbing, this layer can have the following effects. If the load is light and the oxide layer is strongly adhering to the bulk metal, the strength of the junctions between the asperities is weak and wear is low. In this situation, the oxide layer acts as a protective film and the type of wear is known as mild wear.

The oxide layer can be broken up under high normal loads if the oxide layer is brittle and is not strongly adhering to the bulk metal, or if the asperities are rubbing against each other

**TABLE 4.2**

**APPROXIMATE ORDER OF MAGNITUDE FOR WEAR COEFFICIENT $k$ IN AIR**

| UNLUBRICATED | $k$ | LUBRICATED | $k$ |
|---|---|---|---|
| Mild steel on mild steel | $10^{-2}$ to $10^{-3}$ | 52100 steel on 52100 steel | $10^{-7}$ to $10^{-10}$ |
| 60–40 brass on hardened tool steel | $10^{-3}$ | Aluminum bronze on hardened steel | $10^{-8}$ |
| Hardened tool steel on hardened tool steel | $10^{-4}$ | Hardened steel on hardened steel | $10^{-9}$ |
| Polytetrafluoroethylene (PTFE) on tool steel | $10^{-5}$ | | |
| Tungsten carbide on mild steel | $10^{-6}$ | | |

repeatedly, so that the oxide layer breaks up by fatigue. However, if the surfaces are smooth, the oxide layers are more difficult to break up. When this layer is eventually broken up, a wear particle is formed. An upper asperity can then form a strong junction with the lower asperity, which is now unprotected by the oxide layer. The wear rate will now be higher until a fresh oxide layer is formed.

In addition, a surface exposed to the environment is covered with adsorbed layers of gas and usually with other contaminants. Such films, even if they are very thin (5 Å), generally weaken the interfacial bond strength of contacting asperities by separating the metal surfaces. The net effect of these films is to provide a protective layer by reducing adhesion and thus reducing wear. The magnitude of the effect of adsorbed gases and contaminants depends on many factors. For instance, it has been shown repeatedly that even small differences in humidity can have a profound influence on the wear rate.

**Example 4.2:  Adhesive wear in sliding**

The end of a rod made of 60–40 brass is sliding over the unlubricated surface of hardened tool steel with a load of 200 lb. The hardness of brass is 120 HB. What is the distance traveled to produce a wear volume of 0.001 in$^3$ by adhesive wear of the brass rod?

SOLUTION.  The parameters in Eq. (4.9) for adhesive wear are as follows:

$$V = 0.001 \text{ in}^3$$
$$k = 10^{-3} \text{ (from Table 4.2)}$$
$$W = 200 \text{ lb}$$
$$p = 120 \text{ kg/mm}^2 = 170,700 \text{ lb/in}^2.$$

Thus, the distance traveled is

$$L = \frac{3Vp}{kW} = \frac{(3)(0.001)(170,700)}{(10^{-3})(200)} = 2560 \text{ in.} = 213 \text{ ft.}$$

**FIGURE 4.11**   Schematic illustration of abrasive wear in sliding. Longitudinal scratches on a surface usually indicate abrasive wear.

## Abrasive wear

**Abrasive wear** is caused by a hard and rough surface—or a surface containing hard, protruding particles—sliding across a surface. This type of wear removes particles by forming microchips or slivers, thereby producing grooves or scratches on the softer surface (Fig. 4.11). In fact, the abrasive processes that we describe in Chapter 9, such as grinding, ultrasonic machining, and abrasive-jet machining, act in this manner. The difference is that in those operations we control the process parameters to produce desired shapes and surfaces, whereas abrasive wear is unintended and unwanted.

The abrasive wear resistance of pure metals and ceramics is directly proportional to their hardness (Fig. 4.12). Abrasive wear can thus be reduced by increasing the hardness of materials (such as by heat treating and microstructural changes) or by reducing the normal load. Other materials that resist abrasive wear are elastomers and rubbers, because they deform elastically and then recover when abrasive particles cross over their surfaces. The best example is automobile tires, which have long lives even though they are operated on abrasive road surfaces. Even hardened steels would not last long under such conditions.

There are two basic types of abrasive wear: 2-body and 3-body wear. The first type is the basis of **erosive** wear. The abrasive particles (such as sand) are usually carried in a jet of fluid or air, and they remove material from the surface by erosion. The 3-body wear is important in metalworking processes because a lubricant between the die and workpiece may carry with it wear particles (which are generated over a period of time) and cause abrasive wear. Thus, the necessity for proper inspection and filtering of a metalworking fluid is apparent. This type of wear is also significant in the maintenance of various machinery and equipment. There is a possibility that particles (from machining, grinding, the environment, etc.) may contaminate the lubricating system of the machine components and thus scratch and damage the surfaces.

**FIGURE 4.12**   Abrasive wear resistance as a function of the hardness of several classes of materials.

**Corrosive wear.**   **Corrosive wear,** also known as oxidation, or chemical, wear, is caused by chemical or electrochemical reactions between the surface and the environment. The fine corrosive products on the surface constitute the wear particles. When the corrosive layer is destroyed or removed, as by sliding or abrasion, another layer begins to form, and the process of removal and corrosive-layer formation is repeated. Among corrosive media are water, seawater, oxygen, acids and chemicals, and atmospheric hydrogen sulfide and sulfur dioxide. We can reduce corrosive wear by selecting materials that will resist environmental attack, controlling the environment, and reducing operating temperatures to lower the rate of chemical reaction.

**Fatigue wear.**   **Fatigue wear,** also called surface fatigue or surface-fracture wear, is caused when the surface of a material is subjected to cyclic loading, such as in rolling contact in bearings. The wear particles are usually formed by spalling or pitting. Another type of fatigue wear is *thermal fatigue*. Cracks on the surface are generated by thermal stresses from thermal cycling, such as a cool die repeatedly contacting hot workpieces (heat checking). These cracks then join, and the surface begins to spall, producing fatigue wear. This type of wear usually occurs in hot-working and die-casting dies. We can reduce fatigue wear by lowering contact stresses, reducing thermal cycling, and improving the quality of materials by removing impurities, inclusions, and various other flaws that may act as local points for crack initiation.

**Other types of wear.**   Several other types of wear are important in manufacturing processes. **Erosion** is caused by loose abrasive particles abrading a surface. **Fretting corrosion** occurs at interfaces that are subjected to very small movements, such as in machinery. **Impact wear** is the removal of small amounts of materials from a surface by impacting particles. Deburring by vibratory finishing and tumbling (Chapter 9) is an example of this type of wear. In many cases, component wear is the result of a combination of different types of wear. Note in Fig. 4.13, for example, that even in the same forging die, various types of wear take place in different locations. A similar situation exists in cutting tools.

## General observations

It is apparent from the foregoing that wear is a complex phenomenon and that each type of wear is influenced by many factors. Numerous theories have been advanced to explain and interpret the experimental data and observations made regarding wear. Because of the many factors involved, quantifying wear data is difficult. In fact, it is well known that wear data are difficult to reproduce.

The most significant aspects of wear are the nature and geometry of the contacting asperities. The strength of the bond at the asperity junction relative to the cohesive strength of the weaker body is an important factor. When a polymer such as PTFE, that is, *Teflon,* is rubbed against steel, the surface of the polymer (which is much weaker than the steel) becomes embedded with steel wear particles. Also, when you rub gold with your fingers, wear particles of gold remain on the fingers. The mechanisms for this phenomenon are not clear.

The question also arises as to whether friction and wear for a pair of metals are directly related. Although at first we would expect such a relationship, experimental evidence for unlubricated metal pairs indicates that this is not always the case. These results are not surprising, in view of the many factors involved in friction and wear. For instance, ball bearings and races have very little friction, yet they can undergo wear after a large number of cycles.

**FIGURE 4.13**   Types of wear observed in a single die used for hot forging. *Source:* T. A. Dean.

1. Erosion
2. Pitting (lubricated dies only)
3. Thermal fatigue
4. Mechanical fatigue
5. Plastic deformation

## Measuring wear

Several methods can be used to observe and measure wear. The choice of a particular method depends on the accuracy desired and the physical constraints of the system, such as specimen or part size and difficulty of disassembly.

Although not quantitative, the simplest method is visual and tactile (touching) inspection. Measuring dimensional changes, gaging the worn component, profilometry, and weighing are more accurate methods. However, for large workpieces or tools and dies, the weighing method is not accurate because the amount of wear is usually very small compared to the overall weight of the components involved. Performance and noise level can be monitored. Worn machinery components emit more noise than new parts.

Radiography is a method in which wear particles from an irradiated surface are transferred to the mating surface, which is then measured for the amount of radiation. An example is the transfer of wear particles from irradiated cutting tools to the back side of chips. In other situations, the lubricant can be analyzed for wear particles (spectroscopy). This is a precision method and is used widely for applications such as checking jet-engine component wear.

## Wear of plastics and ceramics

The wear behavior of plastics is similar to that of metals. Thus, wear may occur in ways similar to those described earlier. Abrasive wear behavior depends partly on the ability of the polymer to deform and recover elastically, similar to the behavior of elastomers. There is evidence that the parameter describing this behavior may be the ratio of hardness to elastic modulus. Thus, the abrasive wear resistance of the polymer increases as this ratio increases.

Typical polymers with good wear resistance are polyimides, nylons, polycarbonate, polypropylene, acetals, and high-density polyethylene. These polymers are molded or machined to make gears, pulleys, sprockets, and similar mechanical components. Because plastics can be made with a wide variety of compositions, they can also be blended with internal lubricants, such as PTFE, silicon, graphite, and molybdenum disulfide, and with rubber particles that are interspersed within the polymer matrix.

Wear resistance of reinforced plastics depends on the type, amount, and direction of reinforcement in the polymer matrix. Carbon, glass, and aramid fibers all improve wear resistance. Wear takes place when fibers are pulled out of the matrix (fiber pullout). Wear is greatest when the sliding direction is parallel to the fibers because they can be pulled out more easily. Long fibers increase the wear resistance of composites because they are more difficult to pull out, and cracks in the matrix cannot propagate to the surface as easily.

When ceramics slide against metals, wear is caused by small-scale plastic deformation and brittle surface fracture, surface chemical reactions, ploughing, and possibly some fatigue. Metals can be transferred to the oxide-type ceramics surfaces, forming metal oxides. Thus, sliding actually takes place between the metal and metal-oxide surface. Conventional lubricants do not appear to influence ceramics wear significantly.

### 4.4.3 Lubrication

In manufacturing processes, the interface between tools, dies, and workpieces is usually subjected to a wide range of variables. Among the major ones are:

1. Contact pressures, ranging from low values of elastic stresses to multiples of the yield stress of the workpiece material

2. Relative speeds, ranging from very low (such as in some superplastic metal-forming operations) to very high speeds (such as in explosive forming, thin wire drawing, grinding, and high-speed metal cutting)

3. Temperatures, ranging from ambient to almost melting, such as in hot extrusion and squeeze casting

If two surfaces slide against each other under these conditions, with no protective layers at the interface, friction and wear will be high. To reduce friction and wear, surfaces should be held as far apart as possible. This generally is done by metalworking lubricants, which may be solid, semisolid, or liquid in nature. Metalworking lubricants are not only fluids or solids with certain desirable physical properties, such as **viscosity,** but also are *chemicals* that can react with the surfaces of the tools, dies, and workpieces and alter their physical properties.

### Lubrication regimes

Basically, four regimes of lubrication are relevant to metalworking processes.

1. **Thick film:** The surfaces are completely separated by a fluid film having a thickness of about one order of magnitude greater than the surface roughness; thus, there is no metal-to-metal contact. The normal load is light and is supported by the **hydrodynamic fluid film** by the wedge effect caused by the relative velocity of the two bodies and the viscosity of the fluid. The coefficient of friction is very low, usually ranging between 0.001 and 0.02.

2. **Thin film:** As the normal load increases, or as the speed and viscosity of the fluid decrease (caused by, say, a rise in temperature), the film thickness is reduced to about three to five times the surface roughness. There may be some metal-to-metal contact at the higher asperities; this contact increases friction and leads to slightly higher wear.

3. **Mixed lubrication:** In this situation, a significant portion of the load is carried by the metal-to-metal contact of the asperities, and the rest of the load is carried by the pressurized fluid film present in hydrodynamic pockets, such as in the valleys of asperities. The film thickness is less than three times the surface roughness. With proper selection of lubricants, a strongly adhering *boundary* film a few molecules thick can be formed on the surfaces. This film prevents direct metal-to-metal contact and thus reduces wear. Depending on the strength of the boundary film and other parameters, the friction coefficient in mixed lubrication may range up to about 0.4.

4. **Boundary lubrication:** The load here is supported by the contacting surfaces, which are covered with a **boundary layer.** Depending on the boundary film thickness and its strength, the friction coefficient ranges from about 0.1 to 0.4. Wear can be relatively high if the boundary layer is destroyed. Typical **boundary lubricants** are natural oils, fats, fatty acids, and soaps. Boundary films form rapidly on metal surfaces. As the film thickness decreases and metal-to-metal contact takes place, the chemical aspects of surfaces and their roughness become significant. (In thick-film lubrication, the viscosity of the lubricant is the important parameter in controlling friction and wear; chemical aspects are not particularly significant, except as they affect corrosion and staining of metal surfaces.)

   A boundary film can **break down** or it can be removed by being disturbed or rubbed off during sliding, or because of desorption due to high temperatures at the interface. The metal surfaces may thus be deprived of this protective layer. The clean metal surfaces then contact each other and, as a consequence, severe wear and scoring can occur. Thus, the adherence of boundary films is an important aspect in lubrication.

The role of surface roughness, particularly in mixed lubrication, should be recognized. Roughness can serve to create local reservoirs or pockets for lubricants. The lubricant can be trapped in the valleys of the surface and, because these fluids are incompressible, they can support a substantial portion of the normal load. These pockets also supply lubricant to regions where the boundary layer has been destroyed. There is an optimal roughness for lubricant entrapment. In metalworking operations, it is generally desirable for the workpiece, not the die, to have the rougher surface. Otherwise, the workpiece surface will be damaged by the rougher and harder die surface. The recommended surface roughness on most dies is about 0.40 $\mu$m (15 $\mu$in).

## Geometric effects

In addition to surface roughness, the overall *geometry* of the interacting bodies is an important consideration in lubrication. Figure 4.14 shows the typical metalworking processes of drawing, extrusion, and rolling. The movement of the workpiece into the deformation zone must allow a supply of the lubricant to be carried into the die–workpiece interfaces.

Analysis of hydrodynamic lubrication shows that the *inlet angle* is an important parameter. As this angle decreases, more lubricant is entrained; thus lubrication is improved and friction is reduced. With proper selection of parameters, a relatively thick film of lubricant

(a) Drawing or extrusion

(b) Rolling

Die

Roll

Pressure

Inlet zone

Plastic  Rigid

**FIGURE 4.14** Inlet zones for entrainment of lubricants in drawing, extrusion, and rolling. Entrainment of lubricants is important in reducing friction and wear in metalworking operations. *Source:* After W. R. D. Wilson.

can be maintained at the die–workpiece interface. A similar situation exists in the compression of a straight solid cylinder between flat platens (Fig. 4.15). As the approach speed of the platen increases, the lubricant is trapped more and more because of its viscosity; a thick film is thus developed. The *pressure distribution* at the interface is given by the parabolic expression

$$p = \frac{3\eta V(R - r)^2}{t^3}, \tag{4.10}$$

where $\eta$ is the viscosity of the lubricant and $t$ is the film thickness. Thus the pressure increases with increasing viscosity, approach speed, and workpiece diameter; the pressure decreases with increasing film thickness.

(a)

$V$

Pressure distribution

$p$

Workpiece

Fluid film

$t$

$R$

$r$

(b)

**FIGURE 4.15** (a) Pressure distribution at the lubricated interface between a flat die (platen) and a solid cylindrical workpiece. (b) Rough surface produced in compressing an aluminum specimen, due to the high viscosity of the lubricant and high speed of compression. The lubricant is trapped at the die–specimen interface. The surface roughness produced depends on the grain size.

The reduction of friction and wear by entrainment or entrapment of lubricants may not always be desirable. The reason is that with a thick fluid film, the metal surface cannot come into full contact with the die surfaces and hence does not acquire the shiny appearance that may be required on the product. (Note, for instance, the shiny appearance of a piece of copper wire.) A thick film of lubricant generates a grainy dull surface on the workpiece; the degree of roughness depends on the grain size of the material (Fig. 4.15b). In operations such as coining and forging, trapped lubricants are undesirable because they interfere with the forming process and prevent precise shape generation.

### 4.4.4  Metalworking fluids

On the basis of our previous discussions, we can summarize the functions of metalworking fluids as follows:

- Reduce friction, thus reducing force and energy requirements and temperature rise
- Reduce wear, seizure, and galling
- Improve material flow in tools, dies, and molds
- Act as a thermal barrier between the workpiece and tool and die surfaces, thus preventing workpiece cooling in hot-working processes
- Act as a release or parting agent to help in the removal or ejection of parts from dies and molds

Many types of metalworking fluids are now available to fulfill these requirements. Because of their diverse chemistries, properties, and characteristics, the behavior and performance of lubricants can be complex. In this section we describe the general properties of only the most commonly used lubricants.

**Oils** have high film strength on the surface of a metal, as you well know if you have ever tried to clean an oily surface. Although they are very effective in reducing friction and wear, oils have low thermal conductivity and specific heat. Thus they are not effective in conducting away the heat generated by friction and plastic deformation in metalworking operations. Oils also are difficult to remove from component surfaces that subsequently are to be painted or welded, and they are difficult to dispose of. The sources of oils are mineral (petroleum), animal, vegetable, and fish. Oils may be **compounded** with a variety of additives or with other oils to impart special properties.

Metalworking fluids are usually blended with several **additives.** These include oxidation inhibitors, rust preventatives, odor control agents, antiseptics, and foam inhibitors. Important additives in oils are sulfur, chlorine, and phosphorus. Known as **extreme-pressure (EP) additives** and used singly or in combination, they react chemically with metal surfaces and form adherent surface films of metallic sulfides and chlorides. These films have low shear strength and good antiweld properties and thus effectively reduce friction and wear. While EP additives are important in boundary lubrication, these lubricants may preferentially attack the cobalt binder in tungsten carbide tools and dies, causing changes in their surface roughness and integrity.

An **emulsion** is a mixture of two immiscible liquids, usually mixtures of oil and water in various proportions, along with additives. Emulsions, also known as **water-soluble fluids,** are of two types: direct and indirect. In a direct emulsion, mineral oil is dispersed in water as very small droplets. In an indirect emulsion, water droplets are dispersed in the oil. Direct emulsions are important fluids because the presence of water gives them high cooling capacity. They are particularly effective in high-speed metal machining, where temperature rise has detrimental effects on tool life, workpiece surface integrity, and dimensional accuracy.

**Synthetic solutions** are chemical fluids containing inorganic and other chemicals dissolved in water. Various chemical agents are added to impart different properties. **Semisynthetic solutions** are basically synthetic solutions to which small amounts of emulsifiable oils have been added.

**Soaps** are generally reaction products of sodium or potassium salts with fatty acids. Alkali soaps are soluble in water, but other metal soaps are generally insoluble. They are effective boundary lubricants and can also form thick film layers at die–workpiece interfaces, particularly when applied on conversion coatings for cold metalworking applications.

**Greases** are solid or semisolid lubricants and generally consist of soaps, mineral oil, and various additives. They are highly viscous and adhere well to metal surfaces. Although used extensively in machinery, greases have limited use in manufacturing processes.

**Waxes** may be of animal or plant *(paraffin)* origin and have complex structures. Compared to greases, waxes are less "greasy" and are more brittle. They have limited use in metalworking operations, except for copper and, as chlorinated paraffin, for stainless steels and high-temperature alloys.

### Solid lubricants

Because of their unique properties and characteristics, several solid materials are used as lubricants in manufacturing operations. In this section we describe four of the most commonly used **solid lubricants.**

The properties of **graphite** are described in Section 11.14. Graphite is weak in shear along its layers and has a low coefficient of friction in that direction. Thus it can be a good solid lubricant, particularly at elevated temperatures. However, the graphite friction is low only in the presence of air or moisture. In a vacuum or an inert gas atmosphere, friction is very high; in fact, graphite can be quite abrasive in these environments. Graphite may be applied either by rubbing it on surfaces or as a *colloidal* (dispersion of small particles) suspension in liquid carriers, such as water, oil, or alcohols.

Another widely used lamellar solid lubricant, **molybdenum disulfide ($MoS_2$)** is somewhat similar in appearance to graphite. However, unlike graphite it has a high friction coefficient in an ambient environment. Oils are commonly used as carriers for molybdenum disulfide and are used as a lubricant at room temperature. Molybdenum disulfide can also be rubbed onto the surfaces of a workpiece.

Because of their low strength, thin layers of **soft metals** and **polymer coatings** are used as solid lubricants. Suitable metals are lead, indium, cadmium, tin, silver, and polymers such as PTFE, polyethylene, and methacrylates. However, these coatings have limited applications because of their lack of strength under high stresses and at elevated temperatures. Soft metals are used to coat high-strength metals such as steels, stainless steels, and high-temperature alloys. Copper or tin, for example, is chemically deposited on the surface of the metal before it is processed. If the oxide of a particular metal has low friction and is sufficiently thin, the oxide layer can serve as a solid lubricant, particularly at elevated temperatures.

Although a solid material, **glass** becomes viscous at high temperatures and hence can serve as a liquid lubricant. Viscosity is a function of temperature, but not of pressure, and depends on the type of glass. Poor thermal conductivity also makes glass attractive, since it acts as a thermal barrier between hot workpieces and relatively cool dies. Typical glass lubrication applications are in hot extrusion and forging.

**Conversion coatings.**    Lubricants may not always adhere properly to workpiece surfaces, particularly under high normal and shearing stresses. This condition is a special problem in forging, extrusion, and wire drawing of steels, stainless steels, and high-temperature alloys. For these applications, acids transform the workpiece surface by chemically reacting with it, leaving a somewhat rough and spongy surface that acts as a carrier for the lubricant. After treatment, borax or lime is used to remove any excess acid from the surfaces. A liquid lubricant, such as a soap, is then applied to the coated surface. The lubricant film adheres to the surface and cannot be scraped off easily. **Zinc phosphate** conversion coatings are often used on carbon and low-alloy steels. **Oxalate** coatings are used for stainless steels and high-temperature alloys.

## Lubricant selection

Selecting a lubricant for a particular process and workpiece material involves consideration of several factors:

- The particular manufacturing process
- Compatibility of the lubricant with the workpiece and tool and die materials
- Surface preparation required
- Method of lubricant application
- Removal of lubricant after processing
- Contamination of the lubricant by other lubricants, such as those used to lubricate the machinery
- Treatment of waste lubricant
- Storage and maintenance of lubricants
- Biological and ecological considerations
- Costs involved in all aspects listed above

In selecting an oil as a lubricant, we should recognize the importance of its viscosity temperature pressure characteristics. Low viscosity can have a significant detrimental effect on friction and wear. The different functions of a metalworking fluid, whether primarily a lubricant or a coolant, must also be taken into account. Water-base fluids are very effective coolants, but as lubricants are not as effective as oils.

Metalworking fluids should not leave any harmful residues that could interfere with machinery operations. The fluids should not stain or corrode workpiece or equipment. The fluids should be checked periodically for deterioration caused by bacterial growth, accumulation of oxides, chips, and wear debris—and also general degradation and breakdown because of temperature and time. A lubricant may carry wear particles with it and cause damage to the system, so proper inspection and filtering of metalworking fluids are important. These precautions are necessary for all types of machinery, including internal combustion engines and jet engines.

After completion of manufacturing operations, metal surfaces are usually covered with lubricant residues, which should be removed prior to further workpiece processing such as welding or painting. Various cleaning solutions and techniques can be used for this purpose (Section 4.5.2). **Biological and ecological considerations,** with their accompanying legal ramifications, are also important considerations. Potential health hazards may be involved in contacting or inhaling some metalworking fluids. Recycling of waste fluids and their disposal are additional important factors to be considered.

## 4.5 Surface Treatments, Coatings, and Cleaning

After a component is manufactured, all or parts of its surfaces may have to be processed further in order to impart certain properties and characteristics. **Surface treatment** may be necessary to:

- Improve resistance to wear, erosion, and indentation (slideways in machine tools, wear surfaces of machinery, and shafts, rolls, cams, and gears).
- Control friction (sliding surfaces on tools, dies, bearings, and machine ways).
- Reduce adhesion (electrical contacts).
- Improve lubrication (surface modification to retain lubricants).
- Improve corrosion and oxidation resistance (sheet metals for automotive or other outdoor uses, gas-turbine components, and medical devices).
- Improve stiffness and fatigue resistance (bearings and multiple-diameter shafts with fillets).
- Rebuild surfaces on worn components (worn tools, dies, and machine components).
- Improve surface roughness (appearance, dimensional accuracy, and frictional characteristics).
- Impart decorative features, color, or special surface texture.

### 4.5.1 Processes

In **shot peening,** the workpiece surface is hit repeatedly with a large number of cast steel, glass, or ceramic shot (small balls), making overlapping indentations on the surface. This action causes plastic deformation of surfaces, to depths up to 1.25 mm (0.05 in.), using shot sizes ranging from 0.125 mm to 5 mm (0.005 in. to 0.2 in.) in diameter. Because plastic deformation is not uniform throughout a part's thickness, shot peening imparts compressive residual stresses on the surface, thus improving the fatigue life of the component. This process is used extensively on shafts, gears, springs, oil-well drilling equipment, and jet-engine parts (such as turbine and compressor blades).

In **roller burnishing,** also called **surface rolling,** the surface of the component is cold worked by a hard and highly polished roller or series of rollers. This process is used on various flat, cylindrical, or conical surfaces (Fig. 4.16). Roller burnishing improves surface finish by removing scratches, tool marks, and pits. Consequently, corrosion resistance is also improved since corrosive products and residues cannot be entrapped. Internal cylindrical surfaces are burnished by a similar process, called **ballizing** or **ball burnishing.** A smooth ball, slightly larger than the bore diameter, is pushed through the length of the hole.

Roller burnishing is used to improve the mechanical properties of surfaces, as well as the shape and surface finish of components. It can be used either singly or in combination with other finishing processes, such as grinding, honing, and lapping. Soft and ductile, as well as very hard metals, can be roller burnished. Typical applications include hydraulic-system components, seals, valves, spindles, and fillets on shafts.

In **explosive hardening,** surfaces are subjected to high transient pressures by placing a layer of explosive sheet directly on the workpiece surface and detonating it. Large increases in surface hardness can be obtained by this method, with very little change (less than 5 percent) in the shape of the component. Railroad rail surfaces can be hardened by this method.

In **cladding** (clad bonding), metals are bonded with a thin layer of corrosion-resistant metal by applying pressure with rolls or other means. A typical application is cladding of aluminum (*Alclad*), in which a corrosion-resistant layer of aluminum alloy is clad over pure aluminum. Other applications are steels clad with stainless steel or nickel alloys. The cladding material may also be applied through dies, as in cladding steel wire with copper, or by explosives.

In **mechanical plating** (also called mechanical coating, impact plating, or peen plating), fine metal particles are compacted over the workpiece surfaces by impacting them with spherical glass, ceramic, or porcelain beads. The beads are propelled by rotary means. The process is used typically for hardened-steel parts for automobiles, with plating thickness usually less than 0.025 mm (0.001 in.).

We describe traditional methods of **case hardening** (carburizing, carbonitriding, cyaniding, nitriding, flame hardening, and induction hardening) in Section 5.10.3 and summarize them in Table 5.8. In addition to the common heat sources of gas and electricity, laser beams are also used as a heat source in surface hardening of both metals and ceramics. Case hardening, as well as some of the other surface-treatment processes, induce residual stresses on surfaces. The formation of martensite in case hardening causes compressive residual stresses

**FIGURE 4.16** Examples of roller burnishing of (a) a conical surface and (b) a flat surface and the burnishing tools used. *Source:* Cogsdill Tool Products.

on surfaces. Such stresses are desirable because they improve the fatigue life of components by delaying the initiation of fatigue cracks.

In **hard facing,** a relatively thick layer, edge, or point of wear-resistant hard metal is deposited on the surface by any of the welding techniques we describe in Chapter 12. A number of layers are usually deposited (*weld overlay*). Hard coatings of tungsten carbide, chromium, and molybdenum carbide can also be deposited using an electric arc (*spark hardening*). Hard-facing alloys are available as electrodes, rod, wire, and powder. Typical applications for hard facing are valve seats, oil-well drilling tools, and dies for hot metalworking. Worn parts are also hard faced for extended use.

In **thermal spraying,** metal in the form of rod, wire, or powder is melted in a stream of oxyacetylene flame, electric arc, or plasma arc, and the droplets are sprayed on the preheated surface, at speeds up to 100 m/s (20,000 ft/min) with a compressed-air spray gun. All of the surfaces to be sprayed should be cleaned and roughened to improve bond strength. Typical applications for thermal spraying, also called *metallizing,* are steel structures, storage tanks, and tank cars, sprayed with zinc or aluminum up to 0.25 mm (0.010 in.) in thickness.

There are several types of thermal spraying processes, which we list in decreasing order of performance: (1) *Plasma,* either conventional, high-energy, or vacuum; it produces temperatures on the order of 8300°C (15,000°F) and very good bond strength with very low oxide content. (2) *Detonation gun,* in which a controlled explosion takes place using oxyfuel gas mixture; it has a performance similar to plasma. (3) *High-velocity oxyfuel* (HVOF) gas spraying, which has a similarly high performance as above, but is less expensive. (4) *Wire arc,* in which an arc is formed between two consumable wire electrodes; it has good bond strength and is the least expensive process. (5) *Flame wire* spraying, in which the oxyfuel flame melts the wire and deposits it on the surface; its bond is of medium strength, but the process is relatively inexpensive.

Manufactured surfaces can further be modified by secondary operations for technical, functional, optical, or aesthetic reasons. Called **surface texturing,** these additional processes generally consist of (1) etching, using chemicals or sputtering techniques, (2) electric arcs, and (3) atomic oxygen, which reacts with surfaces and produces fine, conelike surface textures. The possible adverse effects of any of these processes on material properties and performance should be considered.

Techniques have also been developed for spraying **ceramic coatings** for high-temperature and electrical-resistance applications, such as to withstand repeated arcing. Powders of hard metals and ceramics are also used as spraying materials. Plasma-arc temperatures may reach 15,000°C (27,000°F), which is much higher than that obtained with flames. Typical applications are nozzles for rocket motors and wear-resistant parts.

## Vapor deposition

**Vapor deposition** is a process in which the substrate (workpiece surface) is subjected to chemical reactions by gases that contain chemical compounds of the materials to be deposited. The coating thickness is usually a few $\mu$m. The deposited materials may consist of metals, alloys, carbides, nitrides, borides, ceramics, or various oxides. The substrate may be metal, plastic, glass, or paper. Typical applications are coating cutting tools, drills, reamers, milling cutters, punches, dies, and wear surfaces.

There are two major deposition processes: physical vapor deposition and chemical vapor deposition. These techniques allow effective control of coating composition, thickness, and porosity. The three basic types of **physical vapor deposition (PVD)** processes are vacuum or arc evaporation (PV/ARC), sputtering, and ion plating. These processes are carried out in a high vacuum at temperatures in the range of 200–500°C (400–900°F). In physical vapor deposition, the particles to be deposited are carried physically to the workpiece, rather than by chemical reactions as in chemical vapor deposition.

In **vacuum evaporation,** the metal to be deposited is evaporated at high temperatures, in a vacuum and is deposited on the substrate, which is usually at room temperature or slightly higher. Uniform coatings can be obtained on complex shapes. In PV/ARC, which was developed recently, the coating material (cathode) is evaporated by a number of arc evaporators, using highly localized electric arcs. The arcs produce a highly reactive plasma consisting of ionized vapor of the coating material. The vapor condenses on the substrate (anode) and coats it. Applications for this process may be functional (oxidation-resistant coatings for high-temperature applications, electronics, and optics) or decorative (hardware, appliances, and jewelry).

In **sputtering,** an electric field ionizes an inert gas (usually argon). The positive ions bombard the coating material (cathode) and cause sputtering (ejecting) of its atoms. These atoms then condense on the workpiece, which is heated to improve bonding. In **reactive sputtering,** the inert gas is replaced by a reactive gas, such as oxygen, in which case the atoms are oxidized and the oxides are deposited. **Radiofrequency (RF) sputtering** is used for nonconductive materials such as electrical insulators and semiconductor devices.

**Ion plating** is a generic term describing the combined processes of sputtering and vacuum evaporation. An electric field causes a glow discharge, generating a plasma. The vaporized atoms in this process are only partially ionized.

**Chemical vapor deposition (CVD)** is a thermochemical process. In a typical application, such as for coating cutting tools with titanium nitride (TiN), the tools are placed on a graphite tray and heated to 950–1050°C (1740–1920°F) at atmospheric pressure in an inert atmosphere. Titanium tetrachloride (a vapor), hydrogen, and nitrogen are then introduced into the chamber. The chemical reactions form titanium nitride on the tool surfaces. For coating with titanium carbide, methane is substituted for the gases. Chemical vapor deposition coatings are usually thicker than those obtained from PVD.

**Ion implantation.** Here, ions are introduced into the surface of the workpiece material. The ions are accelerated in a vacuum to such an extent that they penetrate the substrate to a depth of a few $\mu$m. Ion implantation (not to be confused with ion plating) modifies surface properties by increasing surface hardness and improving friction, wear, and corrosion resistance. This process can be controlled accurately, and the surface can be masked to prevent ion implantation in unwanted places. When used in specific applications, such as semiconductors, this process is called **doping** (meaning alloying with small amounts of various elements).

**Diffusion coating.** This is a process in which an alloying element is diffused into the surface of the substrate, thus altering its properties. Such elements can be supplied in solid, liquid, or gaseous states. This process acquires different names, depending on the diffused element, that describe diffusion processes such as carburizing, nitriding, and boronizing (see Table 5.8).

## Electroplating

In **electroplating,** the workpiece (cathode) is plated with a different metal (anode) while both are suspended in a bath containing a water-base electrolyte solution. Although the plating process involves a number of reactions, basically the metal ions from the anode are discharged under the potential from the external source of electricity, combine with the ions in the solution, and are deposited on the cathode.

All metals can be electroplated, with thicknesses ranging from a few atomic layers to a maximum of about 0.05 mm (0.002 in.). Complex shapes may have varying plating thicknesses. Common plating materials are chromium, nickel, cadmium, copper, zinc, and tin. Chromium plating is carried out by first plating the metal with copper, then with nickel, and finally with chromium. **Hard chromium plating** is done directly on the base metal and has a hardness up to 70 HRC.

Typical electroplating applications are copper plating aluminum wire and phenolic boards for printed circuits, chrome plating hardware, tin plating copper electrical terminals for ease of soldering, and components requiring good appearance and resistance to wear and corrosion. Because they do not develop oxide films, noble metals (such as gold, silver, and platinum) are important electroplating materials for the electronics and jewelry industries. Plastics such as ABS, polypropylene, polysulfone, polycarbonate, polyester, and nylon also can be electroplated. Because they are not electrically conductive, plastics must be preplated by such processes as electroless nickel plating. Parts to be coated may be simple or complex, and size is not a limitation.

A variation of electroplating, **electroforming** is actually a metal fabricating process. Metal is electrodeposited on a mandrel (also called mold or matrix), which is then removed. Thus the coating itself becomes the product. Simple and complex shapes can be produced by electroforming, with wall thicknesses as small as 0.025 mm (0.001 in.). Parts may weigh from a few grams to as much as 270 kg (600 lb). The electroforming process is particularly suited to low production quantities or intricate parts (such as molds, dies, waveguides, nozzles, and bellows) made of nickel, copper, gold, and silver. It is also suitable for aerospace, electronics, and electro-optics applications. Production rates can be increased with multiple mandrels.

**Electroless plating.**   This process is carried out by chemical reactions, without the use of an external source of electricity. The most common application utilizes nickel, although copper is also used. In electroless nickel plating, nickel chloride (a metallic salt) is reduced— with sodium hypophosphite as the reducing agent—to nickel metal, which is then deposited on the workpiece. The hardness of nickel plating ranges between 425 HV and 575 HV, and can be heat treated to 1000 HV. The coating has excellent wear and corrosion resistance.

**Anodizing.**   This is an oxidation process (*anodic oxidation*) in which the workpiece surfaces are converted to a hard and porous oxide layer that provides corrosion resistance and a decorative finish. The workpiece is the anode in an electrolytic cell immersed in an acid bath, resulting in chemical adsorption of oxygen from the bath. Organic dyes of various colors (typically black, red, bronze, gold, and gray) can be used to produce stable, durable surface films. Typical applications for anodizing are aluminum furniture and utensils, architectural shapes, automobile trim, picture frames, keys, and sporting goods. Anodized surfaces also serve as a good base for painting, especially for aluminum, which otherwise is difficult to paint.

**Conversion coating.**   Also called **chemical reaction priming,** in this process a coating forms on metal surfaces as a result of chemical or electrochemical reactions. Various metals, particularly steel, aluminum, and zinc, can be conversion coated. Oxides that naturally form on their surfaces are a form of conversion coating. Phosphates, chromates, and oxalates are used to produce conversion coatings. These coatings are for purposes such as corrosion protection, prepainting, and decorative finish. An important application is in conversion coating of workpieces as a lubricant carrier in cold-forming operations. The two common methods of coating are immersion and spraying. The equipment involved depends on the method of application, the type of product, and considerations of quality.

As the name implies, **coloring** involves processes that alter the color of metals, alloys, and ceramics. It is caused by the conversion of surfaces (by chemical, electrochemical, or thermal processes) into chemical compounds, such as oxides, chromates, and phosphates.

**Hot dipping.**   In this process the workpiece, usually steel or iron, is dipped into a bath of molten metal, such as zinc (for galvanized-steel sheet and plumbing supplies), tin (for tin-plate and tin cans for food containers), aluminum (aluminizing), and terne (lead alloyed with 10–20 percent tin). Hot-dipped coatings on discrete parts or sheet metal provide long-term corrosion resistance to galvanized pipe, plumbing supplies, and many other products. The coating thickness is usually given in terms of coating weight per unit surface area of the sheet, typically 150—900 g/m$^2$ (0.5–3 oz/ft$^2$). Service life depends on the thickness of the zinc coating and the environment to which it is exposed. Various *precoated sheet steels* are used extensively in automobile bodies. Proper draining to remove excess coating materials is an important consideration.

**Porcelain enameling.**   Metals may be coated with a variety of glassy (vitreous) coatings to provide corrosion and electrical resistance and for service at elevated temperatures. These coatings are usually classified as porcelain enamels and generally include enamels and ceramics. The word **enamel** is also used for glossy paints, indicating a smooth, hard coating.

Porcelain enamels are glassy inorganic coatings consisting of various metal oxides. **Enameling** involves fusing the coating material on the substrate by heating them both to 425–1000°C (800–1800°F) to liquefy the oxides. Depending on their composition, enamels have varying resistances to alkali, acids, detergents, cleansers, and water—and come in different colors.

Typical applications for porcelain enameling are household appliances, plumbing fixtures, chemical processing equipment, signs, cookware, and jewelry. Porcelain enamels are also used as protective coatings on jet-engine components. The coating may be applied by dipping, spraying, or electrodeposition, and thicknesses are usually 0.05–0.6 mm (0.002–0.025 in.). Metals that are coated are typically steels, cast iron, and aluminum. Glasses are used as lining for chemical resistance, and the thickness is much greater than in enameling. **Glazing** is the application of glassy coatings on ceramic wares to give them decorative finishes and to make them impervious to moisture.

Metal surfaces may be coated or precoated with a variety of **organic coatings,** films, and laminates to improve appearance, eye appeal, and corrosion resistance. Coatings are applied to the coil stock on continuous lines, with thicknesses generally of 0.0025–0.2 mm (0.0001–0.008 in.). Such coatings have a wide range of properties: flexibility, durability, hardness, resistance to abrasion and chemicals, color, texture, and gloss. Coated sheet metal is subsequently formed into various products, such as TV cabinets, appliance housings, paneling, shelving, residential building siding, gutters, and metal furniture.

More critical applications involve, for example, naval aircraft that are subjected to high humidity, rain, seawater, pollutants (such as from ship exhaust stacks), aviation fuel, deicing fluids, and battery acid, and which are also impacted by particles such as dust, gravel, stones, and deicing salts. For aluminum structures, organic coatings have consisted typically of an epoxy primer and a polyurethane topcoat, with a lifetime of four to six years. Primer performance is very important for coating durability; consequently, much research is being conducted to develop improved coating materials.

Important advances have been made in **diamond coating** of metals, glass, ceramics, and plastics, using various chemical and plasma-assisted vapor deposition processes and ion-beam enhanced deposition techniques. Techniques have also been developed to produce free-standing diamond films on the order of 1 mm (0.040 in.) thick and up to 125 mm (5 in.) in diameter, including smooth and optically clear diamond film (unlike the hazy gray diamond film formerly produced). Development of these techniques, combined with important properties of diamond such as hardness, wear resistance, high thermal conductivity and transparency to ultraviolet light and microwave frequencies, have enabled the production of various aerospace and electronic parts and components.

Examples of diamond-coated products are scratchproof windows (such as for aircraft and missile sensors to protect against sandstorms), sunglasses, cutting tools (such as drills and end mills), calipers, surgical knives, razors, electronic and infrared heat seekers and sensors, light-emitting diodes, diamond-coated speakers for stereo systems, turbine blades, and fuel-injection nozzles. Studies are continuing on growing diamond films on crystalline copper substrate by implantation of carbon ions. An important application is in making computer chips. Diamond can be doped to form $p$- and $n$-type ends on semiconductors to make transistors, and its high thermal conductivity allows closer packing of chips than silicon or gallium-arsenide chips, thus significantly increasing the speed of computers.

### 4.5.2  Cleaning surfaces

We have stressed the importance of surfaces and the influence of deposited or adsorbed layers of various elements and contaminants on surfaces. A clean surface can have both beneficial and detrimental effects. Although an unclean surface may reduce the tendency for adhesion and galling, in general cleanliness is essential for more effective application of metalworking fluids, coating and painting, adhesive bonding, welding, brazing, soldering, reliable functioning of manufactured parts in machinery, food and beverage containers, storage, and in assembly operations.

*Cleaning* involves removal of solid, semisolid, or liquid contaminants from a surface, and it is an important part of manufacturing operations and the economics of production. The word *clean,* or the degree of cleanliness of a surface, is somewhat difficult to define. How, for example, would you test the cleanliness of a dinner plate? Two simple and common tests are based on:

1.  Wiping with a clean cloth and observing any residues on the cloth, as we all have done at one time or another.

2. Observing whether water continuously coats the surface. If water collects as individual droplets, the surface is not clean (*waterbreak test*). Test this phenomenon yourself by wetting dinner plates that have been cleaned to varying degrees.

The type of cleaning process required depends on the type of contaminants to be removed. Contaminants, also called *soils,* may consist of rust, scale, chips and other metallic and nonmetallic debris, metalworking fluids, solid lubricants, pigments, polishing and lapping compounds, and general environmental elements.

Basically there are two types of cleaning methods: mechanical and chemical. **Mechanical methods** consist of physically disturbing the contaminants, as with wire or fiber brushing, dry or wet abrasive blasting, tumbling, and steam jets. Many of these processes are particularly effective in removing rust, scale, and other solid contaminants. Ultrasonic cleaning may also be placed in this category.

**Chemical cleaning** usually involves the removal of oil and grease from surfaces. It consists of one or more of the following processes:

- *Solution.* The soil dissolves in the cleaning solution.

- *Saponification.* A chemical reaction that converts animal or vegetable oils into a soap that is soluble in water.

- *Emulsification.* The cleaning solution reacts with the soil or lubricant deposits and forms an emulsion. The soil and the emulsifier then become suspended in the emulsion.

- *Dispersion.* The concentration of soil on the surface is decreased by surface-active materials in the cleaning solution.

- *Aggregation.* Lubricants are removed from the surface by various agents in the cleaner and collect as large dirt particles.

Some common **cleaning fluids** are used in conjunction with electrochemical processes for more effective cleaning. These fluids include: alkaline solutions, emulsions, solvents, hot vapors, acids, salts, and organic compound mixtures.

Cleaning discrete parts having complex shapes can be difficult. Design engineers should be aware of this difficulty and provide alternative designs, such as avoiding deep blind holes or making several smaller components instead of one large component that may be difficult to clean, and provide appropriate drain holes in the part to be cleaned.

## 4.6   Engineering Metrology and Instrumentation

**Engineering metrology** is defined as the measurement of dimensions: length, thickness, diameter, taper, angle, flatness, profiles, and others. Traditionally, measurements have been made after the part has been produced, which is known as postprocess inspection. The current trend in manufacturing is to make measurements while the part is being produced on the machine, which is known as in-process, on-line, or real-time inspection. Here the term **inspection** means to check the dimensions of what we have produced or are producing and to see whether it complies with the specified dimensional accuracy. Much progress has been made in developing new and automated instruments that are highly precise and sensitive.

An important aspect of metrology in manufacturing processes is **dimensional tolerances,** that is, the permissible variation in the dimensions of a part. Tolerances are important not only for proper functioning of products, but they also have a major economic impact on manufacturing costs. The smaller we make the tolerance, the higher the production costs.

### 4.6.1 Instruments

### Line-graduated instruments

Line-graduated instruments are used for measuring length (linear measurements) or angles (angular measurements). *Graduated* means marked to indicate a certain quantity.

The simplest and most commonly used instrument for making linear measurements is a steel rule (*machinist's rule*), bar, or tape, with fractional or decimal graduations. Lengths are measured directly, to an accuracy that is limited to the nearest division, usually 1 mm or $\frac{1}{64}$ in. Rules may be rigid or flexible and may be equipped with a hook at one end for ease of measuring from an edge. Rule depth gages are similar to rules and slide along a special head.

**Vernier calipers** have a graduated beam and a sliding jaw with a *vernier*. These instruments are also called **caliper gages.** The two jaws of the caliper contact the part being measured, and the dimension is read at the matching graduated lines. Vernier calipers can be used to measure inside or outside lengths. The vernier improves the sensitivity of a simple rule by indicating fractions of the smallest division on the graduated beam, usually to 25 $\mu$m (0.001 in.). Vernier calipers are also equipped with digital readouts, which are easy to read and less subject to human error than reading verniers. Vernier height gages are vernier calipers with setups similar to a depth gage and have similar sensitivity.

**Micrometers.** Commonly used for measuring the thickness and inside or outside diameters of parts, the *micrometer* has a graduated, threaded spindle. Circumferential vernier readings to a sensitivity of 2.5 $\mu$m (0.0001 in.) can be obtained. Micrometers are also available for measuring depths (*micrometer depth gage*) and internal diameters (*inside micrometer*) with the same sensitivity. Micrometers are also equipped with digital readouts to reduce errors in reading. The anvils on micrometers can be equipped with conical or ball contacts. They are used to measure inside recesses, threaded rod diameters, and wall thicknesses of tubes and curved sheets.

**Diffraction gratings** consist of two flat optical glasses with closely spaced parallel lines scribed on their surfaces (Fig. 4.17). The grating on the shorter glass is inclined slightly. As a result, interference fringes develop when it is viewed over the longer glass. The position of these fringes depends on the relative position of the two sets of glasses. With modern

Direction of travel

Reference

¼ line travel

½ line travel

Index grating          Master grating

**FIGURE 4.17** Measurement of very small linear displacements by fringes (moiré fringes). Note the shift in the fringe patterns. Gratings can typically be 40 lines/mm (1000 lines/in.).

equipment, using electronic counters and photoelectric sensors, sensitivities of 2.5 $\mu$m (0.0001 in.) can be obtained with gratings having 40 lines/mm (1000 lines/in.).

Indirect-reading instruments are typically **calipers** and **dividers** without any graduated scales. They are used to transfer the size measured to a direct-reading instrument, such as a rule. After adjusting the legs to contact the part at the desired location, the instrument is held against a graduated rule, and the dimension is read. Because of the experience required to use them and their dependence on graduated scales, the accuracy of this type of indirect measurement is limited. **Telescoping gages** are available for indirect measurement of holes or cavities.

*Angles* are measured in degrees, radians, or minutes and seconds of arc. Because of the geometry involved, angles are usually more difficult to measure than are linear dimensions. A **bevel protractor** is a direct-reading instrument similar to a common protractor, except that it has a movable member. The two blades of the protractor are placed in contact with the part being measured, and the angle is read directly on the vernier scale. The sensitivity of the instrument depends on the graduations of the vernier. Another type of bevel protractor is the **combination square,** which is a steel rule equipped with devices for measuring 45° and 90° angles.

Measuring with a **sine bar** involves placing the part on an inclined bar or plate and adjusting the angle by placing gage blocks on a surface plate. After the part is placed on the sine bar, a dial indicator is used to scan the top surface of the part. **Gage blocks** are added or removed as necessary until the top surface is parallel to the surface plate. The angle on the part is then calculated from geometric relationships. Angles can also be measured using **angle gage blocks.** These are blocks with different tapers that can be assembled in various combinations and used in a manner similar to sine bars. Angles on small parts can be measured through microscopes, with graduated eyepieces, or with optical projectors.

## Comparative length-measuring instruments

Unlike the instruments we have just described, instruments used for measuring comparative lengths, also called *deviation-type* instruments, amplify and measure variations or deviations in distance between two or more surfaces. These instruments compare dimensions, hence the word *comparative*. We describe below the common types of instruments used for making comparative measurements.

**Dial indicators** are simple mechanical devices that convert linear displacements of a pointer to rotation of an indicator on a circular dial. The indicator is set to zero at a certain reference surface, and the instrument or the surface to be measured—either external or internal—is brought into contact with the pointer. The movement of the indicator is read directly on the circular dial—either plus or minus—to accuracies as high as 1 $\mu$m (40 $\mu$in.).

Unlike mechanical systems, **electronic gages** sense the movement of the contacting pointer through changes in the electrical resistance of a strain gage or through inductance or capacitance. The electrical signals are then converted and displayed as linear dimensions. A commonly used electronic gage is the **linear variable differential transformer (LVDT),** used extensively for measuring small displacements. Although they are more expensive than other types, electronic gages have advantages such as ease of operation, rapid response, digital readout, less possibility of human error, versatility, flexibility, and the capability to be integrated into automated systems through microprocessors and computers.

## Measuring straightness, flatness, roundness, and profile

The geometric features of straightness, flatness, roundness, and profile are important aspects of engineering design and manufacturing. For example, piston rods, instrument components, and machine-tool slideways should all meet certain requirements with regard to these characteristics in order to function properly. Consequently, their accurate measurement is essential.

**Straightness** can be checked with straight edges or with dial indicators. **Autocollimators,** resembling a telescope with a light beam that bounces back from the object, are used for accurately measuring small angular deviations on a flat surface. Optical means such as **transits** and **laser beams** are used for aligning individual machine elements in the assembly of machine components.

**Flatness** can be measured by mechanical means, using a surface plate and a dial indicator. This method can be used for measuring perpendicularity, which can also be measured with the use of precision steel squares.

Another method for measuring flatness is by **interferometry,** using an **optical flat.** The device—a glass or fused quartz disk with parallel flat surfaces—is placed on the surface of the workpiece (Fig. 4.18a). When a monochromatic (one wavelength) light beam is aimed at the surface at an angle, the optical flat splits it into two beams, appearing as light and dark bands to the naked eye (Fig. 4.18b). The number of fringes that appear is related to the distance between the surface of the part and the bottom surface of the optical flat (Fig. 4.18c). Consequently, a truly flat workpiece surface (that is, when the angle between the two surfaces

**FIGURE 4.18** (a) Interferometry method for measuring flatness using an optical flat. (b) Fringes on a flat inclined surface. An optical flat resting on a perfectly flat workpiece surface will not split the light beam, and no fringes will be present. (c) Fringes on a surface with two inclinations. Note: the greater the incline, the closer the fringes. (d) Curved fringe patterns indicate curvatures on the workpiece surface. (e) Fringe pattern indicating a scratch on the surface.

in Fig. 4.18a is zero) will not split the light beam and fringes will not appear. When surfaces are not flat, fringes are curved (Fig. 4.18d). The interferometry method is also used for observing surface textures and scratches (Fig. 4.18e) through microscopes for better visibility.

**Roundness** is usually described as deviations from true roundness (mathematically, a circle). The term **out of roundness** is actually more descriptive of the shape of the part. Roundness is very important to the proper functioning of rotating shafts, bearing races, pistons and cylinders, and steel balls in bearings.

The various methods of measuring roundness basically fall into two categories. In the first, the round part is placed on a V-block or between centers and is rotated, with the point of a dial indicator in contact with the surface. After a full rotation of the workpiece, the difference between the maximum and minimum readings on the dial is noted. This difference is called the **total indicator reading** (TIR) or full indicator movement. This method is also used for measuring the straightness (squareness) of shaft end faces. In the second method, called **circular tracing,** the part is placed on a platform, and its roundness is measured by rotating the platform. Conversely, the probe can be rotated around a stationary part to make the measurement.

**Profile** may be measured by several methods. In one method, a surface is compared with a template or profile gage to check shape conformity. Radii or fillets can be measured by this method. Profile may also be measured with a number of dial indicators or similar instruments. Profile-tracing instruments are the latest development.

**Threads** and **gear teeth** have several features with specific dimensions and tolerances. These dimensions must be produced accurately for smooth operation of gears, reducing wear and noise level, and part interchangeability. These features are measured basically by means of thread gages of various designs that compare the thread produced against a standard thread. Some of the gages used are threaded plug gages, screw-pitch gages (similar to radius gages), micrometers with cone-shaped points, and snap gages with anvils in the shape of threads. Gear teeth are measured with instruments that are similar to dial indicators, with calipers and with micrometers using pins or balls of various diameters. Special profile-measuring equipment is also available, including optical projectors.

**Optical projectors,** also called **optical comparators,** were first developed in the 1940s to check the geometry of cutting tools for machining screw threads, but are now used for checking all profiles. The part is mounted on a table, or between centers, and the image is projected on a screen at magnifications up to $100\times$ or higher. Linear and angular measurements are made directly on the screen, which is equipped with reference lines and circles. The screen can be rotated to allow angular measurements as small as 1 min, using verniers.

**Coordinate measuring and layout machines.** Coordinate measuring and layout machines are the latest development in measurement technology. Basically, they consist of a platform on which the workpiece being measured is placed and moved linearly or rotated (Fig. 4.19). A stylus, attached to a head capable of lateral and vertical movements, records all measurements. **Coordinate measuring machines (CMM),** also called **measuring machines,** are versatile in their capability to record measurements of complex profiles with high sensitivity ($0.25\ \mu$m; $10\ \mu$in.).

These machines are built rigidly and are very precise. They are equipped with digital readout or can be linked to computers for on-line inspection of parts. They can be placed close to machine tools for efficient inspection and rapid feedback for correction of processing parameters before the next part is made. They are also being made more rugged to resist environmental effects in manufacturing plants, such as temperature variations, vibration,

**FIGURE 4.19** A coordinate measuring machine, measuring dimensions on an engine block. *Source:* Courtesy of Sheffield Measurement Division, Giddings & Lewis.

and dirt. Dimensions of large parts are measured by **layout machines,** which are equipped with digital readout. These machines are equipped with scribing tools for marking dimensions on large parts, with an accuracy of ± 0.04 mm (0.0016 in.).

## Gages

Thus far we have used the word *gage* to describe some types of measuring instruments, such as a caliper gage, depth gage, telescoping gage, electronic gage, strain gage, and radius gage. The reason is that the words instrument and gage (also spelled gauge) have traditionally been used interchangeably. However, gage has a variety of meanings, such as pressure gage, gage length of a tension-test specimen, and gages for sheet metal, wire, railroad rail, and the bore of shotguns.

**Gage blocks** are individual square, rectangular, or round metal blocks of various sizes, made very precisely from heat-treated and stress-relieved alloy steels or from carbides. Their surfaces are lapped and are flat and parallel within a range of 0.02−0.12 $\mu$m (1−5 $\mu$in.). Gage blocks are available in sets of various sizes, some sets containing almost a hundred gage blocks. The blocks can be assembled in many combinations to obtain desired lengths. Dimensional accuracy can be as high as 0.05 $\mu$m (2 $\mu$in.). Environmental temperature control is important in using gages for high-precision measurements. Although their use requires some skill, gage-block assemblies are commonly utilized in industry as an accurate reference length. Angle blocks are made similarly and are available for angular gaging.

**Fixed gages** are replicas of the shapes of the parts to be measured. **Plug gages** are commonly used for holes. The *GO gage* is smaller than the *NOT GO* (or *NO GO*) *gage* and slides into any hole whose smallest dimension is less than the diameter of the gage. The *NOT GO* gage must not go into the hole. Two gages are required for such measurements, although both may be on the same device, either at opposite ends or in two steps at one end (step-type gage). Plug gages are also available for measuring internal tapers (in which deviations between the gage and the part are indicated by the looseness of the gage), splines, and threads (in which the *GO* gage must screw into the threaded hole).

**Ring gages** are used to measure shafts and similar round parts. Ring thread gages are used to measure external threads. The *GO* and *NOT GO* features on these gages are identified by the type of knurling on the outside diameters of the rings. **Snap gages** are commonly used to measure external dimensions. They are made with adjustable gaging surfaces for use with parts having different dimensions. One of the gaging surfaces may be set at a different gap from the other, thus making a one-unit *GO–NOT GO* gage.

Although fixed gages are easy to use and inexpensive, they only indicate whether a part is too small or too large, compared to an established standard. They do not measure actual dimensions.

There are several types of **pneumatic gages,** also called *air gages.* The gage head has holes through which pressurized air, supplied by a constant-pressure line, escapes. The smaller the gap between the gage and the hole, the more difficult it is for the air to escape, and hence the back pressure is higher. The back pressure, sensed and indicated by a pressure gage, is calibrated to read dimensional variations of holes.

**Microscopes** are optical instruments used to view and measure very fine details, shapes, and dimensions on small and medium-sized tools, dies, and workpieces. The most common and versatile microscope used in tool rooms is the **toolmaker's microscope.** It is equipped with a stage that is movable in two principal directions and can be read to 2.5 $\mu$m (0.0001 in.). Several models of microscopes are available with various features for specialized inspection, including models with digital readout.

The **light section microscope** is used to measure small surface details, such as scratches, and the thickness of deposited films and coatings. A thin light band is applied obliquely to the surface and the reflection is viewed at 90°, showing surface roughness, contours, and other features. Unlike ordinary optical microscopes, the **scanning electron microscope** (**SEM**) has excellent depth of field. As a result, all regions of a complex part are in focus and can be viewed in and photographed to show extremely fine detail. This type of microscope is particularly useful for studying surface textures and fracture patterns. Although expensive, such microscopes are capable of magnifications greater than 100,000×.

### 4.6.2   General characteristics of measuring instruments

The characteristics and quality of measuring instruments are generally described by certain specific terms. These terms, in alphabetical order, are defined as follows:

1. *Accuracy.* The degree of agreement of the measured dimension with its true magnitude.
2. *Amplification.* See Magnification.
3. *Calibration.* Adjusting or setting an instrument to give readings that are accurate within a reference standard.
4. *Drift.* See Stability.

5. *Linearity.* The accuracy of the readings of an instrument over its full working range.

6. *Magnification.* The ratio of instrument output to the input dimension.

7. *Precision.* Degree to which an instrument gives repeated measurement of the same standard.

8. *Repeat accuracy.* Same as accuracy, but repeated many times.

9. *Resolution.* Smallest dimension that can be read on an instrument.

10. *Rule of 10 (Gage Maker's rule).* An instrument or gage should be 10 times more accurate than the dimensional tolerances of the part being measured. Similarly, a factor of 4 is known as the *Mil Standard rule.*

11. *Sensitivity.* Smallest difference in dimension that an instrument can distinguish or detect.

12. *Speed of response.* How rapidly an instrument indicates the measurement, particularly when a number of parts are measured in rapid succession.

13. *Stability.* An instrument's capability to maintain its calibration over a period of time (also called *drift*).

Selection of an appropriate measuring instrument for a particular application depends on the foregoing factors. In addition, the size and type of parts to be measured, the environment (temperature, humidity, dust, and so on), operator skills required, and costs have to be considered in the purchase of such equipment.

### 4.6.3 Automated measurement

With increasing automation in all aspects of manufacturing processes and operations, the need for automated measurement (also called **automated inspection**) has become much more apparent. Flexible manufacturing systems and manufacturing cells have led to the adoption of advanced measurement techniques and systems. In fact, installation and utilization of these systems is now a necessary—not an optional—manufacturing technology.

Traditionally, a batch of parts was manufactured and sent for measurement in a separate quality-control room, and if the parts passed measurement inspection, they were put into inventory. Automated inspection, however, is based on various on-line sensor systems that monitor the dimensions of parts being made and use these measurements to correct the process.

To appreciate the importance of on-line monitoring of dimensions, let's find the answer to the following question: If a machine has been producing a certain part with acceptable dimensions, what factors contribute to subsequent deviation in the dimension of the same part produced by the same machine? The major factors are:

- Static and dynamic deflections of the machine because of vibrations and fluctuating forces, caused by variations such as in the properties and dimensions of the incoming material.

- Deformation of the machine because of thermal effects. These effects include changes in temperatures of the environment, metalworking fluids, and machine bearings and components.

- Wear of tools and dies, which, in turn, affects the dimensions of the parts produced.

As a result of these factors, the dimensions of parts produced will vary, necessitating monitoring of dimensions during production. **In-process** workpiece control is accomplished by special gaging and is used in a variety of applications, such as high-quantity machining and grinding.

## 4.7   Dimensional Tolerances

**Dimensional tolerance** is defined as the permissible or acceptable variation in the dimensions (height, width, depth, diameter, angles) of a part. Tolerances are unavoidable because it is virtually impossible (and unnecessary) to manufacture two parts that have precisely the same dimensions. Furthermore, because close tolerances substantially increase the product cost, a narrow tolerance range is undesirable economically.

Tolerances become important only when a part is to be assembled or mated with another part. Surfaces that are free and not functional do not need close tolerance control. Thus, for example, the accuracies of the holes and the distance between the holes for a connecting rod are far more critical than the rod's width and thickness at various locations along its length. By reviewing the figures throughout this text, you can determine which dimensions and features of the parts illustrated are more critical than others.

To illustrate the importance of dimensional tolerances, let's assemble a simple shaft (axle) and a wheel with a hole, assuming that we want the axle's diameter to be 1 in. (Fig. 4.20). We go to the hardware store and purchase a 1-in. round rod and a wheel with a 1-in. hole. Will the rod fit into the hole without forcing it, or will it be loose in the hole?

The 1-in. dimension is the *nominal* size of the shaft. If we purchase such a rod from different stores or at different times—or select one randomly from a lot of, say, 50 shafts—the chances are that each rod will have a slightly different diameter. Machines may, with the same setup, produce rods of slightly different diameters, depending on a number of factors, such as speed of operation, temperature, lubrication, variations in the incoming material, and similar variables.

If we now specify a range of diameters for both the rod and the hole of the wheel, we can predict correctly the type of fit that we will have after assembly. Certain terminology has been established to clearly define these geometric quantities, such as the ISO system shown in Fig. 4.20. Note that both the shaft and the hole have minimum and maximum diameters, respectively, the difference being the tolerance for each member. A proper engineering drawing should specify these parameters with numerical values, as shown in Fig. 4.21.

**FIGURE 4.20**   Basic size, deviation, and tolerance on a shaft, according to the ISO system.

**FIGURE 4.21**  Various methods of assigning tolerances on a shaft. *Source:* L. E. Doyle.

The range of tolerances obtained in various manufacturing processes is given in Fig. 4.22. There is a general relationship between tolerances and surface finish of parts manufactured by various processes. Note the wide range of tolerances and surface finishes obtained. Also, the larger the part, the greater the obtainable tolerance range becomes. Experience has shown that dimensional inaccuracies of manufactured parts are approximately proportional to the cube root of the size of the part. Thus, doubling the size of a part increases the inaccuracies by $\sqrt[3]{2} = 1.26$ times, or 26 percent.

Several terms are used to describe features of dimensional relationships between mating parts. These terms, in alphabetical order, are defined as follows:

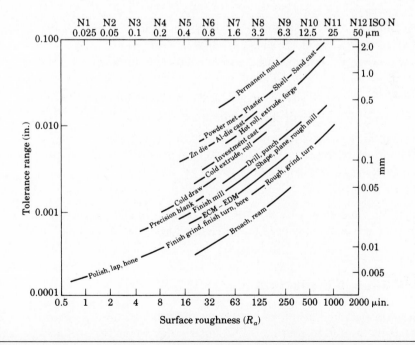

**FIGURE 4.22**  Tolerances and surface roughness obtained in various manufacturing processes. These tolerances apply to a 25 mm (1 in.) workpiece dimension. *Source:* J. A. Schey, Tribology, in *Metalworking, Friction, Lubrication, and Wear* (1983), ASM International, Materials Park, OH.

- *Allowance.* The specified difference in dimensions between mating parts; also called *functional dimension* or *sum dimension.*

- *Basic size.* Dimension from which limits of size are derived, using tolerances and allowances.

- *Bilateral tolerance.* Deviation—plus or minus—from the basic size.

- *Clearance.* The space between mating parts.

- *Datum.* A theoretically exact axis, point, line, or plane.

- *Feature.* Physically identifiable portion of a part, such as hole, slot, pin, or chamfer.

- *Fit.* The amount of clearance or interference between mating parts.

- *Geometric tolerancing.* Tolerances that involve shape features of the part.

- *Hole-basis system.* Tolerances based on a zero line on the hole; also called *standard hole practice* or *basic hole system.*

- *Interference.* Negative clearance.

- *Interference fit.* A fit having limits of size so prescribed that an interference always results when mating parts are assembled.

- *International tolerance grade (IT).* A group of tolerances that vary depending on the basic size, but provide the same relative level of accuracy within a grade.

- *Limit dimensions.* The maximum and minimum dimensions of a part; also called *limits.*

- *Maximum material condition (MMC).* The condition where a feature of size contains the maximum amount of material within the stated limits of size.

- *Nominal size.* Dimension that is used for the purpose of general identification.

- *Positional tolerancing.* A system of specifying the true position, size, and form of the features of a part, including allowable variations.

- *Shaft-basis system.* Tolerances based on a zero line on the shaft; also called *standard shaft practice* or *basic shaft system.*

- *Standard size.* Nominal size in integers and common subdivisions of length.

- *Transition fit.* Fit with small clearance or interference that allows for accurate location of mating parts.

- *Unilateral tolerancing.* Deviation in one direction only from the nominal dimension.

- *Zero line.* Reference line along the basic size from which a range of tolerances and deviations are specified.

Because the dimensions of holes are more difficult to control than those of shafts, the hole-basis system is commonly used for specifying tolerances in shaft and hole assemblies.

## 4.8  Testing and Inspection

Before they are marketed, manufactured parts and products are inspected for several characteristics. This inspection routine is particularly important for products or components whose failure or malfunction has potentially serious implications, such as bodily injury or fatality. Typical examples are cables breaking, switches malfunctioning, brakes failing, grinding wheels breaking, railroad wheels fracturing, turbine blades failing, pressure vessels bursting, and weld or joints failing. In this section we identify and describe the various methods that are commonly used to inspect manufactured products.

*Product quality* has always been one of the most important elements in manufacturing operations, and with increasing domestic and international competition, it has become even more important. Prevention of defects in products and on-line inspection are now major goals in all manufacturing activities. We again emphasize that quality must be built into a product and not merely checked after the product has been made. Close cooperation and communication between design and manufacturing engineers are essential.

### 4.8.1  Nondestructive testing

**Nondestructive testing (NDT)** is carried out in such a way that product integrity and surface texture remain unchanged. These techniques generally require considerable operator skill. Interpreting test results accurately may be difficult because test results can be quite subjective. However, the use of computer graphics and other enhancement techniques has reduced the likelihood of human error in nondestructive testing. We describe below the basic principles of the more commonly used nondestructive testing techniques.

In the **liquid-penetrants technique,** fluids are applied to the surfaces of the part and allowed to penetrate into surface openings, cracks, seams, and porosity. The penetrant can seep into cracks as small as 0.1 $\mu$m (4 $\mu$in.) in width. Two common types of liquids are (1) fluorescent penetrants with various sensitivities, which fluoresce under ultraviolet light; and (2) visible penetrants, using dyes usually red in color, which appear as bright outlines on the surface.

The surface to be inspected is first thoroughly cleaned and dried. The liquid is brushed or sprayed on the surface to be inspected and allowed to remain long enough to seep into surface openings. Excess penetrant is then wiped off or washed away with water or solvent. A developing agent is then added to allow the penetrant to seep back to the surface and spread to the edges of openings, thus magnifying the size of defects. The surface is then inspected for defects, either visually in the case of dye penetrants or with fluorescent lighting. This method is capable of detecting a variety of surface defects and is used extensively. The equipment is simple and easy to use, can be portable, and is less costly to operate than other methods. However, this method can only detect defects that are open to the surface, not internal defects.

The **magnetic-particle inspection technique** consists of placing fine ferromagnetic particles on the surface. The particles can be applied either dry or in a liquid carrier such as water or oil. When the part is magnetized with a magnetic field, a discontinuity (defect) on the surface causes the particles to gather visibly around it. The collected particles generally take the shape and size of the defect. Subsurface defects can also be detected by this method, provided they are not deep. The ferromagnetic particles may be colored with pigments for better visibility on metal surfaces. Wet particles are used for detecting fine discontinuities, such as fatigue cracks.

In **ultrasonic inspection,** an ultrasonic beam travels through the part. An internal defect, such as a crack, interrupts the beam and reflects back a portion of the ultrasonic energy. The amplitude of the energy reflected and the time required for return indicates the presence and location of any flaws in the workpiece. The ultrasonic waves are generated by transducers, called search units or probes, of various types and shapes. They operate on the principle of piezoelectricity, using materials such as quartz, lithium sulfate, and various ceramics. Most inspections are carried out at a frequency range of 1–25 MHz. Couplants are used to transmit the ultrasonic waves from the transducer to the test piece. Typical couplants are water, oil, glycerin, and grease.

The ultrasonic inspection method has high penetrating power and sensitivity. It can be used to inspect flaws in large volumes of material, such as railroad wheels, pressure vessels, and die blocks, from various directions. Accuracy is higher than that of other nondestructive inspection methods. However, this method requires experienced personnel to carry out the inspection and interpret the results correctly.

The **acoustic-emission technique** detects signals (high-frequency stress waves) generated by the workpiece itself during plastic deformation, crack initiation and propagation, phase transformation, and sudden reorientation of grain boundaries. Bubble formation during boiling and friction and wear of sliding interfaces are other sources of acoustic signals. Acoustic-emission inspection is typically performed by stressing elastically the part or structure, such as bending a beam, applying torque to a shaft, and pressurizing a vessel. Acoustic emissions are detected by sensors consisting of piezoelectric ceramic elements. This method is particularly effective for continuous surveillance of load-bearing structures.

The **acoustic-impact technique** consists of tapping the surface of an object and listening to and analyzing the signals to detect discontinuities and flaws. The principle is basically the same as tapping walls, desktops, or countertops in various locations with your fingers or a hammer and listening to the sound emitted. Vitrified grinding wheels are tested in a similar manner (ring test) to detect cracks in the wheel that may not be visible to the naked eye. The acoustic-impact technique can be instrumented and automated and is easy to perform. However, the results depend on part geometry and mass, thus requiring a reference standard to identify flaws.

**Radiography** involves x-ray inspection to detect internal flaws or density and thickness variations in the part. The radiation source is typically an x-ray tube, and a visible permanent image is made on an x-ray film or radiographic paper.

The **eddy-current inspection method** is based on the principle of electromagnetic induction. The part is placed in or adjacent to an electric coil through which alternating current (exciting current) flows at frequencies ranging from 60 Hz to 6 MHz. This current causes eddy currents to flow in the part. Defects in the part impede and change the direction of eddy currents, causing changes in the electromagnetic field. These changes affect the exciting coil (inspection coil), whose voltage is monitored to determine the presence of flaws.

**Thermal inspection** involves observing temperature changes by contact- or noncontact-type heat-sensing devices. Defects in the workpiece, such as cracks, debonded regions in laminated structures, and poor joints, cause a change in temperature distribution. In *thermographic inspection*, materials such as heat-sensitive paints and papers, liquid crystals, and other coatings are applied to the surface. Any changes in their color or appearance indicate defects.

The **holography technique** creates a three-dimensional image of the workpiece, utilizing an optical system. This technique is generally used on simple shapes and highly polished surfaces, and the image is recorded on a photographic film. Its use has been extended to inspection of parts (**holographic interferometry**) having various shapes and surface conditions.

Using double- and multiple-exposure techniques while the part is being subjected to external forces or time-dependent variations, changes in the images reveal defects in the part.

In **acoustic holography,** information on internal defects is obtained directly from the image of the interior of the part. In *liquid-surface acoustical holography,* the workpiece and two ultrasonic transducers (one for the object beam and the other for the reference beam) are immersed in a water-filled tank. The holographic image is then obtained from the ripples in the tank. In **scanning acoustical holography,** only one transducer is used and the hologram is produced by electronic-phase detection. This system is more sensitive, the equipment is usually portable, and very large workpieces can be accommodated by using a water column instead of a tank.

### 4.8.2  Destructive testing

As the name suggests, the part or product tested using **destructive testing** methods no longer maintains its integrity, original shape, or surface texture. Mechanical test methods are all destructive, in that a sample or specimen has to be removed from the product in order to test it. In addition to mechanical testing, other destructive tests include speed testing of grinding wheels to determine their bursting speed and high-pressure testing of pressure vessels to determine their bursting pressure.

Hardness tests leaving large impressions may be regarded as destructive testing. However, microhardness tests may be regarded as nondestructive because of the very small permanent indentations. This distinction is based on the assumption that the material is not notch sensitive. However, most glasses, highly heat-treated metals, and ceramics are notch sensitive; that is, the small indentation produced by the indenter may lower their strength and toughness.

### 4.8.3  Automated inspection

Note that in all the preceding examples we discussed testing parts or products that had already been manufactured. Traditionally, individual parts and assemblies of parts have been manufactured in batches, sent to inspection in quality-control rooms, and if approved, put in inventory. If products do not pass the quality inspection, they are either scrapped or kept on the basis of a certain acceptable deviation from the standard. Obviously, such a system lacks flexibility, requires maintenance an inventory, and inevitably results in some defective parts going through the system. The traditional method actually counts the defects after they occur (**postprocess inspection**), and in no way attempts to prevent defects.

In contrast, one of the important trends in modern manufacturing is **automated inspection.** This method uses a variety of sensor systems that monitor the relevant parameters *during* the manufacturing process (*on-line inspection*). Then, using these measurements, the process automatically corrects itself to produce acceptable parts. Thus, further inspection of the part at another location in the plant is unnecessary. Parts may also be inspected immediately after they are produced (**in-process inspection**).

The use of accurate sensors (see below) and computer-control systems has integrated automated inspection into manufacturing operations. Such a system ensures that no part is moved from one manufacturing process to another (for example, a turning operation on a lathe followed by cylindrical grinding) unless the part is made correctly and meets the standards of the first operation. Automated inspection is flexible and responsive to product design changes. Furthermore, because of automated equipment, less operator skill is required, productivity is increased, and parts have higher quality, reliability, and dimensional accuracy.

### Sensors for automated inspection

Recent advances in *sensor technology* are making on-line or real-time monitoring of manufacturing processes feasible. Directly or indirectly, and with the use of various probes, **sensors** can detect dimensional accuracy, surface roughness, temperature, force, power, vibration, tool wear, and the presence of external or internal defects. Sensors operate on the principles of strain gages, inductance, capacitance, ultrasonics, acoustics, pneumatics, infrared radiation, optics, lasers, and various electronic gages. Sensors may be tactile (touching) or nontactile (see Section 13.8).

Sensors, in turn, are linked to microprocessors and computers for graphic data display. This capability allows rapid on-line adjustment of any processing parameters in order to produce parts that are consistently within specified standards of tolerance and quality. Such systems have already been implemented as standard equipment on many metal-cutting machine tools and grinding machines.

## 4.9   Quality Assurance

**Quality assurance** is the total effort by a manufacturer to ensure that its products conform to a detailed set of specifications and standards. These standards cover several parameters, such as dimensions, surface finish, tolerances, composition, color, and mechanical, physical, and chemical properties. In addition, standards are usually written to ensure proper assembly using interchangeable, defect-free components and a product that performs as intended by its designers.

Quality assurance is the responsibility of everyone involved with design and manufacturing. The often-repeated statement that quality must be built into a product reflects this important concept. Quality cannot be inspected into a finished product. Although product quality has always been important, increased domestic and international competition has caused quality assurance to become even more important. Every aspect of design and manufacturing operations, such as material selection, production, and assembly, is now being analyzed in detail to ensure that quality is truly built into the final product.

If you were in charge of product quality for a manufacturing plant, how would you make sure that the final product is of acceptable quality? The best method is to control materials and processes in such a manner that the products are made correctly in the first place. However, 100 percent inspection is usually too costly to maintain. Therefore, several methods of inspecting smaller, statistically relevant sample lots have been devised. These methods all use statistics to determine the probability of defects occurring in the total production batch.

Inspection involves a series of steps:

1. Inspecting incoming materials to make sure that they meet certain property, dimension, and surface finish and integrity requirements

2. Inspecting individual product components to make sure that they meet specifications

3. Inspecting the product to make sure that individual parts have been assembled properly

4. Testing the product to make sure that it functions as designed and intended

Why must inspections be continued once a product of acceptable quality has been produced? Because there will always be variations in the dimensions and properties of incoming materials, variations in the performance of tools, dies, and machines used in various stages of manufacturing, possibilities of human error, and errors made during assembly of the product. As a result, no two products are ever made exactly alike.

Another important aspect of quality control is the capability to analyze defects and promptly eliminate them or reduce them to acceptable levels. In an even broader sense, quality control involves evaluating the product design and customer satisfaction. The sum total of all these activities is referred to as **total quality management.**

Thus, in order to control quality we have to be able to (1) measure quantitatively the level of quality, and (2) identify all the material and process variables that can be controlled. The level of quality obtained during production can then be established by inspecting the product to determine whether it meets the specifications for tolerances, surface finish, defects, and other characteristics. The identification of material and process variables and their effect on product quality is now possible through the extensive knowledge gained from research and development activities in all aspects of manufacturing.

### 4.9.1 Statistical methods of quality control

**Statistics** deals with the collection, analysis, interpretation, and presentation of large amounts of numerical data. The use of statistical techniques in modern manufacturing operations is necessary because of the large number of material and process variables involved. Those events that occur randomly—that is, without any particular trend or pattern—are called **chance variations.** Those that can be traced to specific causes are called **assignable variations.** The existence of *variability* in production operations has been recognized for centuries, but Eli Whitney (1765–1825) first grasped its full significance when he found that interchangeable parts were indispensable to the mass production of firearms. Modern statistical concepts relevant to manufacturing engineering were first developed in the early 1900s.

To understand **statistical quality control (SQC),** we first need to define some of the terms that are commonly used in this field.

- **Sample size:** The number of parts to be inspected in a sample, whose properties are studied to gain information about the whole population.
- **Random sampling:** Taking a sample from a population or lot in which each item has an equal chance of being included in the sample. Thus, when taking samples from a large bin, the inspector does not take only those that happen to be within reach.
- **Population:** The totality of individual parts of the same design from which samples are taken (also called the **universe**).
- **Lot size:** A subset of population. A lot or several lots can be considered subsets of the population and may be treated as representative of the population.

The sample is inspected for certain characteristics and features, such as tolerances, surface finish, and defects, with the instruments and techniques that we described in this chapter.

These characteristics fall into two categories: those that can be measured quantitatively (method of variables) and those that are qualitative (method of attributes).

The **method of variables** is the quantitative measurement of characteristics such as dimensions, tolerances, surface finish, or physical or mechanical properties. Such measurements are made for each of the units in the group under consideration, and the results are compared against specifications.

The **method of attributes** involves observing the presence or absence of qualitative characteristics, such as external or internal defects in machined, formed, or welded parts or dents in sheet-metal products, for each of the units in the group under consideration. Sample size for attributes-type data is generally larger than for variables-type data.

During the inspection process, measurement results will vary. For example, assume that you are measuring the diameter of turned shafts as they are produced on a lathe, using a micrometer. You soon note that their diameters vary, even though ideally you want all the shafts to be exactly the same size. Let's now turn to consideration of statistical quality-control techniques, which allow us to evaluate these variations and set limits for the acceptance of parts.

If we list the measured diameters of the turned shafts in a given population, we note that one or more parts have the smallest diameter, and one or more have the largest diameter. The majority of the turned shafts have diameters that lie between these extremes. If we group these diameters and plot them, the plot consists of a bar graph representing the number of parts in each diameter group (Fig. 4.23). The bars show a **distribution,** also called a **spread** or **dispersion,** of the shaft-diameter measurements. The bell-shaped curve in Fig. 4.23 is called **frequency distribution,** or the frequency with which parts within each diameter group are being produced.

Data from manufacturing processes often fit curves represented by a mathematically derived *normal distribution curve* (Fig. 4.24). These curves are also called *Gaussian,* developed on the basis of probability. The bell-shaped normal distribution curve fitted to the data in Fig. 4.23 has two important features. First, it shows that most part diameters tend to cluster around an **average** value (**arithmetic mean**). This average is usually designated as $\bar{x}$ and is calculated from the expression

$$\bar{x} = \frac{x_1 + x_2 + x_3 + \cdots + x_n}{n},$$    (4.11)

where the numerator is the sum of all measured values (diameters) and $n$ is the number of measurements (number of shafts).

Diameter of shafts (mm)

**FIGURE 4.23**   A plot of the number of shafts measured and their respective diameters. This type of curve is called a frequency distribution.

**FIGURE 4.24** A normal distribution curve indicating areas within each range of standard deviation. Note: The greater the range, the higher the percentage of parts that fall within it.

The second feature of this curve is its width, indicating the *dispersion* of the diameters measured. The wider the curve, the greater the dispersion. The difference between the largest value and smallest value is called the *range, R*:

$$R = x_{max} - x_{min}. \qquad (4.12)$$

Dispersion is estimated by the **standard deviation**, which is generally denoted as $\sigma$ and is obtained from the expression

$$\sigma = \sqrt{\frac{(x_1 - \bar{x})^2 + (x_2 - \bar{x})^2 + (x_3 - \bar{x})^2 + \cdots + (x_n - \bar{x})^2}{n - 1}}, \qquad (4.13)$$

where $x$ is the measured value for each part. Note from the numerator in Eq. (4.13) that as the curve widens, the standard deviation becomes greater. Also note that $\sigma$ has the unit of linear dimension. In comparing Eqs. (4.12) and (4.13), we note that the range $R$ is a simpler and more convenient measure of dispersion.

Since we know the number of turned parts that fall within each group, we can calculate the percentage of the total population represented by each group. Thus, Fig. 4.24 shows that the diameters of 99.73 percent of the turned shafts fall within the range of $\pm 3\sigma$, 95.46 percent within $\pm 2\sigma$, and 68.26 percent within $\pm 1\sigma$. Thus, only 0.27 percent fall outside the $\pm 3\sigma$ range.

### 4.9.2 Statistical process control

If the number of parts that do not meet set standards (defective parts) increases during a production run, we must be able to determine the cause (incoming materials, machine controls, degradation of metalworking fluids, operator boredom, or others) and take appropriate action. Although this statement at first appears to be self-evident, it was only in the early 1950s that a systematic statistical approach was developed to guide operators in manufacturing plants.

This approach advises the operator to take certain measures and when to take them in order to avoid producing further defective parts. Known as **statistical process control (SPC)**, this technique consists of several elements: control charts and setting control limits, capabilities of the particular manufacturing process, and characteristics of the machinery involved.

## Control charts

The frequency distribution curve in Fig. 4.23 shows a range of shaft diameters being produced that may fall beyond the predetermined design tolerance range. Figure 4.25 shows the same bell-shaped curve, which now includes the specified tolerances for the diameter of turned shafts.

**Control charts** graphically represent the variations of a process over a period of time. They consist of data plotted during production and, typically, there are two plots. The quantity $\bar{x}$ (Fig. 4.26a) is the average for each subset of samples taken and inspected; say, a subset consisting of 5 parts. A sample size of between 2 and 10 parts is sufficiently accurate, provided that sample size is held constant throughout the inspection.

The frequency of sampling depends on the nature of the process. Some processes may require continuous sampling, whereas others may require only one sample per day. Quality-control analysts are best qualified to determine this frequency for a particular situation. Since the measurements in Fig. 4.26a are made consecutively, the abscissa of these control charts also represents time. The solid horizontal line in this figure is the **average of averages (grand average)**, denoted as $\bar{\bar{x}}$, and represents the population mean. The upper and lower horizontal broken lines in these control charts indicate the control limits for the process.

The **control limits** are set on these charts according to statistical-control formulas designed to keep actual production within the usually acceptable $\pm 3\sigma$ range. Thus, for $\bar{x}$,

$$\text{Upper control limit (UCL}_{\bar{x}}) = \bar{x} + 3\sigma = \bar{\bar{x}} + A_2\bar{R} \tag{4.14}$$

and

$$\text{Lower control limit (LCL}_{\bar{x}}) = \bar{x} - 3\sigma = \bar{\bar{x}} - A_2\bar{R}, \tag{4.15}$$

where $A_2$ is read from Table 4.3 and $\bar{R}$ is the average of $R$ values.

These limits are calculated based on the historical production capability of the equipment itself. They are not generally associated with either design tolerance specifications or dimensions. They indicate the limits within which a certain percentage of measured values are normally expected to fall because of the inherent variations of the process itself, upon which the

**FIGURE 4.25** Frequency distribution curve, showing lower and upper specification limits.

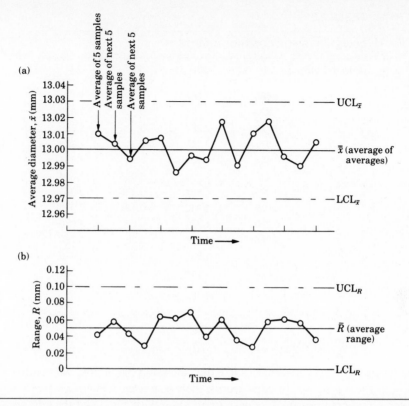

**FIGURE 4.26** Control charts used in statistical quality control. The process shown is in statistical control because all points fall within the lower and upper control limits. In this illustration, sample size is 5 and the number of samples is 15.

limits are based. The major goal of statistical process control is to improve the manufacturing process with the aid of control charts to eliminate assignable causes. The control chart continually indicates progress in this area.

The second control chart (Fig. 4.26b) shows the range $R$ in each subset of samples. The solid horizontal line represents the average of $R$ values in the lot, denoted as $\overline{R}$, and is a measure of the variability of the samples. The upper and lower control limits for $R$ are obtained from the equations

$$\text{UCL}_R = D_4\overline{R} \tag{4.16}$$

and

$$\text{LCL}_R = D_3\overline{R}, \tag{4.17}$$

where the constants $D_4$ and $D_3$ are obtained from Table 4.3. This table also includes the constant $d_2$, which is used in estimating the standard deviation from the equation

**TABLE 4.3**

**CONSTANTS FOR CONTROL CHARTS**

| SAMPLE SIZE | $A_2$ | $D_4$ | $D_3$ | $d_2$ |
|:---:|:---:|:---:|:---:|:---:|
| 2 | 1.880 | 3.267 | 0 | 1.128 |
| 3 | 1.023 | 2.575 | 0 | 1.693 |
| 4 | 0.729 | 2.282 | 0 | 2.059 |
| 5 | 0.577 | 2.115 | 0 | 2.326 |
| 6 | 0.483 | 2.004 | 0 | 2.534 |
| 7 | 0.419 | 1.924 | 0.078 | 2.704 |
| 8 | 0.373 | 1.864 | 0.136 | 2.847 |
| 9 | 0.337 | 1.816 | 0.184 | 2.970 |
| 10 | 0.308 | 1.777 | 0.223 | 3.078 |
| 12 | 0.266 | 1.716 | 0.284 | 3.258 |
| 15 | 0.223 | 1.652 | 0.348 | 3.472 |
| 20 | 0.180 | 1.586 | 0.414 | 3.735 |

$$\sigma = \frac{\overline{R}}{d_2}. \qquad (4.18)$$

When the curve of a control chart is like that shown in Fig. 4.26b, we say that the process is "in good statistical control." In other words, there is no clear discernible trend in the pattern of the curve, the points (measured values) are random with time, and they do not exceed the control limits. However, you can see that curves such as those shown in Fig. 4.27a, b, and c indicate certain trends. For example, note in the middle of curve 4.27a that the diameter of the shafts increases with time. The reason for this increase may be a change in one of the process variables, such as wear of the cutting tool. If, as in curve 4.27b, the trend is toward consistently larger diameters, hovering around the upper control limit, it could mean that the tool settings on the lathe may not be correct and, as a result, the parts being turned are consistently too large. Curve 4.27c shows two distinct trends that may be caused by factors such as a change in the properties of the incoming material or a change in the performance of the cutting fluid (for example, its degradation). These situations place the process "out of control."

Analyzing control-chart patterns and trends requires considerable experience in order to identify the specific cause(s) of an out-of-control situation. Overcontrol of the manufacturing process—that is, setting upper and lower control limits too close to each other (smaller standard deviation range)—is a further reason for out-of-control situations. This is the reason for calculating control limits on process capability rather than on a potentially inapplicable statistic.

It is evident that operator training is critical for successful implementation of SPC on the shop floor. Once process control guidelines are established, operators should, in the interest of efficiency of operation, also have some responsibility to make adjustments in processes that begin to become out of control. The capabilities of individual operators should be taken into account so that they are not overloaded with data input and their proper interpretation. This task is, however, being made easier by a variety of software that is being continually developed. For example, digital readouts on electronic measuring devices are integrated directly

(a)

(b)

(c)

**FIGURE 4.27**  Control charts.
(a) Process begins to become out of
control because of factors such as
tool wear. The tool is changed and
the process is then in statistical control. (b) Process parameters are not
set properly; thus all parts are
around the upper control limit.
(c) Process becomes out of control
because of factors such as a sudden
change in the properties of the incoming material.

into a computer system for real-time SPC. Figure 4.28 shows such a multifunctional computer system in which the output from a digital caliper or micrometer is analyzed with a
miniprocessor in real time and displayed in several ways, such as frequency distribution
curves and control charts.

**Process capability.**  **Process capability** is defined as the limits within which individual measurement values resulting from a particular manufacturing process would normally be expected
to fall when only random variation is present. Process capability tells us that the process can
produce parts within certain limits of precision. Since a manufacturing process involves materials, machinery, and operators, each can be analyzed individually to identify a problem when
process capabilities do not meet part specifications.

**FIGURE 4.28**   Schematic illustration showing integration of digital gages with miniprocessor for real-time data acquisition and SPC/SQC capabilities. Note the examples on the CRT displays, such as frequency distribution and control charts. *Source:* Mitutoyo Corp.

## Example 4.3:   Calculation of control limits and standard deviation

The data in Table 4.4 below show length measurements (in.) taken on a machined workpiece. Sample size is 5 and the sample number is 10; thus, the total number of parts measured is 50. The quantity $\bar{x}$ is the average of five measurements in each sample. We first calculate the average of averages $\bar{\bar{x}}$,

$$\bar{\bar{x}} = \frac{44.296}{10} = 4.430 \text{ in.,}$$

**TABLE 4.4**

| SAMPLE NUMBER | $x_1$ | $x_2$ | $x_3$ | $x_4$ | $x_5$ | $\bar{x}$ | $R$ |
|:---:|:---:|:---:|:---:|:---:|:---:|:---:|:---:|
| 1 | 4.46 | 4.40 | 4.44 | 4.46 | 4.43 | 4.438 | 0.06 |
| 2 | 4.45 | 4.43 | 4.47 | 4.39 | 4.40 | 4.428 | 0.08 |
| 3 | 4.38 | 4.48 | 4.42 | 4.42 | 4.35 | 4.410 | 0.13 |
| 4 | 4.42 | 4.44 | 4.53 | 4.49 | 4.35 | 4.446 | 0.18 |
| 5 | 4.42 | 4.45 | 4.43 | 4.44 | 4.41 | 4.430 | 0.04 |
| 6 | 4.44 | 4.45 | 4.44 | 4.39 | 4.40 | 4.424 | 0.06 |
| 7 | 4.39 | 4.41 | 4.42 | 4.46 | 4.47 | 4.430 | 0.08 |
| 8 | 4.45 | 4.41 | 4.43 | 4.41 | 4.50 | 4.440 | 0.09 |
| 9 | 4.44 | 4.46 | 4.30 | 4.38 | 4.49 | 4.414 | 0.19 |
| 10 | 4.42 | 4.43 | 4.37 | 4.47 | 4.49 | 4.436 | 0.12 |

and the average of $R$ values,

$$\bar{R} = \frac{1.03}{10} = 0.103 \text{ in.}$$

Since the sample size is 5, we determine from Table 4.3 the following constants: $A_2 = 0.577$, $D_4 = 2.115$, and $D_3 = 0$. We can now calculate the control limits using Eqs. (4.14)–(4.17). Thus, for averages we have

$$\text{UCL}_{\bar{x}} = 4.430 + (0.577)(0.103) = 4.489 \text{ in.,}$$
$$\text{LCL}_{\bar{x}} = 4.430 - (0.577)(0.103) = 4.371 \text{ in.,}$$

and for ranges we have

$$\text{UCL}_R = (2.115)(0.103) = 0.218 \text{ in.,}$$
$$\text{LCL}_R = (0)(0.103) = 0 \text{ in.}$$

We can calculate the standard deviation using Eq. (4.18) and a value of $d_2 = 2.326$. Thus,

$$\sigma = \frac{0.103}{2.326} = 0.044 \text{ in.}$$

### 4.9.3   Acceptance sampling and control

**Acceptance sampling** consists of taking only a few random samples from a lot and inspecting them for the purpose of judging whether the entire lot is acceptable or should be rejected or reworked. Developed in the 1920s and used extensively during World War II for military hardware (MIL STD 105), this statistical technique is widely used and valuable. It is particularly useful for inspecting high-production-rate parts, where 100 percent inspection would be too costly.

A variety of acceptance sampling plans have been prepared for both military and national standards, based on an acceptable, predetermined, and limiting percentage of nonconforming parts in the sample. If this percentage is exceeded, the entire lot is rejected, or it is reworked if economically feasible. Note that the actual number of samples (not percentages of the lot that are in the sample) can be significant in acceptance sampling. The greater the number of samples taken from a lot, the greater will be the percentage of nonconforming parts and the lower the probability of acceptance of the lot. **Probability** is defined as the relative occurrence of an event. The probability of acceptance is obtained from various operating characteristic curves, one example of which is shown in Fig. 4.29.

The **acceptance quality level (AQL)** is commonly defined as the level at which there is a 95 percent probability of acceptance of the lot. This percentage would indicate to the manufacturer that 5 percent of the parts in the lot may be rejected by the consumer (called **the producer's risk**). Likewise, the consumer knows that 95 percent of the parts are acceptable (called **the consumer's risk**).

Lots that do not meet the desired quality standards can be salvaged by the manufacturer by a secondary rectifying inspection. In this method, a 100 percent inspection is made of a rejected lot, and the defective parts are removed. This time-consuming and costly process is an incentive for the manufacturer to control the production process.

**FIGURE 4.29** A typical operating-characteristic curve used in acceptance sampling. The higher the percentage of defective parts, the lower is the probability of their being accepted by the consumer. There are several methods of obtaining these curves.

Acceptance sampling requires less time and fewer inspections than do other sampling methods. Consequently, inspection of the parts can be more detailed. Keep in mind, however, that automated inspection techniques are being developed rapidly so that 100 percent inspection of all parts is indeed feasible and can be economical.

### 4.9.4  Total quality control

The **total quality control (TQC)** concept is a management system emphasizing the fact that quality must be designed and built into a product. Defect *prevention*, rather than defect detection, is the major goal. Total quality control is a systems approach in that both management and workers make an integrated effort to manufacture high-quality products consistently. All tasks concerning quality improvements and responsibilities in the organization are clearly identified. The TQC concept also requires 100 percent inspection of parts, usually by automated inspection systems. No defective parts are allowed to continue through the production line.

A related concept that has gained some acceptance is **the quality circle.** This activity consists of regular meetings by groups of workers who discuss how to improve and maintain product quality at all stages of the manufacturing process. Worker involvement and responsibility are emphasized. Comprehensive training is provided so that the worker can become capable of analyzing statistical data, identifying causes of poor quality, and taking immediate action to correct the situation. Putting this concept into practice recognizes the importance of quality assurance as a major company-wide management policy, affecting all personnel and all aspects of production.

### Quality engineering as a philosophy

Many of the quality-control concepts and methods that we have described have been put into larger perspective by certain experts in quality control. Notable among these experts are W. E. Deming, G. Taguchi, and J. M. Juran, whose philosophies of quality and product cost have had a major impact on modern manufacturing.

During World War II, Deming and several others developed new methods of statistical process control in manufacturing plants for wartime industry. The need for statistical control arose from the recognition that there were variations in the performance of machines and people and the quality and dimensions of raw materials. Their efforts, however, involved not only statistical methods of analysis, but a new way of looking at manufacturing operations to improve quality and lower costs. Recognizing the fact that manufacturing organizations are systems of management, workers, machines, and products, Deming emphasizes communication, direct worker involvement, and education in statistics, as well as in modern manufacturing technology. His ideas have been widely accepted in Japan, but only in some segments of U.S. industry.

In the Taguchi methods, high quality and low costs are achieved by combining engineering and statistical methods to optimize product design and manufacturing processes. Loss of quality is defined as the financial loss to society after the product is shipped, with the following results: (1) poor quality leads to customer dissatisfaction; (2) costs are involved in servicing and repairing defective products, some in the field; (3) the manufacturer's credibility is diminished in the marketplace; and (4) the manufacturer eventually loses its share of the market.

The Taguchi methods of quality engineering emphasize the importance of:

- Enhancing cross-functional team interaction. In this interaction, design engineers and process or manufacturing engineers communicate with each other in a common language. They quantify the relationships between design requirements and the manufacturing process.

- Implementing experimental design, in which the factors involved in a process or operation and their interactions are studied simultaneously.

In **experimental design,** the effects of controllable and uncontrollable variables on the product are identified. This approach minimizes variations in product dimensions and properties, bringing the mean to the desired level. The methods used for experimental design are complex, involving the use of fractional factorial design and orthogonal arrays, which reduce the number of experiments required. These methods are also capable of identifying the effect on the product of variables that cannot be controlled (called *noise*), such as changes in environmental conditions.

The use of these methods allows rapid identification of the controlling variables and determination of the best method of process control. These variables are then controlled, without the need for costly new equipment or major modifications to existing equipment. For example, variables affecting tolerances in machining a particular component can be readily identified, and the correct cutting speed, feed, cutting tool, and cutting fluids can be specified.

### 4.9.5 The ISO 9000 standard

With increasing international trade, global manufacturing, and price-sensitive competition, a wide range of industrial and consumer products is now becoming available. Also, customers are increasingly demanding high-quality products and services at low prices, and are looking for suppliers that can respond to this demand consistently and reliably. This strong trend, in turn, has created the need for international conformity and consensus regarding the establishment of quality-control methods and reliability and safety of the products made in different countries and traded internationally.

First issued in 1987 (with a 1994 revision), the ISO 9000 standard (*Quality Management and Quality Assurance Standards*) is a deliberately generic series of quality-system management standards. By 1993, more than 50 countries (industrial as well as developing) had already adopted it as a national industry standard. It has permanently influenced the way manufacturing companies conduct business in world trade and is rapidly becoming the world standard for quality. The ISO 9000 series includes the following standards:

- ISO 9001: Quality systems—Model for quality assurance in design/development, production, installation, and servicing
- ISO 9002: Quality systems—Model for quality assurance in production and installation
- ISO 9003: Quality systems—Model for quality assurance in final inspection and test
- ISO 9004: Quality management and quality system elements—Guidelines

Companies voluntarily register for these standards and are issued certificates. As of 1994, registration may be sought generally for ISO 9001 or 9002, although some companies have registration to ISO 9003. The 9004 standard is simply a guideline and not a model or basis for registration. For certification, a company's plants are visited and audited by accredited and independent third-party teams to certify that the standard's 20 key elements are in place and functioning properly. Depending on the extent to which a company does not meet the requirements of the standard, registration may or may not be recommended at that time. The audit team does not advise or consult with the company on how to fix discrepancies, but merely describes the nature of the noncompliance. Periodic audits are required to maintain certification. The certification process can take from six months to a year or more, and can cost tens of thousands of dollars, depending on the size, number of plants, and product line of the company.

The ISO 9000 standard is not a product certification but a *quality process certification*. Although the 1987 standard makes no reference to continuous quality improvement, the 1994 revision has a section on quality planning and implicitly addresses the need for improvement. Companies establish their own criteria and practices for quality. However, the documented quality system must be in compliance with the ISO 9000 standard; thus, a company cannot write into the system any criterion that opposes the intent of the standard.

Registration means conformity to consistent practices, as specified by the company's own quality system (such as quality in design, development, production, installation, and servicing), including proper documentation of such practice. Thus customers, including government agencies, are assured that specified practices are being followed by the supplier of the product or service (which may or may not be within the same country). In fact, manufacturing companies are themselves assured of such practice regarding their own suppliers who have ISO 9000 registration; they may even demand that their suppliers be registered.

## SUMMARY

- A component's surface affects its properties and characteristics, including friction and wear properties, the effectiveness of lubricants, appearance and geometric features, proclivity to cracking, and thermal and electrical conductivity of contacting bodies. (Section 4.1)
- The surface structure of metal lies above the bulk metal and has usually been plastically deformed and work hardened during prior processing. Surface integrity can be compromised by a variety of defects. Flaws, directionality, roughness, and waviness are measurable quantities used to describe surface texture. (Sections 4.2, 4.3)
- Tribology is the science and technology of interacting surfaces, and encompasses friction, wear, and lubrication. Friction may be desirable or undesirable, depending on the specific

manufacturing circumstances. The adhesion theory of friction is most commonly used to model and predict the frictional behavior between sliding surfaces, since it agrees reasonably well with experimental observations. (Section 4.4)

- Wear is defined as the progressive loss or removal of material from a surface. Wear changes the shape of the workpiece, tool, and die interfaces, and thus affects the process, size, and quality of parts produced. Friction and wear can be reduced by use of lubricants. Four regimes of lubrication are relevant to metalworking processes, and a wide variety of lubricants are available. (Section 4.4)

- Surface treatments are used to impart certain physical and mechanical properties, such as resistance to environmental attack and wear and fatigue resistance. Various techniques employed include mechanical working, heat treatment, and coatings. Cleaning involves removal of solid, semisolid, and liquid contaminants from a surface. (Section 4.5)

- Many parts are made with a high degree of precision, thus requiring measuring instrumentation with various features and characteristics. The selection of a particular measuring instrument depends on factors such as the type of measurement, the environment, and the accuracy of measurement. Great advances have been made in automated measurement, linking them to microprocessors and computers for accurate in-process control of manufacturing operations. (Section 4.6)

- Dimensional tolerances and their specification are important factors in manufacturing, as they not only affect subsequent assembly of parts made and the accuracy and operation of all types of machinery and equipment, but also can significantly influence product cost. (Section 4.7)

- Several nondestructive and destructive testing techniques, each having its own applications, advantages, and limitations, are available for inspection of completed parts and products. The traditional approach has been to inspect the part or product after it is manufactured and to accept a certain number of defective parts. The trend now is toward 100 percent inspection of all parts and products being manufactured, using automated inspection techniques. (Section 4.8)

- Quality assurance is the total effort by a manufacturer to ensure that its products conform to a detailed set of specifications and standards. For defect detection, a variety of statistical quality control and statistical process control methods are usually employed. The total quality control philosophy focuses on defect prevention, rather than defect detection. (Section 4.9)

## SUMMARY OF EQUATIONS

Arithmetic mean value, $R_a = a + b + c + d + \cdots / n$

Root-mean-square average, $R_q = \sqrt{a^2 + b^2 + c^2 + d^2 + \cdots} / n$

Coefficient of adhesion = Tensile force/compressive force applied

Coefficient of friction, $\mu = F/N = \tau/\text{hardness}$

Friction (shear) factor, $m = \tau_i / k$

Adhesive wear, $V = kLW/3p$

Pressure distribution in trapped fluid film, $p = 3\eta V(R - r)^2/t^3$

Arithmetic mean, $x = x_1 + x_2 + x_3 + \ldots + x_n/n$

Range $R = x_{max} - x_{min}$

Standard deviation, $\sigma = \sqrt{\dfrac{(x_1 - \bar{x})^2 + (x_2 - \bar{x})^2 + (x_3 - \bar{x})^2 + \cdots + (x_n - \bar{x})^2}{n - 1}}$

Upper control limit, $\mathrm{UCL}_{\bar{x}} = \bar{\bar{x}} + A_2\bar{R}$

Lower control limit, $\mathrm{LCL}_{\bar{x}} = \bar{\bar{x}} - A_2\bar{R}$

## BIBLIOGRAPHY

### Surfaces, friction, and wear

Arnell, R.D., P.B. Davies, J. Halling, and T.L. Whomes, *Tribology: Principles and Design Applications.* New York: Springer, 1991.

*ASM Handbook,* Vol. 5: *Surface Cleaning, Finishing, and Coating,* 1994; Vol. 18: *Friction, Lubrication, and Wear,* 1992. Materials Park, Ohio: ASM International.

Bayer, R.G., *Mechanical Wear Prediction and Prevention.* New York: Marcel Dekker, 1995.

Bhushan, B., and B. K. Gupta, *Handbook of Tribology: Materials, Coatings, and Surface Treatments.* New York: McGraw-Hill, 1991.

Blau, P.J., *Friction Science and Technology.* New York: Marcel Dekker, 1995.

Halling, J., *Principles of Tribology.* New York: Macmillan, 1975.

Lansdown, A.R., and A.L. Price, *Materials to Resist Wear: A Guide to Their Selection and Use.* Oxford: Pergamon Press, 1986.

Ludema, K.C., *Friction, Wear, Lubrication: A Textbook in Tribology.* Boca Raton, Fla.: CRC Press, 1996.

Moore, D.F., *Principles and Applications of Tribology.* New York: Pergamon, 1975.

Peterson, M.B., and W.O. Winer (eds.), *Wear Control Handbook.* New York: American Society of Mechanical Engineers, 1980.

Rabinowicz, E., *Friction and Wear of Materials,* 2d ed. New York: Wiley, 1995.

Sarkar, A.D., *Wear of Metals.* New York: Pergamon, 1976.

Scott, D. (ed.), *Wear.* Volume 13 in *Treatise on Materials Science and Technology.* New York: Academic Press, 1979.

Williams, J.A., *Engineering Tribology.* New York: Oxford University Press, 1994.

### Tribology in metalworking

Byers, J.P. (ed.), *Metalworking Fluids.* New York: Marcel Dekker, 1994.

Nachtman, E.S., and S. Kalpakjian, *Lubricants and Lubrication in Metalworking Operations.* New York: Marcel Dekker, 1985.

Schey, J.A., *Tribology in Metalworking: Friction, Lubricating and Wear.* Metals Park, Ohio: American Society for Metals, 1983.

### Surface treatments

*ASM Handbook,* Vol. 5: *Surface Engineering.* Materials Park, Ohio: ASM International, 1994.

Budinski, K.G., *Surface Engineering for Wear Resistance.* Englewood Cliffs, New Jersey: Prentice-Hall, 1988.

Rudzki, G.J., *Surface Finishing Systems: Metal and Non-Metal Finishing Handbook-Guide.* Middlesex, UK: Finishing Publications Ltd., 1983.

*Tool and Manufacturing Engineers Handbook,* 4th ed., Vol. 3: *Materials, Finishing and Coating.* Dearborn, Mich.: Society of Manufacturing Engineers, 1985.

## Metrology

Bosch, J.A., ed., *Coordinate Measuring Machines and Systems.* New York: Marcel Dekker, 1995.

Farrago, F.T., and M.A. Curtis, *Handbook of Dimensional Measurement,* 3d ed. New York: Industrial Press, 1994.

Meadows, J.D., *Geometric Dimensioning and Tolerancing.* New York: Marcel Dekker, 1995.

*Tool and Manufacturing Engineers Handbook,* 4th ed., Vol. 4: *Quality Control and Assembly.* Dearborn, Mich.: Society of Manufacturing Engineers, 1987.

Whitehouse, D.J., *Handbook of Surface Metrology.* U.K.: Institute of Physics, 1994.

## Testing, inspection, and quality assurance

*ASM Handbook,* Vol. 17: *Nondestructive Evaluation and Quality Control.* Materials Park, Ohio: ASM International, 1989.

Feigenbaum, A.V., *Total Quality Control,* 3d ed. New York: McGraw-Hill, 1991.

Juran, J.M., and F.M. Gryna, Jr., *Quality Planning and Analysis,* 3d ed. New York: McGraw-Hill, 1993.

Menon, H.G., *TQM in New Product Manufacturing.* New York: McGraw-Hill, 1992.

Robinson, S.L., and R.K. Miller, *Automated Inspection and Quality Assurance.* New York: Marcel Dekker, 1989.

*Tool and Manufacturing Engineering Handbook,* Vol. 4: *Assembly, Testing, and Quality Control.* Dearborn, Mich.: Society of Manufacturing Engineers, 1985.

Wadsworth, H.M., *Handbook of Statistical Control Methods for Engineers and Scientists.* New York: McGraw-Hill, 1990.

Wesner, J.W., J.M. Hiatt, and D.C. Trimble, *Winning With Quality: Applying Quality Principles in Product Development.* Reading, Mass.: Addison-Wesley, 1994.

# QUESTIONS

4.1 Explain what is meant by surface integrity. Why should we be interested in it?

4.2 Why are surface roughness design requirements in engineering so broad? Give appropriate examples.

4.3 We have seen that a surface has various layers. Describe the factors that influence the thickness of these layers.

4.4 What is the consequence of the hardness of oxides of metals being generally much harder than the base metal? Explain.

4.5 What factors would you consider in specifying the lay of a surface?

4.6 Describe the effects of various surface defects (see Section 4.3) on the performance of engineering components in service. How would you go about determining whether or not each of these defects is important for a particular application?

4.7 Explain why the same surface-roughness values do not necessarily represent the same type of surface.

4.8 In using a surface-roughness measuring instrument, how would you go about determining the cut-off value? Give appropriate examples.

4.9 What is the significance of the fact that the stylus path and the actual surface profile are not generally the same?

4.10 Give two examples each in which waviness of a surface would be (1) desirable, and (2) undesirable.

4.11   Explain the significance of surface temperature rise due to friction.

4.12   To what factors would you attribute the fact that the coefficient of friction in hot working is higher than in cold working, as shown in Table 4.1?

4.13   In Section 4.4.1, we note that the values of coefficient of friction can be much higher than unity. Explain why.

4.14   Describe the tribological differences between ordinary machine elements (such as meshing gears, cams in contact with followers, and ball bearings with inner and outer races) and metalworking processes (such as forging, rolling, and extrusion, that involve workpieces in contact with tools and dies).

4.15   Give the reasons why an originally round specimen in a ring-compression test may become oval after deformation.

4.16   Can the temperature rise at a sliding interface exceed the melting point of the metals? Explain.

4.17   Explain clearly the reasoning behind the adhesive-wear formula given by Eq. (4.9).

4.18   How can adhesive wear be reduced? Fatigue wear?

4.19   It has been stated that as the normal load decreases, abrasive wear is reduced. Explain why this is so.

4.20   Why is the abrasive-wear resistance of a material a function of its hardness?

4.21   We have seen that wear can have detrimental effects on engineering components, tools, dies, etc. Can you visualize situations in which wear could be beneficial? Give some examples. (*Hint:* Note that writing with a pencil is a wear process.)

4.22   On the basis of the topics discussed in this chapter, do you think there is a direct correlation between friction and wear of materials? Explain.

4.23   You have undoubtedly replaced parts in various appliances and automobiles because they were worn. Describe the methodology you would follow in determining the type(s) of wear these components have undergone.

4.24   Why is the study of lubrication regimes important?

4.25   Explain why so many different types of metalworking fluids have been developed.

4.26   Differentiate between (1) coolants and lubricants, (2) liquid and solid lubricants, (3) direct and indirect emulsions, and (4) plain and compounded oils.

4.27   Inspect Eq. (4.10) and give a qualitative explanation why three terms are in the numerator and the film thickness is in the denominator.

4.28   Explain why surface treatment of manufactured products may be necessary. Give several examples.

4.29   Which surface treatments are functional and which are decorative? Give examples.

4.30   Give some typical applications of mechanical surface treatment.

4.31   Explain the difference between case hardening and hard facing.

4.32   List several applications for coated sheet metal, including galvanized steel.

4.33   Explain how roller burnishing processes induce residual stresses on the surface of workpieces.

4.34   List several products or components that could not be made properly, or function effectively in service, without implementing the knowledge involved in Sections 4.2 through 4.5.

4.35   Explain the difference between direct- and indirect-reading linear measurements.

4.36   Why have coordinate measuring machines become important instruments in modern manufacturing? Give some examples.

4.37   Give reasons why the control of dimensional tolerances is important.

4.38   Give several engineering examples for the fits described in Section 4.7.

4.39   Explain why a measuring instrument may not have sufficient precision.

4.40   Comment on the differences, if any, between (1) roundness and circularity, (2) roundness and eccentricity, and (3) roundness and cylindricity.

4.41   It has been stated that tolerances for nonmetallic stock, such as plastics, are usually wider than for metals. Explain why. Consider physical and mechanical properties.

4.42 Describe the basic features of nondestructive testing techniques that use electrical energy.

4.43 Identify the nondestructive techniques that are capable of detecting internal flaws and those that detect external flaws only.

4.44 Which of the nondestructive inspection techniques are suitable for nonmetallic materials? Why?

4.45 Why is automated inspection becoming an important part of manufacturing engineering?

4.46 Describe situations in which the use of destructive testing techniques is unavoidable.

4.47 Should products be designed and built for a certain expected life? Explain.

4.48 What are the consequences of setting lower and upper specifications closer to the peak of the curve in Fig. 4.25?

4.49 Identify factors that can cause a process to become out of control. Give several examples.

## PROBLEMS

4.50 Referring to the surface profile in Fig. 4.3, give some numerical values to the vertical distances from the center line. Calculate the $R_a$ and $R_q$ values. Then give another set of values for the same general profile, and calculate the same quantities. Comment on your results.

4.51 Using Fig. 4.8a, make a plot of the coefficient of friction versus the change in internal diameter for a reduction in height of (1) 10, (2) 20, and (3) 50 percent.

4.52 In Example 4.1, assume that the coefficient of friction is 0.15. If all other parameters remain the same, what is the new internal diameter of the ring specimen?

4.53 How would you go about estimating forces required for roller burnishing? (*Hint:* Consider hardness testing.)

4.54 Assume that a steel rule expands by 1 percent because of an increase in environmental temperature. What will be the indicated diameter of a shaft whose actual diameter is 50.00 mm?

4.55 Assume that in Example 4.3, the number of samples was 5 instead of 10. Using the top half of the data given in the table, recalculate control limits and the standard deviation. Compare your re-

sults with the results obtained by using 10 samples.

4.56 Calculate the control limits for averages and ranges for the following: number of samples $= 4, \bar{\bar{x}} = 70, \bar{R} = 7$.

4.57 Calculate the control limits for the following: number of samples $= 3, \bar{\bar{x}} = 36.5, \text{UCL}_R = 4.75$.

4.58 In an inspection with sample size 7 and a sample number of 50, it was found that the average range was 12 and the average of averages was 75. Calculate the control limits for averages and ranges.

4.59 Determine the control limits for the data shown in the table below.

| $x_1$ | $x_2$ | $x_3$ | $x_4$ |
|-------|-------|-------|-------|
| 0.55 | 0.60 | 0.57 | 0.55 |
| 0.59 | 0.55 | 0.60 | 0.58 |
| 0.55 | 0.50 | 0.55 | 0.51 |
| 0.54 | 0.57 | 0.50 | 0.50 |
| 0.58 | 0.58 | 0.60 | 0.56 |
| 0.60 | 0.61 | 0.55 | 0.61 |

4.60  The average of averages of a number of samples of size 9 was determined to be 125. The average range was 17.82 and the standard deviation was 6. The following measurements were taken in a sample: 120, 132, 124, 130, 118, 132, 121, and 127. Is the process in control?

4.61  Assume that you are asked to give a quiz to students on the contents of this chapter. Prepare three quantitative problems and three qualitative questions, and supply the answers.

# 5

## Metal-Casting Processes and Equipment; Heat Treatment

**5.1** Introduction
**5.2** Solidification of Metals
**5.3** Cast Structures
**5.4** Fluid Flow and Heat Transfer
**5.5** Melting Practice and Furnaces
**5.6** Casting Alloys
**5.7** Ingot Casting and Continuous Casting
**5.8** Casting Processes: Expendable Mold
**5.9** Casting Processes: Permanent Mold
**5.10** Processing of Castings; Heat Treatment
**5.11** Casting Design
**5.12** Economics of Casting

## 5.1  Introduction

We can use several methods to shape materials into useful products. Making parts by **casting** molten metal into a mold and letting it solidify is an obvious choice. Indeed, casting is among the oldest methods of net-shape and near-net-shape manufacturing and was first used in about 4000 B.C. to make ornaments, copper arrowheads, and various other objects.

Basically, metal-casting processes involve the introduction of molten metal into a mold cavity where, upon solidification, the metal conforms to the shape of the cavity. The casting process is thus capable of producing intricate shapes in a single piece, including those with internal cavities. Very large, very small, and hollow parts can be produced economically by casting techniques. Typical cast products are engine blocks, cylinder heads, transmission housings, pistons, turbine disks, railroad and automotive wheels, and ornamental artifacts.

Almost all metals can be cast in (or nearly in) the final shape desired, often with only minor finishing required. With appropriate control of material and process parameters, parts can be cast with uniform properties throughout. As with all other manufacturing processes, certain fundamental relationships are essential to the production of good quality and economical castings. Knowledge of these relationships helps us establish proper techniques for mold design and casting practice. The objective is to produce castings that are free from defects and meet requirements for strength, dimensional accuracy, and surface finish.

Fundamentally, the important factors in casting operations are

- Solidification of the metal from its molten state, and accompanying shrinkage (volume change)
- Flow of the molten metal into the mold cavity

- Heat transfer during solidification and cooling of the metal in the mold
- Influence of the type of mold material

We first discuss these aspects of casting, then continue with the description of melting practice and the characteristics of casting alloys. The remainder of the chapter covers ingot and continuous casting, various expendable-mold and permanent-mold casting processes, including their advantages and limitations, processing of castings, heat treatment, casting design, and the economics of casting.

## 5.2  Solidification of Metals

In this section we present an overview of the solidification of metals and alloys. We cover topics that are particularly relevant to casting and essential for our understanding of the structures developed in casting and the structure–property relationships obtained in the casting processes described in this chapter.

**Pure** metals have clearly defined melting or freezing points, and solidification takes place at a constant temperature (Fig. 5.1). When the temperature of the molten metal is reduced to the freezing point, the latent heat of fusion is given off while the temperature remains constant. At the end of this isothermal phase change, solidification is complete and the solid metal cools to room temperature.

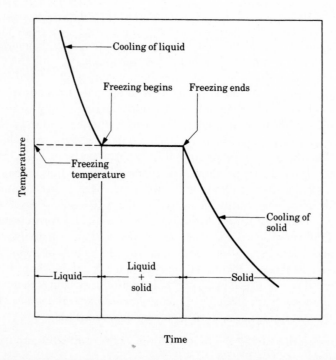

**FIGURE 5.1**  Cooling curve for the solidification of pure metals. Note that freezing takes place at a constant temperature.

Unlike pure metals, **alloys** solidify over a range of temperatures. Solidification begins when the temperature of the molten metal drops below the **liquidus;** it is completed when the temperature reaches the **solidus.** Within this temperature range the alloy is in a mushy or pasty state. Its composition and state are described by the particular alloy's phase diagram.

### 5.2.1 Solid solutions

Two terms are essential in describing alloys: solute and solvent. **Solute** is the minor element (such as salt or sugar) that is added to the **solvent,** which is the major element (such as water). In terms of the elements involved in a metal's crystal structure, the solute is the element (solute atoms) added to the solvent (host atoms). When the particular crystal structure of the solvent is maintained during alloying, the alloy is called a **solid solution.**

**Substitutional solid solutions.**   If the size of the solute atom is similar to that of the solvent atom, the solute atoms can replace solvent atoms and form a **substitutional solid solution** (see Fig. 3.8). An example of this phenomenon is brass, an alloy of zinc and copper, in which zinc (solute atom) is introduced into the lattice of copper (solvent atoms). The properties of brasses can thus be altered over a certain range by controlling the amount of zinc in copper.

**Interstitial solid solutions.**   If the size of the solute atom is much smaller than that of the solvent atom, the solute atom occupies an interstitial position and forms an **interstitial solid solution.** An important example of interstitial solution is steel, an alloy of iron and carbon, in which carbon atoms are present in an interstitial position between iron atoms. As we show in Section 5.10, we can vary the properties of steel over a wide range by controlling the amount of carbon in iron. This is one reason that, in addition to being inexpensive, steel is such a versatile and useful material with a wide variety of properties and applications.

### 5.2.2 Intermetallic compounds

**Intermetallic compounds** are complex structures in which solute atoms are present among solvent atoms in certain specific proportions. Thus some intermetallic compounds have solid solubility. The type of atomic bonds may range from metallic to ionic. Intermetallic compounds are strong, hard, and brittle.

### 5.2.3 Two-phase alloys

A solid solution is one in which two or more elements are soluble in a solid state, forming a single homogeneous solid phase in which the alloying elements are uniformly distributed throughout the solid mass. However, there is a limit to the concentration of solute atoms in a solvent-atom lattice, just as there is a solubility limit to sugar in water. Most alloys consist of two or more solid phases and may be regarded as mechanical mixtures. Thus we call a system with two solid phases a **two-phase system.** Each phase is a homogeneous part of the total mass and has its own characteristics and properties.

A typical example of a two-phase system in metals is lead added to copper in the molten state. After the mixture solidifies, the structure consists of two phases: One phase has a small

(a)

(b)

FIGURE 5.2   (a) Schematic illustration of grains, grain boundaries, and particles dispersed throughout the structure of a two-phase system, such as lead–copper alloy. The grains represent lead in solid solution of copper, and the particles are lead as a second phase.
(b) Schematic illustration of a two-phase system consisting of two sets of grains: dark and light. Dark and light grains have their own compositions and properties, respectively.

amount of lead in solid solution in copper; in the other phase, lead particles, roughly spherical in shape, are *dispersed* throughout the matrix of the primary phase (Fig. 5.2a). This copper–lead alloy has properties that are different from those of either copper or lead alone. Alloying with finely dispersed particles (*second-phase particles*) is an important method of strengthening alloys and controlling their properties. Generally, in two-phase alloys the second-phase particles present obstacles to dislocation movement, which increases the alloy's strength.

Another example of a two-phase alloy is the aggregate structure shown in Fig. 5.2b. This alloy system contains two sets of grains, each with its own composition and properties. The darker grains may, for example, have a different structure than the lighter grains and be brittle, whereas the lighter grains may be ductile.

### 5.2.4   Phase diagrams

A **phase diagram,** also called an *equilibrium diagram* or a *constitutional diagram*, graphically illustrates the relationships among temperature, composition, and the phases present in a particular alloy system. *Equilibrium* means that the state of a system remains constant over an indefinite period of time. *Constitutional* indicates the relationships among structure, composition, and physical makeup of the alloy.

One example of a phase diagram is shown in Fig. 5.3 for the nickel–copper alloy. It is called a **binary phase diagram** because of the two elements (nickel and copper) in the system. The left boundary of this phase diagram (100% Ni) indicates the melting point of nickel, and the right boundary (100% Cu) the melting point of copper. (All percentages in this discussion are by *weight*.)

Note that for a composition of, say, 50% Cu–50% Ni, the alloy begins to solidify at a temperature of 1313°C (2395°F), and solidification is complete at 1249°C (2280°F). Above 1313°C, a homogeneous liquid of 50% Cu–50% Ni exists. When cooled slowly to 1249°C, a homogeneous solid solution of 50% Cu–50% Ni results. However, between the liquidus and solidus curves, say, at a temperature of 1288°C (2350°F), is a two-phase region: a *solid phase* composed of 42% Cu–58% Ni, and a *liquid phase* of 58% Cu–42% Ni. To determine the solid composition, go left horizontally to the solidus curve and read down

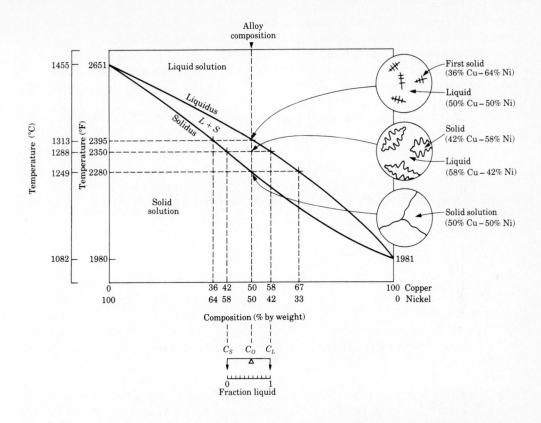

**FIGURE 5.3** Phase diagram for nickel–copper alloy system obtained by slow rate of solidification. Note that pure nickel and pure copper each have one freezing or melting temperature. The top circle on the right depicts the nucleation of crystals. The second circle shows the formation of dendrites. The bottom circle shows the solidified alloy with grain boundaries. See the text for an explanation of the lever rule illustrated in the bottom portion of the figure.

to obtain 42% Cu–58% Ni. Obtain the liquid composition similarly by going to the right to the liquidus curve (58% Cu–42% Ni).

The completely solidified alloy in the phase diagram shown in Fig. 5.3 is a solid solution, because the alloying element (Cu, solute atom) is completely dissolved in the host metal (Ni, solvent atom), and each grain has the same composition. The mechanical properties of solid solutions of Cu–Ni depend on their composition. By increasing the nickel content, the properties of pure copper are improved. The improvements in properties result from pinning (blocking) of dislocations at solute atoms of nickel, which may also be regarded as impurity atoms. As a result, dislocations cannot move as freely, and consequently the strength of the alloy increases.

### Lever rule

The composition of various phases in a phase diagram can be determined by a procedure called the **lever rule**. As shown in the lower portion of Fig. 5.3, we first construct a lever between the solidus and liquidus lines (*tie line*), which is balanced (on the triangle support) at the nominal weight composition $C_0$ of the alloy. The left end of the lever represents the composition $C_S$ of the solid phase and the right end of the composition $C_L$ of the liquid phase. Note from the graduated scale in the figure that we also have indicated the liquid fraction along this tie line, ranging from 0 at the left (fully solid) to 1 at the right (fully liquid).

The lever rule states that the weight fraction of solid is proportional to the distance between $C_0$ and $C_L$:

$$\frac{S}{S + L} = \frac{C_0 - C_L}{C_S - C_L}. \tag{5.1}$$

Likewise, the weight fraction of liquid is proportional to the distance between $C_S$ and $C_0$, hence

$$\frac{L}{S + L} = \frac{C_S - C_0}{C_S - C_L}. \tag{5.2}$$

Note that these quantities are fractions and must be multiplied by 100 to obtain percentages.

From inspection of the tie line in Fig. 5.3—and for a nominal alloy composition of $C_0 = 50\%$ Cu–50% Ni—we see that, because $C_0$ is closer to $C_L$ than it is to $C_S$, the solid phase contains less copper than the liquid phase. Measuring on the phase diagram and using the lever-rule equations, we find that the composition of the solid phase is 42% Cu and of the liquid phase is 58% Cu, as stated in the middle circle at the right in Fig. 5.3.

Note that these calculations refer to copper. If we now reverse the phase diagram in the figure, so that the left boundary is 0% nickel (whereby nickel becomes the alloying element in copper), these calculations give us the compositions of the solid and liquid phases in terms of nickel. The lever rule is also known as the *inverse lever rule* because, as the foregoing equations show, the amount of each phase is proportional to the length of the opposite end of the lever.

### 5.2.5  The iron–carbon system

Steel and cast iron are represented by the *iron–carbon binary system*. Commercially pure iron contains up to 0.008 percent carbon, steels up to 2.11 percent carbon, and cast irons up to 6.67 percent carbon, although most cast irons contain less than 4.5 percent carbon. In this section we discuss the iron–carbon system in order to evaluate and modify the properties of these important materials for specific applications.

The **iron–iron carbide phase diagram** is shown in Fig. 5.4. Although this diagram can be extended to the right—to 100 percent carbon (pure graphite)—the range that is significant to engineering applications is up to 6.67 percent carbon, where cementite forms. Pure iron melts at a temperature of 1538°C (2800°F), as shown at the left in Fig. 5.4. As iron cools, it first forms δ-iron, then γ-iron, and finally α-iron.

**Ferrite.**    Delta ferrite is stable only at very high temperatures and has no practical engineering significance. Alpha ferrite, or simply **ferrite**, is a solid solution of body-centered cubic iron and has a maximum solid solubility of 0.022 percent carbon at a temperature of 727°C (1341°F). Just as there is a solubility limit to the salt in water—with any extra

**FIGURE 5.4**  The iron–iron carbide phase diagram. Because of the importance of steel as an engineering material, this diagram is one of the most important phase diagrams.

amount precipitating as solid salt at the bottom of the container—so there also is a solid solubility limit to carbon in iron.

Ferrite is relatively soft and ductile and is magnetic from room temperature to 768°C (1414°F). Although very little carbon can dissolve interstitially in bcc iron, the amount of carbon can significantly affect the mechanical properties of ferrite. Also, significant amounts of chromium, manganese, nickel, molybdenum, tungsten, and silicon can be contained in iron in solid solution, imparting certain desirable properties.

**Austenite.**  Between 1394°C (2541°F) and 912°C (1674°F) iron undergoes an **allotrophic transformation** from the bcc to fcc structure, becoming what is known as gamma iron, or more commonly, **austenite.** This structure has a solid solubility of up to 2.11 percent carbon at 1148°C (2098°F). Thus the solid solubility of austenite is about two orders of magnitude higher than that of ferrite, with the carbon occupying interstitial positions (Fig. 5.5). Note that the atomic radius of Fe is 0.124 nm and for C 0.071 nm.

Austenite is an important phase in the heat treatment of steels (Section 5.10). It is denser than ferrite, and its single-phase fcc structure is ductile at elevated temperatures; thus it possesses good formability. Large amounts of nickel and manganese can also be dissolved in fcc iron to impart various properties. Austenitic steel is nonmagnetic at high temperatures; austenitic stainless steels are nonmagnetic at room temperature.

**Cementite.**  The right boundary of Fig. 5.4 represents **cementite**, also called **carbide**, which is 100 percent iron carbide ($Fe_3C$), with the carbon content of 6.67 percent. (This carbide should not be confused with carbides used for tool and die materials.) Cementite is a

**FIGURE 5.5**   The unit cell for (a) austenite, (b) ferrite, and (c) martensite. The effect of percentage of carbon (by weight) on the lattice dimensions for martensite is shown in (d). Note the interstitial position of the carbon atoms (see Fig. 3.8) and the increase in dimension $c$ with increasing carbon content. Thus the unit cell of martensite is in the shape of a rectangular prism.

very hard and brittle intermetallic compound and significantly influences the properties of steels. It can be alloyed with elements such as chromium, molybdenum, and manganese.

### 5.2.6   The iron–iron carbide phase diagram

The region of the iron–iron carbide phase diagram that is significant for steels is shown in Fig. 5.6, an enlargement of the lower left portion of Fig. 5.4. Various microstructures can be developed in steels, depending on the carbon content and the method of heat treatment. For example, let's consider iron with a 0.77 percent carbon content being cooled very slowly from a temperature of, say, 1100°C (2012°F) in the austenite phase. The reason for the *slow* cooling rate is to maintain equilibrium; higher rates of cooling are used in heat treating. At 727°C (1341°F) a reaction takes place in which austenite is transformed into alpha ferrite (bcc) and cementite. As the solid solubility of carbon in ferrite is only 0.022 percent, the extra carbon forms cementite.

This reaction is called a **eutectoid** (meaning *eutecticlike*) **reaction**, indicating that at a certain temperature a single solid phase (austenite) is transformed into two other solid phases (ferrite and cementite). The structure of the eutectoid steel is called **pearlite,** because it resembles mother of pearl at low magnifications. The microstructure of pearlite consists of alternate layers (*lamellae*) of ferrite and cementite (Fig. 5.6). Consequently, the mechanical properties of pearlite are intermediate between ferrite (soft and ductile) and cementite (hard and brittle).

In iron with less than 0.77 percent carbon, the microstructure formed consists of a pearlite phase (ferrite and cementite) and a ferrite phase. The ferrite in the pearlite is called *eutectoid ferrite.* The ferrite phase is called *proeutectoid ferrite, pro* meaning before, because it forms at a temperature higher than the eutectoid temperature of 727°C (1341°F). If the carbon content is greater than 0.77 percent, the austenite transforms into pearlite and cementite. The cementite in the pearlite is called *eutectoid cementite,* and the cementite phase

**FIGURE 5.6** Schematic illustration of the microstructures for an iron–carbon alloy of eutectoid composition (0.77 percent carbon) above and below the eutectoid temperature of 727°C (1341°F).

is called *proeutectoid cementite,* because it forms at a temperature higher than the eutectoid temperature.

**Effects of alloying elements in iron.** Although carbon is the basic element that transforms iron into steel, other elements are also added to impart various desirable properties. The effect of these alloying elements on the iron–iron carbide phase diagram is to shift the eutectoid temperature and eutectoid composition (the percentage of carbon in steel at the eutectoid point).

The eutectoid temperature may be raised or lowered from 727°C (1341°F), depending on the particular alloying element. However, alloying elements always lower the eutectoid composition; that is, the carbon content becomes less than 0.77 percent. Lowering the eutectoid temperature means increasing the austenite range. Thus an alloying element such as nickel is known as an *austenite former*. Because nickel has an fcc structure, it probably favors the fcc structure of austenite. Conversely, chromium and molybdenum have the bcc structure, causing these elements to favor the bcc structure of ferrite; these elements are known as *ferrite formers*.

### 5.2.7 Cast irons

The term **cast iron** refers to a family of ferrous alloys composed of iron, carbon (ranging from 2.11 percent to about 4.5 percent), and silicon (up to about 3.5 percent). Cast irons (Section 5.6.1) are usually classified according to their solidification morphology:

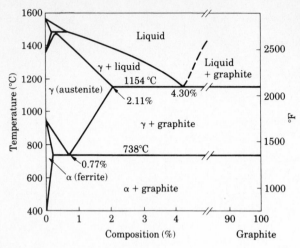

**FIGURE 5.7** Phase diagram for the iron–carbon system with graphite, instead of cementite, as the stable phase. Note that this figure is an extended version of Fig. 5.4.

1. Gray cast iron or gray iron
2. Nodular cast iron, ductile cast iron, or spheroidal graphite cast iron
3. White cast iron
4. Malleable iron
5. Compacted graphite iron

Cast irons are also classified by their structure: ferrite, pearlite, quenched and tempered, or austempered.

The equilibrium diagram relevant to cast irons is shown in Fig. 5.7, in which the right boundary is 100% C, that is, pure graphite. The horizontal liquid + graphite line is at 1154°C (2109°F). Thus cast irons are completely liquid at temperatures lower than those required for liquid steels. Consequently, cast irons have lower melting temperatures, which is why the casting process is so suitable for iron with high carbon content.

Although cementite exists in steels almost indefinitely, it is not completely stable. That is, it is *metastable*, with an extremely low rate of decomposition. However, cementite can be made to decompose into alpha ferrite and graphite. The formation of graphite (*graphitization*) can be controlled, promoted, and accelerated by modifying the composition and the rate of cooling, and by the addition of silicon.

---

### Example 5.1: Determining the amount of phases

Determine the amount of gamma and alpha phases in a 10-kg, 1040 steel casting as it is being cooled slowly to the following temperatures: (a) 900°C, (b) 728°C, and (c) 726°C.

SOLUTION:   (a) Referring to Fig. 5.6, we draw a vertical line at 0.40% C at 900°C. We are in the single-phase austenite region, so percent gamma is 100 (10 kg) and percent alpha is 0. (b) At 728°C, the alloy is in the two-phase gamma–alpha field. When we draw the phase diagram in greater detail, we can find the weight percentages of each phase by the lever rule:

$$\text{Percent alpha} = \left(\frac{C_\gamma - C_0}{C_\gamma - C_\alpha}\right)100 = \left(\frac{0.77 - 0.40}{0.77 - 0.022}\right)100 = 50\%, \text{ or } 5 \text{ kg};$$

$$\text{Percent gamma} = \left(\frac{C_0 - C_\alpha}{C_\gamma - C_\alpha}\right)100 = \left(\frac{0.40 - 0.022}{0.77 - 0.022}\right)100 = 50\%, \text{ or } 5 \text{ kg}.$$

(c) At 726°C, the alloy will be in the two-phase alpha and $Fe_3C$ region. No gamma phase will be present. Again, we use the lever rule to find the amount of alpha present:

$$\text{Percent alpha} = \left(\frac{6.67 - 0.40}{6.67 - 0.022}\right)100 = 94\%, \text{ or } 9.4 \text{ kg}.$$

## 5.3  Cast Structures

The type of *cast structure* developed during solidification of metals and alloys depends not only on the composition of the particular alloy, but also on the rate of heat transfer and the flow of the liquid metal during the casting process. As we describe in other sections in this chapter, the structures developed, in turn, affect the properties of the finished castings.

### 5.3.1  Pure metals

The grain structure of a pure metal solidified in a square mold is shown in Fig. 5.8a. At the mold walls the metal cools rapidly, because the walls are at ambient temperature. As a result of cooling first at the mold walls, the casting develops a solidified *skin*, or *shell*, or fine equiaxed grains. The grains grow in the direction opposite to the heat transfer from the mold. Grains that have a favorable orientation, called *columnar grains,* grow preferentially (Fig. 5.9). Grains that have substantially different orientations are blocked from further growth.

### 5.3.2  Alloys

Pure metals have limited mechanical properties. However, their properties can be enhanced and modified by **alloying.** The vast majority of metals used in engineering applications are some form of an **alloy,** consisting of two or more chemical elements, at least one of which is a metal. Because many consumer and industrial products contain cast alloys, a study of their solidification is essential.

Solidification in alloys begins when the temperature drops below the liquidus, $T_L$, and is complete when it reaches the solidus, $T_S$ (Fig. 5.10). Within this temperature range the alloy is in a mushy or pasty state with **columnar dendrites** (from the Greek *dendron* meaning akin to, and *drys* meaning tree). Note the presence of liquid metal between the dendrite arms.

(a) Chill zone — Columnar zone   (b) Equiaxed zone   (c) Equiaxed structure

**FIGURE 5.8** Schematic illustration of three cast structures of metals solidified in a square mold: (a) pure metals; (b) solid-solution alloys; and (c) structure obtained by heterogenous nucleation of grains using nucleating agents. *Source:* G. W. Form, J. F. Wallace, J. L. Walker, and A. Cibula.

Dendrites have three-dimensional arms and branches (secondary arms) and they eventually interlock, as shown in Fig. 5.11.

The width of the mushy zone, where both liquid and solid phases are present, is an important factor during solidification. We describe this zone in terms of a temperature difference, known as the **freezing range,** as follows:

$$\text{Freezing range} = T_L - T_S. \tag{5.3}$$

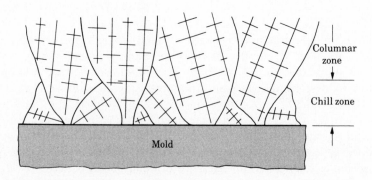

Columnar zone

Chill zone

Mold

**FIGURE 5.9** Development of a preferred texture at a cool mold wall. Note that only favorable oriented grains grow away from the surface of the mold.

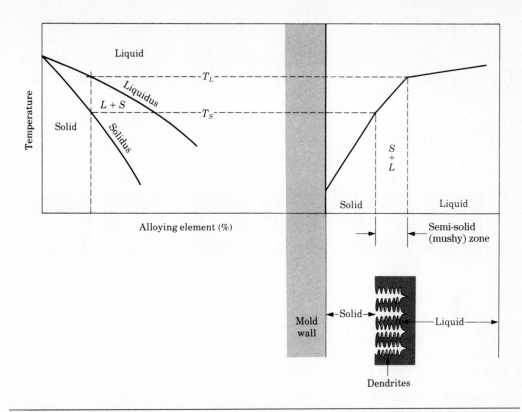

**FIGURE 5.10** Schematic illustration of alloy solidification and temperature distribution in the solidifying metal. Note the formation of dendrites in the semi-solid (mushy) zone.

Figure 5.10 shows that pure metals have a freezing range that approaches zero and that the solidification front moves as a plane front, without forming a mushy zone. Eutectics solidify in a similar manner with an approximately plane front. The type of solidification structure developed depends on the composition of the eutectic. For alloys with a nearly symmetrical phase diagram, the structure is generally lamellar with two or more solid phases present, depending on the alloy system. When the volume fraction of the minor phase of the alloy is less than about 25 percent, the structure generally becomes fibrous. These conditions are particularly important for cast irons.

For alloys, although it is not precise, a short freezing range generally involves a temperature difference of less than $50°C$ ($90°F$), and a long freezing range greater than $110°C$ ($200°F$). Ferrous castings generally have narrow semi-solid (mushy) zones, whereas aluminum and magnesium alloys have wide mushy zones. Consequently, these alloys are in a semi-solid state throughout most of the solidification process.

**Effects of cooling rates.** Slow cooling rates—on the order of $10^2$ K/s—or long local solidification times result in coarse dendritic structures with large spacing between the dendrite arms. For faster cooling rates—on the order of $10^4$ K/s—or short local solidification times, the structure becomes finer with smaller dendrite arm spacing. For still faster cooling rates—on the order of $10^6$ to $10^8$ K/s—the structures developed are *amorphous* (without any

**FIGURE 5.11**    (a) Solidification patterns for gray cast iron in a 180-mm (7-in.)-square casting. Note that after 11 min of cooling, dendrites reach each other, but the casting is still mushy throughout. It takes about two hours for this casting to solidify completely. (b) Solidification of carbon steels in sand and chill (metal) molds. Note the difference in solidification patterns as the carbon content increases. *Source:* H. F. Bishop and W. S. Pellini.

ordered crystalline structure), as described in Section 5.9.8. The structures developed and the resulting grain size, in turn, influence the properties of the casting. As grain size decreases, (1) the strength and ductility of the cast alloy increase (see Hall–Petch equation, Section 3.4.1), (2) microporosity (interdendritic shrinkage voids) in the casting decreases, and (3) the tendency for the casting to crack (*hot tearing*) during solidification decreases. Lack of uniformity in grain size and distribution results in castings with anisotropic properties.

### 5.3.3    Structure–property relationships

Because all castings are expected to possess certain properties to meet design and service requirements, the relationships between these properties and the structures developed during solidification are important aspects of casting. In this section, we describe these relationships in terms of dendrite morphology and the concentration of alloying elements in various regions.

The compositions of dendrites and the liquid metal are given by the phase diagram of the particular alloy. When the alloy is cooled very slowly, each dendrite develops a uniform composition. Under normal cooling encountered in practice, however, *cored dendrites* are formed and have a surface composition different from that at their centers (concentration gradient). The surface has a higher concentration of alloying elements than at the core of the dendrite owing to solute rejection from the core toward the surface during solidification of the dendrite. This phenomenon is known as *microsegregation*. The darker shading in the interdendritic liquid near the dendrite roots in Fig. 5.12 indicates that these regions have a higher solute concentration. Thus microsegregation in these regions is much more pronounced than in others.

In contrast to microsegregation, *macrosegregation* involves differences in composition throughout the casting itself. In situations where the solidifying front moves away from the

**FIGURE 5.12**  Schematic illustration of three basic types of cast structures: (a) columnar dendritic; (b) equiaxed dendritic; and (c) equaxed nondendritic. *Source:* After D. Apelian.

surface of a casting as a plane front (Fig. 5.13), lower melting point constituents in the solidifying alloy are driven toward the center (*normal segregation*). Consequently, such a casting has a higher concentration of alloying elements at its center than at its surfaces. In dendritic structures such as those for solid-solution alloys (Fig. 5.8b) the opposite occurs: The center of the casting has lower concentration of alloying elements (*inverse segregation*). The reason is that liquid metal (having a higher concentration of alloying elements) enters the cavities developed from solidification shrinkage in the dendrite arms, which have solidified sooner. Another form of segregation is the result of gravity (*gravity segregation*), whereby higher density inclusions or compounds sink, and lighter elements (such as antimony in an antimony–lead alloy) float to the surface.

A typical cast structure of a solid-solution alloy with an inner zone of equiaxed grains is shown in Fig. 5.8b. This inner zone can be extended throughout the casting, as shown in Fig. 5.8c, by adding an *inoculant* (nucleating agent) to the alloy. The inoculant induces nucleation of grains throughout the liquid metal (*heterogeneous nucleation*). An example is $TiB_2$ in AlSi alloys.

Because of the presence of thermal gradients in a solidifying mass of liquid metal and because of gravity (hence density differences), convection has a strong influence on the structures developed. Convection promotes the formation of an outer chill zone, refines grain size, and accelerates the transition from columnar to equiaxed grains. The structure shown in Fig. 5.12b can also be obtained by increasing convection within the liquid metal,

**FIGURE 5.13**  Schematic illustration of cast structures in (a) plane front, single phase, and (b) plane front, two phase. *Soure:* After D. Apelian.

whereby dendrite arms separate (*dendrite multiplication*). Conversely, reducing or eliminating convection results in coarser and longer columnar dendritic grains. The dendrite arms are not particularly strong, and they can be broken up by agitation or mechanical vibration in the early stages of solidification (**rheocasting**). This results in finer grain size, with equiaxed nondendritic grains distributed more uniformly throughout the casting (Fig. 5.12c). Convection can be enhanced by the use of mechanical or electromagnetic methods. Experiments are now being conducted during space flights concerning the effects of gravity on the microstructure of castings.

## 5.4 Fluid Flow and Heat Transfer

### 5.4.1 Fluid flow

To emphasize the importance of fluid flow, let's briefly describe a basic gravity casting system, as shown in Fig. 5.14. The molten metal is poured through a **pouring basin** or *cup*. It then flows through the **sprue** and **runners** into the mold cavity. **Risers** serve as reservoirs of molten metal to supply the metal necessary to prevent shrinkage (porosity). Although such a **gating system** appears to be relatively simple, successful casting requires careful design and control of the solidification process to ensure adequate fluid flow during solidification in the system. For example, one of the most important functions of the gating system is to trap contaminants (such as oxides and other inclusions) in the molten metal by having the contaminants adhere to the walls of the gating system, thereby preventing their reaching the actual mold cavity. Furthermore, the proper design of the gating system avoids or minimizes problems such as premature cooling, turbulence, and gas entrapment. Even before it reaches the mold cavity, the molten metal must be handled carefully to avoid forming oxides on molten metal surfaces from exposure to the environment or introduction of impurities into the molten metal.

Two basic principles of fluid flow are relevant to gating design: Bernoulli's theorem and the law of mass continuity.

**FIGURE 5.14** Schematic illustration of a sand mold showing various features.

**Bernoulli's theorem.** Bernoulli's theorem is based on the principle of conservation of energy and relates pressure, velocity, elevation of the fluid at any location in the system, and the frictional losses in a system that is full of liquid:

$$h + \frac{p}{\rho g} + \frac{v^2}{2g} = \text{constant},$$  (5.4)

where $h$ is the elevation above a certain reference plane, $p$ is the pressure at that elevation, $v$ is the velocity of the liquid at that elevation, $\rho$ is the density of the fluid (assuming that it is incompressible), and $g$ is the gravitational constant. Conservation of energy requires that, at a particular location in the system, the following relationship be satisfied:

$$h_1 + \frac{p_1}{\rho g} + \frac{v_1^2}{2g} = h_2 + \frac{p_2}{\rho g} + \frac{v_2^2}{2g} + f,$$  (5.5)

where the subscripts 1 and 2 represent two different elevations, respectively, and $f$ represents the frictional loss in the liquid as it travels downward through the system. The frictional loss includes such factors as energy loss at the liquid–mold wall interfaces and turbulence in the liquid.

**Continuity.**  The continuity law states that, for incompressible liquids and in a system with impermeable walls, the rate of flow is constant:

$$Q = A_1 v_1 = A_2 v_2,$$  (5.6)

where $Q$ is the volumetric rate of flow, such as $m^3/s$, $A$ is the cross-sectional area of the liquid stream, and $v$ is the velocity of the liquid in that cross-sectional location. The subscripts 1 and 2 pertain to two different locations in the system. Thus the flow rate must be maintained anywhere in the system. The permeability of the walls of the system is important, because otherwise some liquid will permeate through (for example, in sand molds) and the flow rate will decrease as the liquid moves through the system. Coatings are often used to inhibit such behavior in sand molds.

**Application.**  An application of the two principles stated is the traditional tapered design of sprues (Fig. 5.14). We can determine the shape of the sprue by using Eqs. (5.5) and (5.6). Assuming that the pressure at the top of the sprue is equal to the pressure at the bottom and that there are no frictional losses, at any point in the sprue the relationship between height and cross-sectional area is given by the parabolic relationship:

$$\frac{A_1}{A_2} = \sqrt{\frac{h_2}{h_1}},$$  (5.7)

where, for example, the subscript 1 denotes the top of the sprue and 2 the bottom. Moving downward from the top, the cross-sectional area of the sprue must decrease. Recall that in a free-falling liquid, such as water from a faucet, the cross-sectional area of the stream decreases as it gains velocity downward. If we design a sprue with a constant cross-sectional area and pour the molten metal, regions may develop where the liquid loses contact with the sprue walls. As a result, *aspiration* may take place whereby air will be sucked in or entrapped in the liquid. On the other hand, in many systems tapered sprues are now replaced by straight-sided sprues with a *choking* mechanism at the bottom, consisting of either a choke core or a runner choke (as shown in Fig. 5.14).

### Flow characteristics

An important consideration in fluid flow in gating systems is the presence of *turbulence*, as opposed to *laminar flow* of fluids. We use the *Reynolds number* to characterize this aspect of fluid flow. The Reynolds number, Re, represents the ratio of the inertia to the viscous forces in fluid flow, and is defined as

$$\text{Re} = \frac{vD\rho}{\eta},\qquad(5.8)$$

where $v$ is the velocity of the liquid, $D$ is the diameter of the channel, and $\rho$ and $\eta$ are the density and viscosity, respectively, of the liquid. The higher this number, the greater is the tendency for turbulent flow. In ordinary gating systems Re ranges from 2000 to 20,000. An Re value of up to 2000 represents laminar flow; between 2000 and 20,000 it is a mixture of laminar and turbulent flow and is generally regarded as harmless in gating systems. However, Re values in excess of 20,000 represent severe turbulence, resulting in an air entrainment and *dross* formation (the scum that forms on the surface of the molten metal) from the reaction of the liquid metal with air and other gases. Techniques for minimizing turbulence generally involve avoidance of sudden changes in flow direction and in the geometry of channel cross-sections in the design of the gating system.

The elimination of dross or slag is another important consideration in fluid flow. It can be achieved by skimming, properly designing pouring basins and gating systems, or using filters. Filters are usually made of ceramics, mica, or fiberglass and their proper location and placement is important for effective filtering of dross and *slag* (nonmetallic products from mutual dissolution of flux and nonmetallic impurities).

### 5.4.2    Fluidity of molten metal

A term commonly used to describe the capability of the molten metal to fill mold cavities is **fluidity**. This term consists of two basic factors: (1) characteristics of the molten metal and (2) casting parameters. The following characteristics of molten metal influence fluidity.

1. *Viscosity*. As viscosity and its sensitivity to temperature (*viscosity index*) increase, fluidity decreases.

2. *Surface tension*. A high surface tension of the liquid metal reduces fluidity. Oxide films developed on the surface of the molten metal thus have a significant adverse effect on fluidity. For example, the oxide film on the surface of pure molten aluminum triples the surface tension.

3. *Inclusions*. As insoluble particles, inclusions can have a significant adverse effect on fluidity. This effect can be verified by observing the viscosity of a liquid such as oil with and without sand particles in it; the latter has higher viscosity.

4. *Solidification pattern of the alloy*. The manner in which solidification occurs, as described in Section 5.3, can influence fluidity. Moreover, fluidity is inversely proportional

to the freezing range (see Eq. 5.3). Thus the shorter the range (as in pure metals and eutectics), the higher the fluidity becomes. Conversely, alloys with long freezing ranges (such as solid-solution alloys) have lower fluidity.

The following casting parameters influence fluidity and also influence the fluid flow and thermal characteristics of the system.

1. *Mold design.* The design and dimensions of components such as the sprue, runners, and risers all influence fluidity.
2. *Mold material and its surface characteristics.* The higher the thermal conductivity of the mold and the rougher its surfaces, the lower the fluidity of the molten metal becomes. Heating the mold improves fluidity, even though it slows down solidification of the metal and the casting develops coarse grains; hence it has less strength.
3. *Degree of superheat.* Defined as the increment of temperature above the melting point of an alloy, superheat improves fluidity by delaying solidification.
4. *Rate of pouring.* The slower the rate of pouring the molten metal into the mold, the lower the fluidity becomes because of the faster local rate of cooling.
5. *Heat transfer.* This factor directly affects the viscosity of the liquid metal.

Although the interrelationships are complex, we use the term **castability** generally to describe the ease with which a metal can be cast to obtain a part with good quality. Obviously, this term includes not only fluidity but casting practices as well.

**Tests for fluidity.**   Although none is accepted universally, several tests have been developed to quantify fluidity. In one such test, the molten metal is made to flow along a channel at room temperature. The distance of metal flow before it solidifies and stops is a measure of its fluidity.

### 5.4.3   Heat transfer

An important consideration in casting is the heat transfer during the complete cycle from pouring to solidification and cooling to room temperature. Heat flow at different locations in the system is a complex phenomenon; it depends on many factors relating to the casting material and the mold and process parameters. For instance, in casting thin sections the metal flow rates must be high enough to avoid premature chilling and solidification. But, the flow rate must not be so fast as to cause excessive turbulence, with its detrimental effects on the properties of the casting.

A typical temperature distribution in the mold–liquid metal interface is shown in Fig. 5.15. Heat from the liquid metal is given off through the mold wall and the surrounding air. The temperature drop at the air–mold and mold–metal interfaces is caused by the presence of boundary layers and imperfect contact at these interfaces. The shape of the curve depends on the thermal properties of the molten metal and the mold.

### 5.4.4   Solidification time

During the early stages of solidification, a thin solidified skin begins to form at the cool mold walls and, as time passes, the skin thickens. With flat mold walls, this thickness is proportional to the square root of time. Thus doubling the time will make the skin $\sqrt{2} = 1.41$ times, or 41 percent, thicker.

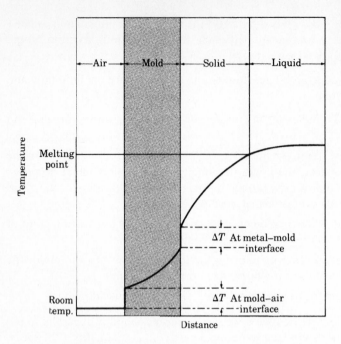

**FIGURE 5.15**  Temperature distribution at the mold wall and liquid–metal interface during solidification of metals in casting.

The **solidification time** is a function of the volume of a casting and its surface area (*Chvorinov's rule*):

$$\text{Solidification time} = C\left(\frac{\text{volume}}{\text{surface area}}\right)^2, \tag{5.9}$$

where $C$ is a constant that reflects mold material, metal properties (including latent heat), and temperature. Thus a large sphere solidifies and cools to ambient temperature at a much slower rate than does a smaller sphere. The reason is that the volume of a sphere is proportional to the cube of its diameter, and the surface area is proportional to the square of its diameter. Similarly, we can show that molten metal in a cube-shaped mold will solidify faster than in a spherical mold of the same volume.

The effects of mold geometry and elapsed time on skin thickness and shape are shown in Fig. 5.16. As illustrated, the unsolidified molten metal has been poured from the mold at dif-

**FIGURE 5.16**  Solidified skin on a steel casting. The remaining molten metal is poured out at the times indicated in the figure. Hollow ornamental and decorative objects are made by a process called slush casting, which is based on this principle. *Source:* H. F. Taylor, J. Wulff, and M. C. Flemings.

ferent time intervals, ranging from 5 s to 6 min. Note that the skin thickness increases with elapsed time but that the skin is thinner at internal angles (location A in the figure) than at external angles (location B). This latter condition is caused by slower cooling at internal angles than at external angles.

---

**Example 5.2:** Solidification times for various shapes

Three pieces being cast have the same volume but different shapes. One is a sphere, one a cube, and the other a cylinder with a height equal to its diameter. Which piece will solidify the fastest and which one the slowest?

SOLUTION.   The volume is unity, so we have from Eq. (5.9):

$$\text{Solidification time} \propto \frac{1}{(\text{surface area})^2}.$$

The respective surface areas are

$$\text{Sphere:} \quad V = \left(\frac{4}{3}\right)\pi r^3, \quad r = \left(\frac{3}{4\pi}\right)^{1/3}, \quad \text{and } A = 4\pi r^2 = 4\pi\left(\frac{3}{4\pi}\right)^{2/3} = 4.84$$

$$\text{Cube:} \quad V = a^3, \quad a = 1, \quad \text{and } A = 6a^2 = 6.$$

$$\text{Cylinder:} \quad V = \pi r^2 h = 2\pi r^3, \quad r = \left(\frac{1}{2\pi}\right)^{1/3}, \quad \text{and}$$

$$A = 2\pi r^2 + 2\pi rh = 6\pi r^2 = 6\pi\left(\frac{1}{2\pi}\right)^{2/3} = 5.54.$$

Thus the respective solidification times $t$ are

$$t_{\text{sphere}} = 0.043C, \quad t_{\text{cube}} = 0.028C, \quad \text{and} \quad t_{\text{cylinder}} = 0.033C.$$

Hence the cube-shaped casting will solidify the fastest and the sphere-shaped casting will solidify the slowest.

---

### 5.4.5  Shrinkage

Metals shrink (contract) during solidification and cooling. Shrinkage, which causes dimensional changes—and, sometimes, cracking—is the result of:

1. Contraction of the molten metal as it cools prior to its solidification;
2. Contraction of the metal during phase change from liquid to solid (latent heat of fusion); and
3. Contraction of the solidified metal (the casting) as its temperature drops to ambient temperature.

The largest amount of shrinkage occurs during cooling of the casting. The amount of contraction for various metals during solidification is shown in Table 5.1. Note that gray cast iron expands. The reason is that graphite has a relatively high specific volume, and when it

**TABLE 5.1**

**SOLIDIFICATION CONTRACTION FOR VARIOUS CAST METALS**

| METAL OR ALLOY | VOLUMETRIC SOLIDIFICATION CONTRACTION (%) | METAL OR ALLOY | VOLUMETRIC SOLIDIFICATION CONTRACTION (%) |
|---|---|---|---|
| Aluminum | 6.6 | 70% Cu–30% Zn | 4.5 |
| Al–4.5% Cu | 6.3 | 90% Cu–10% Al | 4 |
| Al–12% Si | 3.8 | Gray iron | Expansion to 2.5 |
| Carbon steel | 2.5–3 | Magnesium | 4.2 |
| 1% carbon steel | 4 | White iron | 4–5.5 |
| Copper | 4.9 | Zinc | 6.5 |

Note: After R. A. Flinn.

precipitates as graphite flakes during solidification, it causes a net expansion of the metal. Silicon has the same effect in aluminum alloys.

### 5.4.6  Defects

Depending on casting design and method, several defects can develop in castings. Because different names have been used to describe the same defect, the International Committee of Foundry Technical Associations has developed standardized nomenclature consisting of seven basic categories of casting defects:

A. *Metallic projections,* consisting of fins, flash, or massive projections such as swells and rough surfaces.

B. *Cavities,* consisting of rounded or rough internal or exposed cavities, including blowholes, pinholes, and shrinkage cavities (see porosity below).

C. *Discontinuities,* such as cracks, cold or hot tearing, and cold shuts. If the solidifying metal is constrained from shrinking freely, cracking and tearing can occur. Although many factors are involved in tearing, coarse grain size and the presence of low-melting segregates along the grain boundaries (intergranular) increase the tendency for hot tearing. Incomplete castings result from the molten metal being at too low a temperature or pouring the metal too slowly. Cold shut is an interface in a casting that lacks complete fusion because of the meeting of two streams of partially solidified metal.

D. *Defective surface,* such as surface folds, laps, scars, adhering sand layers, and oxide scale.

E. *Incomplete casting,* such as misruns (due to premature solidification), insufficient volume of metal poured, and runout (due to loss of metal from mold after pouring).

F. *Incorrect dimensions or shape*, owing to factors such as improper shrinkage allowance, pattern mounting error, irregular contraction, deformed pattern, or warped casting.

G. *Inclusions,* which form during melting, solidification, and molding. Generally nonmetallic, they are regarded as harmful because they act like stress raisers and reduce the strength of the casting. Hard inclusions (spots) also tend to chip or break tools in machining. They can be filtered out during processing of the molten metal. Inclusions may form during melting because of reaction of the molten metal with the environment (usually oxygen) or the crucible material. Chemical reactions among components in the molten metal may produce inclusions; slags and other foreign material entrapped in the molten metal also become inclusions. Reactions between the metal and the mold material may produce inclusions. Spalling of the mold and core surfaces also produces inclusions, indicating the importance of maintaining melt quality and monitoring the conditions of the molds.

## Porosity

Porosity in a casting may be caused by *shrinkage* or *trapped gases*, or both. Porosity is detrimental to the ductility of a casting and its surface finish, making it permeable and thus affecting pressure tightness of a cast pressure vessel.

Porous regions can develop in castings because of shrinkage of the solidified metal. Thin sections in a casting solidify sooner than thicker regions. As a result, molten metal cannot be fed into the thicker regions that have not yet solidified. Because of contraction, as the surfaces of the thicker regions begin to solidify, porous regions develop at their centers. *Microporosity* can also develop when the liquid metal solidifies and shrinks between dendrites and between dendrite branches.

Porosity caused by shrinkage can be reduced or eliminated by various means. Basically, adequate liquid metal feeding should be provided to all parts of the solidifying casting to avoid porosity. External or internal **chills,** used in sand casting (Fig. 5.17), also are an effective means of reducing shrinkage porosity. The function of chills is to increase the rate of solidification in critical regions. Internal chills are usually made of the same material as the castings.

**FIGURE 5.17** Various types of (a) internal and (b) external chills (dark areas at corners), used in castings to eliminate porosity caused by shrinkage. Chills are placed in regions where there is a larger volume of metal, as shown in (c).

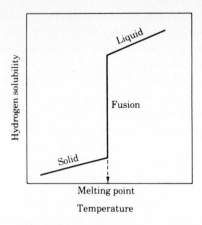

**FIGURE 5.18**  Solubility of hydrogen in aluminum. Note the sharp decrease in solubility as the molten metal begins to solidify.

External chills may be made of the same material or may be iron, copper, or graphite. With alloys, porosity can be reduced or eliminated by making the temperature gradient steep, using, for example, mold materials that have high thermal conductivity. Subjecting the casting to hot isostatic pressing is another method of reducing porosity; however, this is a costly process and is used mainly for critical aircraft components (see also Section 11.3.3).

Liquid metals have much greater *solubility* for gases than do solids (Fig. 5.18). When a metal begins to solidify, the dissolved gases are expelled from the solution. Gases may also result from reactions of the molten metal with the mold materials. Gases either accumulate in regions of existing porosity, such as in interdendritic areas, or they cause microporosity in the casting, particularly in cast iron, aluminum, and copper. Dissolved gases may be removed from the molten metal by flushing or purging with an inert gas or by melting and pouring the metal in a vacuum. If the dissolved gas is oxygen, the molten metal can be *deoxidized*. Steel is usually deoxidized with aluminum or silicon and copper-base alloys with copper alloy containing 15% phosphorus.

Whether microporosity is a result of shrinkage or is caused by gases may be difficult to determine. If the porosity is spherical and has smooth walls, much like the shiny surfaces of holes in Swiss cheese, it is generally from trapped gases. If the walls are rough and angular, porosity is likely from shrinkage between dendrites. Gross or macroporosity is from shrinkage and is usually called **shrinkage cavities.**

## 5.5  Melting Practice and Furnaces

Melting practice is an important aspect of casting operations, because it has a direct bearing on the quality of castings. Furnaces are charged with melting stock consisting of liquid and/or solid metal, alloying elements, and various other materials such as flux and slag-forming constituents. **Fluxes** are inorganic compounds that refine the molten metal by removing dissolved gases and various impurities. Fluxes have several functions, depending on the metal. For example, for aluminum alloys there are cover fluxes (to form a barrier to oxidation), cleaning fluxes, dross-ing fluxes, refining fluxes, and wall-cleaning fluxes (because of the detrimental effect that some fluxes have on furnace linings, particularly induction furnaces). Fluxes may be added manually or can be injected automatically into the molten metal.

Fluxes for aluminum typically consist of chlorides, fluorides, and borates of aluminum, calcium, magnesium, potassium, and sodium. A typical flux for magnesium consists of a composition of magnesium chloride, potassium chloride, barium chloride, and calcium fluo-ride. For copper alloys, there are oxidizing fluxes (which include cupric oxide or manganese dioxide), neutral cover fluxes (borax, boric acid, or glass), reducing fluxes (graphite or char-coal), refining fluxes, and mold fluxes for semicontinuous casting (to prevent oxidation and improve lubricity). For zinc alloys, such as in die casting, typical fluxes contain chlorides of zinc, potassium, and sodium. Fluxes for cast iron typically include sodium carbonate and cal-cium fluoride.

To protect the surface of the molten metal against atmospheric reaction and contamination—and to refine the melt—the pour must be insulated against heat loss. Insulation is usually provided by covering the surface or mixing the melt with compounds that form a **slag.** In casting steels, the composition of the slag includes CaO, $SiO_2$, MnO, and FeO. A small quantity of liquid metal is usually tapped and its composition analyzed. Necessary additions or inoculations are then made prior to pouring the metal into the molds.

The metal charge may be composed of commercially pure **primary** metals, which are remelted scrap. Clean scrapped castings, gates, and risers may also be included in the charge. If the melting points of the alloying elements are sufficiently low, pure alloying elements are added to obtain the desired composition in the melt. If the melting points are too high, the alloying elements do not mix readily with the low-melting-point metals. In this case, **master alloys,** or **hardeners,** are often used. They usually consist of lower-melting point alloys with higher concentrations of one or two of the needed alloying elements. Differences in specific gravity of master alloys should not be great enough to cause segregation in the melt.

## Melting furnaces

The melting furnaces commonly used in foundries are: electric-arc, induction, crucible, and cupolas.

**Electric-arc furnaces** are used extensively in foundries and have advantages such as high rate of melting (thus high production rate), much less pollution than other types of furnaces, and the ability to hold the molten metal for any length of time for alloying purposes.

**Induction furnaces** are especially useful in smaller foundries and produce composition-controlled smaller melts. There are two basic types. The *coreless* induction furnace consists of a crucible completely surrounded with a water-cooled copper coil through which high fre-quency current passes. Because there is a strong electromagnetic stirring action during in-duction heating, this type of furnace has excellent mixing characteristics for alloying and adding new charge of metal. The other type is called a *core* or *channel* furnace which uses low frequency (as low as 60 Hz) and has a coil that surrounds only a small portion of the unit. It is commonly used in nonferrous foundries and is particularly suitable for superheat-ing (heating above normal casting temperature to improve fluidity), holding (keeping the molten metal at a constant temperature for a period of time, thus making it suitable for die-casting applications), and duplexing (using two furnaces, such as melting the metal in one furnace and transferring it to another).

**Crucible furnaces,** which have been used extensively throughout history, are heated with various fuels, such as commercial gases, fuel oil, fossil fuel, as well as electricity. They may be stationary, tilting, or movable. Many ferrous and nonferrous metals are melted in these furnaces.

**Cupolas** are basically refractory-lined vertical steel vessels that are charged with alternating layers of metal, coke, and flux. Although they require major investments and are being replaced by induction furnaces, cupolas operate continuously, have high melting rates, and produce large amounts of molten metal.

**Levitation melting** is a recent development involving magnetic suspension of the molten metal. An induction coil simultaneously heats a solid billet and stirs and confines the metal, thus eliminating the need for a crucible, which could be a source of contamination with oxide inclusions. The molten metal then flows downward into an investment-casting mold placed directly below the coil. Experiments indicate that investment castings made by this method are free of refractory inclusions and gas porosity, and have uniform fine-grained structure.

*Furnace selection* requires careful consideration of several factors that can significantly influence the quality of castings, as well as the economics of casting operations. A variety of furnaces meet requirements for melting and casting metals and alloys in foundries, so proper selection of a furnace generally depends on:

1. Economic considerations, such as initial cost and operating and maintenance costs.
2. The composition and melting point of the alloy to be cast and ease of controlling metal chemistry.
3. Control of the furnace atmosphere to avoid contamination of the metal.
4. Capacity and the rate of melting required.
5. Environmental considerations, such as air pollution and noise.
6. Power supply and its availability and cost of fuels.
7. Ease of superheating the metal.
8. Type of charge material that can be used.

### 5.5.1 Foundries and foundary automation

The casting operations we describe in this chapter are usually carried out in **foundries.** Although casting operations have traditionally involved much manual labor, modern foundries have automated and computer-integrated facilities for all aspects of their operations. They produce a wide variety and sizes of castings at high production rates, at low cost, and with excellent quality control.

Foundry operations initially involve two separate activities. The first is pattern and mold making, which now regularly utilize computer-aided design and manufacturing and **rapid prototyping** techniques (see Section 10.14), minimizing trial and error and improving efficiency. A variety of automated machinery is used to minimize labor costs, which can be significant in the production of castings. The second activity is melting the metals while controlling their composition and impurities.

The rest of the operations, such as pouring into molds carried along conveyors, shakeout, cleaning, heat treatment, and inspection, are also automated. Automation minimizes labor, reduces the possibility of human error, increases the production rate, and can be used to attain higher quality levels. Industrial robots (Section 13.7) are being used extensively in various foundry operations such as removing die castings, cleaning, riser cutting, mold venting, mold spraying, pouring, sorting, inspection, and automatic storage and retrieval systems for cores and patterns using automated guided vehicles (Section 13.6).

The level of automation in foundries is an important economic consideration, particularly in view of the fact that many foundries are small businesses. The degree of automation depends on the type of products made. A die-casting facility or a foundry making parts for the automotive industry may involve production runs in the hundreds of thousands; thus a high level of automation is desirable. Such facilities can afford automation. On the other hand, a jobbing foundry producing short production runs may not be as automated. Furthermore, foundries in general still tend to be somewhat dirty, hot, and labor-intensive operations. It is difficult to find qualified personnel to work in such an environment. Consequently, automation has become increasingly necessary to compensate for the decline in worker availability.

### Safety in foundries

As in all other manufacturing operations, safety in foundries is an important consideration, particularly because of the following factors:

1. Dust from sand and other compounds used in casting, thus requiring proper ventilation and safety equipment for the workers.
2. Fumes from molten metals and lubricants, as well as splashing of the molten metal during transfer or pouring.
3. The presence of fuels for furnaces, the control of their pressure, the proper operation of valves, etc.
4. The presence of water and moisture in crucibles, molds, and other locations, since it rapidly converts to steam, creating severe danger of explosion.
5. Improper handling of fluxes, which are hygroscopic, thus absorbing moisture and creating a danger.
6. Inspection of crucibles, tools, and other equipment for wear, cracks, etc.
7. Inspection of equipment, such as pyrometers, for accuracy and proper calibration.
8. The need for proper personal safety equipment such as gloves, aprons, face shields, and shoes.

## 5.6   Casting Alloys

The general properties of various casting alloys are summarized in Fig. 5.19 and Tables 5.2–5.4.

### 5.6.1   Ferrous alloys

Cast irons represent the largest amount (by weight) of all metals cast. They generally possess several desirable properties, such as wear resistance, hardness, and good machinability. These alloys can easily be cast into complex shapes. The term **cast iron** refers to a family of

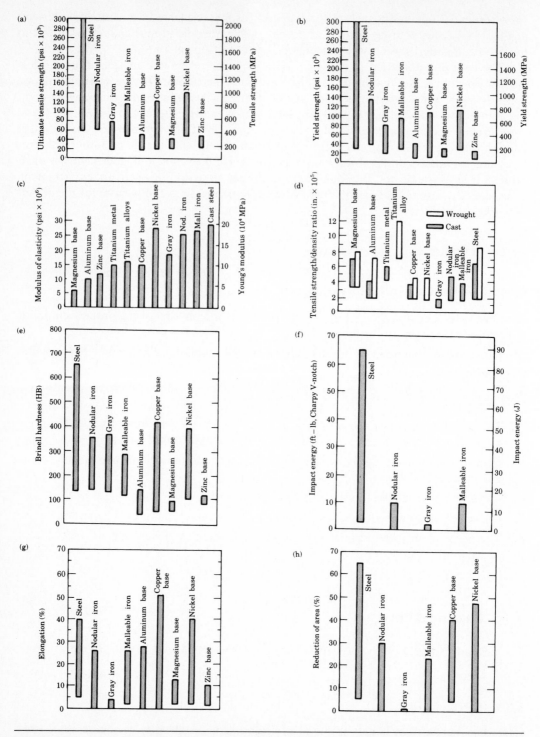

**FIGURE 5.19**   Mechanical properties for various groups of cast alloys. Compare with various tables of properties in Chapter 3. *Source:* Courtesy of Steel Founders' Society of America.

**TABLE 5.2**

**TYPICAL APPLICATIONS FOR CASTINGS AND CASTING CHARACTERISTICS**

| TYPE OF ALLOY | APPLICATION | CAST-ABILITY* | WELD-ABILITY* | MACHIN-ABILITY* |
|---|---|---|---|---|
| Aluminum | Pistons, clutch housings, intake manifolds, engine blocks, heads, cross members, valve bodies, oil pans, suspension components | G–E | F | G–E |
| Copper | Pumps, valves, gear blanks, marine propellers | F–G | F | G–E |
| Gray iron | Engine blocks, gears, brake disks and drums, machine bases | E | D | G |
| Magnesium | Crankcase, transmission housings | G–E | G | E |
| Malleable iron | Farm and construction machinery, heavy-duty bearings, railroad rolling stock | G | D | G |
| Nickel | Gas turbine blades, pump and valve components for chemical plants | F | F | F |
| Nodular iron | Crankshafts, heavy-duty gears | G | D | G |
| Steel (carbon and low alloy) | Die blocks, heavy-duty gear blanks, aircraft undercarriage members, railroad wheels | F | E | F–G |
| Steel (high alloy) | Gas turbine housings, pump and valve components, rock crusher jaws | F | E | F |
| White iron ($Fe_3C$) | Mill liners, shot blasting nozzles, railroad brake shoes, crushers and pulverizers | G | VP | VP |
| Zinc | Door handles, radiator grills, carburetor bodies | E | D | E |

*E, excellent; G, good; F, fair; VP, very poor; D, difficult.

alloys: gray cast iron (gray iron), nodular (ductile or spheroidal) iron, white cast iron, malleable iron, and compacted-graphite iron.

**Gray cast iron.** In this structure, graphite exists largely in the form of flakes (Fig. 5.20a). It is called *gray cast iron*, or *gray iron*, because when broken, the fracture path is along the graphite flakes and thus it has a gray, sooty appearance. These flakes act as stress raisers. Consequently, gray iron has negligible ductility and is weak in tension, although strong in compression, as are other brittle materials. The graphite flakes give this material the capacity to dampen vibrations by the internal friction (thus energy dissipation) caused by these flakes. This capacity makes gray cast iron a suitable and commonly used material for constructing machine tool bases and structures.

Various types of gray cast iron are called ferritic, pearlitic, and martensitic (see Section 5.10). Because of the different structures, each has different properties and applications. In **ferritic gray iron,** also known as fully gray iron, the structure consists of graphite flakes in an alpha ferrite

**TABLE 5.3**

**PROPERTIES AND TYPICAL APPLICATIONS OF CAST IRONS**

| CAST IRON | TYPE | ULTIMATE TENSILE STRENGTH (MPa) | YIELD STRENGTH (MPa) | ELONGA- TION IN 50 mm (%) | TYPICAL APPLICATIONS |
|---|---|---|---|---|---|
| Gray | Ferritic | 170 | 140 | 0.4 | Pipe, sanitary ware |
| | Pearlitic | 275 | 240 | 0.4 | Engine blocks, machine tools |
| | Martensitic | 550 | 550 | 0 | Wearing surfaces |
| Ductile (Nodular) | Ferritic | 415 | 275 | 18 | Pipe, general service |
| | Pearlitic | 550 | 380 | 6 | Crankshafts, highly stressed parts |
| | Tempered martensite | 825 | 620 | 2 | High-strength machine parts, wear resistance |
| Malleable | Ferritic | 365 | 240 | 18 | Hardware, pipe fittings, general engineering service |
| | Pearlitic | 450 | 310 | 10 | Couplings |
| | Tempered martensite | 700 | 550 | 2 | Gears, connecting rods |
| White | Pearlitic | 275 | 275 | 0 | Wear resistance, mill rolls |

**TABLE 5.4**

**TYPICAL PROPERTIES OF NONFERROUS CASTING ALLOYS**

| ALLOY | CONDITION | CASTING METHOD* | UTS (MPa) | YIELD STRESS (MPa) | ELONGATION IN 50 mm (%) | HARDNESS (HB) |
|---|---|---|---|---|---|---|
| *Aluminum* | | | | | | |
| 357 | T6 | S | 345 | 296 | 2.0 | 90 |
| 380 | F | D | 331 | 165 | 3.0 | 80 |
| 390 | F | D | 279 | 241 | 1.0 | 120 |
| *Magnesium* | | | | | | |
| AZ63A | T4 | S, P | 275 | 95 | 12 | — |
| AZ91A | F | D | 230 | 150 | 3 | — |
| QE22A | T6 | S | 275 | 205 | 4 | — |
| *Copper* | | | | | | |
| Brass C83600 | — | S | 255 | 177 | 30 | 60 |
| Bronze C86500 | — | S | 490 | 193 | 30 | 98 |
| Bronze C93700 | — | S | 240 | 124 | 20 | 60 |
| *Zinc* | | | | | | |
| No. 3 | — | D | 283 | — | 10 | 82 |
| No. 5 | — | D | 331 | — | 7 | 91 |
| ZA27 | — | P | 425 | 365 | 1 | 115 |

*S, sand; D, die; P, permanent mold.

**FIGURE 5.20** Microstructure for cast irons. Magnification: 100x. (a) Ferritic gray iron with graphite flakes. (b) Ferritic nodular iron, (ductile iron) with graphite in nodular form. (c) Ferritic malleable iron. This cast iron solidified as white cast iron, with the carbon present as cementite ($Fe_3C$), and was heat treated to graphitize the carbon. *Source:* ASM International, Materials Park, OH.

matrix. **Pearlitic gray iron** has a structure of graphite in a matrix of pearlite. Although still brittle, it is stronger than gray iron. **Martensitic gray iron** is obtained by asutenitizing a pearlitic gray iron, followed by rapid quenching to produce a structure of graphite in a martensite matrix. As a result, this cast iron is very hard.

Castings of gray iron have relatively few shrinkage cavities and little porosity. Typical uses of gray cast iron are for engine blocks, machine bases, electric-motor housings, pipes, and machine wear surfaces. Gray cast irons are specified by a two-digit ASTM designation. Class 20, for example, specifies that the material must have a minimum tensile strength of 140 MPa (20 ksi).

**Ductile iron (nodular iron).** In the structure of ductile iron, graphite is in *nodular* or *spheroid* form (Fig. 5.20b). This shape permits the material to be somewhat ductile and shock resistant. The shape of graphite flakes is changed into nodules (spheres) by small additions of magnesium and/or cerium to the molten metal prior to pouring. Ductile iron can be made ferritic or pearlitic by heat treatment. It can also be heat treated to obtain a structure of tempered martensite.

Typically used for machine parts, pipe, and crankshafts, ductile cast irons are specified by a set of two-digit numbers. Thus class or grade 80–55–06, for example, indicates that the material has a minimum tensile strength of 550 MPa (80 ksi), a minimum yield strength of 380 MPa (55 ksi), and a 6 percent elongation in 50 mm (2 in.).

**White cast iron.** The structure of white cast iron is very hard, wear resistant, and brittle because it contains large amounts of iron carbide instead of graphite. White cast iron is obtained either by rapid cooling of gray iron or by adjusting the material's composition by keeping the carbon and silicon content low. This cast iron is also called white iron, because the lack of graphite gives the fracture surface a white crystalline appearance. Because of its extreme hardness and wear resistance, white cast iron is used mainly for liners for machinery that processes abrasive materials, rolls for rolling mills, and railroad-car brake shoes.

**Malleable iron.**    Malleable iron is obtained by annealing white cast iron in an atmosphere of carbon monoxide and carbon dioxide, between 800°C and 900°C (1470°F and 1650°F) for up to several hours, depending on the size of the part. During this process the cementite decomposes (*dissociates*) into iron and graphite. The graphite exists as *clusters* (Fig. 5.20c) in a ferrite or pearlite matrix and thus has a structure similar to nodular iron. This structure promotes ductility, strength, and shock resistance—hence the term malleable. Typical uses are for railroad equipment and various hardware. Malleable irons are specified by a five-digit designation. Thus 35018, for example, indicates that the yield strength of the material is 240 MPa (35 ksi), and its elongation is 18 percent in 50 mm (2 in.).

**Compacted-graphite iron.**    The compacted-graphite structure has short, thick, and interconnected flakes with undulating surfaces and rounded extremities. The mechanical and physical properties of this cast iron are intermediate between those of flake graphite and nodular graphite cast irons. Its machinability is better than nodular iron. Typical applications are crankcases, ingot molds, cylinder heads, and brake disks. It is particularly suitable for components at elevated temperatures and resists thermal fatigue.

**Magnesium-base alloys.**    Alloys with a magnesium base have good corrosion resistance and moderate strength. These characteristics depend on the particular heat treatment used and on the proper use of protective coating systems.

**Cast steels.**    Because of the high temperatures required to melt cast steels, up to about 1650°C (3000°F), their casting requires considerable knowledge and experience. The high temperatures involved present difficulties in the selection of mold materials—particularly in view of the high reactivity of steels with oxygen—in melting and pouring the metal. Steel castings possess properties that are more uniform (isotropic) than those made by mechanical working processes. Cast steels can be welded; however, welding alters the cast microstructure in the heat-affected zone (see Section 12.5), influencing the strength, ductility, and toughness of the base metal. Subsequent heat treatment must be performed to restore the mechanical properties of the casting. Cast weldments have gained importance where complex configurations, or the size of the casting, may prevent casting the part economically in one place.

**Cast stainless steels.**    Casting of stainless steels involves considerations similar to those for steels in general. Stainless steels generally have a long freezing range and high melting temperatures. They develop various structures, depending on their composition and the process parameters. Cast stainless steels are available in various compositions and can be heat treated and welded. These cast products have high heat and corrosion resistance. Nickel-base casting alloys are used for severely corrosive environments and very high temperature service.

### 5.6.2   Nonferrous alloys

**Aluminum-base alloys.**    Alloys with an aluminum base have a wide range of mechanical properties, mainly because of various hardening mechanisms and heat treatments that can be used with them. Their fluidity depends on oxides and alloying elements in the metal. These alloys have high electrical and thermal conductivity and generally good corrosion resistance to most elements (except alkali). They are nontoxic and lightweight and have good machinability. However, except for alloys with silicon, they generally have low resistance to wear and abrasion. Aluminum-base alloys have many applications, including architectural, decorative, aerospace, and electrical. Increasingly, engine blocks of automobiles are made of aluminum-alloy castings.

**Copper-base alloys.** Alloys with a copper base have the advantages of good electrical and thermal conductivity, corrosion resistance, nontoxicity (unless they contain lead), and wear properties suitable for bearing materials. Mechanical properties and fluidity are influenced by the alloying elements. These alloys also possess good machinability characteristics.

**Zinc-base alloys.** Alloys with a zinc base have good fluidity and sufficient strength for structural applications. These alloys are commonly used in die casting.

**High-temperature alloys.** Having a wide range of properties, high-temperature alloys typically require temperatures of up to 1650°C (3000°F) for casting titanium and superalloys and higher for refractory alloys. Special techniques are used to cast these alloys into parts for jet and rocket engine components. Some of these alloys are more suitable and economical for casting than for shaping by other manufacturing methods such as forging.

## 5.7 Ingot Casting and Continuous Casting

Traditionally, the first step in metal processing is the shaping of the molten metal into a solid form—an **ingot**—for further processing by rolling it into shapes, casting it into semifinished forms, or forging. The shaping process is being rapidly replaced by **continuous casting,** thus improving efficiency and eliminating the need for ingots (see Section 5.7.2). In ingot casting, the molten metal is poured (*teemed*) from the ladle into ingot molds in which the metal solidifies.

The cooled ingots are removed (*stripped*) from the molds and lowered into soaking pits, where they are reheated to a uniform temperature of about 1200°C (2200°F) for subsequent processing by rolling. Ingots may be square, rectangular, or round in cross-section, and their weight ranges from a few hundred pounds to 40 tons.

### 5.7.1 Ferrous alloy ingots

Certain reactions take place during solidification of an ingot, which in turn have an important influence on the quality of the steel produced. For example, significant amounts of oxygen and other gases can dissolve in the molten metal during steelmaking. However, much of these gases are rejected during solidification of the metal, because the solubility limit of gases in the metal decreases sharply as its temperature decreases. The rejected oxygen combines with carbon, forming carbon monoxide, which causes porosity in the solidified ingot. Depending on the amount of gas evolved during solidification, three types of steel ingots can be produced: killed, semi-killed, and rimmed.

**Killed steel** is a fully deoxidized steel; that is, oxygen is removed and thus porosity is eliminated. In the deoxidation process the dissolved oxygen in the molten metal is made to react with elements such as aluminum (most common) that are added to the melt. Other elements such as vanadium, titanium, and zinconium are also used. These elements have an affinity for oxygen and form metallic oxides. If aluminum is used, the product is called aluminum-killed steel. The term *killed* comes from the fact that the steel lies quietly after being produced into the mold.

If sufficiently large, the oxide inclusions in the molten bath float out and adhere to, or are dissolved in, the slag. A fully killed steel is thus free from porosity caused by gases. It is also free from *blowholes* (large spherical holes near the surfaces of the ingot). Consequently, the chemical and mechanical properties of killed steels are relatively uniform throughout the ingot. However, because of metal shrinkage during solidification, an ingot of this type develops a *pipe* at its top. Also called a *shrinkage cavity*, it has a funnel-like appearance. This pipe can comprise a substantial portion of the ingot and has to be cut off and scrapped.

**Semi-killed steel** is a partially deoxidized steel. It contains some porosity, generally in the upper central section of the ingot but has little or no pipe; thus scrap is reduced. Piping in semi-killed is less because it is compensated for by the presence of porosity in that region. Semi-killed steels are economical to produce.

In a **rimmed steel,** which generally has a low carbon content (less than 0.15 percent), the evolved gases are only partially killed or controlled by the addition of elements such as aluminum. The gases form blowholes along the outer rim of the ingot—hence the term *rimmed*. Blowholes are generally not objectionable unless they break through the outer skin of the ingot. Rimmed steels have little or no piping and have a ductile skin with good surface finish.

### 5.7.2 Continuous casting

The traditional method of casting ingots is a batch process; that is, each ingot has to be stripped from its mold after solidification and processed individually. Piping and microstructural and chemical variations also are present throughout the ingot. These problems are alleviated by continuous casting processes, which produce higher quality steels for less cost.

Conceived in the 1860s, **continuous,** or **strand, casting** was first developed for casting non-ferrous metal strip. Now used for steel production, the process yields major improvements in efficiency and productivity and significant reductions in cost. One system for continuous casting is shown schematically in Fig. 5.21. The molten metal in the ladle is cleaned and equalized in temperature by blowing nitrogen gas through it for 5 to 10 minutes. The metal is then poured into a refractory-lined intermediate pouring vessel (*tundish*) where impurities are skimmed off. The tundish holds as much as three tons of metal. The molten metal travels through water-cooled copper molds and begins to solidify as it travels downward along a path supported by rollers (*pinch rolls*).

Before starting the casting process, a solid *starter*, or *dummy bar*, is inserted into the bottom of the mold. The molten metal is then poured and solidifies on the starter bar (Fig. 5.22). The bar is withdrawn at the same rate the metal is poured. The cooling rate is such that the metal develops a solidified skin (shell) to support itself during its travel downward, at speeds typically 25 mm/s (1 in./s). The shell thickness at the exit end of the mold is about 12–18 mm (0.5–0.75 in.). Additional cooling is provided by water sprays along the travel path of the solidifying metal. The molds are generally coated with graphite or similar solid lubricants to reduce friction and adhesion at the mold–metal interfaces. The molds are vibrated to further reduce friction and sticking.

**FIGURE 5.21** Schematic illustration of continuous casting of steel. Because of high efficiency and favorable economics, continuous casting is rapidly replacing traditional steelmaking processes, as well as those for nonferrous metals. *Source:* USX Corporation.

The continuously cast metal may be cut into desired lengths by shearing or torch cutting, or it may be fed directly into a rolling mill for further reductions in thickness and for shape rolling of products such as channels and I-beams. In addition to costing less, continuously cast metals have more uniform compositions and properties than those obtained by ingot casting. Although the thickness of steel strand is usually about 250 mm (10 in.), new developments have reduced this thickness to about 15 mm (0.6 in.) or less. The thinner

**FIGURE 5.22** (a) Method of starting a continuous casting process with a starter bar, and (b) the cross-section of a continuously cast metal as it begins to solidify.

strand reduces the number of rolling operations required and improves the economy of the overall operation.

After they are hot rolled, steel plates or shapes undergo one or more further processes, such as cleaning and pickling by chemicals to remove surface oxides, cold rolling to improve strength and surface finish, annealing, and coating (galvanizing or aluminizing) for resistance to corrosion.

## 5.8 Casting Processes: Expendable Mold

Casting processes (Table 5.5) are generally classified according to mold materials, molding processes, and methods of feeding the mold with the molten metal. The two major categories are expendable-mold and permanent-mold casting (Section 5.9). **Expendable molds** are made of sand, plaster, ceramics, and similar materials, which generally are mixed with various binders or bonding agents. These materials are refractories; that is, they have the capability to withstand the high temperatures of molten metals. After the casting has solidified, the molds in these processes are broken up to remove the casting.

### 5.8.1 Sand casting

The traditional method of expendable-mold casting is in sand molds. Simply stated, **sand casting** consists of placing a pattern (having the shape of the desired casting) in sand to make an imprint, incorporating a gating system, filling the resulting cavity with molten metal, allowing the metal to cool until it solidifies, breaking away the sand mold, and removing the casting. While the origins of sand casting date to ancient times, it is still the most prevalent form of casting. In the United States alone, about 15 million tons of metal are cast by this method each year. Typical parts made by sand casting are machine-tool bases, engine blocks, cylinder heads, and pump housings.

**Sands.** Most sand casting operations use silica sands ($SiO_2$). Sand is the product of the disintegration of rocks over extremely long periods of time. It is inexpensive and is suitable as mold material because of its resistance to high temperatures. There are two general types of sand: *naturally bonded* (*bank* sands) and *synthetic* (*lake* sands). Because its composition can be controlled more accurately, synthetic sand is preferred by most foundries. Several factors are important in the selection of sand for molds. Sand having fine, round grains can be closely packed and forms a smooth mold surface. Good permeability of molds and cores allows gases and steam evolved during casting to escape easily. The mold should have good collapsibility (the casting shrinks while cooling) to avoid defects in the casting, such as hot tearing and cracking. The selection of sand involves certain trade-offs with respect to properties. For example, fine-grained sand enhances mold strength, but the fine grains also lower mold permeability. Sand is typically conditioned before use. *Mulling* machines are used to uniformly mull (mix thoroughly) sand with additives. Clay is used as a cohesive agent to bond sand particles, giving the sand strength.

### Types of sand molds

Sand molds are characterized by the types of sand that comprise them and by the methods used to produce them. There are three basic types of sand molds: green-sand, cold-box, and no-bake molds.

**TABLE 5.5**

**CASTING PROCESSES, THEIR ADVANTAGES AND LIMITATIONS**

| PROCESS | ADVANTAGES | LIMITATIONS |
|---|---|---|
| Sand | Almost any metal is cast; no limit to size, shape or weight; low tooling cost. | Some finishing required; somewhat coarse finish; wide tolerances. |
| Shell mold | Good dimensional accuracy and surface finish; high production rate. | Part size limited; expensive patterns and equipment required. |
| Expandable pattern | Most metals cast with no limit to size; complex shapes. | Patterns have low strength and can be costly for low quantities. |
| Plaster mold | Intricate shapes; good dimensional accuracy and finish; low porosity. | Limited to nonferrous metals; limited size and volume of production; mold making time relatively long. |
| Ceramic mold | Intricate shapes; close tolerance parts; good surface finish. | Limited size. |
| Investment | Intricate shapes; excellent surface finish and accuracy; almost any metal cast. | Part size limited; expensive patterns, molds, and labor. |
| Permanent mold | Good surface finish and dimensional accuracy; low porosity; high production rate. | High mold cost; limited shape and intricacy; not suitable for high-melting-point metals. |
| Die | Excellent dimensional accuracy and surface finish; high production rate. | Die cost is high; part size limited; usually limited to nonferrous metals; long lead time. |
| Centrifugal | Large cylindrical parts with good quality; high production rate. | Equipment is expensive; part shape limited. |

The most common mold material is *green molding sand*. The term *green* refers to the fact that the sand in the mold is moist or damp while the metal is being poured into it. Green molding sand is a mixture of sand, clay, and water. Green-sand molding is the least expensive method of making molds.

In the **skin-dried** method the mold surfaces are dried, either by storing the mold in air or drying it with torches. Skin-dried molds are generally used for large castings because of their higher strength. Sand molds are also oven dried (*baked*) prior to receiving the molten metal. They are stronger than green-sand molds and impart better dimensional accuracy and surface finish to the casting. However, distortion of the mold is greater, the castings are more susceptible to hot tearing because of the lower collapsibility of the mold, and the production rate is slower because of the drying time required.

In the **cold-box mold** process, various organic and inorganic binders are blended into the sand to bond the grains chemically for greater strength. These molds are dimensionally more accurate than green-sand molds but are more expensive.

In the **no-bake mold** process, a synthetic liquid resin is mixed with the sand, and the mixture hardens at room temperature. Because bonding of the mold in this and the cold-box process takes place without heat, they are called *cold-setting processes*.

Major components of sand molds, as depicted in Fig. 5.14, are:

- The mold itself, which is supported by a *flask*. Two-piece molds consist of a *cope* on top and a *drag* on the bottom. The seam between them is the parting line. When more than two pieces are used, the additional parts are called *cheeks*.

- A *pouring basin* or *pouring cup*, into which the molten metal is poured.

- A *sprue*, through which the molten metal flows downward.

- A *gate*, which is located at the base of the sprue. Molds typically contain a system of gates constructed to minimize turbulence in the molten metal and control flow so that metal is supplied at a rate to adequately supply the critical section thickness of the casting. Gating systems often include passageways called *runners*.

- *Risers*, which supply additional metal to the casting as it shrinks during solidification. Figure 5.14 shows two different types of risers: a *blind riser* and an *open riser*.

- *Cores*, which are inserts made from sand. They are placed in the mold to form hollow regions or otherwise define the interior surface of the casting.

- *Vents*, which are placed in molds to carry off gases produced when the molten metal comes into contact with the sand in the molds and core.

## Patterns

Patterns are used to mold the sand mixture into the shape of the casting. They may be made of wood, plastic, or metal. Patterns can also be made directly by *rapid prototyping* techniques (see Section 10.14). The selection of a pattern material depends on the size and shape of the casting, the dimensional accuracy and the quantity of castings required, and the molding process to be used. Because patterns are used repeatedly to make molds, the strength and durability of the material selected for patterns must reflect the number of castings the mold will produce. Patterns may be made of a combination of materials to reduce wear in critical regions. Patterns are usually coated with a **parting agent** to facilitate their removal from the molds.

Patterns can be designed with a variety of features to fit application and economic requirements. **One-piece patterns** are generally used for simpler shapes and low-quantity production. They are generally made of wood and are inexpensive. **Split patterns** are two-piece patterns made so that each part forms a portion of the cavity for the casting. In this way castings having complicated shapes can be produced. **Match-plate patterns** are a popular type of mounted pattern in which two-piece patterns are constructed by securing each half of one or more split patterns to the opposite sides of a single plate (see Fig. 5.23). In such constructions, the gating system can be mounted on the drag side of the pattern. This type of pattern most often is used in conjunction with molding machines and large production runs.

## Cores

For castings with internal cavities or passages, such as in an automotive engine block or a valve body, cores are utilized. Cores are placed in the mold cavity before casting to form the interior surfaces of the casting and are removed from the finished part during shakeout and

Plate

Cope side

Drag side

FIGURE 5.23   A typical metal match-plate pattern used in sand casting.

further processing. Like molds, cores must possess strength, permeability, ability to withstand heat, and collapsibility. Therefore cores are typically made of sand aggregates.

The core is anchored by **core prints.** These are recesses that are added to the pattern to support the core and to provide vents for the escape of gases. A common problem with cores is that for certain casting requirements, as in the case where a recess is required, they may lack sufficient structural support in the cavity. To keep the core from shifting, metal supports, known as **chaplets,** may be used to anchor the core in place.

Cores are generally made in a manner similar to that used in making molds, and the majority are made with shell, no-bake, or cold-box processes. Cores are formed in *core boxes*, which are used much like patterns are used to form sand molds. The sand can be packed into the boxes with sweeps or blown into the box by compressed air from *core blowers*. Core blowers have the advantages of producing uniform cores and operating at a very high production rate.

### Sand-molding machines

The oldest known method of molding, which is still used for simple castings, is to compact the sand by hand hammering or ramming it around the pattern. For most operations, however, the sand mixture is compacted around the pattern by molding machines. These machines eliminate arduous labor, offer higher quality casting by improving the application and distribution of forces, manipulate the mold in a carefully controlled fashion, and increase the rate of production.

Mechanization of the molding process can be further assisted by *jolting* the assembly. The flask, molding sand, and pattern are placed on a pattern plate mounted on an anvil, and jolted upward by air pressure at rapid intervals. The inertial forces compact the sand around the pattern.

In *vertical flaskless molding* the halves of the pattern form a vertical chamber wall against which sand is blown and compacted. Then the mold halves are packed horizontally, with the parting line oriented vertically and moved along a pouring conveyor. This operation is simple and eliminates the need to handle flasks, making potential production rates very high, particularly when other aspects of the operation, such as coring and pouring, are automated.

*Sandslingers* fill the flask uniformly with sand under a stream of high pressure. Sand-throwers are used to fill large flasks and are typically operated by machine. An impeller in the machine throws sand from its blades or cups at such high speeds that the machine not only places the sand but also rams it appropriately.

In *impact molding* the sand is compacted by controlled explosion or instantaneous release of compressed gases. This method produces molds with uniform strength and good permeability.

In *vacuum molding,* also known as the "V" *process,* the pattern is covered tightly by a thin sheet of plastic. A flask is placed over the coated pattern and is filled with sand. A second sheet of plastic is placed on top of the sand, and a vacuum action hardens the sand so that the pattern can be withdrawn. Both halves of the mold are made this way and then assembled. During pouring, the mold remains under a vacuum but the casting cavity does not. When the metal has solidified, the vacuum is turned off and the sand falls away, releasing the casting. Vacuum molding produces castings having very good detail and accuracy; however, it is quite expensive. It is especially well suited for large, relatively flat castings.

### The sand-casting operation

After the mold has been shaped and the cores have been placed in position, the two halves (*cope* and *drag*) are closed, clamped, and weighted down. They are weighted to prevent the separation of the mold sections under the pressure exerted when the molten metal is poured into the mold cavity.

The design of the gating system is important for proper delivery of the molten metal into the mold cavity. Turbulence must be minimized, air and gases must be allowed to escape by vents or other means, and proper temperature gradients must be established and maintained to eliminate shrinkage and porosity. The design of risers is also important for supplying the necessary molten metal during solidification of the casting; the pouring basin may also serve as a riser. After solidification, the casting is shaken out of its mold, and the sand and oxide layers adhering to the casting are removed by vibration (using a shaker) or by sand blasting. The risers and gates are cut off by oxyfuel-gas cutting, sawing, shearing, and abrasive wheels, or they are trimmed in dies.

Almost all commercially used alloys can be sand cast. The surface finish obtained is largely a function of the materials used in making the mold. Dimensional accuracy is not as good as that of other casting processes. However, intricate shapes such as cast-iron engine blocks and very large propellers for ocean liners and impellers can be cast by this process. Sand casting can be economical for relatively small production runs as well as large production runs; equipment costs are generally low. The characteristics of sand casting and other casting processes are given in Table 5.6.

### 5.8.2   Shell-mold casting

The use of *shell-mold casting* has grown significantly, because it can produce many types of castings with close tolerances and good surface finishes at a low cost. In this process, a mounted pattern, made of a ferrous metal or aluminum, is heated to 175–370°C (350–700°F), coated with a parting agent such as silicone, and clamped to a box or chamber containing a fine sand containing a 2.5–4.0 percent thermosetting resin binder, such as phenol-formaldehyde, which coats the sand particles. The sand mixture is blown over the heated pattern, coating it evenly. The assembly is often—but not always—placed in an oven for a short period of time to complete the curing of the resin. The shell hardens around the pattern and is removed from the pattern using built-in ejector pins. Two half-shells are made in this manner and are bonded or clamped together in preparation for pouring.

**TABLE 5.6**

## GENERAL CHARACTERISTICS OF CASTING PROCESSES

| PROCESS | TYPICAL MATERIALS CAST | WEIGHT (kg) | | TYPICAL SURFACE FINISH ($\mu$m, $R_a$) | POR-OSITY* | SHAPE COMPLEX-ITY* | DIMEN-SIONAL ACCURACY* | SECTION THICKNESS (mm) | |
| --- | --- | --- | --- | --- | --- | --- | --- | --- | --- |
| | | MINIMUM | MAXIMUM | | | | | MINIMUM | MAXIMUM |
| Sand | All | 0.05 | No limit | 5–25 | 4 | 1–2 | 3 | 3 | No limit |
| Shell | All | 0.05 | 100 + | 1–3 | 4 | 2–3 | 2 | 2 | — |
| Plaster | Nonferrous (Al, Mg, Zn, Cu) | 0.05 | 50 + | 1–2 | 3 | 1–2 | 2 | 1 | — |
| Investment | All (High melting pt.) | 0.005 | 100 + | 1–3 | 3 | 1 | 1 | 1 | 75 |
| Permanent mold | All | 0.5 | 300 | 2–3 | 2–3 | 3–4 | 1 | 2 | 50 |
| Die | Nonferrous (Al, Mg, Zn, Cu) | < 0.05 | 50 | 1–2 | 1–2 | 3–4 | 1 | 0.5 | 12 |
| Centrifugal | All | — | 5000 + | 2–10 | 1–2 | 3–4 | 3 | 2 | 100 |

*Relative rating: 1 best, 5 worst. Note: These ratings are only general; significant variations can occur, depending on the methods used.

The shells are light and thin, usually 5–10 mm (0.2–0.4 in.), and consequently their thermal characteristics are different from those for thicker molds. The thin shells allow gases to escape during solidification of the metal. The mold is generally used vertically and is supported by surrounding it with steel shot in a cart. The walls of the mold are relatively smooth, offering low resistance to flow of the molten metal and producing castings with sharper corners, thinner sections, and smaller projections than are possible in green-sand molds. With the use of multiple gating systems, several castings can be made in a single mold.

Shell-mold casting may be more economical than other casting processes, depending on various production factors, particularly energy cost. The relatively high cost of metal patterns becomes a smaller factor as the size of production run increases. The high quality of the finished casting can significantly reduce cleaning, machining, and other finishing costs. Complex shapes can be produced with less labor, and the process can be automated fairly easily. Shell-mold casting applications include small mechanical parts requiring high precision, such as gear housings, cylinder heads, and connecting rods. Shell molding is also widely used in producing high-precision molding cores, such as engine-block water jackets.

**Composite molds.**   *Composite molds* are made of two or more different materials and are used in shell molding and other casting processes. They are generally utilized in casting complex shapes, such as impellers for turbines. An example of composite molds is shown in Fig. 5.24. Molding materials commonly used are shells (made as previously described), plaster, sand with binder, metal, and graphite. Composite molds may also include cores and chills to control the rate of solidification in critical areas of castings. Composite molds increase the strength of the mold, improve the dimensional accuracy and surface finish of castings, and may help reduce overall costs and processing time.

**Sodium silicate process.**   The mold material in the *sodium silicate process* is a mixture of sand and 1.5 to 6 percent sodium silicate (waterglass) as the binder for sand. This mixture is packed around the pattern and hardened by blowing carbon dioxide ($CO_2$) gas through it. This process, also known as *silicate-bonded sand* or the *carbon-dioxide process,* was first used in the 1950s and has been developed further, for example, by using various other chemicals

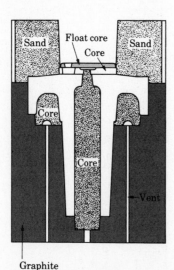

Graphite

**FIGURE 5.24**   Schematic illustration of a semipermanent composite mold. *Source: Steel Castings Handbook*, 5th ed. Steel Founders' Society of America, 1980.

for binders. Cores made by this process reduce the tendency to tear because of their compliance at elevated temperatures.

**Rammed graphite molding.**   Rammed graphite, instead of sand, is used to make molds for casting reactive metals, such as titanium and zirconium. Sand cannot be used because these metals react vigorously with silica. In this specialty method, the molds are packed rather like sand molds, then air dried, baked at 175°C (350°F), fired at 870°C (1600°F), then stored under controlled humidity and temperature. The casting procedures are similar to those for sand molds.

### 5.8.3 Expandable pattern casting (lost foam)

The *expandable pattern casting process* uses a polystyrene pattern, which evaporates upon contact with molten metal to form a cavity for the casting. The process is also known as *evaporative-pattern* or *lost-pattern casting,* and under the trade name *Full-Mold* process. It was formerly known as the expanded polystyrene process, and has become one of the more important casting processes for ferrous and nonferrous metals, particularly for the automotive industry.

First, raw expandable polystyrene (EPS) beads, containing 5 to 8 percent pentane (a volatile hydrocarbon) are placed in a preheated die, usually made of aluminum. The polystyrene expands and takes the shape of the die cavity. Additional heat is applied to fuse and bond the beads together. The die is then cooled and opened, and the polystyrene pattern is removed. Complex patterns may also be made by bonding various individual pattern sections using hot-melt adhesive. In such compound foam-mold assembly, precision of the bonding operation is critically important for maintaining overall dimensional accuracy.

The pattern is then coated with a water-base refractory slurry, dried, and placed in a flask. The flask is filled with loose fine sand, which surrounds and supports the pattern. The sand may be dried or mixed with bonding agents to give it additional strength. The sand is periodically compacted by various means. Then, without removing the polystyrene pattern, the molten metal is poured into the mold. This action immediately vaporizes the pattern (an ablation process) and fills the mold cavity, completely replacing the space previously occupied by the polystyrene pattern. The heat degrades (depolymerizes) the polystyrene and the degradation products are vented into the surrounding sand.

The flow velocity in the mold depends on the rate of degradation of the polymer. Studies have shown that the flow of the molten metal is basically laminar, with Reynolds numbers in the range of 400 to 3000 (Section 5.4.1). The velocity of the molten metal at the metal–polymer pattern front is estimated to be in the range of 0.1–1.0 m/s. The velocity can be controlled by producing patterns with cavities or hollow sections; thus the velocity will increase as the molten metal crosses these hollow regions, similar to pouring in an empty cavity as is done in sand casting. Furthermore, because the polymer requires considerable energy to degrade, large thermal gradients are present at the metal–polymer interface; in other words, the molten metal cools faster than it would if it were poured into a cavity. Consequently, fluidity (see Section 5.4.2) is less than in sand casting. This has important effects on the microstructure throughout the casting and also leads to directional solidification of the metal.

The evaporative-pattern process has a number of advantages over other casting methods:

1.  It is a relatively simple process because there are no parting lines, cores, or riser systems; hence it has a design flexibility.

2.  Inexpensive flasks are sufficient for the process.

3.  Polystyrene is inexpensive and can be easily processed into patterns having very complex shapes, various sizes, and fine surface detail.

4.  The casting requires minimum finishing and cleaning operations.

5.  The process is economical for long production runs. A major fact is the cost to produce the die for expanding the polystyrene beads to make the pattern.

6.  The process can be automated.

Typical applications for this process are cylinder heads, crankshafts, brake components and manifolds for automobiles, and machine bases. The aluminum engine blocks and other components of the General Motors Saturn automobile are made by this process. Recent developments include the use of polymethylmethacrylate (PMMA) and polyalkylene carbonate as pattern materials for ferrous castings. In a modification of the evaporative-pattern process, a polystyrene pattern is surrounded by a ceramic shell (Replicast C-S process). The pattern is burned out prior to pouring the molten metal into the mold. Its principal advantage over investment casting (with its wax patterns) is that carbon pickup into the metal is entirely avoided.

New developments in evaporative-pattern casting include production of metal-matrix composites (Section 11.16). During the process of molding the polymer pattern, it is embedded throughout with fibers or particles, which then become an integral part of the casting. Further studies include modification and grain refinement of the casting by the use of grain refiners and modifier master alloys (Section 5.5) within the pattern while it is being molded.

### 5.8.4  Plaster-mold casting

In the *plaster-mold casting* process the mold is made of plaster of paris (gypsum, or calcium sulfate), with the addition of talc and silica flour to improve strength and control the time required for the plaster to set. These components are mixed with water, and the resulting slurry is poured over the pattern. After the plaster sets, usually within 15 minutes, the pattern is removed and the mold is dried to remove the moisture. The mold halves are then assembled to form the mold cavity and preheated to about 120°C (250°F) for 16 hours. The molten metal is then poured into the mold.

Because plaster molds have very low permeability, gases evolved during solidification of the metal cannot escape. Consequently, the molten metal is poured either in a vacuum or under pressure. Plaster-mold permeability can be increased substantially by the *Antioch* process. The molds are dehydrated in an autoclave (pressurized oven) for 6–12 hours, then rehydrated in air for 14 hours. Another method of increasing permeability is to use foamed plaster, containing trapped air bubbles.

Patterns for plaster molding are generally made of aluminum alloys, thermo-setting plastics, brass, or zinc alloys. Wood patterns are not suitable for making a large number of molds, because the patterns are repeatedly subjected to the water-based plaster slurry which causes the wood to swell. Because there is a limit to the maximum temperature that the plaster mold can withstand, generally about 1200°C (2200°F), plaster-mold casting is used only for aluminum, magnesium, zinc, and some copper-base alloys. The castings have fine detail and good surface finish. Because plaster molds have lower thermal conductivity than other types of molds, the castings cool slowly, yielding more uniform grain structure with less warpage and better mechanical properties.

This process and the ceramic-mold and investment-casting processes are known as *precision casting* because of the high dimensional accuracy and good surface finish obtained. Typical parts made are lock components, gears, valves, fittings, tooling, and ornaments.

### 5.8.5   Ceramic-mold casting

The *ceramic-mold casting* process is similar to the plaster-mold process, with the exception that it uses refractory mold materials suitable for high-temperature applications. The process is also called *cope-and-drag investment casting*. The slurry is a mixture of fine-grained zircon ($ZrSiO_4$), aluminum oxide, and fused silica, which are mixed with bonding agents and poured over the pattern (Fig. 5.25), which has been placed in a flask.

The pattern may be made of wood or metal. After setting, the molds (ceramic facings) are removed, dried, burned off to remove volatile matter, and baked. The molds are clamped firmly and used as all-ceramic molds. In the *Shaw* process, the ceramic facings are backed by fireclay (clay used in making firebricks that resist high temperatures) to give the molds strength. The facings are then assembled into a complete mold, ready to be poured.

The high-temperature resistance of the refractory molding materials allows these molds to be used in casting ferrous and other high-temperature alloys, stainless steels, and tool steels. The castings have good dimensional accuracy and surface finish over a wide range of sizes and intricate shapes, but the process is somewhat expensive. Typical parts made are impellers, cutters for machining, dies for metalworking, and molds for making plastic or rubber components.

### 5.8.6   Investment casting

The *investment-casting* process, also called the *lost-wax* process, was first used during the period 4000–3000 B.C. The pattern is made of wax or a plastic such as polystyrene. The sequences involved in investment casting are shown in Fig. 5.26.

**FIGURE 5.25**   Sequence of operations in making a ceramic mold. *Source: Metals Handbook,* 8th ed., vol. 5, *Forging and Casting* (1970), ASM International, Materials Park, OH.

Mold to make pattern

(a) Injecting wax or plastic pattern

(b) Ejecting pattern — Wax pattern

(c) Pattern assembly (Tree)

(d) Slurry coating

(e) Stucco coating

(f) Completed mold

(g) Autoclaved — Heat — Heat — Molten wax or plastic — Pattern meltout

(h) Molten metal — Pouring

(i) Shakeout

(j) Casting — Pattern

**FIGURE 5.26** Schematic illustration of investment casting (lost-wax process). Castings by this method can be made with very fine detail and from a variety of metals. *Source:* Steel Founders' Society of America.

## Solid mold

The pattern is made by injecting *semisolid* or *liquid* wax or plastic into a metal die in the shape of the pattern. The pattern is then removed and dipped into a slurry of refractory material, such as very fine silica and binders, ethyl silicate, and acids. After this initial coating has dried, the pattern is coated repeatedly to increase its thickness. The term *investment* comes from investing the pattern with the refractory material. Wax patterns require careful handling because they are not strong enough to withstand the forces involved during mold making.

The one-piece mold is dried in air and heated to a temperature of 90–175°C (200–350°F) for about 4 hours, depending on the metal to be cast, to drive off the water of crystallization (chemically combined water). After the metal has been poured and has solidified, the mold is broken up and the casting is removed. A number of patterns can be joined to make one mold called a *tree*, thus increasing process productivity.

Although the labor and materials involved make the lost-wax process costly, it is suitable for casting high-melting-point alloys with a good surface finish and close tolerances. Thus little or no finishing is required, which otherwise would add significantly to the total cost of the casting. This process is capable of producing intricate shapes, with parts weighing from 5 g to 100 kg, from a wide variety of ferrous and nonferrous metals and alloys. Typical parts made are components for office equipment; mechanical components such as gears, cams, valves, and ratchets; and jewelry.

### Ceramic-shell investment casting

A variation of the investment-casting process is *ceramic-shell casting*. It uses the same type of wax or plastic pattern, which is dipped first in a slurry with colloidal silica or ethyl silicate binder, then into a fluidized bed of fine-grained fused silica or zircon flour. The pattern is then dipped into coarse-grain silica to build up additional coatings and thickness to withstand the thermal shock of pouring. This process is economical and is used extensively for precision casting of steels, aluminum, and high-temperature alloys.

If ceramic cores are used in the casting, they are removed by leaching with caustic solutions under high pressure and temperature. The molten metal may be poured in a vacuum to extract evolved gases and reduce oxidation, thus improving the quality of the casting. To further reduce microporosity, the castings made by this and other processes are subjected to hot isostatic pressing. Aluminum castings, for example, are subjected to a gas pressure of up to about 100 MPa (15 ksi) at 500°C (900°F).

---

**Example 5.3:** **Investment–cast superalloy components for gas turbines**

Since the 1960s, investment-cast superalloys have been replacing wrought counterparts in high-performance gas turbines. Much development has been taking place in producing cleaner superalloys (nickel-base and cobalt-base). Improvements have been made in melting and casting techniques, such as by vacuum-induction melting, using microprocessor controls. Impurity and inclusion levels have continually been reduced, thus improving the strength, ductility, and overall reliability of these components. Such control is essential because these parts operate at a temperature only about 50°C (90°F) below the solidus.

The microstructure of an integrally investment-cast, gas-turbine rotor is shown in the upper portion of Fig. 5.27. Note the fine, uniform, equiaxed grain size throughout the cross section. Recent techniques to obtain this result include the use of a nucleant addition to the molten metal, as well as close control of its superheat, pouring techniques, and control of cooling rate of the casting. In contrast, the lower portion of Fig. 5.27 shows the same type of rotor cast conventionally; note the coarse grain structure. This rotor will have inferior properties compared to the fine-grained-rotor. Due to developments in these processes, the proportion of cast parts to other parts in aircraft engines has increased from 20 percent to about 45 percent by weight.

**FIGURE 5.27** Top rotor was investment cast; lower rotor was cast conventionally. *Source: Advanced Materials and Processes,* October 1990, p. 25. ASM International.

### 5.8.7 Vacuum casting

A schematic illustration of the *vacuum-casting process,* or *counter-gravity low-pressure* (CL) *process* (not to be confused with the vacuum-molding process described in Section 5.8.1), is shown in Fig. 5.28. A mixture of fine sand and urethane is molded over metal dies and cured with amine vapor. Then the mold is held with a robot arm and partially immersed into molten metal held in an induction furnace. The metal may be melted in air (CLA process) or in a vacuum (CLV process). The vacuum reduces the air pressure inside the mold to about two-thirds of atmospheric pressure, drawing the molten metal into the mold cavities through a gate in the bottom of the mold. The molten metal in the furnace is at a temperature usually 55°C (100°F) above the liquidus temperature; consequently, it begins to solidify within a fraction of a second. After the mold is filled, it is withdrawn from the molten metal.

**FIGURE 5.28** Schematic illustration of the vacuum-casting process. Note that the mold has a bottom gate. (a) Before and (b) after immersion of the mold into the molten metal. *Source:* From Dr. Robert Blackburn, "Vacuum Casting Goes Commercial," *Advanced Materials and Processes,* February 1990, p. 18. ASM International.

This process is an alternative to investment, shell-mold, and green-sand casting, and is particularly suitable for thin-walled (0.75 mm; 0.03 in.) complex shapes with uniform properties. Carbon and low- and high-alloy steel and stainless steel parts, weighing as much as 70 kg (155 lb), have been vacuum cast by this method. CLA parts are easily made at high volume and relatively low cost. CLV parts usually contain reactive metals, such as aluminum, titanium, zirconium, and hafnium. These parts, which are often in the form of superalloys for gas turbines, may have walls as thin as 0.5 mm (0.02 in.). The process can be automated and production costs are similar to those for green-sand casting.

## 5.9  Casting Processes: Permanent Mold

*Permanent molds,* as the name implies, are used repeatedly and are designed so that the casting can be easily removed and the mold reused. These molds are made of metals that maintain their strength at high temperatures and thus can be used repeatedly. Because metal molds are better heat conductors than expendable molds, the solidifying casting is subjected to a higher rate of cooling, which in turn affects the microstructure and grain size within the casting, as described in Section 5.3.

In permanent-mold casting, two halves of a mold are made from materials such as steel, bronze, refractory metal alloys, or graphite (semipermanent mold). The mold cavity and gating system are machined into the mold and thus become an integral part of it. To produce castings with internal cavities, cores made of metal or sand aggregate are placed in the mold prior to casting. Typical core materials are shell or no-bake cores, gray iron, low-carbon steel, and hot-work die steel. Gray iron is the most commonly used core material, particularly for large molds for aluminum and magnesium castings. Inserts are also used for various parts of the mold.

In order to increase the life of permanent molds, the surfaces of the mold cavity are sometimes coated with a refractory slurry or sprayed with graphite every few castings. These coatings also serve as parting agents and thermal barriers, controlling the rate of cooling of the casting. Mechanical ejectors, such as pins located in various parts of the mold, may be needed for removal of complex castings. Ejectors usually leave small round impressions on castings and gating systems.

The molds are clamped together by mechanical means and heated to facilitate metal flow and reduce thermal damage to the dies. The molten metal is then poured through the gating system. After solidification, the molds are opened and the casting is removed. Special methods of cooling the mold include water or the use of air-cooled fins, similar to those found on motorcycle or lawnmower engines that cool the engine block and cylinder head.

Although the permanent-mold casting operation can be performed manually, the process can be automated for large production runs. This process is used mostly for aluminum, magnesium, and copper alloys because of their generally lower melting points. Steels can also be cast in graphite or heat-resistant metal molds.

This process produces castings with good surface finish, close tolerances, uniform and good mechanical properties and at high production rates. Typical parts made by permanent-mold castings are automobile pistons, cylinder heads, and connecting rods, gear blanks for appliances, and kitchenware. Although equipment costs can be high because of die costs, the process can be mechanized, thus keeping labor costs low. Permanent-mold casting is not economical for small production runs. Furthermore, because of the difficulty in removing the casting from the mold, intricate shapes cannot be cast by this process. However, easily collapsed sand cores can be used and removed from castings to leave intricate internal cavities.

### 5.9.1   Slush casting

We noted in Fig. 5.16 that a solidified skin develops first in a casting and that this skin becomes thicker with time. Hollow castings with thin walls can be made by permanent-mold casting using this principle—a process called *slush casting*. The molten metal is poured into the metal mold, and after the desired thickness of solidified skin is obtained, the mold is inverted or slung, and the remaining liquid metal is poured out. The mold halves are then opened and the casting is removed. The process is suitable for small production runs and is generally used for making ornamental and decorative objects and toys from low-melting-point metals, such as zinc, tin, and lead alloys.

### 5.9.2   Pressure casting

In the two permanent-mold processes that we just described, the molten metal flows into the mold cavity by gravity. In the *pressure-casting* process, also called *pressure pouring* or *low-pressure casting* (Fig. 5.29), the molten metal is forced upward by gas pressure into a graphite or metal mold. The pressure is maintained until the metal has completely solidified in the mold. The molten metal may also be forced upward by a vacuum, which also removes dissolved gases and produces a casting with lower porosity.

**FIGURE 5.29**  The bottom-pressure casting process utilizes graphite molds for the production of steel railroad wheels. *Source:* Griffin Wheel Division of Amsted Industries Incorporated.

### 5.9.3 Die casting

The *die-casting* process, developed in the early 1900s, is a further example of permanent-mold casting. The molten metal is forced into the die cavity at pressures ranging from 0.7 to 700 MPa (0.1 to 100 ksi). The European term *pressure die casting,* or simply die casting, that we describe in this section, is not to be confused with the term *pressure casting* that we described in Section 5.9.2. Typical parts made by die casting are transmission housings, valve bodies, carburetors, motors, business machine and appliance components, hand tools, and toys. The weight of most castings ranges from less than 90 g (3 oz) to about 25 kg (55 lb).

**Hot-chamber process.** The *hot-chamber* process (Fig. 5.30) involves the use of a piston, which traps a certain volume of molten metal and forces it into the die cavity through a gooseneck and nozzle. The pressures range up to 35 MPa (5000 psi), with an average of about 15 MPa (2000 psi). The metal is held under pressure until it solidifies in the die. To improve die life and to aid in rapid metal cooling, thus reducing cycle time, dies are usually cooled by circulating water or oil through various passageways in the die block. Cycle times usually range up to 900 shots (individual injections) per hour for zinc, although very small components such as zipper teeth can be cast at 18,000 shots per hour. Low-melting-point alloys such as zinc, tin, and lead are commonly cast by this process.

**Cold-chamber process.** In the *cold-chamber* process (Fig. 5.31) molten metal is introduced into the injection cylinder (*shot chamber*). The shot chamber is not heated—hence the term *cold* chamber. The metal is forced into the die cavity at pressures usually ranging from 20 MPa to 70 MPa (3 ksi to 10 ksi), although they may be as high as 150 MPa (20 ksi). The machines may be horizontal or vertical; in the latter the shot chamber is vertical and the machine is similar to a vertical press.

High-melting-point alloys of aluminum, magnesium, and copper are normally cast by this method, although other metals (including ferrous metals) can also be cast in this manner.

**FIGURE 5.30** Sequence of steps in die casting of a part in the hot-chamber process. *Source:* Courtesy of *Foundry Management and Technology.*

**FIGURE 5.31** Sequence of operations in die casting of a part in the cold-chamber process. *Source:* Courtesy of *Foundry Management and Technology.*

Molten metal temperatures start at about 600°C (1150°F) for aluminum and magnesium alloys and increase considerably for copper-base and iron-base alloys.

## Process capabilities and machine selection

Because of the high pressures involved, the dies have a tendency to part unless clamped together tightly. Die-casting machines are rated according to the clamping force that can be exerted to keep the dies closed. The capacities of commercially available machines range from about 25 tons to 3000 tons. Other factors involved in the selection of die-casting machines are die size, piston stroke, shot pressure, and cost.

Die-casting dies (Fig. 5.32) may be made single cavity, multiple cavity (several identical cavities), combination cavity (several different cavities), or unit dies (simple small dies that can be combined in two or more units in a master holding die). Dies are usually made of hot-work die steels or mold steels. Die wear increases with the temperature of the molten metal. **Heat-checking** of the dies (surface cracking from repeated heating and cooling of the die) can be a problem. When die materials are selected and maintained properly, dies may last more than half a million shots before die wear becomes significant.

Die design includes taper (draft) to allow the removal of the casting. The sprues and runners may be removed either manually or by using trim dies in a press. The entire die-casting

Single-cavity die    Multiple-cavity die    Combination die    Unit die

**FIGURE 5.32** Various types of cavities in die-casting dies. *Source:* Courtesy of The North American Die Casting Association.

and finishing process can be highly automated. Lubricants (parting agents) are usually applied as thin coatings on die surfaces. Alloys, except magnesium alloys, generally require lubricants. They usually have a water base, with graphite or other compounds in suspension. Because of the high cooling capacity of water, these lubricants also are effective in keeping die temperatures low.

Die casting has the capability for high production rates with good strength, high-quality parts with complex shapes, and good dimensional accuracy and surface detail, thus requiring little or no subsequent machining or finishing operations. Components such as pins, shafts, and fasteners can be cast integrally (**insert molding**), a process similar to putting wooden sticks (the pin) in popsicles (the casting). Ejector marks remain, as do small amounts of flash (thin material squeezed out between the dies) at the die parting line.

In the fabrication of certain parts, die casting can compete favorably with other manufacturing methods, such as sheet-metal stamping and forging, or other casting processes. Additionally, because the molten metal chills rapidly at the die walls, the casting has a fine-grain, hard skin with higher strength than in the center. Consequently, the strength-to-weight ratio of die-cast parts increases with decreasing wall thickness. With good surface finish and dimensional accuracy, die casting can produce bearing surfaces that would normally be machined. Equipment costs, particularly the cost of dies, are somewhat high, but labor costs are generally low because the process is semi- or fully automated. Die casting is economical for large production runs. The properties and typical applications of common die-casting alloys are given in Table 5.7.

**TABLE 5.7**

**PROPERTIES AND TYPICAL APPLICATIONS OF COMMON DIE-CASTING ALLOYS**

| ALLOY | ULTIMATE TENSILE STRENGTH (MPa) | YIELD STRENGTH (MPa) | ELONGA-TION IN 50 mm (%) | APPLICATIONS |
|---|---|---|---|---|
| Aluminum 380 (3.5 Cu–8.5 Si) | 320 | 160 | 2.5 | Appliances, automotive components, electrical motor frames and housings, engine blocks |
| 13 (12 Si) | 300 | 150 | 2.5 | Complex shapes with thin walls, parts requiring strength at elevated temperatures |
| Brass 858 (60 Cu) | 380 | 200 | 15 | Plumbing fixtures, lock hardware, bushings, ornamental castings |
| Magnesium AZ91B (9 Al–0.7 Zn) | 230 | 160 | 3 | Power tools, automotive parts, sporting goods |
| Zinc No. 3 (4 Al) | 280 | — | 10 | Automotive parts, office equipment, household utensils, building hardware, toys |
| 5 (4 Al–1 Cu) | 320 | — | 7 | Appliances, automotive parts, building hardware, business equipment |

Source: Data from The North American Die Casting Association.

**FIGURE 5.33** Schematic illustration of the centrifugal casting process. Pipes, cylinder liners, and similarly shaped parts can be cast by this process.

### 5.9.4 Centrifugal casting

As its name implies, the *centrifugal casting* process utilizes the inertial forces caused by rotation to distribute the molten metal into the mold cavities. This method was first suggested in the early 1800s. There are three types of centrifugal casting: true centrifugal casting, semicentrifugal casting, and centrifuging.

### True centrifugal casting

In *true centrifugal casting*, hollow cylindrical parts, such as pipes, gun barrels, and lampposts, are produced by the technique shown in Fig. 5.33, in which molten metal is poured into a rotating mold. The axis of rotation is usually horizontal but can be vertical for short workpieces. Molds are made of steel, iron, or graphite and may be coated with a refractory lining to increase mold life. Mold surfaces can be shaped so that pipes with various outer shapes, including square or polygonal, can be cast. The inner surface of the casting remains cylindrical because the molten metal is uniformly distributed by centrifugal forces. However, because of density differences, lighter elements such as dross, impurities, and pieces of the refractory lining tend to collect on the inner surface of the casting.

Cylindrical parts ranging from 13 mm (0.5 in.) to 3 m (10 ft) in diameter and 16 m (50 ft) long can be cast centrifugally, with wall thicknesses ranging from 6 mm to 125 mm (0.25 in. to 5 in.). The pressure generated by the centrifugal force is high, as much as 150 gs, and is necessary for casting thick-walled parts. Castings of good quality, dimensional accuracy, and external surface detail are obtained by this process. In addition to pipes, typical parts made are bushings, engine cylinder liners, and bearing rings with or without flanges.

**Semicentrifugal casting.** An example of *semicentrifugal casting* is shown in Fig. 5.34a. This method is used to cast parts with rotational symmetry, such as a wheel with spokes.

**Centrifuging.** In *centrifuging*, also called *centrifuge casting*, mold cavities of any shape are placed at a certain distance from the axis of rotation. The molten metal is poured at the

**FIGURE 5.34** (a) Schematic illustration of the semicentrifugal casting process. Wheels with spokes can be cast by this process. (b) Schematic illustration of casting by centrifuging. The molds are placed at the periphery of the machine, and the molten metal is forced into the molds by centrifugal forces.

center and is forced into the mold by centrifugal forces (Fig. 5.34b). The properties within the castings vary by the distance from the axis of rotation.

### 5.9.5 Squeeze casting

The *squeeze-casting* process, developed in the 1960s, involves solidification of the molten metal under high pressure. Thus it is a combination of casting and forging (Fig. 5.35). The machinery includes a die, punch, and ejector pin. The pressure applied by the punch keeps the entrapped gases in solution (especially hydrogen in aluminum alloys), and the high-pressure contact at the die–metal interface promotes heat transfer, resulting in a fine microstructure with good mechanical properties and limited microporosity.

Parts can be made to near-net shape, with complex shapes and fine surface detail, from both nonferrous and ferrous alloys. Typical products made are automotive wheels and mortar bodies (a short-barreled cannon). The pressures required in squeeze casting are typically higher than pressure die casting but lower than those for hot or cold forging.

Melt metal     Pour molten metal into die     Close die and apply pressure     Eject squeeze casting and charge melt stock and repeat cycle

**FIGURE 5.35** Sequence of operations in the squeeze-casting process. This process combines the advantages of casting and forging.

### 5.9.6   Semisolid metal forming

Also called *semisolid metalworking,* semisolid metal forming was developed in the 1970s. The metal or alloy has a nondentritic, roughly spherical, and fine-grained structure when it enters the die or mold. The alloy exhibits *thixotropic* behavior, that is, its viscosity decreases when agitated; hence this process is also known as **thixoforming** or **thixocasting**. For example, at rest and above its solidus temperature the alloy has the consistency of table butter, but when agitated vigorously its consistency is more like machine oils. This behavior has been utilized in developing technologies combining casting and forging of parts, with cast billets that are forged when 30 to 40 percent liquid. Semisolid metal forming technology was in commercial production by 1981 and is also used in making cast metal-matrix composites.

Another technique for forming in a semisolid state is **rheocasting,** in which a slurry is produced in a mixer and delivered to the mold or die. However, this process has not yet been commercially successful.

---

**Example 5.4:**   Cast automotive-engine components replaced with semisolid forgings

The aluminum rocker-shaft pedestal for a Chrysler 3.5 L 24-valve V–6 engine was originally designed to be produced by permanent-mold casting; however, the resulting part did not have sufficient strength. The function of this part is to attach the valve-actuation mechanism to the cylinder head. Ductile (nodular) iron was considered for this part, but it weighed three times as much. The process was replaced with semisolid forging in which specially cast aluminum slugs were forged when in a thixotropic state, about 50 percent solid and 50 percent liquid. It was found that the forged parts had higher strength, better dimensional accuracy and surface finish, less porosity with better controlled microstructure, and the cost was lower than that for cast parts. The semi-solid forging process was also used to make an aluminum timing-belt tensioner-puller pivot bracket for this engine, redesigned from a ductile iron design. The new bracket had good tolerances, did not require any significant additional machining processes, and cost $2.15 less per piece than the ductile iron brackets. *Source: Advanced Materials & Processes,* June 1993.

---

### 5.9.7   Casting techniques for single-crystal components

We illustrate these techniques by describing the casting of gas turbine blades, which are generally made of nickel-base superalloys. The procedures involved can also be used for other alloys and components.

**Conventional casting of turbine blades.**   The *conventional casting* process involves investment casting using a ceramic mold as shown in Fig. 5.26. The molten metal is poured into the mold and begins to solidify at the ceramic walls. The grain structure developed is poly-crystalline, and the presence of grain boundaries makes this structure susceptible to creep and cracking along those boundaries under the centrifugal forces at elevated temperatures.

**Directionally solidified blades.**   In the *directional solidification* process (Fig. 5.36a), first developed in 1960, the ceramic mold is prepared basically by investment casting techniques and is preheated by radiant heating. The mold is supported by a water-cooled chill plate. After the metal is poured into the mold, the assembly is lowered slowly. Crystals begin to grow at the chill-plate surface and upward. The blade is thus directionally solidified, with

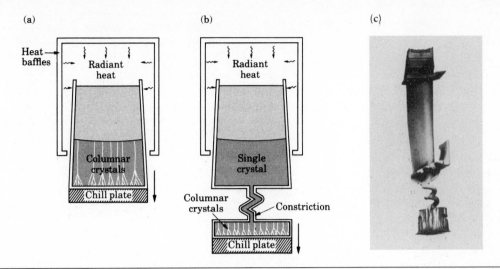

**FIGURE 5.36** Methods of casting turbine blades: (a) directional solidification; (b) method to produce a single-crystal blade; and (c) a single-crystal blade with a constriction portion still attached. *Source:* (a) and (b) Adapted from "Advanced Metal" by B.H. Kear, copyright © 1986 by Scientific American, Inc. All Rights Reserved. (c) *Advanced Materials and Processes,* October 1990, p. 29, ASM International.

longitudinal but no transverse grain boundaries. Consequently, the blade is stronger in the direction of centrifugal forces developed in the gas turbine.

**Single-crystal blades.** In the growing process for single-crystal blades, developed in 1967, the mold is prepared basically by investment casting techniques and has a construction in the shape of a corkscrew (Fig. 5.36b and c), the cross-section of which allows only one crystal through. As the assembly is lowered slowly, a single crystal grows upward through the constriction and begins to grow in the mold. Strict control of the rate of movement is necessary. The solidified mass in the mold is a single-crystal blade. Although more expensive than other blades, the lack of grain boundaries makes these blades resistant to creep and thermal shock. Thus they have a longer and more reliable service life.

**Single-crystal growing.** With the advent of the semiconductor industry, *single-crystal growing* has become a major activity in the manufacture of microelectronic devices. There are basically two methods of crystal growing. In the **crystal-pulling** method, known as the *Czochralski* process (Fig. 5.37a), a seed crystal is dipped into the molten metal and then pulled out slowly, at a rate of about 10 $\mu$m/s, while being rotated about 1 rev/s. The liquid metal begins to solidify on the seed, and the crystal structure of the seed is continued throughout. *Dopants* (alloying elements) may be added to the liquid metal to impart special electrical properties. Single crystals of silicon, germanium, and various other elements are grown by this process. Single-crystal ingots typically 50–150 mm (2–6 in.) in diameter and over 1 m (40 in.) in length have been produced by this technique.

The second technique for crystal growing is the **floating-zone** method (Fig. 5.37b). Starting with a rod of polycrystalline silicon resting on a single crystal, an induction coil heats these two pieces while moving slowly upward. The single crystal grows upward while maintaining its orientation. Thin wafers are then cut from the rod, cleaned, and polished for use in microelectronic device fabrication.

(a)

10 μm/s

~1 rev/s

Seed

Liquid

(b)

20 μm/s

Polycrystalline feed

Induction coil

Single crystal

**FIGURE 5.37** Two methods of crystal growing: (a) crystal pulling (Czochralski process) and (b) floating-zone method. Crystal growing is especially important in the semiconductor industry. *Source:* L. H. Van Vlack, *Materials for Engineering.* (Fig. 18.5.1, p. 513) © 1982 Addison-Wesley Publishing Company, Inc. Reprinted by permission of Addison-Wesley Longman Publishing Co., Inc..

### 5.9.8 Rapid solidification

First developed in the 1960s, *rapid solidification* involves cooling of molten metal at rates as high as $10^6$ K/s, whereby the metal does not have sufficient time to crystallize. These alloys are called **amorphous alloys** or **metallic glasses** because they do not have a long-range crystalline structure. They typically contain iron, nickel, and chromium, which are alloyed with carbon, phosphorus, boron, aluminum, and silicon. Among other effects, rapid solidification results in a significant extension of solid solubility, grain refinement, and reduced microsegregation.

Amorphous alloys exhibit excellent corrosion resistance, good ductility, and high strength. Furthermore, they exhibit very little loss from magnetic hysteresis, high resistance to eddy currents, and high permeability. The latter properties are utilized in making magnetic steel cores for transformers, generators, motors, lamp ballasts, magnetic amplifiers, and linear accelerators, with greatly improved efficiency. Another major application is superalloys of rapidly solidified powders consolidated into near-net shapes for use in aerospace engines.

Gas

Crucible

Induction coil

Melt

Strip

+

Copper disk

**FIGURE 5.38** Schematic illustration of the melt-spinning process to produce thin strips of amorphous metal.

These alloys are produced in the form of wire, ribbon, strip, and powder. In one method, called *melt spinning* (Fig. 5.38), the alloy is melted by induction in a ceramic crucible and propelled under high gas pressure at very high speed against a rotating copper disk (chill block) and chills rapidly (splat cooling).

# 5.10 Processing of Castings; Heat Treatment

The various microstructures developed during metal processing can be modified by **heat treatment** techniques, that is, by controlled heating and cooling of the alloys at various rates. These thermal treatments induce phase transformations that greatly influence mechanical properties such as strength, hardness, ductility, toughness, and wear resistance of the alloys.

The effects of thermal treatment depend primarily on the alloy, its composition and microstructure, the degree of prior cold work, and the rates of heating and cooling during heat treatment. The processes of recovery, recrystallization, and grain growth (see Section 3.6) are examples of thermal treatment, involving changes in the grain structure of the alloy.

## 5.10.1 Heat treating ferrous alloys

In this section we describe the microstructural changes that occur in the iron–carbon system discussed in Section 5.2.5.

**Pearlite.** If the ferrite and cementite lamellae in the pearlite structure (see Section 5.2.6) of the eutectoid steel are thin and closely packed, the microstructure is called **fine pearlite.** If the lamellae are thick and widely spaced, it is called **coarse pearlite.** The difference between the two depends on the rate of cooling through the eutectoid temperature, a reaction in which austenite is transformed into pearlite. If the rate of cooling is relatively high, as in air, fine pearlite is produced; if slow, as in a furnace, coarse pearlite is produced.

The transformation from austenite to pearlite (and for other structures) is best illustrated by Fig. 5.39b and c. These diagrams are called **isothermal transformation (IT) diagrams** or *time–temperature–transformation* (TTT) diagrams. They are constructed from the data in Fig. 5.39a, which shows the percentage of austenite transformed into pearlite as a function of temperature and time. The higher the temperature and/or the longer the time, the greater is the percentage of austenite transformed to pearlite. Note that for each temperature a minimum time is required for the transformation to begin and that some time later all the austenite is transformed to pearlite.

**Spheroidite.** When pearlite is heated to just below the eutectoid temperature and held at that temperature for a period of time, say, for a day at 700°C (1300°F), the cementite lamellae (see Fig. 5.6) transform to *spherical* shapes. Unlike the lamellar shape of cementite, which acts as stress raisers, **spheroidites** (spherical particles) are less conducive to stress concentration because of their rounded shapes. Consequently, this structure has higher toughness and lower hardness than the pearlite structure. In this form it can be cold worked since the ductile ferrite has high toughness, and the spheroidal carbide particles prevent the propagation of cracks within the material.

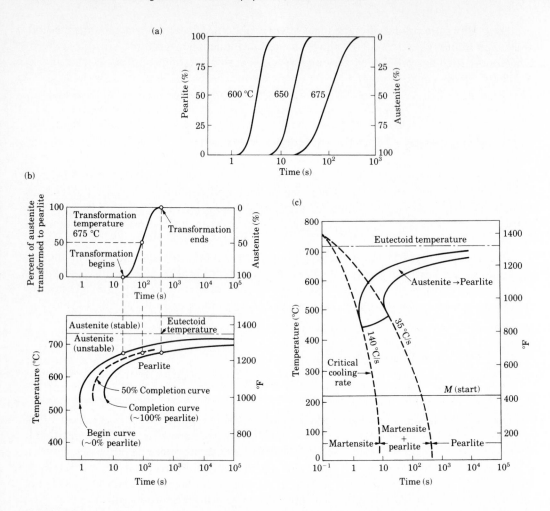

**FIGURE 5.39** (a) Austenite to pearlite transformation of iron–carbon alloys as a function of time and temperature. (b) Isothermal transformation diagram obtained from (a) for a transformation temperature of 675°C (1247°F). (c) Microstructures obtained for a eutectoid iron–carbon alloy as a function of cooling rate. *Source:* ASM International, Materials Park, OH.

**Bainite.** Visible only using electron microscopy, *bainite* has a very fine microstructure, consisting of ferrite and cementite. It can be produced in steels with alloying elements and at cooling rates that are higher than those required for transformation to pearlite. This structure, called *bainitic steel,* is generally stronger and more ductile than pearlitic steel at the same hardness level.

**Martensite.** When austenite is cooled rapidly, as by quenching in water, its fcc structered rectangular prism, which is slightly elongated along one of its principal axes called *martensite.* Because it does not have as many slip systems as a bcc structure, and the carbon is in interstitial positions, martensite is extremely hard and brittle, lacks toughness, and thus has limited use. Martensite transformation takes place almost instantaneously (Fig. 5.39c), because it does not involve the diffusion process, a time-dependent phenomenon that is the mechanism in other transformations.

Transformations involve volume changes because of the different densities of the various phases in the structure. For example, when austenite transforms to martensite, its volume increases (hence its density decreases) by as much as 4 percent. A similar but smaller volume expansion also occurs when austenite transforms to pearlite. These expansions, and thus the thermal gradients present in a quenched part, can cause internal stresses within the body. These stresses may cause parts to crack during heat treatment, as in *quench cracking* of steels caused by rapid cooling during quenching.

**Retained austenite.** If the temperature to which the alloy is quenched is not sufficiently low, only a portion of the structure is transformed to martensite. The rest is *retained austenite*, which is visible as white areas in the structure along with dark needlelike martensite. Retained austenite can cause dimensional instability and cracking and lowers alloy hardness and strength.

**Tempered martensite.** *Tempering* is a heating process that reduces martensite's hardness and improves its toughness. The body-centered tetragonal martensite is heated to an intermediate temperature, where it transforms to a two-phase microstructure, consisting of body-centered cubic alpha ferrite and small particles of cementite. Longer tempering time and higher temperature decrease martensite's hardness. The reason is that the cementite particles coalesce and grow, and the distance between the particles in the soft ferrite matrix increases as the less stable, smaller carbide particles dissolve.

**Hardenability of ferrous alloys.** The capability of an alloy to be hardened by heat treatment is called its **hardenability.** It is a measure of the depth of hardness that can be obtained by heating and subsequent quenching. The term hardenability should not be confused with hardness, which is the resistance of a material to indentation or scratching. Hardenability of ferrous alloys depends on the carbon content, the grain size of the austenite, and the alloying elements present in the material. A test (the **Jominy test**) has been developed in order to determine alloy hardenability.

## Quenching media

Quenching may be carried out in water, brine (saltwater), oils, molten salts, or air. Caustic solutions, polymer solutions, and gases are also used. Because of the differences in the thermal conductivity, specific heat, and heat of vaporization of these media, the rate of cooling of the alloy (*severity of quench*) will also be different.

In relative terms and in decreasing order, the cooling capacity of several quenching media is: agitated brine, 5; still water, 1; still oil, 0.3; cold gas, 0.1; and still air, 0.02. Agitation is also a significant factor in the rate of cooling. In tool steels the quenching medium is specified by a letter (see Table 3.5), such as W for water hardening, O for oil hardening, and A for air hardening. The cooling rate also depends on the surface area-to-volume ratio of the part (see Eq. 5.9). The higher this ratio, the higher is the cooling rate. Thus, for example, a thick plate cools more slowly than a thin plate with the same surface area.

Water is a common medium for rapid cooling. However, the heated metal may form a *vapor blanket* along its surfaces from water-vapor bubbles that form when water boils at the metal–water interface. This blanket creates a barrier to heat conduction because of the lower thermal conductivity of the vapor. Agitating the fluid or the part helps to reduce or eliminate the blanket. Also, water may be sprayed on the part under high pressure. Brine is an effective quenching medium because salt helps to nucleate bubbles at the interfaces, thus improving agitation. However, brine can corrode the part. *Die quenching* is a term used to describe the process of clamping the part to be heat treated to a die, which chills selected regions of the part. In this way cooling rates and warpage can be controlled.

### 5.10.2   Heat treating nonferrous alloys and stainless steels

Nonferrous alloys and some stainless steels generally cannot be heat treated by the techniques used with ferrous alloys. The reason is that nonferrous alloys do not undergo phase transformations as steels do. The hardening and strengthening mechanisms for these alloys are fundamentally different.

Heat-treatable aluminum alloys, copper alloys, and martensitic and precipitation-hardening stainless steels are hardened and strengthened by a process called **precipitation hardening.** This is a technique in which small particles—of a different phase and called *precipitates*—are uniformly dispersed in the matrix of the original phase (see Fig. 5.2a). In this process precipitate forms, because the solid solubility of one element (one component of the alloy) in the other is exceeded.

Three stages are involved in precipitation hardening. We can best describe them by referring to Fig. 5.40, the phase diagram for the aluminum–copper system. For an alloy with the composition of 95.5% Al–4.5% Cu, a single-phase (kappa) substitutional solid-solution of copper (solute) in aluminum (solvent) exists between 500°C and 570°C (930°F and 1060°F). The kappa phase is aluminum rich, with an fcc structure, and is ductile. Below the lower temperature, that is, below the lower solubility curve, two phases are present: kappa

**FIGURE 5.40**   (a) Phase diagram for the aluminum–copper alloy system. (b) Various microstructures obtained during the age-hardening process. *Source:* L. H. Van Vlack, *Materials for Engineering,* (Fig. 5.6.1, p. 123), © 1982, Addison-Wesley Publishing Company, Inc. Reprinted by permission of Addison Wesley Longman, Inc.

and theta (a hard intermetallic compound of $CuAl_2$). This alloy can be heat treated and its properties modified by solution treatment or precipitation.

**Solution treatment.** In *solution treatment* the alloy is heated to within the solid-solution kappa phase, say, 540°C (1000°F), and cooled rapidly, such as by quenching in water. The structure obtained soon after quenching (*A* in Fig. 5.40b), consists only of the single phase kappa. This alloy has moderate strength and considerable ductility.

**Precipitation hardening.** The structure obtained in *A* in Fig. 5.40b can be strengthened by precipitation hardening. The alloy is reheated to an intermediate temperature and held there for a period of time, during which precipitation takes place. The copper atoms diffuse to nucleation sites and combine with aluminum atoms, producing the theta phase, which form as submicroscopic precipitates, shown in *B* by the small dots within the grains of the kappa phase. This structure is stronger than that in *A*, although it is less ductile. The increase in strength is attributed to increased resistance to dislocation movement in the region of the precipitates.

**Aging.** Because the precipitation process is one of time and temperature, it is also called *aging,* and the property improvement is known as *age hardening.* If carried out above room temperature, the process is called **artificial aging.** However, several aluminum alloys harden and become stronger over a period of time at room temperature, by a process known as **natural aging.** Such alloys are first quenched and then, if desired, are formed at room temperature into various shapes and allowed to gain strength and hardness by natural aging. Natural aging can be slowed by refrigerating the quenched alloy.

In the precipitation process, if the reheated alloy is held at that temperature for an extended period of time, the precipitates begin to coalesce and grow. They become larger but fewer, as shown by the larger dots in *C* in Fig. 5.40b. This process is called *overaging,* which makes the alloy softer and weaker. Thus there is an optimal time–temperature relationship in the aging process for obtaining desired properties. Obviously, an aged alloy can be used only up to a certain maximum temperature in service; otherwise it will overage and lose its strength and hardness. Although weaker, an overaged part has better dimensional stability.

**Maraging.** Maraging is a precipitation-hardening treatment for a special group of high-strength iron-base alloys. The word *maraging* is derived from the words martensite and aging. In this process one or more intermetallic compounds are precipitated in a matrix of low-carbon martensite. A typical maraging steel may contain 18 percent nickel, in addition to other elements, and aging is done at 480°C (900°F). Hardening by maraging does not depend on the cooling rate. Thus full uniform hardness can be obtained throughout large parts with minimal distortion. Typical uses of maraging steels are for dies and tooling for casting, molding, forging, and extrusion.

### 5.10.3 Case hardening

The heat treatment processes we have described so far involve microstructural alterations and property changes in the *bulk* of the material or component by *through hardening.* In many situations, however, alteration of only the *surface* properties of a part—hence the term *case hardening*—is desirable. This method is particularly useful for improving resistance to surface indentation, fatigue, and wear. Typical applications for case hardening are gear teeth,

## TABLE 5.8

**OUTLINE OF HEAT TREATMENT PROCESSES FOR SURFACE HARDENING**

| PROCESS | METALS HARDENED | ELEMENT ADDED TO SURFACE | PROCEDURE | GENERAL CHARACTERISTICS | TYPICAL APPLICATIONS |
|---|---|---|---|---|---|
| Carburizing | Low-carbon steel (0.2% C), alloy steels (0.08–0.2% C) | C | Heat steel at 870–950°C (1600–1750°F) in an atmosphere of carbonaceous gases (gas carburizing) or carbon-containing solids (pack carburizing). Then quench. | A hard, high-carbon surface is produced. Hardness 55 to 65 HRC. Case depth < 0.5–1.5 mm (< 0.020 to 0.060 in.) Some distortion of part during heat treatment. | Gears, cams, shafts, bearings, piston pins, sprockets, clutch plates |
| Carbonitriding | Low-carbon steel | C and N | Heat steel at 700–800°C (1300–1600°F) in an atmosphere of carbonaceous gas and ammonia. Then quench in oil. | Surface hardness 55 to 62 HRC. Case depth 0.07 to 0.5 mm (0.003 to 0.020 in.) Less distortion than in carburizing. | Bolts, nuts, gears |
| Cyaniding | Low-carbon steel (0.2% C), alloy steels (0.08–0.2% C) | C and N | Heat steel at 760–845°C (1400–1550°F) in a molten bath of solutions of cyanide (e.g., 30% sodium cyanide) and other salts. | Surface hardness up to 65 HRC. Case depth 0.025 to 0.25 mm (0.001 to 0.010 in.) Some distortion. | Bolts, nuts, screws, small gears |
| Nitriding | Steels (1% Al, 1.5% Cr, 0.3% Mo), alloy steels (Cr, Mo), stainless steels, high-speed tool steels | N | Heat steel at 500–600°C (925–1100°F) in an atmosphere of ammonia gas or mixtures of molten cyanide salts. No further treatment. | Surface hardness up to 1100 HV. Case depth 0.1 to 0.6 mm (0.005 to 0.030 in.) and 0.02 to 0.07 mm (0.001 to 0.003 in.) for high-speed steel. | Gears, shafts, sprockets, valves, cutters boring bars, fuel-injection pump parts |

(continued)

cams, shafts, bearings, fasteners, pins, automotive clutch plates, and tools and dies. Through hardening of these parts would not be desirable, because a hard part lacks the necessary toughness for these applications. A small surface crack can propogate rapidly through the part and cause total failure.

Various surface-hardening processes are available (Table 5.8): **carburizing** (gas, liquid, and pack carburizing), **carbonitriding, cyaniding, nitriding, boronizing,** and **flame** and in-

**TABLE 5.8 (CONTINUED)**

**OUTLINE OF HEAT TREATMENT PROCESSES FOR SURFACE HARDENING**

| PROCESS | METALS HARDENED | ELEMENT ADDED TO SURFACE | PROCEDURE | GENERAL CHARACTERISTICS | TYPICAL APPLICATIONS |
|---|---|---|---|---|---|
| Boronizing | Steels | B | Part is heated using boron-containing gas or solid in contact with part. | Extremely hard and wear resistant surface. Case depth 0.025–0.075 mm (0.001–0.003 in.) | Tool and die steels |
| Flame hardening | Medium-carbon steels, cast irons | None | Surface is heated with an oxyacetylene torch, then quenched with water spray or other quenching methods. | Surface hardness 50 to 60 HRC. Case depth 0.7 to 6 mm (0.030 to 0.25 in.) Little distortion. | Gear and sprocket teeth, axles, crankshafts, piston rods, lathe beds and centers |
| Induction hardening | Same as above | None | Metal part is placed in copper induction coils and is heated by high-frequency current, then quenched. | Same as above. | Same as above. |

duction hardening. Basically, these are heat-treating operations in which the component is heated in an atmosphere containing elements (such as carbon, nitrogen, or boron) that alter the composition, microstructure, and properties of surfaces.

For steels with sufficiently high carbon content, surface hardening takes place without using any of these additional elements. Only the heat-treatment processes described in Section 5.10.1 are needed to alter the microstructures, usually by flame hardening or induction hardening. Laser beams and electron beams are also used effectively to harden both small and large surfaces and also for through hardening of relatively small parts.

Because case hardening is a localized heat treatment, case-hardened parts have a hardness gradient. Typically, the hardness is greatest at the surface and decreases below the surface, the rate of decrease depending on the composition of the metal and the process variables. Surface-hardening techniques can also be used for tempering, thus modifying the properties of surfaces that have been subjected to heat treatment. Various other processes and techniques for surface hardening, such as shot peening and surface rolling, improve wear resistance and various other characteristics (see Section 4.5.1).

**Decarburization** is the phenomenon in which alloys containing carbon lose carbon from their surfaces as a result of heat treatment or hot working in a medium, usually oxygen, that reacts with the carbon. Decarburization is undesirable because it affects the hardenability of the surfaces of the part by lowering the carbon content. It also adversely affects the hardness, strength, and fatigue life of steels by significantly lowering their endurance limits. Decarburization is best avoided by processing the alloy in an inert atmosphere or a vacuum or using neutral salt baths during heat treatment.

### 5.10.4  Annealing

*Annealing* is a general term used to describe the restoration of a cold-worked or heat-treated metal or alloy to its original properties, so as to increase ductility (hence formability), reduce hardness and strength, or modify the microstructure. Annealing is also used to relieve residual stresses in a manufactured part for improved machinability and dimensional stability. (The term annealing also applies to thermal treatment of glasses and similar products and weldments.)

The annealing process involves (1) heating the workpiece to a specific range of temperature, (2) holding it at that temperature for a period of time (soaking), and (3) cooling it slowly. The process may be carried out in an inert or controlled atmosphere or performed at low temperatures to prevent or minimize surface oxidation. Annealing temperatures may be higher than the recrystallization temperature, depending on the degree of cold work (hence stored energy). For example, the recrystallization temperature for copper ranges between 200°C and 300°C (400°F and 600°F), whereas the annealing temperature needed to fully recover the original properties ranges from 260°C to 650°C (500°F to 1200°F), depending on the degree of prior cold work.

**Full annealing** is a term applied to annealing ferrous alloys, generally low- and medium-carbon steels. The steel is heated to above $A_1$ or $A_3$ (Fig. 5.41), and cooling takes place slowly, say, 10°C (20°F) per hour, in a furnace after it is turned off. The structure obtained in full annealing is coarse pearlite, which is soft and ductile and has small uniform grains. Excessive softness in the annealing of steels can be avoided if the entire cooling cycle is carried out in still air. This process is called **normalizing**, in which the part is heated to a temperature of $A_3$ or $A_{cm}$ to transform the structure to austenite. It results in somewhat higher strength and hardness and lower ductility than in full annealing. The structure obtained is fine pearlite with small uniform grains. Normalizing is generally done to refine the grain structure, obtain uniform structure (*homogenization*), decrease residual stresses, and improve machinability.

**Process annealing.**  In *process annealing* (also called *intermediate annealing, subcritical annealing,* or *in-process annealing*) the workpiece is annealed to restore its ductility, part or

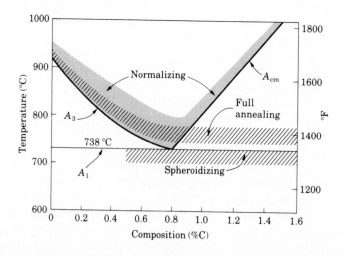

**FIGURE 5.41**  Heat-treating temperature ranges for plain-carbon steels, as indicated on the iron–iron carbide phase diagram. *Source:* ASM International, Materials Park, OH.

all of which may have been exhausted by work hardening during cold working. In this way, the part can be worked further into the final desired shape. If the temperature is high and/or the time of annealing is long, grain growth may result, with adverse effects on formability of annealed parts.

**Stress-relief annealing.** To reduce or eliminate residual stresses, a workpiece is generally subjected to *stress-relief annealing,* or simply *stress relieving.* The temperature and time required for this process depend on the material and the magnitude of residual stresses present. The residual stresses may have been induced during forming, machining, or other shaping processes or caused by volume changes during phase transformations. For steels, the part is heated to below $A_1$, thus avoiding phase transformations. Slow cooling rates, such as in still air, are generally employed. Stress relieving promotes dimensional stability in situations where subsequent relaxing of residual stresses present may cause distortion of the part over a period of time when it is in service. It also reduces the tendency for stress-corrosion cracking.

### 5.10.5 Tempering

If steels are hardened by heat treatment, *tempering* is used in order to reduce brittleness, increase ductility and toughness, and reduce residual stresses. The term tempering is also used for glasses. In tempering, the steel is heated to a specific temperature, depending on composition, and cooled at a prescribed rate. Alloy steels may undergo **temper enbrittlement,** caused by the segregation of impurities along the grain boundaries at temperatures between 480°C and 590°C (900°F and 1100°F).

In **austempering,** the heated steel is quenched from the austenitizing temperature rapidly enough to avoid formation of ferrite or pearlite. It is then held at a certain temperature until isothermal transformation from austenite to bainite is complete. It is then cooled to room temperature, usually in still air, at a moderate rate to avoid thermal gradients within the part. The quenching medium most commonly used is molten salt, at temperatures ranging from 160°C to 750°C (320°F to 1380°F).

Austempering is often substituted for conventional quenching and tempering, either to reduce the tendency for cracking and distortion during quenching or to improve ductility and toughness while maintaining hardness. Because of the relatively short cycle time, this process is economical for many applications. In *modified austempering,* a mixed structure of pearlite and bainite is obtained. The best example of this practice is **patenting,** which provides high ductility and moderately high strength, such as the patented wire used in the wire industry.

In **martempering** *(marquenching),* the steel or cast iron is quenched from the austenitizing temperature into a hot fluid medium, such as hot oil or molten salt. It is held at that temperature until the temperature is uniform throughout the part and then cooled at a moderate rate, such as in air, to avoid temperature gradients within the part. The part is then tempered, because the structure thus obtained is primarily untempered martensite and is not suitable for most applications. Martempered steels have less tendency to crack, distort, and develop residual stresses during heat treatment. In modified martempering the quenching temperature is lower and thus the cooling rate is higher. The process is suitable for steels with lower hardenability.

In **ausforming,** also called **thermomechanical processing,** the steel is formed into desired shapes within controlled ranges of temperature and time to avoid formation of nonmartensitic transformation products. The part is then cooled at various rates to obtain the desired microstructures. Ausformed parts have superior mechanical properties.

### 5.10.6  Design considerations for heat treating

In addition to the metallurgical factors we described, successful heat treating involves design considerations of avoiding problems such as cracking, warping, and nonuniform properties throughout the heat-treated part. The rate of cooling during quenching may not be uniform, particularly with complex shapes of varying cross-sections and thicknesses, thus producing severe temperature gradients. These gradients lead to variations in contraction, which induce thermal stress and may cause the part to crack. Furthermore, nonuniform cooling causes residual stresses in the part, which can lead to stress-corrosion cracking. Thus the method selected and care taken in quenching, as well as the proper choice of quenching media and temperatures, are important considerations.

As a general guideline for part design for heat treating, internal or external sharp corners should be avoided. Otherwise stress concentrations at these corners raise stress levels high enough to cause cracking. Parts should have as nearly uniform thicknesses as possible, or the transition between regions of different thicknesses should be smooth. Parts with holes, grooves, keyways, splines, and unsymmetrical shapes may also be difficult to heat treat because they may crack during quenching. Large surfaces with thin cross-sections are likely to warp. Hot forgings and hot steel-mill products may have a decarburized skin and thus may not successfully respond to heat treatment.

### 5.10.7  Cleaning, finishing, and inspection of castings

After solidification and removal from the mold or die, castings may be subjected to a variety of additional processes. In sand casting, the casting is shaken out of its mold and the sand and oxide layers adhering to the castings are removed by vibration (using a shaker) or by sand blasting. In sand casting and other casting processes—and depending on size—the risers and gates are cut off by oxyfuel-gas cutting, sawing, shearing, and abrasive wheels, or they are trimmed in dies. Castings may be cleaned electrochemically or by pickling with chemicals to remove surface oxides. The surface of castings is important in subsequent machining operations, because machinability can be adversely affected if the castings are not cleaned properly. If regions of the casting have not formed properly or have formed incompletely, the defects may be repaired by welding, thus filling them with weld metal. Sand castings generally have rough, grainy surfaces, although the extent depends on the quality of the mold and the materials used. Finishing operations for castings may involve straightening or forging with dies and machining to obtain final dimensions.

Inspection is an important final step and is done to ensure that the casting meets all design and quality-control requirements. Several methods are available for inspection of castings to determine quality and the presence of any defects. Castings can be inspected visually or optically for surface defects. Subsurface and internal defects are investigated using various nondestructive techniques described in Section 4.8. In destructive testing, test specimens are removed from various sections of a casting and tested for strength, ductility, and other mechanical properties and to determine the presence and location of any defects.

Pressure tightness of cast components (valves, pumps, pipes) is usually determined by sealing the openings in the casting and pressurizing it with water, oil, or air. The casting is then inspected for leaks while the pressure is maintained. Unacceptable or defective castings are melted down for reprocessing. Because of the major economic impact, the types of defects present in castings and their causes must be investigated. Control of all stages during casting, from mold preparation to removal of castings from molds or dies, is important in maintaining good quality.

## 5.11 Casting Design

As in all engineering practice and manufacturing operations, certain guidelines and design principles pertaining to casting have been developed over many years. Although these principles were established primarily through practical experience, analytical methods and computer-aided design and manufacturing techniques are now coming into wider use, improving productivity and the quality of castings. Moreover, careful design can result in significant cost savings.

### 5.11.1 Designing for expendable-mold casting

The following guidelines generally apply to all types of castings. The most significant design considerations are identified and addressed.

**Risers.** A major concern in the design of castings is the size and placement of risers (See Fig. 5.14). Traditionally, riser design has been based on experience and considerations of fluid flow and heat transfer. More recently, commercial computer programs, based on finite-difference mesh control volume algorithms, have become available. These programs also show the importance of obtaining directional solidification (see Fig. 5.9) from the extremities of a casting toward the riser. This can be achieved by proper geometric design of the casting and/or the use of chills or insulators.

**Corners, angles, and section thickness.** Sharp corners, angles, and fillets should be avoided (Fig. 5.42a), because they may cause cracking and tearing during solidification of the metal. Fillet radii should be selected to reduce stress concentrations and to ensure proper liquid-metal flow during the pouring process. If the fillet radii are too large, the volume of the material in those regions is also large and, consequently, the rate of cooling is less.

**FIGURE 5.42** (a) Suggested design modifications to avoid defects in castings. Note that sharp corners are avoided to reduce stress concentrations. (b, c, d) Examples of designs showing the importance of maintaining uniform cross-sections in castings to avoid hot spots and shrinkage cavities.

Section changes in castings should smoothly blend into each other. The location of the largest circle that can be inscribed in a particular region is critical as far as shrinkage cavities are concerned (Fig. 5.42b and c). Because the cooling rate in regions with the larger circles is less, they are called **hot spots**. These regions could develop *shrinkage cavities* and *porosity*. Although they increase the cost of production, metal paddings in the mold can eliminate or minimize hot spots. These paddings act as external chills.

**Flat areas.**    Large flat areas (plain surfaces) should be avoided, as they may warp because of temperature gradients during cooling or develop poor surface finish because of uneven flow of metal during pouring. Flat surfaces can be broken up with ribs and serrations.

**Shrinkage.**    Allowance for shrinkage during solidification should be provided for, so as to avoid cracking of the casting, particularly in permanent molds. Pattern dimensions should also provide for shrinkage of the metal during solidification and cooling. Allowances for shrinkage, also known as *patternmaker's shrinkage allowances,* usually range from about 10 mm/m to 20 mm/m (1/8 in./ft to 1/4 in./ft). Table 5.9 gives the normal shrinkage allowance for metals commonly sand cast.

**Parting lines.**    In general, the parting line should be along a flat plane, rather than contoured. Whenever possible, the parting line should be at the corners or edges of castings, rather than on flat surfaces in the middle of the casting. In this way, the **flash** at the parting line (material running out between the two halves of the mold) will not be as visible. The location of the parting line is important because it influences mold design, ease of molding, number and shape of cores, method of support, and the gating system. Preparation of dry-sand cores requires additional time and cost, so they should be avoided or minimized. We can usually do so by reviewing the design of castings and simplifying them. Two examples of casting design modifications are shown in Fig. 5.43.

**Draft.**    A small *draft* (taper) is provided in sand-mold patterns to enable removal of the pattern without damaging the mold. Depending on the quality of the pattern, draft angles usually range from 0.5° to 2°. The angles on inside surfaces of molds are typically twice this range.

**TABLE 5.9**

**NORMAL SHRINKAGE ALLOWANCE FOR METALS CAST IN SAND MOLDS**

| METAL | PERCENT |
|-------|---------|
| Gray cast iron | 0.83–1.3 |
| White cast iron | 2.1 |
| Malleable cast iron | 0.78–1.0 |
| Aluminum alloys | 1.3 |
| Magnesium alloys | 1.3 |
| Yellow brass | 1.3–1.6 |
| Phosphor bronze | 1.0–1.6 |
| Aluminum bronze | 2.1 |
| High-manganese steel | 2.6 |

(a)

(Outside cores)    (No cores)

(b)

Core

Poor    Good

**FIGURE 5.43** Examples of casting design modifications. *Source: Steel Castings Handbook,* 5th ed. Steel Founders' Society of America, 1980. Used with permission.

They have to be higher than those for outer surfaces, because the casting shrinks inward toward the core.

**Tolerances.** Tolerances depend on the particular casting process, size of the casting, and type of pattern used. Tolerances are smallest within one part of the mold and, because they are cumulative, increase between different parts of the mold. Tolerances should be as wide as possible, within the limits of good part performance; otherwise the cost of the casting increases. In commercial practice, tolerances usually are about $\pm 0.8$ mm (1/32 in.) for small castings and increase with size of castings, say, to 6 mm (1/4 in.) for large castings.

**Allowances.** Because most expendable-mold castings require some additional finishing operations, such as machining, allowances should be made in casting design for these operations. *Machining allowances* are included in pattern dimensions and depend on the type of casting. They increase with the size and section thickness of castings, usually ranging from about 2 mm to 5 mm (0.1 in. to 0.2 in.) for small castings to more than 25 mm (1 in.) for large castings.

### 5.11.2 Designing for permanent-mold casting

The design principles for permanent-mold casting are similar to those for expendable-mold casting. Typical design guidelines and examples for permanent-mold casting are shown schematically in Fig. 5.44 for die casting. Note that the cross-sections have been reduced in order to decrease solidification time and save material. Special considerations are involved in designing and tooling for die casting. Designs may be modified to eliminate the draft for better dimensional accuracy.

### 5.11.3 Modeling of casting processes

Because casting processes involve complex interactions among material and process variables, a quantitative study of these interactions is essential to proper design of castings and the production of high quality castings. In the past such studies have presented major difficulties. However, rapid advances in computers and modeling techniques have led to important innovations in modeling various aspects of casting, including fluid flow, heat transfer, and microstructures developed during solidification under various casting process conditions.

(a)

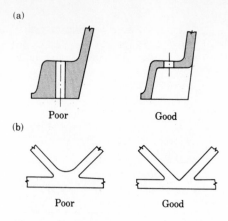

Poor                Good

(b)

Poor                Good

**FIGURE 5.44** Examples of undesirable and desirable design practices for die-cast parts. Note that section-thickness uniformity is maintained throughout the part. *Source:* Courtesy of The North American Die Casting Association.

Modeling of *fluid flow* is based on Bernoulli's and the continuity equations (see Section 5.4). It predicts the behavior of the metal during pouring into the gating system and its travel into the mold cavity, as well as velocity and pressure distributions in the system. Progress has also been made in modeling of *heat transfer* in casting. These studies include investigation of the coupling of fluid flow and heat transfer and the effects of surface conditions, thermal properties of the materials involved, and natural and forced convection on cooling. The capability to calculate isotherms, for example, gives us insight into possible hot spots and subsequent development of shrinkage cavities.

Similar studies are being conducted on modeling the development of *microstructures* in casting. These studies encompass heat flow, temperature gradients, nucleation and growth of crystals, formation of dendritic and equiaxed structures, impingement of grains on each other, and movement of the liquid–solid interface during solidification. Such models are now capable of predicting, for example, the width of the mushy zone (see Fig. 5.10) during solidification and grain size in castings. With the availability of user-friendly computers and advances in computer-aided design and manufacturing (see Chapter 14), modeling techniques are becoming easier to implement. The benefits are increased productivity, improved quality, easier planning and cost estimating, and quicker response to design changes. Several commercial software programs are now available on modeling and casting processes, such as Magmasoft, ProCast, Solidia, and AFSsolid.

## 5.12  Economics of Casting

When looking at various casting processes, we noted that some require more labor than others, some require expensive dies and machinery, and some take a great deal of time to complete. Outlined in Table 5.10, each of these important factors affects to a greater or lesser degree the overall cost of a casting operation.

As we describe in greater detail in Chapter 15, the total cost of a product involves the costs of materials, labor, tooling, and equipment. Preparations for casting a product include

**TABLE 5.10**

### GENERAL COST CHARACTERISTICS OF CASTING PROCESSES

| PROCESS | COST* | | | PRODUCTION RATE† (Pieces/hr) |
|---|---|---|---|---|
| | DIE | EQUIPMENT | LABOR | |
| Sand | L | L | L–M | < 20 |
| Shell | L–M | M–H | L–M | < 50 |
| Plaster | L–M | M | M–H | < 10 |
| Investment | M–H | L–M | H | < 1000 |
| Permanent mold | M | M | L–M | < 60 |
| Die | H | H | L | < 200 |
| Centrifugal | M | H | L–M | < 50 |

*L, low; M, medium; H, high.
†Depends greatly on casting size.

making molds and dies that require raw materials, time, and effort, which we can translate into costs. As Table 5.9 shows, relatively little cost is involved in molds for sand casting. At the other extreme, die-casting dies require expensive materials and a great deal of machining and preparation. In addition to molds and dies, facilities are required for melting and pouring the molten metal into the molds or dies. These facilities include furnaces and related machinery, their costs depending on the level of automation desired. Finally, costs are involved in cleaning and inspecting castings.

The amount of labor required for these operations can vary considerably, depending on the particular process and level of automation. Investment casting, for example, requires a great deal of labor because of the large number of steps involved in this operation. Conversely, operations such as highly automated die casting can maintain high production rates with little labor required.

The cost of equipment per casting (*unit cost*), however, decreases as the number of parts cast increases (Fig. 5.45). Thus sustained high production rates can justify the high cost of

**FIGURE 5.45** Economic comparison of making a part by two different casting processes. Note that because of the high cost of equipment, die casting is economical for large production runs. *Source:* The North American Die Casting Association.

dies and machinery. However, if demand is relatively small, the cost per casting increases rapidly. It then becomes more economical to manufacture the parts by sand casting or by other manufacturing processes. Note that Fig. 5.45 can be expanded to include other casting processes suitable for making the same part.

The two processes (sand and die casting) we compared produce castings with significantly different dimensional and surface-finish characteristics. Thus not all manufacturing decisions are based purely on economic considerations. In fact, parts can usually be made by more than one or two processes. The final decision depends on both economic and technical considerations. We describe competitive aspects of manufacturing processes further in Chapter 15.

---

CASE STUDY:    Engine Crankcase Design and Casting Process Selection

The Subaru brand of automobiles and sport utility vehicles made by the Fuji Heavy Industries Group utilizes a horizontally opposed or *boxer* type of engine configuration. In their most common embodiment, these engines consist of two banks of cylinders, with the pistons from each bank firing in opposite directions. This engine configuration yields excellent specific power and performance within a desirable low-profile package. When designing with flexibility in mind, a common crankcase can be used to produce a number of different engine variants.

The crankcase for a horizontally opposed engine is typically made from two aluminum alloy castings, one for each of the right- and left-hand cylinder banks, bolted together about the crankshaft. When designing these components, the engineer must consider a number of performance-related crankcase features, including:

- Optimum wall thickness and/or reinforcement geometry to prevent deflection under high torque and/or thermal transient loads
- Boss positioning for fastening engine mounts
- Flange dimensions to assure rigid coupling to the transmission
- Internal flow paths to facilitate both oil lubricant and water coolant circulation
- Sealing of the two castings to eliminate any possibility of leaks over the life of the engine
- Volume production of 100,000+ crankcase assemblies per year

An exploded view of the crankcase assembly for a 2.2-liter horizontally opposed engine is shown in Fig. 5.46. Note the required geometric detail of the journal bearings, water coolant jacket around the cylinders, and oil and water coolant recirculation lines, which must be continuous across the two separate crankcase halves. Such geometries are not attainable with sand, gravity-feet, or low-pressure permanent-mold casting processes without extensive post-casting machining and finishing.

In order to create the many bosses, recesses (water jacket), flanges, ports, bearing surfaces, oil galleries, and other crankcase features at a minimum cost, the crankcase engineer typically designs for the high-pressure die-casting process. The high injection pressure (Table 5.11) of the molten metal stream during die fill precludes the use of sand cores. Therefore, designs must

**FIGURE 5.46** EJ22 Open-deck engine block teardown.

be free of internal cavities or undercuts. The added stiffness associated with an undercut or *closed deck* design is not readily achievable with high-pressure die-casting methods.

Each crankcase half is cast in a water-cooled, single-cavity die that rapidly cools the casting (short production cycle time). A long, narrow, drafted steel core is used to create the open-deck water jacket. Cast-iron cylinder liners are loaded as inserts on die mandrels immediately prior to pressure injection of the molten aluminum alloy. This combination of crankcase design features and processing conditions substantially increases productivity.

The additional rigidity of a closed-deck crankcase configuration can be accommodated by selection of a hybrid process known as medium-pressure casting. In this process, the injection speed of the molten alloy during filling of the die is reduced to a point that accommodates the use of sand cores to create the undercut cavities of a closed-deck water jacket. Clean molten metal transfer and reduced die filling turbulence results in less trapped gas and/or entrained deflects. Medium-pressure castings can also be heat treated (T6) and/or welded. Improved crankcase stiffness is achieved at the cost of additional sand cores and associated bake-out and recovery equipment. Cycle time is also slightly increased. The cast features of the medium-pressure cast crankcase are exactly the same as the high-pressure crankcase shown in Fig. 5.46.

In order to produce an entire family of engines with the maximum of common components, Subaru designed the EJ family of engines. The 2.0-liter variants have closed-deck crankcases made by medium-pressure die casting. The 2.2-liter variants have open-deck

**TABLE 5.11**

**CASTING PROCESS COMPARISONS:
OPERATING PARAMETERS**

| | PROCESS | | |
| PARAMETER | LOW PRESSURE | MEDIUM PRESSURE | HIGH PRESSURE |
|---|---|---|---|
| Casting pressure (MPa) | 0.105–.20 | 5.0–30 | 50–100 |
| Injection speed (m/s) | 0.03–0.05 | 0.05–0.25 | 35–60 |
| Die surface temperature (°C) | 300–450 | 150–360 | 120–250 |
| Holding furnace alloy temp (°C) | 700–720 | 680–700 | 650–690 |

crankcases made by high-pressure die casting. By selecting a common alloy and overall design, both crankcases can be machined on the same transfer line. By creating design features in concert with crankcase functional requirements and casting process limitations, Subaru is able to produce an entire family of engines (EJ series) with a minimum of capital equipment and/or tooling investment.

*Source:* James G. Conley, Northwestern University. For details, see T. Yamada, H. Usui, H. Tosa, H. Terauchi, and J. G. Conley, "Experience with the Production of Cast Aluminum Alloy Engine Blocks by Low, Medium, and High Pressure Casting Processes," SAE 1992 Transactions, *Journal of Engines*, Volume 101, Section 3, pp. 1015–1028.

## SUMMARY

- Solidification of pure metals takes place at a clearly defined, constant temperature. Solidification of alloys occurs over a range of temperatures, depending on composition. Phase diagrams are useful tools for identifying the solidification point or points for technologically important materials. (Section 5.2)

- Composition and cooling rates of the melt affect the size and shape of grains and dendrites in solidifying metals and alloy. In turn, the size and structure of grains and dendrites influence properties of the solidified casting. (Section 5.3)

- Most metals shrink during the solidification process, although gray cast iron and some aluminum alloys actually expand. Either way, dimensional changes and sometimes cracking are difficulties that can arise during solidification and cooling. In addition, seven basic categories of casting defects have been identified. (Section 5.3)

- In casting, the molten metal or alloy may flow through a variety of passages, including pouring basins, sprues, runners, risers, and gating systems, before reaching the final mold cavity. Bernoulli's theorem, the continuity law, and the Reynolds number are the analytical tools used in designing an appropriate flow rate and system and in eliminating defects associated with fluid flow. (Section 5.4)

- Heat transfer affects fluid flow and solidification time in a casting process. Solidification time is a function of the volume of a casting and its surface area (Chvorinov's rule). (Section 5.4)

- Melting practices have a direct effect on the quality of castings. Factors that affect melting include: 1) fluxes: inorganic compounds added to the molten metal to remove dissolved gases and various impurities; 2) the type of furnace used; and 3) foundry operations—that is, pattern and mold making, pouring of the melt, removal of cast items from molds, cleaning, heat treatment, and inspection. (Section 5.5)

- Several ferrous and nonferrous casting alloys are available. They have a wide range of properties, casting characteristics, and applications. Because many castings are designed and produced to be assembled with other mechanical components and structures, various other considerations, such as weldability, machinability, and surface conditions, are also important. (Section 5.6)

- The traditional ingot-casting process has been replaced, for many applications, by continuous-casting methods for both ferrous and nonferrous metals. (Section 5.7)

- Casting processes are generally classified as either expendable-mold or permanent-mold casting. The most common expendable-mold methods are sand, shell-mold, plaster, ceramic-mold, and investment casting. Common permanent-mold methods include slush casting, pressure casting, and die casting. Compared to permanent-mold casting, expendable-mold casting usually involves lower mold and equipment costs, but produces lower dimensional accuracy. (Sections 5.8 and 5.9)

- Castings may be subjected to further heat-treatment processes. Various transformations take place in microstructures that have widely varying characteristics and properties, as temperature is reduced at different rates. Important mechanisms of hardening and strengthening involve thermal treatments, including quenching and precipitation hardening. (Section 5.10)

- General principles have been established to aid designers in producing castings that are free from defects and meet tolerance and service requirements. These principles concern shape of casting and various techniques to minimize hot spots that could lead to shrinkage cavities. Because of the large number of variables involved, close control of all parameters is essential, particularly those related to the nature of liquid metal flow into the molds and dies and the rate of cooling in different regions of the mold or die. (Section 5.11)

- Within the limits of good performance, the economic aspects of casting are just as important as technical considerations. Factors affecting the overall cost are the cost of materials, molds, dies, equipment, and labor, each of which varies with the particular casting process. An important parameter is the cost per casting, which, for large production runs, can justify large expenditures for automated machinery. (Section 5.12)

## SUMMARY OF EQUATIONS

Bernoulli's theorem:   $h + \dfrac{p}{\rho g} + \dfrac{v^2}{2g} = \text{constant}$

Continuity equation:    $\dfrac{A_1}{A_2} = \sqrt{\dfrac{h_2}{h_1}}$

Chvorinov's rule:    Solidification time $= C\left(\dfrac{\text{volume}}{\text{surface area}}\right)^2$

## BIBLIOGRAPHY

Allsop, D.F., and D. Kennedy, *Pressure Die Casting–Part II: The Technology of the Casting and the Die.* Oxford: Pergamon, 1983.

*An Introduction to Die Casting.* Des Plaines, Ill.: American Die Casting Institute, 1981.

*ASM Handbook*, Vol. 3: *Alloy Phase Diagrams.* Materials Park, Ohio: ASM International, 1992.

*ASM Handbook*, Vol. 4: *Heat Treating.* Materials Park, Ohio: ASM International, 1991.

*ASM Handbook*, Vol. 15: *Casting.* Materials Park, Ohio: ASM International, 1988.

Bradley, E.F., *High-Performance Castings: A Technical Guide.* Columbus, Ohio: Edison Welding Institute, 1989.

*Case Hardening of Steel.* Metals Park, Ohio: ASM International, 1987.

*Casting Defects Handbook.* Des Plaines, Ill.: American Foundrymen's Society, 1972.

Clegg, A.J., *Precision Casting Processes.* Tarrytown, New York: Pergamon, 1991.

Flemings, M.C., *Solidification Processing.* New York: McGraw-Hill, 1974.

Flinn, R.A., *Fundamentals of Metal Casting.* Reading, Mass.: Addison-Wesley, 1963.

Heine, R.W., C.R. Loper, Jr., and C. Rosenthal, *Principles of Metal Casting,* 2d ed. New York: McGraw-Hill, 1967.

*Investment Casting Handbook.* Chicago, Ill.: Investment Casting Institute, 1979.

Kaye, A., and A.C. Street, *Die Casting Metallurgy.* London: Butterworth, 1982.

Krauss, G., *Steels: Heat Treatment and Processing.* Materials Park, Ohio: ASM International, 1990.

Liebermann, H.H. (ed.), *Rapidly Solidified Alloys.* New York: Marcel Dekker, 1993.

*The Metallurgy of Die Castings.* River Grove, Ill.: Society of Die Casting Engineers, 1986.

Minkoff, I., *Solidification and Cast Structure.* New York: Wiley, 1986.

Powell, G.W., S.-H. Cheng and C.E. Mobley, Jr., *A Fractography Atlas of Casting Alloys.* Columbus, Ohio: Battelle Press, 1992.

*Product Design for Die Casting.* La Grange, Ill.: Diecasting Development Council, 1988.

Rowley, M.T. (ed.), *International Atlas of Casting Defects.* Des Plaines, Ill.: American Foundrymen's Society, 1974.

*Steel Castings Handbook,* 6th ed. Materials Park, Ohio: ASM International, 1995.

Street, A.C., *The Diecasting Book,* 2d ed. Surrey, England: Portcullis Press, 1986.

Szekely, J., *Fluid Flow Phenomena in Metals Processing.* New York: Academic Press, 1979.

Upton, B., *Pressure Die Casting–Part 1: Metals, Machines, Furnaces.* Oxford: Pergamon, 1982.

Walton, C.F., and T.J. Opar (eds.), *Iron Castings Handbook,* 3d ed. Des Plaines, Ill.: Iron Castings Society, 1981.

## QUESTIONS

5.1    Describe the characteristics of (1) an alloy, (2) pearlite, (3) austenite, (4) martensite, and (5) cementite.

5.2    What are the effects of mold materials on fluid flow and heat transfer?

5.3    How does the shape of graphite in cast iron affect its properties?

5.4    Explain the difference between short and long freezing ranges. How are they determined? Why are they important?

5.5 We know that pouring molten metal at a high rate into a mold has certain disadvantages. Are there any disadvantages to pouring it very slowly? Explain.

5.6 Why does porosity have detrimental effects on the mechanical properties of castings? Which physical properties are also affected adversely by porosity?

5.7 A spoked hand wheel is to be cast in gray iron. In order to prevent hot tearing of the spokes, would you insulate the spokes or chill them? Explain.

5.8 Which of the following considerations are important for a riser to function properly? (1) Have a surface area larger than the part being cast. (2) Be kept open to atmospheric pressure. (3) Solidify first. Why?

5.9 Explain why the constant C in Eq. (5.9) depends on mold material, metal properties, and temperature.

5.10 Explain why gray iron undergoes expansion, rather than contraction, during solidification.

5.11 How can you tell whether cavities in a casting are due to porosity or to shrinkage?

5.12 Explain the reasons for hot tearing in castings.

5.13 Would you be concerned about the fact that parts of internal chills are left within the casting? What materials do you think chills should be made of, and why?

5.14 Are external chills as effective as internal chills? Explain.

5.15 Do you think early formation of dendrites in a mold can impede the free flow of molten metal into the mold? Explain.

5.16 Is there any difference in the tendency for shrinkage void formation for metals with short freezing and long freezing ranges, respectively? Explain.

5.17 It has long been observed by foundrymen that low pouring temperatures, i.e., low superheat, promote equiaxed grains over columnar grains. Also, equiaxed grains become finer as the pouring temperature decreases. Explain these phenomena.

5.18 The figures below indicate various defects and discontinuities in cast products. Review each one and offer solutions to avoid them.

5.19 What are the reasons for the large variety of casting processes that have been developed over the years?

5.20 Why can blind risers be smaller than open-top risers?

5.21 Would you recommend preheating the molds in permanent-mold casting? Also, would you remove the casting soon after it has solidified? Explain.

5.22 In a sand-casting operation, what factors determine the time at which you would remove the casting from the mold?

5.23 Explain why the strength-to-weight ratio of all die-cast parts increases with decreasing wall thickness.

5.24 We note that the ductility of some cast alloys is nil (see Fig. 5.19). Do you think this should be a significant concern in engineering applications of castings? Explain.

5.25 The modulus of elasticity E of gray iron varies significantly with its type, such as the ASTM class. Explain why.

5.26 List and explain the considerations for selecting pattern materials.

5.27 Why is the investment-casting process capable of producing fine surface detail on castings?

5.28 Explain why a casting may have a slightly different shape than the pattern used to make the mold.

5.29    Explain why squeeze casting produces parts with better mechanical properties, dimensional accuracy, and surface finish than expendable-mold processes.

5.30    Why are steels more difficult to cast than cast irons?

5.31    What would you do to improve the surface finish in expendable-mold casting processes?

5.32    You have seen that even though die casting produces thin parts, there is a limit to the minimum thickness. Why can't even thinner parts be made by this process?

5.33    What differences, if any, would you expect in the properties of castings made by permanent-mold versus sand-casting methods?

5.34    Which of the casting processes would be suitable for making small toys in large numbers? Why?

5.35    Why are allowances provided for in making patterns? What do they depend on?

5.36    Explain the difference in the importance of drafts in green-sand casting versus permanent-mold casting.

5.37    Make a list of the mold and die materials used in the casting processes described in this chapter. Under each type of material, list the casting processes that are used, and explain why these processes are suitable for that particular mold or die material.

5.38    Explain why carbon is so effective in imparting strength to iron in the form of steel.

5.39    Describe the engineering significance of the existence of a eutectic point in phase diagrams.

5.40    Explain the difference between hardness and hardenability.

5.41    Explain why it may be desirable for castings to be subjected to various heat treatments.

5.42    Describe the differences between case hardening and through hardening insofar as engineering applications are concerned.

## PROBLEMS

5.43    Using Fig. 5.3, estimate the following quantities for a 20% Cu–80% Ni alloy: (1) liquidus temperature, (2) solidus temperature, (3) percentage of nickel in the liquid at 1400°C (2550°F), (4) the major phase at 1400°C, and (5) the ratio of solid to liquid at 1400°C.

5.44    Determine the amount of gamma and alpha phases (Fig. 5.6) in a 5-kg, 1060 steel casting as it is being cooled to the following temperatures: (1) 750°C, (2) 728°C, and (3) 726°C.

5.45    A round casting is 0.1 m in diameter and 0.5 m in length. Another casting of the same metal is elliptical in cross-section, with a major-to-minor axis ratio of 2, and has the same length and cross-sectional area as the round casting. Both pieces are cast under the same conditions. What is the difference in the solidification times of the two castings?

5.46    Derive Eq. (5.7).

5.47    Two halves of a mold (cope and drag) are weighted down to keep them from separating under the pressure exerted by the molten metal (buoyancy). Consider a solid, spherical steel casting, 10 in. in diameter, that is being produced by sand casting. Each flask (see Fig. 5.14) is 30 in. by 30 in. and 15 in. deep. The parting line is at the middle of the part. Estimate the clamping force required.

5.48    Would the position of the parting line in Problem 5.47 influence your answer? Explain.

5.49    Plot the clamping force in Problem 5.47 as a function of increasing diameter of the casting, from 10 in. to 20 in.

5.50    Sketch a graph of specific volume versus temperature for a metal that shrinks as it cools from the liquid state to room temperature. On the graph, mark the area where shrinkage is compensated for by risers.

5.51    A round casting has the same dimensions as in Problem 5.45. Another casting of the same metal is rectangular in cross section, with a width-to-thickness ratio of 2, and has the same length and cross-sectional area as the round casting. Both pieces are cast under the same conditions. What is the difference in the solidification times of the two castings?

5.52   A 50 mm thick square plate and a right circular cylinder with a radius of 100 mm and height of 50 mm each have the same volume. If each is to be cast using a cylindrical riser, will each part require the same size riser to ensure proper feeding? Explain.

5.53   Assume that the top of a round sprue has a diameter of 3 in. and is at a height of 10 in. from the runner. Based on Eq. (5.7), plot the profile of the sprue diameter as a function of its height.

5.54   Estimate the clamping force for a die-casting machine in which the casting is rectangular, with projected dimensions of 100 mm × 200 mm. Would your answer depend on whether or not it is a hot-chamber or cold-chamber process? Explain.

5.55   When designing patterns for casting, patternmakers use special rulers that automatically incorporate solid shrinkage allowances into their designs. Therefore, a 12-in. patternmaker's ruler is longer than a foot. How long is a patternmaker's ruler designed for the making of patterns for (1) aluminum castings and (2) high-manganese steel?

5.56   The blank for the spool shown in the figure below is to be sand cast out of A–319, an aluminum casting alloy. Make

5.57   Repeat Problem 5.56, but assume that the aluminum spool is to be cast using expendable-pattern casting. Explain the important differences between the two patterns.

5.58   In sand casting, it is important that the cope mold half be held down with enough force to keep it from floating when the molten metal is poured in. For the casting shown in the figure below, calculate the minimum amount of weight necessary to keep the cope from floating up as the molten metal is poured in. (*Hint:* The buoyancy force exerted by the molten metal on the cope is related to the

a sketch of the wooden pattern for this part. Include all necessary allowances for shrinkage and machining.

Section A-A

Material: Low-carbon steel
Density: 0.26 lb/in$^3$
All dimensions in inches

0.50 in.

0.45 in.

3.00 in.

4.00 in.

effective height of the metal head above the cope.)

5.59 The original casting design shown in (a) in the figure below was changed to the design shown in (b). The casting is round, with a vertical axis of symmetry. As a functional part, what advantages do you think the new design has over the old one?

(a)

(b)
Ribs or brackets

5.60 An incorrect and a correct design for casting are shown, respectively, in the figure below. Review the changes made and comment on their advantages.

(a) incorrect

Outside core          Outside core

(b) correct

5.61 Three sets of designs for die casting are shown below. Note the changes made

to die design 1 and comment on the reasons.

(a)

(1)

(2)

(b)

(1)

(2)

Parting line

Parting line

(3)

Parting line

(c)

(1)

(2)

5.62 The optimum shape of a riser is spherical to ensure that it cools more slowly than the casting it feeds. Spherically shaped risers, however, are difficult to cast. (1) Sketch the shape of a blind riser that is easy to mold, but also has the smallest possible surface area-to-volume ratio. (2) Compare the solidification time of the riser in part (a) to that of a riser shaped like a right circular cylinder. Assume that the volume of each riser is the same, and that for each

the height is equal to the diameter (see Example 5.2).

5.63 Assume that you are asked to give a quiz to students on the contents of this chapter. Prepare three quantitative problems and three qualitative questions, and supply the answers.

# 6

# Bulk Deformation Processes

**6.1** Introduction
**6.2** Forging
**6.3** Rolling
**6.4** Extrusion
**6.5** Rod, Wire, and Tube Drawing
**6.6** Swaging
**6.7** Die Manufacturing Methods
**6.8** Die Failures

## 6.1   Introduction

Deformation processes in manufacturing are operations that induce shape changes on the workpiece by plastic deformation under forces applied by various tools and dies. Deformation processes can be grouped by temperature, size, and shape of the workpiece and type of operation. For example, using temperature as a criterion, deformation processes may be divided into three basic categories of cold (room temperature), warm, and hot working. Also, deformation processes may be classified by type of operation as primary working or secondary working.

**Primary working** operations are those that take a solid piece of metal (generally from a cast state, such as an ingot) and break it down successively into shapes such as slabs, plates, and billets. Traditionally, primary working processes are forging, rolling, and extrusion. **Secondary working** involves further processing of the products from primary working into final or semifinal products such as bolts, sheet metal parts, and wire.

These classifications are not rigid because some operations belong to both primary and secondary working. For instance, some forgings result in a product as final as those that result from secondary working. Likewise, in extrusion some products are ready for use without any further processing.

The recent trend has been to classify deformation processes according to the size and shape of the workpiece. In these terms, all deformation processes are classified as either bulk deformation or sheet forming.

**Bulk deformation** is the processing of workpieces having a relatively small surface area-to-volume (or surface area-to-thickness) ratio—hence the term *bulk*. In all bulk deformation processing, the thickness or cross-section of the workpiece changes. In **sheet-forming** operations the surface area-to-thickness ratio is relatively large. In general, the material is subjected to shape changes by various dies. Thickness changes are usually undesirable and, in fact, can lead to failure.

**294**

**TABLE 6.1**

**GENERAL CHARACTERISTICS OF BULK DEFORMATION PROCESSES**

| PROCESS | CHARACTERISTICS |
|---|---|
| Forging | Production of discrete parts with a set of dies; some finishing operations usually required; similar parts can be made by casting and powder-metallurgy techniques; usually performed at elevated temperatures; die and equipment costs are high; moderate to high labor cost; moderate to high operator skill. |
| Rolling<br>  Flat | Production of flat plate, sheet, and foil in long lengths, at high speeds, and with good surface finish, especially in cold rolling; requires high capital investment; low to moderate labor cost. |
|   Shape | Production of various structural shapes, such as I-beams, at high speeds; includes thread rolling; requires shaped rolls and expensive equipment; low to moderate labor cost; moderate operator skill. |
| Extrusion | Production of long lengths of solid or hollow products with constant cross-section; usually performed at elevated temperatures; product is then cut into desired lengths; can be competitive with roll forming; cold extrusion has similarities to forging and is used to make discrete products; moderate to high die and equipment cost; low to moderate labor cost; low to moderate operator skill. |
| Drawing | Production of long rod and wire, with round or various cross-sections; smaller cross-sections than extrusions; good surface finish; low to moderate die, equipment, and labor cost; low to moderate operator skill. |

In this chapter we describe four basic bulk deformation processes for metal: forging, rolling, extrusion, and drawing of rod and wire (Table 6.1). We cover bulk deformation processing of plastics and nonmetallic materials in Chapters 10 and 11, respectively.

## 6.2  Forging

**Forging** denotes a family of processes by which plastic deformation of the workpiece is carried out by *compressive forces*. Forging is one of the oldest metalworking operations known, dating back to 5000 B.C., and is used in making parts of widely varying sizes and shapes from a variety of metals. Typical parts made by forging today are crankshafts and connecting rods for engines, turbine disks, gears, wheels, bolt heads, hand tools, and many types of structural components for machinery and transportation equipment.

Forging can be carried out at room temperature (cold working), or at elevated temperatures, called warm and hot forging, depending on the temperature. The temperature range for these categories is given in Table 3.2 in terms of the homologous temperature, $T/T_m$, where $T_m$ is the melting point of the workpiece material on the absolute scale. Note that the homologous recrystallization temperature for metals is about 0.5 (see also Fig. 3.16).

Simple forgings can be made with a heavy hammer and an anvil by techniques used by blacksmiths for centuries. Usually, though, a set of dies and a press are required. The three basic categories of forging are open die, impression die, and closed die.

### 6.2.1   Open-die forging

In its simplest form, **open-die forging** generally involves placing a solid cylindrical workpiece between two flat dies (platens) and reducing its height by compressing it (Fig. 6.1a). This operation is also known as **upsetting.** The die surfaces may be shaped, such as a conical or curved cavity, thereby forming the ends of the cylindrical workpiece during upsetting.

Under ideal conditions, a solid cylinder deforms as shown in Fig. 6.1a. This is known as **homogeneous deformation.** Because volume is constant, any reduction in height increases the diameter of the cylinder. For a specimen that has been reduced in height from $h_0$ to $h_1$, we have

$$\text{Reduction in height} = \frac{h_0 - h_1}{h_0} \times 100\%. \qquad (6.1)$$

From Eqs. (2.1) and (2.9) and using absolute values (as is generally the case with bulk deformation processes), we have

$$e_1 = \frac{h_0 - h_1}{h_0} \qquad (6.2)$$

and

$$\varepsilon_1 = \ln\!\left(\frac{h_0}{h_1}\right). \qquad (6.3)$$

With a relative velocity $v$ between the platens, the specimen is subjected to a strain rate, according to Eqs. (2.17) and (2.18), of

$$\dot{e}_1 = -\frac{v}{h_0} \qquad (6.4)$$

and

$$\dot{\varepsilon}_1 = -\frac{v}{h_1}. \qquad (6.5)$$

If the specimen is reduced in height from $h_0$ to $h_2$, the subscript 1 in the foregoing equations is replaced by subscript 2. Note that the true strain rate $\dot{\varepsilon}$ increases rapidly as the height of the specimen approaches zero. Positions 1 and 2 in Fig. 6.1a may be regarded as instantaneous positions during a continuous upsetting operation.

**FIGURE 6.1** (a) Ideal deformation of a solid cylindrical specimen compressed between flat frictionless dies. This process is known as upsetting. (b) Deformation in upsetting with friction at the die-workpiece interfaces.

Actually, the specimen develops a **barrel** shape, as shown in Fig. 6.1b. Barreling is caused primarily by *frictional* forces at the die-workpiece interfaces that oppose the outward flow of the material at these interfaces. Barreling also occurs in upsetting *hot* workpieces between *cool* dies. The material at and near the interfaces cools rapidly, while the rest of the specimen is relatively hot. Because the strength of the material decreases with temperature, the ends of the specimen show a greater resistance to deformation than does the center. Thus the central portion of the cylinder deforms to a greater extent than do its ends.

In barreling, the material flow within the specimen becomes *nonuniform,* or *inhomogeneous,* as can be seen from the polished and etched section of a barreled cylindrical specimen in Fig. 6.3. Note the roughly triangular stagnant (*dead*) zones at the top and bottom surfaces, which have a wedge effect on the rest of the material, thus causing barreling. This inhomogeneous deformation can also be observed in grid patterns, as shown in Fig. 6.2, which are obtained experimentally using various techniques (Section 6.4.1).

In addition to the single barreling shown in Fig. 6.2, *double barreling* can also be observed. It occurs at large ratios of height-to-cross-sectional area and when friction at the die–workpiece interfaces is high. Under these conditions, the stagnant zone under the platen

**FIGURE 6.2** Schematic illustration of grid deformation in upsetting: (a) original grid pattern; (b) after deformation, without friction; (c) after deformation, with friction. Such deformation patterns can be used to calculate the strains within a deforming body.

**FIGURE 6.3** Grain flow lines in upsetting a solid steel cylinder at elevated temperatures. Note the highly inhomogeneous deformation and barreling. The different shape of the bottom section of the specimen (as compared to the top) results from the hot specimen resting on the lower cool die before deformation proceeded. The bottom surface was chilled; thus it exhibits greater strength and hence deforms less than the top surface. *Source:* J. A. Schey, et al., IIT Research Institute.

is sufficiently far from the midsection of the workpiece to allow the midsection to deform uniformly while the top and bottom portions barrel out—hence the term double barreling.

Barreling caused by friction can be minimized by applying an effective lubricant or ultrasonically vibrating the platens. Also, the use of heated platens or a thermal barrier at interfaces will reduce barreling in hot working.

### Forces and work of deformation under ideal conditions

If friction at the interfaces is zero and the material is perfectly plastic with a yield stress of $Y$, then the normal compressive stress on the cylindrical specimen is uniform at a level $Y$. The force at any height $h_1$ is then

$$F = YA_1, \tag{6.6}$$

where $A_1$ is the cross-sectional area and is obtained from volume constancy. Thus

$$A_1 = \frac{A_0 h_0}{h_1}.$$

The ideal work of deformation is the product of the volume of the specimen and the specific energy $u$ (Eq. 2.58):

$$\text{Work} = \text{Volume} \int_0^{\varepsilon_1} \sigma \, d\varepsilon, \tag{6.7}$$

where $\varepsilon_1$ is obtained from Eq. (6.3).

If the material is strain hardening, with a true stress–true strain curve given by

$$\sigma = K\varepsilon^n,$$

then the expression for force at any stage during deformation becomes

$$F = Y_f A_1, \tag{6.8}$$

where $Y_f$ is the **flow stress** of the material, corresponding to the true strain given by Eq. (6.3) and Fig. 2.42.

The expression for the work done is

$$\text{Work} = (\text{Volume})(\overline{Y})(\varepsilon_1), \tag{6.9}$$

where $\overline{Y}$ is the **average flow stress** and is given by

$$\overline{Y} = \frac{K\displaystyle\int_0^{\varepsilon_1} \varepsilon^n \, d\varepsilon}{\varepsilon_1} = \frac{K\varepsilon_1^n}{n+1}. \tag{6.10}$$

### 6.2.2  Methods of analysis

#### Slab method

The **slab method** is one of the simpler methods of analyzing the stresses and loads in forging and other bulk deformation processes. It requires the selection of an element in the workpiece and identification of all the normal and frictional stresses acting on that element.

**Forging of a rectangular workpiece in plane strain.**   Let's take the case of simple compression with friction (Fig. 6.4), which is the basic deformation in forging. As the flat dies compress the part, it is reduced in thickness and, as the volume remains constant, the part expands laterally. This relative movement at the die–workpiece interfaces causes frictional forces acting in opposition to the movement of the piece. These frictional forces are shown by the horizontal arrows in Fig. 6.4. For simplicity, let's also assume that the deformation is in **plane strain**; that is, the workpiece is not free to flow in the direction perpendicular to this page.

Let's now take an element and indicate all the stresses acting on it (Fig. 6.4b). Note the correct direction of the frictional stresses. Also note the difference in the horizontal stresses acting on the sides of the element; this difference is caused by the presence of frictional stresses on the element. We assume that the lateral stress distribution $\sigma_x$ is uniform along the height $h$.

The next step in this analysis is to balance the horizontal forces on this element, because it must be in static equilibrium. Thus, assuming unit width,

$$(\sigma_x + d\sigma_x)h + 2\mu\sigma_y \, dx - \sigma_x h = 0$$

**FIGURE 6.4** Stresses on an element in plane-strain compression (forging) between flat dies. The stress $\sigma_x$ is assumed to be uniformly distributed along the height $h$ of the element. Identifying the stresses on an element (slab) is the first step in the slab method of analysis for metalworking processes.

or

$$d\sigma_x + \frac{2\mu\sigma_y}{h}\,dx = 0.$$

Note that we have one equation but two unknowns: $\sigma_x$ and $\sigma_y$. We obtain the necessary second equation from the yield criteria as follows. As shown in Fig 6.4c, this element is subjected to triaxial compression (compare this element with that in Fig. 2.39d). Using the distortion-energy criterion for plane strain, we obtain

$$\sigma_y - \sigma_x = \frac{2}{\sqrt{3}}\,Y = Y'. \tag{6.11}$$

Thus

$$d\sigma_y = d\sigma_x.$$

Note that we assumed $\sigma_y$ and $\sigma_x$ to be **principal stresses**. In the strictest sense, $\sigma_y$ cannot be a principal stress because a shear stress is also acting on the same plane. However, this assumption is acceptable for low values of the coefficient of friction $\mu$ and is the standard practice in this method of analysis. Note also that $\sigma_z$ in Fig. 6.4c is derived in a manner similar to that in Eq. (2.44).

We now have two equations that can be solved by noting that

$$\frac{d\sigma_y}{\sigma_y} = -\frac{2\mu}{h}\,dx \quad \text{or} \quad \sigma_y = Ce^{-2\mu x/h}. \tag{6.12}$$

The boundary conditions are such that at $x = a$, $\sigma_x = 0$, and thus $\sigma_y = Y'$ at the edges of the specimen. (All stresses are compressive, so we may ignore negative signs for stresses, which is traditional in such analyses.) Hence the value of C becomes

$$C = Y'e^{2\mu a/h},$$

and thus

$$p = \sigma_y = Y'e^{2\mu(a-x)/h}. \tag{6.13}$$

Also,

$$\sigma_x = \sigma_y - Y' = Y'\left[e^{2\mu(a-x)/h} - 1\right]. \tag{6.14}$$

Equation (6.13) is plotted qualitatively in Fig. 6.5 in dimensionless form. Note that the pressure increases exponentially toward the center of the part and also that it increases with the $a/h$ ratio and increasing friction.

For a strain-hardening material, $Y'$ in Eqs. (6.13) and (6.14) is replaced by $Y'_f$. Because of its shape, the pressure-distribution curve in Fig. 6.5 is referred to as the **friction hill.** Note that the pressure with friction is higher than it is without friction. This result is expected, because the work required to overcome friction must be supplied by the upsetting force.

The area under the pressure curve in Fig. 6.5 is the upsetting **force per unit width** of the specimen. This area can be obtained by integration, but an approximate expression for the *average pressure* $p_{av}$ is

$$p_{av} \simeq Y'\left(1 + \frac{\mu a}{h}\right). \tag{6.15}$$

Again note the significant roles of $a/h$ and friction on the pressure required, especially at high $a/h$ ratios. The *forging force* $F$ is the product of the average pressure and the contact area; that is,

$$F = (p_{av})(2a)(\text{width}). \tag{6.16}$$

Pressure distribution

Workpiece

**FIGURE 6.6** Normal stress (pressure) distribution in the compression of a rectangular workpiece with sliding friction under conditions of plane stress using the *distortion-energy criterion.* Note that the stress at the corners is equal to the uniaxial yield stress of the material, $Y$.

**FIGURE 6.7**   Increase in contact area of a rectangular specimen (top view) compressed between flat dies with friction. Note that the length of the specimen has increased proportionately less than its width. Likewise, a specimen in the shape of a cube acquires the shape of a pancake after deformation with friction.

Note that the expressions for the pressure $p$ are in terms of an instantaneous height $h$. Thus the force at any $h$ during a continuous upsetting operation must be calculated individually.

A *rectangular specimen* can be upset without being constrained on its sides *(plane stress)*. According to the distortion-energy criterion, the normal stress distribution can be given qualitatively by Fig. 6.6. The pressure is $Y$ at the corners because the elements at the corners are in uniaxial compression. A friction hill remains along the edges of the specimen because of friction along the edges.

The lateral expansion (top view) of the edges of a rectangular specimen in actual plane stress upsetting is shown in Fig. 6.7. Whereas the increase in the length of the specimen is 40 percent, the increase in width is 230 percent. The reason is that, as expected, *the material flows in the direction of least resistance*. The width has less frictional resistance than the length. Likewise, after upsetting, a specimen in the shape of an equilateral cube acquires a pancake shape. The diagonal direction expands at a slower rate than the other directions.

### Forging of a solid cylindrical workpiece

Using the slab method of analysis, we can also determine the pressure distribution in forging of a solid cylindrical specimen (Fig. 6.8). We first isolate a segment of angle $d\theta$ in the cylinder of radius $r$ and height $h$, take a small element of radial length $dx$, and place on this element all the normal and frictional stresses acting on it. We then balance the forces on this element in the radial direction. Thus,

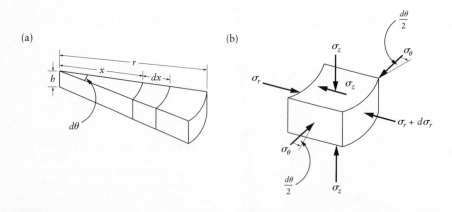

**FIGURE 6.8**   Stresses on an element in forging of a solid cylindrical workpiece between flat dies. Compare with Fig. 6.4. *Source:* (a) After J. F. W. Bishop, J. *Mech Phys. Solids,* vol. 6, 1958, pp. 132–144. (b) Adapted from W. Schroeder and D. A. Webster, *Trans. ASME,* vol. 71, 1949, pp. 289–294.

$$\sigma_r x \cdot d\theta \cdot h + 2\sigma_\theta h \, dx \, \frac{d\theta}{2} - 2\mu\sigma_z x \cdot d\theta \cdot dx - (\sigma_r + d\sigma_r)(x + dx) \cdot d\theta \cdot h = 0.$$

Simplifying this equation, noting that the product $drd\theta$ is very small and hence can be neglected, and dividing by $xhdx$ we obtain

$$\frac{d\sigma_r}{dx} + \frac{\sigma_r - \sigma_\theta}{x} = -\frac{2\mu\sigma_z}{h}.$$

We also note that the circumferential and radial incremental strains are equal by virtue of the fact that

$$d\varepsilon_\theta = \frac{2\pi \, dx}{2\pi x} = \frac{dx}{x} \text{ and } d\varepsilon_r = \frac{dx}{x}.$$

Next we introduce the concept of *proportional straining*, which simply states

$$\frac{d\varepsilon_r}{\sigma_r} = \frac{d\varepsilon_\theta}{\sigma_\theta} = \frac{d\varepsilon_z}{\sigma_z}.$$

Consequently, we have

$$\sigma_r = \sigma_\theta,$$

and thus we have

$$\frac{d\sigma_r}{dx} = -\frac{2\mu\sigma_z}{h}.$$

Also, using the distortion-energy criterion

$$(\sigma_r - \sigma_\theta)^2 + (\sigma_\theta - \sigma_z)^2 + (\sigma_z - \sigma_r)^2 = 2Y^2,$$

we have

$$\sigma_r - \sigma_z = Y,$$

and since the yield stress $Y$ is a constant, we note that

$$d\sigma_r = d\sigma_z.$$

Thus,

$$\frac{d\sigma_z}{dx} = -\frac{2\mu\sigma_z}{h}.$$

Note that this expression is similar to Eq. (6.12) for plane-strain forging. Following the same procedure as before, we obtain the expression for the pressure $p$ at any radius $x$ as

$$p = Ye^{2\mu(r-x)/h}. \tag{6.17}$$

The *average pressure* $p_{av}$ can be given approximately as

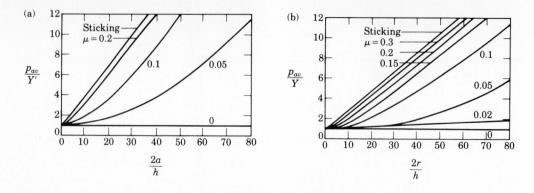

**FIGURE 6.9**  Ratio of average die pressure to yield stress as a function of friction and aspect ratio of the specimen: (a) plane-strain compression; and (b) compression of a solid cylindrical specimen. Note that the yield stress in (b) is $Y$, and not $Y'$ as in plane-strain compression in (a).

$$p_{av} \simeq Y\left(1 + \frac{2\mu r}{3h}\right). \tag{6.18}$$

The *forging force* is thus

$$F = (p_{av})(\pi r^2). \tag{6.19}$$

For strain-hardening materials, $Y$ in Eqs. (6.17) and (6.18) is replaced by the flow stress $Y_f$. In forging operations, the value of the coefficient of friction $\mu$ in Eqs. (6.17) and (6.18) can be estimated to be 0.05 to 0.1 for cold forging, and 0.1 to 0.2 for hot forging. The exact value depends on the effectiveness of the lubricant and can be higher than stated, especially in regions where the workpiece surface is devoid of lubricant.

The effects of friction and the aspect ratio of the specimen ($a/h$ or $r/h$) on the average pressure $p_{av}$ in upsetting are shown in Fig. 6.9. The pressure is given in a dimensionless form and can be regarded as a pressure-multiplying factor. These curves are convenient to use and again show the significance of friction and the aspect ratio of the specimen on upsetting pressure.

**Forging under sticking condition.**  The product of $\mu$ and $p$ is the *frictional stress* (surface shear stress) at the interface at any location $x$ from the center of the specimen. As $p$ increases toward the center, $\mu p$ also increases. However, the value of $\mu p$ cannot be greater than the shear yield stress $k$ of the material. When $\mu p = k$, **sticking** takes place. (In plane strain the value of $k$ is $Y'/2$. See Section 2.11.) Sticking does not necessarily mean adhesion at the interface; it reflects the fact that, relative to the platen surfaces, the material does not move.

For the sticking condition, the normal stress distribution in plane strain can be shown to be

$$p = Y'\left(1 + \frac{a - x}{h}\right). \tag{6.20a}$$

The pressure varies linearly with $x$, as shown in Fig. 6.10. The normal stress distribution for a *cylindrical* specimen under sticking condition can be shown to be

$$p = Y\left(1 + \frac{r - x}{h}\right). \tag{6.20b}$$

The stress distribution again is linear, as shown in Fig. 6.10 for plane strain.

---

**Example 6.1:** Upsetting force

A cylindrical specimen made of annealed 4135 steel is 6 in. in diameter and 4 in. high. It is upset by open-die forging with flat dies to a height of 2 in. at room temperature. Assuming that the coefficient of friction is 0.2, calculate the force required at the end of the stroke. Use the average-pressure formula.

SOLUTION. The average-pressure formula, from Eq. (6.18), is

$$p_{av} \simeq Y\left(1 + \frac{2\mu r}{3h}\right),$$

where $Y$ is replaced by $Y_f$ because the workpiece material is strain hardening. From Table 2.3, $K = 1015$ MPa (147,000 psi) and $n = 0.17$.

The absolute value of the true strain is

$$\varepsilon_1 = \ln\left(\frac{4}{2}\right) = 0.693.$$

Thus

$$Y_f = K\varepsilon_1^n = (147,000)(0.693)^{0.17} = 138,000 \text{ psi}.$$

The final height $h_1$ is 2 in. The radius $r$ at the end of the stroke is found from volume constancy:

$$\frac{\pi 6^2}{4} \cdot 4 = \pi r_1^2 \cdot 2 \quad \text{and} \quad r_1 = 4.24 \text{ in.}$$

Therefore

$$p_{av} \simeq 138,000\left[1 + \frac{(2)(0.2)(4.24)}{(3)(2)}\right] = 177,000 \text{ psi}.$$

The upsetting force then is

$$F = (177,000)\,\pi \cdot (4.24)^2 = 10^7\ \text{lb}.$$

[Note: Figure 6.9b can also be used to solve this problem.]

---

**Example 6.2:**   **Transition from sliding to sticking friction**

In plane-strain upsetting, the frictional stress cannot be higher than the shear yield stress $k$ of the workpiece material. Thus, there may be a distance $x$ in Fig. 6.4 where a transition occurs from sliding to sticking friction. Derive an expression for $x$ in terms of $a$, $h$, and $\mu$ only.

SOLUTION.   The shear stress at the interface due to friction can be expressed as

$$\tau = \mu p.$$

However, we know that the shear stress cannot exceed the yield shear stress $k$ of the material, which, for plane strain, is $Y'/2$. The pressure curve in Fig. 6.5 is given by Eq. (6.13). Thus, in the limit, we have the condition

$$\mu Y' e^{2\mu(a-x)/h} = Y'/2,$$

or

$$2\mu(a - x)/h = \ln(1/2\mu).$$

Hence,

$$x = a - (h/2\mu)\ln(1/2\mu).$$

Note that, as expected, the magnitude of $x$ decreases as $\mu$ decreases. However, the pressures must be sufficiently high to cause sticking, i.e., the $a/h$ ratio must be high. For example, let $a = 10$ mm, and $h = 1$ mm. Then, for $\mu = 0.2$, $x = 7.71$ mm, and for $\mu = 0.4$, $x = 9.72$ mm.

---

**Slip-line analysis**

The **slip-line analysis** method is applied generally to plane-strain conditions. The deforming body is assumed to be rigid, perfectly plastic, and isotropic. The technique consists of the construction of a family of straight or curvilinear lines that intersect each other orthogonally. These lines, known as a **slip-line field**, correspond to the directions of yield stress of the material in shear, $k$.

The network of slip-lines, the construction of which depends largely on intuition and experience and are postulated a priori, must satisfy certain conditions: static equilibrium of forces, yield criteria, and boundary conditions. The slip-line fields must be checked for equilibrium conditions so that, for instance, the slip-lines meet at a free surface at a 45° angle. Also, they must be compatible with the velocity field. This means that the movement of the material must maintain mass continuity.

## Upper-bound technique

In the **upper-bound technique,** the overall deformation zone is divided into a number of smaller zones within which the velocity of a particle is continuous. However, the particle velocities in adjacent zones may be different. As in slip-line analysis, at the boundaries between the zones or between a zone and the die surface, all movement must be such that discontinuities in velocity occur only in the tangential direction.

This technique, which has been used successfully for axisymmetric cases, such as extrusion and rod drawing, is lengthy and gives results similar to those in slip-line analysis. In this analysis, the total power consumed in an operation is the sum of the (1) **ideal** power of deformation, (2) power consumed in **shearing** the material along velocity discontinuities, and (3) power required in overcoming **friction** at the die-workpiece interfaces.

Finally, a velocity field that minimizes the total calculated power is taken as the actual one and is subsequently compared with experimental data.

Theorems have been developed to obtain lower- and upper-bound solutions for the required loads. The lower-bound solution underestimates the load and hence is of no practical value. Conversely, the upper-bound solutions overestimate the load and are therefore of practical interest in metalworking operations.

## Finite-element method

In the **finite-element method,** the deformation zone in an elastic-plastic body is divided into a number of elements interconnected at a finite number of nodal points. The actual velocity distribution is then approximated for each element. A set of simultaneous equations is then developed representing unknown velocity vectors. From the solution of these equations actual velocity distributions and the stresses are calculated.

This technique can incorporate friction conditions at the die-workpiece interfaces and actual properties of the workpiece material. It has been applied to relatively complex geometries in bulk deformation and in sheet-forming problems. Accuracy is influenced by the number and shape of the finite elements, the deformation increment, and the methods of calculation. To ensure accuracy, complex problems require extensive computations. The finite-element method gives a detailed picture of the actual stresses and strain distributions throughout the workpiece.

Figure 6.11 illustrates the results of applying the finite-element method of analysis on a solid cylindrical workpiece in impression-die forging. Only one quarter of the workpiece is shown, with the vertical and horizontal lines being the axes of symmetry. The grid pattern is known as a *mesh*, and its distortion is predicted from theory, aided by the use of computers with the results displayed on the video output screen. Note how the workpiece deforms as it is being forged.

This technique is also capable of predicting microstructural changes in the material during hot-working operations, temperature distribution throughout the workpiece, and the onset of defects without actually running experiments. This knowledge is then utilized in modifying die design to avoid making defective parts. Application of this technique requires inputs such as the stress-strain characteristics of the material, as a function of strain rate and temperature, and frictional and heat transfer characteristics of the die and workpiece.

### Deformation-zone geometry

The observations made concerning Fig. 2.27 are important in estimating and calculating forces in forging and other bulk deformation processes. As shown previously for hardness testing, the compressive stress required for indentation is, ideally, about 3 times the yield stress $Y$ required for uniaxial compression (see Section 2.11). Moreover, (1) the deformation under the indenter is localized, making the overall deformation highly nonuniform; and (2) the deformation zone is relatively small compared to the size of the specimen.

However, in a simple frictionless compression test with flat dies, the top and bottom surfaces of the specimen are always in contact with the dies and the specimen deforms uniformly. Various situations occur between these two extreme examples of specimen-deformation geometry. The *deformation zones* and the pressures required are shown in Fig. 6.12 for the frictionless condition as obtained from **slip-line analysis.** Note that the ratio $h/L$ is the important parameter in determining the inhomogeneity of deformation. The frictionless nature of these examples is important, because friction significantly affects forces, particularly at

**FIGURE 6.12** Die pressures required in *frictionless* plane-strain conditions for a variety of metalworking operations. The geometric relationship between contact area of the dies and workpiece dimensions is an important factor in predicting forces in plastic deformation of materials. *Source:* After W. A. Backofen, *Deformation Processing,* (Fig. 7.1, p. 135), © 1972, Addison-Wesley Publishing Company, Inc. Reprinted by permission of Addison Wesley Longman, Inc.

small values of $h/L$. Deformation-zone geometry depends on the particular process and such parameters as die geometry and percent reduction of the material (Fig. 6.13).

### 6.2.3 Impression-die forging

In **impression-die forging,** the workpiece acquires the shape of the die cavities (impressions) while it is being upset between the closing dies. A typical example is shown in Fig 6.14. Some of the material flows radially outward and forms a **flash.** Because of its high length-to-thickness ratio (equivalent to a high $a/h$ ratio), the flash is subjected to high pressure. These pressures in turn mean high frictional resistance to material flow in the radial direction in the flash gap. Because high friction encourages the filling of the die cavities, the flash has a significant role in the flow of material in impression-die forging.

Furthermore, if the operation is carried out at elevated temperatures, the flash, because of its high surface-to-thickness ratio, cools faster than the bulk of the workpiece. Thus the flash resists deformation more than the bulk does and helps fill the die cavities.

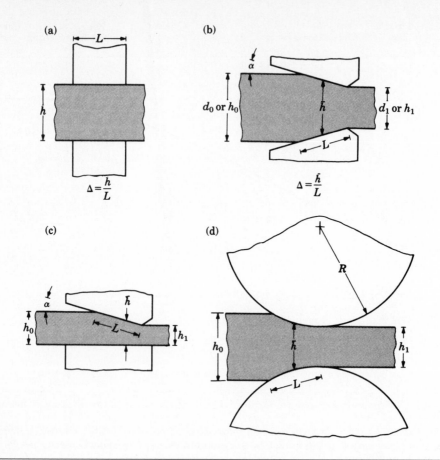

**FIGURE 6.13**   Examples of plastic deformation processes in plane strain showing the $h/L$ ratio. (a) Indenting with flat dies. This operation is similar to cogging, shown in Fig. 6.19. (b) Drawing or extrusion of strip with a wedge-shaped die, described in Sections 6.4 and 6.5. (c) Ironing. See also Fig. 7.57. (d) Rolling, described in Section 6.3. As shown in Fig 6.12, the larger the $h/L$ ratio, the higher the die pressure becomes. In actual processing, however, the smaller this ratio, the greater is the effect of friction at the die-workpiece interfaces. The reason is that contact area, hence friction, increases with decreasing $h/L$ ratio.

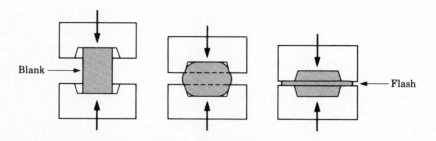

**FIGURE 6.14**   Schematic illustration of stages in impression-die forging. Note the formation of flash, or excess material that is subsequently trimmed off.

**TABLE 6.2**

**RANGE OF $K_p$ VALUES FOR EQ. (6.21) IN IMPRESSION-DIE FORGING**

| | |
|---|---|
| Simple shapes, without flash | 3–5 |
| Simple shapes, with flash | 5–8 |
| Complex shapes, with flash | 8–12 |

Because of the more complex shapes involved, accurate calculation of forces in impression-die forging is difficult. Depending on its position, each element within the workpiece is generally subject to different strains and strain rates. Consequently, the level of strength that the material exhibits at each location depends not only on the strain and strain rate, but also on the exponents $n$ and $m$ of the workpiece material in Eqs. (2.11) and (2.16).

Because of the many difficulties involved in calculating forces in impression-die forging, certain **pressure-multiplying factors** $K_p$ have been recommended, as shown in Table 6.2. These factors are to be used with the expression

$$F = (K_p)(Y_f)(A), \tag{6.21}$$

where $F$ is the forging load, $A$ is the projected area of the forging (including the flash), and $Y_f$ is the flow stress of the material at the strain and strain rate to which the material is subjected.

A typical impression-die forging load as a function of the stroke of the die is shown in Fig. 6.15. For this axisymmetric workpiece, the force increases gradually as the cavity is filled (Fig. 6.15b). The force then increases rapidly as flash forms. In order to obtain the final dimensions and details on the forged part, the dies must close further, with an even steeper rise in the forging load.

Note that the flash has a finite contact length with the die called **land** (see Fig. 6.29). The land ensures that the flash generates sufficient resistance of the outward flow of the material to aid in die filling without contributing excessively to the forging load.

### Closed-die forging

Although not quite correct, the example shown in Fig. 6.14 is also referred to as **closed-die forging.** In true closed-die forging no flash is formed and the workpiece is completely surrounded by the dies. In impression-die forging any excess metal in the die cavity is formed into a flash, but this is not the case in closed-die forging. Thus proper control of the volume of material is essential to obtain a forging of desired dimensions. Undersized blanks in closed-die forging prevent the complete filling of the die; oversized blanks may cause premature die failure or jamming of the dies.

**Precision forging,** or **flashless forging,** and similar operations where the part formed is close to the final dimensions of the desired component, are also known as **near-net-shape** production. Any excess material is subsequently removed by various machining processes.

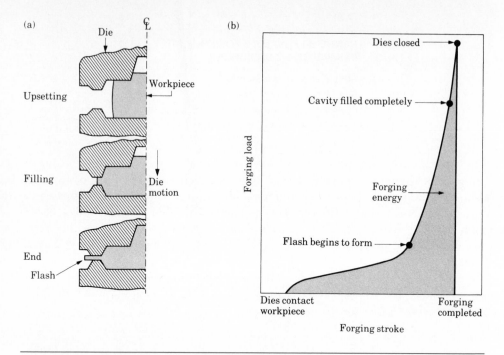

**FIGURE 6.15** Typical load-stroke curve for closed-die forging. Note the sharp increase in load after the flash begins to form. In hot-forging operations, the flash requires high levels of stress because it is thin, that is, small *h,* and cooler than the bulk of the forging. *Source:* After T. Altan.

In precision forging, special dies are made and machined to greater accuracy than in ordinary impression-die forging. Precision forging requires higher capacity forging equipment than do other forging processes. Aluminum and magnesium alloys are particularly suitable for precision forging because of the low forging loads and temperatures required. Also, they result in little die wear and produce a good surface finish. Steels and other alloys are more difficult to precision forge. The choice between conventional forging and precision forging requires an economic analysis. Precision forging requires special dies. However, much less machining is involved, because the part is closer to the desired final shape.

In the **isothermal forging** process, also known as **hot-die forging,** the dies are heated to the same temperature as the hot blank. In this way, cooling of the workpiece is eliminated, the low flow stress of the material is maintained, and material flow within the die cavities is improved. The dies are generally made of nickel alloys, and complex parts with good dimensional accuracy can be forged in one stroke in hydraulic presses. Isothermal forging generally is expensive; however, it can be economical for intricate forgings of expensive materials, provided that the quantity required is large enough to justify die costs.

**Orbital forging** is a relatively new process in which the die moves along an orbital path (Fig. 6.16) and forges or forms the part incrementally. Thus the forces are lower, the operation is more quiet, and part flexibility can be achieved.

**FIGURE 6.16** Schematic illustration of the orbital forging process. Note that the die is in contact only with a portion of the workpiece surface. This process is also called rotary forging, swing forging, or rocking-die forging and can be used for forming bevel gears, wheels, and bearing rings.

### 6.2.4  Miscellaneous forging operations

**Coining.**   Another example of closed-die forging is the minting of *coins,* where the slug is shaped in a completely closed cavity. The pressures required can be as great as five to six times the flow stress of the material in order to produce the fine details of a coin or a medallion. Several coining operations may be necessary in order to obtain full detail on some parts.

The coining process is also used with forgings and other products to improve surface finish and impart the desired dimensional accuracy (**sizing**) of the products. The pressures are large, and usually very little change in shape takes place. Lubricants cannot be tolerated in coining, because they can be trapped in die cavities and prevent reproduction of fine die surface details.

**Heading.**   Basically an upsetting operation, **heading** is performed at the end of a rod (usually round) to form a shape with a larger cross-section. Typical examples are the heads of bolts, screws, and nails (Fig 6.17). An important aspect of the upsetting process during heading is the tendency for buckling if the length-to-diameter ratio is too high. Heading is done on machines called **headers,** which are usually highly automated horizontal machines with high production rates. The operation can be carried out cold, warm, or hot. Heading forces can be estimated reliably using the slab method of analysis described in Section 6.2.2.

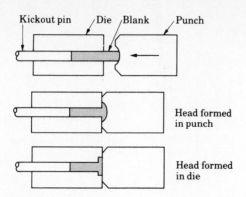

**Piercing.**    In **piercing** a punch indents the surface of a workpiece to produce a cavity or an impression (Fig. 6.18). The workpiece may be confined in a die cavity or it may be unconstrained. The piercing force depends on the punch cross-sectional area and its tip geometry, the flow stress of the material, and the friction at the interfaces. Punch pressures may range from three to five times the flow stress of the material. The term piercing is also used for the process of cutting of holes with a punch and die.

**Hubbing.**    In **hubbing** a hardened punch with a particular tip geometry is pressed into the surface of a block. This process is used to produce a die cavity, which is then used for subsequent forming operations. The cavity is usually shallow; for deeper cavities some material may be removed from the surface by machining prior to hubbing. The pressure required to generate a cavity by hubbing is approximately three times the ultimate tensile strength (UTS) of the material of the block. Thus the force required is

$$\text{Hubbing force} = 3\,(\text{UTS})(A) \tag{6.22}$$

where $A$ is the projected area of the impression. Note that the factor 3 is in agreement with the observations made with regard to hardness of materials in Section 2.6.8.

**Cogging.**    In **cogging,** also called *drawing out,* the thickness of a bar is reduced by successive steps at certain intervals (Fig. 6.19). A long section of a bar can be reduced in thickness without large forces, because the contact area is small.

**FIGURE 6.18** Examples of piercing operations.

**FIGURE 6.19** Schematic illustration of a cogging operation on a rectangular bar. With simple tools, the thickness and cross-section of a bar can be reduced by multiple cogging operations. Note the barreling after cogging. Blacksmiths use a similar procedure to reduce the thickness of parts in small increments by heating the workpiece and hammering it numerous times.

**Fullering and edging.** The operations of **fullering** and **edging** are performed—usually on bar stock—to distribute the material in certain regions of a forging (see Fig. 6.27). The parts are then formed into final shapes by other forging processes.

**Roll forging.** In **roll forging** the cross-sectional area of a bar is reduced and altered in shape by passing it through a pair of rolls with grooves of various shapes (Fig. 6.20). This operation may also be used to produce a part that is basically the final product, such as tapered shafts, tapered leaf springs, table knives, and numerous tools. Roll forging is also used as a preliminary forming operation, followed by other forging processes. Typical products are crankshafts and other automotive components.

**Skew rolling.** A process similar to roll forging is **skew rolling** (Fig. 6.21), which is used for making ball bearings. Round wire or rod stock is fed into the roll gap and spherical blanks are formed continuously by the rotating rolls. (Another method of forming blanks for ball bearings is by cutting pieces from a round bar and upsetting them, as shown in Fig. 6.22. The balls are then ground and polished in special machinery.)

### 6.2.5 Defects

Besides *surface cracking,* other defects in forgings can be caused by the material flow patterns in the die cavity. As shown in Fig. 6.23, excess material in the web of a forging can buckle

**FIGURE 6.20** Schematic illustration of a roll forging (cross rolling) operation. Tapered leaf springs and knives can also be made by this process with specially designed rolls. *Source:* After J. Holub, *Machinery,* vol. 102, Jan 16, 1963.

**FIGURE 6.21** Production of steel balls for bearings by the skew rolling process. Balls for bearings can also be made by the forging process shown in Fig. 6.22.

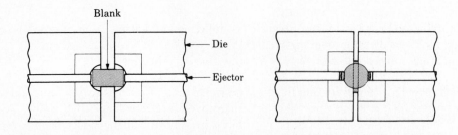

**FIGURE 6.22** Production of steel balls by upsetting of a cylindrical blank. Note the formation of flash. The balls are subsequently ground and polished for use in ball bearings and other mechanical components.

during forging and develop **laps.** If the web is thick, the excess material flows past the already forged portions and develops internal cracks (Fig. 6.24). These examples indicate the importance of properly distributing material and controlling the flow in the die cavity.

The various radii in the die cavity can significantly affect formation of defects. In Fig. 6.25, the material follows a large corner radius better than a small radius. With small radii, the material can fold over itself and produce a lap, called **cold shut.**

Blocked forging     Begin finishing     Web buckles     Laps in finished forging

**FIGURE 6.23** Laps formed by buckling of the web during forging. The solution to this problem is to increase the initial web thickness to avoid buckling.

**FIGURE 6.24** Internal defects produced in a forging because of an oversized billet. The die cavities are filled prematurely and the material at the center of the part flows past the filled regions as deformation continues.

During the service life of the forged component, such defects can lead to fatigue failure and other problems, such as excessive corrosion and wear. The importance of inspecting forgings before they are put into service, particularly for critical applications, is obvious. (For inspection techniques, see Section 4.8.)

Although it may not be considered a flaw, another important aspect of quality in a forging is the **grain flow pattern.** At times the grain flow lines reach a surface perpendicularly, exposing the grain boundaries directly to the environment. They are known as **end grains.** In service they can be attacked by the environment and develop a rough surface and act as stress raisers. For critical components, end grains in forgings can be avoided by proper selection of the original workpiece orientation in the die cavity and by control of material flow.

Because the metal flows in various directions in a forging—and because of temperature variations—the properties of a forging are generally anisotropic. As Fig. 6.26 shows, strength and ductility vary significantly in test specimens taken from different locations and orientations in a forged part.

**FIGURE 6.25** Effect of fillet radius on defect formation in forging. Small fillets (right side of drawings) cause the defects. *Source:* ALCOA.

**FIGURE 6.26** Mechanical properties of five tensile-test specimens taken at various locations and directions in an AZ61 magnesium alloy forging. Note the anisotropy of properties caused by inhomogeneous deformation during forging. *Source:* After S. M. Jablonski, Modern Metals, vol. 16, 1963, pp. 62-70.

| Specimen location | Yield strength, psi | Ultimate tensile strength, psi | Elongation, % |
|---|---|---|---|
| 1. | 26,500 | 41,200 | 20.0 |
| 2. | 23,000 | 39,400 | 17.7 |
| 3. | 13,800 | 38,000 | 18.6 |
| 4. | 25,400 | 40,400 | 16.2 |
| 5. | 17,100 | 37,500 | 14.0 |

### 6.2.6   Forgeability

The **forgeability** of a metal can be defined as its capability to undergo deformation by forging without cracking. This definition can be expanded to include the flow strength of the metal. Thus a material with good forgeability is one that can be shaped with low forces without cracking. A number of tests have been developed to measure forgeability, although none is universally accepted. We describe the more commonly used tests.

**Upsetting test.**   A standard test is to *upset* a solid cylindrical specimen and observe any cracking on the barreled surfaces (Fig. 3.20d). The greater the reduction in height prior to cracking, the greater is the forgeability of the metal. These cracks are caused by **secondary tensile stresses** on the barreled surfaces. They are called secondary because no external tensile stress is applied to the material, and the stresses develop because of changes in the material's shape.

Because barreling is a result of friction at the die—workpiece interfaces, friction has a marked effect on cracking in upsetting. As friction increases the specimen cracks at a lower reduction in height. Of interest also is whether the crack is at 45 degrees or vertical; the angle depends on the sign of the axial stress $\sigma_z$ on the barreled surface. If this stress is negative (compressive) the crack is at 45 degrees; if positive (tensile) it is longitudinal. Depending on the notch sensitivity of the material, any surface defects will also affect the results by causing premature cracking. A typical surface defect is a **seam:** a longitudinal scratch, a string of inclusions, or a fold from prior working of the material.

Upsetting tests can be performed at various temperatures and strain rates. An optimal range for these parameters can then be specified for forging a particular material. Such tests can serve only as guidelines, since in actual forging the metal is subjected to a different state of stress as compared with a simple upsetting operation.

**Hot-twist test.**   A round specimen is twisted continuously until it fails. The test is performed at various temperatures, and the number of turns that each specimen undergoes before failure is observed. The optimal forging temperature is then determined. The hot-twist test is particularly useful in determining the forgeability of steels, although upsetting tests can also be used for that purpose. Small changes in the composition of or impurities in the metal can have a significant effect on forgeability.

**Effect of hydrostatic pressure on forgeability.** As described in Section 2.2.8, hydrostatic pressure has a significant beneficial effect on the ductility of metals and nonmetallic materials. Experiments have indicated that the results of room-temperature forgeability tests, outlined above, are improved (that is, cracking takes place at higher strain levels) if the tests are conducted in an environment of high *hydrostatic pressure*. Although such a test would be difficult to perform at elevated temperatures, techniques have been developed to forge metals in a high compressive environment in order to take advantage of this phenomenon. The pressure-transmitting medium is usually a low-strength ductile metal.

**Forgeability of various metals.** Based on the results of various tests and observations, the forgeability of several metals and alloys has been determined. It is based on such considerations as the ductility and strength of the metals, forging temperature required, frictional behavior, and quality of the forgings obtained.

### 6.2.7 Dies

Proper design of forging dies and selection of die materials require considerable experience and knowledge of the strength and ductility of the workpiece material, its sensitivity to strain rate and temperature, and its frictional characteristics. Die *distortion* under high forging loads can also be an important consideration in die design, particularly if close tolerances are required.

The most important rule in die design is that the workpiece material flow *in the direction of least resistance*. Thus common practice is to distribute the material so that it can properly fill die cavities. An example is the forging of a connecting rod, as shown in Fig. 6.27. Starting with a round bar stock, the bar is first preformed (**intermediate shape**) by some of the techniques previously described and formed into the final shape in two additional forging operations, and then trimmed. Note how proper distribution of the material fills the die cavities to forge the part successfully.

Another example of intermediate shapes is known in Fig. 6.28. Selection of these shapes requires experience and involves calculating cross-sectional areas at each location in the forging. The dies used for intermediate forging operations are called **blocker dies.** Care in calculating

Blank

Edging

Blocking

Finishing

Trimming

**FIGURE 6.27** Stages in forging a connecting rod for an internal combustion engine. Note the amount of flash that is necessary to fill the die cavities properly.

1. Roll forged (first pass)

2. Roll forged (second pass)

3. Blocked in closed dies

4. Finish forged in closed dies

5. Trimmed (before twisting)

**FIGURE 6.28** Intermediate stages in forging a crankshaft. These intermediate stages are important for distributing the material and filling the die cavities properly.

the volume of the material required to properly fill the die cavities is essential. Computer techniques are now being developed to expedite these calculations. The terminology used in die design is shown in Fig. 6.29. The significance of various parameters are as follows.

The **parting line** is where the two dies meet. For simple symmetrical shapes the parting line is a straight line at the center of the forging; for more complex shapes the line may be offset and may not be in a single plane. Selection of the proper location for the parting line is based on the shape of the part, metal flow, balance of forces, and the flash.

The significance of **flash** is described in Section 6.2.3. After sufficiently constraining lateral flow (by the length of the land), the flash is allowed to flow into a **gutter.** Thus the extra flash does not increase the forging load unnecessarily. A general guideline for flash clearance (between the dies) is 3 percent of the maximum thickness of the forging. The length of the land is usually five times the flash clearance.

**Draft angles** are necessary in almost all forgings to facilitate the removal of the part from the die. Draft angles usually range between 3° and 10°. Because the forging shrinks in its radial direction (as well as in other directions) as it cools, internal draft angles are made larger than external ones. Thus internal angles are about 7° to 10° and external angles about 3° to 5°.

**FIGURE 6.29** Standard terminology for various features of a typical forging die.

**TABLE 6.3**

**FORGING TEMPERATURE RANGES FOR VARIOUS METALS**

|  | °C | °F |  | °C | °F |
|---|---|---|---|---|---|
| Aluminum alloys | 400–450 | 750–850 | Titanium alloys | 750–795 | 1400–1800 |
| Copper alloys | 625–950 | 1150–1750 | Refractory alloys | 975–1650 | 1800–3000 |
| Alloy steels | 925–1250 | 1700–2300 |  |  |  |

Proper selection of the **radii** for corners and fillets ensures smooth flow of the metal in the die cavity and improves die life. Small radii are generally not desirable because of their adverse effect on metal flow and their tendency to wear rapidly from stress concentration and thermal cycling. Small radii in fillets can cause fatigue cracking in dies.

**Die materials.** Because most forgings, particularly large ones, are performed at elevated temperatures, die materials generally must have strength and toughness at elevated temperatures, hardenability, mechanical and thermal shock resistance, and wear resistance—particularly, abrasive wear resistance because of the presence of scale on heated forgings.

Selection of die materials depends on die size, composition and properties of the workpiece, complexity of its shape, forging temperature, type of operation, cost of die material, the number of forgings required, and the heat transfer and distortion characteristics of the die material. Common die materials are tool and die steels, which contain chromium, nickel, molybdenum, and vanadium.

**Forging temperatures and lubrication.** The range of temperatures for hot forging are given in Table 6.3.

*Lubrication* in forging plays an important role. It affects friction and wear and, consequently, the flow of metal into the die cavities. Lubricants can also serve as a *thermal barrier* between the hot forging and the relatively cool dies, thus slowing the rate of cooling of the workpiece. Another important role of a lubricant is to serve as a *parting agent*, that is, to prevent the forging from sticking to the dies. Lubricants can significantly affect the wear pattern in forging dies. A wide variety of metal working fluids can be used in forging. For hot forging, graphite, molybdenum disulfide, and (sometimes) glass are commonly used as lubricants. For cold forging, mineral oils and soaps are common lubricants (see Table 6.4).

### 6.2.8 Equipment

Forging equipment of various designs, capacities, speeds, and speed-stroke characteristics is available (Fig. 6.30).

**Hydraulic presses.** Hydraulic presses have a constant low speed and are **load limited**. Large amounts of energy can be transmitted to the workpiece by a constant load available throughout the stroke. Ram speed can be varied during the stroke. These presses are used for both open-die and closed-die forging operations. The largest hydraulic press in existence has a capacity of 670 MN (75,000 tons).

**TABLE 6.4**

**LUBRICANTS COMMONLY USED IN METALWORKING OPERATIONS (SEE ALSO CHAPTER 4)**

| MATERIAL AND ALLOY | TEMPERATURE | FORGING | ROLLING | ROD AND EXTRUSION | WIRE DRAWING | SHEET METALWORKING |
|---|---|---|---|---|---|---|
| Aluminum | Cold | FA + MO, S | FA + MO, MO | D, G, MS, L, S | MO, E, Wa, FA + MO | FO + MO, E, L, Wa, S |
| | Hot | MO + G, MS | FA + CO, FA + E | D, G, PI | — | G |
| Beryllium | Hot | MO + G, J | G | MS, G, J | G | — |
| Copper | Cold | S, E, T | E, MO | S, T, L, Wa, G, MS | FO + E + S, MO, Wa | FO + E, FO + MO, S + Wa |
| | Hot | G | D, E | D, MO + G | — | G |
| Lead | Cold | FO + MO | FA + MO, EM | D, S, T, E, FO | FO | FO + MO |
| | Hot | | | D | | |
| Magnesium | Cold | FA + MO, CC + S | D, FA + MO | D, T | — | FO + MO |
| | Hot | W + G, MO + G | MO + FA + EM, D, G | D, PI | — | G + MO, S + Wa, T + G |
| Nickel | Cold | — | MO + CL | CC + S | CC + S, MO + CL | E, MO + EP, CL, CC + S |
| | Hot | MO + G, W + G, Gl | W, E | Gl, J | — | G, MS |
| Refractory | Hot | Gl, G, MS | G + MS | J + Gl | G + MS, PI | MS, G |
| Steels (carbon and low alloy) | Cold | EP + MO, CC + S | E, CO, MO | CC + S, T, W + MS | S, CC + S | E, S, Wa, FO + MO, PI, CSN |
| | Hot | MO + G, Sa, Gl | W, MO, CO, G + E | Gl, G | — | G |
| Steels (stainless) | Cold | CL + MO, CC + S | CL + EM, CL + MO | CC + S, CP, MO | CC + S, CL + MO | FO+MO, Wa, PI, EP + MO, S |
| | Hot | MO + G, Gl | D, W, E | Gl | – | G |
| Titanium | Cold | S, MO | MO, CC + FO, G | CC + G, CC + S | CC + PI | CL, S, PI, Wa |
| | Hot | W + G, Gl, MS | PI, CC, G, MS | J + Gl, Gl | — | G, MS |

CC  Conversion coatings
CI  Chlorinated paraffin
CO  Compounded oil
CP  Copper plate
CSN Chemicals and synthetics

D   Dry
E   Emulsion
EP  Extreme pressure
FA  Fatty acid

FO  Fatty oil
G   Graphite
Gl  Glass
J   Canned or jacketed

L   Lanolin
MO  Mineral oil
MS  Molybdenum disulfide

PI  Polymer
S   Soap
Sa  Salts

T   Tallow
W   Water
Wa  Wax

**FIGURE 6.30** Schematic illustration of various types of presses used in metalworking. The choice of the press is an important factor in the overall operation.

**Mechanical presses.** Mechanical presses are **stroke limited.** They are basically crank or eccentric types, with speeds varying from a maximum at the center of the stroke to zero at the bottom. The force available depends on the stroke position and becomes extremely large at the bottom dead center. Thus proper setup is essential to avoid breaking the dies or other equipment. The largest mechanical press has a capacity of 107 MN (12,000 tons).

**Screw presses.** Screw presses derive their energy from a flywheel. The forging load is transmitted through a vertical screw. These presses are **energy limited** and can be used for many forging operations. They are particularly suitable for producing small quantities, for parts requiring precision (such as turbine blades), and for control of ram speed. The largest screw press has a capacity of 280 MN (31,500 tons).

**Hammers.** Hammers derive their energy from the potential energy of the ram, which is then converted to kinetic energy; thus they are **energy limited.** The speeds are high; therefore the low forming times minimize cooling of the hot forging, allowing the forging of complex shapes, particularly with thin and deep recesses. To complete the forging, several blows may have to be made on the part. In **power hammers** the ram is accelerated in the downstroke by steam or air, in addition to gravity. The highest energy available in power hammers is 1150 kJ (850,000 ft-lb).

**Counterblow hammers.** Counterblow hammers have two rams that simultaneously approach each other to forge the part. They are generally of the mechanical-pneumatic or mechanical-hydraulic type. These machines transmit less vibration to the foundation. The largest counterblow hammer has a capacity of 1200 kJ (900,000 ft-lb).

**Equipment selection.** Selection of forging equipment depends on the size and complexity of the forging, strength of the material and its sensitivity to strain rate, and degree of deformation required. Generally, presses are usually preferred for aluminum, magnesium, beryllium, bronze, and brass; hammers are preferred for copper, steels, titanium, and refractory alloys. Production rate is also a consideration in equipment selection. The number of strokes per minute ranges from a few for hydraulic presses to as many as 300 for power hammers.

## 6.3 Rolling

**Rolling** is the process of reducing the thickness or changing the cross-section of a long workpiece by compressive forces applied through a set of **rolls** (Fig. 6.31) similar to rolling dough with a rolling pin to reduce its thickness. Rolling, which accounts for about 90 percent of all

**FIGURE 6.31** Schematic outline of various flat-and shape-rolling processes. *Source:* American Iron and Steel Institute.

**FIGURE 6.32** Changes in the grain structure of cast or large-grain wrought metals during hot rolling. Hot rolling is an effective way to reduce grain size in metals for improved strength and ductility. Cast structures of ingots or continuous castings are converted to a wrought structure by hot working.

metals produced by metalworking processes, was first developed in the late 1500s. The basic operation is flat rolling, or simply rolling, where the rolled products are flat plate and sheet.

**Plates,** which are generally regarded as having a thickness greater than 6 mm (1/4 in.), are used for structural applications such as ship hulls, boilers, bridges, girders, machine structures, and nuclear vessels. Plates can be as much as 0.3 m (12 in.) thick for the supports for large boilers, 150 mm (6 in.) for reactor vessels, and 100–125 mm (4–5 in.) for battleships and tanks.

**Sheets** are generally less than 6 mm thick and are used for automobile bodies, appliances, containers for food and beverages, and kitchen and office equipment. Commercial aircraft fuselages are usually made of about 1 mm (0.040 in.) thick aluminum-alloy sheet, while beverage cans are made of 0.15 mm (0.006 in.) aluminum sheet. Aluminum **foil** used to wrap candy and cigarettes has a thickness of 0.008 mm (0.0003 in.). Sheets are provided as flat pieces or as strip in coils to manufacturing facilities for further processing into products.

Traditionally, the initial material form for rolling is an ingot. However, this practice is now being rapidly replaced by continuous casting and rolling, with much higher efficiency and lower cost. Rolling is first carried out at elevated temperatures (hot rolling), wherein the coarse-grained, brittle, and porous structure of the ingot or continuously cast metal is broken down into a **wrought structure,** with finer grain size (Fig. 6.32).

### 6.3.1  Mechanics of flat rolling

The flat rolling process is shown schematically in Fig. 6.33. A strip of thickness $h_0$ enters the roll gap and is reduced to $h_f$ by the rotating (powered) rolls. The surface speed of the roll is $V_r$. To keep the volume rate of metal flow constant, the velocity of the strip must increase as it moves through the roll gap (similar to fluid flow through a converging channel). At the exit of the roll gap, the velocity of the strip is $V_f$ (Fig. 6.34). Because $V_r$ is constant along the roll gap, sliding occurs between the roll and the strip.

At one point along the arc of contact the two velocities are the same. It is known as the **neutral point** or **no-slip point.** To the left of this point, the roll moves faster than the workpiece, and to the right the workpiece moves faster than the roll. Because of friction at the interfaces, the frictional forces—which oppose motion—act on the strip surfaces, as shown in Fig. 6.34.

Note the similarity between Figs. 6.34 and 6.1b. The friction forces oppose each other at the neutral point. In upsetting, these forces are equal to each other because of the symmetry

(Top roll removed)

**FIGURE 6.33** Schematic illustration of the flat-rolling process. A greater volume of metal is formed by rolling than by any other metalworking process.

of the operation. In rolling, the frictional force on the left of the neutral point must be greater than the force on the right. This difference yields a net frictional force to the right, which makes the rolling operation possible by pulling the strip into the roll gap. Furthermore, the net frictional force and the surface velocity of the roll must be in the same direction in order to supply work to the system. Thus the neutral point should be toward the exit in order to satisfy these requirements.

**Forward slip** in rolling is defined in terms of the exit velocity of the strip $V_f$ and the surface speed of the roll $V_r$ as

$$\text{Forward slip} = \frac{V_f - V_r}{V_r} \tag{6.23}$$

and is a measure of the relative velocities involved.

### Roll pressure distribution

Although the deformation zone in the roll gap is subjected to a state of stress similar to that in upsetting, the calculation of forces and stress distribution in flat rolling is more involved because of the curved surface of contact. In addition, the material at the exit is strain hardened, so the flow stress at the exit is higher than that at the entry.

The stresses on an element in the entry and exit zones, respectively, are shown in Fig. 6.35. Note that the only difference between the two elements is the direction of the friction force. Using the **slab method** of analysis for **plane strain** (described in Section 6.2.2), we may analyze the stresses in rolling as follows.

**FIGURE 6.34** Relative velocity distribution between roll and strip surfaces. Note the difference in the direction of frictional forces. The arrows represent the frictional forces acting on the strip.

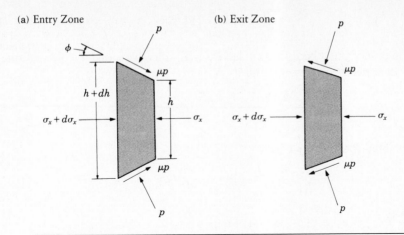

**FIGURE 6.35** Stresses on an element in rolling: (a) entry zone and (b) exit zone.

From the equilibrium of the horizontal forces on the element shown in Fig. 6.35, we have

$$(\sigma_x + d\sigma_x)(h + dh) - 2pR\,d\phi\,\sin\phi - \sigma_x h \pm 2\mu pR\,d\phi\,\cos\phi = 0.$$

When we simplify, neglecting second-order terms, this expression reduces to

$$\frac{d(\sigma_x h)}{d\phi} = 2pR(\sin\phi \mp \mu\cos\phi).$$

In rolling, the angle $\alpha$ is only a few degrees; hence we may assume that $\sin\phi = \phi$ and $\cos\phi = 1$. Thus

$$\frac{d(\sigma_x h)}{d\phi} = 2pR(\phi \mp \mu). \tag{6.24}$$

Because the angles involved are small, we assume that $p$ is a *principal stress*, $\sigma_x$ being the other principal stress. The relationship between these two stresses and the flow stress $Y_f$ of the material is given by Eq. (2.45) for plane strain, or

$$p - \sigma_x = \frac{2}{\sqrt{3}}\,Y_f = Y_f'. \tag{6.25}$$

For a strain-hardening material, the flow stress $Y_f$ in these expressions corresponds to the strain that the material has undergone at that particular location in the roll gap.

Rewriting Eq. (6.24),

$$\frac{d\big[(p - Y'_f)h\big]}{d\phi} = 2pR(\phi \mp \mu),$$

or

$$\frac{d}{d\phi}\left[Y'_f\left(\frac{p}{Y'_f} - 1\right)h\right] = 2pR(\phi \mp \mu)$$

$$Y'_f h \frac{d}{d\phi}\left(\frac{p}{Y'_f}\right) + \left(\frac{p}{Y'_f} - 1\right)\frac{d}{d\phi}(Y'_f h) = 2pR(\phi \mp \mu)$$

The second term is very small because as $h$ decreases $Y'_f$ increases, due to cold working, thus making the product nearly a constant, and its derivative thus becoming zero. We now have

$$\frac{(d/d\phi)(p/Y'_f)}{p/Y'_f} = \frac{2R}{h}(\phi \mp \mu). \tag{6.26a}$$

Letting $h_f$ be the final thickness, we have

$$h = h_f + 2R(1 - \cos\phi),$$

or, approximately,

$$h = h_f + R\phi^2. \tag{6.26b}$$

Substituting this in Eq. (6.26a) and integrating, we have

$$\ln\frac{p}{Y'_f} = \ln\frac{h}{R} \mp 2\mu\sqrt{\frac{R}{h_f}}\tan^{-1}\sqrt{\frac{R}{h_f}}\,\phi + \ln C,$$

or

$$p = C\,Y'_f \frac{h}{R}\,e^{\mp\mu H}.$$

where

$$H = 2\sqrt{\frac{R}{h_f}}\tan^{-1}\left(\sqrt{\frac{R}{h_f}}\,\phi\right). \tag{6.27}$$

At entry, $\phi = \alpha$; hence, $H = H_0$ with $\phi$ replaced by $\alpha$. At exit, $\phi = 0$ hence; $H = H_f = 0$. Also, at entry and exit, $p = Y'_f$. Hence, in the **entry zone**,

$$C = \frac{R}{h_f}e^{\mu H_i}$$

and

$$p = Y'_f \frac{h}{h_0}e^{\mu(H_0 - H)} \tag{6.28}$$

In the **exit zone**, we have

$$C = \frac{R}{h_f}$$

and

$$p = Y'_f \frac{h}{h_f}e^{\mu H}. \tag{6.29}$$

**FIGURE 6.36** Pressure distribution in the roll gap as a function of the coefficient of friction. Note that, as friction increases, the neutral point shifts toward the entry. Without friction, the rolls slip and the neutral point shifts completely to the exit.

**FIGURE 6.37** Pressure distribution in the roll gap as a function of reduction in thickness. Note the increase in the area under the curves with increasing reduction in thickness, thus increasing the roll-separating force.

Note that the pressure $p$ is a function of $h$ and the angular position $\phi$ along the arc of contact. These expressions also indicate that the pressure increases with increasing strength of the material, increasing coefficient of friction, and increasing $R/h_f$ ratio. The $R/h_f$ ratio in rolling is equivalent of the $a/h$ ratio in upsetting (see Section 6.2.2).

The dimensionless theoretical pressure distribution in the roll gap is shown in Fig. 6.36. Note the similarity of this curve to Fig 6.5 (**friction hill**). Also note that the neutral point shifts toward the exit as friction decreases, which is to be expected. When friction approaches zero, the rolls begin to slip instead of pulling the strip in, so the neutral point must approach the exit. (This situation is similar to the spinning of automobile wheels when accelerating on wet pavement.)

The effect of reduction in thickness of the strip on the pressure distribution is shown in Fig. 6.37. As reduction increases, the length of contact in the roll gap increases, which in turn increases the peak pressure. The curves shown are theoretical; actual pressure distributions, as determined experimentally, have smoother curves with their peaks rounded off.

**Determination of the neutral point.** We determine the neutral point simply by equating Eqs. (6.28) and (6.29). Thus at the neutral point,

$$\frac{h_0}{h_f} = \frac{e^{\mu H_0}}{e^{2\mu H_n}} = e^{\mu(H_0 - 2H_n)}$$

**FIGURE 6.38** Pressure distribution as a function of front and back tension. Note the shifting of the neutral point and the reduction in the area under the curves with increasing tension.

or

$$H_n = \frac{1}{2}\left(H_0 - \frac{1}{\mu}\ln\frac{h_0}{h_f}\right).$$

(6.30)

Substituting Eq. (6.30) into Eq. (6.27), we obtain

$$\phi_n = \sqrt{\frac{h_f}{R}}\tan\left(\sqrt{\frac{h_f}{R}}\cdot\frac{H_n}{2}\right).$$

(6.31)

### Front and back tension

The roll force $F$ can be reduced by various means, such as lower friction, smaller roll radii, smaller reductions, and higher workpiece temperatures. A particularly effective method is to reduce the apparent compressive yield stress of the material by applying longitudinal *tension*. Recall from our discussion of yield criteria (Section 2.11) that if tension is applied to a strip (Fig. 6.38), the yield stress normal to the strip surface drops and hence the roll pressure will decrease.

Tensions in rolling can be applied either at the entry (**back tension,** $\sigma_b$) or at the exit of the strip (**front tension** $\sigma_f$) or both. We then modify Eqs. (6.28) and (6.29) to include the effect of tension for the entry and exit zones, respectively, as follows:

$$p = (Y_f' - \sigma_b)\frac{h}{h_0}e^{\mu(H_0-H)}$$

(6.32)

and

$$p = (Y_f' - \sigma_f)\frac{h}{h_f}e^{\mu H}.$$

(6.33)

Depending on the relative magnitudes of the tensions applied, the neutral point may shift, as shown in Fig. 6.38. This shift affects the pressure distribution, torque, and power requirements in rolling. Front tension is controlled by the torque on the coiler (**delivery reel**) around which the rolled sheet is coiled. The back tension is controlled by a braking system in the uncoiler (**payoff reel**). Special instrumentation is available for these controls. Tensions are particularly important in rolling thin, high-strength materials because they require high roll forces.

**Example 6.3:** Back tension required to cause roll slip

It is to be expected that in rolling a strip, the rolls will begin to slip if the back tension $\sigma_b$ is too high. Derive an expression for the magnitude of back tension required to make the rolls begin to slip.

SOLUTION.  Slipping of the rolls means that the neutral point has moved all the way to the exit in the roll gap. Thus, the whole contact area becomes the entry zone and Eq. (6.32) is applicable. Also, we know that when $\phi = 0$, $H = 0$. Thus, the pressure at the exit is

$$p_{\phi=0} = (Y_f' - \sigma_b)\left(\frac{h_f}{h_0}\right)e^{\mu H_0}.$$

However, we also know that at the exit, the pressure is equal to $Y_f'$. Rearranging this equation, we obtain

$$\sigma_b = Y_f'\left[1 - \left(\frac{h_0}{h_f}\right)(e^{-\mu H_0})\right],$$

where $H_0$ is obtained from Eq. (6.29) with the condition $\phi = \alpha$. Since all the quantities are known, we can thus calculate the magnitude of the back tension.

---

### Roll forces

The area under these pressure-contact length curves multiplied by the strip width, $w$, is the **roll force**, $F$, on the strip. This area can be obtained graphically after the points are plotted. This force can also be calculated from the expression

$$F = \int_0^{\phi_n} wpR\,d\phi + \int_{\phi_n}^{\alpha} wpR\,d\phi. \tag{6.34}$$

A simpler method of calculating the roll force is to multiply the contact area with an average contact stress:

$$F = Lwp_{\text{av}}, \tag{6.35}$$

where $L$ is the *arc of contact*. We can approximate the dimension $L$ as the projected area:

$$L = \sqrt{R\Delta h}, \tag{6.36}$$

where $R$ is the roll radius and $\Delta h$ is the difference between the original and final thicknesses of the strip (**draft**).

The estimation of $p_{\text{av}}$ depends on the $h/L$ ratio, where $h$ is now the *average* thickness of the strip in the roll gap (see Figs. 6.12 and 6.13). For large $h/L$ ratios (small reductions and/or small roll diameters), the rolls act as indenters in a hardness test. Friction is not significant, and $p_{\text{av}}$ is obtained from Fig. 6.12, where $a = L/2$. Small $h/L$ ratios (large reductions and/or large roll diameters) are equivalent to high $a/h$ ratios. Thus friction is predominant, and $p_{\text{av}}$ is obtained from Eq. (6.15). For strain-hardening materials the appropriate flow stresses must be calculated.

As a rough approximation and for low frictional conditions, we can simplify Eq. (6.35) as follows:

$$F = Lw\overline{Y}', \tag{6.37}$$

where $\overline{Y}'$ is the **average flow stress** in plane strain (see Fig. 2.42) of the material in the roll gap.

For higher frictional conditions, we may write an expression similar to Eq. (6.15), or

$$F = Lw\overline{Y}'\left(1 + \frac{\mu L}{2h_{av}}\right) \tag{6.38}$$

**Roll torque and power**

The **roll torque** $T$ for each roll can be calculated analytically from the expression

$$T = \int_{\phi_n}^{\alpha} w\mu p R^2 d\phi - \int_{0}^{\phi_n} w\mu p R^2 d\phi \tag{6.39}$$
$$\quad\quad \text{(entry)} \quad\quad\quad \text{(exit)}$$

Note that the negative sign indicates the change in direction of the friction force at the neutral point. Thus, if the frictional forces are equal to each other, the torque is zero.

The torque in rolling can also be estimated by assuming that the roll-separating force $F$ acts in the *middle* of the arc of contact, that is, a moment arm of $0.5L$, and that $F$ is perpendicular to the plane of the strip. (Whereas $0.5L$ is a good estimate for hot rolling, $0.4L$ is a better estimate for cold rolling.)

The **torque per roll** then is

$$T = \frac{FL}{2}.$$

The **power required per roll** is

$$\text{Power} = T\omega, \tag{6.40}$$

where $\omega = 2\pi N$ and $N$ is the revolutions per minute of the roll. Consequently, the power per roll is

$$\text{Power} = \frac{\pi F L N}{60,000} \text{ kW}, \tag{6.41}$$

where $F$ is in newtons, $L$ is in meters, and $N$ is the rpm of the roll. We can also express the power as

$$\text{Power} = \frac{\pi F L N}{33,000} \text{ hp}, \tag{6.42}$$

where $F$ is in lb and $L$ is in ft.

---

**Example 6.4:   Power required in rolling**

A 9 in. wide 6061-O aluminum strip is rolled from a thickness of 1.00 in. to 0.80 in. If the roll radius is 12 in. and the roll rpm is 100, estimate the horsepower required for this operation.

SOLUTION. The power needed for a set of *two* rolls is given by Eq. (6.42):

$$\text{Power} = \frac{2\pi FLN}{33,000} \text{ hp,}$$

where $F$ is given by Eq. (6.37) and $L$ by Eq. (6.36). Thus

$$F = Lw\overline{Y}' \quad \text{and} \quad L = \sqrt{R\Delta h}.$$

We now have

$$L = \sqrt{(12)(1.0 - 0.8)} = 1.55 \text{ in.} = 0.13 \text{ ft} \quad \text{and} \quad w = 9 \text{ in.}$$

For 6061-O aluminum, $K = 30,000$ psi and $n = 0.2$ (from Table 2.3). The true strain in this operation is

$$\varepsilon_1 = \ln\left(\frac{1.0}{0.8}\right) = 0.223.$$

Thus, from Eq. (6.10),

$$\overline{Y} = \frac{(30,000)(0.223)^{0.2}}{1.2} = 18,500 \text{ psi,} \quad \text{and} \quad \overline{Y}' = (1.15)(18,500) = 21,275 \text{ psi.}$$

Therefore

$$F = (1.55)(9)(21,275) = 297,000 \text{ lb,}$$

and

$$\text{Power} = \frac{(2\pi)(297,000)(0.13)(100)}{33,000} = 735 \text{ hp.}$$

---

**Forces in hot rolling.** Because ingots and slabs are usually hot rolled, the calculation of forces and torque in hot rolling is important. However, there are two major difficulties. One is the proper estimation of the coefficient of friction $\mu$ at elevated temperatures. The other is the strain-rate sensitivity of metals at high temperatures.

The *average strain rate* in flat rolling can be obtained by dividing the strain by the time required for an element to undergo this strain in the roll gap. The time can be approximated as $L/V_r$, so

$$\dot{\varepsilon} = \frac{V_r}{L} \ln\left(\frac{h_0}{h_f}\right). \tag{6.43}$$

The flow stress $Y_f$ of the material corresponding to this strain rate must first be obtained and then substituted into the proper equations. These calculations are approximate. Variations in $\mu$ and the temperature within the strip in hot rolling further contribute to the difficulties in calculating forces accurately.

## Friction

In rolling, although the rolls cannot pull the strip into the roll gap without some friction, forces and power requirements rise with increasing friction. In *cold* rolling, the coefficient of friction $\mu$ usually ranges between 0.02 and 0.3, depending on the materials and lubricants used. The low ranges for the coefficient of friction are obtained with effective lubricants and regimes approaching hydrodynamic lubrication, such as in cold rolling of aluminum at high speeds. In *hot* rolling, the coefficient of friction may range from about 0.2, with effective lubrication, to as high as 0.7, indicating sticking, which usually occurs with steels, stainless steels, and high-temperature alloys.

The maximum possible **draft,** that is, $h_0 - h_f$, in flat rolling is a function of friction and roll radius:

$$\Delta h_{max} = \mu^2 R. \tag{6.44}$$

Hence the higher the friction and the larger the roll radius, the greater is the maximum draft. As expected, the draft is zero when there is no friction. The maximum value of the angle $\alpha$ (**angle of acceptance**) in Fig. 6.34 is geometrically related to Eq. (6.44). From the simple model of a block sliding down an inclined plane, we can see that

$$\alpha_{max} = \tan^{-1}\mu. \tag{6.45}$$

If $\alpha_{max}$ is larger than this value, the rolls begin to slip, because the friction is not high enough to pull the material through the roll gap.

## Roll deflection and flattening

Roll forces tend to bend the rolls, as shown in Fig. 6.39a, with the result that the strip is thicker at its center than at its edges (**crown**). The usual method of avoiding this problem is to grind the rolls so that their diameter at the center is slightly larger than at the edges. This is known as **camber.** In rolling sheet metals the camber is generally less than 0.25 mm (0.01 in.) on the radius.

(a)

(b)

**FIGURE 6.39** (a) Bending of straight cylindrical rolls because of the roll-separating force. (b) Bending of rolls, ground with camber, that produce a sheet of uniform thickness during rolling.

When properly designed, such rolls produce flat strips as shown in Fig. 6.39b. However, a particular camber is correct only for a certain load and width of strip. In hot rolling, uneven temperature distribution in the roll can also cause its diameter to vary along the length of the roll. In fact, camber can be controlled by varying the location of the coolant (lubricant) on the rolls in hot rolling.

Forces also tend to **flatten** the rolls elastically, much like the flattening of tires on automobiles. Flattening of the rolls increases roll radius and hence yields a larger contact area for the same reduction in thickness. Thus the *roll force F* increases.

The new (distorted) roll radius $R'$ is

$$R' = R\left(1 + \frac{CF'}{h_0 - h_f}\right). \tag{6.46}$$

In Eq. (6.46), $C$ is $2.3 \times 10^{-2}$ mm$^2$/kN ($1.6 \times 10^{-4}$ in$^2$/klb) for steel rolls and $4.57 \times 10^{-2}$ ($3.15 \times 10^{-4}$) for cast iron rolls. $F'$ is the **roll force per unit width** of strip, kN/mm (klb/in.). The higher the elastic modulus of the roll material, the less the roll distorts. Note from Eq. (6.46) that reducing the roll force, such as by using an effective lubricant or taking smaller reductions, also reduces roll flattening.

From Eq. (6.46) we cannot determine $R'$ directly, because it is a function of the roll force $F$, which itself is a function of the roll radius. The solution is obtained by trial and error. (Note that $R$ in all previous equations should be replaced by $R'$ when significant roll flattening occurs.)

### Spreading

Although rolling plates and sheets with high width-to-thickness ratios is essentially a process of plane strain, with smaller ratios, such as a square cross-section, the width increases considerably during rolling. This increase in width is known as **spreading** (Fig. 6.40). Spreading decreases with increasing width-to-thickness ratios of the entering material, decreasing friction, and increasing ratios of roll radius-to-strip thickness.

### 6.3.2 Defects

Successful rolling practice requires balancing many factors, including material properties, process variables, and lubrication. Defects may be on the surfaces of the rolled plates and sheets, or they may be structural defects within the material. **Surface defects** may result from inclusions and impurities in the material, scale, rust, dirt, roll marks, and other causes related to the prior treatment and working of the material. In hot rolling blooms, billets, and slabs, the surface is usually preconditioned by various means, such as by torch (**scarfing**), to remove scale.

**Structural defects** are those that distort or affect the integrity of the rolled product. Some typical defects are shown in Fig. 6.41. Wavy edges are caused by bending of the rolls; the edges of the strip are thinner than the center. Because the edges elongate more than the center and are restrained from expanding freely, they buckle. The cracks shown in Fig. 6.41b and c are usually caused by low ductility and barreling. **Alligatoring** is a complex phenomenon resulting from inhomogeneous deformation of the material during rolling or defects in the original cast ingot, such as piping.

**FIGURE 6.40** Increase in the width (spreading) of a strip in flat rolling. Spreading can be similarly observed when dough is rolled with a rolling pin.

**FIGURE 6.41** Schematic illustration of typical defects in flat rolling: (a) wavy edges; (b) zipper cracks in center of strip; (c) edge cracks; (d) alligatoring.

**Residual stresses.**    Residual stresses can be generated in rolled sheets and plates due to inhomogeneous plastic deformation in the roll gap. Small-diameter rolls or small reductions tend to work the metal plastically at its surfaces (similarly to shot peening or roller burnishing). This working generates compressive residual stresses on the surfaces and tensile stresses in the bulk (Fig. 6.42a). Large-diameter rolls and high reductions, however, tend to deform the bulk to a greater extent than the surfaces. The reason is the frictional constraint at the surfaces along the arc of contact. This situation generates residual stresses that are opposite to the previous case, as shown in Fig. 6.42b.

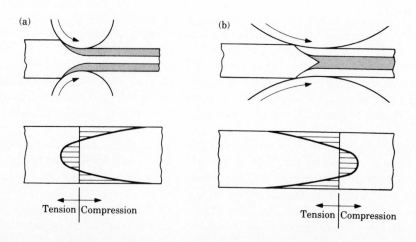

**FIGURE 6.42**    The effect of roll radius on the type of residual stresses developed in flat rolling: (a) small rolls, or small reduction; and (b) large rolls, or large reduction in thickness.

### 6.3.3 Flat-rolling practice

The initial breaking down of an ingot (or continuous casting) by hot rolling converts the coarse-grained, brittle, and porous cast structure to a wrought structure (see Fig. 6.32). This structure has finer grains and enhanced ductility resulting from breaking up brittle grain boundaries and closing up internal defects, such as porosity. Traditional methods of rolling ingots are now being rapidly replaced by continuous casting. Temperature ranges for hot rolling are similar to those for forging (see Table 6.3).

The product of the first hot-rolling operation is called a **bloom** or **slab** (see Fig. 6.31). A bloom usually has a square cross-section, at least 150 mm (6 in.) on the side; a slab is usually rectangular in cross-section. Blooms are processed further by shape rolling into structural shapes, such as I-beams and railroad rails. Slabs are rolled into plates and sheet. **Billets** are usually square, with a cross-sectional area smaller than blooms, and are rolled into various shapes, such as round rods and bars, using shaped rolls. (Hot-rolled round rods are used as the starting material for rod and wire drawing and are called **wire rods.**)

In hot rolling blooms, billets, and slabs, the surface of the material is usually **conditioned** (prepared for a subsequent operation) prior to rolling. Conditioning is done by various means, such as with a torch (scarfing) to remove heavy scale, or by rough grinding to smoothen surfaces. Prior to cold rolling, the scale developed during hot rolling or other defects may be removed by pickling with acids or by mechanical means, such as blasting with water, or by grinding.

**Pack rolling** is a flat-rolling operation in which two or more layers of metal are rolled together, thus improving productivity. Aluminum foil, for example, is pack rolled in two layers. One side of aluminum foil is matte and the other side shiny. The foil-to-foil side has a matte and satiny finish, and the foil-to-roll side is shiny and bright because it has been in contact with the polished roll.

Mild steel, when stretched during sheet-forming operations, undergoes **yield-point elongation.** This phenomenon causes surface irregularities called **stretcher strains** or **Lueder's bands** (see also Section 7.2). To avoid this situation, the sheet metal is subjected to a light pass of 0.5–1.5 percent reduction, known as **temper rolling** (*skin pass*).

A rolled sheet may not be sufficiently flat as it leaves the roll gap because of variations in the material or in the processing parameters during rolling. To improve flatness, the strip is then passed through a series of **leveling rolls.** Each roll is usually driven separately with individual electric motors. The strip is flexed in opposite directions as it passes through the sets of rollers. Several different roller arrangements are used (Fig. 6.43).

**FIGURE 6.43** Schematic illustration of methods of roller leveling. These processes are used to flatten rolled sheets.

Sheet thickness is identified by a **gage number**; the smaller the number, the thicker the sheet. There are different numbering systems for different sheet metals. Rolled sheets of copper and brass are also identified by thickness changes.

**Lubrication.** Ferrous alloys are usually hot rolled without a lubricant, although graphite may be used. Aqueous solutions are used to cool the rolls and break up the scale on the workpiece. Nonferrous alloys are hot rolled with a variety of compounded oils, emulsions, and fatty acids. Cold rolling is done with low-viscosity lubricants, including mineral oils, emulsions, paraffin, and fatty oils (see Table 6.4).

## Equipment

A wide variety of rolling equipment is available with a number of roll arrangements, the major types of which are shown in Fig. 6.44. **Two-high** or **three-high** (developed in the mid 1800s) rolling mills are used for initial breakdown passes on cast ingots with roll diameters ranging from 600 to 1400 mm (24 to 55 in.). Small-diameter rolls are preferable, because the smaller the roll radius the lower the roll force. However, small rolls deflect under roll forces and have to be supported by other rolls (Fig. 6.44c and d).

The range of dimensions in rolling is extremely wide. The rolled product can be as wide as 5000 mm (200 in.) and as thin as 0.0025 mm (0.0001 in.) obtained in a cluster mill. The **cluster mill (Sendzimir)** is particularly suitable for cold rolling thin strips of high-strength metals. One arrangement for the rolls is shown in Fig. 6.45. The **work roll** (smallest roll) can be as small as 6 mm (1/4 in.) in diameter and is usually made of tungsten carbide for rigidity, strength, and wear resistance.

Stiffness in rolling mills is important for controlling dimensions. Mills can be highly automated with rolling speeds as high as 25 m/s (5000 ft/min). Flat rolling can also be carried out with front tension only, using idling rolls (**Steckel rolling**). For this case, the torque $T$ on the roll is zero, assuming frictionless bearings.

Requirements for roll materials are mainly strength and resistance to wear. Three common roll materials are cast iron, cast steel, and forged steel. For hot rolling, roll surfaces are generally rough and may even have notches or grooves in order to pull the metal through the roll gap at high reductions. Rolls for cold rolling are ground to a fine finish and, for special applications, are also polished.

**FIGURE 6.44** Schematic illustration of various roll arrangements: (a) 2 high; (b) 3 high; (c) 4 high; (d) cluster; (e) tandem rolling, with three stands; and (f) planetary.

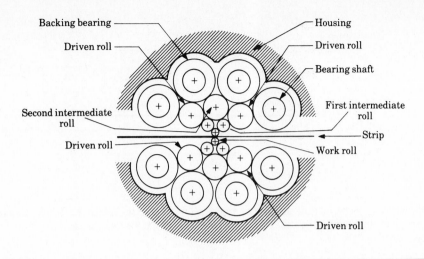

**FIGURE 6.45** Schematic illustration of a cluster (*Sendzimir*) mill. These mills are very rigid and are used in rolling thin sheets of high-strength materials, with good control of dimensions.

### 6.3.4 Miscellaneous rolling operations

**Shape rolling.** Straight structural shapes such as bars of various cross-sections, channel sections, I-beams, and railroad rails are rolled by passing the stock through a number of pairs of specially designed rollers (Fig. 6.46). These processes were first developed in the late 1700s. The original material is usually a bloom (Fig. 6.31). Designing a series of rolls (**roll-pass design**) requires experience in order to avoid defects and hold tolerances, although some

Stage 1 — Blooming rolls

Stage 2 — Edging rolls

Stage 3 — Roughing horizontal and vertical rolls

Stage 4 — Intermediate horizontal and vertical rolls

Stage 5 — Edging rolls

Stage 6 — Finishing horizontal and vertical rolls

**FIGURE 6.46** Stages in shape rolling of an H-section part. Various other structural sections, such as channels and I-beams are also rolled by this process.

**FIGURE 6.47**   (a) Schematic illustration of a ring-rolling operation. Thickness reduction results in an increase in the part diameter. (b) Examples of cross-sections that can be formed by ring rolling.

of these defects may also be due to the material rolled. The material elongates as it is reduced in cross-section. However, for a shape such as a channel the reduction is different in various locations within the section. Thus elongation is not uniform, which can cause the product to warp or crack. Airfoil shapes can also be produced by shape-rolling techniques.

**Ring rolling.**   In ring rolling, a small-diameter, thick ring is expanded into a larger-diameter, thinner ring. The ring is placed between two rolls, one of which is driven (Fig. 6.47). The thickness is reduced by bringing the rolls closer as they rotate. The reduction in thickness is compensated for by an increase in the diameter of the ring. A great variety of cross-sections can be ring rolled with shaped rolls. This process can be carried out at room or elevated temperatures, depending on the size and strength of the product. The advantages of ring rolling, compared with other processes for making the same part, are short production runs, material savings, close tolerances, and favorable grain flow direction. Typical applications of ring rolling are large rings for rockets and turbines, gearwheel rims, ball and roller bearing races, flanges and reinforcing rings for pipes, and pressure vessels.

**Thread and gear rolling.**   In the cold-forming thread and gear rolling process, threads are formed on round rods or workpieces by passing them between reciprocating or rotating dies (Fig. 6.48a). Typical products are screws, bolts, and similar threaded parts. With flat dies the threads are rolled on the rod or wire with each stroke of the reciprocating die. Production rates are very high but depend on the diameter of the product. With small diameters the rates can be as high as eight pieces per second and with larger diameters (as much as 25 mm [1 in.]) about one per second. Two- or three-roller thread-rolling machines are also available. In another design (Fig. 6.48b) threads are formed using a rotary die with production rates as high as 80 pieces per second.

The thread-rolling process generates threads without any metal loss and with greater strength because of cold working. The surface finish is very smooth, and the process induces compressive residual stresses on the surfaces, which improves fatigue life. The product is superior to that made by thread cutting and is used in the production of almost all externally threaded fasteners.

**FIGURE 6.48** Thread-rolling processes: (a) flat dies and (b) two-roller dies. These processes are used extensively in making threaded fasteners at high rates of production.

Because of volume constancy in plastic deformation, a rolled thread requires a smaller diameter round stock to produce the same major diameter as a machined thread (Fig. 6.49). Also, whereas machining removes material by cutting through the grain flow lines of the material, rolled threads have a grain flow pattern that improves the strength of the thread because of cold working. Thread rolling can also be carried out internally with a fluteless forming tap. The process is similar to external thread rolling and produces accurate threads with good strength. In all thread-rolling processes it is essential that the material have sufficient ductility and that the rod or wire be of proper size. Lubrication is also important for good surface finish and to minimize defects.

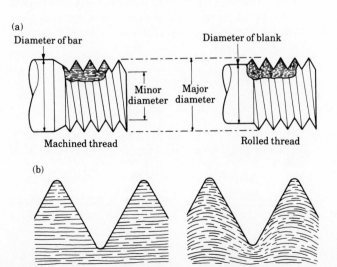

**FIGURE 6.49** (a) Schematic illustration of machined and rolled threads. Note the size of the blank diameter for the rolled thread. (b) Grain flow lines in machined and rolled threads. Unlike machined threads, which are cut through the grains of the metal, rolled threads follow the grains and are stronger because of cold working.

**FIGURE 6.50** Cavity formation by secondary tensile stresses in a solid round bar and its utilization in the rotary tube piercing process. This is the principle of the Mannesmann mill for seamless tubemaking. The mandrel is held in place by the long rod, although techniques have been developed in which the mandrel remains in place without the rod.

Spur and helical gears are also produced by cold-rolling processes similar to thread rolling. The process may be carried out on solid cylindrical blanks or on precut gears. Helical gears can also be made by a direct extrusion process, using specially shaped dies. Cold rolling of gears has many applications in automatic transmissions and power tools.

**Rotary tube piercing.** A hot-working process, rotary tube piercing is used to make long, thick-walled seamless tubing, as shown in Fig. 6.50. The process is based on the principle that when a round bar is subjected to radial compression in the manner shown in Fig. 6.50a, tensile stresses develop at the center of the rod. When subjected to cycling compressive stresses, as shown in Fig. 6.50b, a cavity begins to form at the center of the rod.

The rotary tube piercing (**Mannesmann process,** developed in the 1880s) process is carried out by an arrangement of rotating rolls, as shown in Fig. 6.50c. The axes of the rolls are skewed in order to pull the round bar through the rolls by the longitudinal force component of their rotary action. A mandrel assists the operation by expanding the hole and sizing the inside diameter of the tube. Because of the severe deformation that the metal undergoes in this process, high-quality, defect-free bars must be used.

## 6.4 Extrusion

In the basic extrusion process, developed in the late 1700s for lead pipe, a round billet is placed in a chamber and forced through a die opening by a ram. The die may be round or of various other shapes. There are four basic types of extrusion: direct, indirect, hydrostatic, and impact (Figs. 6.51 and 6.52).

**Direct extrusion** (*forward* extrusion) is similar to forcing the paste through the opening of a toothpaste tube. The billet slides relative to the container wall; the wall friction increases the ram force considerably.

In **indirect extrusion** (*reverse, inverted,* or *backward* extrusion), the die moves toward the billet; thus, except at the die, there is no relative motion at the billet-container interface.

**FIGURE 6.51** Types of extrusion. (a) direct; (b) indirect; (c) hydrostatic; (d) impact.

In **hydrostatic extrusion,** the chamber is filled with a fluid that transmits the pressure to the billet, which is then extruded through the die. There is no friction along the container walls.

**Impact extrusion** is a form of indirect extrusion and is particularly suitable for hollow shapes.

Extrusion processes can be carried out hot or cold. Because a chamber is involved, each billet is extruded individually and hence extrusion is basically a *batch* process. **Cladding** by extrusion can also be carried out with coaxial billets (such as copper clad with silver), provided that the flow stresses of the two metals are similar.

**FIGURE 6.52** Extrusions and examples of products made by sectioning off extrusions. *Source*: Kaiser Aluminum.

**FIGURE 6.53** Schematic illustration of three different types of metal flow in direct extrusion.

### 6.4.1 Metal flow in extrusion

Because the billet is forced through a die, with a substantial reduction in its cross-section, the *metal flow pattern* in extrusion is an important factor in the overall process. A common technique for investigating the flow pattern is to halve the round billet lengthwise and mark one face with a grid pattern. The two halves are then placed together in the container (they may also be fastened together or *brazed* to keep the two halves intact) and extruded. They are then taken apart and inspected.

Figure 6.53 shows three typical results in direct extrusion with square dies. The conditions under which these different flow patterns are obtained are as follows.

1.  The most homogeneous flow pattern is obtained when there is no friction at the billet-container-die interfaces (Fig. 6.53a). This type of flow occurs when the lubricant is very effective or with indirect extrusion.

2.  When friction along all interfaces is high, a **dead-metal zone** develops (Fig. 6.53b). Note the high-shear area as the material flows into the die exit, somewhat like a funnel. This configuration may indicate that the billet surfaces (with their oxide layer and lubricant) could enter this high-shear zone and be extruded, causing defects in the extruded product.

3.  The high-shear zone extends farther back (Fig. 6.53c). This extension can result from high container-wall friction, which retards the flow of the billet, or materials in which the flow stress drops rapidly with increasing temperature. In hot working, the material near the container walls cools rapidly and hence increases in strength. Thus the material in the central regions flows toward the die more easily than that at the outer regions. As a result, a large dead-metal zone forms and the flow is inhomogeneous. This flow pattern leads to a defect known as a **pipe** or **extrusion defect**.

Thus the two factors that greatly influence metal flow in extrusion are the frictional conditions at billet-container-die interfaces and thermal gradients in the billet.

### 6.4.2 Mechanics of extrusion

The ram force in direct extrusion can be calculated in the following ways for different situations.

#### Ideal deformation

The **extrusion ratio** $R$ is defined as

$$R = \frac{A_o}{A_f}, \tag{6.47}$$

where $A_o$ is the billet cross-sectional area and $A_f$ is the area of the extruded product (Fig. 6.54). The absolute value of the true strain then is

$$\varepsilon_1 = \ln\!\left(\frac{A_o}{A_f}\right) = \ln\!\left(\frac{L_f}{L_o}\right) = \ln R, \qquad (6.48)$$

where $L_o$ and $L_f$ are the lengths of the billet and the extruded product, respectively.

For a perfectly plastic material with a yield stress $Y$, the energy dissipated in plastic deformation per unit volume, $u$, is

$$u = Y\varepsilon_1. \qquad (6.49)$$

Hence the work done on the billet is

$$\text{Work} = (A_o)(L_o)(u). \qquad (6.50)$$

This work is supplied by the ram force $F$, which travels a distance $L_o$. Thus

$$\text{Work} = FL_o = pA_oL_o, \qquad (6.51)$$

where $p$ is the **extrusion pressure** at the ram. Equating the work of plastic deformation to the external work done, we find that

$$p = u = Y\ln\!\left(\frac{A_o}{A_f}\right) = (Y)(\ln R). \qquad (6.52)$$

For strain-hardening materials, $Y$ should be replaced by the *average flow stress* $\overline{Y}$.

Note that Eq. (6.52) is the same as the area under the true stress-true strain curve for the material.

### Ideal deformation and friction

Equation (6.52) pertains to ideal deformation without any friction. Based on the slab method of analysis, when friction at the die-billet interface is included (but not the container-wall friction) and for small die angles, the pressure $p$ is

$$p = Y\!\left(1 + \frac{\tan \alpha}{\mu}\right)\!\left[(R)^{\mu \cot \alpha} - 1\right]. \qquad (6.53)$$

An estimate of $p$ can also be obtained as follows. Referring to Fig. 6.54, the total power input $P$ with a ram velocity of $V_o$ is

**FIGURE 6.54**  Variables in the direct-extrusion process.

$$P_{\text{input}} = pV_o(\pi D_o^2/4).$$

The power due to plastic work of deformation is the product of volume rate of flow and the energy per unit volume. Thus,

$$P_{\text{plastic}} = V_o(\pi D_o^2/4)(Y)[\ln(D_o/D_f)^2].$$

Let us now assume that, because of the dead zone, material flow in the die region takes place along a $45°$ "die angle" and that the frictional stress is equal to the shear yield stress $k = Y/2$ of the material. The power dissipated due to friction along the die angle can then be calculated as

$$P_{\text{friction}} = (V_o)(\pi D_o^2/\sqrt{2})(Y/2) \ln (D_o/D_f).$$

Equating the power input to the sum of the plastic deformation and frictional powers, we obtain

$$p/Y = 3.41 \ln(D_o/D_f) = 1.7 \ln(D_o/D_f)^2,$$

or

$$p = 1.7 \, Y \ln R. \tag{6.54}$$

In this analysis, we neglected the force required to overcome friction at the billet-container interface. If we assume that the frictional stress is equal to the shear yield stress of the material, $k$, we can obtain the additional ram pressure required due to friction, $p_f$, as

$$(p_f)\left(\frac{\pi D_o^2}{4}\right) = \pi D_o kL$$

or

$$p_f = k \frac{4L}{D_o} = Y \frac{2L}{D_o}, \tag{6.55}$$

where $L$ is the length of the billet remaining in the container. Thus Eq. (6.54) becomes

$$p = Y\left(1.7 \ln R + \frac{2L}{D_o}\right). \tag{6.56}$$

For strain-hardening materials, $Y$ in these expressions should be replaced by $\overline{Y}$.

Note that as the billet is extruded farther, $L$ decreases and thus the ram force decreases (Fig. 6.55), whereas in indirect extrusion the ram force is not a function of billet length.

**FIGURE 6.55** Schematic illustration of typical extrusion pressure as a function of ram travel: (a) direct extrusion and (b) indirect extrusion. The pressure in direct extrusion is higher because of frictional resistance in the chamber as the billet moves toward the die.

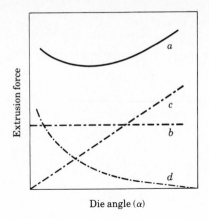

Die angle ($\alpha$)

**FIGURE 6.56** Schematic illustration of extrusion force as a function of die angle: (a) total force; (b) ideal force; (c) force required for redundant deformation; and (d) force required to overcome friction. Note that there is an optimum die angle where the total extrusion force is a minimum.

**Actual forces.** The derivation of analytical expressions, including friction, die angle, and redundant work due to inhomogeneous deformation of the material, can be difficult. Furthermore, there are difficulties in estimating the coefficient of friction, the flow stress of the material, and the redundant work in a particular operation. Consequently, a convenient *empirical* formula has been developed:

$$p = Y(a + b \ln R), \tag{6.57}$$

where $a$ and $b$ are constants determined experimentally. Approximate values for $a$ and $b$ are 0.8 and 1.2 to 1.5, respectively. Again, note that for strain-hardening materials $Y$ is replaced by $\overline{Y}$.

**Optimum die angle.** The die angle has an important effect on forces in extrusion. Its relationship to work is as follows.

1. The **ideal work** of deformation is independent of the die angle (Fig. 6.56), because it is a function only of the extrusion ratio.
2. The **frictional work** increases with decreasing die angle because the length of contact at the billet-die interface increases, thus requiring more work.
3. The **redundant work** caused by inhomogeneous deformation increases with die angle.

Because the total ram force is the sum of these three components, there is an angle where this force is a *minimum* (Fig. 6.56). Unless the behavior of each component as a function of the die angle is known, determination of this *optimum angle* is difficult.

---

**Example 6.5: Strain rate in extrusion**

Determine the true strain rate in extruding a round billet of radius $r_0$ as a function of distance $x$ from die entry.

SOLUTION. From geometry in the die gap, we note that

$$\tan \alpha = (r_o - r)/x,$$

or

$$r = r_o - x\tan\alpha.$$

We can define incremental true strain as

$$d\varepsilon = dA/A,$$

where $A = \pi r^2$. Therefore, $dA = 2\pi r dr$, and hence

$$d\varepsilon = 2dr/r, \text{ where } dr = -\tan\alpha dx.$$

We also know that

$$\dot{\varepsilon} = d\varepsilon/dt = -(2/r)(\tan\alpha)(dx/dt).$$

However, $dx/dt = V$, which is the velocity of the material at any location $x$ in the die. Thus,

$$\dot{\varepsilon} = -2\tan\alpha\,(V)/2.$$

From constancy of flow rate, we can write

$$V = r_o^2 V_o / r^2,$$

and hence,

$$\dot{\varepsilon} = -2\tan\alpha\,(r_o^2)(V_o)/r^3 = -2\tan\alpha\,(r_o^2)(V_o)/(r_o - x\tan\alpha)^3.$$

The negative sign is due to the fact that we have defined true strain in terms of the cross sectional area, which decreases as $x$ increases.

---

**Forces in hot extrusion.**   Because of the strain-rate sensitivity of metals at elevated temperatures, forces in hot extrusion are difficult to calculate. The average true strain rate $\bar{\dot{\varepsilon}}$ is

$$\bar{\dot{\varepsilon}} = \frac{6V_o D_o^2 \tan\alpha}{D_o^3 - D_f^3} = \ln R, \tag{6.58}$$

where $V_o$ is the ram velocity. Note from this equation that for high extrusion ratios $(D_o \gg D_f)$ and for $\alpha = 45°$, as may be the case with a square die (thus developing a dead zone) and poor lubrication, the strain rate reduces to

$$\bar{\dot{\varepsilon}} = \frac{6V_o}{D_o} \ln R. \tag{6.59}$$

The effect of ram speed and temperature on extrusion pressure is shown in Fig. 6.57. As expected, pressure increases rapidly with ram speed, especially at elevated temperatures. As extrusion speed increases, the rate of work done per unit time also increases. Because work is converted into heat, the heat generated at high speeds may not be dissipated fast enough.

**FIGURE 6.57** Schematic illustration of the effect of temperature and ram speed on extrusion pressure. Compare with Fig. 2.11.

The subsequent rise in temperature can cause **incipient melting** of the workpiece material and cause defects. Circumferential surface cracks caused by **hot shortness** may also develop; in extrusion this is known as **speed cracking.** These problems can be eliminated by reducing the extrusion speed.

A convenient parameter that is used to estimate forces in extrusion is an experimentally determined **extrusion constant** $K_e$, which includes various factors and can be determined from

$$p = K_e \ln R. \tag{6.60}$$

Figure 6.58 gives some typical values of $K_e$ for various materials.

**FIGURE 6.58** Extrusion constant $K_e$ for various materials as a function of temperature. Note the ranges of temperature for various materials. (a) 1100 aluminum; (b) copper; (c) 70–30 brass; (d) beryllium; (e) cold-rolled steel; (f) stainless steel; (g) molybdenum; and (h) chromium. *Source:* After P. Loewenstein, ASTME Paper SP63-89.

**Example 6.6:**   Force in hot extrusion

A copper billet 5 in. in diameter and 10 in. long is extruded at 1500°F at a speed of 10 in./s. Using square dies and assuming poor lubrication, estimate the force required in this operation if the final diameter is 2 in.

SOLUTION.   The extrusion ratio is

$$R = \frac{5^2}{2^2} = 6.25.$$

The average true strain rate, from Eq. (6.59), is

$$\dot{\bar{\varepsilon}} = \frac{(6)(10)}{(5)} \ln 6.25 = 22/s.$$

From Table 2.5 let's assume that $C = 19,000$ psi (an average value) and $m = 0.06$. Then

$$\sigma = C\dot{\varepsilon}^m = (19,000)(22)^{0.06} = 22,870 \text{ psi}.$$

Assuming that $\overline{Y} = \sigma$, then from Eq. (6.56),

$$p = \overline{Y}\left(1.7 \ln R + \frac{2L}{D_o}\right) = (22,870)\left[(1.7)(1.83) + \frac{(2)(10)}{(5)}\right] = 162,630 \text{ psi}.$$

Hence

$$F = (p)(A_o) = (162,630)\frac{(\pi)(5)^2}{4} = 3.2 \times 10^6 \text{ lb}.$$

### 6.4.3   Miscellaneous extrusion processes

### Cold extrusion

*Cold extrusion* is a general term often used to denote a combination of processes, such as direct and indirect extrusion and forging (Fig. 6.59). Many materials can be extruded into various configurations, with the billet either at room temperature or at a temperature of a few hundred degrees.

Cold extrusion has gained wide acceptance in industry because of the following advantages over hot extrusion:

1.   Improved mechanical properties resulting from strain hardening, provided that the heat generated by plastic deformation and friction does not recrystallize the extruded metal
2.   Good control of tolerances, thus requiring a minimum of machining operations
3.   Improved surface finish, provided that lubrication is effective
4.   Lack of oxide layers
5.   High production rates and relatively low cost

**FIGURE 6.59** Examples of cold extrusion. Arrows indicate the direction of material flow. These parts may also be considered as forgings.

However, the stresses on tooling in cold extrusion are very high, especially with steel workpieces. The stress levels on tooling are on the order of the hardness of the material, that is, at least three times its flow stress. The design of tooling and selection of appropriate tool materials are therefore crucial to success in cold extrusion. The hardness of tooling usually ranges between 60 and 65 HRC for the punch and 58 to 62 HRC for the die. Punches are a critical component; they must have sufficient strength, toughness, wear, and fatigue resistance.

Lubrication also is crucial, especially with steels, because of the generation of new surfaces and the possibility of seizure between the metal and the tooling caused by the breakdown of lubrication. The most effective lubrication is phosphate conversion coatings on the workpiece and soap (or wax in some cases) as the lubricant (see Table 6.4). Temperature rise in cold extrusion is an important factor, especially at high extrusion ratios. The temperature may be sufficiently high to initiate and complete the recrystallization process of the cold-worked metal, thus reducing the advantages of cold working.

### Impact extrusion

A process often included in the category of cold extrusion, *impact extrusion* is similar to indirect extrusion (Fig. 6.51d). The punch descends at a high speed and strikes the *blank (slug)*, extruding it upward. The thickness of the extruded tubular section is a function of the clearance between the punch and the die cavity.

The impact-extrusion process usually produces tubular sections having wall thicknesses that are small in relation to their diameters. This ratio can be as small as 0.005. The concentricity between the punch and the blank is important for uniform wall thickness. A typical example of impact extrusion is the production of collapsible tubes, such as for toothpaste (Fig. 6.60a). The punch travel is determined by the setting of the press. A variety of nonferrous metals are impact extruded in this manner into various shapes (Fig. 6.60b), using vertical presses at production rates as high as two parts per second.

### Hydrostatic extrusion

In *hydrostatic extrusion* the pressure required for extrusion is supplied through a fluid medium surrounding the billet (Fig. 6.51c). Consequently, there is no container-wall friction. Pressures are usually about 1400 MPa (200 ksi). The high pressure in the chamber transmits some of the fluid to the die surfaces, thus significantly reducing friction and forces (Fig. 6.61). Hydrostatic extrusion, which was developed in the early 1950s, has

**FIGURE 6.60** (a) Impact extrusion of a collapsible tube (Hooker process). (b) Two examples of products made by impact extrusion. These parts may also be made by casting, forging, and machining, depending on the dimensions and materials involved and the properties desired. Economic considerations are also important in final process selection.

been improved by extruding the part into a second pressurized chamber, which is under lower pressure (**fluid-to-fluid extrusion**). This operation reduces the defects in the extruded product.

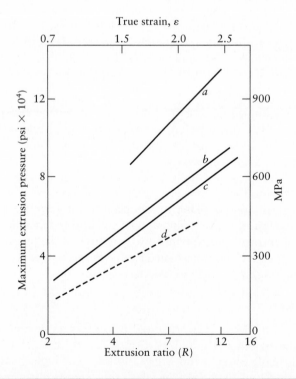

**FIGURE 6.61** Extrusion pressure as a function of the extrusion ratio for an aluminum alloy. (a) Direct extrusion, $\alpha$ = 90 deg. (b) Hydrostatic extrusion, $\alpha$ = 45 deg. (c) Hydrostatic extrusion, $\alpha$ = 22.5 deg. (d) Ideal homogeneous deformation, calculated. *Source:* After H. Ll., D. Pugh and K. Ashcroft.

Because the hydrostatic pressure increases the ductility of the material, brittle materials can be extruded successfully by this method. However, the main reasons for this success appear to be low friction and use of low die angles and high extrusion ratios. Most commercial hydrostatic-extrusion operations use ductile materials. However, a variety of metals and polymers, solid shapes, tubes and other hollow shapes, as well as honeycomb and clad profiles, have been extruded successfully.

Hydrostatic extrusion is usually carried out at room temperature, typically using vegetable oils as the fluid, particularly castor oil because it is a good lubricant and its viscosity is not influenced significantly by pressure. For elevated-temperature extrusion, waxes, polymers, and glass are used as the fluid. These materials also serve as thermal insulators and help maintain the billet temperature during extrusion. In spite of the success obtained, hydrostatic extrusion has had limited industrial applications, largely because of the somewhat complex nature of tooling, the experience required in working with high pressures and design of specialized equipment, and the long cycle times required.

### 6.4.4  Defects

There are three principal defects in extrusion. They are surface cracking, extrusion defect, and internal cracking.

### Surface cracking

If the extrusion temperature, friction, or extrusion speed is too high, surface temperatures rise significantly and can lead to surface cracking and tearing (**fir-tree cracking** or **speed cracking**). These cracks are intergranular, are usually the result of hot shortness, and occur especially with aluminum, magnesium, and zinc alloys. This defect is also observed with other metals, such as molybdenum alloys. This situation can be avoided by using lower temperatures and speeds.

However, surface cracking may also occur at low temperatures and has been attributed to periodic sticking of the extruded product along the die land. When the product being extruded sticks to the die land, the extrusion pressure increases rapidly. Shortly thereafter, the product moves forward again and the pressure is released. The cycle is then repeated.

### Extrusion defect

The type of metal flow observed in Fig. 6.53c tends to draw surface oxides and impurities toward the center of the billet, much like a funnel. This defect is known as **extrusion defect, pipe, tailpipe,** or **fishtailing.** A considerable portion of the material can be rendered useless as an extruded product because of it—as much as one-third the length of the extrusion.

This defect can be reduced by modifying the flow pattern to a less inhomogeneous one, such as by controlling friction and minimizing temperature gradients. Another method is to machine the surface of the billet prior to extrusion to eliminate scale and impurities. The extrusion defect can also be avoided by using a dummy block (Fig. 6.51a) that is smaller in diameter than the container, thus leaving a thin shell along the container wall as extrusion progresses.

### Internal cracking

The center of an extruded product can develop cracks (variously known as **centerburst, center cracking, arrowhead fracture,** or **chevron cracking**), as shown in Fig. 6.62. These cracks are attributed to a state of *hydrostatic tensile stress* (also called *secondary tensile stresses*) at

**FIGURE 6.62** Chevron cracking in round steel bars during extrusion. Unless the part is inspected properly, such internal defects may remain undetected and possibly cause failure of the part in service.

the centerline of the deformation zone in the die. This situation is similar to the necked region in a uniaxial tensile-test specimen.

The major variables affecting hydrostatic tension are the die angle, extrusion ratio (reduction in cross-sectional area), and friction. We can understand them best by observing the extent of inhomogeneous deformation in extrusion. Experimental results indicate that, for the same reduction, as the die angle becomes larger, the deformation across the part becomes more inhomogeneous.

In addition to the die angle, another factor in internal cracking is the die contact length. The smaller the die angle, the longer is the contact length. This situation is similar to a hardness test with a flat indenter. The size and depth of the deformation zone increases with increasing contact length. This condition is illustrated in Fig. 6.63.

As shown in Fig. 6.12, an important parameter is the $h/L$ ratio: the higher this ratio, the more inhomogeneous the deformation. High ratios mean small reductions and large die angles. Inhomogeneous deformation indicates that the center of the billet is not in a fully plastic state (it is rigid). The reason is that the plastic deformation zones under the die contact lengths do not reach each other (Fig 6.63). Likewise, small reductions and high die angles retard the flow of the material at the surfaces, while the central portions are freer to move through the die.

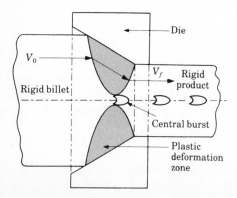

**FIGURE 6.63** Deformation zone in extrusion showing rigid and plastic zones. Note that the plastic zones do not meet, leading to chevron cracking. The same observations are also made in drawing round bars through conical dies, and drawing flat sheet plate through wedge-shaped dies. *Source:* After B. Avitzur.

The high $h/L$ ratios described generate hydrostatic tensile stresses in the center of the billet, causing the type of defects shown in Fig. 6.62. These defects form more readily in materials with impurities, inclusions, and voids, because they serve as nucleation sites for defect formation. As for the role of friction, high friction in extrusion apparently delays formation of these cracks.

Such cracks have also been observed in tube extrusion and in spinning of tubes. The cracks appear on the inside surfaces for the reasons already given. In summary, the tendency for center cracking increases with increasing die angles and levels of impurities, and decreases with increasing extrusion ratio. These observations are also valid for drawing of rod and wire (see Section 6.5).

### 6.4.5  Extrusion practice

Numerous materials can be extruded to a wide variety of cross-sectional shapes and dimensions. Extrusion ratios can range from about 10 to 100 or over. Ram speeds may be up to 0.5 m/s (100 ft/min). Generally, slower speeds are preferable for aluminum, magnesium, and copper and higher speeds for steels, titanium, and refractory alloys. Presses for hot extrusion are generally hydraulic and horizontal and are usually vertical for cold extrusion.

**Hot extrusion.**    In addition to the strain-rate sensitivity of the material at elevated temperatures, hot extrusion requires other special considerations. Cooling of the billet in the container can result in highly inhomogeneous deformation. Furthermore, because the billet is heated prior to extrusion, it is covered with an oxide layer (unless heated in an inert atmosphere). The resulting different frictional properties can affect the flow of the material and may produce an extrusion covered with an oxide layer. In order to avoid this problem, the diameter of the dummy block ahead of the ram (Fig. 6.51a) is made a little smaller than that of the container. A thin cylindrical shell (**skull**), consisting mainly of the oxidized layer, is thus left in the container, and the extruded product is free of oxides. Temperature ranges for hot extrusion are similar to those for forging (see Table 6.3).

**Lubrication.**    Lubrication is important in hot extrusion. For steels, stainless steels, and high-temperature materials, glass is an excellent lubricant. It maintains its viscosity at elevated temperatures, has good wetting characteristics, and acts as a thermal barrier between the billet and the container and the die, thus minimizing cooling. A circular glass pad is usually placed at the die entrance. This pad softens and melts away slowly as extrusion progresses and forms an optimal die geometry. The viscosity-temperature index of the glass is an important factor in this application. Solid lubricants such as graphite and molybdenum disulfide are also used in hot extrusion. Nonferrous metals are usually extruded without a lubricant, although graphite may be used (see Table 6.4).

For materials that have a tendency to stick to the container and the die, the billet can be enclosed in a **jacket** of a softer metal such as copper or mild steel (**canning**). Besides acting as a low-friction interface, canning prevents contamination of the billet by the environment or the billet material from contaminating the environment if the material is toxic or radioactive. The canning technique is also used for processing metal powders.

**Dies.**    Die materials for hot extrusion are usually hot-work die steels. To extend die life, coatings may be applied to the dies. Designing dies requires considerable experience because of the variety of products extruded. Dies with angles of 90° (square dies or **shear dies**) can

**FIGURE 6.64**   Extrusion of a seamless tube. The hole in the billet may be prepunched or pierced, or it may be generated during extrusion.

also be used in extrusion of nonferrous metals, especially aluminum. Tubing is also extruded with a ram fitted with a mandrel (Fig. 6.64). For billets with a pierced hole, the mandrel may be attached to the ram. If the billet is solid, it must first be pierced in the container by the mandrel.

The complexity of an extrusion is a function of the ratio of the perimeter to the cross-sectional area of the part, known as the **shape factor.** Thus a solid, round extrusion is the simplest shape. Hollow shapes (Fig. 6.65a) can also be extruded by **welding-chamber methods** using various special dies known as **spider, porthole,** and **bridge dies** (Fig. 6.65b). The metal flows around the arms of the die into strands, which are then rewelded under the high die pressures at the exit. This process is suitable only for aluminum and some of its alloys

**FIGURE 6.65**   (a) An extruded 6063-T6 aluminum ladder lock for aluminum extension ladders. This part is 8 mm (5/16 in.) thick and is sawed from the extrusion (see Fig. 6.52). (b) Components of various dies for extruding intricate hollow shapes. *Source:* for (b): K. Laue and H. Stenger, *Extrusion–Processes, Machinery, Tooling.* ASM International, Metals Park, Ohio, 1981. Used with permission.

because of their capacity for pressure welding. Lubricants cannot be used as they prevent rewelding during extrusion.

**Equipment**   The basic equipment for extrusion is a hydraulic press, usually horizontal. These presses are designed for a variety of extrusion operations. Crank-type mechanical presses are also used for cold extrusion and piercing and for mass production of steel tubing. The largest hydraulic press for extrusion has a capacity of 160 MN (16,000 tons).

## 6.5   Rod, Wire, and Tube Drawing

Drawing is an operation in which the cross-sectional area of a bar or tube is reduced by pulling it through a converging die (Fig. 6.66). The die opening may be any shape. Wire drawing involves smaller diameter materials than rod drawing, with sizes as small as 0.025 mm (0.001 in.). The drawing process, which was an established art by the 11th century, is somewhat similar to extrusion, except that in drawing the bar is under tension, whereas in extrusion it is under compression.

*Rod and wire drawing* are usually finishing processes. The product is either used as produced or is further processed into other shapes, usually by bending or machining. Rods are used for various applications, such as small pistons, tension-carrying structural members, shafts, and spindles, and as the raw material for fasteners such as bolts and screws. Wire and wire products have a wide range of applications, such as electrical and electronic equipment and wiring, cables, springs, musical instruments, fencing, bailing, wire baskets, and shopping carts.

### 6.5.1   Mechanics of rod and wire drawing

The major variables in the drawing process are reduction in cross-sectional area, die angle, and friction. These variables are illustrated in Fig. 6.66.

**Ideal deformation**

For a round rod or wire, the **drawing stress** $\sigma_d$ for the simplest case of ideal deformation (no friction or redundant work) can be obtained by the same approach as that in extrusion. Thus

$$\sigma_d = Y \ln \left( \frac{A_o}{A_f} \right). \tag{6.61}$$

**FIGURE 6.66**   Variables in drawing round rod wire.

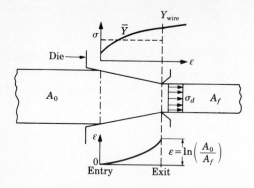

**FIGURE 6.67** Variation in strain and flow stress in the deformation zone in drawing. Note that the strain increases rapidly toward the exit. The reason is that when the exit diameter is zero, the true strain reaches infinity. The point $Y_{wire}$ represents the yield stress of the wire.

Note that this expression is the same as Eq. (6.52) and that it also represents the energy per unit volume $u$. For strain-hardening materials $Y$ is replaced by an average flow stress $\overline{Y}$ in the deformation zone (Fig. 6.67). Thus for a material that exhibits the true stress-true strain behavior of

$$\sigma = K\varepsilon^n,$$

the average flow stress $\overline{Y}$ is

$$\overline{Y} = \frac{K\varepsilon_1^n}{n+1}.$$

The drawing force $F$ then is

$$F = \overline{Y}A_f \ln\left(\frac{A_o}{A_f}\right). \tag{6.62}$$

Note that the greater the reduction in cross-sectional area and the stronger the material, the higher the drawing force.

### Ideal deformation and friction

In drawing, friction increases the drawing force, because work has to be supplied to overcome that friction. Using the slab method of analysis and on the basis of Fig. 6.68, we equate the forces in the horizontal direction as follows:

$$(\sigma_x + d\sigma_x)\frac{\pi}{4}(D + dD)^2 - \sigma_x\frac{\pi}{4}D^2 + p\,\frac{\pi D\,dx}{\cos\alpha}\sin\alpha + \mu p\,\frac{\pi D\,dx}{\cos\alpha}\cos\alpha = 0.$$

Simplifying and ignoring the second-order terms, we obtain

$$D d\sigma_x + 2\sigma_x dD + 2p\left(1 + \frac{\mu}{\tan\alpha}\right)dD = 0.$$

Note that we have two unknowns ($\sigma_x$ and $p$) but only one equation. We obtain the second equation from yield criteria, recognizing that this is an axisymmetric case and that both the maximum shear-stress and the distortion-energy criterion give the relationship

$$\sigma_x + p = Y. \tag{6.63}$$

**FIGURE 6.68** Stresses acting on an element in drawing of a solid cylindrical rod or wire through a conical converging die.

Letting $\mu/\tan\alpha = B$ and using the above relationships, we now obtain

$$\frac{dD}{D} = \frac{d\sigma_x}{2B\sigma_x - 2Y(1 + B)}.$$

Integrating this equation between the limits $D_f$ and $D_o$—and by noting that at $D = D_o$, $\sigma_x = 0$, and at $D = D_f$, $\sigma_x = \sigma_d$—yields

$$\sigma_d = Y\frac{1 + B}{B}\left[1 - \left(\frac{D_f}{D_o}\right)^{2B}\right]$$

or

$$\sigma_d = Y\frac{1 + B}{B}\left[1 - \left(\frac{A_f}{A_o}\right)^{B}\right]$$

$$= Y\left(1 + \frac{\tan\alpha}{\mu}\right)\left[1 - \left(\frac{A_f}{A_o}\right)^{\mu\cot\alpha}\right]. \tag{6.64}$$

For a strain-hardening material, the yield stress $Y$ is replaced by the average flow stress $\overline{Y}$, as shown in Fig. 6.67. Investigations have shown that, even though Eq. (6.64) does not include the redundant work, it is in good agreement with experimental data for small die angles and for a wide range of reductions.

**Redundant work of deformation**

Depending on the die angle and reduction, the material in drawing undergoes *inhomogeneous deformation,* much as it does in extrusion. Thus the redundant work of deformation also has to be included in the expression for the drawing stress. One such expression is:

$$\sigma_d = \overline{Y}\left\{\left(1 + \frac{\tan\alpha}{\mu}\right)\left[1 - \left(\frac{A_f}{A_o}\right)^{\mu\cot\alpha}\right] + \frac{4}{3\sqrt{3}}\alpha^2\left(\frac{1 - r}{r}\right)\right\}, \tag{6.65}$$

where $r$ is the fractional reduction of area and $\alpha$ is the die angle in radians.

The first term in Eq. (6.65) represents the ideal and frictional work components. The second term represents the redundant-work component, which, as expected, is a function of the die angle. The larger the angle is, the greater the inhomogeneous deformation and hence the greater the redundant work.

Another expression for the drawing stress for small die angles (including all three components of work) is

$$\sigma_d = \overline{Y}\left[\left(1 + \frac{\mu}{\alpha}\right)\ln\left(\frac{A_o}{A_f}\right) + \frac{2}{3}\alpha\right]. \tag{6.66}$$

The last term in this expression is the redundant-work component, whereby this work increases linearly with the die angle, as shown in Fig. 6.56.

Because redundant deformation is a function of the $h/L$ ratio (see Figs. 6.12 and 6.13), an **inhomogeneity factor** $\Phi$ can replace the last term in Eq. (6.66). For drawing of round sections this factor is, approximately,

$$\Phi = 1 + 0.12\left(\frac{h}{L}\right). \tag{6.67}$$

A simple expression for the drawing stress then is

$$\sigma_d = \Phi\overline{Y}\left(1 + \frac{\mu}{\alpha}\right)\ln\left(\frac{A_o}{A_f}\right). \tag{6.68}$$

Equations (6.64), (6.65), (6.66), and (6.68), while not always agreeing with experimental data, approximate reasonably well the stresses required for wire drawing. More important, they identify the effects of the various parameters involved. With good lubrication, the value of $\mu$ in these equations ranges from about 0.03 to 0.1.

## Die pressure

The *die pressure* $p$ at any diameter along the die contact length can be conveniently obtained from

$$p = Y_f - \sigma, \tag{6.69}$$

where $\sigma$ is the tensile stress in the deformation zone at any diameter. (Thus $\sigma$ is equal to $\sigma_d$ at the exit and zero at entry). And $Y_f$ is the flow stress of the material at any diameter corresponding to that strain.

Equation (6.69) is based on yield criteria for an element subjected to the stress system shown in Fig. 2.35b, where the compressive stresses in the two principal directions are both equal to $p$. Note that as the tensile stress increases toward the exit, the die pressure drops toward the exit. Tensile stress and die pressure distributions are shown qualitatively in Fig. 6.69.

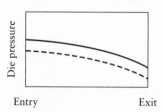

**FIGURE 6.69**   Variation in the drawing stress and die contact pressure along the deformation zone. Note that as the drawing stress increases, the die pressure decreases. This condition can be observed from the yield criteria, described in Section 2.11. Note the effect of back tension.

**Example 6.7:** Power and die pressure in rod drawing

A round rod of annealed 302 stainless steel is being drawn from a diameter of 10 mm to one of 8 mm at a speed of 0.5 m/s. Assume that the frictional and redundant work together constitute 40% of the ideal work of deformation. (1) Calculate the power required in this operation. (2) Calculate the die pressure at the die exit.

SOLUTION.

1. The true strain in this operation is

$$\varepsilon_1 = \ln\left(\frac{10^2}{8^2}\right) = 0.446.$$

From Table 2.3, $K = 1300$ MPa and $n = 0.30$. Thus

$$\overline{Y} = \frac{K\varepsilon_1^n}{n+1} = \frac{(1300)(0.446)^{0.30}}{1.30} = 785 \text{ MPa.}$$

From Eq. (6.62)

$$F = \overline{Y}A_f \ln\left(\frac{A_o}{A_f}\right),$$

where

$$A_f = \frac{(\pi)(0.008)^2}{4} = 5 \times 10^{-5} \text{ m}^2.$$

Hence

$$F = (785)(5 \times 10^{-5})(0.446) = 0.0175 \text{ MN,}$$
$$\text{Power} = (F)(V_f) = (0.0175)(0.5)$$
$$= 0.00875 \text{ MN} \cdot \text{m/s}$$
$$= 0.00875 \text{ MW} = 8.75 \text{ kW,}$$

and

$$\text{Actual power} = (1.4)(8.75) = 12.25 \text{ kW.}$$

2. From Eq. (6.69)

$$p = Y_f - \sigma,$$

where $Y_f$ represents the flow stress of the material at the exit. Thus

$$Y_f = K\varepsilon_1^n = (1300)(0.446)^{0.30} = 1020 \text{ MPa.}$$

In this equation $\sigma$ is the drawing stress $\sigma_d$. Hence, using the actual force,

$$\sigma_d = \frac{F}{A_f} = \frac{(1.4)(0.0175)}{0.00005} = 490 \text{ MPa}.$$

Therefore the die pressure at the exit is

$$p = 1020 - 490 = 530 \text{ MPa}.$$

---

**Drawing at elevated temperatures.**   At elevated temperatures, the flow stress of metals is a function of the strain rate. The *average true strain rate* $\dot{\bar{\varepsilon}}$ in the deformation zone in drawing is

$$\dot{\bar{\varepsilon}} = \frac{6V_o}{D_o} \ln \left( \frac{A_o}{A_f} \right). \tag{6.70}$$

[Note that this expression is the same as Eq. (6.59).] For a particular drawing operation at an elevated temperature, we first calculate the average strain rate. From that we determine the flow stress and the average flow stress $\bar{Y}$ of the material.

**Optimum die angle.**   Because of the various effects of the die angle on the three work components (ideal, friction, and redundant work), there is an *optimum die angle* where the extrusion force is a minimum. Drawing involves a similar type of deformation, so there is an optimum die angle in rod and wire drawing also, as shown in Fig. 6.56. Figure 6.70 shows a typical example from an experimental study. The optimum die angle for the minimum force increases with reduction. Note that the optimum angles in drawing are rather small.

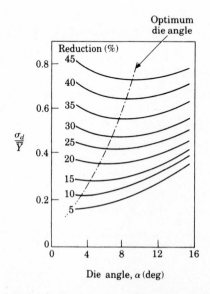

**FIGURE 6.70**   The effect of reduction in cross-sectional area on the optimum die angle increases with reduction. *Source:* After J. G. Wistreich.

**Maximum reduction per pass.** With greater reduction the drawing stress increases, but obviously the magnitude of the drawing stress has a limit. If it reaches the yield stress of the material, the material will simply continue to yield further as it leaves the die. This result is unacceptable because the product will undergo further deformation. Thus the maximum possible drawing stress can equal only the yield stress of the exiting material.

In the ideal case of a perfectly plastic material with a yield stress $Y$, the limiting condition is

$$\sigma_d = Y \ln \left(\frac{A_o}{A_f}\right) = Y, \tag{6.71}$$

or

$$\ln \left(\frac{A_o}{A_f}\right) = 1.$$

Hence

$$\frac{A_o}{A_f} = e$$

and therefore

$$\text{Maximum reduction per pass} = \frac{A_o - A_f}{A_o} = 1 - \frac{1}{e} = 0.63 = 63\%. \tag{6.72}$$

The effects of friction and die angle on maximum reduction per pass are similar to those shown in Fig. 6.56. Because both friction and redundant work increase the drawing stress, the maximum reduction per pass will be lower than the ideal. In other words, the exiting material must have a larger cross-sectional area to sustain the higher drawing forces.

With strain hardening, the exiting material is stronger than the rest, and the maximum reduction per pass increases.

---

**Example 6.8:  Maximum reduction per pass for a strain-hardening material**

Obtain an expression for the maximum reduction per pass for a material with a true stress–true strain curve of $\sigma = K\varepsilon^n$. Ignore friction and redundant work.

SOLUTION.  From Eq. (6.61) for this material,

$$\sigma_d = \overline{Y} \ln \left(\frac{A_o}{A_f}\right) = \overline{Y}\varepsilon_1,$$

where

$$\overline{Y} = \frac{K\varepsilon_1^n}{n + 1}$$

and $\sigma_d$, for this problem, can have a maximum value equal to the flow stress at $\varepsilon_1$, or

$$\sigma_d = Y_f = K\varepsilon_1^n.$$

Hence we can write Eq. (6.71) as

$$K\varepsilon_1^n = \frac{K\varepsilon_1^n}{n + 1}\varepsilon_1, \quad \text{or} \quad \varepsilon_1 = n + 1.$$

With $\varepsilon_1 = \ln(A_o/A_f)$ and maximum reduction $= (A_o - A_f)/A_o$, these expressions reduce to

$$\text{Maximum reduction per pass} = 1 - e^{-(n+1)}. \tag{6.73}$$

Note that when $n = 0$ (perfectly plastic material) this expression reduces to Eq. (6.72). As $n$ increases, the maximum reduction per pass increases.

---

**Drawing of flat strip.**  Whereas drawing of round sections is axisymmetric, flat-strip drawing with high width-to-thickness ratios can be regarded as a **plane-strain** problem. The dies are wedge-shaped and thus the process is somewhat similar to rolling of wide strips. There is little or no change in the width of the strip during drawing. This process, although not of any industrial significance, is the fundamental deformation mechanism in ironing, as described in Section 7.12.

The treatment of this subject, as far as forces and maximum reductions are concerned, is similar to that for round sections. The drawing stress for the ideal condition is

$$\sigma_d = Y' \ln\left(\frac{t_o}{t_f}\right), \tag{6.74}$$

where $Y'$ is the yield stress of the material in plane strain and $t_o$ and $t_f$ are the original and final thicknesses of the strip, respectively. The effects of friction and redundant deformation in strip drawing are similar to those for round sections.

We obtain the maximum reduction per pass by equating the drawing stress (Eq. 6.74) to the *uniaxial* yield stress of the material, because the drawn strip is subjected only to simple tension. Thus

$$\sigma_d = Y' \ln\left(\frac{t_o}{t_f}\right) = Y, \quad \ln\left(\frac{t_o}{t_f}\right) = \frac{Y}{Y'} = \frac{\sqrt{3}}{2} \quad \text{and} \quad \frac{t_o}{t_f} = e^{\sqrt{3}/2}.$$

We can then show that

$$\text{Maximum reduction per pass} = 1 - \frac{1}{e^{\sqrt{3}/2}} = 0.58 = 58\%. \tag{6.75}$$

**Drawing of tubes.**  Tubes produced by extrusion or other processes can be reduced in thickness or diameter (**tube sinking**) by the tube-drawing processes illustrated in Fig. 6.71. A variety of mandrels are used to produce different shapes. Shape changes can also be imparted by using dies and mandrels with various profiles. Drawing forces, die pressures, and maximum reduction per pass in tube drawing can be calculated by methods similar to those described for round rods.

### 6.5.2 Defects

Defects in drawing are similar to those observed in extrusion, especially **center cracking.** The factors influencing these internal cracks are the same, namely, the tendency for cracking in-

(a)          (b)          (c)

**FIGURE 6.71** Various methods of tube drawing

creases with increasing die angle and decreasing reduction per pass, and with friction and the presence of inclusions in the material.

Another type of defect is the formation of **seams,** which are longitudinal scratches or folds in the material. Such defects can open up during subsequent forming operations by up-setting, heading, thread rolling, or bending of the rod or wire. Various surface defects can also result from improper selection of process parameters and lubrication.

Because of inhomogeneous deformation, a cold-drawn rod, wire, or tube usually contains residual stresses. Typically, as shown in Fig. 6.72, a wide range of residual stresses are

**FIGURE 6.72** Residual stesses in cold-drawn AISI 1045 carbon steel round rod: $T$ = transverse, $L$ = longitudinal, and $R$ = radial directions. *Source:* After E. S. Nachtman.

present within the rod in three principal directions. However, for very light reductions, the surface residual stresses are compressive. Light reductions are equivalent to shot peening or surface rolling, thus improving fatigue life. In addition to fatigue, residual stresses can also be significant in stress-corrosion cracking over a period of time, and in warping of the component when a layer is subsequently removed, as by machining or grinding.

### 6.5.3   Drawing practice

Successful drawing operations require careful selection of process parameters and consideration of many factors. A typical die design for drawing, with its characteristic features, is shown in Fig. 6.73. Die angles usually range from 6° to 15°. The purpose of the land is to size and set the final diameter of the product. Also, when the die is reground after use, the land maintains the exit dimension of the die opening. Reductions in cross-sectional area per pass range from about 10 percent to 45 percent; usually, the smaller the cross-section the smaller is the reduction per pass. Reductions per pass greater than 45 percent may result in breakdown of lubrication and deterioration of surface finish. Light reductions may also be made (**sizing pass**) on rods to improve surface finish and dimensional accuracy.

A rod or wire is fed into the die by first **pointing** it by swaging (forming the tip of the rod into a conical shape, Section 6.6). After the rod or wire is placed in the die, the tip is clamped into the jaws of the wire-drawing machine and the rod or wire is drawn continuously through the die. In most wire drawing operations the wire passes through a series of dies (**tandem drawing**). In order to avoid excessive tension in the exiting wire, it is wound one or two turns around a capstan between each pair of dies. The speed of the capstan is adjusted so that it supplies not only tension but also a small back tension to the wire entering the next die. Back tension reduces the die pressure and extends die life.

Drawing speeds depend on the material and cross-sectional area. They may be as low as 0.15 m/s (30 ft/min) for heavy sections, to as high as 50 m/s (10,000 ft/min) for very fine wire. Temperature can rise substantially at high drawing speeds.

Rods and tubes that are not sufficiently straight (or if supplied in a coiled form) can be straightened by passing them through pairs of rolls placed at different axes. The rolls subject the material to a series of bending and unbending operations, similar to the method shown in Fig. 6.43 for rolling.

Large sections can be drawn at elevated temperatures. In cold drawing, because of strain hardening, intermediate annealing between passes may be necessary to maintain sufficient ductility. Steel wires for springs and musical instruments are made by a heat-treatment process that precedes or follows the drawing operation (**patenting**). These wires have ultimate tensile strengths as high as 4800 MPa (700,000 psi), with tensile reduction of area of about 20 percent.

### Dies

Die materials are usually alloy tool steels, carbides, or diamond. A diamond die (used for drawing of fine wires) may be a single crystal or a polycrystalline diamond in a metal matrix. Carbide and diamond dies are made as inserts or nibs, which are then supported in a steel casing.

FIGURE 6.73 Terminology for a typical die for drawing round rod or wire.

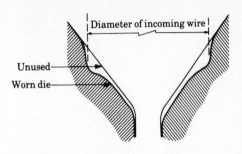

FIGURE 6.74 Schematic illustration of typical wear pattern in a wire-drawing die.

In addition to rigid dies, a set of idling rolls is also used in drawing of rods or bars of various shapes. This arrangement is known as **Turk's head** and is more versatile than ordinary dies since the rolls can be adjusted to various positions for different products.

A typical wear pattern on a drawing die is shown in Fig. 6.74. Note that die wear is highest at the entry. Although the die pressure is highest in this region and may be partially responsible for wear, other factors are involved. They are variation in the diameter of the entering wire and vibration (thus subjecting the entry contact zone to fluctuating stresses) and the presence of abrasive scale on the surface of the entering wire.

## Lubrication

Proper lubrication is essential in rod, tube, and wire drawing, regardless of whether the process is dry or wet drawing. In **dry drawing** the surface of the wire is coated with various lubricants, depending on its strength and frictional characteristics. A common lubricant used is soap. The rod to be drawn is first surface treated by **pickling.** This treatment removes the surface scale that could lead to surface defects and, being quite abrasive, would considerably reduce die life. The soap is picked up by the wire as it goes through a box filled with soap powder (stuffing box).

With high-strength materials, such as steels, stainless steels, and high-temperature alloys, the surface of the rod or wire may be coated either with a softer metal or with a **conversion coating.** Copper or tin can be chemically deposited on the surface of the metal. This thin layer of softer metal acts as a solid lubricant during drawing. Conversion coatings may consist of sulfate or oxalate coatings on the rod, which are then typically coated with soap, as a lubricant. Polymers are also used as solid lubricants, such as in drawing of titanium.

In **wet drawing**, the dies and rod are completely immersed in a lubricant. Typical lubricants are oils and emulsions containing fatty or chlorinated additives, and various chemical compounds.

**Equipment.** Two types of equipment are used in drawing. A **draw bench** is similar to a long horizontal tensile testing machine. It is used for single draws of straight, large cross-section rods and tubes for lengths up to 30 m (100 ft) with a hydraulic or chain-drive mechanism. Rod and wire of smaller cross-section are drawn by a **bull block (capstan),** which is

**FIGURE 6.75** Schematic illustration of the swaging process: (a) side view and (b) front view. (c) Schematic illustration of roller arrangement, curvature on the four radial hammers (that give motion to the dies), and the radial movement of a hammer as it rotates over the rolls.

**FIGURE 6.76** Reduction of outer and inner diameters of tubes by swaging. (a) Free sinking without a mandrel. The ends of solid bars and wire are tapered (pointing) by this process in order to feed the material into the conical die. (b) Sinking on a mandrel. Coaxial tubes of different materials can also be swaged in one operation.

(a)                                                                                    (b)

**FIGURE 6.77**   (a) Typical cross-sections produced by swaging constant-wall-thickness tube blanks on shaped mandrels. Rifling of small gun barrels can also be made by swaging, using a specially shaped mandrel. The formed tube is then removed by slipping it out of the mandrel. (b)These parts can also be made by swaging.

a rotating drum around which the wire is wrapped. The tension in this setup provides the force required to draw the wire.

## 6.6  Swaging

In swaging, also known as **rotary swaging** or **radial forging,** a solid rod or a tube is reduced in diameter by the reciprocating radial movement of two or four dies (Fig. 6.75). The die movements are generally obtained by means of a set of rollers in a cage. The internal diameter and the thickness of the tube can be controlled with or without mandrels as shown in Fig. 6.76. Mandrels can also be made with longitudinal grooves (similar in appearance to a splined shaft); thus, internally shaped tubes can be swaged (Fig. 6.77a). The rifling in gun barrels is made by swaging a tube over a mandrel with spiral grooves. Externally shaped parts can also be made by swaging (Fig. 6.77b).

The swaging process is usually limited to a diameter of about 50 mm (2 in.), although special machinery has been built to swage gun barrels of larger diameter. The length of the product in this process is limited only by the length of the mandrel (if needed). Die angles are usually only a few degrees and may be compound—the die may have more than one angle, for more favorable material flow during swaging. Lubricants are used for improved surface finish and longer die life. The process is generally carried out at room temperature. Parts produced by swaging have improved mechanical properties and good dimensional accuracy.

## 6.7  Die Manufacturing Methods

Various manufacturing methods, either singly or in combination, are used in making dies. These processes include casting, forging, machining, grinding, electrical, and electrochemical methods of die sinking and finishing operations, such as honing and polishing. The process of hubbing, either cold or hot, may also be used to make small dies with shallow cavities. Dies are usually heat treated for greater strength and wear resistance. If necessary, their surface profile and finish are improved by finish grinding and polishing, either by hand or with programmable industrial robots.

The *choice* of a die manufacturing method depends on the particular manufacturing process in which the die is to be used and the size and shape of the die. Cost often dictates the process selected because, as we have shown, tool and die costs can be significant in manufacturing operations. For example, the cost of a set of dies for automotive body panels may run $2 million. Even small and relatively simple dies can cost hundreds of dollars. On the other hand, because a large number of parts are usually made from the same die, die cost per piece is generally only a small portion of a part's manufacturing cost.

Dies may be classified as *male* and *female*. Dies may also be classified by their size. Small dies generally are those that have a surface area of $10^3 - 10^4$ mm$^2$ ($2 - 15$ in$^2$), whereas large dies have surface areas of 1 m$^2$ (9 ft$^2$) and larger, such as those used for pressworking automotive body panels.

Dies of various sizes and shapes can be *cast* from steels, cast irons, and nonferrous alloys. The processes used may be sand casting for large dies weighing many tons to shell molding for small dies. Cast steels are generally preferred as die materials because of their strength and toughness and the ease with which their composition, grain size, and properties can be controlled and modified.

Depending on how they solidify, cast dies, unlike those made of wrought metal by *forging* or *rolling,* may not have directional properties. Thus they exhibit the same properties on all working surfaces. However, because of shrinkage, control of dimension and accuracy in cast dies can be difficult, compared to machined dies.

Most commonly, dies are *machined* from forged *die blocks* by processes such as milling, turning, grinding, electrical and electrochemical machining, and polishing. Typically, a die for hot-working operations is machined by milling on an automatic copy machine that traces a master model (*pattern*), or on computer-controlled machine tools that use various software. The patterns may be made of wood, epoxy resins, aluminum, or gypsum. Conventional machining can be difficult for high-strength and wear-resistant die materials that are hard or are heat treated. These operations can be very time consuming even with copy milling machines. As a result, nontraditional machining processes are used extensively, particularly for small or medium-sized dies. These processes are generally faster and more economical, and the dies usually do not require additional finishing. Diamond dies for drawing fine wire are manufactured by producing holes with a thin rotating needle, coated with diamond dust suspended in oil.

To obtain improved hardness, wear resistance, and strength, die steels are usually *heat treated*. Improper heat treatment is one of the most common causes of die failure. Particularly important are the condition and composition of the die surfaces. Thus the proper selection of temperatures and atmospheres for heat treatment, quenching media, quenching practice, tempering procedures, and handling is important. Dies may be subjected to various surface treatments for improved frictional and wear characteristics (see Chapter 4).

After heat treatment, tools and dies are subjected to grinding, polishing, or chemical and electrical machining processes to obtain the desired surface finish and dimensional accuracy. Heat treatment may distort dies because of microstructural changes and uneven thermal cycling. The grinding process, if not controlled properly, can cause surface damage by excessive heat and induce harmful tensile residual stresses on the surface of the die, thus reducing its fatigue life. Scratches on a die's structure can act as stress raisers. Likewise, commonly used die-making processes, such as electrical-discharge machining, can cause surface damage and cracks, unless the process parameters are carefully controlled.

## 6.8   Die Failures

Failure of dies in manufacturing operations generally result from one or more of the following causes: improper design, defective material, improper heat treatment and finishing operations, overheating and heat checking (cracking caused by temperature cycling), excessive wear, overloading, misuse, and improper handling. We describe in the following paragraphs some of the important factors leading to die failure. These factors apply to dies made of die steels, but many are also applicable to other die materials. The proper design of dies is as important as the proper selection of die materials. In order to withstand the forces in manufacturing processes, a die must have proper cross sections and clearances. Sharp corners, radii, and fillets, as well as sudden changes in cross-section, act as stress raisers and can have detrimental effects on die life. Dies may be made in segments and prestressed during assembly for improved strength.

The proper handling, installation, assembly, and aligning of dies are important. *Overloading* of tools and dies can cause premature failure. A common cause of failure of cold-extrusion dies, for example, is the failure of the operator to remove a formed part from the die before loading it with another blank.

In spite of their hardness and resistance to abrasion, die materials such as carbides and diamond are susceptible to cracking and chipping from *impact forces* or *thermal stresses* caused by temperature gradients within the die. Thus surface preparation and finishing are important. Even metalworking fluids can adversely affect tool and die materials. Sulfur and chlorine additives in lubricants and coolants, for example, can leach away the cobalt binder in tungsten carbide and lower its strength and toughness.

Even if they are manufactured properly, dies are subjected to stresses and temperature during their use, which cause wear and hence shape changes. Die *wear* is important because when die shape changes, the parts, in turn, have improper dimensions. Thus the economics of the manufacturing operation is adversely affected.

During use, dies may also undergo *heat checking* from thermal cycling. To reduce heat checking (which has the appearance of parched land) and eventual die breakage in hot-working operations, dies are usually preheated to temperatures of about 150°C to 250°C (300°F to 500°F). Cracked or worn dies may be repaired by welding and metal-deposition techniques, including lasers. Dies may be designed and constructed with inserts that can be replaced when worn or cracked. The proper design and placement of these inserts is important, because otherwise they can crack.

Die failure and fracture in manufacturing plants can be hazardous to employees. It is not unusual for a set of dies resting on the floor or a shelf to suddenly disintegrate because of the highly stressed condition of its components. The broken pieces are propelled at high speed and can cause serious injury or death. Therefore highly stressed dies and tooling should always be surrounded by metal shielding. These shields should be properly designed and sufficiently strong to contain the fractured pieces in the event of die failure.

## CASE STUDY:    Development of Isothermally Forged Jet-Engine Blades

Fan and compressor blades for jet engines (see Fig. 3.1) have been forged to near-net shape for many years. However, blades on advanced aircraft engines are made from difficult-to-forge alloys and have very thin airfoils with small tolerances for thickness and geometric deviations. Fan blade design for the General Electric F–100 aircraft engine, for example, includes thinner airfoils, sharper edge radii, closer tolerances, and use of Ti–8Al–1Mo–1V titanium alloy with high strength-to-density and stiffness-to-density ratios (see Section 3.11.6). Engine performance is excellent, but manufacturing costs are high because, traditionally, such blades are machined from oversized forgings, typically requiring 1.47 kg (3.25 lb) of titanium alloy to produce a 0.45 kg (1 lb) final product, thus contributing significantly to the cost of raw material and manufacture.

In a more advanced manufacturing process, a fully contoured but slightly oversized forged blade is brought to finished airfoil dimensional and surface finish requirements by isothermally sizing the blade in a hot die. Although this approach had proven cost effective by eliminating a significant amount of manually controlled finishing operations, the next step was to combine conventional press forging and isothermal sizing into a single isothermal forging operation (see Section 6.2.3).

**Process development.**    The capability of isothermal forging to achieve substantial metal flow in titanium alloys using hot dies and low strain rates had previously been demonstrated. In addition, studies on die materials had identified IN–100, a highly alloyed nickel-base material, as an acceptable die material for titanium forging. However, there were many concerns, such as: (1) Can an acceptable surface finish be obtained in precision-forged titanium-alloy blades? (2) Will the IN–100 dies withstand the high temperatures and pressures required to forge this alloy? (3) Can finished blades be removed from the die without distortion? (4) Will finished blades meet specified metallurgical characteristics? (5) Will the dies be serviceable over long periods without deformation, distortion, or washout? (6) Can die-temperature uniformity be maintained under production conditions? (7) Is the process economically viable?

The development of a forging lubrication system for use on hot dies was a major accomplishment. Earlier development programs had used glass lubricants (see Sections 4.4.4 and 6.2.7, and Table 6.4). However, as expected, these materials built up in die cavities, causing forged parts to stick in the die; but several proprietary lubricant systems were developed to overcome these problems. The system consisted of a vitreous (glasslike) hydrodynamic component, a graphite boundary film component, and boron nitride (included primarily as a parting agent). The IN–100 dies were heated by electrical-resistance cartridge heating elements placed through the width of the die blocks, and the temperature distribution in the die was monitored using several thermocouples. Lubricant accumulation in the die cavities was minimal and acceptable surface condition was maintained by occasional wire brushing.

**Preform design.**    Blade preforms for the developmental effort were produced by conventional mechanical-press forging procedures. A series of forging trials established the preferred preform configuration and mass distribution (see, for example, Figs. 6.27 and 6.28). The preforms exceeded finish thickness by approximately 1.5 mm (0.060 in.) on the airfoil surfaces and were 6.60 mm (0.260 in.) oversize on the root envelope. Initially, blades were forged in a single isothermal squeeze to final net size. Although a few blades were successfully forged by this procedure, the process was judged as suboptimal for production purposes. Problems arose because the isothermal lubrication system was not adequate for the large amount of metal movement required. In order to ensure the production of quality blades, a two-squeeze technique was adopted. It was determined that in production, it would be preferable to conventionally forge the preforms to within 0.75 mm (0.030 in.) of the final thickness. Starting with a thinner preform, a single isothermal squeeze would produce blades with net dimensions and acceptable surface finish. Using isothermal forging, the starting material requirement was reduced to 0.83 kg (1.83 lb) from 1.47 kg for the traditional forging, i.e., a reduction of almost 44 percent.

**Forging conditions.** More than 300 blades were forged using the double-squeeze isothermal forging process. Forging temperature was maintained at 960°C (1760°F) ± 20°F for most of these forgings. Between the first and second squeeze, the blades were cleaned, given an acid treatment to remove any surface oxidation, and recoated with the lubricant. The second squeeze was identical to the first except for a shorter dwell time (1 min) at peak pressure. The die blocks were inspected after almost six hundred pressing operations and found to be still in good condition. Some plastic deformation of the die was evident in localized regions, but because of the excess stock in the root, these regions were not detrimental. Some fine cracks developed in the die surface early in the forging trials, and the cause was attributed to possible porosity in the cast dies.

**Heat treatment and mechanical properties.** The blades were solution treated and aged after forging (see section 5.10.2). Solution treatment was in a vacuum at 930°C (1700°F) for 1 hour, followed by forced argon cooling. The aging cycle consisted of 8 hours at 565°C (1050°F). The mechanical properties of the isothermally forged and heat-treated blades were determined on specimens taken from the center portion of the airfoil, between the root and mid-span. Properties were well above specification, and blade microstructure was entirely satisfactory.

Two alternative thermomechanical treatments were also investigated. In the first, the blades were processed through the solution anneal heat treatment prior to, rather than after, isothermal forging. The second approach was to assume that the isothermal forging operation itself provided a solution anneal, i.e., the blades were aged only. Microstructures and mechanical properties of the blades that were solution annealed before isothermal forging were found to be equivalent to those of the production-lot blades. The aged-only blades appeared to be significantly stronger but slightly less ductile than production-lot blades and had similar microstructures.

**Blades twisting.** When removed from the isothermal dies, the forged blades were excellent from a bow standpoint, but were significantly undertwisted and well outside permissible angular tolerances. The undertwisted condition worsened when the blades were given a solution anneal heat treatment. This behavior appeared to be reasonably reproducible, that is, it would be possible to design an extra increment of twist into the tooling to compensate for the tendency of the blade to unwind. Because of the undertwisted condition of the blades, a warm straightening operation was added to the manufacturing process (such operations are common practice in conventional fan-blade manufacture). Straightening was carried out at 590°C (1100°F) in overtwisted tool-steel dies.

**Machining.** The blades were machined conventionally (see Chapter 8), with the root completely machined and the mid-span elements rough machined. Before finish machining the close-tolerance mid-span notch faces, fan blades are customarily given a stress-relief heat treatment. When this procedure was followed with the isothermally forged blades, they unwound to an uncommon degree. This distortion forced more than half the blades outside tolerance specifications for airfoil twist and mid-span back contour. This degree of blade distortion could not be corrected using manual techniques; however, investigation of straightening alternatives was not required, either. Instead, the blades were given an intense glass-bead peening (see Section 4.5.1), which is normally the last operation. The surface stresses introduced during peening caused the airfoils to wind up to approximately the same degree they had unwound during stress relieving. With the bulk of the blades back within bow and twist tolerances, the remaining mid-span machining operations could be completed. A localized glass-bead peening, restricted to the mid-span region, completed this stage of the processing.

As a result of this study, reproducible isothermal forging procedures were established for titanium blades that met all metallurgical requirements. The process can reduce manufacturing costs, particularly for larger blades, through manufacturing time savings and improved material utilization. Further process refinements involve isothermal forging of these blades in a single squeeze and eliminating the post-forging solution anneal treatment.

*Source:* After J.D. Snow and C.R. Cook, TRW, Inc, in *ManTech Journal,* Vol.5, No.3, 1980.

## SUMMARY

- Bulk deformation processes, which consist of forging, rolling, extrusion, and drawing, involve major changes in dimensions of the workpiece which are subsequently subjected to further processing. The bulk properties as well as surface characteristics of the material are important. (Section 6.1)

- Forging denotes a family of processes by which deformation of the workpiece is carried out by compressive forces. Forging is capable of producing a wide variety of parts with favorable characteristics of strength, roughness, dimensional accuracy, and reliability in service. Among important factors in forging are friction, material flow in dies, heat transfer, and lubrication. Various defects can develop, depending on die geometry, quality of the billet materials, and preform. Several forging machines are available with various characteristics and capabilities. (Section 6.2)

- Methods of analysis of the stresses and loads in forging and other bulk-deformation processes include the slab method, slip-line analysis, the upper-bound technique, and the finite-element method. (Section 6.2)

- Rolling is the process of reducing the thickness or changing the cross-section of a long workpiece by compressive forces applied through a set of rolls. Rolling is a continuous operation and involves several material and process variables, including roll size relative to material thickness, reduction per pass, speed, lubrication, and temperature. Rolled products include plate, sheet, foil, rod, pipe and tube, as well as shape-rolled products such as l-beams, structural shapes, and bars of various cross-section. A variety of roll configurations are available and are used for rolling these products. (Section 6.3)

- Extrusion is the process of forcing a material through a die, much like squeezing toothpaste from a tube. The extrusion process is capable of producing lengths of solid or hollow sections with constant cross-sectional area. Important factors are die design, extrusion ratio, lubrication, billet temperature, and extrusion speed. Cold extrusion, which is a combination of extrusion and forging operations, is capable of producing parts economically and with good mechanical properties. (Section 6.4)

- Rod, wire, and tube drawing involve pulling the material through one or more dies. Proper die design and selection of materials and lubricants are essential to obtaining products of high quality with good surface finish and dimensional accuracy. The major process variables in the mechanics of drawing are die angle, friction, and reduction per pass. (Section 6.5)

- In swaging, a solid rod or a tube is reduced in diameter by the reciprocating radial movement of two or four dies. Swaging is suitable for producing short or long lengths of bar or tubing with various internal or external profiles. (Section 6.6)

- Because die failure has a major economic impact, die design, die material selection, and die manufacturing methods are of major importance. A wide variety of die materials and manufacturing methods are available, including nontraditional material-removal processes and subsequent treatment and finishing operations. (Sections 6.7, 6.8)

## SUMMARY OF EQUATIONS

### Forging

Pressure in plane-strain compression, $p = Y' e^{2\mu(a-x)/h}$

Average pressure in plane-strain compression, $p \simeq Y'\left(1 + \dfrac{\mu a}{h}\right)$

Pressure in axisymmetric compression, $p = Y e^{2\mu(r-x)/h}$

Average pressure in axisymmetric compression, $p = Y\left(1 + \dfrac{2\mu r}{3h}\right)$

Pressure in plane-strain compression, sticking, $p = Y'\left(1 + \dfrac{a-x}{h}\right)$

Pressure in axisymmetric compression, sticking, $p = Y\left(1 + \dfrac{r-x}{h}\right)$

### Rolling

Roll pressure in entry zone, $p = Y'_f \dfrac{h}{h_o} e^{\mu(H_0 - H)}$

    with back tension, $p = (Y'_f - \sigma_b)\dfrac{h}{h_o} e^{\mu(H_0 - H)}$

Roll pressure in exit zone, $p = Y'_f \dfrac{h}{h_f} e^{\mu H}$

    with front tension, $p = (Y'_f - \sigma_f)\dfrac{h}{h_f} e^{\mu H}$

Parameter $H = 2\sqrt{\dfrac{R}{h_f}} \tan^{-1}\left(\sqrt{\dfrac{R}{h_f}} \cdot \phi\right)$

Roll force, $F = \displaystyle\int_{o}^{\phi_u} w\, p\, R\, d\phi + \int_{\phi_n}^{\alpha} w\, p\, R\, d\phi$

Roll force, approximate, $F = Lw\overline{Y}'\left(1 + \dfrac{\mu L}{2h_{av}}\right)$

Roll torque, $T = \displaystyle\int_{\phi_u}^{\alpha} w\mu\, p\, R^2 d\phi - \int_{o}^{\phi_u} w\mu\, p\, R^2 d\phi$

Roll contact length, approximate, $L = \sqrt{R\Delta h}$

Power per roll $= \pi FLN/60{,}000$ kW

$\qquad\qquad\quad = \pi FLN/33{,}000$ hp

Maximum draft, $\Delta h_{max} = \mu^2 R$

Maximum angle of acceptance, $\alpha_{max} = \tan^{-1}\mu$

### Extrusion

Extrusion ratio, $R = A_o/A_f$

Extrusion pressure, ideal, $p = Y \ln R$

Extrusion pressure, with friction, $p = Y\left(1 + \dfrac{\tan\alpha}{\mu}\right)\left(R^{\mu\cot\alpha} - 1\right)$

### Rod and wire drawing

Drawing stress, ideal, $\sigma_d = Y \ln\left(\dfrac{A_o}{A_f}\right)$

Drawing stress, with friction, $\sigma_d = Y\left(1 + \dfrac{\tan\alpha}{\mu}\right)\left[1 - \left(\dfrac{A_f}{A_o}\right)^{\mu\cot\alpha}\right]$

Die pressure, $p = Y_f - \sigma$

## BIBLIOGRAPHY

### General Texts

Altan, T., S.-I. Oh, and H. Gegel, *Metal Forming–Fundamentals and Applications*. Metals Park, Ohio: American Society for Metals, 1983.

Avitzur, B., *Handbook of Metal-Forming Processes*. New York: Wiley, 1983.

Blazynski, T.Z., *Plasticity and Modern Metal-Forming Technology*. New York: Elsevier, 1989.

Boer, C.R., N. Rebelo, H. Rydstad, and G. Schroder, *Process Modelling of Metal Forming and Thermomechanical Treatment*. New York: Springer-Verlag, 1986.

Dieter, G.E., *Mechanical Metallurgy*, 3d ed. New York: McGraw-Hill, 1986.

Hosford, W.F., and R.M. Caddell, *Metal Forming, Mechanics and Metallurgy*, 2d ed. Englewood Cliffs, New Jersey: Prentice-Hall, 1993.

Johnson, W., and P.B. Mellor, *Engineering Plasticity.* New York: Van Nostrand Reinhold, 1973.
Kobayashi, S., S.-I. Oh, and T. Altan, *Metal Forming and the Finite-Element Method.* New York: Oxford, 1989.
Lange, K. (ed.), *Handbook of Metal Forming.* New York: McGraw-Hill, 1985.
Rowe, G.W., *Elements of Metalworking Theory.* London: Edward Arnold, 1979.
————, *Principles of Industrial Metalworking Processes.* London: Edward Arnold, 1977.

## Forging

Altan, T., S.-I. Oh, and H.C. Gegel, *Metal Forming–Fundamentals and Applications.* Metals Park, Ohio: American Society for Metals, 1983.
*ASM Handbook,* Vol. 14: *Forming and Forging.* Materials Park, Ohio: ASM International, 1988.
Byrer, T.G. (ed.), *Forging Handbook.* Metals Park, Ohio: American Society for Metals, 1985.
*Forging Design Handbook.* Metals Park, Ohio: American Society for Metals, 1972.
Lange, K. (ed.), *Handbook of Metal Forming.* New York: McGraw-Hill, 1985.
*Open Die Forging Manual,* 3d ed. Cleveland: Forging Industry Association, 1982.
Sabroff, A.M., F.W. Boulger, and H.J. Henning, *Forging Materials and Practices.* New York: Reinhold, 1968.

## Rolling

Lange, K. (ed.), *Handbook of Metal Forming.* New York: McGraw-Hill, 1985.
Larke, E.C., *The Rolling of Strip, Sheet, and Plate,* 2d ed. London: Chapman and Hall, 1963.
Roberts, W.L., *Cold Rolling of Steel.* New York: Marcel Dekker, 1978.
————, *Hot Rolling of Steel.* New York: Marcel Dekker, 1983.
Underwood, L.R., *The Rolling of Metals,* Vol. 1. New York: Wiley, 1950.
Wusatowski, Z., *Fundamentals of Rolling.* New York: Pergamon, 1969.

## Extrusion

*ASM Handbook,* Vol. 14: *Forming and Forging.* Materials Park, Ohio: ASM International, 1988.
Alexander, J.M., and B. Lengyel, *Hydrostatic Extrusion.* London: Mills and Boon, 1971.
Everhart, J.E., *Impact and Cold Extrusion of Metals.* New York: Chemical Publishing Company, 1964.
Inoue, N., and M. Nishihara (eds.), *Hydrostatic Extrusion: Theory and Applications.* New York: Elsevier, 1985.
Lange, K. (ed.), *Handbook of Metal Forming.* New York: McGraw-Hill, 1985.
Laue, K., and H. Stenger, *Extrusion—Processes, Machinery, Tooling.* Metals Park, Ohio: American Society for Metals, 1981.
Michaeli, W., *Extrusion Dies.* New York: Macmillan, 1984.
Pearson, C.E., and R.N. Parkins, *The Extrusion of Metals,* 2d ed. New York: Wiley, 1961.

## Drawing

Bernhoeft, C.P., *The Fundamentals of Wire Drawing.* London: The Wire Industry Ltd., 1962.
Lange, K. (ed.), *Handbook of Metal Forming.* New York: McGraw-Hill, 1985.
*Nonferrous Wire Handbook,* 2 vols. Branford, Conn.: The Wire Association International, Inc., 1977 and 1981.
*Steel Wire Handbook,* Vol. 1, 1968; Vol. 2, 1969; Vol. 3, 1972; Vol. 4, 1980. Guilford, Conn.: Wire Association International.

# QUESTIONS

## Forging

6.1 How can you tell whether a certain part is forged or cast? Describe the features that you would investigate to arrive at a conclusion.

6.2 Why is the control of blank volume important in closed-die forging?

6.3 What are the advantages and limitations of a cogging operation? Die inserts in forging?

6.4 Explain why there are so many different kinds of forging machines.

6.5 Devise an experimental method whereby you can measure the force required for forging only the flash in impression-die forging (see Fig. 6.15a).

6.6 A manufacturer is successfully hot forging a certain part using material supplied by Company A. A new supply of material is obtained from Company B, with the same nominal composition of the major alloying elements as the previous material. However, it is found that the new forgings are cracking even though the same procedure is followed as before. What is the probable reason?

6.7 Explain why there might be a change in the density of a forged product as compared to that of the blank.

6.8 Describe the role of surface oxide layers on a blank as it is being deformed by impression-die forging.

6.9 Since glass is a good lubricant for hot extrusion, would you use glass for impression-die forging also? Explain.

6.10 Describe and explain the factors that influence spread in cogging operations on square billets.

6.11 In deriving the formula for forging, we had to use the yield criteria described in Chapter 2. Explain why.

6.12 Why are end grains generally undesirable in forged products? Give examples.

## Rolling

6.13 Rolling reduces the thickness of plates and sheets. However, it is possible to reduce the thickness by simply stretching the material. Would this be a feasible process? Explain.

6.14 In Fig. 6.36 explain why the neutral point moves towards the roll-gap entry as friction increases.

6.15 Explain how a cast structure is converted into a wrought structure by hot rolling.

6.16 We described three factors that influence spreading in rolling: strip width-to-thickness ratio, friction, and ratio of roll radius to strip thickness. Explain how these factors affect spreading.

6.17 Explain how the residual stress patterns in Fig. 6.42 become reversed when roll radius or reduction per pass is changed.

6.18 Explain how you would go about applying front and back tensions to sheet metals during rolling.

6.19 We noted that rolls tend to flatten under roll forces. Which property(ies) of the roll material can be increased to reduce flattening?

6.20 Describe the methods by which roll flattening can be reduced.

6.21 Explain the technical and economic reasons for taking larger rather than smaller reductions per pass in flat rolling.

6.22 Surface roughness in hot-rolled products is higher than in cold-rolled products. Explain why.

6.23 List and explain the methods that can be used to reduce the roll-separating force.

6.24 Explain the advantages and limitations of using small-diameter rolls on flat rolling.

6.25 Is it possible to adjust back and front tensions simultaneously so that the neutral point remains in the same position as in ordinary rolling without tensions? Explain.

## Extrusion

6.26 The extrusion ratio, die geometry, extrusion speed, and billet temperature all affect the extrusion pressure. Explain why.

6.27 How would you go about avoiding centerburst defects in extrusion? Explain why your methods would be effective.

6.28 Assume that you are reducing the diameter of two round rods, one by simple tension and the other by indirect extrusion. Which method will require more force? Why?

6.29 How would you go about making a stepped extrusion that has increasingly larger cross-sections along its length? Is it possible? Would your process be economical and suitable for high production runs? Explain.

6.30 We have stated that the temperature ranges for extruding various metals are similar to those for forging (see Table 6.3). Describe the consequences of extruding at a temperature (1) below and (2) above these ranges.

6.31 We note from Eq. (6.52) that, for low values of extrusion ratio such as $R = 2$, the ideal extrusion pressure $p$ can be lower than the yield stress $Y$ of the material. Explain whether or not this is logical.

6.32 Referring to Eq. (6.59), explain why the three variables in that equation should influence the strain rate in extrusion.

## Drawing

6.33 What changes would you expect in the strength, hardness, ductility, and anisotropy of annealed metals after being cold drawn through dies? Why?

6.34 We have seen that in rod and wire drawing, the maximum die pressure is at the die entry. Why?

6.35 Describe the conditions under which wet drawing and dry drawing, respectively, are desirable.

6.36 Name the important process variables in drawing and explain how they affect the drawing process.

6.37 Assume that a rod drawing operation can be carried out either in one pass or in two passes in tandem. If all die angles are the same and the total reduction is the same, will the drawing forces be different? Explain.

6.38 In Fig. 6.66, assume that the reduction in the cross-section is taking place by pushing the rod through the die instead of pulling it. Assuming that the material is perfectly plastic, sketch the die pressure distribution and explain your reasoning for the following situations: (1) frictionless, (2) with friction, and (3) frictionless but with front tension.

6.39 In deriving Eq. (6.72), no mention is made of the ductility of the original material being drawn. Explain why.

6.40 Explain why the die pressure in drawing decreases toward the exit.

6.41 What is the value of the die pressure at the exit when a drawing operation is being carried out at the maximum reduction per pass?

6.42 Explain why the maximum reduction per pass in drawing should increase with increasing strain-hardening exponent $n$.

6.43 Explain why the maximum reduction per pass is higher in drawing round bars (63 percent) as compared to plane-strain drawing of flat sheet or plate (58 percent).

6.44 If in deriving Eq. (6.71) we include friction, will the maximum reduction per pass be the same, higher, or lower than 63 percent? Explain.

6.45 Explain what effects back tension has on the die pressure in wire or rod drawing, and why.

6.46 Explain why the inhomogeneity factor $\phi$ in rod and wire drawing depends on the ratio $h/L$ in Fig. 6.12.

6.47 Describe the reasons why the swaging process was developed.

## General

6.48 From the topics covered in this chapter, list and explain specifically two examples where (1) friction is desirable and (2) friction is not desirable.

6.49 Take any three topics from Chapter 2 and with a specific example for each, show their relevance to the topics covered in this chapter.

6.50 Take any three topics from Chapter 3 and with a specific example for each, show their relevance to the topics covered in this chapter.

6.51 List and explain the reasons why there are so many different types of die materials used for the processes described in this chapter.

6.52   Explain the reasons why we should be interested in residual stresses developed in workpieces made by the forming processes described in this chapter.

6.53   Make a summary of the types of defects found in the processes described in this chapter. Indicate, for each, methods of reducing or eliminating the defects.

## PROBLEMS

### Forging

6.54   In the free-body diagram in Fig. 6.4b, we have shown the incremental stress $d\sigma_x$ on the element pointing to the left. Yet it would appear that, because of the direction of frictional stresses $\mu p$, it should point to the right in order to balance the horizontal forces. Show that the same answer for the forging pressure is obtained regardless of the direction of this incremental stress.

6.55   Plot the force vs. reduction-in-height curve in open-die forging of a cylindrical, annealed copper specimen 1 in. high and 1 in. in diameter, up to a reduction of 75 percent for the case of (1) no friction between the flat dies and the specimen, (2) $\mu = 0.1$, and (3) $\mu = 0.2$. Ignore barreling. Use average pressure formulas.

6.56   Provide the answers to Problem 6.55 using Fig. 6.9b.

6.57   Calculate the work done for each case in Problem 6.55.

6.58   Determine the temperature rise in the specimen for each case in Problem 6.55, assuming that the process is adiabatic and the temperature is uniform throughout the specimen.

6.59   To determine forgeability, a hot-twist test is performed on a round bar 25 mm in diameter and 300 mm long. It is found that it underwent 200 turns before it fractured. Calculate the shear strain at the outer surface of the bar at fracture.

6.60   Derive an expression for the average pressure in plane-strain compression with sticking friction.

6.61   What is the value of $\mu$ when, for plane-strain compression, the forging load with sliding friction is equal to the load with sticking friction? Use average pressure formulas.

6.62   We note that in cylindrical upsetting, the frictional stress cannot be greater than the shear yield stress $k$ of the material. Thus, there may be a distance $x$ in Fig. 6.8 where a transition occurs from sliding to sticking friction. Derive an expression for $x$ in terms of $r$, $h$, and $\mu$ only.

6.63   Assume that the workpiece is being pushed to the right by a lateral force $F$ while being compressed between flat dies (see figure below). (1) Make a sketch of the die pressure distribution for the condition where $F$ is not large enough to slide the workpiece to the right. (2) Make a similar sketch except that $F$ is now large enough so that the workpiece is sliding to the right while being compressed.

6.64   For the sticking example in Fig. 6.10, derive an expression for the lateral force $F$ required to slide the workpiece to the

right while being compressed between flat dies.

6.65   Two solid cylindrical specimens $A$ and $B$, made of a perfectly plastic material, are being forged with friction and isothermally at room temperature to a reduction in height of 50 percent. Specimen *A* is 2 in. high and 1 sq. in. in cross-sectional area, and specimen *B* is 1 in. high and 2 sq. in. in cross-section. Will the work done be the same for the two specimens? Explain.

6.66 In Fig. 6.6, does the pressure distribution along the four edges of the workpiece depend on the particular yield criterion used? Explain.

6.67 Under what conditions would you have a normal pressure distribution in forging a workpiece as shown below? Explain.

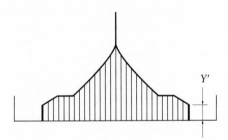

6.68 Derive the average die pressure formula given by Eq. (6.15). (*Hint:* Obtain the volume under the friction hill over the surface by integration and divide it by the cross-sectional area of the workpiece.)

6.69 Take two solid cylindrical specimens of equal diameter but different heights and compress them (frictionless) to the same percent reduction in height. Show that the final diameters will be the same.

6.70 A rectangular workpiece has the following original dimensions: $2a = 100$ mm, $h = 25$ mm, and width = 20 mm. The metal has a strength coefficient of 400 MPa and a strain-hardening exponent of 0.5. It is being forged in plane strain with $\mu = 0.3$. Calculate the force required when the height is reduced by 20 percent. Do not use average pressure formulas.

6.71 Assume that in upsetting a solid cylindrical specimen between two flat dies and with friction, the dies are rotated at opposite directions to each other. How, if at all, will the forging force change from that for non-rotating dies? (*Hint:* Note that the dies will now require torque because of the change in the direction of frictional forces at the die–workpiece interfaces.)

6.72 A solid cylindrical specimen, made of a perfectly plastic material, is being upset between flat dies with no friction. The process is being carried out by a falling weight, as in a drop hammer. The downward velocity of the hammer is at a maximum when it first contacts the workpiece, and becomes zero when the hammer stops at a certain height of the specimen. Establish quantitative relationships between workpiece height and velocity and make a qualitative sketch of the velocity profile of the hammer. (*Hint:* The loss in the kinetic energy of the hammer is the plastic work of deformation; thus there is a direct relationship between workpiece height and velocity.)

6.73 How would you go about estimating the swaging force acting on each die?

6.74 We note that in swaging a tube, its thickness increases. How would you go about calculating the thickness change without measuring it? You are allowed to make any other measurements.

## Rolling

6.75 In Example 6.4, what is the velocity of the strip leaving the rolls?

6.76 With appropriate sketches, explain the changes that occur in the roll pressure distribution if one of the rolls is idling, i.e., power is shut off to that roll.

6.77 It can be shown that it is possible to determine $\mu$ in flat rolling without measuring torque or forces. By inspecting rolling equations, describe an experimental procedure to do so. Note that you are allowed to measure any quantity other than torque or forces.

6.78 Derive a relationship between back tension $\sigma_b$ and front tension $\sigma_f$ in rolling such that when both tensions are increased, the neutral point remains in the same position.

6.79 Take an element at the center of the deformation zone in flat rolling. Assuming that all the stresses acting on this element are principal stresses, place these stresses qualitatively and state whether they are tension or compression. Explain why. Is it possible for all these three principal stresses to be equal to each other in magnitude? Explain.

6.80 It has been stated that in rolling a flat strip, the roll-separating force is reduced about twice as effectively by back tension as it is by front tension. With appropriate sketches, explain why this is so. (*Hint:* Note the position of the neutral point.)

6.81 We can see that in rolling a strip, the rolls will begin to slip if the back tension $\sigma_b$ is too high. Derive an analytical expression for the magnitude of the back tension in order to make the powered rolls begin to slip. Use the same terminology as in the text.

6.82 Prove Eq. (6.44).

6.83. In Steckel rolling, the rolls are idling and thus there is no net torque, assuming frictionless bearings. Where then is the energy coming from to supply the work of deformation in rolling? Explain with appropriate sketches and state the conditions that have to be satisfied.

6.84 Derive an expression for the tension required in Steckel rolling of a flat sheet, without friction, for a workpiece whose true stress–true strain curve is given by $\sigma = a + b\varepsilon$.

6.85 Make a neat sketch of the roll-pressure distribution in ordinary rolling with powered rolls. (1) Assume now that the power to both rolls is shut off and rolling is taking place by front tension only, i.e., Steckel rolling. Superimpose on your diagram the new roll-pressure distribution, explaining your reasoning clearly. (2) After completing part (1), further assume that the roll bearings are getting rusty and deprived of lubrication while rolling is still taking place by front tension only. Superimpose a third roll-pressure distribution diagram for this condition, explaining your reasoning.

6.86 Derive Eq. (6.26b), based on the equation preceding it. Comment on how different the $h$ values are as the angle $\phi$ increases.

6.87 In Fig. 6.37, assume that $L = 2L_2$. Is the roll force $F$ for $L$ twice or more than twice the roll force for $L_2$? Explain.

6.88 A flat-rolling operation is being carried out where $h_o = 0.125$ in., $h_f = 0.100$ in., $w_o = 10$ in., $R = 10$ in., $\mu = 0.1$, and the average flow stress of the material is 40,000 psi. Estimate the roll force and the torque. Include roll flattening.

## Extrusion

6.89 Derive an expression for the true strain rate $\dot{\varepsilon}$ in the deformation zone in extrusion of a square billet, in terms of the entering velocity $V_o$, die angle $\alpha$, initial billet dimension $t_o$, and distance $x$ from entry.

6.90 Calculate the force required in direct extrusion of 1100-O aluminum from a diameter of 6 in. to 2 in. Assume that the redundant work is 40 percent of the ideal work of deformation and the friction work is 25 percent of the total work of deformation.

6.91 Prove Eq. (6.58).

6.92 Referring to Example 6.6, calculate the theoretical temperature rise in the extruded material, assuming that there is no heat loss. (See Chapter 3 for physical properties.)

6.93 Using the same approach as that shown in Section 6.5 for wire drawing, show that the extrusion $\sigma_d$ pressure is given by the following expression:

$$p = Y\left(1 + \frac{\tan\alpha}{\mu}\right)\left[1 - \left(\frac{A_o}{A_f}\right)^{\mu\cot\alpha}\right],$$

where $A_o$ and $A_f$ are the original and final workpiece diameters, respectively.

## Drawing

6.94 Calculate the power required in Example 6.7 if the workpiece material is annealed 70–30 brass.

6.95 Using Eq. (6.64), make a plot similar to Fig. 6.70 for the following conditions: $K = 70$ MPa, $n = 0.25$, $\mu = 0.04$.

6.96 Using the same approach as that in Section 6.5 for wire drawing, show that the drawing stress $\sigma_d$ in plane-strain drawing of a flat sheet or plate is given by the expression

$$\sigma_d = Y'\left(1 + \frac{\tan \alpha}{\mu}\right)\left[1 - \left(\frac{h_f}{h_o}\right)^{\mu\cot\alpha}\right],$$

where $h_o$ and $h_f$ are the original and final thickness, respectively, of the workpiece.

6.97 Derive an analytical expression for the die pressure in wire drawing, without friction or redundant work, as a function of the instantaneous diameter in the deformation zone.

6.98 A linearly strain-hardening material with a true stress–true strain curve $\sigma = 5,000 + 20,000\varepsilon$ psi is being drawn into a wire. If the original diameter of the wire is 0.2 in., what is the minimum possible diameter at the die exit? Assume that there is no redundant work and the frictional work is 25 percent of the ideal work of deformation. (*Hint:* The yield stress of the exiting wire is that point on the true stress–true strain curve corresponding to the total strain that the material has undergone.)

6.99 In Fig. 6.72, assume that the longitudinal residual stress at the center is $-80,000$ psi. Using the distortion-energy criterion, calculate the minimum yield stress that this particular steel must have in order to sustain these levels of residual stresses.

6.100 What is the answer to Problem 6.99 for the maximum-shear-stress criterion?

6.101 Derive an expression for the die-separating force in frictionless wire drawing of a per-

fectly plastic material. Use the same terminology as in the text.

6.102 A material with a true stress–true strain curve $\sigma = 10,000\varepsilon^{0.5}$ is used in wire drawing. Assuming that the friction and redundant work comprise a total of 50 percent of the ideal work of deformation, calculate the maximum reduction in cross-sectional area per pass that is possible.

6.103 Derive an expression for the maximum reduction per pass for a material of $\sigma = K\varepsilon^n$, assuming that the friction and redundant work contribute a total of 20 percent to the ideal work of deformation.

6.104 Prove that the true strain rate $\dot{\varepsilon}$ in drawing or extrusion in plane strain with wedge-shaped dies is given by the expression

$$\dot{\varepsilon} = -2\tan\alpha V_o t_o / (t_o - 2x\tan\alpha)^2,$$

where $\alpha$ is the die angle, $t_o$ is the original thickness, and $x$ is the distance from die entry. (*Hint:* Note that $d\varepsilon = dA/A$.)

6.105 In drawing a strain-hardening material with $n = 0.2$, what should be the percentage of friction plus redundant work, in terms of ideal work, so that the maximum reduction per pass is 63 percent?

6.106 Assume that you are asked to give a quiz to students on the contents of this chapter. Prepare five

# 7

# Sheet-Metal Forming Processes

**7.1** Introduction

**7.2** Sheet-Metal Characteristics

**7.3** Shearing

**7.4** Bending of Sheet and Plate

**7.5** Stretch Forming

**7.6** Bulging

**7.7** Rubber Forming

**7.8** Spinning

**7.9** High-Energy-Rate Forming

**7.10** Superplastic Forming

**7.11** Various Forming Methods

**7.12** Deep Drawing

**7.13** Formability of Sheet Metals

**7.14** Economics of Sheet-Metal Forming

## 7.1 Introduction

In Chapter 6 we noted that workpieces could be characterized by the ratio of their surface area to their volume or thickness. **Sheet forming,** unlike bulk deformation processes, involves workpieces with a high ratio of surface area to thickness. A sheet thicker than 6 mm (1/4 in.) is generally called a **plate.** Although relatively thick plates, such as those used in boilers, bridges, ships, and nuclear power plants, may have smaller ratios of surface area to thickness, sheet forming usually involves relatively thin materials.

The products made by sheet-forming processes include a large variety of shapes and sizes, ranging from simple bends to double curvatures with shallow or deep recesses. Typical examples are metal desks, appliance bodies, hubcaps, aircraft panels, beverage cans, car bodies, and kitchen utensils. Thus sheet forming, which is also called **pressworking,** is among the most important of metalworking processes. It dates back to as early as 5000 B.C. when household utensils, jewelry, and other objects were made by hammering and stamping metals such as gold, silver, and copper.

Sheet metal is produced by a rolling process, as described in Section 6.3. If the sheet is thin, it is generally coiled; if thick, it is available as flat sheets or plates, which may have been decoiled and flattened.

Before a sheet-metal part is formed, a blank of suitable dimensions is first removed from a large sheet. Removal is usually done by a shearing process, as described in Section 7.3. However, there are several other methods for cutting sheet, particularly plates. The sheet or plate may be cut with a band saw, which is a chip removal process. Flame cutting is another common method, particularly for thick steel plates, and is widely used in shipbuilding and heavy structural components. Laser cutting has also become an important process and is used with computer-controlled equipment to cut a variety of shapes consistently. In friction sawing, a disk or blade rubs against the sheet or plate at high surface speeds. Because shearing is the most commonly used method, we focus on it as a method of cutting sheet metal.

## 7.2   Sheet-Metal Characteristics

Sheet metals are generally characterized by a high ratio of surface area to thickness. Forming of sheet metals generally is carried out by tensile forces in the plane of the sheet; otherwise, the application of compressive forces could lead to buckling, folding, and wrinkling of the sheet.

In bulk deformation processes such as forging, rolling, extrusion, and wire drawing (Chapter 6), the thickness or the lateral dimensions of the workpiece are intentionally changed. However, in most sheet-forming processes any thickness change is caused by stretching of the sheet under tensile stresses (Poisson's ratio). Thickness decreases generally should be avoided, as they can lead to necking and failure.

Because the basic mechanisms of all sheet-forming processes are stretching and bending, certain factors significantly influence the overall operation. The major ones are: elongation, yield-point elongation, anisotropy, grain size, residual stresses, springback, and wrinkling.

### 7.2.1   Elongation

Although sheet-forming operations rarely involve simple uniaxial stretching, the observations made in regard to simple tensile testing can be useful in understanding the behavior of sheet metals. Recall from Fig. 2.2 that a specimen subjected to tension first undergoes **uniform elongation** (which corresponds to the UTS, after which necking begins). This elongation is then followed by additional nonuniform elongation until fracture takes place (**postuniform elongation**). The material is being stretched in sheet forming, so high uniform elongation is desirable for good formability.

In Section 2.2.4 we showed that, for a material having a true stress–true strain curve that can be represented by the equation

$$\sigma = K\varepsilon^n, \tag{7.1}$$

the strain at which necking begins (**instability**) is given by

$$\varepsilon = n. \tag{7.2}$$

Hence the true uniform strain in a simple stretching operation (uniaxial tension) is numerically equal to the strain-hardening exponent $n$. A large $n$ indicates large uniform elongation and thus it is desirable for sheet forming.

Necking of a sheet specimen generally takes place at an angle $\phi$ to the direction of tension, as shown in Fig. 7.1a (**localized necking**). For an isotropic sheet specimen in simple tension, the Mohr's circle for this situation is constructed as follows (Fig. 7.1b). The strain $\varepsilon_1$ is the longitudinal strain, and $\varepsilon_2$ and $\varepsilon_1$ are the lateral strains. Poisson's ratio in the plastic range

**FIGURE 7.1** (a) Localized necking in a sheet specimen under tension. (b) Determination of the angle of neck from Mohr's circle for strain. (c) Schematic illustrations for diffuse and localized necking, respectively. (d) Localized necking in an aluminum strip stretched in tension. Note the double neck.

is 0.5, so the lateral strains have the value $-\varepsilon_1/2$. The narrow neck band in Fig. 7.1(a) is in **plane strain** along its length because it is constrained by the material above and below the neck.

The angle $\phi$ can now be determined from the Mohr's circle by a rotation (either clockwise or counterclockwise) of $2\phi$ from the $\varepsilon_1$ position (Fig. 7.1b). This angle is about 110°; thus the angle $\phi$ is about 55°. Note that although the length of the neck band remains essentially constant during the test, its thickness decreases (because of volume constancy), and the specimen eventually fractures. The angle $\phi$ will be different for materials that are anisotropic in the plane of the sheet.

Whether necking is *localized* or **diffuse** (Fig. 7.1c) depends on the strain rate sensitivity $m$ of the material, as given by the equation

$$\sigma = C\dot{\varepsilon}^m. \tag{7.3}$$

As described in Section 2.2.7, the higher the value of $m$, the more diffuse the neck becomes. An example of localized necking on an aluminum strip in tension is shown in Fig. 7.1d. Note the double localized neck, that is, $\phi$ can be in the clockwise and counterclockwise position in Fig. 7.1a.

In addition to uniform elongation, the **total elongation** gage length [such as 50 mm (2 in.)] of a tension-test specimen is also a significant factor in formability of sheet metals. Note that total elongation is the sum of uniform elongation and postuniform elongation. Uniform elongation is governed by the strain-hardening exponent $n$, whereas postuniform elongation is governed by the strain-rate–sensitivity index $m$. The higher the $m$ value, the more diffuse the neck and hence the greater the postuniform elongation becomes before fracture . Thus the total elongation of the material increases with increasing values of both $n$ and $m$.

### 7.2.2 Yield-point elongation

Low-carbon steels exhibit a behavior called **yield-point elongation**, with upper and lower yield points, as shown in Fig. 7.2a. This behavior indicates that, after the material yields, it stretches farther in certain regions in the specimen with no increase in the lower yield point, while other regions have not yet yielded. When the overall elongation reaches the yield-point elongation, the entire specimen has been deformed uniformly. The magnitude of the yield-point elongation depends on the strain rate (with higher rates, the elongation generally increases) and the grain size of the sheet metal. As grain size decreases, yield-point elongation increases. Yield-point elongation is usually on the order of a few percent.

This behavior of low-carbon steels produces **Lueder's bands** (also called *stretcher strain marks* or *worms*) on the sheet, as shown in Fig. 7.2b. These are elongated depressions on the surface of the sheet and can be objectionable in the final product because of surface appearance. They can also cause difficulty in subsequent coating and painting operations. Stretcher strain marks on the curved bottom of steel cans for common household products are shown in Fig. 7.2c; aluminum cans do not exhibit this behavior.

The usual method of avoiding this problem is to eliminate or to reduce yield-point elongation by reducing the thickness of the sheet 0.5 to 1.5 percent by cold rolling, known as **temper rolling** or **skin rolling**. Because of strain aging, yield-point elongation reappears after a few days at room temperature or after a few hours at higher temperatures. Thus the sheet

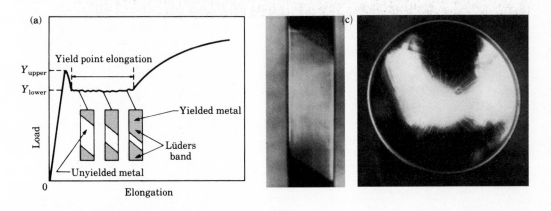

**FIGURE 7.2** (a) Yield point elongation and Lueder's bands in tension testing. (b) Lueder's bands in annealed low-carbon steel sheet. (c) Stretcher strains at the bottom of a steel can for household products. *Source:* (b) Reprinted with permission from Caterpillar, Inc.

metal should be formed within a certain period of time (from one to three weeks for rimmed steel, Section 5.7) to avoid reappearance of stretcher strains.

### 7.2.3   Anisotropy

Another important factor influencing sheet-metal forming is **anisotropy**, or **directionality**, of the sheet metal. Anisotropy is acquired during the thermomechanical processing history of the sheet. Recall from Section 3.5 that there are two types of anisotropy: **crystallographic anisotropy** (from preferred grain orientation) and **mechanical fibering** (from alignment of impurities, inclusions, voids, and the like, throughout the thickness of the sheet during processing).

Anisotropy may be present not only in the plane of the sheet, but also in its thickness direction. The former is called **planar** anisotropy and the latter **normal** or **plastic anisotropy**.

### 7.2.4   Grain size

**Grain size** of sheet metal is important for two reasons: first, because of its effect on the mechanical properties of the material, and second, because of its effect on the surface appearance of the formed part. The coarser the grain, the rougher the surface appears (**orange peel**). An ASTM grain size of No. 7 or finer is preferred for general sheet-metal forming (see section 3.4.1).

### 7.2.5   Residual stresses

Residual stresses can be present in sheet metal parts because of the nonuniform deformation of the sheet during forming. When disturbed, as by removing a portion of it, the part may distort. Tensile residual stresses on surfaces can also lead to **stress–corrosion cracking** of sheet-metal parts (Fig. 7.3) unless they are properly stress relieved.

**FIGURE 7.3**   Stress–corrosion cracking in a deep-drawn brass part for a light fixture. The cracks developed over a period of time. Brass and austenitic (300 series) stainless steels are among metals that are susceptible to stress–corrosion cracking.

### 7.2.6  Springback

Sheet-metal parts, because they are generally thin and are subjected to relatively small strains, are likely to experience considerable **springback**. This is particularly significant in bending and other sheet-forming operations where the bend radius-to-thickness ratio is high, such as in automotive parts.

### 7.2.7  Wrinkling

Although in sheet forming the metal is generally subjected to tensile stresses, the method of forming may be such that compressive stresses are developed in the plane of the sheet. An example is the **wrinkling** of the flange in deep drawing because of circumferential *compression*. Other terms used to describe similar phenomena are **buckling, folding,** and **collapsing**. The tendency for wrinkling increases with the unsupported or unconstrained length or surface area of the sheet metal, decreasing thickness, and nonuniformity of the thickness of the sheet. Lubricants that are trapped or distributed nonuniformly between the surface of the sheet and the dies can contribute to the initiation of wrinkling.

### 7.2.8  Coated sheet metal

Sheet metals, especially steel, *precoated* with a variety of organic coatings, films, and laminates are available. Coated sheet metals are used primarily for appearance and eye appeal and also offer corrosion resistance. Coatings are applied to the coil stock on continuous lines. Coating thicknesses generally range from 0.0025 to 0.2 mm (0.0001 to 0.008 in.) on flat surfaces. Coatings are available with a wide range of properties, such as flexibility, durability, hardness, resistance to abrasion and chemicals, color, texture, and gloss. Coated sheet metal is subsequently formed into products such as TV cabinets, appliance housings, paneling, shelving, residential siding, and metal furniture.

Zinc is used extensively as a coating on sheet steel (*galvanized,* see Sections 3.9.7 and 4.5) to protect it from corrosion, particularly in the automotive industry. Galvanizing of steel can be done by hot dipping, electrogalvanizing, or galvannealing processes.

## 7.3  Shearing

The **shearing** process involves cutting sheet metal by subjecting it to shear stresses, usually between a **punch** and a **die** much like a paper punch (Fig. 7.4). The punch and die may be any shape; for example, they may be circular or straight blades similar to a pair of scissors. The major variables in the shearing process are the punch force $P$, the speed of the punch, lubrication, edge condition and materials of the punch and die, their corner radii, and the clearance between punch and die.

The overall features of typical sheared edges for the two sheared surfaces (the slug and the sheet) are shown in Fig. 7.5. Note that the edges are neither smooth nor perpendicular to the plane of the sheet. The **clearance** $c$ (Fig. 7.4) is the major factor determining the shape and quality of the sheared edge. As shown in Fig. 7.6a, as clearance increases, the edges become

**FIGURE 7.4**  Schematic illustration of the shearing process with a punch and die. This is a common method of producing various openings in sheet metals.

rougher and the zone of deformation becomes larger. The material is pulled into the clearance area, and the edges of the sheared zone become more and more rounded. In fact, if the clearance is too large, the sheet metal is bent and subjected to tensile stresses, instead of un-

**FIGURE 7.5**  Characteristic features of a (a) punched hole and (b) the punched slug. Note that the slug has been sealed down as compared with the hole.

**FIGURE 7.6** (a) Effect of clearance c between the punch and die on the deformation zone in shearing. As clearance increases, the material tends to be pulled into the die, rather than being sheared. In practice, clearances usually range between 2 and 10 percent of the thickness of the sheet. (b) Microhardness (HV) contours for a 6.4-mm (0.25-in.)-thick AISI 1020 hot-rolled steel in the sheared region. *Source:* After H. P. Weaver and K. J. Weinmann.

dergoing a shearing deformation. As Fig. 7.6b shows, sheared edges can undergo severe cold working because of the high strains involved. This result, in turn, can adversely affect the formability of the sheet during subsequent operations.

Note the formation of a **burr** in Fig. 7.5. Burr height increases with increasing clearance and increasing ductility of the metal. Tools with dull edges are also a major factor in burr formation. The height, shape, and size of the burr can significantly affect many subsequent forming operations, such as in flanging, where a burr can lead to cracks.

Observing the shearing mechanism reveals that shearing usually starts with the formation of cracks on both the top and bottom edges of the workpiece (A and B in Fig. 7.4). These cracks eventually meet, and complete separation takes place. The *rough fracture surface* in the slug in Fig. 7.5 is caused by these cracks. The *smooth, shiny, and burnished surface* is from the contact and rubbing of the sheared edge against the walls of the die. In the slug shown, the burnished surface is in the lower region, because this is the section that rubs against the die wall. On the other hand, inspection of the sheared surface on the sheet itself reveals that the burnished surface is on the upper region in the sheared edge and results from rubbing against the punch.

The ratio of the burnished to rough areas on the sheared edge increases with increasing ductility of the sheet metal; it decreases with increasing material thickness and clearance. The punch travel required to complete the shearing process depends on the maximum shear strain that the material can undergo before fracture. Thus a brittle metal, or one that is highly cold worked, requires little travel of the punch to complete shearing.

Note from Fig. 7.7 that the deformation zone is subjected to high shear strains. The width of this zone depends on the rate of shearing, that is, punch speed. With increasing punch speed, the heat generated by plastic deformation is confined to a smaller zone (approaching a narrow adiabatic zone), and the sheared surface is smoother (Fig. 7.7c).

The **punch force** P is basically the product of the shear strength of the sheet metal and the cross-sectional area being sheared. However, friction between the punch and the workpiece can increase this force substantially. Because the sheared zone is subjected to cracks, plastic deformation, and friction, the punch force-stroke curves can have various shapes. One typical

(a)

(c)

FIGURE 7.7   Deformation and temperature rise in the shearing zone. The temperature was measured by thermocouples. Punching at (a) slow speed and (b) high speed. Note that the deformation is confined to a narrow zone in high-speed shearing and that the temperature is higher than in slow-speed shearing. *Source:* After N. Yanagihara, H. Saito, and T. Nakagawa. Numbers above the figures indicate punch penetration. (c) Fracture zone in shearing with static and dynamic loading.

FIGURE 7.8   Typical punch-penetration curve in shearing. The area under the curve is the work done in shearing. The shape of the curve depends on process parameters and material properties.

curve for a ductile material is shown in Fig. 7.8. The area under this curve is the *total work* done in shearing.

An approximate empirical formula for estimating the **maximum punch force** $P_{max}$ is

$$P_{max} = 0.7(\text{UTS})(t)(L), \qquad (7.4)$$

where UTS is the *ultimate tensile strength* of the sheet metal, $t$ is its thickness, and $L$ is the total length of the sheared edge. Thus for a round hole of diameter $D$, we have $L = \pi D$.

**Example 7.1:**   Calculation of maximum punch force

Estimate the force required in punching a 1-in. (25-mm) diameter hole through a $\frac{1}{8}$-in. (3.2-mm) thick annealed titanium-alloy Ti–6Al–4V sheet at room temperature.

SOLUTION.   The force is estimated from Eq. (7.4), and the UTS for this alloy is found from Table 3.15 to be 1000 MPa, or 140,000 psi. Thus

$$P = 0.7(\tfrac{1}{8})(\pi)(1)(140{,}000) = 38{,}500 \text{ lb}$$
$$= 19.25 \text{ tons} = 0.17 \text{ MN.}$$

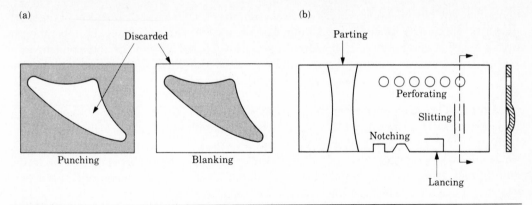

**FIGURE 7.9** (a) Punching (piercing) and blanking. (b) Examples of various shearing operations on sheet metal.

### 7.3.1 Shearing operations

Various operations based on the shearing process are performed. We first define two terms. In **punching**, the sheared slug is discarded (Fig. 7.9a). In **blanking**, the slug is the part and the rest is scrap.

**Die cutting.** **Die cutting** consists of the following operations (Fig. 7.9b): (1) **perforating**, or punching a number of holes in a sheet; (2) **parting**, or shearing the sheet into two or more pieces usually when the adjacent blanks do not have a matching contour; (3) **notching**, or removing pieces or various shapes from the edges; and (4) **lancing**, or leaving a tab without removing any material. Parts produced by these processes have various uses, particularly in assembly with other components.

**Fine blanking.** Very smooth and square edges can be produced by **fine blanking** (Fig. 7.10a). One basic die design is shown in Fig. 7.10b. A V-shaped stinger, or impingement, locks the sheet metal tightly in place and prevents the type of distortion of the material shown

**FIGURE 7.10** (a) Comparison of sheared edges by conventional (left) and fine-blanking (right) techniques. (b) Schematic illustration of setup for fine blanking. *Source:* Feintool U.S. Operations.

in Fig. 7.6. The fine-blanking process involves clearances on the order of 1 percent of the sheet thickness, which may range from 0.13 mm to 13 mm (0.005 in. to 0.5 in.). The operation is usually carried out on triple-action hydraulic presses where the movements of the punch, pressure pad, and die are controlled individually. Fine blanking usually involves parts having holes, which are punched simultaneously with blanking.

**Slitting.**    Shearing operations can be carried out with a pair of circular blades (**slitting**), similar to those in a can opener (Fig. 7.11). The blades follow either a straight line or a circular or curved path. A slit edge normally has a burr, which may be removed by rolling. There are two types of slitting equipment. In the *driven* type, the blades are powered. In the *pull-through* type, the strip is pulled through idling blades. Slitting operations, if not performed properly, may cause various distortions of the sheared part.

**Steel rules.**    Soft metals, paper, leather, and rubber can be blanked with **steel-rule dies**. Such a die consists of a thin strip of hardened steel bent to the shape to be sheared (similar to a cookie cutter) and held on its edge on a flat wooden base. The die is pressed against the sheet, which rests on a flat surface, and cuts the sheet to the shape of the steel rule.

**Nibbling.**    In **nibbling**, a machine called a nibbler moves a straight punch up and down rapidly into a die. The sheet metal is fed through the gap, and a number of overlapping holes are made. This operation is similar to making a large elongated hole by successively punching holes with a paper punch. Sheets can be cut along any desired path by manual control. The process is economical for small production runs since no special dies are required.

**Scrap in shearing.**    The amount of *scrap* produced, or **trim loss**, in shearing operations can be significant. On large stampings it can be as high as 30 percent. A significant factor in manufacturing cost, scrap can be reduced significantly by proper arrangement of the shapes on the sheet to be cut (*nesting* of blanks). Computer-aided design techniques have been developed to minimize scrap in shearing operations.

### 7.3.2  Shearing dies

Because the formability of the sheared part can be influenced by the quality of its sheared edges, clearance control is important. In practice, clearances usually range between 2 percent

Driven cutter

Workpiece

Clearance

**FIGURE 7.11**    Slitting with rotary knives. This process is similar to opening cans.

**FIGURE 7.12** Schematic illustrations of shaving on a sheared edge. (a) Shaving a sheared edge. (b) Shearing and shaving, combined in one stroke.

and 8 percent of the sheet's thickness. The thicker the sheet is, the larger the clearance (as much as 10 percent). However, the smaller the clearance, the better is the quality of the edge. In a process called **shaving** (Fig. 7.12), the extra material from a rough sheared edge is trimmed by cutting.

**Punch and die shapes.**   Note in Fig. 7.4 that the surfaces of the punch and die are flat. Thus the punch force builds up rapidly during shearing because the entire thickness is sheared at the same time. The area being sheared at any moment can be controlled by beveling the punch and die surfaces (Fig. 7.13). The geometry is similar to that of a paper punch, which you can see by looking closely at the tip. This geometry is particularly suitable for shearing thick blanks because it reduces the total shearing force. It also reduces the operation's noise level.

**Compound dies.**   Several operations on the same strip may be performed in one stroke with a **compound die** in one station. Although they yield higher productivity, these operations are usually limited to relatively simple shearing because they are somewhat slow, and the dies are more expensive than those for individual shearing operations.

**Progressive dies.**   Parts requiring multiple operations, such as punching, bending, and blanking, are made at high production rates in **progressive dies**. The sheet metal is fed through a coil strip, and a different operation is performed at the same station with each

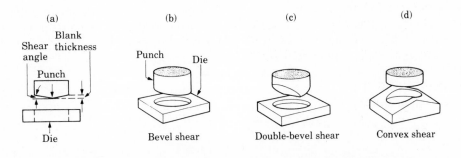

**FIGURE 7.13** Examples of the use of shear angles on punches and dies.

(a)                                                                  (b)

**FIGURE 7.14**  (a) Schematic illustration of making a washer in a progressive die. (b) Forming of the top piece of an aerosol spray can in a progressive die. Note that the part is attached to the strip until the last operation is completed.

stroke of a series of punches (Fig. 7.14a). An example of a part made in progressive dies is shown in Fig. 7.14b.

**Transfer dies.**   In a **transfer die** setup, the sheet metal undergoes different operations at different stations, which are arranged along a straight line or a circular path. After each operation, the part is transferred to the next station for additional operations.

**Tool and die materials.**   Tool and die materials for shearing are generally tool steels and, for high production rates, carbides (see Table 3.7). Lubrication is important for reducing tool and die wear and improving edge quality.

## 7.4   Bending of Sheet and Plate

One of the most common metalworking operations is **bending**. This process is used not only to form parts such as flanges, curls, seams, and corrugations, but also to impart stiffness to the part by increasing its moment of inertia (note, for instance, that a long strip of metal is much less rigid when flat than when it is formed into the shape of a gutter). In this section, we describe the mechanics and practices of bending operations.

The terminology used in bending is shown in Fig. 7.15. In bending, the outer fibers of the material are in tension and the inner fibers are in compression. Theoretically, the strains at the outer and inner fibers are equal in magnitude and are given by the equation

$$e_0 = e_i = \frac{1}{(2R/T) + 1},\qquad(7.5)$$

where $R$ is the bend radius and $T$ is the sheet thickness.

**FIGURE 7.15** Bending terminology. The bend radius is measured to the inner surface of the bend. Note that the length of the bend is the width of the sheet. Also note that the bend angle and the bend radius (sharpness of the bend) are two different variables.

Experimental evidence indicates that, although Eq. (7.5) holds reasonably well for the inner fiber strain $e_i$, actual values of $e_o$ are considerably higher than $e_i$. The reason is the shifting of the neutral axis toward the inner surface. The length of bend (Fig. 7.15) is smaller in the outer region than in the inner region (as can easily be observed when you bend a rectangular eraser). The difference between the outer and inner strains increases with decreasing $R/T$ ratio, that is, sharper bends.

### 7.4.1 Minimum bend radius

It is apparent from Eq. (7.5) that, as the $R/T$ ratio decreases, the tensile strain at the outer fiber increases and that the material may crack after a certain strain is reached. The radius $R$ at which a crack appears on the outer surface of the bend is called the **minimum bend radius**. The minimum radius to which a part can be bent safely is normally expressed in terms of its thickness, such as $2T$, $3T$, $4T$, and so on. Thus a $3T$ bend radius indicates that the smallest radius to which the sheet can be bent without cracking is 3 times its thickness. The minimum bend radius for various materials has been determined experimentally and is available in various handbooks. Some typical results are given in Table 7.1.

**TABLE 7.1**

**MINIMUM BEND RADIUS FOR VARIOUS MATERIALS AT ROOM TEMPERATURE**

| MATERIAL | CONDITION | |
|---|---|---|
| | SOFT | HARD |
| Aluminum alloys | 0 | 6T |
| Beryllium copper | 0 | 4T |
| Brass, low-leaded | 0 | 2T |
| Magnesium | 5T | 13T |
| Steels | | |
|   austenitic stainless | 0.5T | 6T |
|   low-carbon, low-alloy and HSLA | 0.5T | 4T |
| Titanium | 0.7T | 3T |
| Titanium alloys | 2.6T | 4T |

Studies have also been conducted to establish a relationship between the minimum $R/T$ ratio and a given mechanical property of the material. One such analysis is based on the following assumptions.

1. The true strain at cracking on the outer fiber in bending is equal to the true strain at fracture $\varepsilon_f$ of that material in a simple tension test.
2. The material is homogeneous and isotropic.
3. The sheet is bent in a state of plane stress, that is, its $L/T$ ratio is small.

The true strain at fracture in tension is

$$\varepsilon_f = \ln\left(\frac{A_0}{A_f}\right) = \ln\left(\frac{100}{100 - r}\right),$$

where $r$ is the percent reduction of area in a tension test. From Section 2.2.2,

$$\varepsilon_0 = \ln(1 + e_0) = \ln\left(1 + \frac{1}{(2R/T) + 1}\right) = \ln\left(\frac{R + T}{R + (T/2)}\right).$$

Equating the two expressions and simplifying, we obtain

$$\text{Minimum } \frac{R}{T} = \frac{50}{r} - 1. \tag{7.6}$$

For the experimental data shown in Fig. 7.16, the curve that best fits the data is

$$\text{Minimum } \frac{R}{T} = \frac{60}{r} - 1. \tag{7.7}$$

The $R/T$ ratio approaches zero (complete bendability, that is, the material can be folded over itself) at a tensile reduction of area of 50 percent. Interestingly, this is the same value obtained in spinnability of metals (described in Section 7.8), where a material with 50 percent reduction of area is found to be completely spinnable. A curve similar to that in Fig. 7.16 also is obtained when the data are plotted against the total elongation (percent) of the material.

The bendability of a metal may be increased by increasing its tensile reduction of area, either by *heating* or by the application of *hydrostatic pressure*. Other techniques may also be employed to increase the compressive environment in bending, such as applying compressive

**FIGURE 7.16** Relationship between the ratio of bend radius-to-sheet thickness and tensile reduction of area for various materials. Note that sheet metal with a reduction of area of about 50 percent can be bent and flattened over itself without cracking. *Source:* After J. Datsko and C. T. Yang, *J. Eng. Ind.,* vol. 82, 1960, pp. 309–314.

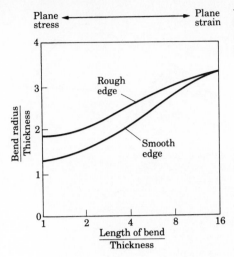

**FIGURE 7.17** The effect of length of bend and edge condition on bend radius-thickness ratio of 7075-T aluminum. *Source:* After G. Sachs and G. Espey.

forces in the plane of the sheet during bending to minimize tensile stresses in the outer fibers of the bend area.

As the length of the bend increases, the state of stress at the outer fibers changes from uniaxial stress to a biaxial stress. The reason is that $L$ tends to become smaller due to stretching of the outer fibers, but it is constrained by the material around the bend area. Biaxial stretching tends to reduce ductility, that is, strain to fracture. Thus as the length $L$ increases, the minimum bend radius increases (Fig. 7.17). However, at a length of about $10T$, the minimum bend radius increases no further, and a plane-strain condition is fully developed.

As the $R/T$ ratio decreases, narrow sheets (smaller length of bend) crack at the edges and wider sheets crack at the center, where the biaxial stress is the highest. Bendability also depends on the edge condition of the sheet being bent. Because rough edges are points of stress concentration, bendability decreases as edge roughness increases.

Another significant factor in edge cracking is the amount and shape of *inclusions* in the sheet metal. Inclusions in the form of **stringers** are more detrimental than globular-shaped inclusions. Thus anisotropy of the sheet is also important in bendability. As depicted in Fig. 7.18, cold rolling of sheets produces anisotropy because of alignment of impurities, inclusions, and voids (mechanical fibering). Thus transverse ductility is reduced, as shown in Fig. 7.18c. In bending such a sheet, caution should be used in cutting the blank in the proper direction from the rolled sheet, although this may not always be possible in practice.

Another important factor in bendability is the amount of cold working that the edges undergo during shearing (Fig. 7.6b). Removal of the cold-worked regions, such as by shaving, machining, or heat treating, greatly improves resistance to edge cracking during bending.

### 7.4.2 Springback

Because all materials have a finite modulus of elasticity, plastic deformation is followed by **elastic recovery** upon removal of the load. In bending, this recovery is known as **springback**. As shown in Fig. 7.19, the final bend angle after springback is smaller and the final bend radius is larger than before. (This can be easily observed by bending a piece of wire.)

**FIGURE 7.18**   (a) and (b) The effect of elongated inclusions (stringers) on cracking as a function of the direction of bending with respect to the original rolling direction of the sheet. This example shows the importance of the direction of cutting from large sheets in workpieces that are subsequently bent to make a product. (c) Cracks on the outer radius of an aluminum strip bent to an angle of 90°.

Springback occurs not only in flat sheets or plate, but also in bars, rod, and wire of any cross-section.

A quantity characterizing springback is the **springback factor** $K_s$, which is determined as follows. As the bend allowance (Fig. 7.15) is the same before and after bending, the relationship obtained for pure bending is

$$\text{Bend allowance} = \left(R_i + \frac{T}{2}\right)\alpha_i = \left(R_f + \frac{T}{2}\right)\alpha_f. \tag{7.8}$$

From this relationship, $K_s$ is defined as

$$K_s = \frac{\alpha_f}{\alpha_i} = \frac{(2R_i/T) + 1}{(2R_f/T) + 1}, \tag{7.9}$$

where $R_i$ and $R_f$ are the initial and final bend radii, respectively.

We note from this expression that the springback factor $K_s$ depends only on the $R/T$ ratio. A springback factor of $K_s = 1$ indicates no springback, and $K_s = 0$ indicates complete elastic recovery (Fig. 7.20), as in the leaf spring of an automobile.

Recall that the amount of elastic recovery depends on the stress level and the modulus of elasticity $E$ of the material. Thus elastic recovery increases with the stress level and with de-

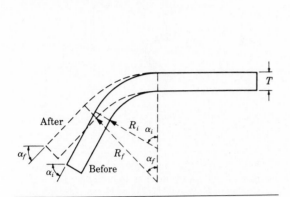

**FIGURE 7.19** Terminology for springback in bending. Springback is caused by the elastic recovery of the material upon unloading. In this example, the material tends to recover toward its originally flat shape. However, there are situations where the material bends farther upon unloading (negative springback), as shown in Fig. 7.21.

**FIGURE 7.20** Springback factor $K_s$ for various materials: (a) 2024–0 and 7075–0 aluminum; (b) austenitic stainless steels; (c) 2024-T aluminum; (d) 1/4 hard austenitic stainless steels; (e) 1/2 hard to full-hard austenitic stainless steels. *Source:* After G. Sachs, *Principles and Methods of Sheet-Metal Fabricating.* Reinhold, 1951, p. 100.

creasing elastic modulus. Based on this observation, an approximate formula has been developed to estimate springback:

$$\frac{R_i}{R_f} = 4\left(\frac{R_i Y}{ET}\right)^3 - 3\left(\frac{R_i Y}{ET}\right) + 1, \tag{7.10}$$

where $Y$ is the yield stress of the material at 0.2 percent offset (see Fig. 2.2b).

### Negative springback

The springback observed in Fig. 7.19 is called positive springback. Under certain conditions negative springback is also possible. In other words, the bend angle becomes larger after the bend has been completed and the load is removed. This phenomenon is generally associated with V-die bending (Fig. 7.21).

The development of negative springback can be explained by observing the sequence of deformation in Fig. 7.21. If we remove the piece at stage (b), it will undergo positive springback. At stage (c) the ends of the part are touching the male punch. Note that between stages (c) and (d) the part is actually being bent in the direction opposite to that between (a) and (b). Note also the lack of conformity of the punch radius and the inner radius of the part in (b) and (c). In (d), however, the radii are the same.

Upon unloading, the part in stage (d) will spring back inwardly because it is being *unbent* from stage (c) both at the tip of the punch and in the arms of the part. This inward (negative) springback can be greater than the positive springback that would result upon unloading from stage (c) because of the large strains that the material is undergoing in the bend area in stage (b).

**FIGURE 7.21**  Schematic illustration of the stages in bending round wire in a V-die. This type of bending can lead to negative springback, which does not occur in air bending (shown in Fig. 7.26a). *Source:* After K. S. Turke and S. Kalpakjian, *Proc. NAMRC III,* 1975, pp. 246–262.

Some experimental results with round wires using three different materials are shown in Fig. 7.22. Note that positive springback increases with increasing $R/D$ ratio, where $D$ is the wire diameter. (This result is similar to the $R/T$ effect in Fig. 7.20). However, springback becomes negative at low $R/D$ ratios because low $R/D$ ratios indicate less conformity at the punch radius-wire interface in Fig. 7.21.

**FIGURE 7.22**  Range of positive and negative springback for various materials (with the same modulus of elasticity) as a function of the ratio of bend radius to wire diameter. *Source:* After K. S. Turke and S. Kalpakjian, *Proc. NAMRC III,* 1975, pp. 246–262.

**FIGURE 7.23** Methods of reducing or eliminating springback in bending operations. *Source:* V. Cupka, T. Nakagawa, and H. Tyamoto.

**Compensation for springback.**

In practice, springback is usually compensated for by using various techniques:

1.  **Overbending** the part in the die (Figs. 7.23a and b). Overbending can also be achieved by the **rotary bending** technique shown in Fig. 7.23e. The upper die has a cylindrical rocker (with an angle of $< 90°$) and is free to rotate; as it travels downward, the sheet is clamped and bent by the rocker over the lower die (die anvil). A relief angle in the lower die allows overbending of the sheet at the end of the stroke, thus compensating for springback.

2.  **Coining** the bend region by subjecting it to high localized compressive stresses between the tip of the punch and the die surface (Figs. 7.23c and d), known as **bottoming.**

3.  **Stretch bending,** in which the part is subjected to tension while being bent. The bending moment required to make the sheet deform plastically will be reduced as the combined tension (due to bending of the outer fibers and the applied tension) in the sheet increases. Thus springback, which is the result of nonuniform stresses due to bending, will also decrease. This technique is used to limit springback in stretch forming of shallow automotive bodies (see Section 7.5).

4.  Because springback decreases as yield stress decreases—all other parameters being the same—bending may also be carried out at elevated temperatures to reduce springback.

**Example 7.2:**   **Estimating springback**

A 20-gage steel sheet is bent to a radius of 0.5 in. Assuming that its yield stress is 40,000 psi, calculate the radius of the part after it is bent.

SOLUTION.   The appropriate formula is Eq. (7.10), where

$$R_i = 0.5 \text{ in.}, Y = 40,000 \text{ psi}, E = 29 \times 10^6 \text{ psi, and } T = 0.0359 \text{ in.},$$

for 20-gage steel, as found in handbooks. Thus

$$\frac{R_i Y}{ET} = \frac{(0.5)(40,000)}{(29 \times 10^6)(0.0359)} = 0.0192,$$

and

$$\frac{R_i}{R_f} = 4(0.0192)^3 - 3(0.0192) + 1 = 0.942.$$

Hence

$$R_f = \frac{0.5}{0.942} = 0.531 \text{ in.}$$

### 7.4.3 Forces

Bending forces can be estimated by assuming that the process is one of the simple bending of a rectangular beam. Thus the bending force is a function of the strength of the material, the length and thickness of the part, and the die opening $W$, as shown in Fig. 7.24. Excluding friction, the general expression for the **maximum bending force** $P_{max}$ is

$$P_{max} = k \frac{(\text{UTS})LT^2}{W}, \tag{7.11}$$

where the factor $k$ ranges from about 1.2 to 1.33 for a $V$ die. The $k$ values for wiping and $U$ dies are 0.25 and 2 times, respectively, that of $k$ for $V$ dies. The effect of various factors on bending force, such as friction, is included in the factor $k$. Equation (7.11) applies well to situations in which the punch radius and sheet thickness are small compared to die opening $W$.

The bending force is also a function of punch travel. It increases from zero to a maximum and may decrease as the bend is completed. The force then increases sharply as the punch bottoms in the case of die bending. In *air bending*, or *free bending* (see Fig. 7.26a), the force does not increase again after it begins to decrease.

**FIGURE 7.24**    Common die-bending operations, showing the die-opening dimension $W$ used in calculating bending forces [see Eq. (7.11)].

(a) Channel forming   (b) Joggle

(c) Hemming (flattening)   (d) Two-stage lock seam   (e) Offset forming

**FIGURE 7.25**   Schematic illustrations of various bending operations in a press brake.

### 7.4.4   Common bending operations

In this section we describe common bending operations. Some of these processes are performed on discrete sheet-metal parts; others are done continuously, as in roll forming of coiled sheet stock.

**Press brake forming.**   Sheet metal or plate can be bent easily with simple fixtures using a press. Long—7 m (20 ft) or more—and relatively narrow pieces are usually bent in a **press brake**. This machine utilizes long dies in a mechanical or hydraulic press and is suitable for small production runs. The tooling is simple and adaptable to a wide variety of shapes (Fig. 7.25). Die materials may range from hardwood, for low-strength materials and small production runs, to carbides. For most applications, carbon-steel or gray-iron dies are generally used.

**Other bending operations.**   Sheet metal may be bent by a variety of processes (Fig. 7.26). In **air bending** (*free bending*), there is only one die. Bending in a **4-slide machine** allows making tubular parts, such as bushings of various sizes. Bending of sheet metals can also be carried out with a pair of rolls, the larger one of which is flexible and made of *polyurethane*. The upper roll pushes the sheet into the flexible lower roll and imparts a curvature to the sheet, the shape of which depends on the degree of indentation into the flexible roll. Thus, by controlling this depth, it is possible to form a sheet with a variety of curvatures on it. Plates are bent by the **roll-bending** process shown in Fig. 7.26d. Adjusting the distance between the three rolls produces various curvatures.

**Beading.**   In **beading**, the edge of the sheet metal is bent into the cavity of a die (Fig. 7.27). The bead gives stiffness to the part by increasing the moment of inertia of the edges. Also, it improves the appearance of the part and eliminates exposed sharp edges.

**FIGURE 7.26**   Examples of various bending operations.

**Flanging.**   Flanging is a process of bending the edges of sheet metals. In **shrink flanging** (Fig. 7.28a), the flange is subjected to compressive hoop stresses which, if excessive, cause the flange edges to wrinkle. The wrinkling tendency increases with decreasing radius of curvature of the flange. In **stretch flanging**, the flange edges are subjected to tensile stresses and, if excessive, can lead to cracking at the edges.

In the **dimpling** operation (Fig. 7.28b), a hole is first punched and then expanded into a flange, or a shaped punch pierces the sheet metal and expands the hole. Flanges may be produced by **piercing** with a shaped punch (Fig. 7.28c). The ends of tubes are flanged by a similar process (Fig. 7.28d). When the angle of bend is less than 90°, as in fittings with conical ends, the process is called **flaring**.

The condition of the edges is important in these operations. Stretching the material causes high tensile stresses at the edges, which may lead to cracking and tearing of the flange. As the ratio of flange to hole diameters increases, the strains increase proportionately. The rougher the edge, the greater will be the tendency for cracking. Sheared or punched edges may be shaved with a sharp tool (see Fig. 7.12) to improve the surface finish of the edge and reduce the tendency for cracking.

**FIGURE 7.27**   (a) Bead forming with a single die. (b) Bead forming with two dies in a press brake.

FIGURE 7.28 Various flanging operations. (a) Flanges on flat sheet. (b) Dimpling. (c) Piercing sheet metal to form a flange. In this operation a hole does not have to be prepunched before the punch descends. Note, however, the rough edges along the circumference of the flange. (d) Flanging of a tube. Note thinning of the edges of the flange.

**Hemming.** In the **hemming** process (also called **flattening**), the edge of the sheet is folded over itself (Fig. 7.25c). Hemming increases the stiffness of the part, improves its appearance, and eliminates sharp edges. *Seaming* involves joining two edges of sheet metal by hemming (Fig. 7.25d). Double seams are made by a similar process, using specially shaped rollers, for watertight and airtight joints, such as in food and beverage containers.

FIGURE 7.29 The roll-forming process.

Stage 1    Stage 2    Stage 3    Stage 4    Stage 5    Stage 6    Stage 7

**FIGURE 7.30**  Stages in roll forming of sheet-metal door frame. In stage 6, the rolls may be shaped as in *A* or *B*. *Source:* G. Oehler.

**Roll forming.**    For bending continuous lengths of sheet metal and for large production runs, **roll forming** is used (*contour roll forming* or *cold roll forming*). The metal strip is bent in stages by passing it through a series of rolls (Fig. 7.29). Typical products are channels, gutters, siding, panels, frames (Fig. 7.30), and pipes and tubing with lock seams. The length of the part is limited only by the amount of material supplied from the coiled stock. The parts are usually sheared and stacked continuously. The sheet thickness typically ranges from about 0.125 mm to 20 mm (0.005 in. to 0.75 in.). Forming speeds are generally below 1.5 m/s (300 ft/min), although they can be much higher for special applications.

Proper design and sequencing of the rolls, which usually are mechanically driven, requires considerable experience. Tolerances, springback, and tearing and buckling of the strip have to be considered. The rolls are generally made of carbon steel or gray iron and may be chromium plated—for better surface finish of the product and wear resistance of the rolls. Lubricants may be used to improve roll life and surface finish and to cool the rolls and the workpiece.

**Tube bending.**    Bending and forming tubes and other hollow sections require special tooling to avoid buckling and folding. The oldest and simplest method of bending a tube or

(a) Stretch bending

(b) Draw bending

(c) Compression bending

(d) Mandrels for tube bending

Plug    Balls    Laminated    Cable

**FIGURE 7.31**  Methods of bending tubes. Internal mandrels, or filling tubes with particulate materials such as sand, are often necessary to prevent collapsing of the tubes during bending. Solid rods and structural shapes are also bent by these techniques.

**FIGURE 7.32** A method of forming a tube with sharp angles using axial compressive forces. Compressive stresses are beneficial in forming operations because they delay fracture. Note that the tube is supported internally with rubber or fluid to avoid collapsing during forming. *Source:* After J. L. Remmerswaal and A. Verkaik, *Int. Conf. Manufacturing Technology,* ASME, 1967, pp. 1171–1181.

pipe is to pack the inside with loose particles, commonly sand, and bend the part in a suitable fixture. This technique prevents the tube from buckling. After the tube has been bent, the sand is shaken out. Tubes can also be plugged with various flexible internal mandrels, as shown in Fig. 7.31, which also illustrates various bending techniques and fixtures for tubes and sections. A relatively thick tube having a large bend radius can be bent without filling it with particulates or using plugs.

The beneficial effect of forming metals with highly compressive forces is demonstrated in the bending of a tube with relatively sharp corners (Fig. 7.32). In this operation, the tube is subjected to longitudinal compressive stresses, which reduce the stresses in the outer fibers in the bend area, thus improving the bendability of the material.

## 7.5  Stretch Forming

In **stretch forming,** the sheet metal is clamped around its edges and stretched over a die or form block, which moves upward, downward, or sideways, depending on the particular machine (Fig. 7.33). Stretch forming is used primarily to make aircraft wingskin panels, automobile door panels, and window frames. Although this process is generally used for low-volume productions, it is versatile and economical. Aluminum skins for the Boeing 767 and 757 aircraft are made by stretch forming, with a tensile force of 9 MN (2 million lb). The rectangular sheets are 12 m × 2.5 m × 6.4 mm (40 ft × 8.3 ft × 0.25 in.).

In most operations, the blank is a rectangular sheet, clamped along its narrower edges and stretched lengthwise, thus allowing the material to shrink in width. Controlling the amount of stretching is important to avoid tearing. Stretch forming cannot produce parts with sharp contours or re-entrant corners (depressions of the surface of the die).

Dies for stretch forming are generally made of zinc alloys, steel, plastics, or wood. Most applications require little or no lubrication. Various accessory equipment can be used in conjunction with stretch forming, including additional forming with both male and female dies while the part is in tension.

**FIGURE 7.33**  Schematic illustration of a stretch-forming process. Aluminum skins for aircraft can be made by this process. *Source:* Cyril Bath Co.

---

## Example 7.3:  Work done in stretch forming

A 15 in. long workpiece (see figure below) is stretched by a force $F$ until $\alpha = 20°$. The original cross-sectional area is 0.5 in². The material has a curve $\sigma = 100,000\ \varepsilon^{0.3}$. Find the total work done, ignoring end effects or bending. What is $\alpha_{max}$ before necking begins?

SOLUTION.   This situation is equivalent to stretching a workpiece from 15 in. to a length of $a + b$. For $\alpha = 20°$, we find that the final length $L_f = 16.8$ in. The true strain then is

$$\varepsilon = \ln\left(\frac{L_f}{L_0}\right) = \ln\left(\frac{16.8}{15}\right) = 0.114.$$

The work done per unit volume is

$$u = \int_0^{0.114} \sigma \cdot d\varepsilon = 10^5 \int_0^{0.114} \varepsilon^{0.3} d\varepsilon = 10^5 \frac{\varepsilon^{1.3}}{1.3}\bigg]_0^{0.114} = 4570 \text{ in} \cdot \text{lb/in}^3.$$

The volume of the workpiece is

$$V = (15)(0.5) = 7.5 \text{ in}^3.$$

Hence

$$\text{Work} = (u)(V) = 34,275 \text{ in.} \cdot \text{lb.}$$

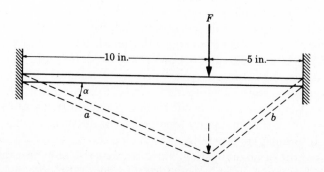

The necking limit for uniaxial tension is given by Eq. (7.2), the problem being similar to a stretch-forming operation. Thus

$$L_{max} = L_0 e^n = 15e^{0.3} = 20.2 \text{ in.}$$

Hence $a + b = 20.2$ in., and from similarities in triangles, we obtain

$$a^2 - 10^2 = b^2 - 5^2, \quad \text{or} \quad a^2 = b^2 + 75.$$

Hence $a = 12$ in. and $b = 8.2$ in. Therefore

$$\cos \alpha = \frac{10}{12} = 0.833 \quad \text{or} \quad \alpha_{max} = 33.6°.$$

## 7.6 Bulging

The basic forming process of **bulging** involves placing a tubular, conical, or curvilinear part in a split-female die and expanding it with, say, a polyurethane plug (Fig. 7.34a). The punch is then retracted, the plug returns to its original shape, and the part is removed by opening the dies. Typical products are coffee or water pitchers, barrels, and beads on drums. For parts with complex shapes, the plug, instead of being cylindrical, may be shaped in order to apply greater pressure at critical points. The major advantage of using polyurethane plugs is that they are very resistant to abrasion, wear, and lubricants, and they do not damage the surface finish of the part being formed.

**FIGURE 7.34** (a) Bulging of a tubular part with a flexible plug. Water pitchers can be made by this method. (b) Production of fittings for plumbing by expanding tubular blanks with internal pressure. The bottom of the piece is then punched out to produce a "T." *Source:* J. A. Schey, *Introduction to Manufacturing Processes,* 2d ed. New York: McGraw-Hill Publishing Company, 1987. Reproduced by permission of the McGraw-Hill Companies. (c) Manufacturing of bellows.

*Hydraulic pressure* may be used in bulging operations, although they require sealing and need hydraulic controls (Fig. 7.34b). **Segmented dies** that are expanded and retracted mechanically may also be used. These dies are relatively inexpensive and can be used for large production runs. Formability in bulging processes can be enhanced by the application of longitudinal compressive stresses.

*Bellows* are manufactured by a bulging process, as shown in Fig. 7.34c. After the tube is bulged at several equidistant locations, it is compressed axially to collapse the bulged regions, thus forming bellows. The tube material must be able to undergo the large strains involved during the collapsing process.

## 7.7 Rubber Forming

In the processes described in the preceding sections, we noted that the dies are made of solid materials. However, in **rubber forming** one of the dies in a set can be made of flexible material, such as a rubber or polyurethane membrane. Polyurethanes are used widely because of their resistance to abrasion, resistance to cutting by burrs or sharp edges of the sheet metal, and long fatigue life.

In bending and embossing sheet metal, as shown in Fig. 7.35, the female die has been replaced with a rubber pad. Note that the outer surface of the sheet is protected from damage or scratches because it is not in contact with a hard metal surface during forming. Pressures are usually on the order of 10 MPa (1500 psi).

In the **hydroform** or **fluid-forming process** (Fig. 7.36), the pressure over the rubber membrane is controlled throughout the forming cycle, with maximum pressures of up to 100 MPa (15,000 psi). This procedure allows close control of the part during forming to prevent wrinkling or tearing. Deeper draws are obtained than in conventional deep drawing because the pressure around the rubber membrane forces the cup against the punch. The friction at the punch-cup interface reduces the longitudinal tensile stresses in the cup and thus delays frac-

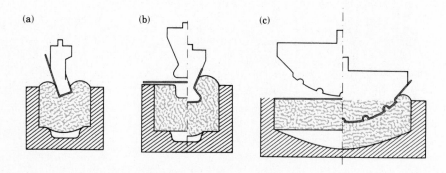

**FIGURE 7.35** Examples of bending and embossing sheet metal with a metal punch and flexible pad serving as the female die. *Source:* Polyurethane Products Corporation (630) 543-6700.

(a)  (b)  (c)  (d)

Pressure control
valve

Forming cavity
(oil filled)

Rubber
diaphragm

Draw ring
Punch

Blank

Part

**FIGURE 7.36** The hydroform, or fluid forming, process. Note that unlike the ordinary deep-drawing process, the dome pressure forces the cup walls against the punch. The cup travels with the punch and thus deep drawability is improved.

ture. The control of frictional conditions in rubber forming as well as other sheet-forming operations can be a critical factor in making parts successfully. Selection of proper lubricants and application methods is important.

When selected properly, rubber forming processes have the advantages of low tooling cost, flexibility and ease of operation, low die wear, no damage to the surface of the sheet, and capability to form complex shapes. Parts can also be formed with laminated sheets of various nonmetallic materials or coatings.

# 7.8 Spinning

**Spinning** involves the forming of axisymmetric parts over a rotating mandrel with the use of rigid tools or rollers. There are three basic types of spinning processes: conventional (or manual), shear, and tube spinning. The equipment for all these processes is basically similar to a lathe with various special features.

## 7.8.1 Conventional spinning

In the **conventional spinning** process, a circular blank of flat or preformed sheet metal is held against a rotating mandrel while a rigid tool deforms and shapes the material over the mandrel (Fig. 7.37a). The tools may be actuated either manually or by a hydraulic mechanism. The operation involves a sequence of passes (Fig. 7.38) and requires considerable skill. This process has been found to be quite difficult to study analytically.

Typical shapes made by conventional spinning processes are shown in Fig. 7.39. This process is particularly suitable for conical and curvilinear shapes, which would otherwise be difficult or uneconomical to form by other methods. Part diameters may range up to 6 m (20 ft). Although most spinning is performed at room temperature, thick parts or metals with low ductility or high strength require spinning at elevated temperatures. Tooling costs in spinning are relatively low. However, because the operation requires multiple passes to form the final part, it is economical for small production runs only.

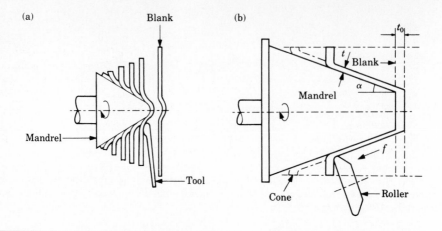

**FIGURE 7.37** Schematic illustration of spinning processes: (a) conventional spinning and (b) shear spinning. Note that in shear spinning the diameter of the spun part, unlike conventional spinning, is the same as that of the blank. The quantity $f$ is the feed (mm/rev or in./rev).

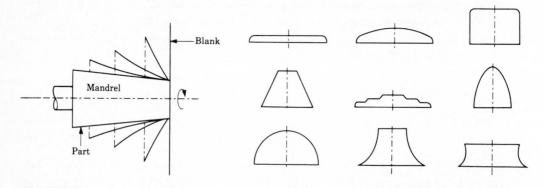

**FIGURE 7.38** Stages in conventional spinning of a tubular component from a flat, circular metal disk. This operation requires considerable skill to prevent the part from collapsing or buckling during spinning.

**FIGURE 7.39** Typical shapes produced by the conventional spinning process. Circular marks on the external surfaces of components usually indicate that the parts have been made by spinning. Examples are aluminum kitchen utensils and light reflectors.

### 7.8.2  Shear spinning

In the **shear spinning** process (also known as **power spinning, flow turning, hydrospinning,** or **spin forging**), an axisymmetrical conical or curvilinear shape is generated by holding the diameter of the part constant (see Fig. 7.37b). Parts typically made by this process are rocket-motor casings and missile nose cones. Although a single roller can be used, two rollers are desirable to balance the radial forces acting on the mandrel. Large parts, up to about 3 m (10 ft), can be spun to close tolerances by shear spinning. Little material is wasted, and the operation is completed in a relatively short time. If carried out at room temperature, the part has a higher yield strength than the original material, but lower ductility and toughness. A great variety of shapes can be spun with relatively simple tooling, generally made of tool steel. Because of the large plastic deformation involved, the process generates considerable heat. It is usually carried away by a coolant-type fluid applied during spinning.

**FIGURE 7.40** Schematic illustration of a shear-spinnability test. As the roller advances, the part thickness is reduced. The reduction in thickness at fracture is called the maximum spinning reduction per pass. *Source:* After R. L. Kegg, *J. Eng. Ind.,* vol. 83, 1961, pp. 119–124.

In shear spinning over a conical mandrel, the thickness $t$ of the spun part is simply

$$t = t_0 \sin \alpha. \tag{7.12}$$

The force primarily responsible for supplying energy is the tangential force, $F_t$. For an ideal situation in the shear spinning of a cone, this force is

$$F_t = u t_0 f \sin \alpha, \tag{7.13}$$

where $u$ is the specific energy of deformation, Eq. (2.59) and $f$ is the feed. It is the area under the true stress-true strain curve corresponding to a true strain related to the shear strain by

$$\varepsilon = \frac{\gamma}{\sqrt{3}} = \frac{\cot \alpha}{\sqrt{3}} \tag{7.14}$$

(see Section 2.11.7). Actual forces can be as much as 50 percent higher than those given by Eq. (7.13) because of factors such as redundant work and friction.

An important factor in shear spinning is the **spinnability** of the metal, that is, the smallest thickness to which a part can be spun without fracture. To determine spinnability, a simple test method has been developed (Fig. 7.40), in which a circular blank is spun over an ellipsoid mandrel, according to Eq. (7.12). As the thickness is eventually reduced to zero, all materials will fail at some thickness $t_f$.

Data for various materials are given in Fig. 7.41. The maximum spinning reduction in thickness is defined as

$$\text{Maximum reduction} = \frac{t_0 - t_f}{t_0} \times 100\% \tag{7.15}$$

and is plotted against the tensile reduction of area of the material. Note that if a metal has a tensile reduction of area of about 50 percent or greater, it can be reduced in thickness by 80 percent by spinning in one pass. For less ductile materials the spinnability is lower. Process variables such as feed and speed do not appear to have a significant influence on spinnability.

**FIGURE 7.41** Experimental data showing the relationship between maximum spinning reduction per pass and the tensile reduction of area of the original material. Note that once a material has about 50 percent reduction of area in a tension test, further increase in the ductility of the original material does not improve its spinnability. *Source:* S. Kalpakjian, *J. Eng. Ind.,* vol. 86, 1964, pp. 49–54.

Figures 7.16 and 7.41 show that both maximum bendability and spinnability coincide with a tensile reduction of area of about 50 percent. Any further increase in the ductility of the original material does not improve its formability. For materials with low ductility, spinnability can usually be improved by forming at elevated temperatures.

### 7.8.3 Tube spinning

In the **tube-spinning** process, tubes are reduced in thickness by spinning them on a mandrel using rollers. The operation may be carried out externally or internally (Fig. 7.42). Also, the part may be spun forward or backward, similar to a drawing or a backward extrusion

**FIGURE 7.42** Examples of external and internal tube spinning and the variables involved.

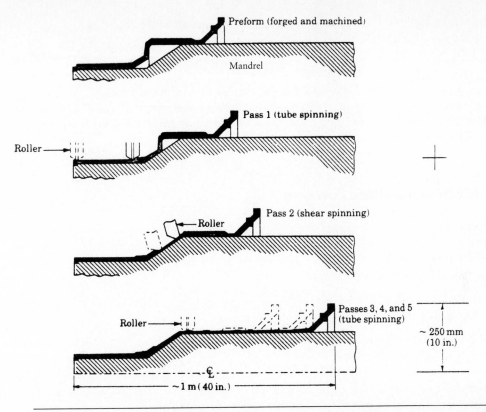

Preform (forged and machined)

Mandrel

Pass 1 (tube spinning)

Roller

Pass 2 (shear spinning)

Roller

Roller

Passes 3, 4, and 5 (tube spinning)

~ 250 mm (10 in.)

~1 m (40 in.)

**FIGURE 7.43** Stages in tube and shear spinning of a compressor shaft for the jet engine of the supersonic Concorde aircraft. Economic analysis indicated that the best method of manufacturing this part was to spin a preformed (forged and machined) tubular blank.

process. In either case, the reduction in wall thickness results in a longer tube. Various internal and external profiles can be generated by changing the path of the roller during its travel along the mandrel. Tube spinning can be used for making pressure vessels, automotive components, and rocket and missile parts. It can also be combined with shear spinning, as producing the compressor shaft for an aircraft engine shown in Fig. 7.43.

Spinnability in this process is determined by a test method similar to that for shear spinning (Fig. 7.44). Maximum reduction per pass in tube spinning is related to the tensile reduction of area of the material (see Fig. 7.41), with results very similar to shear spinning. As in shear spinning, ductile metals fail in tension after the reduction in thickness has taken place, whereas the less ductile metals fail in the deformation zone under the roller.

Roller path

Workpiece

Roller

Mandrel

$\alpha$

$t_0$

$t_f$

**FIGURE 7.44** Schematic illustration of a tube-spinnability test. As the roller advances, the wall thickness is reduced and the part eventually fractures. *Source:* S. Kalpakjian, *J. Eng. Ind.,* vol. 86, 1964, pp. 49–54.

The *ideal tangential force* $F_t$ in forward tube spinning can be expressed as

$$F_t = \overline{Y} \Delta t f, \tag{7.16}$$

where $\Delta t$ is $(t_0 - t)$ (see Fig. 7.42) and $\overline{Y}$ is the average flow stress of the material. For backward spinning, the ideal tangential force is roughly twice that given by Eq. (7.16). Friction and redundant work of deformation cause the actual forces to be about twice those obtained by Eq. (7.16).

## 7.9 High-Energy-Rate Forming

Sheet-metal-forming processes that use chemical, electrical, or magnetic sources of energy are described in this section. They are called **high-energy-rate** processes because the energy is released in a very short time. The basic operations are explosive forming, electrohydraulic forming, and magnetic-pulse forming.

### 7.9.1 Explosive forming

The most common **explosive forming** process (first proposed in the early 1900s) is shown in Fig. 7.45. The workpiece is clamped over a die, the air in the die cavity is evacuated, and the workpiece is lowered into a tank filled with water. An explosive charge is then placed at a certain height and detonated. The rapid conversion of the explosive into gas generates a shock wave. The pressure of this wave is sufficiently high to form the metal.

The peak pressure $p$, generated in water, is given by the expression

$$p = K \left( \frac{\sqrt[3]{W}}{R} \right)^a \tag{7.17}$$

where $p$ is the peak pressure in pounds per square inch; $K$ is a constant that depends on the explosive, e.g., 21,600 for TNT; $W$ is the weight of the explosive in pounds; $R$ is the distance of the explosive from the workpiece (standoff) in feet; and $a$ is a constant, generally taken as 1.15.

**FIGURE 7.45** Schematic illustration of the explosive forming process. Although explosives are generally used for destructive purposes, their energy can be controlled and utilized in forming large parts that would otherwise be difficult or expensive to produce by other methods.

**FIGURE 7.46** Influence of the standoff distance and type of energy-transmitting medium on the peak pressure obtained using 1.8 kg (4 lb) of TNT. To be effective, the pressure-transmitting medium should have high density and low compressibility. In practice, water is a commonly used medium.

**FIGURE 7.47** Schematic illustration of the confined method of explosive bulging of tubes. Thin-walled tubes of nonferrous metals can be formed to close tolerances by this process.

An important factor in determining peak pressure is the **compressibility** of the energy-transmitting medium (say, water) and its **acoustic impedance**, that is, the product of mass density and sound velocity in the medium. The lower the compressibility and the higher the density of the medium, the higher the peak pressure rises (Fig. 7.46). The distance between the water level and the explosive should not be too small; otherwise, the energy is dissipated rapidly to the environment. Detonation speeds are typically 6700 m/s (22,000 ft/s), and the speed at which the metal is formed is estimated to be on the order of 30 to 200 m/s (100 to 600 ft/s).

A great variety of shapes can be formed explosively, provided that the material is sufficiently ductile at high strain rates. Depending on the number of parts to be formed, dies may be made of aluminum alloys, steel, ductile iron, zinc alloys, reinforced concrete, wood, plastics, and composite materials. The final properties of parts made by this process are basically the same as those made by conventional methods. Safety is an important aspect in these operations.

Needing only one die and being versatile, with no limit to the size of the workpiece, explosive forming is particularly suitable for small-quantity production runs of large parts. Steel plates 25 mm (1 in.) thick and 3.6 m (12 ft) in diameter have been formed by this method. Tubes of 25 mm (1 in.) wall thickness have also been bulged by explosive forming techniques. Another explosive forming method, which uses a cartridge as the source of energy, is shown in Fig. 7.47; no other energy-transmitting medium is used. The process can be used for bulging and expanding of thin-walled tubes. Die wear and die failure can be a significant problem in this operation.

## Example 7.4:  Peak pressure in explosive forming

Calculate the peak pressure in water for 0.1 lb of TNT at a standoff of 1 ft. Is this pressure sufficiently high for forming sheet metals?

SOLUTION.   Using Eq. (7.17), we find that

$$p = (21{,}600)\left(\frac{\sqrt[3]{0.1}}{1}\right)^{1.15} \simeq 9000 \text{ psi.}$$

This pressure would be sufficiently high to form sheet metals. Note in Example 2.7, for instance, that the pressure required to expand a thin-walled spherical shell of a material similar to soft aluminum alloys is only 400 psi. Also note that a process such as hydroform (Section 7.7) has a maximum hydraulic pressure in the dome of about 100 MPa (15,000 psi). Other rubber-forming processes use pressures from about 1500 to 7500 psi. Thus the pressure obtained in this problem is sufficient for most sheet-forming processes.

### 7.9.2   Electrohydraulic forming

In the **electrohydraulic forming** process, also called **underwater-spark** or **electric-discharge forming**, the source of energy is a spark from electrodes connected with a thin wire (Fig. 7.48). The energy is stored in a bank of condensers charged with direct current. The rapid discharge of this energy through the electrodes generates a shock wave, which then forms the part. This process is essentially similar to explosive forming, except that it utilizes a lower level of energy and is used with smaller workpieces. It is also a safer operation.

### 7.9.3   Magnetic-pulse forming

In the **magnetic-pulse forming** process, the energy stored in a capacitor bank is discharged rapidly through a magnetic coil. In a typical example, a ring-shaped coil is placed over a tubular workpiece to be formed (collapsed) over another solid piece to make an integral part (Fig. 7.49). The magnetic field produced by the coil crosses the metal tube, generating eddy currents in the tube. This current, in turn, produces its own magnetic field. The forces produced by the two magnetic fields oppose each other; thus there is a repelling force between the coil and the tube. The high forces generated collapse the tube over the inner piece.

The higher the electrical conductivity of the workpiece, the higher is the magnetic force. The metal does not have to have any special magnetic properties. Magnetic-pulse forming is used for a variety of operations, such as swaging of thin-walled tubes over rods, cables, plugs, etc., and for bulging and flaring. Flat coils are also used for forming of flat sheet metal, as for embossing and shallow drawing operations.

**FIGURE 7.48**   Schematic illustration of the electrohydraulic forming process.

## 7.10  Superplastic Forming

The *superplastic behavior* of certain very fine-grained alloys (normally less than 10 to 15 μm), where very large elongations (up to 2000 percent) are obtained at certain temperatures and low strain rates, was described in Section 2.2.7. These alloys, such as Zn–Al and titanium, can be formed into complex shapes with common metalworking or polymer processing techniques. The high ductility and relatively low strength of superplastic alloys present the following advantages:

1. Lower strength of tooling, because of the low strength of the material at forming temperatures, hence, lower tooling costs.

2. Forming of complex shapes in one piece, with fine detail and close tolerances, and elimination of secondary operations.

3. Weight and material savings because of formability of the materials.

4. Little or no residual stresses in the formed parts.

The limitations of superplastic forming are:

1. The material must not be superplastic at service temperatures.

2. Because of the extreme strain-rate sensitivity of the superplastic material, it must be formed at sufficiently low rates (typically at strain rates of $10^{-4}$ to $10^{-2}/s$). Forming times range anywhere from a few seconds to several hours. Thus cycle times are much longer than conventional forming processes. Superplastic forming is therefore a batch-forming process.

Superplastic alloys (particularly Zn–22Al and Ti–6Al–4V) can be formed by bulk deformation processes, such as compression molding, closed-die forging, coining, hubbing, and extrusion. Sheet forming of these materials can also be done using operations such as thermoforming, vacuum forming, and blow molding.

An important development is the ability to fabricate sheet-metal structures by combining diffusion bonding with superplastic forming. Typical structures in which flat sheets are diffusion bonded and formed are shown in Fig. 7.50. After diffusion bonding of selected

**FIGURE 7.50** Two types of structures made by diffusion bonding and super-plastic forming of sheet metal. Such structures have a high stiffness-to-weight ratio. *Source:* Rockwell International Corp.

locations of the sheets, the unbonded regions (*stop-off*) are expanded into a mold by air pressure. These structures are thin and have high stiffness-to-weight ratios. Hence they are particularly important in aircraft and aerospace applications.

This process improves productivity by eliminating mechanical fasteners and produces parts with good dimensional accuracy and low residual stresses. The technology is now well advanced for titanium structures (typically Ti–6Al–4V alloy) for aerospace applications. Structures made of 7475-T6 aluminum alloy are also being developed using this technique. Other metals for superplastic forming are Inconel 100 and Incoloy 718 nickel alloys and iron-base, high-carbon alloys. Commonly used die materials in superplastic forming are low-alloy steels, cast tool steels, ceramics, graphite, and plaster of paris. Selection depends on the forming temperature and strength of the superplastic alloy.

## 7.11 • Various Forming Methods

**Peen forming** is used to produce curvatures on thin sheet metals by **shot peening** one surface of the sheet. Peening is done with cast-iron or steel shot, discharged either from a rotating wheel or with an air blast from a nozzle. Peen forming is used by the aircraft industry to generate smooth and complex curvatures on aircraft wing skins (Fig. 7.51). Cast-steel shot about 2.5 mm (0.1 in.) in diameter at speeds of 60 m/s (200 ft/s) has been used to form wing panels 25 m (80 ft) long. For heavy sections, shot diameters as large as 6 mm (1/4 in.) may be used.

In peen forming, the surface of the sheet is subjected to compressive stresses, which tend to expand the surface layer. Since the material below the peened surface remains rigid, the surface expansion causes the sheet to develop a curvature. The process also induces compressive surface residual stresses, with improved fatigue strength of the sheet. The peen forming process is also used for straightening twisted or bent parts. Out-of-round rings, for example, can be straightened by this method.

**FIGURE 7.51** Peen-forming machine to form a large sheet-metal part, such as an aircraft-skin panel. The sheet is stationary and the machine traverses it. *Source: Metal Improvement Company.*

**Gas mixtures** in a closed container have been utilized as an energy source. When ignited, the pressures generated are sufficient to form parts. The principle is similar to the generation of pressure in an internal combustion engine.

**Liquefied gases**, such as liquid nitrogen, may be used to develop pressures high enough to form sheet metals. When allowed to reach room temperature in a closed container, liquefied nitrogen becomes gaseous and expands, developing the necessary pressure to form the part.

### 7.11.1 Manufacturing honeycomb materials

There are two principal methods of manufacturing honeycomb materials. In the **expansion** process (Fig. 7.52a)—the most common method—sheets are cut from a coil and an adhesive

**FIGURE 7.52** Methods of making honeycomb materials: (a) expansion process and (b) corrugation process. *Source: Materials Engineering.* Reprinted with permission. (c) Making a honeycomb sandwich.

is applied at intervals (node lines). The sheets are stacked and cured in an oven, whereby strong bonds develop at the adhesive joints. The block is then cut into slices of desired dimension and stretched to produce a honeycomb structure. This procedure is similar to expanding folded paper structures into the shape of decorative objects.

In the **corrugation** process (Fig. 7.52b), the sheet passes through a pair of specially designed rolls, which produce corrugated sheets that are then cut into desired lengths. Again, adhesive is applied to the node lines, and the block is cured. Note that no expansion process is involved. The honeycomb material is then made into a sandwich structure (Fig. 7.52c). Face sheets are subsequently joined with adhesives to the top and bottom surfaces.

Honeycomb structures are most commonly made of 3000-series aluminum, but are also made of titanium, stainless steels, and nickel alloys. Because of their light weight and high resistance to bending forces, these structures are used for aircraft and aerospace components, as well as buildings and transportation equipment.

## 7.12  Deep Drawing

In **deep drawing**, a flat sheet-metal blank is formed into a cylindrical or box-shaped part by means of a punch that presses the blank into the die cavity (Fig. 7.53a). Although the process is generally called deep drawing, meaning deep parts, the basic operation also produces parts that are shallow or have moderate depth. Deep drawing, first developed in the 1700s, has been studied extensively and has become an important metalworking process. Typical parts produced are beverage cans, pots and pans, containers of all shapes and sizes, sinks, and automobile panels.

The basic parameters in deep drawing a cylindrical cup are shown in Fig. 7.53b. A circular sheet blank, with a diameter $D_o$ and thickness $t_o$, is placed over a die opening with a corner radius of $R_d$. The blank is held in place with a **blankholder,** or **hold-down ring,** with a certain force. A punch, with a diameter of $D_p$ and a corner radius $R_p$, moves downward and pushes the blank into the die cavity, thus forming a cup. The significant independent variables in deep drawing are:

1. Properties of the sheet metal
2. The ratio of blank diameter to punch diameter
3. Sheet thickness
4. The clearance between the punch and the die
5. Punch and die corner radii
6. Blankholder force
7. Friction and lubrication at the punch, die, and workpiece interfaces
8. Speed of the punch

At an intermediate stage during the deep drawing operation, the workpiece is subjected to the states of stress shown in Fig. 7.54. On element $A$ in the blank, the radial tensile stress is due to the blank being pulled into the cavity, and the compressive stress normal to the element is due to the blankholder pressure. With a free-body diagram of the blank along its diameter, the radial tensile stresses lead to compressive hoop stresses on element $A$ because of

(a)

Punch

Pressure plate

Blankholder

Die

Blank

Spring stripper ring

**FIGURE 7.53**   (a) Schematic illustration of the deep-drawing process. This is the first step in the basic process by which aluminum beverage cans are produced today. The stripper ring facilitates the removal of the formed cup from the punch. (b) Variables in deep drawing of a cylindrical cup. Only the punch force in this illustration is a dependent variable; all others are independent variables, including the blankholder force.

(b)

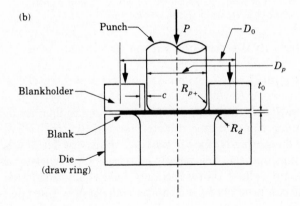

Punch — $P$ — $D_0$

$D_p$

Blankholder

$c$   $R_{p+}$   $t_0$

Blank

$R_d$

Die (draw ring)

(a)                    (b)

$A$

$B$

**FIGURE 7.54**   Deformation of elements in the flange (a) and the cup wall (b) in deep drawing of a cylindrical cup.

**FIGURE 7.55** Examples of drawing operations: (a) pure drawing and (b) pure stretching. The bead prevents the sheet metal from flowing freely into the die cavity. (c) Possibility of wrinkling in the unsupported region of a sheet in drawing. *Source:* After W. F. Hosford and R. M. Caddell.

reduction in the circumferential direction. (It is these hoop stresses that tend to cause the flange to wrinkle during drawing, thus requiring a blankholder under a certain force.) Under this state of stress, element $A$ contracts in the hoop direction and elongates in the radial direction.

The cup wall, which is already formed, is subjected principally to a longitudinal tensile stress, as shown in element $B$. The punch transmits the drawing force $P$ (see Fig. 7.53b) through the walls of the cup and to the flange that is being drawn into the cavity. The tensile hoop stress on element $B$ is caused by the cup being held tightly on the punch because of its contraction under tensile stresses in the cup wall. (A thin-walled tube, when subjected to tension, becomes smaller in a diameter, as can be shown from the generalized flow rule equations; see Eq. (2.43).) Thus element $B$ tends to elongate in the longitudinal (axial) direction with no change in its width, because it is constrained by the rigid punch.

An important aspect of this operation is determining how much **stretching** and how much **pure drawing** is taking place (Fig. 7.55). Note that either with a high blankholder force or with the use of **draw beads** (Figs. 7.55b and 7.56) the blank can be prevented from flowing freely into the die cavity. The deformation of the sheet metal takes place mainly under the punch and the sheet begins to stretch, eventually resulting in necking and tearing. Whether necking is localized or diffused depends on (1) the strain rate sensitivity exponent $m$ of the sheet metal (the higher the $m$ value the more diffuse the neck), (2) geometry of the punch, and (3) lubrication.

Conversely, a low blankholder force will allow the blank to flow freely into the die cavity (pure drawing), whereby the blank diameter is reduced as drawing progresses. The deformation of the sheet is mainly in the flange, and the cup wall is subjected only to elastic stresses. However, these stresses increase with increasing $D_o/D_p$ ratio and can eventually lead to failure when the cup wall cannot support the load required to draw in the flange (Fig. 7.55a). Also note that, in pure drawing, element $A$ in Fig. 7.54 tends to increase in thickness as it moves toward the die cavity because it is being reduced in diameter.

(a)    (b)    (c)

**FIGURE 7.56**    (a) Schematic illustration of a draw bead. (b) Metal flow during drawing of a box-shaped part, using beads to control the movement of the material. (c) Deformation of circular grids in drawing (see Section 7.13). *Source:* After S. Keeler.

In drawing operations, the length of the *unsupported* wall is significant in that it can lead to **wrinkling.** As shown in Fig. 7.55c, element *A* in the sheet is being pulled into the die cavity as the punch descends. Note, however, that the blank is becoming smaller in diameter and the circumference at the element is also becoming smaller as the element moves to position *A'*. Thus, at this position, the element is being subjected to circumferential compressive strains and is unsupported by any tooling, unlike an element between the blankholder and the die surface. Because the sheet is thin and cannot support compressive strains to any significant extent, it will tend to wrinkle in the unsupported region. This situation is particularly common in pure drawing, whereas it is less so as the process becomes more pure stretching and the blank diameter remains basically the same.

**Ironing.**    If the thickness of the sheet as it enters the die cavity is greater than the clearance between the punch and the die, the thickness will be reduced. This effect is known as **ironing;** it produces a cup with constant wall thickness (Fig. 7.57). Thus the smaller the clearance,

**FIGURE 7.57**    Schematic illustration of the ironing process. Note that the cup wall is thinner than its bottom. All beverage cans without seams (known as 2-piece cans) are ironed, generally in three steps, after being drawn into a cup by deep drawing. (Cans with separate tops and bottoms are known as 3-piece cans.)

the greater is the ironing. Obviously, because of volume constancy, an ironed cup will be longer than one produced with a large clearance.

### 7.12.1   Deep drawability (limiting drawing ratio)

The **limiting drawing ratio (LDR)** is defined as the maximum ratio of blank diameter to punch diameter that can be drawn without failure, or $D_o/D_p$. Many attempts have been made to correlate this ratio with various mechanical properties of the sheet metal.

In an ordinary deep drawing process, failure generally occurs by thinning in the cup wall under high longitudinal tensile stresses. By observing the movement of the material into the die cavity (Fig. 7.54), we note that the material should be capable of undergoing a reduction in width (by being reduced in diameter), yet it should resist thinning under the longitudinal tensile stresses in the cup wall.

The ratio of width $(w)$ to thickness $(t)$ strain (Fig. 7.58) is

$$R = \frac{\varepsilon_w}{\varepsilon_t} = \frac{\ln\left(\dfrac{w_o}{w_f}\right)}{\ln\left(\dfrac{t_o}{t_f}\right)}, \tag{7.18}$$

where $R$ is known as the **normal anisotropy** of the sheet metal (also known as **plastic anisotropy** or **strain ratio**). The subscripts $o$ and $f$ refer to the original and final dimensions, respectively. An $R$ value of one indicates that the width and thickness strains are equal to each other; that is, the material is isotropic.

As sheet metals are generally thin compared to their surface area, error in the measurement of small thickness is possible. Equation (7.18) can be modified, based on volume constancy, to

**FIGURE 7.58**   Definition of normal anisotropy ratio $R$ in terms of width and thickness strains in a tensile-test specimen cut from a rolled sheet. Note that the specimen can be cut in different directions with respect to the length, or rolling direction, of the sheet.

$$R = \frac{\ln\left(\dfrac{w_o}{w_f}\right)}{\ln\left(\dfrac{w_f \ell_f}{w_o \ell_o}\right)}, \tag{7.19}$$

where $\ell$ refers to the gage length of the sheet specimen. To calculate $R$, the final length and width in a test specimen are generally measured at an elongation of 15 percent to 20 percent, or for materials with lower ductility, below the elongation where necking begins.

Rolled sheets generally have **planar anisotropy**. Thus the $R$ value of a specimen cut from a rolled sheet (Fig. 7.58) will depend on its orientation with respect to the rolling direction of the sheet. In this case, an *average R value*, $\overline{R}$, is calculated as follows:

$$\overline{R} = \frac{R_0 + 2R_{45} + R_{90}}{4}, \tag{7.20}$$

where the subscripts 0, 45, and 90 refer to angular orientation (in degrees) of the test specimen with respect to the rolling direction of the sheet. Thus an isotropic material has an $\overline{R}$ value of 1. Some typical $\overline{R}$ values are given in Table 7.2. Although hexagonal close-packed metals usually have high $\overline{R}$ values, the low value for zinc is the result of its high $c/a$ ratio in the crystal lattice (see Fig. 3.2c). The $\overline{R}$ value also depends on the grain size and texture of the sheet metal. For cold-rolled steels, for example, $\overline{R}$ increases as grain size increases, (or ASTM number decreases, Fig. 7.59.) For hot-rolled sheet steels, $\overline{R}$ is approximately 1, because the texture developed has a random orientation.

Experiments with sheet steels have shown that $\overline{R}$ is related to the average modulus of elasticity, $\overline{E}$, as shown in Fig. 7.60. Thus a high value of $\overline{R}$ is associated with a high value of $\overline{E}$. The average modulus, $\overline{E}$, is determined in the same manner as $\overline{R}$. Techniques and equipment are now available to measure the modulus of elasticity of sheet specimens by observing their natural frequencies.

The direct relationship between $\overline{R}$ and LDR, as determined experimentally, is shown in Fig. 7.61. In spite of its scatter, no other mechanical property of sheet metal indicates as consistent a relationship to LDR as $\overline{R}$ does. For an isotropic material and based on ideal deformation, the maximum LDR is equal to $e = 2.718$.

**TABLE 7.2**

**TYPICAL RANGE OF AVERAGE NORMAL ANISOTROPY RATIO $\overline{R}$ FOR VARIOUS SHEET METALS**

| | |
|---|---|
| Zinc alloys | 0.4–0.6 |
| Hot-rolled steel | 0.8–1.0 |
| Cold-rolled rimmed steel | 1.0–1.4 |
| Cold-rolled aluminum-killed steel | 1.4–1.8 |
| Aluminum alloys | 0.6–0.8 |
| Copper and brass | 0.6–0.9 |
| Titanium alloys ($\alpha$) | 3.0–5.0 |
| Stainless steels | 0.9–1.2 |
| High-strength low-alloy steels | 0.9–1.2 |

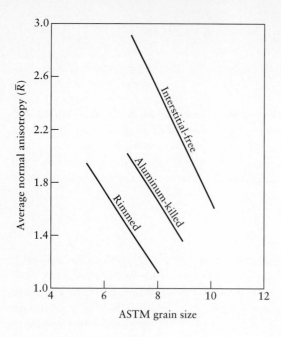

**FIGURE 7.59** Effect of grain size on the average normal anisotropy for various low-carbon steels. *Source:* After D.J. Blickwede.

The **planar anisotropy** of a sheet, $\Delta R$, can also be defined in terms of directional $R$ values as follows:

$$\Delta R = \frac{R_0 - 2R_{45} + R_{90}}{2}, \tag{7.21}$$

**FIGURE 7.60** Relationship between average normal anisotropy $\bar{R}$ and the average modulus of elasticity $\bar{E}$ for steel sheet. *Source*: After P. R. Mould and T. R. Johnson, Jr., *Sheet Met. Ind.,* vol. 50, 1973, p. 328.

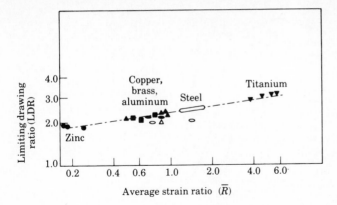

**FIGURE 7.61** Effect of average normal anisotropy $\bar{R}$ on limiting drawing ratio (LDR) for a variety of sheet metals. Zinc has a high $c/a$ ratio (see Fig. 3.2c), whereas titanium has a low ratio. *Source:* After M. Atkinson, *Sheet Met. Ind.,* vol 44, 1967, p. 167.

which is the difference between the average of the $R$ values in the $0°$ and $90°$ directions to rolling and the $R$ value at $45°$. And $\Delta R$ is related to $\Delta E$, where $\Delta E$ is determined in the same manner as $\Delta R$ is in Eq. (7.21), with $R$ replaced by $E$.

### Earing

Planar anisotropy causes **ears** to form in drawn cups (Fig. 7.62), producing a wavy edge. The number of ears produced may be four, six, or eight. The height of the ears increases with increasing $\Delta R$. When $\Delta R = 0$, no ears form. Ears are objectionable because they have to be trimmed off, wasting material. The earing length in a typical drawn aluminum can is about 1–2 percent of the length of the can.

Deep drawability is thus enhanced with a high $\bar{R}$ and a low $\Delta R$. Generally, however, sheet metals with a high $\bar{R}$ also have a high $\Delta R$. Attempts are being made to develop textures in sheet metals to improve drawability. The controlling parameters in processing metals have been found to be: alloying elements, additives, processing temperatures, annealing cycles after processing, thickness reduction in rolling, and cross rolling (biaxial) of plates in processing them into sheets.

**FIGURE 7.62** Earing in a drawn steel cup, caused by the planar anisotropy of the sheet metal.

**FIGURE 7.63** Schematic illustration of the variation of punch force with stroke in deep drawing. Note that ironing does not begin until after the punch has traveled a certain distance and the cup is formed partially. Arrows indicate the start of ironing.

**Example 7.5:** Estimating the limiting drawing ratio

Estimate the limiting drawing ratio (LDR) that you would expect from a sheet metal, which, when stretched by 23% in length, decreases in thickness by 10%.

SOLUTION.   From volume constancy of the test specimen, we have

$$w_o t_o \ell_o = w_f t_f \ell_f \quad \text{or} \quad \frac{w_f t_f \ell_f}{w_o t_o \ell_o} = 1.$$

From the information given,

$$\frac{\ell_f - \ell_o}{\ell_o} = 0.23 \quad \text{or} \quad \frac{\ell_f}{\ell_o} = 1.23$$

and

$$\frac{t_f - t_o}{t_o} = -0.10 \quad \text{or} \quad \frac{t_f}{t_o} = 0.90.$$

Hence

$$\frac{w_f}{w_o} = 0.903.$$

From Eq. (7.18), we obtain

$$R = \frac{\ln\left(\dfrac{w_o}{w_f}\right)}{\ln\left(\dfrac{t_o}{t_f}\right)} = \frac{\ln 1.107}{\ln 1.111} = 0.965.$$

If the sheet has planar isotropy, then $R = \bar{R}$ and from Fig. 7.61 we estimate that

$$\text{LDR} = 2.4.$$

The **maximum punch force** $P_{max}$ supplies the work required in deep drawing. The punch force supplies the work which as in other deformation processes, consists of ideal work of deformation, redundant work, friction work and, when present, the work required for ironing (Fig. 7.63). Because of the many variables involved in this operation and because deep drawing is not a steady-state process, calculating the punch force is difficult. Of the various expressions that have been developed, one simple and very approximate formula for the punch force is

$$P_{max} = \pi D_p t_o (\text{UTS}) \left( \frac{D_o}{D_p} - 0.7 \right). \qquad (7.22)$$

Note that Eq. (7.22) does not include friction, the punch and die corner radii, or the blankholder force. However, this empirical equation makes rough provision for these factors. The punch force is supported basically by the cup wall. If this force is excessive, **tearing** occurs, as shown in Fig. 7.55a. Note that the cup was drawn to a considerable depth before failure occurred. This result can be expected from Fig. 7.63, where we observe that the punch force does not reach a maximum until after the punch has traveled a certain distance. The punch corner radius and die radius (if they are greater than 10 times the sheet thickness) do not affect the maximum punch force significantly.

---

**Example 7.6:** Estimating cup diameter and earing

A steel sheet has $R$ values of 0.9, 1.3, and 1.9 for the $0°, 45°$, and $90°$ directions to rolling, respectively. If a round blank is 100 mm in diameter, estimate the smallest cup diameter to which it can be drawn. Will ears form?

SOLUTION.   Substituting these values in Eq. (7.20), we obtain

$$\bar{R} = [0.9 + (2)(1.3) + 1.9]/4 = 1.35.$$

The limiting drawing ratio (LDR) is defined as the maximum ratio of blank diameter to punch diameter that can be drawn without failure, i.e., $D_o/D_p$. From Fig. 7.61, we estimate the LDR for this steel to be approximately 2.5.

To determine whether or not earing will occur, we substitute the $R$ values in Eq. (7.21). Thus,

$$\Delta R = [0.9 - (2)(1.3) + 1.9]/2 = 0.1.$$

We know that ears will not form if $\Delta R = 0$, and since this is not the case here, ears will form in deep drawing this material.

---

### 7.12.2  Drawing practice

The **blankholder pressure** is generally 0.7 percent to 1.0 percent of the sum of the yield and ultimate tensile strength of the sheet metal. Too high a blankholder force increases the punch load (because of friction) and leads to tearing of the cup wall. However, if the force is too low, wrinkling occurs in the flange.

(a)          (b)

**FIGURE 7.64**   Effect of die and punch radii in deep drawing on fracture of a cylindrical cup. (a) Die radius too small. The die radius should generally be 5 to 10 times the sheet thickness. (b) Punch corner radius too small. Because friction between the cup and the punch aids in the drawing operation, excessive lubrication of the punch is detrimental to drawability.

**Clearances** generally are 7–14 percent greater than the original sheet thickness. The choice of clearance depends on the thickening of the cup wall (which is a function of the drawing ratio). As the clearance decreases, ironing increases. If the clearance is too small, the blank may simply be pierced and sheared by the punch.

The **corner radius** of the punch and the die are important. If they are too small they can cause fracture at the corners (Fig. 7.64). If they are too large, the unsupported area wrinkles. Wrinkling in this region (and from the flange area) causes a defect on the cup wall called **puckering**.

**FIGURE 7.65**   Reducing the diameter of drawn cups by redrawing operations: (a) conventional redrawing and (b) reverse redrawing. Small-diameter deep containers undergo many drawing and redrawing operations.

(a)                          (b)                          (c)

**FIGURE 7.66**  Deep drawing without a blankholder, using a *tractrix* die profile. The tractrix is a special curve, the construction for which can be found in texts on analytical geometry or in handbooks.

**Draw beads** (see Fig. 7.56) are useful in controlling the flow of the blank into the die cavity. They are essential in drawing box-shaped and nonsymmetric parts. Draw beads also help in reducing the blankholder forces required because of the stiffness imparted to the flange by the bent regions at the beads. Proper design and location of draw beads requires considerable experience.

**Redrawing.**    Containers or shells that are too difficult to draw in one operation are generally **redrawn**, as shown in Fig. 7.65a. Another process is **reverse drawing**, shown in Fig. 7.63b, where the metal is subjected to bending in the direction opposite to its original bending configuration. This reversal in bending results in **strain softening** and is another example of the Bauschinger effect (described in Section 2.3.2). This operation requires lower forces, and the material behaves in a more ductile manner.

**Drawing without a blankholder.**    Deep drawing may be carried out *without* a blankholder, provided that the sheet metal is thick enough to prevent wrinkling. The dies are specially contoured for this operation, an example of which is shown in Fig. 7.66. An approximate limit for drawing without a blankholder is given by

$$D_o - D_p < 5t_o. \tag{7.23}$$

It is apparent from Eq. (7.23) that the process can be used with thin materials for shallow draws. Although the punch stroke is greater, a major advantage of this process is the reduced cost of tooling and equipment.

**Various operations.**    Numerous types of parts are also made with shallow or moderate depths. Some involve drawing or stretching, or a combination of these operations. Parts may be embossed with male and female dies or by other means. **Embossing** involves a number of very shallow draws (made with matching dies) on a sheet. The process is used principally for decorative purposes.

**Lubrication.**    Lubrication in deep drawing is important in lowering forces, increasing drawability, reducing tooling wear, and reducing part defects. In general, lubrication of the punch should be minimized, as friction between the punch and the cup improves drawability.

For general applications, commonly used lubricants are mineral oils, soap solutions, and heavy-duty emulsions. For more difficult applications, coatings, wax, and solid lubricants are used (see Table 6.4).

**Tooling and equipment.**    The most commonly used tool materials for deep drawing are tool steels and alloyed cast iron. Other materials, including carbides and plastics, may also be used, depending on the particular application (see Table 3.7). A double-action mechanical press is generally used for deep drawing, although hydraulic presses are also used. The double-action press controls the punch and blankholder independently and forms the part at a constant speed. Punch speeds generally range between 0.1 and 0.3 m/s (20 and 60 ft/min). Speed is generally not important in drawability, although lower speeds are used for high-strength metals.

Since, as we have seen, the blankholder force controls the flow of the sheet metal within the die, new presses have been designed with *variable blankholder force*. In these presses, the blankholder force can be varied with punch stroke. Further control of metal flow is achieved by the use of separate die cushions that allow the local blankholder force to be applied at the proper location on the sheet.

# 7.13   Formability of Sheet Metals

The formability of sheet metals has been of great interest because of its technological and economic significance. Sheet-metal **formability** is generally defined as the ability of the metal to undergo the desired shape changes without failures such as necking and tearing. Three factors have a major influence on formability: properties of the sheet metal, lubrication at various interfaces between the sheet and dies and tooling, and the characteristics of the equipment and tools and dies used in forming the sheet. Several techniques have been developed to test the formability of sheet metals. Formability may also be predicted by modeling the forming processes, with several inputs, including the crystallographic texture of the sheet metal.

### 7.13.1   Testing for formability

**Tensile tests.**    The tensile test (see Section 2.2) is the most basic and common test used to evaluate formability. Important properties are determined, such as total elongation of the sheet specimen at fracture, the strain hardening exponent $n$, the planar anisotropy, and the normal anisotropy $R$ of the sheet, as we shall describe below.

**Cupping tests.**    Because sheet forming is basically a biaxial stretching process, the earliest tests developed to determine or to predict formability were cupping tests, such as the **Erichsen** and **Olsen** tests (*stretching*) and the **Swift** and **Fukui** tests *(drawing)*. A typical cupping test is the Erichsen test, shown in Fig. 7.67. The sheet-metal specimen is clamped over a circular flat die with a load of 1000 kg. A 20-mm–diameter steel ball is then hydraulically pushed into the sheet metal until a crack appears on the stretched specimen or the punch force reaches a maximum. The distance $d$, in millimeters, is the Erichsen number. The greater the value of $d$, the greater is the formability of the sheet. Cupping tests measure the capability of the material to be stretched before fracturing and are relatively easy to perform. However, they do not simulate the exact conditions of actual forming operations, because the stretching under the ball is axisymmetric.

**FIGURE 7.67** The Erichsen cupping test to determine formability of sheet metal. The sheet metal under the steel ball is subjected to biaxial stretching. The greater the distance $d$ before failure, the greater is the formability of the material. Ball diameter = 20 mm.

**Bulge test.** Equal (balanced) biaxial stretching of sheet metals is also performed in the **bulge test**, which has been used extensively to simulate sheet-forming operations. In this test, a circular blank is clamped at its periphery and is bulged by *hydraulic* pressure, thus replacing the punch. The process is one of pure stretch forming, and no friction is involved, as in using a punch. This test can be used to provide effective stress–effective strain curves for biaxial loading under frictionless conditions. The biaxial bulge limit is also a sensitive measure of sheet cleanliness (quality).

## Forming-limit diagrams

An important development in testing formability of sheet metals is the construction of **forming-limit diagrams (FLD)**. In these tests, the sheet blank is marked with a grid pattern of circles or

**FIGURE 7.68** Schematic illustration of the punch–stretch test on sheet specimens with different widths and clamped at the edges. The narrower the specimen, the more uniaxial is the stretching. A large square specimen stretches biaxially under the hemispherical punch (see also Fig. 7.69).

**FIGURE 7.69** Bulge test results on steel sheets of various widths. The first specimen (left) stretched farther before cracking than the last specimen. From left to right, the state of stress changes from uniaxial to biaxial stretching. *Source:* Courtesy of R. W. Thompson, Inland Steel Research Laboratories.

similar patterns, using chemical etching or photoprinting techniques. The blank is then stretched over a punch (Fig. 7.68), and the deformation of the circles is observed and measured in regions where failure occurs. Note that the sheet is fixed by a drawbead that prevents the sheet from being drawn into the die as is the case in the Erichsen test. For improved accuracy of measurement, the circles on the sheet are made as small as possible and the lines as narrow as possible.

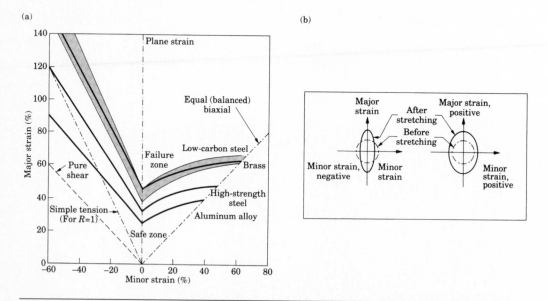

**FIGURE 7.70** (a) Forming-limit diagram (FLD) for various sheet metals. The major strain is always positive. The region above the curves is the failure zone; hence the state of strain in forming must be such that it falls below the curve for a particular material; $R$ is the normal anisotropy. (b) Note the definition of positive and negative minor strains. If the area of the deformed circle is larger than the area of the original circle, the sheet is thinner than the original, because the volume remains constant during plastic deformation. *Source:* After S.S. Hecker and A.K. Ghosh.

Because lubricants can significantly influence the test results, lubrication conditions should be stated in the test report. In this manner, different lubricants can be evaluated for the same sheet metal, or different metals can be evaluated for the same lubricant. This test can also be run with an unlubricated punch.

In order to vary the deformation strain path, the specimens are prepared with varying widths (Fig. 7.68). Thus a square specimen produces equal (balanced) biaxial stretching under the punch, whereas a specimen with a small width approaches the state of a simple tension test. After a series of such tests is performed on a particular sheet metal (Fig. 7.69), the boundaries between failed and safe regions are plotted on the forming-limit diagram (Fig. 7.70). Another example of a failed part with round and square grid patterns is shown in Fig. 7.71.

In the forming-limit diagram the **major** and **minor strains** (engineering) are obtained as follows. Note in Fig. 7.70b that after stretching the original circle has deformed into an ellipse. The major axis of the ellipse represents the major direction and magnitude of stretching. The major strain plotted in Fig. 7.70a is the *engineering strain* (percent) in this direction. Likewise, the minor axis of the ellipse represents the magnitude of the stretching *or* shrinking in the transverse direction. Thus the minor strain can be negative or positive. (The major strain is always positive, because forming sheet metal takes place by stretching at least in one direction.) These diagrams may also be plotted in terms of true major and minor strains.

As an example, if we draw a circle in the center and on the surface of a sheet-metal tensile-test specimen and then stretch it plastically, the specimen becomes narrower; thus the minor strain is negative. Because of volume constancy in the plastic range (see Eq. 2.48) and assuming that the normal anisotropy is $R = 1$, thus the width and thickness strains are equal, then $\varepsilon_w = -0.5\,\varepsilon_\ell$. As a model, this phenomenon can be demonstrated easily by experimenting with a wide rubber band (hence it will be only in the elastic range), placing a circle on it and stretching it.

However, if we place a circle on a spherical rubber balloon and inflate it, the minor strain is positive and equal in magnitude to the major strain. The circle simply becomes a larger circle.

**FIGURE 7.71** An example of using grid marks (circular and square) to determine the magnitude and direction of surface strains in sheet-metal forming. Note that the crack (tear) is generally perpendicular to the major (positive) strain. *Source:* After S.P. Keeler.

By observing the difference in surface area between the original circle and the ellipse, we can also determine whether the thickness of the sheet has changed. If the area of the ellipse is larger than the original circle, the sheet has become thinner (because volume remains constant in plastic deformation).

*Friction* at the punch–metal interface can also be important to test results. With well-lubricated interfaces, the strains are more uniformly distributed over the punch. Depending on the notch sensitivity of the sheet metal, surface scratches, deep gouges, and blemishes can reduce formability and cause premature tearing and failure during testing and in actual forming operations. With effective lubrication the coefficient of friction in sheet-metal forming generally ranges from about 0.05 to 0.1 for cold forming, and from 0.1 to 0.2 for elevated temperature forming.

From Fig. 7.70a, we observe, as expected, that different materials have different forming-limit diagrams. The higher the curve, the better is the formability of the material. Also note that for the same minor strain, say 20 percent, a compressive minor strain is associated with a higher major strain before failure than is a tensile (positive) minor strain. In other words, it is desirable for the minor strain to be negative, that is, shrinking in the minor direction. Special tools have been designed for forming sheet metals that take advantage of the beneficial effect of negative minor strains on extending formability (see Fig. 7.32).

The effect of sheet-metal thickness on forming-limit diagrams is to raise the curves in Fig. 7.70a. Thus the thicker the sheet, the higher its formability curve, hence the more formable it is. In actual forming operations, however, a thick blank may not bend as easily around small radii. The possible effect of the rate of deformation on forming-limit diagrams should also be assessed for each material. Extensive studies are being carried out to develop new test methods to predict the behavior of metals in sheet-forming operations.

---

**Example 7.7:    Estimating diameter of expansion**

A thin-walled spherical shell made of the aluminum alloy shown in Fig. 7.70(a) is being expanded by internal pressure. If the original diameter is 200 mm, what is the maximum diameter to which it can be expanded safely?

SOLUTION.    Because the material is being stretched in a state of equal (balanced) biaxial tension, we find from Fig. 7.70a that the maximum allowable engineering strain is about 40%. Thus

$$e = \frac{\pi D_f - \pi D_o}{\pi D_o} = \frac{D_f - 200}{200} = 0.40.$$

Hence

$$D_f = 280 \text{ mm}.$$

---

**Example 7.8:    Strains in the sheet-metal body of an automobile**

Computer programs have been developed that compute the major and minor strains and their orientations from the measured distortions of grid patterns placed on the surface of

**FIGURE 7.72** Major and minor strains in various regions of an automobile body.

sheet metal. One application for an automobile body is shown in Fig. 7.72 for various panels. Note that the trunk lid and the roof are subjected mainly to plane strain (see Fig. 7.70a), whereas the front door and front fender are subjected to biaxial strains. The numbers in the strain paths indicate the frequency of occurrence. *Source.* After T. J. Nihill and W. R. Thorpe.

**Limiting dome height test.** Recall that in the test illustrated in Fig. 7.68, the major and minor strains were measured and plotted on the forming-limit diagram. In the **limiting dome height (LDH)** test, performed with similarly prepared specimens, the *height* of the dome at failure of the sheet—or when the punch force reaches a maximum—is measured, as is done in the test shown in Fig. 7.67. Because the specimens are clamped, the LDH test indicates the capability of the material to stretch without failure. It has been shown that high LDH values are related to sheet-metal properties such as high $n$ and $m$ values, as well as high total elongation (percent) of the sheet metal.

### 7.13.2 Dent resistance of sheet-metal parts

A dent is a small, but permanent, biaxial deformation. In certain applications involving sheet-metal parts, such as automotive body panels, appliances, and office furniture, an important consideration is the dent resistance of the sheet-metal panel. The factors significant in dent resistance are the yield stress $Y$, thickness $t$, and the shape of the panel. **Dent resistance** is determined by a combination of material and geometrical parameters, and can be defined as

$$\text{Dent resistance} = \alpha \frac{Y^2 t^4}{S},\qquad (7.24)$$

where $S$ is the panel stiffness, which in turn is defined as

$$S = (E)(t^a)(\text{shape}),\qquad (7.25)$$

where the value of $a$ ranges from 1 to 2 for most panels. As for shape, the smaller the curvature (hence the flatter the panel), the greater the dent resistance is because of its flexibility. Thus dent resistance increases with increasing strength and thickness and decreases with increasing elastic modulus $E$ and stiffness and decreasing curvature. Note that for $a = 2$, dent resistance is proportional to $t^2$.

Dents are usually caused by *dynamic* forces, such as those of falling objects or other objects that hit the sheet-metal panel. In typical automotive panels, for instance, impact velocities range up to 45 m/s (150 ft/s). Thus the dynamic yield stress (high strain rate), rather than the static yield stress, is the significant strength parameter. Denting under quasistatic forces could also be important.

For materials in which yield stress increases with strain rate, denting requires higher energy levels than under static conditions. Furthermore, dynamic forces tend to cause more *localized* dents than static forces, which tend to spread the dented area. Because a portion of

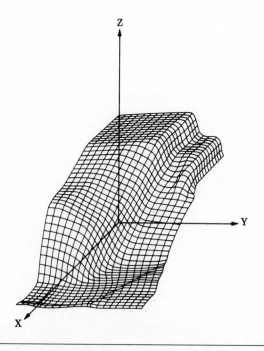

**FIGURE 7.73** Deformation of a square mesh in computer simulation of forming a sheet-metal part. *Source:* J.L Duncan, R. Sowerby, and E. Chu.

the energy goes into elastic deformation, the modulus of resilience of the sheet metal is an additional factor to be considered.

### 7.13.3 Modeling of sheet-forming processes

In Section 6.2.2 we outlined the techniques for studying bulk deformation processes and described *modeling* of impression-die forging using the finite-element method (see Fig 6.11). Such mathematical modeling is also being applied to sheet-forming processes. Deformation of an originally square mesh on the surface of a sheet-metal part is shown in Fig. 7.73, as developed by computer-aided modeling. The ultimate goal of such simulation techniques is rapid analysis of stresses, strains, flow patterns, and springback as functions of parameters such as material characteristics, friction, anisotropy, deformation speed, and temperature.

Through such interactive analysis, we can determine the optimum tool and die geometry to make a certain part, thus reducing or eliminating costly die tryouts. In addition, these simulation techniques can be used to determine blank size and shape (including intermediate shapes), press characteristics, and process parameters to optimize the forming operation. The application of computer modeling, although requiring powerful computers and extensive software, has already proven to be a cost-effective tool in sheet-metal forming, particularly in the automotive industry.

---

**Example 7.9:** Forming of laser-welded sheet metals

It will be noted that in the forming processes we have described thus far, the sheet blank is made of one piece and one thickness, usually blanked from a large sheet. A recent important

g 60/60 (45/45) Hot galvanized alloy steel sheet.
Zinc amount: 60/60 (45/45) g/m$^2$.

m20/20      Double-layered iron-zinc alloy electroplated steel sheet. Zinc amount 20/20 g/m$^2$.

---

**FIGURE 7.74** Production of an outer side panel by laser welding and stamping. *Source:* After M. Geiger and T. Nakagawa.

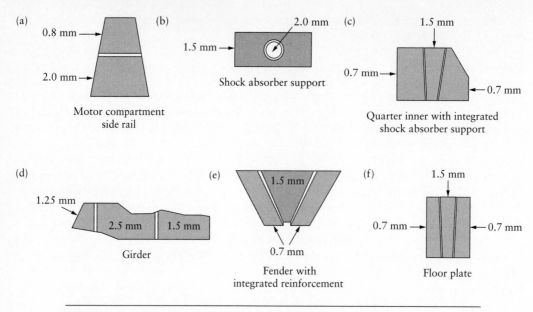

**FIGURE 7.75**  Examples of laser welded and stamped automotive body components. *Source:* After M. Geiger and T. Nakagawa.

trend involves laser welding (see Section 12.4) of pieces of sheet metal of different shapes and thicknesses; the welded assembly is subsequently formed into a final shape. One example of this is the production of an automobile outer side panel, shown in Fig 7.74. Note that five different pieces are first blanked; four of these are 1 mm thick and one 0.8 mm thick. These pieces are butt laser welded and then stamped into the final shape. Thus, the blanks can be tailored for a particular application, including not only shape and thickness, but also different quality sheets and with or without coatings. Laser welding techniques are now highly developed and, as a consequence, weld joints are very strong. This trend of welding and forming of sheet-metal pieces allows significant flexibility in product design, structural stiffness and crash behavior, formability, use of different materials in one component, and weight savings, as well as cost reduction in materials, scrap, equipment, assembly, and labor.

There is a growing number of applications for this type of production in U.S. and Japanese automotive companies (Fig. 7.75). The various components shown in the figure utilize the advantages outlined above. For example, note in part (b) that the required strength and stiffness for supporting the shock absorber is achieved by welding a round piece on the surface of the large sheet. The sheet thickness in these components varies, depending on its location and contribution to characteristics such as stiffness and strength, while allowing significant weight savings.

In *laser-assisted forming,* certain operations are being carried out using lasers as a localized heat source, thus reducing the flow stress of the sheet at specific locations, improving formability, and increasing process flexibility. Although still in experimental stages, examples are straightening, bending, embossing, and forming complex tubular or flat components. The possible adverse effects of such localized heating on product integrity are also being studied.

*Source:* After M. Geiger and T. Nakagawa.

**FIGURE 7.76**   Cost comparison for manufacturing a round sheet-metal container by conventional spinning and deep drawing. Note that for small quantities spinning is more economical.

## 7.14   Economics of Sheet-Metal Forming

Sheet-metal forming involves economic considerations similar to those for the other processes that we have discussed. Sheet-forming operations compete with each other, as well as with other processes, more than other processes do. We have noted that sheet-forming operations are versatile and that a number of different processes can be used to produce the same part. For example, a cup-shaped part can be formed by deep drawing, spinning, rubber forming, or explosive forming. Similarly, a cup can be formed by impact extrusion or casting, or by fabricating it from different pieces.

The part shown in Fig. 7.76 can be made either by deep drawing or by conventional spinning. However, the die costs for the two processes are significantly different. Deep-drawing dies have many components and cost much more than the relatively simple mandrels and tools employed in spinning. Consequently, the die cost per part in drawing will be high if few parts are needed. On the other hand, this part can be formed by deep drawing in a much shorter time (seconds) than by spinning (minutes), even if the latter operation is automated. Furthermore, spinning requires more skilled labor. Considering all these factors, we find that the *breakeven point* is at about 700 parts. Thus deep drawing is more economical for quantities greater than that.

---

CASE STUDY:   FORMING OF A GENERALIZED AUTOMOTIVE FRONT FENDER

Whenever an automobile is designed, dies must be made for producing the new shapes. A particular automobile part such as a front fender, for example, is similar from one car to another; experience gained with previous parts is helpful in designing new dies. This experience is not infallible, however, and a great deal of trial and error is required unless numerical simulation is used at the outset.

In the following example, a fictitious automaker* followed a trial-and-error procedure in its research laboratory. It applied the method to a simplified and scale model of the die shape in order to arrive at a design close enough to the final one to eliminate many die try-out steps with the more expensive, final tool shapes and sizes. While the real part may be more than 1200 mm in length, the scale part is approximately 300 mm in length. This smaller scale allows the use of a research press rather than tying up a production press, and allows the use of much less expensive tools. In order to simplify matters further, the "fender arch" was removed from the scale part without sacrificing much of the fidelity at the corner regions, where most of the forming problems occur.

Figure 7.77a shows the original scale panel with the wheel arch; the same panel without the arch is shown in Fig. 7.77b. The original was formed of aluminum-killed, deep-drawing

**FIGURE 7.77** (a) Scale fender with wheel arch, AKDQ steel. (b) Scale fender without wheel arch, HSLA steel. (c) Scale fender schematic (Trial #1), (d) Forming performance (Trial #2), (e) Scale fender schematic (Trial #3), (f) Forming performance (Trial #3)

steel and the modified panel was formed from high-strength, low-alloy steel, hence the different attainable draw depths. The original tooling trial utilized a uniform draw bead around the periphery of the die cavity (Fig. 7.77c). This draw bead is used to restrain the flow of sheet material into the die, thus forcing it to stretch rather than simply bend and translate. The automaker knew from experience that such restraint and stretching are usually required to avoid severe wrinkling (such as is seen with an aluminum-foil pie plate), which is unacceptable to customers. However, this design proved to be impossible to form without splitting at the corners of the punch (Fig. 7.77d).

It was reasoned, by experience, that removal of the draw bead would allow material to draw more freely into the die, thus reducing the tendency to split at the corners. In fact, Trial #2, which removed the draw bead, did eliminate all splitting, but this problem was replaced by severe wrinkling at all of the corners. It became clear, again according to experience, that it would be necessary to restrain the sheet drawing into the die cavity along the straight edges of the punch while allowing the corner regions to draw more easily. Therefore, Trial #3 added draw beads back along the sides while keeping them off the corner areas. Some trial and error (Trial #3a, #3b, #3c, etc.) found the optimal extent of the draw beads along the sides, as shown in Fig. 7.77e.

The result of this Trial #3 is shown in Fig. 7.77f. By careful placement of the draw beads and selection of the binder force (in production, this adjustment would be accomplished by changing the contours of the draw bead itself so that the exact value of restraint could be tailored to optimize forming performance), it was found that it is possible to eliminate all fractures or all severe wrinkling, but not both simultaneously. The result for a typical binder force is shown in Fig. 7.78 although it is possible to eliminate either the crack or the wrinkle.

**FIGURE 7.78** Wrinkle behavior of scale fender forming with constant and step-down binder force control (Trial #4).

Trial #3 would have been the usual stopping point for many industrial trials. Once the draw beads have been distributed optimally without obtaining acceptable forming performance, the following choices would be available to the production engineer:

1. Substitute a more formable and costly material (for example, interstitial-free instead of drawing-quality steel, or drawing-quality steel in place of rimmed steel). With the more formable material in place, it may be possible to increase the restraining force of the draw beads, thus reducing wrinkling without causing the split failures.

2. Substitute a thicker (and hence more expensive and heavier) material. The thicker material will in general resist bending and thus wrinkling more effectively (and there may be some improvement in formability, although this is controversial) thus, it may be possible to reduce wrinkling at the same draw bead restraining force.

3. Relax the design standards. While splits are almost never acceptable, it may be possible to allow some minor wrinkling. This, however, will require extensive discussions between product and production engineers.

4. Modify the part shape. This is usually a last resort because not only does it involve both product and production engineering approval, but it may also have an impact on other adjacent parts in the automobile. However, such changes are sometimes necessary.

As it turned out, the research department and automotive company were particularly progressive and tried an innovative solution that required none of the drawbacks of the usual choices listed above. They embarked on a program to control the binder force during the press cycle, using production presses, and they worked further in the laboratory (Trial #4) with the scale fender to find an improved binder control strategy.

In particular, they first tried a simple step-down strategy, maintaining a high (294 kN) binder force, up just before the usual fracture point. At a punch travel of 40 mm (out of a total of 48 mm), this binder force was reduced to 98 kN. No fractures were observed and the wrinkle height was reduced dramatically, compared with any constant binder force capable of forming the part without splitting. (The optimal constant binder force was found to be approximately 196 kN.) These results are shown in Fig. 7.78.

This first trial allowed the part to be made successfully. However, the changes to a full-size press require considerable capital expenditure, thus only a single press (for difficult-to-form parts such as this one) has been modified and used so far. The principal difficulty in adopting this strategy for more presses lies mainly in the lack of experience in choosing a blankholder force signature for a new part design. As simulation of sheet-forming operations is used more extensively, modification of presses to take advantage of blankholder control is likely to become a cost-saving option for many parts.

*Source:* R.H Wagoner, Ohio State University

*For more about this actual part design, see the following references:
Yuji Hishida and Robert H. Wagoner, "Experimental Analysis of Blank Holding Force Control in Sheet Forming" (SAE Paper No. 930285). Sheet Metal and Stamping Symposium. SAE SP-944. Warrendale, PA. 1993, pp. 93-100.
Y. Hishida and R. H. Wagoner, "Analysis of Blank Holding Force Control Forming Using Hydraulic Forming Simulator," *Advanced Technology of Plasticity*, 1993, pp. 1740-1746.

# SUMMARY

- Sheet-metal forming processes involve workpieces that have a high ratio of surface area to thickness. In these processes, tensile stresses are usually applied in the plane of the sheet through various punches and dies. The material is generally prevented from being reduced in thickness in order to avoid necking and subsequent tearing and fracture. Material properties and behavior important to these processes include elongation, yield-point elongation, anisotropy, and grain size. (Sections 7.1, 7.2)

- Blanks for forming operations are generally sheared from large sheets using various processes. The quality of the sheared edge is important in subsequent processing, and important process parameters are the clearance between the dies, sharpness, and lubrication. (Section 7.3)

- Bending of sheet, plate, and tubing is a commonly performed operation, and important parameters are the minimum bend radius (without causing cracking in the bend area) and springback. Several techniques are available for bending and for minimizing springback. (Section 7.4)

- Discrete sheet-metal bodies can be made by stretch forming, bulging, and rubber forming, whereby the workpiece is shaped by stretching in different directions. In rubber forming, in contrast to the other two methods, one of the dies can be made of a flexible material such as rubber or polyurethane. (Sections 7.5, 7.6, 7.7)

- Spinning involves the forming of axisymmetric parts over a rotating mandrel using rigid tools or rollers, and comprises conventional, shear, and tube-spinning processes. Various external and internal profiles can be produced. (Section 7.8)

- High-energy-rate forming processes use chemical, electrical, or magnetic means of energy. They include explosive forming of usually large sheets, electrohydraulic forming, and magnetic-pulse forming of relatively small parts. (Section 7.9)

- Superplastic forming methods utilize the superplastic behavior of materials and are capable of producing complex shapes with a high strength-to-weight ratio. They are particularly effective when combined with diffusion bonding of two or more sheets prior to forming. (Section 7.10)

- Various other sheet-forming processes include peen forming based on the principle of shot peening, use of gases, and manufacturing honeycomb structures by either the expansion or corrugation techniques. (Section 7.11)

- Deep drawing, which utilizes a punch and a die, is one of the most important sheet-forming processes because it can produce container-shaped parts with high depth-to-diameter ratios. Deep drawability has been found to depend on the normal anisotropy of the sheet metal. Earing also occurs in deep drawing because of the planar anisotropy of the metal. Buckling and wrinkling in drawing can be significant problems; techniques have been developed to minimize or eliminate these problems through careful design of tools and dies and reducing the unsupported length of the sheet metal. (Section 7.12)

- Tests have been developed to determine the formability of sheet metals in forming processes. The most comprehensive test is the punch–stretch (bulge) test that allows the construction of a forming-limit diagram, indicating safe and failure regions as a function of the planar strains involved. Each type of metal, its condition, and its thickness has its own curve. Limiting-dome height tests have also been developed for determining formability. (Section 7.13)

## SUMMARY OF EQUATIONS

Maximum punch force in shearing, $P_{max} = 0.7(\text{UTS})(t)(L)$

Strain in bending, $e_o = e_i = \dfrac{1}{(2R/T) + 1}$

Minimum $R/T = \dfrac{50}{r} - 1$

Springback, $R_i/R_f = 4\left(\dfrac{R_iY}{ET}\right)^3 - 3\left(\dfrac{R_iY}{ET}\right) + 1$

Maximum bending force, $P_{max} = \dfrac{k\,(\text{UTS})LT^2}{w}$

Normal anisotropy, $R = \dfrac{\varepsilon_w}{\varepsilon_t}$

average, $\overline{R} = \dfrac{R_0 + 2R_{45} + R_{90}}{4}$

Planar anisotropy, $\Delta R = \dfrac{R_0 - 2R_{45} + R_{90}}{2}$

Maximum punch force in drawing, $P_{max} = \pi D_p t_o(\text{UTS})\left(\dfrac{D_o}{D_p} - 0.7\right)$

Pressure in explosive forming, $p = K\left(\dfrac{\sqrt[3]{W}}{R}\right)^a$

## BIBLIOGRAPHY

*ASM Handbook,* Vol. 14: *Forming and Forging.* Materials Park, Ohio: ASM International, 1988.
*Fundamentals of Tool Design,* 3d ed. Dearborn, Mich.: Society of Manufacturing Engineers, 1991.
Hosford, W.F., and R.M. Caddell, *Metal Forming, Mechanics and Metallurgy,* 2d ed. Englewood Cliffs, New Jersey: Prentice-Hall, 1993.
Koistinen, D.P., and N.M. Wang (eds.), *Mechanics of Sheet Metal Forming.* New York: Plenum, 1978.
Morgan, E., *Tinplate and Modern Canmaking Technology.* Oxford, England: Pergamon, 1985.
Pacquin, J.R., and R.E. Crowley, *Die Design Fundamentals.* New York: Industrial Press, 1987.
Sachs, G., *Principles and Methods of Sheet Metal Fabricating,* 2d ed. New York: Reinhold, 1966.
*Tool and Manufacturing Engineers Handbook,* 4th ed., Vol. 2: *Forming.* Dearborn, Mich.: Society of Manufacturing Engineers, 1984.

## QUESTIONS

7.1    Take any three topics from Chapter 2 and, with specific examples for each, show their relevance to the topics covered in this chapter.

7.2    Do the same as for Question 7.1, but for Chapter 3.

7.3    Describe the (1) similarities and (2) dissimilarities between the bulk deformation processes described in Chapter 6 and the sheet-metal forming processes described in this chapter.

7.4 Discuss the material and process variables that influence the shape of the punch force vs. stroke curve for shearing, such as that shown in Fig. 7.8, including its height and width.

7.5 In preparing a large number of blanks for sheet-forming operations, the blanks are spaced (*nesting*) such that material waste is minimized. Make some sketches and describe the considerations involved in this procedure, with particular emphasis on subsequent forming operations.

7.6 Describe your observations concerning Figs. 7.6 and 7.7.

7.7 Inspect a common paper punch and comment on the shape of the punch end, as compared to those shown in Fig. 7.13.

7.8 Explain how you would go about estimating the temperature rise in a shearing operation, as shown in Fig. 7.7.

7.9 As a practicing engineer in manufacturing, would you be interested in the shape of the curve in Fig. 7.8? Explain.

7.10 Do you think the presence of burrs can be beneficial in certain applications? Give specific examples.

7.11 Give the reasons why there are so many different types of die materials used for the processes described in this chapter.

7.12 Describe the difference between compound, progressive, and transfer dies.

7.13 It has been stated that the quality of the sheared edges can influence the formability of sheet metals. Explain why.

7.14 Explain why and how various factors influence springback in bending of sheet metals.

7.15 Can the hardness of a sheet metal have an effect on its springback in bending? Explain.

7.16 We note in Fig. 7.17 that the state of stress shifts from plane stress to plane strain as the length of bend to sheet thickness increases. Explain why.

7.17 Inspect Fig. 7.20 and describe those material properties that have an effect on the relative position of the curves.

7.18 In Table 7.1, we note that hard material conditions have higher $R/T$ ratios than soft ones. Explain why.

7.19 Why do tubes buckle when bent? Experiment with a soda straw.

7.20 If you wish to bend a piece of, say, copper tubing without collapsing it, you could go to a hardware store and buy a helical springlike gadget. Visit a hardware store, find out what this gadget is, and explain why you can use it to successfully bend tubing.

7.21 Based on Fig. 7.24, sketch the shape of a U-die to produce channel-shaped bends.

7.22 Explain why negative springback does not occur in air bending of sheet metals.

7.23 Give examples of products where the presence of beads is beneficial.

7.24 Explain why cupping tests do not always predict the behavior of sheet metals in actual forming operations.

7.25 Assume that you are carrying out a sheet-forming operation and you find that the material is not sufficiently ductile. Make suggestions to improve ductility.

7.26 Referring to Fig. 7.41, explain why a further increase in the ductility of the material (beyond about 50 percent tensile reduction of area) does not improve the maximum spinning reduction per pass.

7.27 Many missile components are made by spinning. What other methods would you use if spinning processes were not available?

7.28 In deep drawing of a cylindrical cup, is it always necessary that there be tensile circumferential stresses on the element in the cup wall (see Fig. 7.54b)?

7.29 When comparing the hydroform process with the deep-drawing process, it is stated that deeper draws are possible in the former method. With appropriate sketches, explain why.

7.30 We note in Fig. 7.54a that element $A$ in the flange is subjected to compressive circumferential (hoop) stresses. Using a free-body diagram, explain why.

7.31 From the topics covered in this chapter, list and explain specifically two examples where (1) friction is desirable, and (2) friction is not desirable.

7.32 Explain why increasing the value of normal anisotropy $R$ improves the deep drawability of sheet metals.

7.33 What is the reason for the negative sign in the numerator of Eq. (7.21)?

7.34 If you had a choice whereby you could control the state of strain in a sheet-forming operation, would you rather work on the left or the right side of the forming-limit diagram? Explain.

7.35 Comment on the effect of punch lubricating on the limiting drawing ratio in deep drawing.

7.36 Comment on the size of the circles used on sheet-metal surfaces for determining their formability. Are square patterns, as seen in Fig. 7.71, useful? Explain.

7.37 Make a list of the independent variables that influence the punch force in deep drawing of a cylindrical cup, and explain why and how they influence the force.

7.38 Explain why the simple tension line in the forming-limit diagram in Fig. 7.70 states that it is for $R = 1$, where $R$ is the normal anisotropy.

7.39 Explain the reasons for developing forming-limit diagrams. Do you have any criticisms of such diagrams? Explain.

7.40 Explain the reasoning behind Eq. (7.20) for normal anisotropy and Eq. (7.21) for planar anisotropy, respectively.

7.41 Describe why earing occurs. How would you avoid it?

7.42 What is the significance of the size of the grid patterns used in the study of sheet-metal formability?

7.43 We have stated that the thicker the sheet metal, the higher the curve in the forming-limit diagram. Explain why.

7.44 Inspect the earing shown in Fig. 7.62 and estimate the direction in which the blank was cut.

7.45 Describe the factors that influence the size and length of beads in sheet-forming operations.

7.46 We have seen that the strength of metals depends on their grain size. Would you then expect that strength influences the $R$ value of sheet metals? Explain.

7.47 A general rule for dimensional relationships for successful drawing without a blankholder is given in Eq. (7.23). Explain what happens if this limit is exceeded.

7.48 Explain why the three broken lines (simple tension, plane strain, and equal biaxial stretching) in Fig. 7.70a have those particular slopes.

7.49 Think of a specific part and explain which of the processes from Chapters 6 and 7, respectively, can be used to make that part. Then make a final choice, explaining your reasoning.

7.50 We have seen that bendability and spinnability have a common aspect as far as workpiece material properties are concerned. Describe this common aspect.

7.51 Explain the reasons why such a wide variety of sheet-forming processes have been developed over the years.

7.52 Make a summary of the type of defects found in sheet-metal forming processes; include brief comments on the reasons for these defects.

7.53 Which of the processes described in this chapter utilize one die only? What are the advantages?

7.54 Make a survey of some of the products you are familiar with that are made of sheet metal. Discuss the process(es) by which you think they were made.

7.55 It has been suggested that deep drawability can be increased by (1) heating the flange and (2) chilling the punch, both by some suitable means. Comment on how these methods could improve drawability.

7.56 Offer designs whereby the suggestions given in Question 7.55 can be implemented. Would the production rate affect your designs? Explain.

## PROBLEMS

7.57 Referring to Eq. (7.5), it is stated that actual values of $e_o$ are considerably higher than $e_i$ due to the shifting of the neutral axis during bending. With an appropriate sketch, explain this phenomenon.

7.58 Referring to Eq. (7.11) we note that the bending force is a function of $T$ squared. Why? (*Hint*: Consider bending-moment equations in mechanics of solids.)

7.59 Calculate the minimum tensile true fracture strain that a sheet metal should have in order to be bent to the following $R/T$ ratios: (1) $2T$, (2) $3T$, and (3) $5T$. See Table 7.1.

7.60 Estimate the maximum bending force required for a 1/8 in. thick and 10 in. wide Ti–5Al–2.5Sn titanium alloy in a V-die with a width of 8 in.

7.61 In Example 7.3, calculate the work done by the force-distance method.

7.62 What would be the answer to Example 7.3 if the tip of the force $F$ is fixed to the strip by some means, thus maintaining its lateral position? (*Hint:* Note that the left portion of the strip will now be strained more than the right portion.)

7.63 Calculate the magnitude of the force $F$ in Example 7.3 for $\alpha = 15°$.

7.64 How would the force in Example 7.3 vary if the workpiece material were made of a perfectly plastic material?

7.65 Calculate the press force needed in punching 5052–0 aluminum foil, 0.1 mm thick, in the shape of a square hole 30 mm on each side.

7.66 A straight bead is being formed on a 1-mm thick aluminum sheet in a 20-mm diameter die, as shown in the figure below (see also Fig. 7.27a). Let $Y = 120$ MPa. Considering springback, calculate the outside diameter of the bead after it is formed and unloaded from the die.

7.67 Inspect Eq. (7.10) and substituting some numerical values, show whether the first term can be neglected without significant error in calculating springback.

7.68 In Example 7.4, calculate the amount of TNT required to develop a pressure of 15,000 psi on the workpiece surface.

7.69 Estimate the limiting drawing ratio ($LDR$) for the materials listed in Table 7.2.

7.70 For the same material and thickness as in Problem 7.60, estimate the force required for deep drawing, with a blank diameter of 12 in. and punch diameter of 9 in.

7.71 A cup is being drawn from a sheet metal having a strain ratio of 3. Estimate the maximum ratio of cup height-to-cup diameter that can be drawn successfully in a single draw. Assume that the sheet thickness throughout the cup remains the same as the original blank thickness.

7.72 Obtain an expression for the curve in Fig. 7.61 in terms of $LDR$ and the average strain ratio $\bar{R}$ (*Hint:* See Fig. 2.7b.)

7.73 A steel sheet has $R$ values of 1.0, 1.5, and 2.0 for the $0°$, $45°$, and $90°$ directions to rolling, respectively. If a round blank is 200 mm in diameter, estimate the smallest cup diameter to which it can be drawn.

7.74 In Problem 7.73, explain whether ears will form, and if so, why.

7.75 Show that in pure drawing of a perfectly plastic material, the limiting drawing ratio is 2.718. (*Hint:* Assume that element $A$ in Fig. 7.54 is being drawn or pulled through a wedge-shaped die, from a diameter of $D_o$ to $D_p$, with the tensile radial stress shown on the element; then determine maximum drawability as done in wire or rod drawing.)

7.76 A 1 in. thick isotropic sheet metal is inscribed with a circle 5 mm in diameter. It is then stretched uniaxially by 25 percent. Calculate (1) the final dimensions of the circle, and (2) the thickness of the sheet at this location.

7.77 Make a search of the literature and obtain the equation for a tractrix curve, as used in Fig. 7.66.

7.78 In Example 7.3, assume that the stretching is done by two equal forces $F$, each at 5 in. from the ends of the workpiece. (1) Calculate the magnitude of this force for $\alpha = 10°$. (2) If we want stretching to be done up to $\alpha_{max} = 50°$ without necking, what should be the minimum value of $n$ of the material?

7.79  Derive Eq. (7.5).

7.80  Estimate the maximum power in shear spinning a 0.5 in. thick annealed 304 stainless steel plate, 15 in. in diameter, on a conical mandrel of $\alpha = 45°$. The man-

drel rotates at 100 rpm and the feed is $f = 0.1$ in./rev.

7.81  Assume that you are asked to give a quiz to students on the contents of this chapter. Prepare five quantitative problems and five qualitative questions, and supply the answers.

# 8

## Material-Removal Processes: Cutting

**8.1** Introduction
**8.2** Mechanics of Chip Formation
**8.3** Tool Wear
**8.4** Surface Finish and Integrity
**8.5** Machinability
**8.6** Cutting-Tool Materials
**8.7** Cutting Fluids
**8.8** Cutting Processes and Machine Tools for Producing Round Shapes
**8.9** Cutting Processes and Machine Tools for Producing Various Shapes
**8.10** Machining and Turning Centers
**8.11** Vibration and Chatter
**8.12** Machine-Tool Structures
**8.13** Machining Economics

## 8.1 Introduction

Parts manufactured by casting, forming, and various shaping processes often require further operations before the product is ready for use. Also, in many engineering applications, parts have to be interchangeable in order to function properly and reliably during their expected service lives, thus requiring control of dimensional accuracy. Often, critical choices have to be made about the extent of shaping and forming versus the extent of machining to be done on a workpiece to produce an acceptable part.

Although **machining** is the broad term used to describe *removal* of material from a workpiece, it covers several processes, which we usually divide into the following categories:

- Cutting, generally involving single-point or multipoint cutting tools, each with a clearly defined geometry.
- Abrasive processes, such as grinding.
- Nontraditional machining processes, utilizing electrical, chemical, and optical sources of energy.

We can summarize why material-removal processes are desirable or even necessary in manufacturing operations as follows:

1. Closer dimensional accuracy may be required than is available from casting, forming, or shaping processes alone. In a forged crankshaft, for example, the bearing surfaces and the holes cannot be produced with good dimensional accuracy and surface finish by forming and shaping processes alone.

2. Parts may have external and internal profiles, as well as sharp corners and flatness, that cannot be produced by forming and shaping processes.

3. Some parts are heat treated for improved hardness and wear resistance. Since heat-treated parts may undergo distortion and surface discoloration, they generally require additional finishing operations, such as grinding, to obtain the desired final dimensions and surface finish.

4. Special surface characteristics or texture that cannot be produced by other means may be required on all or part of the surfaces of the product. Copper mirrors with very high reflectivity, for example, are made by machining with a diamond cutting tool.

5. Machining the part may be more economical than manufacturing it by other processes, particularly if the number of parts desired is relatively small.

Against these advantages, material-removal processes have certain limitations:

1. Removal processes inevitably waste material and generally require more energy, capital, and labor than forming and shaping operations. Thus they should be avoided whenever possible.

2. Removing a volume of material from a workpiece generally takes longer than to shape it by other processes.

3. Unless carried out properly, material-removal processes can have adverse effects on the surface quality and properties of the product.

In spite of these limitations, material-removal processes and machines are indispensable to manufacturing technology. Ever since lathes were introduced in the 1700s, these processes have developed continuously. We now have available a variety of computer-controlled machines, as well as new techniques using electrical, chemical, and optical energy sources. The machines on which material-removal operations are performed are generally called **machine tools.** Their construction and characteristics influence greatly these operations and product quality. It is important to view machining, as well as all manufacturing operations, as a *system* consisting of the workpiece, the tool, and the machine. Cutting operations cannot be carried out efficiently and economically without a knowledge of the interactions among these elements.

## 8.2   Mechanics of Chip Formation

Cutting processes, such as turning on a lathe, drilling, and milling (Fig. 8.1) remove material from the surface of the workpiece thereby producing **chips**. The basic mechanics of chip formation is essentially the same for all these operations, which we represent by the two-dimensional model in Fig. 8.2. In this model a tool moves along the workpiece at a certain velocity $V$ and a depth of cut $t_o$. A chip is produced ahead of the tool by shearing the material continuously along the shear plane.

The major *independent variables* (those that we can change directly) in the cutting process are:

• Tool material and its condition
• Tool shape, surface finish, and sharpness

(a) Straight turning

(b) Cutting off

**FIGURE 8.1** Examples of cutting processes.

(c) Slab milling

(d) End milling

- Workpiece material, condition, and temperature
- Cutting conditions, such as speed and depth of cut
- Use of a cutting fluid
- The characteristics of the machine tool, such as its stiffness and damping
- Workholding, fixturing, etc.

*Dependent variables* are those that are influenced by changes in the independent variables and are:

- Type of chip produced
- Force and energy dissipated in the cutting process
- Temperature rise in the workpiece, the chip, and the tool
- Wear and failure of the tool
- The surface finish produced on the workpiece after machining

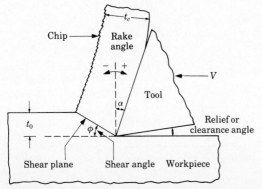

**FIGURE 8.2** Schematic illustration of a two-dimensional cutting process (also called orthogonal cutting).

Ask yourself the following questions. If, for example, the surface finish of the workpiece being cut is poor and unacceptable, which of the independent variables do you change first? The angle of the tool? If so, do you increase it or decrease it? If the tool wears and becomes dull rapidly, do you change the cutting speed, the depth of cut, or the tool material? If the cutting tool begins to vibrate, what should be done to eliminate this vibration? In this section we will describe the mechanics of chip formation in order to establish the effects of various parameters on the overall cutting process.

Although almost all cutting processes are three-dimensional in nature, the model shown in Fig. 8.2 is very useful in studying the basic mechanics of cutting. In this model, known as **orthogonal cutting**, the tool has a **rake angle** of $\alpha$ (positive, as shown in the figure) and a **relief**, or *clearance, angle*. Note that the sum of the rake, relief, and included angles of the tool is $90°$.

Microscopic examinations have revealed that chips are produced by the shearing process shown in Fig. 8.3a, and that shearing takes place along a **shear plane** making an angle $\phi$, called the **shear angle**, with the surface of the workpiece. Below the shear plane the workpiece is undeformed, and above it is the chip, already formed and moving up the face of the tool as cutting progresses. Because of the relative velocity, there is friction between the chip and the rake face of the tool. Note also that this shearing process is like cards in a deck sliding against each other.

The thickness of the chip, $t_c$, can be determined by knowing $t_o$, $\alpha$, and $\phi$. The ratio of $t_o$ to $t_c$ is known as the **cutting ratio**, $r$, which we can express as

$$r = \frac{t_o}{t_c} = \frac{\sin \phi}{\cos (\phi - \alpha)}.$$

(8.1)

The chip thickness is always greater than the depth of cut; hence the value of $r$ is less than unity. The reciprocal of $r$ is known as the *chip compression ratio* and is a measure of how thick the chip has become compared to the depth of cut. Thus the chip compression ratio is always greater than unity.

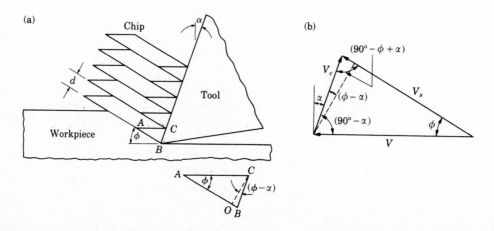

**(a)**

**(b)**

**FIGURE 8.3** (a) Schematic illustration of the basic mechanism of chip formation in cutting. (b) Velocity diagram in the cutting zone.

On the basis of Fig. 8.3a, we can express the **shear strain,** $\gamma$, that the material undergoes as

$$\gamma = \frac{AB}{OC} = \frac{AO}{OC} + \frac{OB}{OC}, \tag{8.2}$$

or

$$\gamma = \cot \phi + \tan (\phi - \alpha). \tag{8.3}$$

Thus large shear strains are associated with low shear angles and low or negative rake angles. Shear strains of 5 or larger have been observed in actual cutting operations. Compared to forming and shaping processes, therefore, the material undergoes greater deformation during cutting. Also, deformation in cutting takes place within a very narrow deformation zone; that is, the dimension $d = OC$ is very small.

From Fig. 8.2 we note that since chip thickness $t_c$ is greater than the depth of cut $t_o$, the *velocity of the chip,* $V_c$, has to be lower than the cutting speed, $V$. Since mass continuity has to be maintained, we have

$$Vt_o = V_c t_c \quad \text{or} \quad V_c = Vr. \tag{8.4}$$

Hence

$$V_c = V \frac{\sin \phi}{\cos (\phi - \alpha)}. \tag{8.5}$$

We can construct a velocity diagram (Fig. 8.3b) and from trigonometric relationships obtain the following relationships:

$$\frac{V}{\cos (\phi - \alpha)} = \frac{V_s}{\cos \alpha} = \frac{V_c}{\sin \phi}, \tag{8.6}$$

where $V_s$ is the velocity at which shearing takes place in the shear plane. The **shear strain rate** is the ratio of $V_s$ to the thickness $d$ of the sheared element (shear zone), or

$$\dot{\gamma} = \frac{V_s}{d}. \tag{8.7}$$

Experimental evidence indicates that $d$ is on the order of $10^{-2}$ to $10^{-3}$ mm ($10^{-3}$ to $10^{-4}$ in). This means that, even at low cutting speeds, the shear strain rate is very high, on the order of $10^3$ to $10^6$/s. A knowledge of the shear strain rate is essential because of its effects on the strength and ductility of the material, and the chip morphology (chip form).

## 8.2.1 Types of chips

When we observe actual chip formation under different metal-cutting conditions, we find significant deviations from the ideal model shown in Figs. 8.2 and 8.3a. Some types of metal chips commonly observed in practice are shown schematically in Fig. 8.4, with micrographs in Fig. 8.5. Because the type of chips produced significantly influences the surface finish produced and the overall cutting operation, we will discuss the type of chips in the following order:

**FIGURE 8.4** Basic types of chips produced in metal cutting: (a) continuous chip with narrow, straight primary shear zone; (b) secondary shear zone at the chip–tool interface; (c) continuous chip with large primary shear zone; (d) continuous chip with built-up edge; (e) segmented or nonhomogeneous chip; and (f) discontinuous chip. *Source:* After M. C. Shaw.

1. Continuous
2. Built-up edge
3. Serrated
4. Discontinuous

Let's first note that a chip has two surfaces: one that is in contact with the tool face (rake face), and the other from the original surface of the workpiece. The tool side of the chip surface is shiny, or *burnished* (Fig. 8.6), which is caused by rubbing of the chip as it climbs up the tool face. The other surface of the chip does not come into contact with any solid body. This surface has a jagged, steplike appearance (Fig. 8.5a), which is caused by the shearing mechanism of chip formation.

**Continuous chips.** *Continuous chips* are usually formed at high cutting speeds and/or high rake angles (Figs. 8.4a and 8.5a). The deformation of the material takes place along a narrow shear zone, called the **primary shear zone**. Continuous chips may develop a **secondary shear zone** at the tool–chip interface (Figs. 8.4b and 8.5b), caused by friction. The secondary zone becomes deeper as tool–chip friction increases.

In continuous chips, deformation may also take place along a wide primary shear zone with curved boundaries (Fig. 8.4c). Note that the lower boundary is below the machined surface, which subjects the machined surface to distortion, as depicted by the distorted vertical lines. This situation occurs particularly in machining soft metals at low speeds and low rake angles. It can produce poor surface finish and induce residual surface stresses, which may be detrimental to the properties of the machined part.

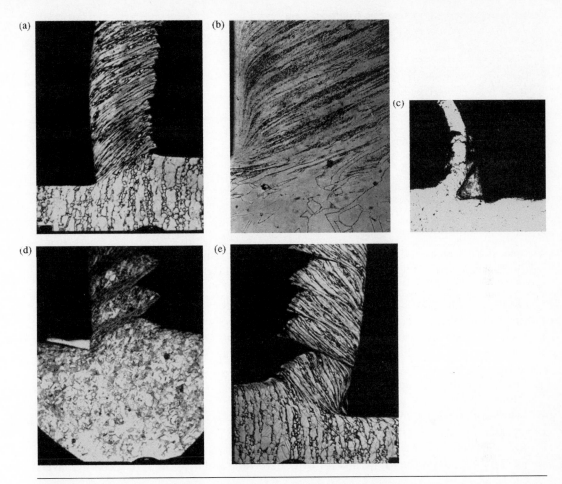

**FIGURE 8.5** Photographs of basic types of chips obtained in cutting metals. (a) Continuous chip in cutting 60–40 brass at 3.4 m/s (670 ft/min) at a depth of 0.2 mm (0.008 in). (b) Secondary shear zone in cutting copper at 1.7 m/s (330 ft/min). (c) Built-up edge in cutting sintered tungsten. (d) Inhomogeneous (serrated) chip in cutting 321 stainless steel at 1.25 m/s (250 ft/min) at a depth of cut of 0.2 mm (0.008 in). (e) Discontinuous chip in cutting 60–40 brass at 0.25 m/s (50 ft/min). Note the small built-up edge at the root of the chip. *Source:* Parts (a), (b), (d), (e) courtesy of P.K. Wright, University of California, Berkeley; part (c) A.J. Moser and S. Kalpakjian.

Although they generally produce good surface finish, continuous chips are not always desirable, particularly in automated machine tools. They tend to get tangled around the toolholder, and the operation has to be stopped to clear away the chips. This problem can be alleviated with chip breakers (see below).

As a result of strain hardening (caused by the shear strains to which it is subjected), the chip usually becomes harder, less ductile, and stronger than the original workpiece material. The increase in hardness and strength of the chip depends on the shear strain (Table 8.1). As

**FIGURE 8.6** Shiny (burnished) surface on the tool side of a continuous chip produced in turning.

the rake angle decreases, the shear strain increases and the chip becomes stronger and harder. With increasing strain, the material tends to behave like a rigid, perfectly plastic body.

**Built-up edge chips.** A *built-up edge* (BUE) may form at the tip of the tool during cutting (Figs. 8.4d and 8.5c). This edge consists of layers of material from the workpiece that are gradually deposited on the tool (hence the term *built-up*). As it becomes larger, the BUE becomes unstable and eventually breaks up. Part of the BUE material is carried away by the tool side of the chip; the rest is deposited randomly on the workpiece surface. The process of BUE formation and destruction is repeated continuously during the cutting operation.

The built-up edge is commonly observed in practice. It is one of the factors that most adversely affects surface finish in cutting, as Figs. 8.5c and 8.7 show. A built-up edge, in effect, changes the geometry of cutting. Note, for example, the large tip radius of the BUE and the rough surface finish produced. Because of work hardening and deposition of successive layers

**TABLE 8.1**

**STRENGTH AND HARDNESS OF CHIPS IN TURNING MILD STEEL**

| Rake angle, $\alpha$ | | 45° | 35° | 27° | 10° | 10° | 10° |
|---|---|---|---|---|---|---|---|
| Feed ($f$); mm/rev | | 0.30 | 0.30 | 0.20 | 0.20 | 0.20 | 0.20 |
| Cutting speed ($V$), m/min | | 50 | 50 | 168 | 168 | 168 | 76 |
| Cutting fluid | | Soluble oil | None | None | Soluble oil | None | None |
| Shear strain, $\gamma$ | | 1.1 | 1.7 | 2.1 | 2.9 | 3.1 | 4.0 |
| Tensile strength (UTS), kg/mm² | 46 | 75 | 84 | 91 | 92 | 93 | 95 |
| Vickers hardness number (HV) | 209 | 272 | 289 | 302 | 320 | 314 | 325 |
| HV/UTS | 4.5 | 3.6 | 3.4 | 3.3 | 3.5 | 3.4 | 3.4 |

Source: After K. Nakayama.

**FIGURE 8.7**   (a) Hardness distribution in the cutting zone for 3115 steel. Note that some regions in the built-up edge are as much as three times harder than the bulk metal. (b) Surface finish in turning 5130 steel with a built-up edge. (c) Surface finish on 1018 steel in face milling. Magnifications: 15×. *Source:* Courtesy of Institute of Advanced Manufacturing Sciences, Inc.

of material, BUE hardness increases significantly (Fig. 8.7a). Although BUE is generally undesirable, a thin, stable BUE is usually regarded as desirable because it protects the tool's surface.

The exact mechanism of formation of the BUE is not yet clearly understood. However, investigations have identified two distinct mechanisms that contribute to formation. One is the adhesion of the workpiece material to the rake face of the tool. The strength of this bond depends on the affinity of the workpiece and tool materials. The other mechanism is the growth of the adhered metal layers to form a BUE.

One of the important factors in forming the BUE is the propensity for strain hardening of the workpiece material; the higher the strain-hardening exponent, the greater the tendency is for BUE formation. Experimental evidence indicates that as the cutting speed increases, the BUE decreases or is eliminated. It is not clear whether the speed, as such, plays a significant role (such as by increasing the strain rate), or whether the effect results from the increase in temperature with speed. The latter appears to contradict the observation that higher temperatures improve adhesion, because as speed increases, temperature increases but BUE decreases. In addition to the factors outlined, the tendency for BUE formation can be reduced by decreasing the depth of cut, increasing the rake angle, using a tool with a small tip radius, and using an effective cutting fluid.

**Serrated chips.** *Serrated chips* (also called **segmented** or *nonhomogeneous* chips) are semicontinuous chips, with zones of low and high shear strain (Figs. 8.4e and 8.5d). Metals with low thermal conductivity and strength that decreases sharply with temperature, such as titanium, exhibit this behavior. The chips have a sawtoothlike appearance.

**Discontinuous chips.** *Discontinuous chips* consist of segments that may be firmly or loosely attached to each other (Figs. 8.4f and 8.5e). Discontinuous chips usually form under the following conditions: (1) brittle workpiece materials, because they do not have the capacity to undergo the high shear strains developed in cutting; (2) materials that contain hard inclusions and impurities; (3) very low or very high cutting speeds; (4) large depths of cut and low rake angles; (5) low stiffness of the machine tool; and (6) lack of an effective cutting fluid.

Impurities and hard particles act as nucleation sites for cracks, thereby producing discontinuous chips. A large depth of cut increases the probability of such defects being present in the cutting zone. Higher cutting speeds mean higher temperatures; hence the material is more ductile and has less tendency to form discontinuous chips. Another factor in discontinuous-chip formation is the magnitude of the compressive stresses on the shear plane. Recall from Fig. 2.22 that the maximum shear strain at fracture increases with increasing compressive stress. If the normal stress is not sufficiently high, the material is unable to undergo the shear strain required to form a continuous chip.

Because of the discontinuous nature of chip formation, forces continually vary during cutting. Consequently, the stiffness of the cutting-tool holder and the machine tool is important in cutting with discontinuous-chip as well as serrated-chip formation. If not stiff enough the machine tool may begin to vibrate and chatter. This, in turn, adversely affects the surface finish and dimensional accuracy of the machined component and may damage or cause excessive wear of the cutting tool.

**Chip formation in nonmetallic materials.** Much of the discussion thus far for metals is also generally applicable to nonmetallic materials. A variety of chips are obtained in cutting thermoplastics, depending on the type of polymer and process parameters such as depth of cut, tool geometry, and cutting speed. Because they are brittle, thermosetting plastics and ceramics generally produce discontinuous chips.

**Chip curl.** *Chip curl* (Figs. 8.6 and 8.8a) is common to all cutting operations with metals and nonmetallic materials such as plastics and wood. The reasons for chip curl are still not clearly understood. Among the possible factors contributing to it are the distribution of stresses in the primary and secondary shear zones, thermal effects, and the work-hardening characteristics of the workpiece material. Process variables also affect chip curl. Generally, the radius of curvature decreases (the chip becomes curlier) with decreasing depth of cut, increasing rake angle, and decreasing friction at the tool–chip interface. The use of cutting fluids and various additives in the workpiece material also influence chip curl.

**Chip breakers.** As we stated earlier, long, continuous chips are undesirable. They tend to become entangled, and interfere with cutting operations, and can be a safety hazard. This

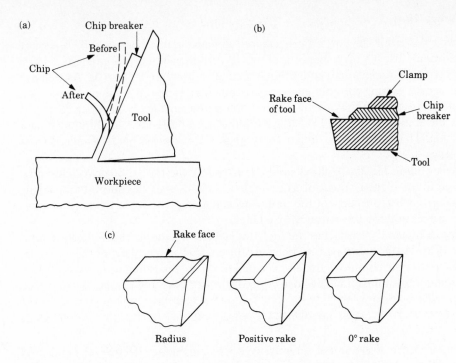

**FIGURE 8.8**
(a) Schematic illustration of the action of a chip breaker. Note that the chip breaker decreases the radius of curvature of the chip. (b) Chip breaker clamped on the rake face of a cutting tool. (c) Grooves in cutting tools acting as chip breakers.

situation is especially troublesome in high-speed automated machinery. The usual procedure to avoid it is to break the chip intermittently with a *chip breaker*. The chip breaker can be a piece of metal clamped to the rake face of the tool (Fig. 8.8b), or it can be an integral part of the tool (Fig. 8.8c).

Chips can also be broken by changing the tool geometry, thus controlling chip flow, as in the turning operations shown in Fig. 8.9. Various cutting tools with chip breaker features are available. In interrupted cutting operations, such as milling, chip breakers are generally not necessary, since the chips already have finite lengths resulting from the intermittent nature of the operation.

**FIGURE 8.9** Various chips produced in turning: (a) tightly curled chip; (b) chip hits workpiece and breaks; (c) continuous chip moving away from workpiece; and (d) chip hits tool shank and breaks off. *Source:* G. Boothroyd, *Fundamentals of Metal Machining and Machine Tools.*

### 8.2.2 Oblique cutting

Thus far we have described the cutting process two-dimensionally. However, the majority of cutting operations involve tool shapes that are three-dimensional (**oblique**). The basic difference between two-dimensional and oblique cutting is shown in Fig. 8.10a. As we have shown, in orthogonal cutting the tool edge is perpendicular to the movement of the tool, and the chip slides directly up the face of the tool. In oblique cutting, the cutting edge is at an angle $i$, the **inclination angle** (Fig. 8.10b). Note the lateral direction of chip movement in oblique cutting. This situation is similar to an angled snow-plow blade, which throws the snow sideways.

Note that the chip in Fig. 8.10a flows up the rake face of the tool at angle $\alpha_c$ (**chip flow angle**), measured in the plane of the tool face. Angle $\alpha_n$ is known as the **normal rake angle**, which is a basic geometric property of the tool. This is the angle between the normal $oz$ to the workpiece surface and the line $oa$ on the tool face.

The workpiece material approaches the tool at a velocity $V$, and leaves the surface (as a chip) with a velocity $V_c$. We calculate the effective rake angle $\alpha_e$ in the plane of these two velocities. Assuming that the chip flow angle $\alpha_c$ is equal to the inclination angle $i$ (which is experimentally found to be approximately correct), the effective rake angle $\alpha_e$ is

$$\alpha_e = \sin^{-1}(\sin^2 i + \cos^2 i \sin \alpha_n). \tag{8.8}$$

Since we can measure both $i$ and $\alpha_n$ directly, we can calculate the effective rake angle. As $i$ increases, the effective rake angle increases, and the chip becomes thinner and longer. The effect of the inclination angle on chip shape is shown in Fig. 8.10c.

A typical single-point turning tool used on a lathe is shown in Fig. 8.11. Note the various angles involved, each of which has to be selected properly for efficient cutting. We discuss various three-dimensional cutting tools in greater detail in Sections 8.8 and 8.9. These include tools for drilling, tapping, milling, planing, shaping, broaching, sawing, and filing.

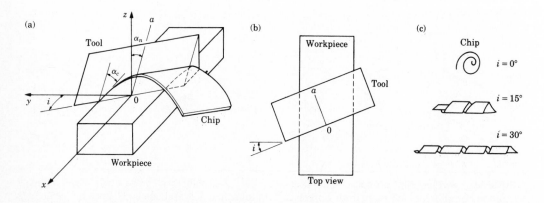

**FIGURE 8.10**   (a) Schematic illustration of cutting with an oblique tool. (b) Top view showing the inclination angle $i$. (c) Types of chips produced with different inclination angles.

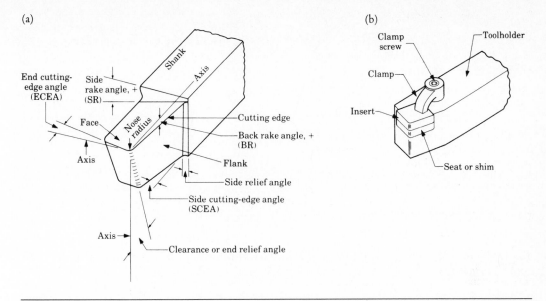

**FIGURE 8.11** (a) Schematic illustration of a right-hand cutting tool. Although these tools have traditionally been produced from solid tool-steel bars, they have been largely replaced by carbide or other inserts of various shapes and sizes, as shown in (b).

**Shaving and skiving.** Thin layers of material can be removed from straight or curved surfaces by a process similar to the use of a plane to shave wood. *Shaving* is particularly useful in improving the surface finish and dimensional accuracy of sheared parts and punched slugs. Another application of shaving is in finishing gears with a cutter that has the shape of the gear tooth. Parts that are long or have a combination of shapes are shaved by *skiving*. The skiving tool moves tangentially across the length of the workpiece.

### 8.2.3 Cutting forces

Knowledge of the forces and power involved in cutting operations is important for the following reasons:

- Power requirements have to be determined so that a motor of suitable capacity can be installed in the machine tool.

- Data on forces are necessary for the proper design of machine tools for cutting operations that avoid excessive distortion of the machine elements and maintain desired tolerances for the machined part.

- Whether the workpiece can withstand the cutting forces without excessive distortion has to be determined in advance.

The forces acting on the tool in orthogonal cutting are shown in Fig. 8.12. The **cutting force**, $F_c$, acts in the direction of the cutting speed $V$ and supplies the energy required for cutting. The **thrust force**, $F_t$, acts in the direction normal to the cutting velocity, that is, perpendicular to the workpiece. These two forces produce the **resultant force**, $R$. Note that the resultant force can be

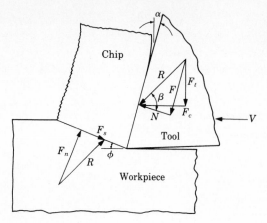

**FIGURE 8.12**    Forces acting on a cutting tool in two-dimensional cutting.

resolved into two components on the tool face: a **friction force**, $F$, along the tool–chip interface, and a **normal force**, $N$, perpendicular to it. From Fig. 8.12, we can show the force $F$ to be

$$F = R \sin \beta, \tag{8.9}$$

and the force $N$ as

$$N = R \cos \beta. \tag{8.10}$$

Note that the resultant force is balanced by an equal and opposite force along the shear plane and is resolved into a *shear force, $F_s$,* and a *normal force, $F_n$.*

The ratio of $F$ to $N$ is the *coefficient of friction, $\mu$,* at the tool–chip interface, and the angle $\beta$ is the **friction angle**. We can express $\mu$ as

$$\mu = \tan \beta = \frac{F_t + F_c \tan \alpha}{F_c - F_t \tan \alpha}. \tag{8.11}$$

The coefficient of friction in metal cutting generally ranges from about 0.5 to 2.0, thus indicating that the chip encounters considerable frictional resistance while climbing up the face of the tool.

Although the magnitude of forces in actual cutting operations is generally on the order of a few hundred newtons, the local stresses in the cutting zone and the pressures on the tool are very high because the contact areas are very small. The chip–tool contact length (Fig. 8.2), for example, is typically on the order of 1 mm (0.04 in.). Thus the tool is subjected to very high stresses, which lead to wear and sometimes chipping and fracture of the tool.

**Thrust force.**   Knowing the thrust force in cutting is important. The tool holder and the machine tool must be stiff enough to minimize deflections caused by this force. For example, if the thrust force is too high or if the machine tool is not sufficiently stiff, the tool will be pushed away from the surface being machined. This movement will, in turn, reduce the depth of cut, causing lack of dimensional accuracy in the machined part.

Referring to Fig. 8.12, note that the cutting force must always be in the direction of cutting in order to supply energy to the system. Although the thrust force does not contribute to the total work done, knowing its magnitude is important. For instance, if the tool holder is not sufficiently stiff, the thrust force will simply deflect the tool and reduce the depth of cut. Thus, tolerances cannot be held.

Also note from Fig. 8.12 that the thrust force is downward. We will now show that this force can act upward. First, we express the thrust force as

$$F_t = R \sin (\beta - \alpha) \tag{8.12}$$

or

$$F_t = F_c \tan (\beta - \alpha). \tag{8.13}$$

Because the magnitude of $F_c$ is always positive (say, as shown in Fig. 8.12), the sign of $F_t$ can be either positive or negative, depending on the relative magnitudes of $\beta$ and $\alpha$. When $\beta > \alpha$, the sign of $F_t$ is positive (downward); when $\beta < \alpha$, it is negative (upward). Thus an upward thrust force at high rake angles and/or with low friction at the tool–chip interface is possible. We can visualize the situation in Fig. 8.12: When $\mu = 0$, then $\beta = 0$ and the resultant force $R$ coincides with the force $N$. In this case, $R$ will have an upward thrust force component. Also note that for $\alpha = 0$ and $\beta = 0$, the thrust force is zero.

These observations have been verified experimentally, as shown in Fig. 8.13. The influence of the depth of cut is obvious: As $t_o$ increases, $R$ must also increase so that $F_c$ will also increase. This action supplies the additional energy required to remove the extra material produced by the increased depth of cut. The change in direction and magnitude of

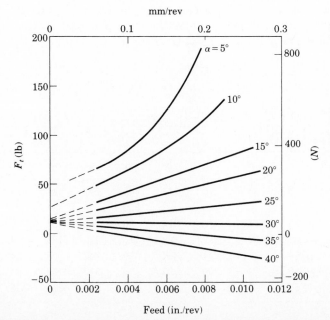

**FIGURE 8.13**  Thrust force as a function of rake angle and feed in orthogonal cutting of AISI 1112 cold-rolled steel. Note that at high rake angles the thrust force is negative. A negative thrust force has important implications in the design of machine tools and in controlling the stability of the cutting process. *Source:* S. Kobayashi and E. G. Thomsen, *J. Eng. Ind.*, vol. 81, 1959, pp. 251–262.

**TABLE 8.2**

**ORTHOGONAL CUTTING DATA FOR 4130 STEEL***

| $\alpha$ | $\phi$ | $\gamma$ | $\mu$ | $\beta$ | $F_c$ (lb) | $F_t$ (lb) | $u_t\left(\dfrac{\text{in.} - \text{lb}}{\text{in}^3} \times 10^3\right)$ | $u_s$ | $u_f$ | $\dfrac{u_f}{u_t}$ (%) |
|---|---|---|---|---|---|---|---|---|---|---|
| 25° | 20.9° | 2.55 | 1.46 | 56 | 380 | 224 | 320 | 209 | 111 | 35 |
| 35 | 31.6 | 1.56 | 1.53 | 57 | 254 | 102 | 214 | 112 | 102 | 48 |
| 40 | 35.7 | 1.32 | 1.54 | 57 | 232 | 71 | 195 | 94 | 101 | 52 |
| 45 | 41.9 | 1.06 | 1.83 | 62 | 232 | 68 | 195 | 75 | 120 | 62 |

*$t_0 = 0.0025$ in.; $w = 0.475$ in.; $V = 90$ ft/min; tool: high-speed steel.
Source: After E. G. Thomsen.

the thrust force can play a significant role. Within a certain range of operating conditions, it leads to instability problems in machining, particularly if the machine tool is not stiff enough. However, such high rake angles are rarely used in machining most metals.

**Observations on forces in the cutting zone.** In addition to being a function of the strength of the workpiece material, forces in cutting are also influenced by other variables. Extensive data are available, and one set, Table 8.2, shows that the cutting force $F_c$ increases with increasing depth of cut, decreasing rake angle, and decreasing speed. From analysis of the data in Table 8.3, the effect of cutting speed can be attributed to the fact that as speed decreases, the shear angle decreases and the friction coefficient increases. Both effects increase the cutting force.

Another factor that can significantly influence the cutting force is the tip radius of the tool. The larger the radius (the duller the tool), the greater is the force. Experimental evidence indicates that, for depths of cut on the order of five times the tip radius or higher, the effect of dullness on the cutting forces is negligible. (See also Section 8.4).

**Stresses.** We can analyze the stresses in the shear plane and at the tool–chip interface by assuming that they are uniformly distributed. The forces in the shear plane can be resolved into shear and normal forces and stresses. The average *shear stress* in the shear plane is

$$\tau = \frac{F_s}{A_s} \tag{8.14}$$

**TABLE 8.3**

**ORTHOGONAL CUTTING DATA FOR 9445 STEEL***

| $\alpha$ | V (ft/min) | $\phi$ | $\gamma$ | $\mu$ | $\beta$ | $F_c$ | $F_t$ | $u_t$ | $u_s$ | $u_f$ | $\dfrac{u_f}{u_t}$ (%) |
|---|---|---|---|---|---|---|---|---|---|---|---|
| +10 | 197 | 17 | 3.4 | 1.05 | 46 | 370 | 273 | 400 | 292 | 108 | 27 |
| | 400 | 19 | 3.1 | 1.11 | 48 | 360 | 283 | 390 | 266 | 124 | 32 |
| | 642 | 21.5 | 2.7 | 0.95 | 44 | 329 | 217 | 356 | 249 | 107 | 30 |
| | 1186 | 25 | 2.4 | 0.81 | 39 | 303 | 168 | 328 | 225 | 103 | 31 |
| -10 | 400 | 16.5 | 3.9 | 0.64 | 33 | 416 | 385 | 450 | 342 | 108 | 24 |
| | 637 | 19 | 3.5 | 0.58 | 30 | 384 | 326 | 415 | 312 | 103 | 25 |
| | 1160 | 22 | 3.1 | 0.51 | 27 | 356 | 263 | 385 | 289 | 96 | 25 |

*$t_0 = 0.037$ in.; $W = 0.25$ in.; tool: cemented carbide.
Source: After M. E. Merchant.

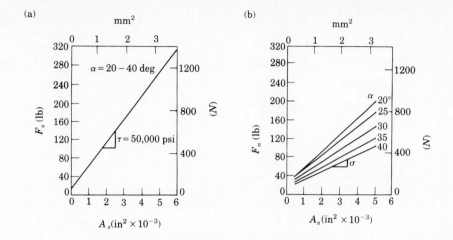

**FIGURE 8.14** Shear force and normal force as a function of shear plane area and rake angle for 85–15 brass. Note that the shear stress in the shear plane is constant, regardless of the magnitude of the normal stress. Thus normal stress has no effect on the shear flow stress of the material. *Source:* After S. Kobayashi and E. G. Thomsen, *J. Eng. Ind.,* vol. 81, 1959, pp. 251–262.

and the average *normal stress* is

$$\sigma = \frac{F_n}{A_s},$$ (8.15)

where $A_s$ is the area of the shear plane, namely,

$$A_s = \frac{wt_o}{\sin \phi}.$$

Figure 8.14 gives some data pertaining to these average stresses. The rake angle is a parameter, and the shear plane area is increased by increasing the depth of cut. We can draw the following conclusions from these curves:

1. The shear stress on the shear plane is independent of the rake angle.
2. The normal stress on the shear plane decreases with increasing rake angle.
3. Consequently, the normal stress in the shear plane has no effect on the magnitude of the shear stress. This phenomenon has also been verified by other mechanical tests. (However, normal stress strongly influences the magnitude of the shear strain in the shear zone. The maximum shear strain to fracture increases with the normal compressive stress.)

Determining the stresses on the rake face of the tool presents considerable difficulties. One problem is accurately determining the length of contact at the tool–chip interface. This length increases with decreasing shear angle, indicating that the contact length is a function of rake angle, cutting speed, and friction at the tool–chip interface. Another problem is that the stresses are not uniformly distributed on the rake face.

Photoelastic studies have determined that the actual stress distribution is qualitatively shown in Fig. 8.15. Note that the stress normal to the tool face is maximal at its tip and decreases rapidly toward the end of the contact length. The shear stress has a similar trend,

Stresses on tool face

Tool tip

**FIGURE 8.15** Schematic illustration of the distribution of normal and shear stresses at the tool–chip interface (rake face). Note that, whereas the normal stress increases continuously toward the tip of the tool, the shear stress reaches a maximum and remains at that value (known as *sticking*; see Section 4.4.1).

except that it levels off at approximately the center of the contact length. This behavior indicates that *sticking* is taking place, whereby the shear stress has reached the shear yield strength of the material. Such a sticking region has been observed on some chips. (Sticking regions also can be observed in various other metal-forming processes.)

### 8.2.4 Shear-angle relationships

Because the shear angle and the shear zone have great significance in the mechanics of cutting, a great deal of effort has been expended to determine the relationship of the shear angle to material properties and process variables. One of the earliest analyses (by M. E. Merchant) is based on the assumption that the shear angle adjusts itself so that the cutting force is a minimum, or so that the maximum shear stress occurs in the shear plane. From the force diagram in Fig. 8.12, the following relationships can be obtained:

$$F_c = R \cos (\beta - \alpha) \tag{8.16}$$

and

$$F_s = R \cos (\phi + \beta - \alpha). \tag{8.17}$$

The area of the shear plane is

$$A_s = \frac{w t_o}{\sin \phi}, \tag{8.18}$$

where $w$ is the width of the cut. Thus the shear stress in the shear plane is

$$\tau = \frac{F_s}{A_s} = \frac{F_c \sec (\beta - \alpha) \cos (\phi + \beta - \alpha) \sin \phi}{w t_o}. \tag{8.19}$$

If we assume that $\beta$ is independent of $\phi$, we can find the shear angle corresponding to the maximum shear stress by differentiating Eq. (8.19) with respect to $\phi$ and equating it to zero:

$$\frac{d\tau}{d\phi} = \cos (\phi + \beta - \alpha) \cos \phi - \sin (\phi + \beta - \alpha) \sin \phi = 0.$$

Thus

$$\tan (\phi + \beta - \alpha) = \cot \phi = \tan (90° - \phi)$$

or

$$\phi = 45° + \frac{\alpha}{2} - \frac{\beta}{2}. \tag{8.20}$$

Note that Eq. (8.20) indicates that, as the rake angle decreases and/or as the friction at the tool–chip interface increases, the shear angle decreases and the chip is thus thicker. This result is to be expected, because decreasing $\alpha$ and increasing $\beta$ tend to lower the shear angle.

A second method of determining $\phi$ is based on slip-line analysis (by E. H. Lee and B. W. Shaffer) and on the following assumptions.

1.   The shear plane $AB$ in Fig. 8.16a is a plane of maximum shear stress.
2.   There are no shear or normal stresses along $AC$, as there should not be since the chip beyond $AC$ is unconstrained.
3.   The material is rigid, perfectly plastic.

The appropriate Mohr's circle construction for this shear zone is shown in Fig. 8.16b. The angle $\eta$ is given by

$$\eta = 45° - \beta.$$

From geometry,

$$\eta = \phi - \alpha,$$

**FIGURE 8.16**   (a) Slip-line field in orthogonal cutting. Line $AB$ is the shear plane along which the shear stress is assumed to be a maximum. (b) Mohr's circle for stresses in the cutting zone. Point $CB$ on the circle represents the normal and shear stresses at the tool–chip interface, respectively.

**FIGURE 8.17**    Comparison of experimental and theoretical shear-angle relationships. More recent analytical studies have resulted in better agreement with experimental data.

so the expression for the shear angle is

$$\phi = 45° + \alpha - \beta. \tag{8.21}$$

Note that this expression is similar to Eq. (8.20) and indicates the same trends, although numerically it gives different values.

In another study (by T. Sata and M. Mizuno), the following simple relationships were proposed and could serve as a guide for estimating the shear angle for practical purposes:

$$\phi = \alpha \quad \text{for} \quad \alpha > 15°, \tag{8.22}$$

$$\phi = 15° \quad \text{for} \quad \alpha < 15°. \tag{8.23}$$

A number of other expressions based on various models and different assumptions have been obtained for the shear angle. However, many of these expressions do not agree well with experimental data over a wide range of conditions (Fig. 8.17), largely because shear rarely occurs in a thin plane. However, the shear angle always decreases with increasing $(\beta - \alpha)$ as shown in Fig. 8.18. Recent and more comprehensive studies appear to accurately predict the shear angle analytically, especially for continuous chips.

**FIGURE 8.18**    Relation between the shear angle and friction angle for various alloys and cutting speeds. *Source:* S. Kobayashi, et al., *Trans. ASME, J. Eng. Ind.,* vol. 82, 1960, pp. 333–347.

### 8.2.5 Specific energy

Refer to Fig. 8.12 and note that the **total power** input in cutting is

$$\text{Power} = F_c V.$$

If we let the width of the cut be $w$, then the **total energy per unit volume** of material removed (**specific energy**), $u_t$, is

$$u_t = \frac{F_c V}{w t_o V} = \frac{F_c}{w t_o}. \tag{8.24}$$

In other words, $u_t$ is simply the ratio of the cutting force to the projected area of the cut. Figures 8.3 and 8.12 also show that the power required to overcome friction at the tool–chip interface is the product of $F$ and $V_c$ or in terms of frictional specific energy, $u_f$,

$$u_f = \frac{F V_c}{w t_o V} = \frac{F r}{w t_o} = \frac{(F_c \sin \alpha + F_t \cos \alpha) r}{w t_o}. \tag{8.25}$$

Likewise, the power required for shearing along the shear plane is the product of $F_s$ and $V_s$. Hence the specific energy for shear, $u_s$, is

$$u_s = \frac{F_s V_s}{w t_o V}. \tag{8.26}$$

The total specific energy, $u_t$, is the sum of the two, or

$$u_t = u_f + u_s. \tag{8.27}$$

There are two other sources of energy in cutting. One is the *surface energy* resulting from the formation of new surfaces when a layer of material is removed by cutting. However, this source represents a very small amount of energy compared to the shear and frictional energies involved. The other source is the energy associated with the *momentum change* as the metal crosses the shear plane. (This source is similar to the forces involved in a turbine blade from momentum changes of the fluid or gas.) Although in ordinary metal-cutting operations the momentum energy is negligible, this energy can be significant at very high cutting speeds [above 125 m/s (25,000 ft/min) or so].

Experimental data on specific energies are given in Tables 8.2 and 8.3. Note that, as the rake angle increases, the frictional specific energy remains more or less constant, whereas the shear specific energy is reduced rapidly. Thus the ratio $u_f / u_t$ increases considerably as $\alpha$ increases. This trend can also be predicted by obtaining an expression for the ratio as follows:

$$\frac{u_f}{u_t} = \frac{F V_c}{F_c V} = \frac{R \sin \beta}{R \cos (\beta - \alpha)} \cdot \frac{V r}{V} = \frac{\sin \beta}{\cos (\beta - \alpha)} \cdot \frac{\sin \phi}{\cos (\phi - \alpha)}. \tag{8.28}$$

Experimental observations have revealed that as $\alpha$ increases, both $\beta$ and $\phi$ increase. Thus inspection of Eq. (8.28) indicates that the ratio $u_f / u_t$ should also increase with $\alpha$. Obviously, $u_f$ and $u_s$ are related. Although $u_f$ is not required for the cutting action to take place, it affects the magnitude of $u_s$. The reason is that, as friction increases, the shear angle decreases. A decreasing shear angle, in turn, increases the magnitude of $u_s$.

**TABLE 8.4**

**APPROXIMATE ENERGY REQUIREMENTS IN CUTTING OPERATIONS (at drive motor, corrected for 80% efficiency; multiply by 1.25 for dull tools).**

| MATERIAL | SPECIFIC ENERGY | |
|---|---|---|
| | $W \cdot s/mm^3$ | $hp \cdot min/in.^3$ |
| Aluminum alloys | 0.4–1.1 | 0.15–0.4 |
| Cast irons | 1.6–5.5 | 0.6–2.0 |
| Copper alloys | 1.4–3.3 | 0.5–1.2 |
| High-temperature alloys | 3.3–8.5 | 1.2–3.1 |
| Magnesium alloys | 0.4–0.6 | 0.15–0.2 |
| Nickel alloys | 4.9–6.8 | 1.8–2.5 |
| Refractory alloys | 3.8–9.6 | 1.1–3.5 |
| Stainless steels | 3.0–5.2 | 1.1–1.9 |
| Steels | 2.7–9.3 | 1.0–3.4 |
| Titanium alloys | 3.0–4.1 | 1.1–1.5 |

The calculation of all the parameters involved in these specific energies in cutting presents considerable difficulties. Good theoretical computations are available, but they are difficult to perform. The reliable prediction of cutting forces and energies is therefore still based largely on experimental data (Table 8.4). These data should serve as a useful guide. The wide range of values in Table 8.4 can be attributed to differences in strength within each material group and other variables, such as friction and operating conditions.

---

**Example 8.1:** Frictional energy in cutting

We are carrying out an orthogonal cutting process in which $t_o$ = 0.005 in., $V$ = 400 ft/min, $\alpha$ = 10°, and the width of cut = 0.25 in. We observe that $t_c$ = 0.009 in., $F_c$ = 125 lb, and $F_t$ = 50 lb. We want to calculate the percentage of the total energy that goes into overcoming friction at the tool–chip interface.

SOLUTION. We can express the percentage as:

$$\frac{\text{Friction energy}}{\text{Total energy}} = \frac{FV_c}{F_c V} = \frac{Fr}{F_c},$$

where

$$r = \frac{t_o}{t_c} = \frac{5}{9} = 0.555, \qquad F = R \sin \beta, \qquad F_c = R \cos(\beta - \alpha),$$

and

$$R = \sqrt{F_t^2 + F_c^2} = \sqrt{50^2 + 125^2} = 135 \text{ lb.}$$

Thus

$$125 = 135 \cos(\beta - 10),$$

from which

$$\beta = 32° \text{ and } F = 135 \sin 32° = 71.5 \text{ lb.}$$

Hence

$$\text{Percentage} = \frac{(71.5)(0.555)}{125} = 0.32 = 32\%.$$

---

**Example 8.2:  Comparison of forming and machining energies**

You are given two pieces of annealed 304 stainless steel rods, each 0.500 in. in diameter and 6 in. long. You are asked to reduce the diameters to 0.480 in., one piece by pulling it in tension and the other by machining it on a lathe in one pass. Calculate the respective amounts of work involved and explain the difference in the energies dissipated.

SOLUTION.   The work done in pulling the rod is

$$W_{\text{tension}} = (u)(\text{volume}),$$

where

$$u = \int_0^{\varepsilon_1} \sigma \, d\varepsilon.$$

The strain is found from

$$\varepsilon_1 = \ln\left(\frac{0.500}{0.480}\right)^2 = 0.0816.$$

From Table 2.3, the following values for $K$ and $n$ are obtained for this material:

$$K = 1275 \text{ MPa} = 185,000 \text{ psi} \quad \text{and} \quad n = 0.45.$$

Thus

$$u = \frac{K\varepsilon_1^{n+1}}{n+1} = \frac{(185,000)(0.0816)^{1.45}}{1.45} = 3370 \text{ in.} \cdot \text{lb/in}^3$$

and

$$W_{\text{tension}} = (3370)(\pi)(0.25)^2(6) = 3970 \text{ in.} \cdot \text{lb.}$$

From Table 8.4, an average value for the specific energy in machining stainless steels is obtained as 1.5 hp $\cdot$ min/in$^3$. The volume of material machined is

$$\frac{\pi}{4}\left[(0.5)^2 - (0.480)^2\right](6) = 0.092 \text{ in}^3.$$

The specific energy, in appropriate units, is

$$\text{Specific energy} = (1.5)(33,000)(12) = 594,000 \text{ in.} \cdot \text{lb/in}^3.$$

Thus, the machining work is

$$W_{\text{mach}} = (594,000)(0.092) = 54,650 \text{ in.} \cdot \text{lb}.$$

Note that the work in machining is about 14 times that for tension. The difference between the energies is that tension requires very little straining and involves no friction. Machining involves friction, and the material removed—even though small in volume—has undergone much higher strains than the bulk material in tension. Assuming an average shear strain of 3 (from Tables 8.2 and 8.3), which is equivalent to an effective strain of 1.7 (Eq. 2.58), we note that in machining the material is subjected to a strain 21 times that in tension. These differences explain why machining consumes much more energy than reducing the diameter of this rod by stretching. However, as the diameter of the rod decreases, the difference between the two energies becomes smaller, assuming that the same depth of material is to be removed.

### 8.2.6  Temperature

As in all metalworking operations, the energy dissipated in cutting operations is converted into heat, which in turn, raises the temperature in the cutting zone. Knowledge of the temperature rise in cutting is important because it

- adversely affects the strength, hardness, and wear resistance of the cutting tool;
- causes dimensional changes in the part being machined, making control of dimensional accuracy difficult; and
- can induce thermal damage to the machined surface, adversely affecting its properties.

Because of the work done in shearing and in overcoming friction on the rake face of the tool, the main sources of heat generation are the primary shear zone and the tool–chip interface. Additionally, if the tool is dull or worn, heat is also generated by the tool tip rubbing against the machined surface.

Various studies have been made of temperatures in cutting, based on heat transfer and dimensional analysis using experimental data. Although Fig. 8.19 shows that there are considerable temperature gradients in the cutting zone, a simple but approximate expression for the mean temperature for orthogonal cutting is

$$T = \frac{1.2Y_f}{\rho c} \sqrt[3]{\frac{Vt_o}{K}}, \tag{8.29}$$

where $T$ is the mean temperature of the tool–chip interface (°F); $Y_f$ is the flow stress of the material (psi); $V$ is the cutting speed (in./s); $t_o$ is the depth of cut (in.); $\rho c$ is the volumetric specific heat of the workpiece (in. $\cdot$ lb/in$^3$ $\cdot$ °F); and $K$ is the thermal diffusivity of the workpiece (in$^2$/s). Thermal diffusivity is the ratio of thermal conductivity to volumetric specific heat.

**FIGURE 8.19** Typical temperature distribution in the cutting zone. Note that the maximum temperature is about halfway up the face of the tool, and that there is a steep temperature gradient across the thickness of the chip. Some chips may become red hot, causing safety hazards to the operator and thus requiring safety guards. *Source:* After G. Vieregge, *Werkstatt und Betrieb,* No. 11, 1953, p. 696.

Equation (8.29) indicates that temperature increases with the strength of the workpiece material, cutting speed, and depth of cut. Also, the lower the thermal conductivity of the workpiece material, the higher the temperature. Thermal properties of the tool are relatively unimportant compared to those of the workpiece. Because some of the parameters in Eq. (8.29) depend on temperature, using appropriate values that are compatible with the predicted temperature range is important.

Based on Eq. (8.29), an expression for the mean temperature $T$ in turning (on a lathe) is

$$T \propto V^a f^b, \tag{8.30}$$

where $a$ and $b$ are constants, $V$ is the cutting speed, and $f$ is the feed of the tool (Fig. 8.20).

Approximate values for $a$ and $b$ are

| TOOL | $a$ | $b$ |
|---|---|---|
| Carbide | 0.2 | 0.125 |
| High-speed steel | 0.5 | 0.375 |

**FIGURE 8.20** Terminology in a turning operation on a lathe, where $f$ is the feed (in./rev or mm/rev) and $d$ is the depth of cut. Note that feed in turning is equivalent to the depth of cut in orthogonal cutting (Fig. 8.2), and the depth of cut in turning is equivalent to width of cut in orthogonal cutting.

**FIGURE 8.21** Temperature distribution in turning: (a) flank temperature for tool shape (see Fig. 8.52); (b) tool-chip interface temperature. Note that the rake face temperature is higher than at the flank surface. *Source:* After B. T. Chao and K. J. Trigger, *J. Eng. Ind.,* vol. 83, 1961, pp. 496–504.

The depth of cut has negligible influence on the mean temperature for depths exceeding twice the tip radius of the tool. Results from one experimental measurement of temperature (using thermocouples) are shown in Fig. 8.21. Note that the maximum temperature is at a point away from the tip of the tool and that it increases with cutting speed.

The temperature generated in the shear plane is a function of the specific energy for shear, $u_s$, and the specific heat of the material. Hence temperature rise is highest in cutting materials with high strength and low specific heat, as Eq. (2.65) shows. The temperature rise at the tool–chip interface is a function of the coefficient of friction. Flank wear (see Section 8.3) is also a source of heat because of rubbing of the tool on the machined surface.

Cutting speed has a major influence on temperature. As the speed increases, there is little time for the heat to be dissipated and hence temperature rises. The chip is a good heat sink, in that it carries away most of the heat generated. (This action of the chip is similar to *ablation,*

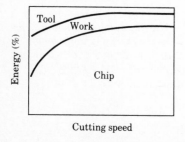

**FIGURE 8.22** Typical energy distribution as a function of cutting speed. Note that most of the cutting energy is carried away by the chip (in the form of heat), particularly as speed increases. For dimensional accuracy during cutting, it is important not to allow the workpiece temperature to rise significantly.

where layers of metal melt away from a surface, thus carrying away the heat.) As cutting speed increases, a larger proportion of the heat is carried away by the chip, as shown in Fig. 8.22.

**Techniques for measuring temperature.** Temperatures and their distribution in the cutting zone may be determined from thermocouples embedded in the tool and/or the workpiece. This technique has been used successfully, although it involves considerable effort. A simpler technique for determining the average temperature is by the thermal emf (electromotive force) at the tool–chip interface, which acts as a hot junction between two different (tool and chip) materials. Infrared radiation from the cutting zone may also be monitored with a radiation pyrometer. However, this technique indicates only surface temperatures, and the accuracy of the results depends on the emissivity of the surfaces, which is difficult to determine accurately.

## 8.3   Tool Wear

We have shown that tool surfaces are subjected to forces, temperature, and sliding—all conditions that induce wear. Because of its effects on the quality of the machined surface and the economics of machining, tool wear is one of the most important and complex aspects of machining operations. Although cutting speed is an independent variable, the forces and temperatures generated are dependent variables and are functions of numerous parameters. Similarly, wear depends on tool and workpiece materials (their physical, mechanical, and chemical properties), tool geometry, cutting fluid properties, and various other operating parameters. The types of wear on a tool depend on the relative roles of these variables. Analytical studies of tool wear present considerable difficulties, and therefore our knowledge of wear is based largely on experimental data.

The basic wear behavior of a tool is shown in Fig. 8.23 (for a two-dimensional cut) and Figs. 8.24 and 8.25 (for a three-dimensional cut, where the position of the tool is shown in Fig. 8.20). The various regions of wear are identified as *flank wear, crater wear, nose wear,* and *chipping* of the cutting edge.

The tool profile can be altered by these various wear and fracture processes, and this change is bound to influence the cutting operation. In addition to these wear processes, plastic deformation of the tool can also take place to some extent, such as in softer tools at elevated temperatures. Gross chipping of the tool is called *catastrophic failure,* whereas wear is generally a gradual process. Because of the complex interactions between material and process variables, the interpretation of these wear patterns and study of the wear mechanisms must be carried out cautiously.

**FIGURE 8.23** Schematic illustration of crater and flank wear on a cutting tool. Wear of tools is one of the most important aspects in cutting and has a major influence on the economics of the overall machining operation.

**FIGURE 8.24** (a) Types of wear observed in cutting tools. The thermal cracks shown are usually observed in interrupted cutting operations, such as in milling. (b) Catastrophic failure of tools. (c) Features of tool wear in a turning operation. The *VB* indicates average flank wear. *Source:* (a) and (b) After V. C. Venkatesh. (c) International Organization for Standardization (ISO).

### 8.3.1   Flank wear

*Flank* wear has been studied extensively and is generally attributed to:

1.   Sliding of the tool along the machined surface causing adhesive and/or abrasive wear, depending on the materials involved; and

2.   Temperature, because of its influence on tool material properties.

(a)          (b)

**FIGURE 8.25** (a) Crater wear on rake face (top surface) and flank wear (side front view) on a carbide insert used in cutting sintered tungsten on a lathe. (b) Crater wear and chipping (lower left side) on a carbide insert used in machining tungsten. *Source:* A. J. Moser and S. Kalpakjian.

Following a classic and extensive study by F. W. Taylor (published in 1907), the following relationship was established for cutting various steels:

$$VT^n = C, \tag{8.31}$$

where $V$ is the cutting speed, $T$ is the time that it takes to develop a flank wear land ($VB$ in Fig. 8.24c) of certain dimensions, $n$ is an exponent that depends on cutting conditions, and $C$ is a constant. Equation (8.31) is a simple version of the many relationships developed by Taylor among the variables involved. Each combination of workpiece and tool material and each cutting condition has its own $n$ value and a different constant $C$.

**Tool-life curves.** *Tool-life curves* are plots of experimental data obtained in cutting tests (Fig. 8.26). Note the rapid decrease in tool life as cutting speed increases and the strong influence of the condition of the workpiece material on tool life. Also note the large difference in tool life for different workpiece microstructures. Heat treatment is important largely because of increasing workpiece hardness. For example, ferrite has a hardness of about 100 HB, pearlite 200 HB, and martensite 300–500 HB. Impurities and hard constituents in the material are also important considerations because they reduce tool life by their abrasive action.

We usually plot tool-life curves on log–log paper, from which we can easily determine the exponent $n$ (Fig. 8.27). The range of $n$ values that have been determined experimentally is given in Table 8.5. These curves are usually linear over a certain range of cutting speeds but are rarely so over a wide range. Moreover, the exponent $n$ can indeed become negative at low cutting speeds. Thus tool-life curves may actually reach a maximum and then curve downward. Therefore, caution should be exercised when using tool-life equations beyond the range of cutting speeds for which they are applicable.

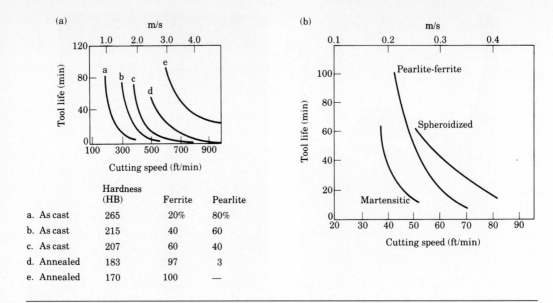

**FIGURE 8.26** Effect of workpiece microstructure on tool life in turning. Tool life is given in terms of the time (in minutes) required to reach a flank wear land of a specified dimension. (a) Ductile cast iron. (b) Steels, with identical hardness. Note the rapid decrease in tool life as the cutting speed increases.

Because of the influence of temperature on the physical and mechanical properties of materials, we would expect that wear is strongly influenced by temperature. Experimental investigations have shown that there is indeed a direct relation between flank wear and temperature generated during cutting (Fig. 8.28). Although cutting speed has been found to be the most significant process variable in tool life, depth of cut and feed rate are also important. Thus Eq. (8.31) can be modified as follows:

**FIGURE 8.27** Tool-life curves for a variety of cutting-tool materials. The negative inverse of the slope of these curves is the exponent *n* in tool-life equations.

**TABLE 8.5**

**RANGE OF *n* VALUES FOR VARIOUS CUTTING TOOLS**

| | |
|---|---|
| High-speed steels | 0.08–02 |
| Cast alloys | 0.1–0.15 |
| Carbides | 0.2–0.5 |
| Ceramics | 0.5–0.7 |

$$VT^n d^x f^y = C, \tag{8.32}$$

where $d$ is the depth of cut and $f$ is the feed rate (in mm/rev or in./rev) in turning.

The exponents $x$ and $y$ must be determined experimentally for each cutting condition. For example, taking $n = 0.15$, $x = 0.15$, and $y = 0.6$ as typical values encountered in practice, we see that cutting speed, feed rate, and depth of cut are of decreasing order of importance.

Equation (8.32) can be rewritten as

$$T = C^{1/n} V^{-1/n} d^{-x/n} f^{-y/n} \tag{8.33}$$

or

$$T \simeq C^7 V^{-7} d^{-1} f^{-4}. \tag{8.34}$$

For a constant tool life, the following observations can be made from Eq.(8.34):

1. If the feed rate or the depth of cut is increased, the cutting speed must be decreased, and vice versa.

2. Depending on the exponents, a reduction in speed can then result in an increase in the volume of the material removed because of the increased feed rate and/or depth of cut.

**FIGURE 8.28** Relationship between measured temperature during cutting and tool life (flank wear). Note that high cutting temperatures severely reduce tool life. See also Eq. (8.30). *Source:* After H. Takeyama and Y. Murata.

Work material: Heat-resistant alloy
Tool material: Tungsten carbide
Tool life criterion: 0.024 in. (0.6 mm) flank wear

**TABLE 8.6**

**ALLOWABLE AVERAGE WEAR LAND (*VB*) FOR CUTTING TOOLS IN VARIOUS OPERATIONS**

| | ALLOWABLE WEAR LAND (mm) | |
|---|---|---|
| OPERATION | HIGH-SPEED STEELS | CARBIDES |
| Turning | 1.5 | 0.4 |
| Face milling | 1.5 | 0.4 |
| End milling | 0.3 | 0.3 |
| Drilling | 0.4 | 0.4 |
| Reaming | 0.15 | 0.15 |

**Allowable wear land.**    The *allowable wear land* (VB) for various conditions is given in Table 8.6. For improved dimensional accuracy and surface finish, the allowable wear land may be made smaller than the values given in Table 8.6. The recommended cutting speed for a high-speed-steel tool is generally the one that gives a tool life of 60–120 min and for carbide tools 30–60 min.

**Optimum cutting speed.**    We have shown that as cutting speed increases, tool life is rapidly reduced. On the other hand, if cutting speeds are low, tool life is long but the rate at which material is removed is also low. Thus there is an optimum cutting speed.

---

**Example 8.3:**   **Effect of cutting speed on material removal**

We can appreciate the effect of cutting speed on the volume of metal removed between tool re-sharpenings or replacements by analyzing Fig. 8.26a. Assume that we are machining the material in the "a" condition, that is, as cast with a hardness of 265 HB. If our cutting speed is 1 m/s (200 ft/min), we note that tool life is about 40 min. Hence the tool travels a distance of (1 m/s) (60 s/min)(40 min)  = 2400 m before it is resharpened or replaced. If we change the cutting speed to 2 m/s, tool life is about 5 min. Hence the tool travels $(2)(60)(5) = 600$ m.

Since the volume of material removed is directly proportional to the distance the tool has traveled, we see that by decreasing the cutting speed, we can remove more material between tool changes. Note, however, that the lower the cutting speed the longer is the time required to machine a part. These variables have important economic impact, and we will discuss them further in Section 8.13.

---

**Example 8.4:**   **Increasing tool life by reducing the cutting speed**

Using the Taylor Eq. (8.31) for tool life and letting $n = 0.5$ and $C = 400$, calculate the percentage increase in tool life when the cutting speed is reduced by 50 percent.

SOLUTION.    Since $n = 0.5$, we can rewrite the Taylor equation as $V\sqrt{T} = 400$. Letting $V_1$ be the initial speed and $V_2$ the reduced speed, we note that for this problem $V_2 = 0.5\, V_1$. Because $C$ is a constant, we have the relationship

$$0.5\, V_1\sqrt{T_2} = V_1\sqrt{T_1}.$$

Simplifying this expression, we find that $T_2/T_1 = 1/0.25 = 4.0$. This indicates that the tool life change is $(T_2 - T_1)/T_1 = (T_2/T_1) - 1 = 4 - 1 = 3$, or that it is increased by 300 percent. Note that the reduction in cutting speed has resulted in a major increase in tool life. You will also note that, in this problem, the magnitude of $C$ is not relevant.

### 8.3.2 Crater wear

The most significant factors in *crater* wear are temperature and the degree of chemical affinity between the tool and the workpiece. The factors affecting flank wear also influence crater wear. We have shown that the rake face of the tool is subjected to high levels of stress and temperature, in addition to sliding at relatively high speeds. Peak temperatures, for instance, can be on the order of $1100°C$ ($2000°F$) (Fig. 8.21b).

Interestingly, the location of maximum crater wear generally coincides with the location of maximum temperature. Experimental evidence indicates a direct relation between crater-wear rate and tool–chip interface temperature (Fig. 8.29). Note the sharp increase in crater wear after a certain temperature range has been reached.

The cross-section of the tool–chip interface in cutting steel at high speeds is shown in Fig. 8.30. Note the location of the crater-wear pattern and the discoloration of the tool (loss of temper) as a result of high temperatures. Note also how well the discoloration profile agrees with the temperature profile shown in Fig. 8.19.

The effect of temperature on crater wear has been described in terms of a **diffusion** mechanism (the movement of atoms across the tool–chip interface). Diffusion depends on the tool–workpiece material combination and on temperature, pressure, and time. As these

°C

Average tool–chip interface temperature (°F)

**FIGURE 8.29** Relation between crater wear rate and average tool-chip interface temperature in turning: (a) high-speed steel tool; (b) C-1 carbide; (c) C-5 carbide. Note that crater wear increases rapidly within a narrow range of temperature. *Source:* After K.J. Trigger and B.T. Chao.

**FIGURE 8.30** Interface of chip (left) and rake face of tool (right) and crater wear in cutting AISI 1004 steel at 3 m/s (585 ft/min). Discoloration of the tool indicates high temperature (loss of temper). Note how the crater-wear pattern coincides with the discoloration pattern. Compare this pattern with the temperature distribution shown in Fig. 8.19 *Source:* Courtesy of P. K Wright, University of California, Berkeley.

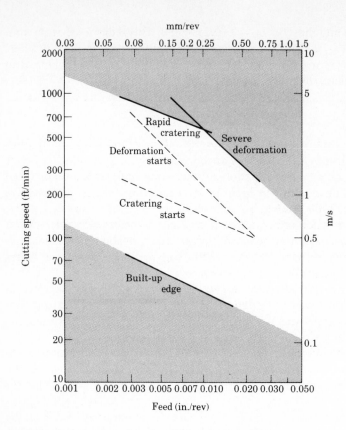

**FIGURE 8.31** Plastic deformation and wear of a carbide tool in cutting austenitic stainless steel, as a function of cutting speed and feed. Note that deformation and crater wear take place at higher speeds, indicating higher temperatures. Charts like this are useful in determining the optimum range of feed and speed in cutting operations. *Source:* After P. A. Dearnley and E. M. Trent.

quantities increase, the diffusion rate increases. Unless these factors are favorable, crater wear will not take place by diffusion. Wear will then be caused by other factors, such as those outlined for flank wear. High temperatures may also cause softening and plastic deformation of the tool because of the decrease in its yield strength with temperature. This type of deformation is generally observed in machining high-strength materials.

The various patterns of wear for a carbide tool in cutting austenitic stainless steel are shown in Fig. 8.31. A BUE forms at low speeds. As the cutting speed is increased, crater wear and plastic deformation of the cutting edge of the tool take place owing to increased temperature. The complex interaction of all the parameters involved prevents generalization about the wear behavior of cutting tools. Each particular set of parameters results in a specific type of wear on the cutting tool.

### 8.3.3 Chipping

The term *chipping* is used to describe the breaking away of a piece from the cutting edge of the tool. The chipped pieces may be very small (microchipping or macrochipping), or they may involve relatively large fragments (gross chipping or fracture). Unlike wear, which is a more gradual process, chipping is a phenomenon that results in a sudden loss of tool material. Two main causes of chipping are mechanical shock (impact by interrupted cutting, such as in milling) and thermal fatigue. Chipping may occur in a region in the tool where a small crack or defect already exists. Thermal cracks (Fig. 8.24a) are caused by the thermal cycling of the tool in interrupted cutting. Thermal cracks are generally perpendicular to the cutting edge.

High positive rake angles can contribute to chipping because of the small included angle of the tool tip. Crater wear may also progress toward the tool tip and weaken it, thus causing chipping. Chipping or fracture can be reduced by selecting tool materials with high-impact and thermal shock resistance.

### 8.3.4 General observations

Because of the many factors involved, including the characteristics of the machine tool, the wear behavior of cutting tools varies significantly. In addition to the wear processes we have already described, other phenomena also occur in tool wear (Fig. 8.24). For example, as a result of the high temperatures generated during cutting, tools may soften and undergo plastic deformation because their yield strength decreases. This type of deformation generally occurs in machining high-strength metals and alloys. Thus tools must maintain their strength and hardness at the elevated temperatures encountered in cutting.

The wear groove or notch on cutting tools (Fig. 8.24) has been attributed to the fact that this region is the boundary where the chip is no longer in contact with the tool. This boundary, or **depth-of-cut line**, oscillates because of inherent variations in the cutting operation and accelerates the wear process. Furthermore, this region is in contact with the machined surface from the previous cut. Since a machined surface may develop a thin work-hardened layer, this contact could contribute to the formation of the wear groove.

Because they are hard and abrasive, scale and oxide layers on a workpiece surface increase wear. In such cases, the depth of cut $d$ should be greater than the thickness of the oxide film or the work-hardened layer. Thus the value of $d$ in Fig. 8.20 should be greater than the thickness of the scale on the workpiece. In other words, light cuts should not be taken on rusted workpieces.

### 8.3.5 Tool condition monitoring

Techniques typically fall into two general categories: direct and indirect. The **direct** method for observing the condition of a cutting tool involves optical measurement of wear, such as by periodically observing changes in the tool profile. This is the most common and reliable technique and is done using a microscope (*toolmakers' microscope*). This procedure, however, requires that the cutting operation be stopped. With rapidly increasing use of computer-controlled machine tools and implementation of automated manufacturing, the reliable and repeatable performance of cutting tools has become an important consideration. Once programmed properly, most modern machine tools now operate with little direct supervision by a machine operator. Consequently, the failure of a cutting tool can have serious effects on the quality of the machined part as well as the efficiency and economics of the overall machining operation. It is thus essential to continuously and indirectly monitor the condition of the cutting tool, i.e., wear, chipping, or gross failure. Most state-of-the-art machine controls are now equipped with tool-condition monitoring systems.

**Indirect** methods of measuring wear involve the correlation of the tool condition with process variables such as forces, power, temperature rise, surface finish, and vibrations. One important development is the **acoustic emission technique,** which utilizes a piezoelectric transducer attached to a tool holder. The transducer picks up signals that are acoustic emissions

**FIGURE 8.32**   Relationship between mean flank wear, maximum crater wear, and acoustic emission (noise generated during cutting) as a function of machining time. This technique is being developed as a means for monitoring wear rate in various cutting processes without interrupting the operation. *Source:* After M. S. Lan and D. A. Dornfeld, *Proc. NAMRC-X,* 1982, pp. 305–311.

(typically above 100 kHz) resulting from the stress waves generated during cutting. By analyzing the signals, tool wear and chipping can be monitored (Fig. 8.32). The acoustic emission technique is particularly effective in precision machining operations where, because of the small amounts of material removed, cutting forces are low.

A similar indirect tool–condition monitoring system consists of **transducers** that are installed in original machine tools, or are retrofitted on existing machines. They continually monitor torque and forces in various directions in a variety of machining processes. The signals are preamplified and a microprocessor analyzes the content of the signals and interprets it. The system is capable of differentiating the signals that come from tool breakage, tool wear, a missing tool, overloading of the machine, or colliding machine components such as spindles. It can also automatically compensate for tool wear.

The design of the transducers must be such that they are nonintrusive to the machining operation, accurate and repeatable in signal detection, resistant to abuse and shop-floor environment, and cost effective. Continued progress is being made in the development of such sensors, including the use of infrared and fiber optic techniques for temperature measurement in machining operations.

## 8.4   Surface Finish and Integrity

Surface finish influences not only the dimensional accuracy of machined parts, but also their properties. Whereas **surface finish** describes the *geometric* features of surfaces, **surface integrity** pertains to properties such as fatigue life and corrosion resistance, which are influ-

**FIGURE 8.33** Range of surface roughnesses obtained in various machining processes. Note the wide range within each group. (See also Fig. 9.31).

enced strongly by the type of surface produced. Factors influencing surface integrity are temperatures generated during processing, residual stresses, metallurgical (phase) transformations, and surface plastic deformation, tearing, and cracking. Surface roughness for machining and other processes is given in Fig. 8.33.

The built-up edge, with its significant effect on tool profile, has the greatest influence on surface roughness. Figure 8.34 shows surfaces obtained in two different cutting operations. Note the considerable damage to the surfaces from BUE. Ceramic and diamond tools generally produce better surface finish than other tools, largely because of their much lower tendency to form BUE.

A tool that is not sharp has a large radius along its edges (see Fig. 8.24c), just as a dull pencil or knife does. Figure 8.35 illustrates the relationship between the radius of the cutting edge and depth of cut in orthogonal cutting. Note that at small depths of cut the rake angle can effectively become negative, and the tool may simply ride over the workpiece surface and not remove chips. You can simulate this behavior by trying to scrape the surface along the length of a stick of butter with a dull knife; you will note that you will not be able to remove a layer

(a)                                                    (b)

**FIGURE 8.34**   Surfaces produced on steel by cutting, as observed with a scanning electron microscope: (a) turned surface and (b) surface produced by shaping. *Source:* J. T. Black and S. Ramalingam.

of butter at very small depths, whereas if the knife is sharp, you will be able to do so. If this radius, not to be confused with radius $R$ in Fig. 8.24c, is large in relation to the depth of cut, the tool will rub over the machined surface. Rubbing in cutting generates heat and induces surface residual stresses, which in turn, may cause surface damage such as tearing and cracking. Thus, the depth of cut should generally be greater than the radius on the cutting edge.

In turning, as in other cutting operations, the tool leaves a spiral profile—**feed marks**—on the machined surface as it moves across the workpiece. The higher the feed $f$ and the smaller the radius $R$, the more prominent these marks. Although not significant in rough machining operations, these marks are important in finish machining.

We describe vibration and chatter in some detail in Section 8.11. For now, we should recognize that if the tool vibrates or chatters during cutting, it will adversely affect surface finish. The reason is that a vibrating tool changes the dimensions of the cut periodically. Excessive chatter can also cause chipping and premature failure of the more brittle cutting tools such as ceramics and diamond.

**FIGURE 8.35**   Schematic illustration of a dull tool in orthogonal cutting (exaggerated). Note that at small depths of cut, the rake angle can effectively become negative. The tool may simply ride over the workpiece surface, burnishing it.

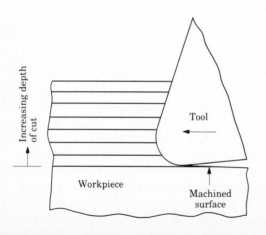

# 8.5 Machinability

We usually define **machinability** of a material in terms of three factors: (1) surface finish and integrity of the machined part, (2) tool life obtained, and (3) force and power requirements. Thus good machinability indicates *good surface finish* and integrity, *long tool life,* and *low force and power* requirements. An additional parameter is chip curl. As we stated earlier, long, thin curled chips, if not broken up, can severely interfere with the cutting operation by becoming entangled in the cutting zone. Thus the type of chip a material produces is also a factor in its machinability.

Because of the complex nature of cutting operations, establishing relationships to define quantitatively the machinability of a material is difficult. In manufacturing plants, tool life and surface roughness are generally considered to be the most important factors in machinability. Although not used much any more, approximate **machinability ratings** are available.

Machinability ratings are usually based on a tool life of $T = 60$ min. The standard is AISI 1112 steel, which is given a rating of 100. Thus for a tool life of 60 min, this steel should be machined at a cutting speed of 100 ft/min (0.5 m/s). Higher speeds will reduce tool life, and lower speeds will increase it.

For example, 3140 steel has a machinability rating of 55. This means that when it is machined at a cutting speed of 55 ft/min (0.275 m/s), tool life will be 60 min. Nickel has a rating of 200, indicating that it should be machined at 200 ft/min (1 m/s) to obtain a tool life of 60 min. Machinability ratings for various materials are as follows: free-cutting brass 300, 2011 wrought aluminum 200, pearlitic gray iron 70, Inconel 30, and precipitation-hardening 17–7 steel 20. As the hardness of a material increases, its machinability ratings decrease proportionately.

## 8.5.1 Machinability of steels

Because steels are among the most important engineering materials, their machinability has been studied extensively. The machinability of steels has been improved mainly by adding lead and sulfur to obtain so-called **free-machining steels.**

**Leaded steels.**   Lead is added to the molten steel and takes the form of dispersed fine lead particles. Lead is insoluble in iron, copper, and aluminum and their alloys. Thus during cutting, the lead particles are sheared and smeared over the tool–chip interface. Because of its low shear strength, the lead acts as a solid lubricant.

This behavior has been verified by the presence of high concentrations of lead on the tool–side face of chips in machining leaded steels. In addition to this effect, lead probably lowers the shear stress in the primary shear zone, thus reducing cutting forces and power consumption. Lead may be used in either nonsulfurized or resulfurized steels. However, because of environmental concerns, there is now a trend toward eliminating the use of leaded steels. Lead in steels may be replaced by bismuth. Leaded steels are identified by the letter $L$ between the second and third numerals; thus, for example, 10L45. However, in stainless steels, similar use of the letter $L$ in their identification means low carbon, which improves their corrosion resistance.

**Resulfurized steels.**   Sulfur in steels forms manganese sulfide inclusions (second-phase particles), which act as stress raisers in the primary shear zone (Fig. 8.36). As a result, the chips produced are small and break up easily, thus improving machinability. The shape,

(a)                                    (b)                                    (c)

**FIGURE 8.36** Photomicrographs showing various types of inclusions in low-carbon, resulfurized free-machining steels. (a) Manganese sulfide inclusions in AISI 1215 steel. (b) Manganese sulfide inclusions and glassy manganese silicate-type oxide (dark) in AISI 1215 steel. (c) Manganese sulfide with lead particle as tails in AISI 12L14 steel. *Source:* Courtesy of Inland Steel Company.

orientation, distribution, and concentration of these inclusions significantly influence machinability. Elements such as tellurium and selenium (both chemically similar to sulfur) in resulfurized steels act as inclusion modifiers.

**Calcium-deoxidized steels.**   An important development is calcium-deoxidized steels, in which oxide flakes of calcium aluminosilicate (CaO, $SiO_2$, and $Al_2O_3$) are formed. These flakes, in turn, reduce the strength of the secondary shear zone, thus decreasing tool–chip interface friction and wear, hence temperature. Consequently, these steels undergo less crater wear, especially at high cutting speeds.

**Effects of other elements in steels on machinability.**   The presence of aluminum and silicon in steels is always harmful because they combine with oxygen and form aluminum oxide and silicates. These compounds are hard and abrasive, thus increasing tool wear and reducing machinability.

Carbon, manganese, and phosphorus have various effects on the machinability of steels, depending on their composition. As the carbon content increases, machinability decreases. However, plain low-carbon steels (less than 0.15% C) can produce poor surface finish by forming a built-up edge. Cast steels are more abrasive, although their machinability is similar to wrought steels. Tool and die steels are very difficult to machine and usually require annealing prior to machining. Machinability of most steels is generally improved by cold working, which reduces the tendency for built-up edge formation.

Other alloying elements, such as nickel, chromium, molybdenum, and vanadium, which improve the properties of steels, generally reduce machinability. The effect of boron is negligible. The role of gaseous elements such as oxygen, hydrogen, and nitrogen has not been clearly established. Any effect that they may have would depend on the presence and quantity of other alloying elements.

In selecting various elements to improve machinability, we should consider the possible detrimental effects of these elements on the properties and strength of the machined part in service. At elevated temperatures, for example, lead causes embrittlement of steels (hot shortness), although at room temperature it has no effect on mechanical properties. Sulfur can severely reduce hot workability of steels, because of the presence of iron sulfide, unless sufficient manganese is present to prevent the formation of iron sulfide. At room temperature, the mechanical properties of resulfurized steels depend on the orientation of the deformed manganese sulfide inclusions (anisotropy).

**Stainless steels.** Austenitic (300 series) steels are generally difficult to machine. Chatter could be a problem, thus requiring machine tools with high stiffness. However, ferritic stainless steels (also 300 series) have good machinability. Martensitic (400 series) steels are abrasive and tend to form built-up edge, and require tool materials with high hot hardness and crater-wear resistance. Precipitation-hardening stainless steels are strong and abrasive, requiring hard and abrasion-resistant tool materials.

### 8.5.2 Machinability of various other metals

Aluminum is generally easy to machine. However, the softer grades tend to form built-up edge. High cutting speeds, high rake angles, and high relief angles are recommended. Wrought alloys with high silicon content and cast aluminum alloys may be abrasive and hence require harder tool materials. Dimensional control may be a problem in machining aluminum since it has a low elastic modulus and a relatively high thermal coefficient of expansion.

*Beryllium* is similar to cast irons but is more abrasive and toxic. Hence it requires machining in a controlled environment.

*Gray cast irons* are generally machinable but are abrasive. Free carbides in castings reduce their machinability and cause tool chipping or fracture, thus requiring tools with high toughness. Nodular and malleable irons are machinable with hard tool materials.

*Cobalt-base alloys* are abrasive and highly work hardening. They require sharp and abrasion-resistant tool materials and low feeds and speeds.

*Wrought copper* can be difficult to machine because of built-up edge formation, although cast copper alloys are easy to machine. Brasses are easy to machine, especially with the addition of lead (leaded free-machining brass). Bronzes are more difficult to machine than brass.

*Magnesium* is very easy to machine, has good surface finish, and prolongs tool life. However, care should be exercised because of its high rate of oxidation and the danger of fire.

*Molybdenum* is ductile and work hardening. Hence it can produce poor surface finish, thus requiring sharp tools.

*Nickel-base alloys* are work hardening, abrasive, and strong at high temperatures. Their machinability is similar to that of stainless steels.

*Tantalum* is very work hardening, ductile, and soft. Hence it produces a poor surface finish; tool wear is high.

*Titanium* and its alloys have poor thermal conductivity (lowest of all metals), thus causing significant temperature rise and built-up edge. Hence it can be difficult to machine.

*Tungsten* is brittle, strong, and very abrasive, hence its machinability is low. Machinability improves greatly at elevated temperature.

*Zirconium* has good machinability. However, it requires a coolant-type cutting fluid because of the danger of explosion and fire.

### 8.5.3 Machinability of various materials

*Graphite* is abrasive. It requires hard, abrasion-resistant, sharp tools. *Thermoplastics* generally have low thermal conductivity, low elastic modulus, and low softening temperature. Consequently, machining them requires tools with positive rake to reduce cutting forces, large relief angles, small depths of cut and feed, relatively high speeds, and proper support of the workpiece. Tools should be sharp. External cooling of the cutting zone may be necessary to keep the chips from becoming "gummy" and sticking to the tools. Cooling can usually be done with a jet of air, vapor mist, or water-soluble oils. Residual stresses may develop during machining. To relieve residual stresses, machined parts can be annealed at temperatures rang-ing from 80°C to 160°C (175°F to 315°F) for a period of time and then cooled slowly and uniformly to room temperature.

*Thermosetting plastics* are brittle and sensitive to thermal gradients during cutting. Their machinability is generally similar to that of thermoplastics.

Because of the fibers present, *reinforced plastics* are very abrasive and are difficult to ma-chine. Fiber tearing and pulling is a significant problem. Furthermore, machining of these materials requires careful removal of machining debris to avoid contact with and inhaling of fibers. Other *composites,* such as metal-matrix and ceramic-matrix composites, can be diffi-cult to machine depending on the properties of the individual components, i.e., reinforcing fibers or whiskers and the matrix material.

### 8.5.4 Thermally assisted machining

Materials that are difficult to machine at room temperature can be machined more easily at elevated temperatures, thus lowering cutting forces and increasing tool life. In **thermally as-sisted machining** (*hot machining*), the source of heat is a torch, high-energy beam (such as laser or electron beam), or plasma arc, focused to an area just ahead of the cutting tool. Most applications in hot machining are in turning and milling. Heating and maintaining a uniform temperature distribution within the workpiece may be difficult to control. Also, the original microstructure of the workpiece may be adversely affected. Except in isolated cases, ther-mally assisted machining offers no significant advantage over machining at room tempera-ture with the use of appropriate cutting tools and fluids. Some success has been obtained in laser-assisted machining of silicon-nitride ceramics and Stellite.

### 8.5.5 Automachining

*Automachining* is a term that describes cutting of metals using single-point tools made of the workpiece material. The cutting tool is not subjected to any additional treatment or coatings. When used on a lathe, it produces continuous chips and good surface finish on materials such as 2011-T3, 2024-T3, and 301 stainless steel. The surface finish is fair in automachining of mild steel and poor in copper. The mechanism by which automachining is possible has been determined experimentally to be the formation of a BUE. The BUE tip acts as the cutting edge; the chip does not contact the original tool at its rake face during cutting. As the hardness of the BUE can be considerably higher than that of the workpiece material (see Fig. 8.7a), cutting is actually performed by the harder BUE. Although it is not likely to be of industrial significance, automachining is nevertheless an interesting method of metal cutting.

## 8.6   Cutting-Tool Materials

Cutting-tool materials and their proper selection are among the most important factors in machining operations, as are mold and die materials for forming and shaping processes. We have noted previously that the tool is subjected to high temperatures, contact stresses, and rubbing on the workpiece surface, as well as by the chip climbing up the rake face of the tool.

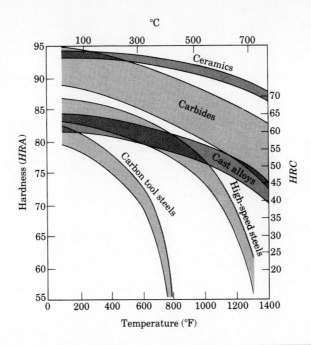

**FIGURE 8.37** Hardness of various cutting-tool materials as a function of temperature (hot hardness). The wide range in each group of materials results from the variety of tool compositions and treatments available for that group.

Consequently, a cutting tool must have the following characteristics in order to produce good-quality and economical parts:

- **Hardness,** particularly at elevated temperatures (hot hardness), so that the hardness and strength of the tool are maintained at the temperatures encountered in cutting operations (Fig. 8.37).
- **Toughness,** so that impact forces on the tool in interrupted cutting operations, such as milling or turning a splined shaft, do not chip or fracture the tool.
- **Wear resistance,** so that an acceptable tool life is obtained before the tool is resharpened or replaced.
- **Chemical stability** or **inertness** with respect to the workpiece material, so that any adverse reactions contributing to tool wear are avoided.

Various cutting-tool materials having a wide range of properties are available (Tables 8.7 and 8.8). Tool materials are usually divided into the following general categories, which are listed in the approximate chronological order in which they were developed and implemented:

1. Carbon and medium-alloy steels
2. High-speed steels
3. Cast-cobalt alloys
4. Carbides
5. Coated tools

**TABLE 8.7**

**TYPICAL PROPERTIES OF TOOL MATERIALS**

| PROPERTY | HIGH-SPEED STEELS | CAST ALLOYS | CARBIDES | | CERAMICS | CUBIC BORON NITRIDE | SINGLE-CRYSTAL DIAMOND* |
| | | | WC | TiC | | | |
|---|---|---|---|---|---|---|---|
| Hardness | 83–86 HRA | 82–84 HRA 46–62 HRC | 90–95 HRA 1800–2400 HK | 91–93 HRA 1800–3200 HK | 91–95 HRA 2000–3000 HK | 4000–5000 HK | 7000–8000 HK |
| Compressive strength | | | | | | | |
| MPa | 4100–4500 | 1500–2300 | 4100–5850 | 3100–3850 | 2750–4500 | 6900 | 6900 |
| psi $\times 10^3$ | 600–650 | 220–335 | 600–850 | 450–560 | 400–650 | 1000 | 1000 |
| Transverse rupture strength | | | | | | | |
| MPa | 2400–4800 | 1380–2050 | 1050–2600 | 1380–1900 | 345–950 | 700 | 1350 |
| psi $\times 10^3$ | 350–700 | 200–300 | 150–375 | 200–275 | 50–135 | 105 | 200 |
| Impact strength | | | | | | | |
| J | 1.35–8 | 0.34–1.25 | 0.34–1.35 | 0.79–1.24 | < 0.1 | < 0.5 | < 0.2 |
| in.-lb | 12–70 | 3–11 | 3–12 | 7–11 | < 1 | < 5 | < 2 |
| Modulus of elasticity | | | | | | | |
| GPa | 200 | — | 520–690 | 310–450 | 310–410 | 850 | 820–1050 |
| psi $\times 10^5$ | 30 | — | 75–100 | 45–65 | 45–60 | 125 | 120–150 |
| Density | | | | | | | |
| kg/m³ | 8600 | 8000–8700 | 10,000–15,000 | 5500–5800 | 4000–4500 | 3500 | 3500 |
| lb/in.³ | 0.31 | 0.29–0.31 | 0.36–0.54 | 0.2–0.22 | 0.14–0.16 | 0.13 | 0.13 |
| Volume of hard phase, % | 7–15 | 10–20 | 70–90 | — | 100 | 95 | 95 |
| Melting or decomposition temperature | | | | | | | |
| °C | 1300 | — | 1400 | 1400 | 2000 | 1300 | 700 |
| °F | 2370 | — | 2550 | 2550 | 3600 | 2400 | 1300 |
| Thermal conductivity, W/m K | 30–50 | — | 42–125 | 17 | 29 | 13 | 500–2000 |
| Coefficient of thermal expansion, $\times 10^{-6}$/°C | 12 | — | 4–6.5 | 7.5–9 | 6–8.5 | 4.8 | 1.5–4.8 |

*The values for polycrystalline diamond are generally lower, except impact strength, which is higher.

**TABLE 8.8**

**GENERAL CHARACTERISTICS OF CUTTING-TOOL MATERIALS. THESE TOOL MATERIALS HAVE A WIDE RANGE OF COMPOSITIONS AND PROPERTIES; THUS OVERLAPPING CHARACTERISTICS EXIST IN MANY CATEGORIES OF TOOL MATERIALS**

| | CARBON AND LOW- TO MEDIUM- ALLOY STEELS | HIGH-SPEED STEELS | CAST- COBALT ALLOYS | UNCOATED CARBIDES | COATED CARBIDES | CERAMICS | POLYCRYSTALLINE CUBIC BORON NITRIDE | DIAMOND |
|---|---|---|---|---|---|---|---|---|
| Hot hardness | | | Increasing | | | | | |
| Toughness | | | Increasing | | | | | |
| Impact strength | | | Increasing | | | | | |
| Wear resistance | | | Increasing | | | | | |
| Chipping resistance | | | Increasing | | | | | |
| Cutting Speed | | | Increasing | | | | | |
| Thermal-shock resistance | | | Increasing | | | | | |
| Tool material cost | | | Increasing | | | | | |
| Depth of cut | Light to medium | Light to heavy | Light to heavy | Light to heavy | Light to heavy | Light to heavy | Light to heavy | Very light for single-crystal diamond |
| Finish obtainable | Rough | Rough | Rough | Good | Good | Very good | Very good | Excellent |
| Method of processing | Wrought | Wrought, cast, HIP* sintering | Cast and HIP sintering | Cold pressing and sintering | CVD or PVD† | Cold pressing and sintering or HIP sintering | High-pressure, high-temperature sintering | High-pressure, high-temperature sintering |
| Fabrication | Machining and grinding | Machining and grinding | Grinding | Grinding | Grinding | Grinding | Grinding and polishing | Grinding and polishing |

Source: R. Komanduri, *Kirk-Othmer Encyclopedia of Chemical Technology*, 3d ed. New York: Wiley. Copyright © 1978, John Wiley. Reprinted by permission of John Wiley & Sons, Inc.
*Hot-isostatic pressing.
†Chemical-vapor deposition, physical-vapor deposition.

6.   Alumina-base ceramics

7.   Cubic boron nitride

8.   Silicon-nitride-base ceramics

9.   Diamond

10.  Whisker-reinforced tool materials

Many of these materials are also used for dies and molds. In this section, we present the characteristics, applications, and limitations of these tool materials in machining operations. We discuss characteristics such as hot hardness, toughness, impact strength, wear resistance, thermal shock resistance, and costs, as well as the range of cutting speeds and depth of cut for optimum performance. We show that proper selection of tools is a critical factor in the quality of surfaces produced and the economics of machining.

### 8.6.1   Carbon and medium-alloy steels

*Carbon steels* are the oldest of tool materials and have been used widely for drills, taps, broaches, and reamers since the 1880s. Low-alloy and medium-alloy steels were developed later for similar applications but with longer tool life. Although inexpensive and easily shaped and sharpened, these steels do not have sufficient hot hardness and wear resistance for cutting at high speeds where, as you have seen, the temperature rises significantly. Note in Fig. 8.37, for example, how rapidly the hardness of carbon steels decreases as the temperature increases. Consequently, the use of these steels is limited to very low-speed cutting operations.

### 8.6.2   High-speed steels

High-speed steel (HSS) tools are so named because they were developed to cut at higher speeds. First produced in the early 1900s, **high-speed steels** are the most highly alloyed of the tool steels. They can be hardened to various depths, have good wear resistance, and are relatively inexpensive. Because of their high toughness and resistance to fracture, high-speed steels are especially suitable for high positive-rake-angle tools (small included angle), interrupted cuts, and for machine tools with low stiffness that are subject to vibration and chatter.

There are two basic types of high-speed steels: *molybdenum* (M series) and *tungsten* (T series). The M series contains up to about 10 percent molybdenum, with chromium, vanadium, tungsten, and cobalt as alloying elements. The T series contains 12 to 18 percent tungsten, with chromium, vanadium, and cobalt as alloying elements. The M series generally has higher abrasion resistance than the T series, undergoes less distortion during heat treating, and is less expensive. Consequently, 95 percent of all high-speed steel tools produced in the United States are made of M-series steels.

High-speed-steel tools are available in wrought, cast, and sintered (powder-metallurgy) forms. They can be *coated* for improved performance (see Section 8.6.5). High-speed-steel tools may also be subjected to surface treatments, such as case hardening for improved hardness and wear resistance and steam treatment at elevated temperatures to develop a black oxide layer for improved performance in machining such as reducing built-up edge formation. High-speed steels account for the largest tonnage of tool materials used today, followed by various die steels and carbides. They are used in a wide variety of cutting operations requiring complex tool shapes such as drills, reamers, taps, and gear cutters. Their basic limitation is the relatively low cutting speeds when compared to carbide tools.

### 8.6.3 Cast-cobalt alloys

Introduced in 1915, **cast-cobalt alloys** have the following ranges of composition: 38 to 53 percent cobalt, 30 to 33 percent chromium, and 10 to 20 percent tungsten. Because of their high hardness, typically 58 to 64 HRC, they have good wear resistance and maintain their hardness at elevated temperatures.

Commonly known as *Stellite* tools, these alloys are cast and ground into relatively simple tool shapes. They are not as tough as high-speed steels and are sensitive to impact forces. Consequently, they are less suitable than high-speed steels for interrupted cutting operations. These tools are now used only for special applications that involve deep, continuous roughing operations at relatively high feeds and speeds—as much as twice the rates possible with high-speed steels.

### 8.6.4 Carbides

The three groups of tool materials just described (alloy steels, high-speed steels, and cast alloys) have the necessary toughness, impact strength, and thermal shock resistance but have important limitations, such as strength and hardness, particularly hot hardness. Consequently, they cannot be used as effectively where high cutting speeds, hence high temperatures, are involved; thus tool life can be short.

To meet the challenge of higher speeds for higher production rates, **carbides** (also known as *cemented* or *sintered carbides*) were introduced in the 1930s. Because of their high hardness over a wide range of temperatures (see Fig. 8.37), high elastic modulus and thermal conductivity, and low thermal expansion, carbides are among the most important, versatile, and cost-effective tool and die materials for a wide range of applications. The two basic groups of carbides used for machining operations are tungsten carbide and titanium carbide. In order to differentiate them from coated tools (see Section 8.6.5), plain carbide tools are usually referred to as **uncoated** carbides.

#### Tungsten carbide

*Tungsten carbide* (WC) is a composite material consisting of tungsten-carbide particles bonded together in a cobalt matrix, hence they are also called *cemented* carbides. These tools are manufactured by powder-metallurgy techniques, in which WC particles are combined together with cobalt in a mixer, resulting in a cobalt matrix surrounding the WC particles. These particles, which are 1–5 $\mu$m (40–200 $\mu$in.) in size, are then pressed and sintered (hence also called *sintered* carbides) into the desired insert shapes (see Example 11.5). WC is frequently compounded with carbides of titanium and niobium to impart special properties to the carbide.

The amount of cobalt significantly affects the properties of carbide tools (Fig. 8.38). As the cobalt content increases, the strength, hardness, and wear resistance of WC decrease, while its toughness increases because of the higher toughness of cobalt. Tungsten carbide tools are generally used for cutting steels, cast irons, and abrasive nonferrous materials, and are rapidly replacing HSS tools due to their better performance.

#### Titanium carbide

*Titanium carbide* (TiC) has higher wear resistance than tungsten carbide but is not as tough. With a nickel-molybdenum alloy as the matrix, TiC is suitable for machining hard materials, mainly steels and cast irons, and for cutting at speeds higher than those for tungsten carbide.

**FIGURE 8.38**   Effect of cobalt content in tungsten-carbide tools on mechanical properties. Note that hardness is directly related to compressive strength and hence, inversely, to wear.

## Inserts

We have seen that carbon-steel and high-speed-steel tools are formed in one piece and ground to various shapes (see Fig. 8.52); other such tools include drills and milling cutters. After the cutting edge wears, the tool has to be removed from its holder and reground. Although a supply of sharp or resharpened tools is usually available from tool rooms, tool-changing operations are time-consuming and inefficient. The need for a more effective method has led to the development of **inserts,** which are individual cutting tools with a number of cutting edges (Fig. 8.39). A square insert, for example, has eight cutting points and a triangular insert has six.

Inserts are usually clamped on the tool *shank* with various locking mechanisms (Fig 8.40). Although not as commonly used, inserts may be *brazed* to the tool shank. However, because of the difference in thermal expansion between the insert and the tool-shank materials, brazing must be done carefully to avoid cracking or warping. Clamping is the preferred method because each insert has a number of cutting edges, and after one edge is worn, it is *indexed* (rotated in its holder) to present another cutting edge. In addition to these examples, a wide variety of other toolholders are available for specific applications, including those with quick insertion and removal features.

Carbide inserts are available in a variety of shapes, such as square, triangle, diamond, and round. The strength of the cutting edge of an insert depends on its shape. The smaller the angle (Fig. 8.41), the lower the strength of the edge. In order to further improve edge strength and prevent chipping, all insert edges are usually honed, chamfered, or produced with a negative land (Fig. 8.42). Most inserts are honed to a radius of about 0.025 mm (0.001 in.).

**FIGURE 8.39**  Typical carbide inserts with various shapes and chip-breaker features. Round inserts are also available. The holes in the inserts are standardized for interchangeability. *Source:* Courtesy of Kyocera Engineered Ceramics, Inc., and *Manufacturing Engineering,* Society of Manufacturing Engineers.

We described *chip breaker* features on cutting tools in Fig. 8.8 for the purpose of controlling chip flow during machining and eliminating long chips. Most carbide inserts are now available with a wide range of chip-breaker features, typical examples of which can be seen in Fig. 8.39. The selection of a particular chip-breaker feature on a tool depends on the feed

**FIGURE 8.40**  Methods of attaching inserts to tool shank: (a) clamping; and (b) wing lockpins. (c) Examples of inserts attached to tool shank with threadless lockpins, which are secured with side screws. *Source:* Courtesy of GTE Valenite. (d) Brazed insert on a tool shank.

**FIGURE 8.41**   Relative edge strength of different shapes of inserts. Strength refers to the cutting edge shown by the angles.

**FIGURE 8.42**   Edge preparation of inserts to improve edge strength and resistance to chipping and fracture.

and depth of cut of the operation as well as the workpiece material and whether it is a roughing or finishing cut.

   **Stiffness** of the machine tool is of major importance in using carbide tools (see Section 8.11). Light feeds, low speeds, and chatter are detrimental because they tend to damage the tool's cutting edge. Light feeds, for example, concentrate the forces and temperature closer to the edges of the tools—hence the tendency for the edges to chip off. Low cutting speeds tend to encourage cold welding of the chip to the tool. Cutting fluids, if used to minimize the heating and cooling of the tool in interrupted cutting operations, should be applied continuously and in large quantities.

## Classification of carbide tools

The American National Standards Institute (ANSI) has for a number of years classified carbide tools as shown in Table 8.9, ranging from grades C–1 to C–8. With rapidly increasing trends in global manufacturing and with increasingly wider use of the ISO (International Organization for Standardization) standards, there has been, for some time, a major effort to standardize carbide classification.

   Carbide grades are now being classified using the symbols P, M, and K, as shown in Table 8.10, for a range of applications. Table 8.9 includes the approximate equivalent ISO standards for the carbides listed. However, because of the wide variety of carbide compositions available and the broad range of machining applications and workpiece materials involved, efforts for ISO classification continue to be a difficult task. This is especially true when comparing ISO grades with the traditional C grades.

**TABLE 8.9**

**CLASSIFICATION OF TUNGSTEN CARBIDES ACCORDING TO MACHINING APPLICATIONS. SEE ALSO VARIOUS SECTIONS FOR CUTTING TOOL RECOMMENDATIONS.**

| ISO STANDARD | ANSI CLASSIFICATION NUMBER (GRADE) | MATERIALS TO BE MACHINED | MACHINING OPERATION | TYPE OF CARBIDE | CHARACTERISTICS OF | |
|---|---|---|---|---|---|---|
| | | | | | CUT | CARBIDE |
| K30-K40 | C–1 | Cast iron, nonferrous metals, and nonmetallic materials requiring abrasion resistance | Roughing | Wear-resistant grades; generally straight WC–Co with varying grain sizes | Increasing cutting speed ↓ Increasing feed rate ↑ | Increasing hardness and wear resistance ↓ ↑ Increasing strength and binder content |
| K20 | C–2 | | General purpose | | | |
| K10 | C–3 | | Light finishing | | | |
| K01 | C–4 | | Precision finishing | | | |
| P30-P50 | C–5 | Steels and steel alloys requiring crater and deformation resistance | Roughing | Crater-resistant grades; various WC–Co composi-tions with TiC and/or TaC alloys | Increasing cutting speed ↓ Increasing feed rate ↑ | Increasing hardness and wear resistance ↓ ↑ Increasing strength and binder content |
| P20 | C–6 | | General purpose | | | |
| P10 | C–7 | | Light finishing | | | |
| P01 | C–8 | | Precision finishing | | | |

Note: The ISO and ANSI comparisons are approximate.

**TABLE 8.10**

**ISO CLASSIFICATION OF CARBIDE CUTTING TOOLS ACCORDING TO USE**

| SYMBOL | WORKPIECE MATERIAL | COLOR CODE | DESIGNATION IN INCREASING ORDER OF WEAR RESISTANCE AND DECREASING ORDER OF TOUGHNESS IN EACH CATEGORY, IN INCREMENTS OF 5 |
|---|---|---|---|
| P | Ferrous metals with long chips | Blue | P01, P05 through P50 |
| M | Ferrous metals with long or short chips; nonferrous metals | Yellow | M10 through M40 |
| K | Ferrous metals with short chips; nonferrous metals; nonmetallic materials | Red | K01, K10 through K40 |

## 8.6.5 Coated tools

As described in Chapter 3, new alloys and engineered materials have been developed contin-uously, particularly since the 1960s. These materials have high strength and toughness but are generally abrasive and are highly reactive chemically with tool materials. The difficulty of ma-chining these materials efficiently—and the need for improving the performance in machining the more common engineering materials—has led to important developments in **coated tools.**

Because of their unique properties, coated tools can be used at high cutting speeds, thus reducing the time required for machining operations and costs. Figure 8.43 shows that cutting time has been reduced by a factor of more than 100 since 1900. Coated tools are now being used for cutting operations, with tool life as much as 10 times that of uncoated tools.

### Coating materials

Coating materials commonly used are titanium nitride (TiN), titanium carbide (TiC), titanium carbonitride (TiCN), and aluminum oxide ($Al_2O_3$). Generally in the thickness range of 2–10 $\mu$m (80–400 $\mu$in.), these coatings are applied on cutting tools and inserts by *chemical-vapor deposition* (CVD) and *physical-vapor deposition* (PVD) techniques, described in Section 4.5. The CVD process is the most commonly used method for carbide tools with multiphase and ceramic coatings. The PVD-coated carbides with TiN coatings, on the other hand, have higher cutting-edge strength, lower friction, lower tendency to form a built-up edge, and are smoother and more uniform in thickness, which is generally in the range of 2–4 $\mu$m (80–160 $\mu$in.).

Coatings for cutting tools, as well as for dies, should have the following general characteristics:

- High hardness at elevated temperatures
- Chemical stability and inertness to the workpiece material
- Low thermal conductivity
- Good bonding to the substrate to prevent flaking or spalling
- Little or no porosity

The effectiveness of coatings, in turn, is enhanced by hardness, toughness, and high thermal conductivity of the substrate, which may be carbide or high-speed steel. Honing (see Section 9.8) of the cutting edges is an important procedure to maintain the strength of the coating. Otherwise the coating may chip off at sharp edges and corners.

**FIGURE 8.44** Wear patterns on high-speed-steel uncoated and titanium-nitride coated tools. Note that flank wear is lower for the coated tool.

**Titanium nitride.**    Titanium-nitride coatings have low coefficient of friction, high hardness, resistance to high temperature, and good adhesion to the substrate. Consequently, they greatly improve the life of high-speed-steel tools, as well as the lives of carbide tools, drills, and cutters. Titanium-nitride coated tools, which are gold in color, perform well at higher cutting speeds and feeds.

Flank wear is significantly lower than for uncoated tools (Fig 8.44), and flank surfaces can be reground after use, since regrinding does not remove the coating on the rake face of the tool. However, coated tools do not perform as well at low cutting speeds because the coating can be worn off by chip adhesion. Hence the use of appropriate cutting fluids to discourage adhesion is important.

**Titanium carbide.**    Titanium-carbide coatings on tungsten-carbide inserts have high flank-wear resistance in machining abrasive materials.

**Ceramics.**    Their resistance to high temperature, chemical inertness, low thermal conductivity, and resistance to flank and crater wear make ceramics suitable coatings for tools. The most commonly used ceramic coating is aluminum oxide ($Al_2O_3$). However, because they are very stable (not chemically reactive), oxide coatings generally bond weakly with the substrate.

**Multiphase coatings.**    The desirable properties of the coatings just described can be combined and optimized with the use of **multiphase coatings** (Fig. 8.45). Carbide tools are now available with two or three layers of such coatings and are particularly effective in machining cast irons and steels.

The first layer over the substrate is TiC, followed by $Al_2O_3$, and then TiN. The first layer should bond well with the substrate; the outer layer should resist wear and have low thermal conductivity; the intermediate layer should bond well and be compatible with both layers. Typical applications of multiple-coated tools are:

1. High-speed, continuous cutting: TiC/$Al_2O_3$.
2. Heavy-duty, continuous cutting: TiC/$Al_2O_3$/TiN.
3. Light, interrupted cutting: TiC/TiC+TiN/TiN.

A new development in coatings is *alternating multiphase layers*, with layers that are thinner than in regular multiphase coatings (Fig. 8.45). The thickness of these layers is in the

**FIGURE 8.45** Multiphase coatings on a tungsten-carbide substrate. Three alternating layers of aluminum oxide are separated by very thin layers of titanium nitride. Inserts with as many as thirteen layers of coatings have been made. Coating thicknesses are typically in the range of 2–10 $\mu$m. *Source:* Courtesy of Kennametal, Inc., and *Manufacturing Engineering,* Society of Manufacturing Engineers.

range of 2–10 $\mu$m. The reason for using thinner coatings is that coating hardness increases with decreasing grain size, a phenomenon that is similar to the increase in strength of metals with decreasing grain size. Thus thinner layers are harder than thicker layers.

### Diamond coatings

A recent development concerns the use of diamond polycrystalline as coatings for cutting tools, particularly tungsten-carbide and silicon-nitride inserts. Problems exist regarding adherence of the diamond film to the substrate and the difference in thermal expansion between diamond and substrate materials (see Table 8.7).

Thin-film *diamond coated* inserts are now available, as are thick-film diamond brazed-tip cutting tools. Thin films are deposited on substrates by PVD and CVD techniques. Thick films are obtained by growing a large sheet of pure diamond, which is then laser cut to shape and brazed to a carbide shank. Diamond-coated tools are particularly effective in machining abrasive materials, such as aluminum alloys containing silicon, fiber-reinforced and metal-matrix composite materials, and graphite. Improvements in tool life of as much as tenfold have been obtained over other coated tools.

### New coating materials

While titanium nitride coatings made by chemical-vapor deposition are still common, advances are being made in developing and testing new coating materials. *Titanium carbonitride* (TiCN) and *titanium aluminum nitride* (TiAlN) have been found to be effective in cutting stainless steels. TiCN, which is deposited by physical-vapor deposition techniques, is harder and tougher than TiN and can be used on carbides and high-speed steel tools. TiAlN is effective in machining aerospace alloys.

*Chromium-based* coatings such as chromium carbide (CrC) have been found to be effective in machining softer metals that tend to adhere to the cutting tool, such as aluminum, copper, and titanium. Other new materials include *zirconium nitride* (ZrN) and *hafnium nitride* (HfN). Considerable experimental data is required before these coatings and their behavior can be fully assessed for proper applications in machining.

### Ion implantation

In this process ions are introduced into the surface of the cutting tool, improving its surface properties (see Section 4.5). The process does not change the dimensions of tools. *Nitrogen-ion* implanted carbide tools have been used successfully on alloy steels and stainless steels. *Xenon-ion* implantation of tools is also under development.

### 8.6.6 Alumina-base ceramics

**Ceramic** tool materials, introduced in the early 1950s, consist primarily of fine-grained, high-purity *aluminum oxide*. They are cold pressed into insert shapes under high pressure, sintered at high temperature, and called *white*, or *cold-pressed*, ceramics. Additions of titanium carbide and zirconium oxide help improve properties such as toughness and thermal-shock resistance.

**Alumina-base ceramic** tools have very high abrasion resistance and hot hardness (Fig. 8.46). Chemically, they are more stable than high-speed steels and carbides. Thus they have less tendency to adhere to metals during cutting and hence less tendency to form a built-up edge. Consequently, good surface finish is obtained with ceramic tools in cutting cast irons and steels. However, ceramics lack toughness, resulting in premature tool failure by chipping or catastrophic failure (see Fig. 8.24).

Ceramic inserts are available in shapes similar to carbide inserts. They are effective in very high speed, uninterrupted cutting operations, such as finishing or semifinishing by turning. To reduce thermal shock, cutting should be performed either dry or with a copious

**FIGURE 8.46** Ranges of properties for various groups of tool materials (see also various tables in this chapter).

amount of cutting fluid applied in a steady stream. Improper or intermittent applications of fluid can cause thermal shock and fracture of the ceramic tool.

Ceramic tool shape and setup are important. Negative rake angles, hence large included angles, are generally preferred in order to avoid chipping due to poor tensile strength. Tool failure can be reduced by increasing the stiffness and damping capacity of machine tools, mountings, and workholding devices thus reducing vibration and chatter (see Section 8.11).

**Cermets.**   *Black,* or *hot-pressed,* ceramics (carboxides) were introduced in the 1960s. They typically contain 70 percent aluminum oxide and 30 percent titanium carbide, and are also called *cermets* (from ceramic and metal). Other cermets contain molybdenum carbide, niobium carbide, and tantalum carbide. Their brittleness and high cost have been a problem when cermets are used as cutting tools. However, further refinements have resulted in improved strength, toughness, and reliability. Although cermets can be coated, the benefits of coatings are somewhat controversial as the wear-resistance improvement appears to be marginal.

### 8.6.7   Cubic boron nitride

Next to diamond, **cubic boron nitride** (CBN) is the hardest material presently available. The CBN cutting tool was introduced in 1962. It is made by bonding a 0.5–1-mm (0.02–0.04-in.) layer of *polycrystalline cubic boron nitride* to a carbide substrate by sintering under pressure (Fig. 8.47). While the carbide provides shock resistance, the CBN layer provides very high wear resistance and cutting-edge strength (Fig. 8.48). Cubic boron nitride tools are also made

**FIGURE 8.47**   Construction of polycrystalline cubic boron nitride or diamond layer on a tungsten-carbide insert.

**FIGURE 8.48**   Inserts with polycrystalline cubic boron nitride tips (top row) and solid polycrystalline CBN inserts (bottom row). *Source:* Courtesy of Valenite, Inc.

in small sizes without a substrate. Because CBN tools are brittle, stiffness of the machine tool and fixturing is important to avoid vibration and chatter.

At elevated temperatures, CBN is chemically inert to iron and nickel and its resistance to oxidation is high. It is therefore particularly suitable for cutting hardened ferrous and high-temperature alloys (see Hard Turning, Section 8.8.2). Cubic boron nitride is also used as an abrasive (see Chapter 9).

### 8.6.8 Silicon-nitride-base ceramics

Developed in the 1970s, **silicon-nitride** (SiN)-**base ceramic** tool materials consist of silicon nitride with various additions of aluminum oxide, yttrium oxide, and titanium carbide. These tools have high toughness, hot hardness, and good thermal-shock resistance.

An example of an SiN-base material is **sialon,** so called after the elements silicon, aluminum, oxygen, and nitrogen in its composition. It has higher thermal-shock resistance than silicon nitride and is recommended for machining cast irons and nickel-base superalloys at intermediate cutting speeds. Because of chemical affinity, SiN-base tools are not suitable for machining steels.

### 8.6.9 Diamond

The hardest substance of all known materials is diamond (see also Section 11.14). It has low friction, high wear resistance, and the ability to maintain a sharp cutting edge. It is used when good surface finish and dimensional accuracy are required, particularly with soft nonferrous alloys and abrasive nonmetallic materials. **Single-crystal** diamonds of various carats are used for special applications, such as machining copper-front precision optical mirrors for the Strategic Defense Initiative (SDI) program.

Because diamond is brittle, tool shape and its sharpness are important. Low rake angles (large included angles) are normally used to provide a strong cutting edge. Special attention should be given to proper mounting and crystal orientation in order to obtain optimum tool use. Diamond wear may occur by microchipping, caused by thermal stresses and oxidation, and transformation to carbon, caused by the heat generated during cutting.

Single-crystal (single-point) diamond tools have been largely replaced by **polycrystalline diamond** tools (*compacts*), which are also used as wire-drawing dies for fine wire. These materials consist of very small synthetic crystals, fused by a high-pressure, high-temperature process to a thickness of about 0.5–1 mm (0.02–0.04 in.) and bonded to a carbide substrate, similar to CBN tools (see Fig. 8.47). The random orientation of the diamond crystals prevents the propagation of cracks through the structure, thus significantly improving its toughness.

Diamond tools can be used satisfactorily at almost any speed, but are suitable mostly for light, uninterrupted finishing cuts. In order to minimize tool fracture, the single-crystal diamond must be resharpened as soon as it becomes dull. Because of its strong chemical affinity, diamond is not recommended for machining plain-carbon steels and titanium, nickel, and cobalt-base alloys. Diamond is also used as an abrasive in grinding and polishing operations (see Chapter 9) and as coatings (see Section 4.5).

### 8.6.10 Whisker-reinforced tool materials

In order to further improve the performance and wear resistance of cutting tools, particularly in machining new workpiece materials and composites that are under continued development, progress is being made in new tool materials with enhanced properties such as (1) high fracture toughness, (2) resistance to thermal shock, (3) cutting-edge strength, and (4) hot hardness.

Recent developments consist of using *whiskers* as reinforcing fibers in composite cutting-tool materials. Examples of **whisker-reinforced** materials include silicon-nitride base tools reinforced with silicon-carbide (SiC) whiskers and aluminum-oxide base tools reinforced with silicon-carbide whiskers, sometimes with the addition of *zirconium oxide* ($ZrO_2$). However, the low inertness of silicon carbide to ferrous metals makes SiC-reinforced tools unsuitable for machining irons and steels.

### 8.6.11  Cutting-tool reconditioning

When tools, particularly high-speed steels, become worn, they are **reconditioned** (resharpened) for further use. They are usually ground on tool and cutter grinders in toolrooms having special fixtures. The reconditioning may be carried out either by hand, which requires considerable operator skill, or on computer-controlled tool and cutter grinders. Other methods (see Chapter 9) may also be used to recondition tools and cutters. Reconditioning may also involve recoating used tools with titanium nitride.

Consistency and precision in reconditioning are important. Resharpened tools should be inspected for their shape and surface finish. As described earlier, inserts are now recycled after use. Whether tools should be reconditioned or recycled depends on the relative costs involved. Skilled labor is costly, as are computer-controlled grinders. Thus an additional consideration is the possible recycling of tool materials and the costs involved, since many contain expensive materials of strategic importance, such as tungsten and cobalt.

## 8.7   Cutting Fluids

Also called lubricants and coolants, **cutting fluids** are used extensively in machining operations to:

- Reduce friction and wear, thus improving tool life and surface finish.
- Reduce forces and energy consumption.
- Cool the cutting zone, thus reducing workpiece temperature and distortion and improving tool life.
- Wash away the chips.
- Protect the newly machined surfaces from environmental corrosion.

A cutting fluid can interchangeably be a *coolant* and a *lubricant*. Its effectiveness in cutting operations depends on a number of factors, such as the method of application, temperature, cutting speed, and type of machining operation. As we have shown, temperature increases as cutting speed increases. Thus cooling of the cutting zone is of major importance at high cutting speeds. Water is an excellent coolant; however, it causes rusting of workpieces and machine components and is a poor lubricant. On the other hand, if the speed is low, such

| Cutting operation | Operation severity | Cutting fluid activity |
|---|---|---|
| | High | High |
| Broaching (internal) | | |
| Tapping | | |
| Broaching (surface) | | |
| Form and thread grinding | | |
| Threading (general) | | |
| Gear shaping (rotary) | | |
| Thread rolling (rotary) | | |
| Gear shaping | | |
| Reaming | | |
| Deep drilling | | |
| Drilling, boring | | |
| Thread rolling (rack) | | |
| Hobbing, gear (spline) | | |
| Milling | | |
| Turning | | |
| Band and hack sawing | | |

**FIGURE 8.49** Relative severity of machining operations. The higher the severity, the greater is the need for an effective cutting fluid. *Source:* Mobil Oil Corporation.

as in broaching or tapping, lubrication—not cooling—is the important factor. Lubrication reduces the tendency for built-up edge formation and thus improves surface finish.

The relative severity of various machining operations is shown qualitatively in Fig. 8.49, which also includes the relative cutting speeds employed. Severity is defined as the magnitude of temperatures and forces encountered, the tendency for built-up edge formation, and the ease with which chips are disposed of from the cutting zone. Note how important cutting-fluid effectiveness is as severity increases.

There are situations, however, in which the use of cutting fluids can be detrimental. In interrupted cutting operations, such as milling, the cooling action of the cutting fluid increases the extent of alternate heating and cooling (thermal cycling) to which the cutter teeth are subjected. This condition can lead to thermal cracks (thermal fatigue or thermal shock). Cutting fluids may also cause the chip to become curlier, thus concentrating the stresses near the tool tip. These stresses, in turn, concentrate the heat closer to the tip and reduce tool life.

### 8.7.1 Cutting-fluid action

Although the basic lubrication mechanisms in metalworking operations are described in greater detail in Section 4.4.3, we briefly discuss here the mechanisms by which cutting fluids influence machining operations. In view of the high contact pressures and relative sliding at the tool–chip interface, how does a cutting fluid penetrate this interface to influence the cutting process?

It appears that the fluid is drawn into the tool–chip interface by the *capillary action* of the interlocking network of surface asperities. (Fig. 8.50, see also Section 4.4.1). Studies have shown that the cutting fluid gains access to the interface by seeping from the sides of the chip. Because of the small size of this capillary network, the cutting fluid should have a small molecular size and proper wetting (surface tension) characteristics.

Chip

Tool

Cutting fluid

0.0001 in.

0.0025 mm

---

### Example 8.5:   Effect of cutting fluids on machining

A machining operation is being carried out with a cutting fluid that is effective as a lubricant. Explain the changes in the mechanics of the cutting operation and total energy consumption if the fluid is shut off.

SOLUTION.    Since the cutting fluid is a good lubricant, when the fluid is shut off, the friction at the tool-chip interface will increase. The following chain of events then takes place:

1.  Fluid is shut off.
2.  Friction at the tool–chip interface increases.
3.  The shear angle decreases.
4.  The shear strain increases.
5.  The chip is thicker.
6.  A built-up edge is likely to form.

As a consequence:

1.  The shear energy in the primary zone increases.
2.  The friction energy in the secondary zone increases.
3.  Hence the total energy increases.
4.  Surface finish is likely to deteriorate.
5.  The temperature in the cutting zone increases; hence tool wear increases.
6.  Tolerances may be difficult to maintain because of the increased temperature and expansion of the workpiece during machining.

### 8.7.2 Types of cutting fluids

There are generally four types of cutting fluids commonly used in machining operations: *oils, emulsions, semisynthesis,* and *synthetics.* The general characteristics of these fluids are described in Section 4.4.4. Cutting fluid recommendations for specific machining operations are given throughout this chapter.

### 8.7.3 Methods of application

The most common method of applying cutting fluid is **flood cooling.** Flow rates range from 10 L/min (3 gal/min) for single-point tools, to 225 L/min (60 gal/min) per cutter for multiple-tooth cutters, such as in milling. In operations such as gun drilling and end milling, fluid pressures of 700–14,000 kPa (100–2000 psi) are used to wash away the chips.

**Mist cooling** is another method of applying cutting fluids and is used particularly with water-base fluids. Although it requires venting (to prevent inhaling of fluid particles by the machine operator and others nearby) and has limited cooling capacity, mist cooling supplies fluid to otherwise inaccessible areas and provides better visibility of the workpiece being machined. It is particularly effective in grinding operations and at air pressures of 70–600 kPa (10–80 psi).

With increasing speed and power of machine tools, heat generation in machining operations has become a significant factor. New developments include the use of *high-pressure* refrigerated coolant systems to improve the rate of heat removal from the cutting zone. High pressures are also used in delivering the cutting fluid to the cutting zone by specially designed nozzles that aim a powerful jet of fluid to the zone. This action breaks up the chips (thus acting as a chip breaker) in situations where the chips produced would otherwise be long and continuous, and thus interfere with the cutting operation.

### 8.7.4 Effects of cutting fluids

The selection of a cutting fluid should also include other considerations, such as effects on workpiece material and machine tools and biological and environmental factors.

**Effects on workpiece material.** When selecting a cutting fluid, you should consider whether the machined component will be subjected to environmental attack and high service stresses, thus possibly leading to stress-corrosion cracking. This consideration is particularly important for cutting fluids having sulfur and chlorine additives. For example, fluids containing chlorine should not be used when machining titanium. Additional considerations are staining of the workpiece by cutting fluids, especially on copper and aluminum. Machined parts should be cleaned and washed in order to remove any cutting-fluid residue.

**Effects on machine tools.** Just as a cutting fluid may adversely affect the workpiece material, it can similarly affect the machine tool and its various components, such as slideways and bearings. The choice of fluid must therefore include consideration of its compatibility with the machine-member materials.

**Biological and environmental effects.** Because the machine-tool operator is usually in close proximity to cutting fluids, the effects of operator contact with fluids should be of primary concern. Fumes, smoke, and odors from cutting fluids can cause severe skin reactions, as well as respiratory problems. Considerable progress has been made in ensuring the safe use of cutting fluids in manufacturing plants. Additionally, the effect on the external environment, particularly with regard to degradation of the fluid, is important.

Cutting fluids, as well as other metalworking fluids used in manufacturing operations, may undergo chemical changes as they are used over time due to environmental effects as well as contamination from various sources, including chips and metal fines produced during machining, and tramp oil. These changes involve the growth of microbes (bacteria, molds, and yeast), particularly in the presence of water, and adversely affect the properties and characteristics of the fluids, thus influencing the machining operation.

**Clarification, recycling, and disposal.**   Several techniques are available for clarifying used cutting fluids such as settling, skimming, centrifuging, and filtering. Recycling involves treatment of the fluids with various additives, agents, biocides, and deodorizers, as well as water treatment (for water-based fluids). Disposal practices must comply with federal, state, and local laws and regulations.

## 8.8   Cutting Processes and Machine Tools for Producing Round Shapes

In this section we describe processes that produce parts that are basically round in shape. Typical products made include parts as small as miniature screws for eyeglass-frame hinges and as large as shafts, pistons, cylinders, gun barrels, and turbines for hydroelectric power plants. These processes are usually performed by turning the workpiece on a lathe. *Turning* means that the part is rotating while it is being machined. The starting material is usually a workpiece that has been made by other processes, such as casting, shaping, forging, extrusion, and drawing. Turning processes are versatile and capable of producing a wide variety of shapes, as we outline in Fig. 8.51:

- **Turning** straight, conical, curved, or grooved workpieces, such as shafts, spindles, pins, handles, and various machine components.
- **Facing,** to produce a flat surface at the end of the part, such as parts that are attached to other components, or to produce grooves for O-ring seats.
- Producing various shapes by **form tools,** such as for functional purposes or for appearance.
- **Boring,** to enlarge a hole made by a previous process or in a tubular workpiece or to produce internal grooves.
- **Drilling,** to produce a hole, which may be followed by boring to improve its accuracy and surface finish.
- **Parting,** also called **cutting off,** to cut a piece from the end of a part, as in making slugs or blanks for additional processing into discrete products.
- **Threading,** to produce external and internal threads in workpieces.
- **Knurling,** to produce a regularly shaped roughness on cylindrical surfaces, as in making knobs.

These operations may be performed at various rotational speeds of the workpiece, depths of cut, $d$, and feed, $f$, depending on the workpiece and tool materials, the surface fin-

**FIGURE 8.51** Various cutting operations that can be performed on a lathe.

ish and dimensional accuracy required, and the capacity of the machine tool. Cutting speeds are usually in the range of 0.15–4 m/s (30–800 ft/min). *Roughing cuts,* which are performed for large-scale material removal, usually involve depths of cut greater than 0.5 mm (0.02 in.) and feeds on the order of 0.2–2 mm/rev (0.008–0.08 in./rev). *Finishing cuts* usually involve lower depths of cut and feed.

## 8.8.1 Turning parameters

The majority of turning operations involve simple single-point cutting tools. The geometry of a typical right-hand cutting tool for turning is shown in Fig. 8.52. Such tools are described by a standardized nomenclature. Each group of tool and workpiece materials has an optimum set of tool angles, which was developed largely through experience (Table 8.11).

**Tool geometry.** The various tool angles used have important functions in cutting operations. **Rake angles** are important in controlling the direction of chip flow and the strength of the tool tip. Positive angles improve the cutting operation by reducing forces and temperatures.

(a)

End view

Top view

End section A–A

Side view

| Tool Signature | Dimensions | Abbreviation |
|---|---|---|
| 8 . . . . . . . . . . . . . . | Back rake angle | BR |
| 14 . . . . . . . . . . . . . . | Side rake angle | SR |
| 6 . . . . . . . . . . . . . . | End relief angle | ER |
| 12 . . . . . . . . . . . . . . | End clearance angle | . . . |
| 6 . . . . . . . . . . . . . . | Side relief angle | SRF |
| 12 . . . . . . . . . . . . . . | Side clearance angle | . . . |
| 20 . . . . . . . . . . . . . . | End cutting-edge angle | ECEA |
| 15 . . . . . . . . . . . . . . | Side cutting-edge angle | SCEA |
| 1/8 . . . . . . . . . . . . . . | Nose radius | NR |

(b)

**FIGURE 8.52**   (a) Designations and symbols for a right-hand cutting tool; solid high-speed-steel tools have a similar designation. Right-hand means that the tool travels from right to left as shown in Fig. 8.20. (b) Square insert in a right-hand toolholder for a turning operation. A wide variety of toolholders are available for holding inserts at various angles. Thus, the angles shown in (a) can be achieved easily by selecting an appropriate insert and toolholder. *Source:* Kennametal, Inc.

**TABLE 8.11**

**GENERAL RECOMMENDATIONS FOR TURNING TOOL ANGLES**

| MATERIAL | HIGH-SPEED STEEL | | | | | CARBIDE (INSERTS) | | | | |
|---|---|---|---|---|---|---|---|---|---|---|
| | BACK RAKE | SIDE RAKE | END RELIEF | SIDE RELIEF | SIDE AND END CUTTING EDGE | BACK RAKE | SIDE RAKE | END RELIEF | SIDE RELIEF | SIDE AND END CUTTING EDGE |
| Aluminum and magnesium alloys | 20 | 15 | 12 | 10 | 5 | 0 | 5 | 5 | 5 | 15 |
| Copper alloys | 5 | 10 | 8 | 8 | 5 | 0 | 5 | 5 | 5 | 15 |
| Steels | 10 | 12 | 5 | 5 | 15 | −5 | −5 | 5 | 5 | 15 |
| Stainless steels | 5 | 8–10 | 5 | 5 | 15 | −5–0 | −5–5 | 5 | 5 | 15 |
| High-temperature alloys | 0 | 10 | 5 | 5 | 15 | 5 | 0 | 5 | 5 | 45 |
| Refractory alloys | 0 | 20 | 5 | 5 | 5 | 0 | 0 | 5 | 5 | 15 |
| Titanium alloys | 0 | 5 | 5 | 5 | 15 | −5 | −5 | 5 | 5 | 5 |
| Cast irons | 5 | 10 | 5 | 5 | 15 | −5 | −5 | 5 | 5 | 15 |
| Thermoplastics | 0 | 0 | 20–30 | 15–20 | 10 | 0 | 0 | 20–30 | 15–20 | 10 |
| Thermosets | 0 | 0 | 20–30 | 15–20 | 10 | 0 | 15 | 5 | 5 | 15 |

However, positive angles produce a small included angle of the tool tip (see Fig. 8.2). Depending on the toughness of the tool material, this may cause premature tool chipping and failure. **Side rake angle** is more important than **back rake angle,** although the latter usually controls the direction of chip flow.

**Relief angles** control interference and rubbing at the tool-workpiece interface. If the relief angle is too large, the tool may chip off; if too small, flank wear may be excessive. **Cutting edge angles** affect chip formation, tool strength, and cutting forces to various degrees. **Nose radius** affects surface finish and tool-tip strength. Also, the sharper the radius, the rougher will be the surface finish of the workpiece and the lower will be the strength of the tool. However, large nose radii can lead to tool chatter.

**Material removal rate.**   The *material removal rate* (MRR) is the volume of material removed per unit time, such as $mm^3/min$ or $in^3/min$. Referring to Fig. 8.53a, note that for each revolution of the workpiece we remove a ring-shaped layer of material whose cross-sectional area is the product of the distance the tool travels in one revolution (the feed, $f$) and the depth of cut $d$. The volume of this ring is the product of the cross-sectional area, that is, $(f)(d)$, and the average circumference of the ring, that is, $\pi D_{avg}$, where $D_{avg} = (D_o + D_f)/2$. For light cuts on large-diameter workpieces, we can replace the average diameter by $D_o$. The rotational speed of the workpiece is $N$.

Thus, the material removal rate per revolution will be $\pi D_{avg}df$. Since we have $N$ revolutions per minute, the removal rate is

$$\text{MRR} = \pi D_{avg} d f N. \tag{8.35}$$

We can check the dimensional accuracy of this equation by substituting dimensions into the right-hand side; thus, (mm)(mm)(mm/rev)(rev/min) = $mm^3/min$, which indicates volume rate of removal. Similarly, the cutting time $t$ for a workpiece of length $l$ can be calculated by

(a)                                                                      (b)

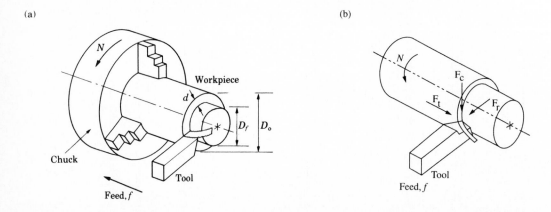

**FIGURE 8.53**   (a) Schematic illustration of a turning operation showing depth of cut, $d$, and feed, $f$. Cutting speed is the surface speed of the workpiece at the tool tip. (b) Forces acting on a cutting tool in turning. $F_c$ is the cutting force; $F_t$ is the thrust or feed force (in the direction of feed); and $F_r$ is the radial force that tends to push the tool away from the workpiece being machined. Compare this figure with Fig. 8.12 for a two-dimensional cutting operation.

noting that the tool travels at a feed rate of $fN =$ (mm/rev)(rev/min) = mm/min. Since the distance traveled is $l$ mm, the cutting time is

$$t = \frac{l}{fN}. \tag{8.36}$$

This time does not include the time required for *tool approach* and *retraction*. Because the time spent in noncutting cycles of a machining operation is nonproductive and affects the overall economics, the time involved in approaching and retracting tools to and from the workpiece is an important consideration. Machine tools are now being designed and built to minimize this time—one method is first by rapid traverse of the tools, then a slower movement as the tool engages the workpiece.

**Forces in turning.**    The three forces acting on a cutting tool are shown in Fig. 8.53b. The *cutting force* $F_c$ acts downward on the tool tip and tends to deflect the tool downward. This is the force that supplies the energy required for the cutting operation. As we can see from the upcoming example on material removal rate, this force can be calculated from the energy per unit volume described in Section 8.2.5 and by using Table 8.4. The *thrust force* $F_t$ acts in the longitudinal direction; this force is also called the *feed force* because it is in the feed direction. The *radial force* $F_r$ is in the radial direction and tends to push the tool away from the workpiece. Forces $F_t$ and $F_r$ are difficult to calculate because of the many factors involved in the cutting process, so they are determined experimentally. These forces are important in the design of machine tools as well as in the deflection of tools for precision machining operations.

**Tool materials, feeds, and cutting speeds.**    The general characteristics of cutting-tool materials were described in Section 8.6. As a general guideline in turning operations, a broad range of applicable cutting speeds and feeds for these tool materials is given in Fig. 8.54.

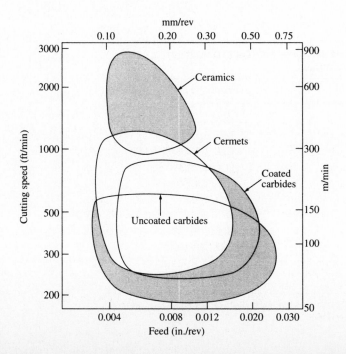

**FIGURE 8.54**  The range of applicable cutting speeds and feeds for a variety of tool materials. *Source:* Valenite, Inc.

**TABLE 8.12**

**APPROXIMATE RANGE OF RECOMMENDED CUTTING SPEEDS FOR TURNING OPERATIONS**

| WORKPIECE MATERIAL | CUTTING SPEED | |
| --- | --- | --- |
| | m/min | ft/min |
| Aluminum alloys | 200–1000 | 650–3300 |
| Cast iron, gray | 60–900 | 200–3000 |
| Copper alloys | 50–700 | 160–2300 |
| High-temperature alloys | 20–400 | 65–1300 |
| Steels | 50–500 | 160–1600 |
| Stainless steels | 50–300 | 160–1000 |
| Thermoplastics and themosets | 90–240 | 300–800 |
| Titanium alloys | 10–100 | 30–330 |
| Tungsten alloys | 60–150 | 200–500 |

Note: (a) These speeds are for carbides and ceramic cutting tools. Speeds for high-speed-steel tools are lower than indicated. The higher ranges are for coated carbides and cermets. Speeds for diamond tools are considerably higher than those indicated.
(b) Depths of cut, $d$, are generally in the range of 0.5–12 mm (0.02–0.5 in.).
(c) Feeds, $f$, are generally in the range of 0.15–1 mm/rev (0.006–0.040 in./rev).

Specific recommendations for turning cutting speeds for various workpiece materials and cutting tools are given in Table 8.12. In this table, **roughing** means cutting material without much consideration to tolerances and surface finish. **Finishing** means cutting to obtain final dimensions with acceptable tolerances and surface finish.

**TABLE 8.13**

**GENERAL RECOMMENDATIONS FOR CUTTING FLUIDS FOR MACHINING**

| MATERIAL | TYPE OF FLUID |
| --- | --- |
| Aluminum | D, MO, E, MO + FO, CSN |
| Beryllium | MO, E, CSN |
| Copper | D, E, CSN, MO + FO |
| Magnesium | D, MO, MO + FO |
| Nickel | MO, E, CSN |
| Refractory alloys | MO, E, EP |
| Steels (carbon and low alloy) | D, MO, E, CSN, EP |
| Steels (stainless) | D, MO, E, CSN |
| Titanium | CSN, EP, MO |
| Zinc | D, MO, E, CSN |
| Zirconium | D, E, CSN |

Note: CSN, chemicals and synthetics; D, dry; E, emulsion; EP, extreme pressure; FO, fatty oil; and MO, mineral oil.

**TABLE 8.14**

**SUMMARY OF TURNING PARAMETERS AND FORMULAS**

$N$ = Rotational speed of the workpiece, rpm

$f$ = Feed, mm/rev or in./rev

$v$ = Feed rate, or linear speed of the tool along workpiece length, mm/min or in./min $= fN$

$V$ = Surface speed of workpiece, m/min or ft/min

    $= \pi D_0 N$ (for maximum speed)

    $= \pi D_{avg} N$ (for average speed)

$l$ = Length of cut, mm or in.

$D_0$ = Original diameter of workpiece, mm or in.

$D_f$ = Final diameter of workpiece, mm or in.

$D_{avg}$ = Average diameter of workpiece, mm or in.

    $= (D_0 + D_f)/2$

$d$ = Depth of cut, mm or in.

  $= (D_0 - D_f)/2$

$t$ = Cutting time, s or min

  $= l/fN$

MRR $= \text{mm}^3/\text{min or in.}^3/\text{min}$

    $= \pi D_{avg} d f N$

Torque $= \text{N} \cdot \text{m or lb} \cdot \text{ft}$

      $= (F_c)(D_{avg}/2)$

Power $= \text{kW or hp}$

      $= (\text{Torque})(\omega)$, where $\omega = 2\pi N$ radians/min

Note: The units given are those that are commonly used; however, appropriate units must be used and checked in the formulas.

**Cutting fluids.** Although many metals and nonmetallic materials can be machined without a cutting fluid, in many cases the application of a cutting fluid can improve the operation significantly. General recommendations for cutting fluids for various workpiece materials are given in Table 8.13.

## Example 8.6: Material removal rate and cutting force in turning

A 6 in. long, $\frac{1}{2}$ in. diameter 304 stainless steel rod is being reduced in diameter to 0.480 in. by turning on a lathe. The spindle rotates at $N = 400$ rpm, and the tool is traveling at an axial speed of 8 in./min. Calculate the cutting speed, material removal rate, time to cut, power dissipated, and cutting force. (See also Table 8.14.)

SOLUTION. The cutting speed is the tangential speed of the workpiece. The maximum cutting speed is at the outer diameter $D_o$ and is obtained from the expression

$$V = \pi D_o N.$$

Thus,

$$V = (\pi)(0.500)(400) = 628 \text{ in./min} = 52 \text{ ft/min.}$$

The cutting speed at the machined diameter is

$$V = (\pi)(0.480)(400) = 603 \text{ in./min} = 50 \text{ ft/min.}$$

From the information given, we note that the depth of cut is

$$d = (0.500 - 0.480)/2 = 0.010 \text{ in.,}$$

and the feed is

$$f = 8/400 = 0.02 \text{ in./rev.}$$

Thus, according to Eq. (8.35), the material removal rate is

$$\text{MRR} = (\pi)(0.490)(0.010)(0.02)(400) = 0.123 \text{ in}^3/\text{min.}$$

The actual time to cut, according to Eq. (8.36), is

$$t = 6/(0.02)(400) = 0.75 \text{ min.}$$

We can calculate the power required by referring to Table 8.4 and taking an average value for stainless steel as 4 W·s/mm$^3$ = 4/2.73 = 1.47 hp·min/in$^3$. Thus power dissipated is

$$\text{Power} = (1.47)(0.123) = 0.181 \text{ hp,}$$

and since 1 hp = 396,000 in.-lb/min, the power dissipated is 71,700 in.-lb/min. The cutting force $F_c$ is the tangential force exerted by the tool. Since power is the product of torque $T$ and rotational speed in radians per unit time, we have

$$T = (71,700)/(400)(2\pi) = 29 \text{ lb-in.}$$

Since $T = (F_c)(D_{\text{avg}}/2)$, we have

$$F_c = (29)/(0.490/2) = 118 \text{ lb.}$$

---

### 8.8.2  Lathes and lathe operations

**Lathes** are generally considered to be the oldest machine tools. Although woodworking lathes were first developed during the period 1000–1 B.C., metalworking lathes with lead screws were not built until the late 1700s. The most common lathe, shown schematically in Fig. 8.55, was originally called an *engine lathe* because it was powered with overhead pulleys and belts from nearby engines. Today these lathes are equipped with individual electric motors. Although simple and versatile, an engine lathe requires a skilled machinist because all controls are manipulated by hand. Consequently, it is inefficient for repetitive operations and

**FIGURE 8.55** Schematic illustration of the components of a lathe.

for large production runs. However, various types of automation can be added to improve efficiency, as described in the rest of this section.

**Lathe components.** Lathes are equipped with a variety of components and accessories (Fig. 8.55). The *bed* supports all the other major components of the lathe. The *carriage* or *carriage assembly*, which slides along the ways, consists of an assembly of the cross-slide, tool post, and apron. The cutting tool is mounted on the *tool post*, usually with a *compound rest* that swivels for tool positioning and adjustments. The *headstock*, which is fixed to the bed, is equipped with motors, pulleys, and V-belts that supply power to the *spindle* at various rotational speeds. Headstocks have a hollow spindle to which workholding devices, such as *chucks* and *collets*, are attached. The *tailstock*, which can slide along the ways and be clamped at any position, supports the other end of the workpiece. The *feed rod*, which is powered by a set of gears from the headstock, rotates during operation of the lathe and provides movement to the carriage and the cross-slide. The *lead screw*, which is used for cutting threads accurately, is engaged with the carriage by closing a split nut around the lead screw.

A lathe is usually specified by its *swing*, that is, the maximum diameter of the workpiece that can be machined, by the maximum distance between the headstock and tailstock centers,

Chuck

Round turret for
OD operations

CNC unit

End turret for
ID operations

Tailstock

**FIGURE 8.56**    A computer numerical control lathe. Note the two turrets on this machine.
*Source:* Jones & Lamson.

and by the length of the bed. A variety of lathe designs are available, including *bench lathes, toolroom lathes, engine lathes, gap lathes,* and *special-purpose lathes.*

*Tracer lathes* are machine tools and attachments that are capable of turning parts with various contours. Also called *duplicating lathes* or *contouring lathes,* the cutting tool follows a path that duplicates the contour of a template through a hydraulic or electrical system. *Automatic lathes,* also called *chucking machines,* or *chuckers,* which are usually vertical and do not have tailstocks, are used for machining individual pieces of regular or irregular shapes. *Turret lathes* are capable of performing multiple cutting operations on the same workpiece, such as turning, boring, drilling, thread cutting, and facing. Several cutting tools are mounted on the hexagonal *main turret.* The lathe usually has a *square turret* on the cross-slide, with as many as four cutting tools mounted on it. Computer-controlled lathes (Fig. 8.56) are the most advanced, with their movement and control actuated by computer numerical controls.

**Computer-controlled lathes.**    In the most advanced lathes, movement and control of the machine and its components are actuated by **computer numerical controls (CNC)**. The features of such a lathe are shown in Fig. 8.56. These lathes are usually equipped with one or more turrets. Each turret is equipped with a variety of tools and performs several operations on different surfaces of the workpiece. These machines are highly automated, the operations are repetitive and maintain the desired accuracy, and less-skilled labor is required (after the machine is set up). They are suitable for low to medium volumes of production. The details of computer controls are presented in Chapters 13 and 14 (see also Section 8.10).

**Example 8.7:    Typical parts made on computer numerical control turning machine tools**

The capabilities of CNC turning machine tools are illustrated in Fig. 8.57. Material and number of cutting tools used and machining times are indicated for each part. Although not

(a) Housing base

67.4 mm
(2.654″)

87.9 mm
(3.462″)

98.4 mm
(3.875″)

85.7 mm (3.375″)
32 threads per in.

**Material: Titanium Alloy**
**Number of tools: 7**
**Total machining time**
**(two operations):**
**5.25 minutes**

(b) Inner bearing race

235.6 mm
(9.275″)

78.5 mm
(3.092″)

Material: 52100
Alloy Steel
Number of tools: 4
Total machining time
(two operations):
6.32 minutes

(c) Tube reducer

50.8 mm
(2″)

23.8 mm
(0.938″)

53.2 mm
(2.094″)

Material: 1020 Carbon Steel
Number of tools: 8
Total machining time
(two operations):
5.41 minutes

**FIGURE 8.57**

as effectively or consistently, these parts can also be made on manual or turret lathes.
*Source:* Monarch Machine Tool Company.

## Turning process capabilities

Relative *production rates* in turning, as well as other cutting operations that are discussed in the rest of this chapter, are shown in Table 8.15. These rates have an important bearing on productivity in machining operations. Note that there are major differences in the production rate among these processes. These differences are not only due to the inherent characteristics of the processes and machine tools, but are also due to various other factors such as setup times and the types and sizes of the workpieces involved. The proper selection of a process and the machine tool for a particular product is essential for minimizing production costs, as is also discussed in Section 8.13 and Chapter 15.

As stated above, the ratings in Table 8.15 are relative and there can be significant variations in specialized applications. For example, heat-treated high-carbon cast steel rolls (for rolling mills, Section 6.3) can be machined on special lathes at material removal rates as high as 6000 cm$^3$/min (370 in$^3$/min) using multiple cermet tools. The important factor in this operation (also called *high removal rate machining*) is the very high rigidity of the machine tool (to avoid tool breakage due to chatter; see Section 8.11) and its high power, which can be up to 450 kW (600 hp).

The surface finish and dimensional accuracy obtained in turning and related operations depend on factors such as the characteristics and condition of the machine tool, stiffness, vibration and chatter, process parameters, tool geometry and wear, cutting fluids, machinability of the workpiece material, and operator skill. As a result, a wide range of surface finishes can be obtained, as shown in Fig. 8.33 (see also Fig. 9.31).

**High-speed machining.**   With increasing demands for higher productivity and lower production costs, investigations have been carried out since the late 1950s to increase the material

**TABLE 8.15**

**TYPICAL PRODUCTION RATES FOR VARIOUS CUTTING OPERATIONS**

| OPERATION | RATE |
|---|---|
| Turning | |
|     Engine lathe | Very low to low |
|     Tracer-lathe | Low to medium |
|     Turret lathe | Low to medium |
|     Computer-control lathe | Low to medium |
|     Single-spindle chuckers | Medium to high |
|     Multiple-spindle chuckers | High to very high |
| Boring | Very low |
| Drilling | Low to medium |
| Milling | Low to medium |
| Planing | Very low |
| Gear cutting | Low to medium |
| Broaching | Medium to high |
| Sawing | Very low to low |

Note: Production rates indicated are relative: Very low is about one or more parts per hour; medium is approximately 100 parts per hour; very high is 1000 or more parts per hour.

removal rate in machining, particularly for applications in the aerospace and automotive industries. As we have seen, one obvious possibility is to increase the surface cutting speed.

The term *high speed* is relative. As shown in Table 8.12, general recommendations for typical applications state that aluminum alloys, for example, should be machined at much higher speeds than titanium alloys or stainless steels. As a general guide, however, an approximate range of speeds may be defined as follows: High speed: 600–1800 m/min (2000–6000 ft/min); very high speed: 1800–18,000 m/min (6000–60,000 ft/min); ultrahigh speed: greater than 18,000 m/min.

Much research and development work has been carried out on **high-speed machining** (turning, as well as milling, boring, and drilling) of aluminum alloys, titanium alloys, steels, and superalloys. Considerable data have been collected regarding the effect of high speeds on the type of chips produced, cutting forces, temperatures generated, tool wear, surface finish, and the economics of the process. These studies have indicated that high-speed machining can be economical for certain applications, and consequently, it is now implemented for machining aircraft turbine components and automotive engines with five to ten times the productivity of traditional machining. Important factors in these operations are the selection of an appropriate cutting tool, the power of the machine tools and their stiffness, the stiffness of toolholders and the workholding devices, spindle design for high power and high rotational speeds, the inertia of the machine-tool components, fast feed drives, and the level of automation.

It is important to note, however, that high-speed machining should be considered basically for situations in which *cutting time* is a significant portion of the floor-to-floor time of the operation. Other factors such as *noncutting time* and labor costs are important considerations in the overall assessment of the benefits of high-speed machining for a particular application.

**Ultraprecision machining.** Beginning with the 1960s, increasing demands have been made on precision manufacturing of components for computer, electronics, nuclear energy, and defense applications. Examples are optical mirrors, computer memory disks, and drums for photocopying machines, with surface finish in the tens of nanometer ($10^{-9}$ m or 0.001 $\mu$m; 0.04 $\mu$in.) range and form accuracies in the $\mu$m and sub-$\mu$m range. The cutting tool for these **ultraprecision machining** applications is exclusively a single-crystal diamond (hence the process is also called *diamond turning*), with a polished cutting-edge having a radius as small as in the tens of nanometer range. Wear of the diamond can be a significant problem, and recent advances include cryogenic diamond turning in which the tooling system is cooled by liquid nitrogen to a temperature of about $-120°$C ($-184°$F).

The materials for ultraprecision machining to date include copper alloys, aluminum alloys, silver, gold, electroless nickel, infrared materials, and plastics (acrylics). The depths of cut involved are in the nanometer range. In this range, hard and brittle materials produce continuous chips (known as *ductile-regime cutting;* see also ductile-regime grinding in Section 9.5.3); deeper cuts produce discontinuous chips.

The machine tools for these applications are built with very high precision and high machine, spindle, and workholding-device stiffness. These ultraprecision machines, parts of which are made of structural materials with low thermal expansion and good dimensional stability, are located in a dust-free environment (clean rooms) where the temperature is controlled within a fraction of one degree. Vibrations from external and internal (machine) sources are also avoided as much as possible. Feed and position controls are made by laser metrology, and the machines are equipped with highly advanced computer control systems and with thermal and geometric error compensating features.

**Hard turning.** We have seen that as the hardness of the workpiece increases, its machinability decreases accordingly and tool wear and fracture, as well as surface finish and integrity, can become a significant problem. As described in Chapter 9, there are several other mechanical (particularly grinding) processes and nonmechanical methods of removing material economically from hard or hardened metals. However, it is still possible to apply traditional cutting processes to hard metals and alloys by selecting an appropriate tool material and machine tools with high stiffness.

One common example is finish machining of heat-treated steel machine and automotive components using polycrystalline cubic boron nitride (PCBN) cutting tools. Called **hard turning,** this process produces machined parts with good dimensional accuracy, surface finish, and surface integrity. It can compete successfully with grinding the same components, from both technical and economic aspects. A comparative example of hard turning versus grinding is presented in Example 9.4.

**Cutting screw threads.** Threads are produced primarily by forming (*thread rolling*), which constitutes the largest quantity of threaded parts produced, or cutting. Casting threaded parts is also possible, although dimensional accuracy and production rate are not as high as those obtained in the other processes. Figure 8.51k shows that turning operations are capable of producing threads on round bar stock. When threads are produced externally or internally by cutting with a lathe-type tool, the process is called *thread cutting* or *threading*.

When cut internally with a special threaded tool (*tap*), the process is called *tapping*. External threads may also be cut with a die or by milling. Although it adds considerably to cost, threads may be ground for improved accuracy and surface finish.

Automatic screw machines are designed for high-production-rate machining of screws and similar threaded parts. Because they are capable of producing other components, they are now called *automatic bar machines*. All operations on these machines are performed automatically, with tools attached to a special turret. The bar stock is fed forward automatically after each screw or part is machined to finished dimensions and cut off. These machines may be equipped with single or multiple spindles, and capacities range from 3 to 150 mm ($\frac{1}{8}$ to 6 in.) diameter bar stock.

### 8.8.3   Boring and boring machines

The basic boring operation, as carried out on a lathe, is shown in Fig. 8.51h. **Boring** consists of producing circular internal profiles in hollow workpieces or on a hole made by drilling or another process and is carried out using cutting tools that are similar to those used in turning. However, because the boring bar has to reach the full length of the bore, tool deflection and hence maintenance of dimensional accuracy can be a significant problem. The boring bar must be sufficiently stiff—that is, a material with high elastic modulus, such as carbides—in order to minimize deflection and avoid vibration and chatter. Boring bars have been designed with capabilities for damping vibrations.

Although boring operations on relatively small workpieces can be carried out on a lathe, **boring mills** are used for large workpieces. These machines are either vertical or horizontal, and are capable of performing operations such as turning, facing, grooving, and chamfering. A *vertical* boring machine (Fig. 8.58) is similar to a lathe but with a vertical axis of workpiece rotation. In *horizontal* boring machines the workpiece is mounted on a table that can move horizontally both axially and radially. The cutting tool is mounted on a spindle that rotates in the headstock, which is capable of both vertical and longitudinal motions. Drills, reamers, taps, and milling cutters can also be mounted on the spindle.

**FIGURE 8.58**  Schematic illustration of the components of a vertical boring mill.

Cross-rail

Tool head

Workpiece

Work table

Bed

Column

**FIGURE 8.59** (a) Standard chisel-point drill indicating various features. (b) Crankshaft-point drill.

### 8.8.4 Drilling, reaming, and tapping

One of the most common of all machining processes is drilling. **Drills** usually have a high length-to-diameter ratio (Fig. 8.59 and Table 8.16) and thus are capable of producing deep holes. However, they are somewhat flexible, depending on their diameter, and should be used with care in order to drill holes accurately and to prevent the drill from breaking. Moreover, the chips are produced within the workpiece, and the chips have to move in a direction opposite to the axial movement of the drill. Consequently, chip disposal and the effectiveness of cutting fluids can present significant problems.

The most common drill is the standard-point twist drill (Fig. 8.59). The main features of the drill point are a *point angle, lip-relief angle, chisel-edge angle,* and *helix angle.* The geometry of the drill tip is such that the normal rake angle and velocity of the cutting edge vary with the distance from the center of the drill. Other types of drills (Fig. 8.60) include the *step* drill, *core* drill, *counterboring* and *countersinking* drills, *center* drill, and *spade* drill. *Crankshaft* drills have good centering ability, and because chips tend to break up easily, these drills are suitable for drilling deep holes. *Gun drilling* requires a special drill and is used for drilling deep holes; hole depth-to-diameter ratios can be 300 or higher.

In the *trepanning* technique a cutting tool produces a hole by removing a disk-shaped piece (*core*), usually from flat plates. Thus a hole is produced without reducing all the material

**TABLE 8.16**

**GENERAL RECOMMENDATIONS FOR DRILL GEOMETRY FOR HIGH-SPEED STEEL TWIST DRILLS**

| WORKPIECE MATERIAL | POINT ANGLE | LIP-RELIEF ANGLE | CHISEL-EDGE ANGLE | HELIX ANGLE | POINT |
|---|---|---|---|---|---|
| Aluminum alloys | 90–118 | 12–15 | 125–135 | 24–48 | Standard |
| Magnesium alloys | 70–118 | 12–15 | 120–135 | 30–45 | Standard |
| Copper alloys | 118 | 12–15 | 125–135 | 10–30 | Standard |
| Steels | 118 | 10–15 | 125–135 | 24–32 | Standard |
| High-strength steels | 118–135 | 7–10 | 125–135 | 24–32 | Crankshaft |
| Stainless steels, low strength | 118 | 10–12 | 125–135 | 24–32 | Standard |
| Stainless steels, high strength | 118–135 | 7–10 | 120–130 | 24–32 | Crankshaft |
| High-temp. alloys | 118–135 | 9–12 | 125–135 | 15–30 | Crankshaft |
| Refractory alloys | 118 | 7–10 | 125–135 | 24–32 | Standard |
| Titanium alloys | 118–135 | 7–10 | 125–135 | 15–32 | Crankshaft |
| Cast irons | 118 | 8–12 | 125–135 | 24–32 | Standard |
| Plastics | 60–90 | 7 | 120–135 | 29 | Standard |

to chips. The process can be used to make disks up to 150 mm (6 in.) in diameter from flat sheet or plate.

**Material-removal rate.**    The *material-removal rate* (MRR) in drilling is the ratio of volume removed to time. Thus

$$\text{MRR} = \frac{\pi D^2}{4} fN, \tag{8.37}$$

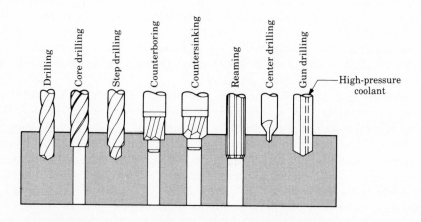

**FIGURE 8.60**    Various types of drills and drilling operations.

where $D$ is the drill diameter, $f$ is the feed, and $N$ is the rpm of the drill. See Table 8.17 for recommendations.

The **thrust force** in drilling is the force that acts in the direction of the hole axis. If this force is excessive, it can cause the drill to break or bend. The thrust force depends on factors such as the strength of the workpiece material, feed, rotational speed, cutting fluids, drill diameter, and drill geometry. Although some attempts have been made, accurate calculation of the thrust force on the drill has proven to be difficult. Experimental data are available as an aid in the design and use of drills and drilling equipment. Thrust forces in drilling range from a few newtons for small drills to as high as 100 kN (22.5 klb) in drilling high-strength materials with large drills. Similarly, drill torque can range as high as 4000 N·m (3000 ft-lb).

The torque during drilling also is difficult to calculate. It can be obtained by using the data in Table 8.4 by noting that power dissipated during drilling is the product of torque and rotational speed. Thus by first calculating the MRR, we can calculate the torque on the drill. **Drill life**, as well as tap life, is usually measured by the number of holes drilled before the drill becomes dull and forces increase.

*Drilling machines* are used for drilling holes, tapping, reaming, and other general-purpose, small-diameter boring operations. Drilling machines are generally vertical, the most common type of which is a **drill press**. The workpiece is placed on an adjustable table, either by clamping it directly into the slots and holes on the table or by using a vise, which in turn can be clamped to the table. The workpiece should be properly clamped, for both safety and accuracy, because the drilling torque can be high enough to rotate the workpiece. The drill is lowered manually by hand wheel or by power feed at preset rates. Manual feeding requires some skill in judging the appropriate feed rate. In order to maintain proper cutting speeds at the cutting edges of drills, the spindle speed on drilling machines has to be adjustable to accommodate different sizes of drills. Adjustments are made by means of pulleys, gear boxes,

**TABLE 8.17**

**GENERAL RECOMMENDATIONS FOR SPEEDS AND FEEDS IN DRILLING**

| WORKPIECE MATERIAL | SURFACE SPEED | | FEED, mm/rev (in./rev) DRILL DIAMETER | | RPM | |
|---|---|---|---|---|---|---|
| | m/min | ft/min | 1.5 mm (0.060 in.) | 12.5 mm (0.5 in.) | 1.5 mm | 12.5 mm |
| Aluminum alloys | 30–120 | 100–400 | 0.025 (0.001) | 0.30 (0.012) | 6400–25,000 | 800–3000 |
| Magnesium alloys | 45–120 | 150–400 | 0.025 (0.001) | 0.30 (0.012) | 9600–25,000 | 1100–3000 |
| Copper alloys | 15–60 | 50–200 | 0.025 (0.001) | 0.25 (0.010) | 3200–12,000 | 400–1500 |
| Steels | 20–30 | 60–100 | 0.025 (0.001) | 0.30 (0.012) | 4300–6400 | 500–800 |
| Stainless steels | 10–20 | 40–60 | 0.025 (0.001) | 0.18 (0.007) | 2100–4300 | 250–500 |
| Titanium alloys | 6–20 | 20–60 | 0.010 (0.0004) | 0.15 (0.006) | 1300–4300 | 150–500 |
| Cast irons | 20–60 | 60–200 | 0.025 (0.001) | 0.30 (0.012) | 4300–12,000 | 500–1500 |
| Thermoplastics | 30–60 | 100–200 | 0.025 (0.001) | 0.13 (0.005) | 6400–12,000 | 800–1500 |
| Thermosets | 20–60 | 60–200 | 0.025 (0.001) | 0.10 (0.004) | 4300–12,000 | 500–1500 |

Note: As hole depth increases, speeds and feeds should be reduced. Selection of speeds and feeds also depends on the specific surface finish required.

**FIGURE 8.61** Terminology for a helical reamer.

or variable-speed motors. Drill presses are usually designated by the largest workpiece diameter that can be accommodated on the table. Sizes typically range from 150 mm (6 in.) to 1250 mm (50 in.).

## Reaming and reamers

**Reaming** is an operation to make an existing hole dimensionally more accurate than can be obtained by drilling alone and to improve its surface finish. The most accurate holes are produced by the following sequence of operations: centering, drilling, boring, and reaming. For even better accuracy and surface finish, holes may be internally ground and honed.

A **reamer** (Fig. 8.61) is a multiple-cutting-edge tool with straight or helically fluted edges that removes very little material. The shanks may be straight or tapered, as in drills. The basic types of reamers are *hand* and *machine (chucking)* reamers. Other types are *rose* with cutting edges having wide margins and no relief, *fluted, shell, expansion,* and *adjustable* reamers.

## Tapping and taps

Internal threads in workpieces can be produced by tapping. A **tap** is basically a threading tool with multiple cutting teeth (Fig. 8.62). Taps are generally available with three or four flutes.

**FIGURE 8.62** Terminology for a tap.

**FIGURE 8.63** Typical parts and shapes produced by the cutting processes described in Section 8.9.

Three-fluted taps are stronger because of the larger amount of material available in the flute. *Tapered* taps are designed to reduce the torque required for tapping through holes, *bottoming* taps for tapping blind holes to their full depth, and *collapsible* taps for large diameter holes. After tapping is completed, the tap is mechanically collapsed and removed from the hole, without having to rotate it. Tap sizes range up to 100 mm (4 in.).

## 8.9 Cutting Processes and Machine Tools for Producing Various Shapes

Several cutting processes and machine tools (Fig. 8.63) are capable of producing complex shapes with the use of multitooth, as well as single-point, cutting tools. We begin with one of the most versatile processes, called milling, in which a multitooth cutter rotates along various axes with respect to the workpiece. We then cover planing, shaping, and broaching, in which flat and shaped surfaces are produced, and the tool or workpiece travels along a straight path. Next, we describe sawing processes, which are generally used to prepare blanks, usually from rods, flat plate, and sheet, for further operations by forming, machining, and welding. We briefly cover filing, which is used to remove small amounts of material, usually from edges and corners.

### 8.9.1 Milling operations

**Milling** includes a number of versatile machining operations, which are capable of producing a variety of configurations using a *milling cutter*, a multitooth tool that produces a number of chips in one revolution. Parts such as the one shown in Fig. 8.63 can be machined efficiently and repeatedly using various milling cutters.

## Slab milling

In **slab milling**, also called *peripheral* milling, the axis of cutter rotation is parallel to the workpiece surface to be machined (Fig. 8.64a). The cutter, generally made of high-speed steel, has a number of teeth along its circumference, each tooth acting like a single-point cutting tool called a *plain mill*.

Cutters used in slab milling may have *straight* or *helical* teeth producing orthogonal or oblique cutting action. The helical teeth on the cutter are shown in Fig. 8.1c. In *conventional milling*, also called *up milling*, the maximum chip thickness is at the end of the cut (Fig. 8.64). Advantages are that tooth engagement is not a function of workpiece surface characteristics, and contamination or scale on the surface does not affect tool life. This is the dominant method of milling. The cutting process is smooth, provided that the cutter teeth are sharp, but there is a tendency for the tool to chatter. Also, the workpiece has a tendency to be pulled upward, thus proper clamping is important.

In *climb milling*, also called *down milling*, cutting starts with the chip at its thickest location. The advantage is that the downward component of cutting forces holds the workpiece in place, particularly for slender parts. However, because of the resulting high impact forces when the teeth engage the workpiece, this operation must have a rigid setup, and backlash must be eliminated in the table feed mechanism. Climb milling is not suitable for machining workpieces having surface scale, such as hot-worked metals, forgings, and castings. The scale is hard and abrasive and causes excessive wear and damage to the cutter teeth, thus reducing tool life. Climb milling is recommended, in general, for maximum cutter life when computer numerical controlled (CNC) machine tools are used. A typical application is in finishing cuts on aluminum.

The cutting speed $V$ in milling is the peripheral speed of the cutter, or

$$V = \pi D N, \tag{8.38}$$

where $D$ is the cutter diameter and $N$ is the rotational speed of the cutter (Fig. 8.64b).

Note that the thickness of the chip in slab milling varies along its length because of the relative longitudinal motion between cutter and workpiece. For a straight-tooth cutter we can determine the approximate *undeformed chip thickness $t_c$* (*chip depth of cut*) from the equation

**FIGURE 8.64**    (a) Schematic illustration of conventional milling and climb milling. (b) Slab milling operation, showing depth of cut $d$, feed per tooth $f$, chip depth of cut $t_c$, and workpiece speed $v$. (c) Schematic illustration of cutter travel distance $t_c$ to reach full depth of cut.

$$t_c = 2f\sqrt{\frac{d}{D}}, \tag{8.39}$$

where $f$ is the feed per tooth of the cutter, measured along the workpiece surface (that is, the distance the workpiece travels per tooth of the cutter, in mm/tooth or in./tooth) and $d$ is the depth of cut. As the value of $t_c$ becomes greater, the force on the cutter tooth increases.

Feed per tooth is determined from the equation

$$f = v/Nn, \tag{8.40}$$

where $v$ is the linear speed (feed rate) of the workpiece and $n$ is the number of teeth on the cutter periphery. We can check the dimensional accuracy of this equation by using appropriate units for the individual terms; thus, (mm/tooth) = (m/min) $(10^3$ mm/m)/(rev/min)(number of teeth/rev), which checks out correctly. The cutting time $t$ is given by the expression

$$t = (l + l_c)/v, \tag{8.41}$$

where $l$ is the length of the workpiece (Fig. 8.64c) and $l_c$ is the extent of the cutter's first contact with the workpiece. Based on the assumption that $l_c \ll l$ (although this is not generally so), the *material removal rate* is

$$\text{MRR} = lwd/t = wdv, \tag{8.42}$$

where $w$ is the width of the cut, which for a workpiece narrower than the length of the cutter, is the same as the width of the workpiece. The distances that the cutter travels in noncutting cycles of the milling operation are important economic considerations and should be minimized.

Although we can calculate the *power requirement* in slab milling, the *forces* acting on the cutter (tangential, radial, and axial; see also Fig. 8.53b) are difficult to calculate because of the many variables involved, particularly tool geometry. These forces can be measured experimentally for a variety of conditions. From the power, however, we can calculate the *torque* on the cutter spindle (see example below). Although the torque is the product of the tangential force on the cutter and its radius, the tangential force per tooth will depend on how many teeth are engaged during the cut.

---

**Example 8.8:** Calculation of material removal rate, power required, torque, and cutting time in slab milling

A slab-milling operation is being carried out on a 12 in. long, 4 in. wide annealed mild-steel block at a feed $f = 0.01$ in./tooth and depth of cut $d = \frac{1}{8}$ in. The cutter is $D = 2$ in. in diameter, has 20 straight teeth, rotates at $N = 100$ rpm, and is wider than the block to be machined. Calculate the material removal rate, estimate the power and torque required for this operation, and calculate the cutting time.

SOLUTION. From the information given, we can calculate the linear speed $v$ of the workpiece from Eq. (8.40). Thus,

$$v = fNn = (0.01)(100)(20) = 20 \text{ in./min.}$$

Using Eq. (8.42), the material removal rate is

$$\text{MRR} = (4)(\tfrac{1}{8})(20) = 10 \text{ in}^3/\text{min}.$$

Since the workpiece is annealed mild steel, let's estimate unit power from Table 8.4 as 1.1 hp·min/in.$^3$. Hence the power required in this operation can be estimated to be

$$\text{Power} = (1.1)(10) = 11 \text{ hp}.$$

We can also calculate the torque acting on the cutter spindle by noting that power is the product of torque and the spindle rotational speed, in radians per unit time. Thus,

$$\text{Torque} = \text{Power} / \text{Rotational speed}$$
$$= (11 \text{ hp})(33,000 \text{ lb-ft/min. hp})/(100 \text{ rpm})(2\pi)$$
$$= 578 \text{ lb-ft}.$$

The cutting time is given by Eq. (8.41) in which the quantity $l_c$ can be shown, from simple geometric relationships and for $D \gg d$, to approximate

$$l_c = \sqrt{Dd} = \sqrt{(2)(\tfrac{1}{8})} = 0.5 \text{ in. in.}$$

Thus, the cutting time is

$$t = (12 + 0.5)/20 = 0.625 \text{ min} = 37.5 \text{ s}.$$

---

### Face milling

In **face milling** the cutter is mounted on a spindle having an axis of rotation perpendicular to the workpiece surface and removes material in the manner shown in Fig. 8.65a. The cutter rotates at a rotational speed $N$ and the workpiece moves along a straight path at a linear speed $v$. When the cutter rotates as shown in Fig. 8.65b, the operation is *climb* milling; when it rotates in the opposite direction (Fig. 8.65c), the operation is *conventional* milling. The cutting tools are usually carbide or high-speed-steel inserts and are mounted on the cutter body.

Because of the relative motion between the cutting teeth and the workpiece, a face-milling cutter leaves *feed marks* on the machined surface, much as in turning operations. Surface roughness depends on insert corner geometry and feed per tooth.

The terminology for a face-milling cutter and the various angles are shown in Fig. 8.66. The side view is shown in Fig. 8.67, where we note that, as in turning operations, the *lead angle* of the insert in face milling has a direct influence on the *undeformed chip thickness*. As the lead angle (positive as shown) increases, the undeformed chip thickness (hence chip thickness) decreases and the length of contact increases. The range of lead angles for most face-milling cutters is from 0° to 45°, thus influencing the axial and tangential forces. Note that the cross-sectional area of the undeformed chip remains constant. The lead angle also influences forces in milling. We can see that as the lead angle decreases, there is less and less of a vertical force (axial force on the cutter spindle) component.

(a)   Insert

**FIGURE 8.65**   Face-milling operation showing (a) action of an insert in face milling; (b) climb milling; (c) conventional milling; (d) dimensions in face milling. The width of cut *w* is not necessarily the same as the cutter radius. *Source:* Courtesy of The Ingersoll Cutting Tool Company.

A wide variety of milling cutters are available. The cutter diameter should be chosen in such a manner that it will not interfere with fixtures and other components in the setup. In a typical face milling operation, the ratio of the cutter diameter $D$ to the width of cut $w$ should be no less than 3:2.

The relationship of cutter diameter and insert angles and their position relative to the surface to be milled is important, in that it will determine the angle at which an insert *enters*

**FIGURE 8.66**   Terminology for a face-milling cutter.

**FIGURE 8.67** The effect of lead angle on the undeformed chip thickness in face milling. Note that as the lead angle increases, the undeformed chip thickness (hence chip thickness) decreases, but the length of contact (hence chip width) increases. The insert must be sufficiently large to accommodate the contact length increase.

and *exits* the workpiece. Note in Fig. 8.65b for climb milling that if the insert has zero axial and radial rake angles (see Fig. 8.66), the rake face of the insert engages the workpiece directly. However, as we see in Fig. 8.68a and b, the same insert engages the workpiece at different angles depending on the relative positions of the cutter and the workpiece. In illustration 8.68a the tip of the insert makes the first contact, and hence there is a possibility for the cutting edge to chip off. In illustration 8.68b, on the other hand, the first contacts

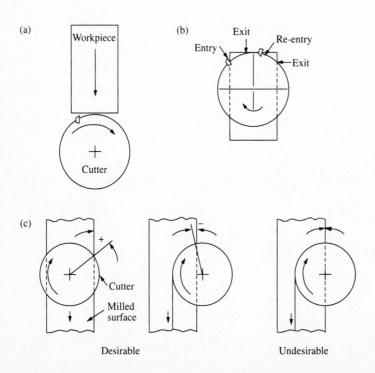

**FIGURE 8.68** (a) Relative position of the cutter and insert as it first engages the workpiece in face milling, (b) insert positions toward the end of cut, and (c) examples of exit angles of insert, showing desirable (positive or negative angle) and undesirable (zero angle) positions. In all figures, the cutter spindle is perpendicular to the page.

**TABLE 8.18**

**SUMMARY OF MILLING PARAMETERS AND FORMULAS**

$N$ = Rotational speed of the milling cutter, rpm

$f$ = Feed, mm/tooth or in./tooth

$D$ = Cutter diameter, mm or in.

$n$ = Number of teeth on cutter

$v$ = Linear speed of the workpiece or feed rate, mm/min or in./min

$V$ = Surface speed of cutter, m/min or ft./min

   = $v/Nn$

$l$ = Length of cut, mm or in.

$t$ = Cutting time, s or min

   = $(l + l_c)/v$, where $l_c$ = extent of the cutter's first contact with workpiece

MRR = mm³/min or in.³/min

     = $wdv$, where $w$ is the width of cut

Torque = N · m or lb · ft

     = $(F_c)(D/2)$

Power = kW or hp

     = (Torque)$(\omega)$, where $\omega = 2\pi N$ radians/min.

Note: The units given are those that are commonly used; however, appropriate units must be used in the formulas.

(ie., at entry, reentry, and the two exits) are at an angle and away from the tip of the insert. There is, therefore, less tendency for the insert to fail because the force on the insert increases and decreases more slowly. Note from Fig. 8.66 that the radial and axial rake angles will also have an effect on this situation.

Figure 8.68c shows the exit angles for various cutter positions. Note that in the first two examples, the insert exits the workpiece at an angle whereby the force on the insert diminishes to zero at a slower rate (desirable) than in the third example, where the insert exits suddenly (undesirable).

**Example 8.9:** Calculation of material removal rate, power required, and cutting time in face milling

Refer to Fig. 8.65 and assume that $D = 150$ mm, $w = 60$ mm, $l = 500$ mm, $d = 3$ mm, $v = 0.6$ m/min, and $N = 100$ rpm. The cutter has 10 inserts, and the workpiece material is a high-strength aluminum alloy. Calculate the material removal rate, cutting time, and feed per tooth, and estimate the power required. (See also Table 8.18.)

SOLUTION. The cross-section of the cut is $(w)(d) = (60)(3) = 180$ mm². Noting that the workpiece speed $v$ is 0.6 m/min = 600 mm/min, we calculate the material removal rate as

$$\text{MRR} = (180)(600) = 108,000 \text{ mm}^3/\text{min}.$$

The cutting time is given by

$$t = (l + 2l_c)/v.$$

We note from Fig. 8.65 that for this problem $l_c = D/2$, or 75 mm. Thus the cutting time is

$$t = (500 + 150)/(600/60) = 65 \text{ s} = 1.08 \text{ min}.$$

We obtain the feed per tooth from Eq. (8.40). Noting that $N = 100$ rpm $= 1.67$ rev/s, we find

$$f = 10/(1.67)(10) = 0.6 \text{ mm/tooth}.$$

For this material we estimate the unit power from Table 8.4 to be 1.1 W·s/mm$^3$. Hence we estimate the power to be

$$\text{Power} = (1.1)(1800) = 1980 \text{ W} = 1.98 \text{ kW}.$$

### End milling

The cutter in **end milling** is shown in Fig. 8.1d and has either straight or tapered shanks for smaller and larger cutter sizes, respectively. The cutter usually rotates on an axis perpendicular to the workpiece, although it can be tilted to machine tapered surfaces. Flat surfaces as well as various profiles can be produced by end milling. The end faces of some end mills have cutting teeth, and these can be used as a drill to start a cavity. End mills are also available with hemispherical ends for producing curved surfaces, as in making dies. *Hollow end mills* have internal cutting teeth and are used for machining the cylindrical surface of solid round workpieces, as in preparing stock with accurate diameters for automatic machines. End mills are made of high-speed steels or have carbide inserts.

We described high-speed machining and its applications in Section 8.8.2. One of the more common applications is *high-speed milling* using an end mill and with the same general provisions regarding the stiffness of machines, workholding devices, etc. that have been described. A typical application is milling aluminum-alloy aerospace components and honeycomb structures, with spindle speeds on the order of 20,000 rpm. Chip collection and disposal can be a significant problem in these operations.

### Other milling operations and cutters

Several other types of milling operations and cutters are used to machine various surfaces. In *straddle milling,* two or more cutters are mounted on an arbor and are used to machine two parallel surfaces on the workpiece (Fig. 8.69a). *Form milling* is used to produce curved profiles, with cutters that have specially shaped teeth (Fig. 8.69b). Such cutters are also used for cutting gear teeth (see Section 8.9.7).

*Circular cutters* are used for slotting and slitting. The teeth may be staggered slightly, as in a sawblade (see Section 8.9.5), to provide clearance for the cutter in making deep slots. *Slitting saws* are relatively thin, usually less than 5 mm ($^3/_{16}$ in.). *T-slot cutters* are used to mill T-slots, such as those in machine-tool work tables for clamping workpieces. A slot is first

(a) Straddle milling
(b) Form milling

**FIGURE 8.69**  Cutters for (a) straddle milling and (b) form milling.

milled with an end mill. The T-slot cutter then cuts the complete profile of the slot in one pass. *Key seat cutters* are used to make the semicylindrical key seats (Woodruff) for shafts. *Angle milling cutters* with a single angle or double angles are used to produce tapered surfaces with various angles.

*Shell mills* are hollow inside and are mounted on a shank, thus allowing the same shank to be used for different sized cutters. The use of shell mills is similar to that of end mills. Milling with a single cutting tooth mounted on a high-speed spindle is known as *fly cutting* and is generally used in simple face-milling and boring operations. The tool can be shaped as a single-point cutting tool and can be placed in various radial positions on the spindle.

**Toolholders.**    Milling cutters are classified as arbor cutters or shank cutters. *Arbor cutters* are mounted on an arbor, such as those for slab, face, straddle, and form milling. In *shank-type cutters,* the cutter and the shank are one piece. The most common examples of shank cutters are end mills. Whereas small end mills have straight shanks, larger end mills have tapered shanks for better clamping to resist the higher forces and torque involved. Cutters with straight shanks are mounted in collet chucks or special end mill holders; those with tapered shanks are mounted in tapered tool holders. In addition to mechanical means, hydraulic tool holders and arbors are also available. The stiffness of cutters and tool holders is important for surface quality and reducing vibration and chatter in milling operations. Conventional tapered tool holders have a tendency to wear and bellmouth under the radial forces in milling.

**Milling process capabilities.**    In addition to various characteristics of milling processes that have been described, milling capabilities include parameters such as production rate, surface finish, tolerances, and cost considerations. The conventional ranges of feeds and cutting speeds for milling are given in Table 8.19. For *cutting fluid* recommendations, see Table 8.13. *Cutting speeds,* which as in turning operations vary over a wide range depending on workpiece material, cutting tool material, and process parameters, range from 30 m/min (90 ft/min) to 3000 m/min (10,000 ft/min).

Because of their capabilities to perform a variety of cutting operations, milling machines (first built in 1876) are among the most versatile and useful machine tools. A wide selection of milling machines with numerous features is available. Used for general-purpose milling operations, **column-and-knee-type** machines are the most common milling machines. The spindle to which the milling cutter is attached may be *horizontal* (Fig. 8.70) for slab milling or *vertical* for face and end milling, boring, and drilling operations (Fig. 8.71). The various components are moved manually or by power. The basic components of these machines are:

**TABLE 8.19**

**APPROXIMATE RANGE OF RECOMMENDED CUTTING SPEEDS FOR MILLING OPERATIONS**

| | CUTTING SPEED | |
|---|---|---|
| WORKPIECE MATERIAL | m/min | ft/min |
| Aluminum alloys | 300–3000 | 1000–10,000 |
| Cast iron, gray | 90–1300 | 300–4200 |
| Copper alloys | 90–1000 | 300–3300 |
| High-temperature alloys | 30–550 | 100–1800 |
| Steels | 60–450 | 200–1500 |
| Stainless steels | 90–500 | 300–1600 |
| Thermoplastics and thermosets | 90–1400 | 300–4500 |
| Titanium alloys | 40–150 | 130–500 |

Note: (a) these speeds are for carbides, ceramic, cermets, and diamond cutting tools. Speeds for high-speed-steel tools are lower than indicated.
(b) Depths of cut, $d$, are generally in the range of 1–8 mm (0.04–0.3 in.).
(c) Feeds per tooth, $f$, are generally in the range of 0.08–0.46 mm/rev (0.003–0.018 in./rev).

1. *Work table*, on which the workpiece is clamped, using the T-slots. The table moves longitudinally with respect to the saddle.

2. *Saddle*, which supports the table and can move transversely.

3. *Knee*, which supports the saddle and gives the table vertical movement for adjusting the depth of cut.

4. *Overarm* in horizontal machines, which is adjustable to accommodate different arbor lengths.

5. *Head*, which contains the spindle and cutter holders. In vertical machines the head may be fixed or vertically adjustable and can be swiveled in a vertical plane on the column for milling tapered surfaces.

In **bed-type** machines, the work table is mounted directly on the bed, which replaces the knee, and can move only longitudinally. These milling machines are not as versatile as others, but have high stiffness and are used for high-production-rate work. The spindles may be

**FIGURE 8.70** Schematic illustration of a horizontal-spindle column-and-knee-type milling machine. *Source:* G. Boothroyd, *Fundamentals of Machining and Machine Tools.*

Head

Work table

Column

Workpiece

Saddle

Knee

Base

**FIGURE 8.71** Schematic illustration of a vertical-spindle column-and-knee-type milling machine. *Source*: G. Boothroyd, *Fundamentals of Machining and Machine Tools*.

horizontal or vertical and of duplex or triplex types, that is, with two or three spindles for simultaneously machining two or three workpiece surfaces.

Several other types of milling machines are available. *Planer-type* machines, which are similar to bed-type machines, are equipped with several heads and cutters to mill various surfaces. They are used for heavy workpieces and are more efficient than planers (see Section 8.9.2) when used for similar purposes. *Rotary-table* machines are similar to vertical milling machines and are equipped with one or more heads for face-milling operations. Using tracer fingers, *duplicating* machines (copy milling machines) reproduce parts from a master model. They are used in the automotive and aerospace industries for machining complex parts and dies (die sinking). Various milling-machine components are being replaced rapidly with *computer numerical control* (CNC) machines. These machine tools are versatile and capable of milling, drilling, boring, and tapping with repetitive accuracy. Other developments include *profile* milling machines, with five-axis movements.

### 8.9.2 Planing and planers

**Planing** is a relatively simple cutting process by which flat surfaces, as well as various cross-sections with grooves and notches, are produced along the length of the workpiece. Planing is usually done on large workpieces—as large as 25 m × 15 m (75 ft × 45 ft).

In a **planer**, the workpiece is mounted on a table that travels along a straight path. A horizontal cross-rail, which can be moved vertically along the ways in the column, is equipped with one or more tool heads. The cutting tools are attached to the heads, and machining is done along a straight path. Because of the reciprocating motion of the workpiece, elapsed noncutting time during the return stroke is significant both in planing and in shaping. Consequently, these operations are not efficient or economical, except for low-quantity production. Efficiency of the operation can be improved by equipping planers with tool holders and tools that cut in both directions of table travel.

### 8.9.3 Shaping and shapers

**Shaping** is used to machine parts much like planing does, except that the parts are smaller. Cutting by shaping is basically the same as in planing. In a *horizontal shaper,* the tool travels along a straight path, and the workpiece is stationary. The cutting tool is attached to the tool

**FIGURE 8.72** (a) Typical parts made by internal broaching. (b) Parts made by surface broaching. Heavy lines indicate broached surfaces. *Source:* General Broach and Engineering Company.

head, which is mounted on the ram. The ram has a reciprocating motion, and in most machines cutting is done during the forward movement of the ram (*push cut*); in others it is done during the return stroke of the ram (*draw cut*).

Vertical shapers (*slotters*) are used for machining notches, keyways, and dies. Shapers are also capable of producing complex shapes, such as cutting a helical impeller, in which the workpiece is rotated during the cut through a master cam. Because of low production rate, shapers are generally used in toolrooms, job shops, and repair work.

### 8.9.4 Broaching and broaching machines

The **broaching** operation is similar to shaping with multiple teeth and is used to machine internal and external surfaces, such as holes of circular, square, or irregular section, keyways, teeth of internal gears, multiple spline holes, and flat surfaces (Fig. 8.72). A **broach** is, in effect, a long multitooth cutting tool (Fig. 8.73) with successively deeper cuts. Thus the total depth of material removed in one stroke is the sum of the depths of cut of each tooth. A broach can remove material as deep as 6 mm (0.25 in.) in one stroke. Broaching can produce parts with good surface finish and dimensional accuracy; hence it competes favorably with other processes, such as boring, milling, shaping, and reaming, to produce similar shapes. Although broaches can be expensive, the cost is justified because of their use for high-quantity production runs.

The terminology for a broach is given in Fig. 8.73b. The rake (hook) angle depends on the material cut, as in turning and other cutting operations, and usually ranges between $0°$ and $20°$. The clearance angle is usually $1°–4°$; finishing teeth have smaller angles. Too small

**FIGURE 8.73** (a) Cutting action of a broach, showing various features. (b) Terminology for a broach.

a clearance angle causes rubbing of the teeth against the broached surface. The pitch of the teeth depends on factors such as length of the workpiece (length of cut), tooth strength, and size and shape of chips. The tooth depth and pitch must be sufficiently large to accommodate the chips produced during broaching, particularly for long workpieces, but at least two teeth should be in contact with the workpiece at all times. The following formula may be used to obtain the pitch for a broach to cut a surface of length $l$:

$$\text{Pitch} = k \sqrt{l}, \tag{8.43}$$

where $k = 1.76$ for $l$ in mm, and $k = 0.35$ for $l$ in inches. An average pitch for small broaches is in the range of 3.2–6.4 mm (0.125–0.25 in.) and for large ones in the range of 12.7–25 mm (0.5–1 in.). The cut per tooth depends on the workpiece material and the surface finish desired. It is usually in the range of 0.025–0.075 mm (0.001–0.003 in.) for medium-size broaches and can be larger than 0.25 mm (0.01 in.) for larger broaches.

Broaches are available with various tooth profiles, including some that have chip breakers. Broaches are also made round with circular cutting teeth and are used to enlarge holes (Fig. 8.74). Note that the cutting teeth on the broach have three regions: roughing, semifinishing, and finishing. A variety of broaches are made for producing various internal and external shapes. Irregular internal shapes are usually broached by starting with a round hole in the workpiece, produced by drilling or boring.

A recent advance in broaching technology is **turn broaching** of crankshafts. The crankshaft rotates between centers and the broach, which is equipped with multiple inserts and passes tangentially across the part. The process is thus a combination of broaching and skiving. Straight as well as circular broaches have been used successfully for such applications. Machines that broach a number of crankshafts simultaneously have been built. Main bearings for engines are also broached in this way.

Broaching machines either pull or push the broaches and are made horizontal or vertical. *Push broaches* are usually shorter, generally in the range of 150–350 mm (6–14 in.). *Pull broaches* tend to straighten the hole, whereas pushing permits the broach to follow any irregularity of the leader hole. Horizontal machines are capable of longer strokes. Broaching

**FIGURE 8.74** Terminology for a pull-type internal broach used for enlarging long holes.

machines are relatively simple in construction, have only linear motions, and are usually actuated hydraulically, although some are moved by crank, screw, or rack. Many styles of broaching machines are manufactured, some with multiple heads, allowing a variety of shapes and parts to be produced, including helical splines and rifled gun barrels. Sizes range from machines for making needlelike parts to those used for broaching gun barrels. The force required to pull or push the broach depends on the strength of the workpiece material, total depth of cut, and width of cut. Tooth profile and cutting fluids also affect this force. Capacities of broaching machines are as high as 0.9 MN (100 tons) of pulling force.

### 8.9.5   Sawing and saws

**Sawing** is a cutting operation in which the cutting tool is a blade (*saw*) having a series of small teeth, with each tooth removing a small amount of material. This process is used for all metallic and nonmetallic materials that are machinable by other cutting processes and is capable of producing various shapes. The width of cut (**kerf**) in sawing is usually narrow, and thus sawing wastes little material.

Typical saw-tooth and saw-blade configurations are shown in Fig. 8.75. Tooth spacing is usually in the range of 0.08–1.25 teeth per mm (2–32 per in.). A wide variety of tooth forms and spacing and blade thicknesses, widths, and sizes are available. Blades are made from carbon and high-speed steels. Carbide-steel or high-speed-steel tipped steel blades are used for sawing harder materials (Fig. 8.76).

In order to prevent the saw from binding and rubbing during cutting, the teeth are alternately set in opposite directions so that the kerf is wider than the blade (Fig. 8.75b). At least two or three teeth should always be engaged with the workpiece in order to prevent snagging (catching of the saw tooth on the workpiece). This is why sawing thin materials satisfactorily is difficult. The thinner the stock, the finer the saw teeth should be—and the greater the number of teeth per unit length. Cutting speed in sawing usually ranges up to 1.5 m/s (300 ft/min), with lower speeds for high-strength metals. Cutting fluids are generally used to improve the quality of cut and the life of the saw.

**FIGURE 8.75**   (a) Terminology for saw teeth. (b) Types of saw teeth, staggered to provide clearance for the saw blade to prevent binding during sawing.

**FIGURE 8.76** (a) High-speed-steel teeth welded on steel blade. (b) Carbide inserts brazed to blade teeth.

*Hacksaws* have straight blades and reciprocating motions; they may be manually or power operated. *Circular* saws (*cold* saws) are generally used for high-production-rate sawing of large cross-sections. *Band* saws have long, flexible, continuous blades and have continuous cutting action. Blades and high-strength wire can be coated with diamond powder (*diamond-edged* blades and *diamond* saws). They are suitable for sawing hard metallic, nonmetallic, and composite materials.

**Friction sawing** is a process in which a mild-steel blade or disk rubs against the workpiece at speeds of up to 125 m/s (25,000 ft/min). The frictional energy is converted into heat, which rapidly softens a narrow zone in the workpiece. The action of the blade or disk, which is sometimes provided with teeth or notches, pulls and ejects the softened metal from the cutting zone. The heat generated in the workpiece produces a heat-affected zone on the cut surfaces; thus properties can be adversely affected by the sawing process. Because only a small portion of the blade is engaged with the workpiece at any time, the blade cools rapidly as it passes through the air.

The friction-sawing process is suitable for hard ferrous metals and reinforced plastics but not for nonferrous metals because they have a tendency to stick to the blade. Disks for friction sawing as large as 1.8 m (6 ft) in diameter are used to cut off large steel sections in rolling mills. Friction sawing can also be used to remove flash from castings.

### 8.9.6 Filing

**Filing** is small-scale removal of material from a surface, corner, or hole. First developed in about 1000 B.C., files are usually made of hardened steels and are available in a variety of cross-sections, including flat, round, half round, square, and triangular. Files have many tooth forms and grades of coarseness, such as smooth-cut, second-cut, and bastard-cut. Although filing is usually done by hand, various machines with automatic features are available for high production rates, with files reciprocating at up to 500 strokes/min. *Band files* consist of file segments, each about 75 mm (3 in.) long, that are riveted to flexible steel bands and used in a manner similar to band saws. Disk type files are also available.

*Rotary files* and **burs** are available and are used for special applications. These cutters are usually conical, cylindrical, or spherical in shape and have various tooth profiles. Similar to that of reamers, their cutting action removes small amounts of material. The rotational speeds range from 1500 rpm for cutting steel with large burs to as high as 45,000 rpm for magnesium with small burs.

**FIGURE 8.77**   Producing gear teeth on a blank by form cutting.

### 8.9.7   Gear manufacturing by cutting

Because of their capability for transmitting motion and power, gears are among the most important of all mechanical components. Gears can be manufactured by casting, forging, extrusion, drawing, thread rolling, powder metallurgy, and blanking (for making thin gears such as those used in watches and small clocks). Nonmetallic gears can be made by injection molding and casting.

In **form cutting** the cutting tool is similar to a form-milling cutter in the shape of the space between the gear teeth (Fig. 8.77). The gear-tooth shape is reproduced by cutting the gear blank around its periphery. The cutter travels axially along the length of the gear tooth at the appropriate depth to produce the gear tooth. After each tooth is cut, the cutter is withdrawn, the gear blank is rotated (indexed), and the cutter proceeds to cut another tooth. The process continues until all teeth are cut. Each cutter is designed to cut a range of tooth numbers. The precision of the form-cut tooth profile depends on the accuracy of the cutter and the machine and its stiffness.

Form cutting can be done on milling machines, with the cutter mounted on an arbor and the gear blank mounted in a dividing head. Because the cutter has a fixed geometry, form cutting can be used only to produce gear teeth that have constant width, that is, on spur or helical gears but not on bevel gears. Internal gears and gear teeth on straight surfaces, such as in rack and pinion, are form cut with a shaped cutter, using a machine similar to a shaper.

**Broaching** can also be used to produce gear teeth and is particularly applicable to internal teeth. The process is rapid and produces fine surface finish with high dimensional accuracy. However, because broaches are expensive—and a separate broach is required for each size of gear—this method is suitable mainly for high-quantity production.

In **gear generating**, the tool may be a (1) pinion-shaped cutter, (2) rack-shaped straight cutter, or (3) hob. The pinion-shaped cutter can be considered as one of the gears in a conjugate pair and the other as the gear blank (Fig. 8.78a). This type of cutter is used in gear generating on machines called **gear shapers** (Fig. 8.78b). The cutter has an axis parallel to that of the gear blank and rotates slowly with the blank at the same pitch-circle velocity with an axial reciprocating motion. A train of gears provides the required relative motion between the cutter shaft and the gear-blank shaft. Cutting may take place either at the downstroke or upstroke of the machine. Because the clearance required for cutter travel is small, gear shaping is suitable for gears that are located close to obstructing surfaces such as flanges (as in the gear blank in Fig. 8.78b). The process can be used for low-quantity as well as high-quantity production.

**FIGURE 8.78**  (a) Schematic illustration of gear generating with a pinion-shaped gear cutter. (b) Schematic illustration of gear generating in a gear shaper using a pinion-shaped cutter. Note that the cutter reciprocates vertically. (c) Gear generating with rack-shaped cutter. (d) Schematic illustration of three views of gear cutting with a hob. *Source:* After E. P. DeGarmo and Society of Manufacturing Engineers.

On a **rack shaper** the generating tool is a segment of a rack (Fig. 8.78c), which reciprocates parallel to the axis of the gear blank. Because it is not practical to have more than 6–12 teeth on a rack cutter, the cutter must be disengaged at suitable intervals and returned to the starting point, the gear blank meanwhile remaining fixed. A gear-cutting **hob** is basically a worm, or screw, that has been made into a gear-generating tool by machining a series of longitudinal slots or gashes into it to form cutting teeth (Fig. 8.78d). When hobbing a spur gear, the angle between the hob and gear blank axes is 90° minus the lead angle at the hob threads. All motions in hobbing are rotary, and the hob and gear blank rotate continuously as in two gears meshing until all teeth are cut.

Straight *bevel gears* are generally roughed out in one cut with a form cutter on machines that index automatically. The gear is then finished to the proper shape on a gear generator. The generating method is analogous to the rack-generating method that we described. The cutters reciprocate across the face of the bevel gear as does the tool on a shaper. The machines that are used to make spiral bevel gears operate on essentially the same principle. The spiral cutter is basically a face-milling cutter that has a number of straight-sided cutting blades protruding from its periphery.

As produced by any of the processes described, the surface finish and dimensional accuracy of gear teeth may not be accurate enough for certain applications. Several *finishing* operations are also performed, including shaving, burnishing, grinding, honing and lapping.

Modern gear-manufacturing machines are *computer controlled*. Multi-axis computer-controlled machines have capabilities of machining many types of gears using indexable cutters.

## 8.10   Machining and Turning Centers

In descriptions of individual machining processes and machine tools in the preceding chapters, it was noted that each machine, regardless of how highly it is automated, is designed to perform basically one type of operation. Previous chapters have also shown that in manufacturing operations most parts require a number of different machining operations on their various surfaces.

Note, for example, that the parts shown in Fig. 8.79 have a variety of features and that all surfaces on these parts require different types of machining operations, such as milling, facing, boring, drilling, reaming, and threading, to certain specified dimensional tolerances and surface finish. None of the processes and machine tools described thus far could, individually, produce these parts.

(a)          (b)          (c)          (d)

**FIGURE 8.79**   Examples of parts that may be machined on machining centers, using various processes such as turning, facing, milling, drilling, boring, reaming, and threading. These processes would ordinarily require a variety of machine tools for such parts to be machined. *Source:* Toyoda Machinery.

Tools (cutters)
Tool storage
Tool interchange arm
Traveling column
Spindle
Spindle carrier
Computer numerical control panel
Index table
Pallets
Bed

**FIGURE 8.80**  A horizontal-spindle machining center, equipped with an automatic tool changer. Tool magazines can store 200 cutting tools. *Source:* Courtesy of Cincinnati Milacron, Inc.

Traditionally, these operations would be performed by moving the part from one machine tool to another until all machining is completed. This is a viable manufacturing method that can be highly automated and is the principle behind *transfer lines,* consisting of numerous machine tools arranged in a sequence (described in Section 13.2.2). The workpiece, such as an engine block, moves from station to station, with a particular machining operation performed at each station. The part is then transferred to the next machine for another operation, and so on. Transfer lines are commonly used in high-volume or mass production.

There are situations and products, however, where such lines are not feasible or economical, particularly when types of products to be machined change rapidly. An important concept, developed in the late 1950s, is machining centers. A **machining center** is a computer-controlled machine tool capable of performing a variety of cutting operations on different surfaces and different directions on a workpiece (Fig. 8.80). In general, the workpiece is stationary and the cutting tools revolve, such as in milling and drilling operations. The development of machining centers is intimately related to advances in *computer control* of machine tools, details of which are described in Chapter 14. Recall that as an example of advances in modern lathes, Fig. 8. 16 illustrated a numerically controlled lathe (*turning center*) with two turrets carrying several cutting tools for operations such as turning, facing, boring, and threading.

The workpiece in a machining center is placed on a **pallet** or *module* that can be oriented in three principal directions as well as rotated around one or more axes on the pallet (Fig. 8.80). Thus, after a particular cutting operation has been completed, the workpiece does not have to be moved to another machine for additional operations, say, drilling, reaming, and tapping. In other words, the tools and the machine are brought to the workpiece. After all the cutting operations have been completed, the pallet automatically moves away with the finished workpiece, and another pallet containing another workpiece to be machined is brought into position by **automatic pallet changers** (Fig. 8.81). All movements are guided by computer control and pallet-changing cycle times are on the order of 10–30 seconds. Pallet stations are available with multiple pallets serving the machining center. The machines can also be equipped with various automatic parts, such as loading and unloading devices.

(a)

(b)

**FIGURE 8.81**  (a) Schematic top view of a horizontal-spindle machining center showing the pallet pool, set-up station for a pallet, pallet carrier, and an active pallet in operation (shown directly below the spindle of the machine). (b) Schematic illustration of two machining centers with a common pallet pool. Various other arrangements are possible in such systems. *Source:* Hitachi Seiki USA Inc. Los Angeles office, 17023 S. Edwards Rd., Cerritos, CA 90703 (310) 926-4500.

The machining center is equipped with a programmable **automatic tool changer**. Depending on the design, up to 200 cutting tools can be stored in a magazine, drum, or chain (tool storage); auxiliary tool storage is available on some special machining centers for many more tools. The cutting tools are automatically selected with random access for the shortest route to the machine spindle. The **tool-exchange arm** shown in Fig. 8.82 is a common design. It swings around to pick up a particular tool (each tool having its own toolholder) and places it in the spindle. Tools are identified by coded tags, bar codes, or memory chips applied directly to the toolholders. Tool changing times are typically on the order of a few seconds.

Machining centers may be equipped with a **tool- and/or part-checking station** that feeds information to the computer-numerical control to compensate for any variations in tool settings or tool wear. **Touch probes** are available which can be automatically loaded into a tool-

**FIGURE 8.82**  Swing-around type tool changer. Other designs are also available, such as that shown at top of Fig. 8.80. *Source:* Cincinnati Milacron, Inc.

holder for determining reference surfaces of the workpiece, for selection of tool setting and for on-line inspection of parts being machined. Several surfaces can be contacted, and their relative positions are determined and stored in the database of the computer software. The data are then used for programming tool paths and compensating for tool length and diameter, including compensation for tool wear in more advanced designs.

### 8.10.1 Types of machining and turning centers

Although there are various designs of machining centers, the two basic types are vertical spindle and horizontal spindle; many machines have the capability of using both axes. The maximum dimensions that the cutting tools can reach around a workpiece in a machining center are known as the *work envelope;* this term was first used in connection with industrial robots (see Section 13.7.2).

**Vertical-spindle machining centers,** or *vertical machining centers,* are suitable for performing various machining operations on flat surfaces with deep cavities, such as in mold and die making. Because the thrust forces in vertical machining are directed downward, these machines have high stiffness and produce parts with good dimensional accuracy. They are generally less expensive than horizontal-spindle machines.

**Horizontal-spindle machining centers,** or *horizontal machining centers,* (see Fig. 8.80) are suitable for large as well as tall workpieces requiring machining on a number of their surfaces. The pallet can be rotated on different axes to various angular positions. Another category of horizontal-spindle machines is **turning centers,** which are computer-controlled lathes with several features. A three-turret computer numerical-controlled turning center is shown in Fig. 8.83. This machine is designed with two horizontal spindles and three turrets

1st Turret head
2nd Turret head
#1 Spindle head
#3 Turret head
#2 Spindle head

**FIGURE 8.83**
Schematic illustration of a three-turret, two-spindle computer numerical-controlled turning center. *Source:* Hitachi Seiki USA Inc. Los Angeles office, 17023 S. Edwards Rd., Cerritos, CA 90703 (310) 926-4500

which are equipped with a variety of cutting tools to perform several operations on a rotating workpiece.

**Universal machining centers** are more recent developments and are equipped with both vertical and horizontal spindles. They have a variety of features and are capable of machining all surfaces of a workpiece (vertical, horizontal, and diagonal), hence the term universal.

### 8.10.2   Characteristics and capabilities of machining centers

The major characteristics of machining centers are:

- They are capable of handling a variety of part sizes and shapes efficiently, economically, and with repetitively high dimensional accuracy and tolerances on the order of ± 0.0025 mm (0.0001 in.).

- The machines are versatile, having as many as six axes of linear and angular movements and the capability of quick changeover from one type of product to another. Thus the need for a variety of machine tools and floor space is reduced significantly.

- The time required for loading and unloading workpieces, changing tools, gaging, and troubleshooting is reduced, thus improving productivity, reducing labor (particularly skilled labor) requirements, and minimizing machining costs.

- Machining centers are highly automated and relatively compact, so that one operator can attend to two or more machines at the same time.

- The machines are equipped with tool-condition monitoring devices for detecting tool breakage and wear, as well as probes for compensation for tool wear and for tool positioning.

- In-process and post-process gaging and inspection of machined workpieces have become features of machining centers.

Machining centers are available in a wide variety of sizes and features, and their cost ranges from about $50,000 to $1 million and higher. Typical capacities range up to 75 kW (100 hp) and maximum spindle speeds are usually in the range of 4000–8000 rpm, but some are as high as 75,000 rpm for special applications using small-diameter cutters. Some pallets are capable of supporting workpieces weighing as much as 7,000 kg (15,000 lb), although higher capacities are available for special applications.

Many machines are now being constructed on a *modular* basis, so that various peripheral equipment and accessories can be installed and modified as the demand for and the type of products change. Product changes are a major consideration in computer-integrated manufacturing.

Because of the high productivity of machining centers, large amounts of chips are produced and must be collected and disposed of properly. Several designs are available for chip collection.

**Machine-tool selection.**   Machining centers can require significant capital expenditures. Thus, to be cost effective, they have to be used for at least two shifts per day. Consequently, there must be sufficient and continued demand for products made in machining centers to justify this purchase. However, because of their inherent versatility, machining centers can be used to produce a very wide range of products, particularly in just-in-time manufacturing (see Section 14.11).

The selection of a particular type and size of machining center depends on several factors, among which are:

- Type of products, their size, and shape complexity
- Type of machining operations to be performed and type and number of tools required
- Dimensional accuracy required
- Production rate required

Although versatility is the key factor in selecting machining centers, these considerations must be weighed carefully against the high capital investment required for machining centers and compared to that of manufacturing the same products by using a number of the more traditional machine tools.

---

**Example 8.10:** Machining outer bearing races on a turning center

Outer bearing races (see Fig. 8.84) are machined on a turning center. The starting material is hot-rolled 52100 steel tube 91 mm (3.592 in.) OD and 75.5 mm (2.976 in.) ID. The cutting speed is 95 m/min (313 ft/min) for all operations. All tools are carbide, including the cutoff tool (last operation) which is 3.18 mm ($\frac{1}{8}$ in.) instead of 4.76 mm ($\frac{3}{16}$ in.) for the high-speed steel cutoff tool that was formerly used. The material saved by this change is significant because the width of the race is small. The turning center was able to machine these races at high speeds and with repeatable tolerances of $\pm$ 0.025 mm (0.001 in.)

1 Finish turning of outside diameter

2 Boring and grooving on outside diameter

3 Internal grooving with a radius-form tool

4 Finish boring internal groove and rough boring of internal diameter

5 Internal grooving with form tool and chamfering

6 Cutting off finished part; inclined bar picks up bearing race

**FIGURE 8.84**

*Source:* McGill Manufacturing Company.

## 8.11   Vibration and Chatter

In describing cutting processes and machine tools, we pointed out the importance of machine stiffness in controlling dimensional accuracy and surface finish of parts. In this section we describe the adverse effects of *low stiffness* on product quality and machining operations. Low stiffness affects the level of vibration and chatter in tools and machines. If uncontrolled, vibration and chatter can result in the following:

- Poor surface finish (right central region in Fig. 8.85)
- Loss of dimensional accuracy of the workpiece
- Premature wear, chipping, and failure of the cutting tool, which is crucial with brittle tool materials, such as ceramics, some carbides, and diamond
- Damage to machine-tool components from excessive vibrations
- Objectionable noise generated, particularly if it is of high frequency, such as the squeal heard in turning brass on a lathe

Extensive studies have shown that vibration and chatter in cutting are complex phenomena. Cutting operations cause two basic types of vibration: forced vibration and self-excited vibration.

**Forced vibration** is generally caused by some periodic applied force present in the machine tool, such as from gear drives, imbalance of the machine-tool components, misalignment, or motors and pumps. In processes such as milling or turning a splined shaft or a shaft with a keyway, forced vibrations are caused by the periodic engagement of the cutting tool with and exit from the workpiece surface.

The basic solution to forced vibrations is to isolate or remove the forcing element. If the forcing frequency is at or near the natural frequency of a component of the machine-tool system, one of the frequencies may be raised or lowered. The amplitude of vibration can be reduced by increasing the stiffness or damping the system (see the discussion below). Although changing the cutting-process parameters generally does not appear to greatly influence forced vibrations, changing the cutting speed and the tool geometry can be helpful.

Generally called **chatter, self-excited vibration** is caused by the interaction of the chip-removal process and the structure of the machine tool. Self-excited vibrations usually have a

**FIGURE 8.85**   Chatter marks (right of center of photograph) on the surface of a turned part. *Source:* General Electric Company.

very high amplitude. Chatter typically begins with a disturbance in the cutting zone. Such disturbances include lack of homogeneity in the workpiece material or its surface condition, changes in type of chips produced, or a change in frictional conditions at the tool-chip interface as also influenced by cutting fluids and their effectiveness.

The most important type of self-excited vibration in machining is **regenerative chatter**. It is caused when a tool cuts a surface that has a roughness or disturbances from the previous cut. Because the depth of cut varies, the resulting variations in the cutting force subject the tool to vibrations and the process continues repeatedly—hence the term regenerative. You may observe this type of vibration while driving over a rough road (the so-called washboard effect).

Self-excited vibrations can generally be controlled by increasing the dynamic stiffness of the system and by damping. We define **dynamic stiffness** as the ratio of the amplitude of the force applied to the amplitude of vibration. Since a machine tool has different stiffnesses at different frequencies, changes in cutting parameters, such as cutting speed, can influence chatter.

### 8.11.1 Damping

We define *damping* as the rate at which vibrations decay. Damping is an important factor in controlling machine-tool vibration and chatter.

**Internal damping of structural materials.** Internal damping results from the energy loss in materials during vibration. Thus, for example, steel has less damping than gray cast iron, and composite materials have more damping than gray iron (Fig. 8.86). You can observe this difference in the damping capacity of materials by striking them with a gavel and listening to the sound. Try, for example, striking pieces of steel, concrete, and wood.

**Joints in the machine-tool structure.** Although less significant than internal damping, bolted joints in the structure of a machine tool are also a source of damping. Because friction dissipates energy, small relative movements along dry (unlubricated) joints dissipate energy and thus improve damping. In joints where oil is present, the internal friction of the oil layers also dissipates energy, thus contributing to damping. Note, for example, that stirring oil with a stick requires force; thus energy is dissipated in stirring it.

**FIGURE 8.86** Relative damping capacity of gray cast iron and epoxy-granite composite material. The vertical scale is the amplitude of vibration and the horizontal scale is time. *Source:* Cincinnati Milacron, Inc.

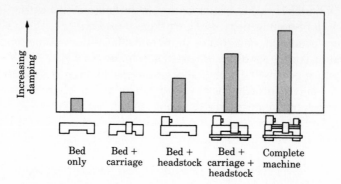

**FIGURE 8.87**    Damping of vibrations as a function of the number of components on a lathe. Joints dissipate energy; thus the greater the number of joints, the higher the damping will be. *Source:* J. Peters.

In describing the machine tools for various cutting operations, we noted that all machines consist of a number of large and small components, assembled into a structure. Consequently, this type of damping is cumulative, owing to the presence of a number of joints in a machine tool. Note in Fig. 8.87 how damping increases as the number of components on a lathe and their contact area increase. Thus the more joints, the greater is the amount of energy dissipated and the higher is the damping.

**External damping.**    External damping is accomplished with external dampers, which are similar to shock absorbers on automobiles. Special vibration absorbers have been developed and installed on machine tools for this purpose.

### 8.11.2    Factors influencing chatter

Studies have indicated that the tendency for a particular workpiece to chatter during cutting is proportional to the cutting forces and the depth and width of cut. Consequently, because cutting forces increase with strength (hardness), the tendency to chatter generally increases as the hardness of the workpiece material increases. Thus aluminum and magnesium alloys have less tendency to chatter than do martensitic and precipitation-hardening stainless steels, nickel alloys, and high-temperature and refractory alloys.

An important factor in chatter is the type of chip produced during cutting operations. As we have shown, continuous chips involve steady cutting forces. Consequently, they generally do not cause chatter. Discontinuous chips and serrated chips, on the other hand, may do so. These chips are produced periodically, and the resulting variations in force during cutting can cause chatter.

## 8.12    Machine-Tool Structures

In this section we discuss the *material* and *design* aspects of machine tools as structures that have certain desired characteristics. Today's markets have stringent requirements for high quality and precision in manufactured products, often made with difficult-to-machine materials, with precise specifications concerning dimensional accuracy and surface finish and integrity. Consequently, the design and construction of machine tools are important aspects of manufacturing engineering.

The proper design of machine-tool structures requires a thorough knowledge of the materials available for construction, their forms and properties, the dynamics of the particular machining process, and the forces involved. *Stiffness* and *damping* are important factors in machine-tool structures. Stiffness involves both the dimensions of the structural components and the elastic modulus of the materials used. Damping, as you have seen, involves the type of materials used, as well as the number and nature of the joints in the structure.

### 8.12.1  Materials and design

Traditionally, the base and some of the major components of machine tools have been made of gray or nodular cast iron, which has the advantages of low cost and good damping capacity, but are heavy. Lightweight designs are desirable because of ease of transportation, higher natural frequencies, and lower inertial forces of moving members. Lightweight designs and design flexibility require fabrication processes such as (1) mechanical fastening (bolts and nuts) of individual components and (2) welding. However, this approach to fabrication increases labor and material costs because of the preparations involved.

Wrought steels are the likely choice for such lightweight structures because of their low cost, availability in various section sizes and shapes (such as channels, angles, and tubes), desirable mechanical properties, and favorable characteristics (such as formability, machinability, and weldability). Tubes, for example, have high stiffness-to-weight ratios. On the other hand, the benefit of the higher damping capacity of castings and composites isn't available with steels.

In addition to stiffness, another factor that contributes to lack of precision of the machine tool is the thermal expansion of its components, causing distortion. The source of heat may be internal, such as bearings, ways, motors, and heat generated from the cutting zone, or external, such as nearby furnaces, heaters, sunlight, and fluctuations in cutting fluid and ambient temperatures.

Equally important in machine-tool precision are foundations, their mass, and how they are installed in a plant. For example, in one grinder for grinding 2.75 m (9 ft) diameter marine-propulsion gears with high precision, the concrete foundation is 6.7 m (22 ft deep). The large mass of concrete and the machine base reduces the amplitude of vibrations and its adverse effects.

### 8.12.2  Recent developments

Significant developments concerning the materials used for machine-tool bases and components have taken place. These developments are summarized in the following paragraphs.

**Concrete.**  Concrete has been used for machine-tool bases since the early 1970s. Compared to cast iron, concrete is less expensive, its curing time is about three weeks (much shorter than for making castings), and it has good damping capacity. However, concrete is brittle and thus not suitable for applications involving impact loads. Concrete can also be poured into cast-iron base structures to increase their mass and improve machine-tool damping capacity. Filling the cavities of bases with loose sand is also an effective means of improving damping capacity.

**Acrylic concretes.**   A mixture of concrete and polymer (polymethylmethacrylate), acrylic concretes can easily be cast into desired shapes for machine bases and various components. They were first introduced in the 1980s, and several new compositions are being developed. These materials can also be used for sandwich construction with cast irons, thus combining the advantages of each type of material.

**Granite-epoxy composite.**   A castable composite has been developed, with a composition of about 93 percent crushed granite and 7 percent epoxy binder. First used in precision grinders in the early 1980s, this composite material has several favorable properties: (1) good castability, which allows design versatility in machine tools; (2) high stiffness-to-weight ratio; (3) thermal stability; (4) resistance to environmental degradation; and (5) good damping capacity (see Fig. 8.86).

## 8.13  Machining Economics

Besides technical considerations, the economics of metal removal processes is very important. In machining a certain part, we may want to determine the parameters that will give us either the minimum cost per part or the maximum production rate.

The **total cost per piece** is composed of four items:

$$C_p = C_m + C_s + C_l + C_t, \tag{8.44}$$

where $C_p$ is the cost (say, in dollars) per piece; $C_m$ is the machining cost; $C_s$ is the cost of setting up for machining, such as mounting the cutter and fixtures and preparing the machine for the particular operation; $C_l$ is the cost of loading, unloading, and machine handling; and $C_t$ is the tooling cost, which includes tool changing, regrinding, and depreciation of the cutter.

The **machining cost** is

$$C_m = T_m(L_m + B_m), \tag{8.45}$$

where $T_m$ is the machining time per piece; $L_m$ is the labor cost of production operator per hour; and $B_m$ is the burden rate, or overhead charge, of the machine, including depreciation, maintenance, indirect labor, and the like.

The **setup cost** $C_s$ is a fixed figure in dollars per piece. The $C_l$ cost is

$$C_l = T_l(L_m + B_m), \tag{8.46}$$

where $T_l$ is the time involved in loading and unloading the part, changing speeds, feed rates, and so on.

The **tooling cost** is

$$C_t = \frac{1}{N_p}\left[T_c(L_m + B_m) + T_g(L_g + B_g) + D_c\right], \tag{8.47}$$

where $N_p$ is the number of parts machined per tool grind; $T_c$ is the time required to change the tool; $T_g$ is the time required to grind the tool; $L_g$ is the labor cost of tool grinder operator per hour; $B_g$ is the burden rate of tool grinder per hour; and $D_c$ is the depreciation of the tool in dollars per grind.

The **time** needed to produce one part is

$$T_p = T_l + T_m + \frac{T_c}{N_p},\tag{8.48}$$

where $T_m$ has to be calculated for a particular operation. For example, let's consider a turning operation. The machining time is

$$T_m = \frac{L}{fN} = \frac{\pi LD}{fV},\tag{8.49}$$

where $L$ is the length of cut, $f$ is the feed, $N$ is the rpm of the workpiece, $D$ is the workpiece diameter, and $V$ is the cutting speed. Note that appropriate units must be used in all these equations.

From the tool-life equation we have

$$VT^n = C.$$

Hence

$$T = \left(\frac{C}{V}\right)^{1/n},\tag{8.50}$$

where $T$ is the time, in minutes, required to reach a flank wear of certain dimension, after which the tool has to be reground or changed. The number of pieces per tool grind is thus simply

$$N_p = \frac{T}{T_m}.\tag{8.51}$$

The combination of Eqs. (8.49), (8.50), and (8.51) gives

$$N_p = \frac{fC^{1/n}}{\pi LDV^{(1/n)-1}}.\tag{8.52}$$

We can now define the **cost per piece** $C_p$ in Eq. (8.44) in terms of several variables. To find the optimum cutting speed and also the optimum tool life for **minimum cost**, we have to differentiate $C_p$ with respect to $V$ and set it to zero. Thus

$$\frac{\partial C_p}{\partial V} = 0.\tag{8.53}$$

We then find that the optimum cutting speed $V_o$ is

$$V_o = \frac{C(L_m + B_m)^n}{\{[(1/n) - 1][T_c(L_m + B_m) + T_g(L_g + B_g) + D_c]\}^n},\tag{8.54}$$

and the optimum tool life $T_o$ is

$$T_o = [(1/n) - 1]\frac{T_c(L_m + B_m) + T_g(L_g + B_g) + D_c}{L_m + B_m}.\tag{8.55}$$

(a)

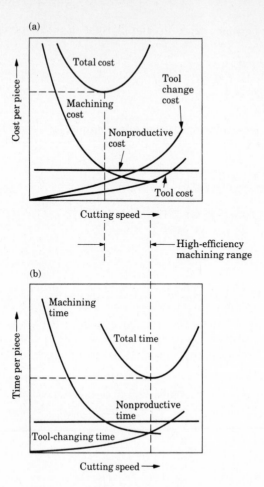

**FIGURE 8.88** Graphs showing (a) cost per piece and (b) time per piece in machining. Note the optimum speeds for both cost and time. The range between the two is known as the *high-efficiency machining range.*

To find the optimum cutting speed and also the optimum tool life for **maximum production**, we have to differentiate $T_p$ with respect to $V$ and set it to zero. Thus

$$\frac{\partial T_p}{\partial V} = 0. \tag{8.56}$$

We find that the **optimum cutting speed** $V_o$ now becomes

$$V_o = \frac{C}{\{[(1/n) - 1]T_c\}^n}, \tag{8.57}$$

and the **optimum tool life** $T_o$ is

$$T_o = [(1/n) - 1]T_c. \tag{8.58}$$

A qualitative plot of minimum cost per piece and the minimum time per piece, that is, the maximum production rate, is given in Fig. 8.88. The cost of a machined surface also depends on the finish required (see Section 9.17); the machining cost increases rapidly with finer surface finish.

This analysis indicates the importance of identifying all relevant parameters in a machining operation, determining various cost factors, obtaining relevant tool-life curves for the particular operation, and properly measuring the various time intervals involved in the overall operation. The importance of obtaining accurate data is clearly shown in Fig. 8.88: Small changes in cutting speed can have a significant effect on the minimum cost or time per piece.

## CASE STUDY: Machining of Various Complex Parts

In Example 8.7 in Section 8.8.2, we illustrated the capabilities of CNC machining by showing three different parts; note, however, that these are axisymmetric components. We will now further illustrate the capabilities of CNC turning by referring to Fig 8.89 below, showing three additional but more complex parts: pump shaft, crankshaft, and a tubular part with internal rope thread. As in most operations, machining of these parts consisted of roughing and finishing cuts.

1. *Pump shaft* (Fig 8.89a and b). This part, as well as a wide variety of similar parts with external and internal features including camshafts, was produced on a CNC lathe with two turrets, similar in construction to those shown in Figs. 8.56 and 8.83; each turret can hold as many as eight tools. Note that to produce this particular shape, the upper turret is programmed in such a manner that its radial movement is synchronized with shaft rotation. The spindle turning angle is monitored directly; a processor performs a high-speed calculation and the CNC issues a command to the cam turret in terms of that angle. It has absolute position feedback using a high-accuracy scale system; the CNC compares the actual value with the commanded one and performs an automatic compensation using a built-in learning function. The turret has a lightweight design for smooth operation (by reducing inertial forces).

**FIGURE 8.89**

**TABLE 8.20**

| OPERATION | CUTTING SPEED | | DEPTH OF CUT | FEED | TOOL |
|---|---|---|---|---|---|
| **(a)** | | | | | |
| OD roughing | 1150 rpm | 160 m/min (525 fpm) | 3 mm (0.12 in.) | 0.3 mm/rev (0.012 ipr) | K10 (C3) |
| OD finishing | 1750 | 250 (820) | 0.2 (0.008) | 0.15 (0.0059) | K10 (C3) |
| Lead roughing | 300 | 45 (148) | 3 (0.12) | 0.15 (0.0059) | K10 (C3) |
| Lead finishing | 300 | 45 (148) | 0.1 (0.004) | 0.15 (0.0059) | Diamond compact |
| **(b)** | | | | | |
| Eccentric roughing | 200 rpm | 5-11 m/min (16-36 fpm) | 1.5 mm (0.059 in) | 0.2 mm/rev (0.008 ipr) | K10 (C3) |
| Eccentric finishing | 200 | 5-11 (16-36) | 0.1 (0.004) | 0.05 (0.0020) | K10 (C3) |
| **(c)** | | | | | |
| Thread roughing | 800 rpm | 70 m/min (230 fpm) | 1.6 mm (0.063 in.) | 0.15 mm/rev (0.0059 ipr) | Coated carbide |
| Thread finishing | 800 | 70 (230) | 0.1 (0.004) | 0.15 (0.0059) | Cermet |

This shaft may be made of aluminum or stainless steel. The machining parameters for aluminum are shown in Table 8.20; these may be compared with the data in Table 8.12, which gives only a broad approximate range as a guideline. The inserts were K10 (C3) uncoated carbide and compacted (polycrystalline) diamond (see Section 8.6). The OD machine refers to the two straight cylindrical ends of the part. The total machining time for an aluminum shaft was 24 min. and for stainless steel 55 min. Note from Table 8.12 that the cutting speed for stainless steel is considerably lower than that for aluminum, thus machining time is longer.

2. *Crankshaft* (Fig. 8.89c). This part is made of ductile (nodular) cast iron (see Section 5.6.1). The machining parameters are shown in Table 8.20b. The insert was K10 carbide. The machining time was 25 min.; note that this time is on the same order of magnitude as that for the first part.

3. *Tubular part with internal rope thread* (Fig. 8.89d). Made of 304 stainless steel, this part was machined under the conditions given in Table 8.20c. The starting blank was a straight tubular piece, similar to a bushing. The cutting tools were coated carbide and cermet. The boring bar was made of tungsten carbide to increase stiffness (see Section 8.8.3) and thus improve dimensional accuracy, which for the threaded portion was $\pm 0.05$ mm (0.002 in.) and with a surface finish of $R_a = 2.5\mu$m (100$\mu$in.). The machining time was 1.5 min. Note that this time is much shorter than that for the two examples given above; the reason is that this part is shorter, less material is removed, it does not have the eccentric features of the first two parts (thus the radial movement of the cutting tool is not a function of the angular position of the part), and the cutting speed is higher.

*Source:* Based on technical literature by Okuma Corp.

# SUMMARY

- Material-removal processes are often necessary in order to impart the desired dimensional accuracy, geometric features, and surface finish characteristics to components, particularly those with complex shapes that cannot be produced economically or properly by other shaping techniques. On the other hand, these processes waste some material in the form of chips, generally take longer, and may have adverse effects on the surfaces produced. (Section 8.1)

- Important process variables are tool shape and material, cutting conditions such as speed, feed, and depth of cut, use of cutting fluids, and the characteristics of the machine tool and the workpiece material. Parameters influenced by these variables are forces and power consumption, tool wear, surface finish and integrity, temperature, and dimensional accuracy of the workpiece. Commonly observed chip types are continuous, built-up edge, discontinuous, and serrated. (Section 8.2)

- Temperature rise is an important consideration, as it can have adverse effects on tool life as well as the dimensional accuracy and surface integrity of the machined part. (Section 8.2)

- Cutting-tool wear depends primarily on workpiece and tool-material characteristics, cutting speed, and cutting fluids. Feeds, depth of cut, and machine-tool characteristics also have an effect. Two major types of wear are flank wear and crater wear. (Section 8.3)

- Surface finish of machined components is an important factor, as it can adversely affect product integrity. Important variables are the geometry and condition of the cutting tool, type of chip produced, and process variables. (Section 8.4)

- Machinability is usually defined in terms of surface finish, tool life, and force and power requirements. Machinability of materials depends not only on their intrinsic properties and microstructure, but also on proper selection and control of process variables. (Section 8.5)

- A wide variety of cutting-tool materials have been developed over the past century, the most common being high-speed steels, carbides, ceramics, and cubic boron nitride. They have a broad range of mechanical and physical properties, such as hot hardness, toughness, chemical stability and inertness, and resistance to chipping and wear. Various tool coatings have also been developed, resulting in major improvements in tool life. The selection of appropriate tool materials depends not only on the material to be machined, but also on process parameters and the characteristics of the machine tool. (Section 8.6)

- Cutting fluids are important in machining operations and can reduce friction and forces and power requirements, and improve tool life. Generally, slower operations with high tool pressures require a fluid with good lubricating characteristics, whereas in high-speed operations, where temperature rise can be significant, fluids with cooling capacity are preferred. Selection should include considerations of the possible adverse effects of cutting fluids on machined products, machine tools, personnel, and the environment. (Section 8.7)

- Cutting processes that produce external and internal circular profiles are turning, boring, drilling, tapping, and thread cutting. Because of the three-dimensional nature of these operations, chip movement and its control are important considerations, since otherwise they interfere with the operation. Chip removal can be a significant problem, especially in drilling and tapping, and can lead to tool breakage. Optimization of each process requires an understanding of the interrelationships among design parameters, such as part shape, dimensional accuracy, and surface finish, and process parameters, such as speed, feed, depth of cut, tool material and shape, cutting fluids, and sequence of operations to be performed. (Section 8.8)

- Among newer developments are high-speed machining, ultraprecision machining, and hard turning. These developments have helped reduce machining costs and have produced parts with exceptional surface finish and dimensional accuracy for critical applications. (Section 8.8)

- Various complex shapes can be machined by slab, face, and end milling, broaching, and sawing. These processes utilize multitooth tools and cutters at various axes with respect to the workpiece. The machine tools are now mostly computer controlled, have various features and attachments, and possess considerable flexibility in operation. Gears are manufactured by some of these techniques. (Section 8.9)

- Because of their versatility and capability of performing a variety of cutting operations on small and large workpieces, machining and turning centers are among the most important developments in machine tools. Their selection depends on factors such as part complexity, the number and type of cutting operations to be performed, number of cutting tools needed, dimensional accuracy, and production rate required. (Section 8.10)

- Vibration and chatter in machining are important considerations for workpiece dimensional accuracy and surface finish, as well as tool life. Stiffness and damping capacity of machine tools are important factors in controlling vibration and chatter. New materials are being developed and used for constructing machine-tool structures. (Sections 8.11 and 8.12)

- The economics of machining processes depend on nonproductive, machining, tool-change, and tool costs. Optimum cutting speeds can be determined for minimum machining time per piece and minimum cost per piece, respectively. (Section 8.13)

## SUMMARY OF EQUATIONS

Cutting ratio, $r = \dfrac{t_o}{t_c} = \dfrac{\sin\phi}{\cos(\phi - \alpha)}$

Shear strain, $\gamma = \cot\phi + \tan(\phi - \alpha)$

Velocity relationships: $\dfrac{V}{\cos(\phi - \alpha)} = \dfrac{V_s}{\cos\alpha} = \dfrac{V_c}{\sin\phi}$

Friction force, $F = R\sin\beta$

Normal force, $N = R\cos\beta$

Coefficient of friction, $\mu = \tan\beta = \dfrac{F_t + F_c\tan\alpha}{F_c - F_t\tan\alpha}$

Thrust force, $F_t = R\sin(\beta - \alpha) = F_c\tan(\beta - \alpha)$

Shear-angle relationships, $\phi = 45° + \dfrac{\alpha}{2} - \dfrac{\beta}{2}$

$\phi = 45° + \alpha - \beta$

Total cutting power $= F_c V$

Specific energy, $u_t = \dfrac{F_c}{w\, t_o}$

Frictional specific energy, $u_f = \dfrac{F\, r}{w\, t_o}$

Shear specific energy, $u_s = \dfrac{F_s V_s}{w\, t_o V}$

Mean temperature, $T = \dfrac{1.2\, Y_f}{\rho c} \sqrt[3]{\dfrac{V\, t_o}{K}}$

$T \propto V^a f^b$

Tool life, $VT^n = C$

Material removal rate, MRR, in turning $= \pi D_{\mathrm{avg}}\, df N$

in drilling $= \pi(D^2/4)f N$

in milling $= wdv$

## BIBLIOGRAPHY

### Fundamentals

Armarego, E.J.A., and R.H. Brown, *The Machining of Metals*. Englewood Cliffs, New Jersey: Prentice-Hall, 1969.

*ASM Handbook*, Vol. 16: *Machining*. Materials Park, Ohio: ASM International, 1989.

Boothroyd, G., and W.A. Knight, *Fundamentals of Machining and Machine Tools*, 2d ed. New York: Marcel Dekker, 1989.

DeVries, W.R., *Analysis of Material Removal Processes*. New York: Springer, 1992.

Mills, B., and A.H. Redford, *Machinability of Engineering Materials*. London: Applied Science Publishers, 1983.

Oxley, P.L.B., *Mechanics of Machining: An Analytical Approach to Assessing Machinability*. New York: Wiley, 1989.

Schneider, A.F., *Mechanical Deburring and Surface Finishing Technology*. New York: Marcel Dekker, 1990.

Shaw, M.C., *Metal Cutting Principles*. New York: Oxford, 1984.

Stephenson, D., and J.S. Agapiou, *Metal Cutting: Theory and Practice*. New York: Marcel Dekker, 1996.

*Tool and Manufacturing Engineers Handbook*, 4th ed., Vol. 1: *Machining*. Dearborn, Mich.: Society of Manufacturing Engineers, 1983.

Trent, E.M., *Metal Cutting*, 4th ed. London: Butterworth Heinemann, 1991.

Venkatesh, V.C., and H. Chandrasekaran, *Experimental Techniques in Metal Cutting*, rev. ed. New Delhi: Prentice-Hall, 1987.

## Cutting tools

*ASM Handbook,* Vol. 1: *Properties and Selection: Irons, Steels, and High-Performance Alloys.* Materials Park, Ohio: ASM International, 1990.
*ASM Handbook,* Vol. 16: *Machining.* Materials Park, Ohio: ASM International, 1989.
*ASM Specialty Handbook–Tool Materials.* Materials Park, Ohio: ASM International, 1995.
Komanduri, R., *"Tool Materials,"* in Kirk-Othmer *Encyclopedia of Chemical Technology,* 3d ed., v. 23. New York: Wiley, 1978.
*Machinery's Handbook.* New York: Industrial Press, revised periodically.
*Machining Data Handbook,* 3d ed., 2 vols. Cincinnati: Machinability Data Center, 1980.
*Metal Cutting Tool Handbook,* 7th ed. New York: Industrial Press, 1989.
Roberts, G.A., and R.A. Cary, *Tool Steels,* 4th ed. Metals Park, Ohio: American Society for Metals, 1980.
*Tool and Manufacturing Engineers Handbook,* 4th ed., Vol. 1: *Machining.* Dearborn, Mich.: Society of Manufacturing Engineers, 1983.
Trent, E.M., *Metal Cutting,* 2d ed. Stoneham, Mass.: Butterworth, 1984.

## Cutting fluids

*ASM Handbook,* Vol. 16: *Machining.* Materials Park, Ohio: ASM International, 1989.
Byers, J.P. (ed.), *Metalworking Fluids.* New York: Marcel Dekker, 1994.
Nachtman, E.S., and S. Kalpakjian, *Lubricants and Lubrication in Metalworking Operations.* New York: Marcel Dekker, 1985.
Schey, J.A., *Tribology in Metalworking—Friction, Wear and Lubrication.* Metals Park, Ohio: American Society for Metals, 1983.
Sluhan, C. (ed.), *Cutting and Grinding Fluids: Selection and Application.* Dearborn, Mich.: Society of Manufacturing Engineers, 1992.

## Machining processes and machine tools

*ASM Handbook,* Vol. 16: *Machining.* Materials Park, Ohio: ASM International, 1989.
*Handbook of High-Speed Machining Technology.* New York: Chapman & Hall, 1985.
Juneja, B.L., *Fundamentals of Metal Cutting and Machine Tools.* New York: Wiley, 1987.
*Machining Data Handbook,* 2 vols., 3d ed. Cincinnati: Machinability Data Center, 1980.
*Tool and Manufacturing Engineering Handbook,* 4th ed., Vol. 1, *Machining.* Dearborn, Mich.: Society of Manufacturing Engineers, 1983.
Townsend, D.P., *Dudley's Gear Handbook: The Design, Manufacturing, and Application of Gears,* 2d ed. New York: McGraw-Hill, 1991.
Walsh, R.A., *McGraw-Hill Machining and Metalworking Handbook.* New York: McGraw-Hill, 1994.
Weck, M., *Handbook of Machine Tools,* 4 vols. New York: Wiley, 1984.

## QUESTIONS

8.1 In metal cutting, it has been observed that the shear strains obtained are higher than those calculated from the properties of the material. To what factors would you attribute this difference? Explain.

8.2 Describe the effects of material properties and process variables on serrated chip formation.

8.3 Explain why the cutting force $F_c$ increases with increasing depth of cut and decreasing rake angle.

8.4 What are the effects of performing a cutting operation with a dull tool tip?

8.5 Describe the trends that you observe in Table 8.1.

8.6 To what factors would you attribute the large difference in the specific energies within each group of materials in Table 8.4?

8.7 Describe the effects of cutting fluids on chip formation. Explain why and how they influence the cutting operation.

8.8 Under what conditions would you discourage the use of cutting fluids?

8.9 Describe the trends that you observe in Tables 8.2 and 8.3.

8.10 Can you offer an explanation as to why the maximum temperature in cutting is located at about the middle of the tool–chip interface? Explain. Remember that there are two principal sources of heat: the shear plane and the tool–chip interface.

8.11 State whether or not the following statements are correct, explaining your reasons: (1) For the same shear angle, there are two rake angles that give the same cutting ratio. (2) For the same depth of cut and rake angle, the type of cutting fluid used has no influence on chip thickness. (3) If the cutting speed, shear angle, and rake angle are known, the chip velocity can be calculated. (4) The chip becomes thinner as the rake angle increases. (5) The function of a chip breaker is to decrease the curvature of the chip.

8.12 It is generally undesirable to allow temperatures to rise too high in cutting operations. Explain why.

8.13 Explain why the same tool life may be obtained at two different cutting speeds.

8.14 Inspect Tables 8.7 and 8.8 and identify tool materials that would not be particularly suitable for interrupted cutting operations. Explain.

8.15 Explain the disadvantages of a cutting operation in which the type of chip produced is discontinuous.

8.16 We note that tool life can be almost infinite at low cutting speeds. Would you recommend that all machining be done at low speeds? Explain.

8.17 How would you explain the effect of cobalt content on the properties of carbides as shown in Fig. 8.38?

8.18 Explain why studying the types of chips produced is important in understanding machining operations.

8.19 How would you expect the cutting force to vary in the case of serrated chip formation? Explain.

8.20 Wood is a highly anisotropic material (it is orthotropic). Explain the effects of cutting wood at different angles to the grain direction on chip formation.

8.21 Describe the advantages of oblique cutting.

8.22 Explain why it is possible to remove more material between tool resharpenings by lowering the cutting speed.

8.23 Explain the significance of Eq. (8.8).

8.24 Explain why there are so many different types of cutting-tool materials.

8.25 How would you go about measuring the hot hardness of cutting tools?

8.26 Describe the reasons for making cutting tools with multiple coatings of different materials.

8.27 Explain the advantages and limitations of inserts. Why were they developed?

8.28 Make a list of alloying elements used in high-speed steels. Explain why they are so effective in cutting tools.

8.29 What are the purposes of chamfers on cutting tools?

8.30 Why does temperature have such an important effect on cutting-tool performance?

8.31 Ceramic and cermet cutting tools have certain advantages over carbide tools. Why then are they not replacing carbide tools to a greater extent?

8.32 Why are chemical stability and inertness important in cutting tools?

8.33 What precautions would you take in cutting with brittle tool materials?

8.34 Why do cutting fluids have different effects at different cutting speeds? Is the control of cutting-fluid temperature important? If so, why?

8.35 Which of the two materials, diamond or cubic boron nitride, is more suitable for cutting steels? Why?

8.36 Describe the properties that the substrate for multiple-coated tools should have.

8.37 List and explain the considerations involved in whether a cutting tool should be reconditioned or discarded.

8.38 List the parameters that influence the temperature in metal cutting, and explain why and how they do so.

8.39 List and explain factors that contribute to poor surface finish.

8.40 Explain the functions of different angles on a single-point lathe cutting tool. How does the chip thickness vary as the side cutting-edge angle is increased?

8.41 We note that the helix angle for drills is different for different groups of materials. Why?

8.42 A turning operation is being carried out on a long round bar at a constant depth of cut. Explain what changes, if any, may occur in the machined diameter from one end of the bar to the other. Give reasons for any changes that may occur.

8.43 Describe the relative characteristics of climb milling and up milling.

8.44 In Fig. 8.76a, high-speed-steel cutting teeth are welded to a steel blade. Would you recommend that the whole blade be made of high-speed steel? Explain your reasons.

8.45 Explain the technical requirements that led to the development of machining centers. What are the distinctive features of these centers? Why do their spindle speeds vary over a wide range? Are there cutting operations that cannot be performed on these centers? Explain.

8.46 Describe the adverse effects of vibrations and chatter in machining.

8.47 Make a list of components of machine tools that could be made of ceramics and explain why ceramics would be suitable.

8.48 In Fig. 8.13, why do the thrust forces start at a finite value when the feed is zero?

8.49 Do you think temperature rise in cutting is related to the hardness of the workpiece? Explain.

8.50 Describe the effects of tool wear on the workpiece and on the machining operation in general.

8.51 In Fig. 8.31, explain why the curves go downward as feed increases.

8.52 Explain whether or not it is desirable to have a high or low $n$ value in the Taylor tool-life equation.

8.53 Explain why the cutting force is a function of the cutting speed, feed, and rake angle.

8.54 Devise a means whereby you can perform an orthogonal cutting operation with a round workpiece on a lathe.

8.55 Cutting tools are sometimes designed so that the tool–chip contact length is controlled (see, for example, B.T. Chao and K.J. Trigger, *Trans. ASME, J. Eng. Ind.*, v. 81, 1959, 139–151). Discuss the effect of this on the mechanics of orthogonal cutting.

8.56 It is stated that it is possible for the $n$ value in the Taylor tool-life equation to be negative. Explain how.

8.57 Assume that you are asked to estimate the cutting force in slab milling with a straight-tooth cutter, without running a test. Describe the procedure that you would follow.

8.58 Explain the possible reasons why a knife cuts better when it is moved back and forth. Consider factors such as the material cut, friction, and the dimensions of the knife.

8.59 What are the effects of lowering the friction at the tool–chip interface, say with an effective cutting fluid, on the mechanics of cutting operations? Explain, giving several examples.

8.60 Why is it not always advisable to increase cutting speed in order to increase production rate? Explain.

8.61 It has been stated that the shear strain rate in metal cutting is high. Why?

8.62 We note from the exponents in Eq. (8.30) that the cutting speed has a greater influence on temperature than the feed. Why?

8.63 Comment on the location of various phenomena indicated in Fig. 8.31.

8.64 Describe the consequences of exceeding the allowable wear land (Table 8.6) for cutting tools.

8.65 Explain whether having a high or low $n$ value in the Taylor tool-life equation is desirable. How about the value of C?

8.66 The tool-life curve for ceramic tools in Fig. 8.27 is to the right of those for other tool materials. Why?

8.67 In Fig. 8.22, we note that the percentage of the energy carried away by the chip increases with cutting speed. Why?

8.68 How would you go about measuring the effectiveness of cutting fluids? Explain.

8.69 Describe the conditions that are critical in utilizing the capabilities of diamond and cubic boron nitride cutting tools.

8.70 In Table 8.7, the last two properties listed can be important to the life of the cutting tool. Why? Which of the properties listed are, in your opinion, the least important in machining? Explain.

8.71 In Fig. 8.37, we note that all tool materials have a wide range of hardness for a particular temperature, especially carbides. Why?

8.72 List and discuss the considerations involved in whether a used cutting tool should be recycled, and if so, how would you proceed.

8.73 As you can see, there is a wide range of tool materials available and used successfully today, yet much research and development is presently being carried out on these materials. Why?

8.74 Drilling, boring, and reaming of larger holes is generally more accurate than just drilling and reaming. Why?

8.75 A badly oxidized and uneven round bar is being turned on a lathe. Would you recommend a relatively small or large depth of cut, *d*? Explain your reasons.

8.76 Does the force or torque in drilling change as the hole depth increases? Explain.

8.77 Explain the advantages and limitations of producing threads by forming and cutting, respectively.

8.78 Describe your observations regarding the contents of Tables 8.11, 8.12, 8.16, 8.17, and 8.19.

8.79 The footnote to Table 8.17 states that as the hole depth increases, speeds and feeds should be reduced. Why?

8.80 List and explain factors that contribute to poor surface finish in machining operations.

8.81 Describe your observations regarding Fig. 8.49. Why is tapping so high in operational severity?

8.82 Explain whether the feed marks left on the workpiece by a face-milling cutter are segments of a true circle.

8.83 What determines the selection of the number of teeth on a milling cutter? See, for example, Figs. 8.64 and 8.66.

8.84 Explain the technical requirements that led to the development of machining and turning centers. Why do their spindle speeds vary over a wide range?

8.85 In addition to the number of components shown in Fig. 8.87, what other factors influence the rate at which damping increases in a machine tool? Explain.

8.86 Why is thermal expansion of machine-tool components important?

8.87 Would using the machining processes described in this chapter be difficult on various nonmetallic or rubberlike materials? Explain your thoughts, commenting on the influence of various physical and mechanical properties of workpiece materials, the cutting forces involved, parts geometries, fixturing required, etc.

8.88 The accompanying illustration shows a part that is to be machined from a rectangular blank. Suggest the type of operations required and their sequence, and the machine tool(s) required.

Stepped cavity

Drilled and tapped holes

# PROBLEMS

8.89 Prove Eq. (8.1).

8.90 Using Eq. (8.3), make a plot of the shear strain $\gamma$ vs. the shear angle $\phi$, with the rake angle $\alpha$ as a parameter. Describe your observations.

8.91 Assume that in orthogonal cutting, the rake angle is $10°$ and the coefficient of friction is 0.5. Determine the percentage change in chip thickness when the friction is doubled.

8.92 Derive Eq. (8.11).

8.93 Determine the shear angle in Example 8.1. Is this an exact calculation or an estimate? Explain.

8.94 Taking carbide as an example and using Eq. (8.30), determine how much the feed should be changed in order to keep the mean temperature constant when the cutting speed is doubled.

8.95 With appropriate diagrams, show how the use of a cutting fluid can change the magnitude of the thrust force $F_t$ in orthogonal cutting.

8.96 An 8 in. diameter stainless steel bar is being turned on a lathe at 500 rpm and a depth of cut $d = 0.1$ in. If the power of the motor is 2 hp, what is the maximum feed that you can have before the motor stalls?

8.97 Using the Taylor equation for tool wear and letting $n = 0.5$ and $C = 400$, calculate the percentage increase in tool life if the cutting speed is reduced by (1) 20% and (2) 50%.

8.98 Determine the $n$ and $C$ values for the four tool materials shown in Fig. 8.27.

8.99 Using Eq. (8.30) and referring to Fig. 8.21a, estimate the magnitude of the coefficient $a$.

8.100 Estimate the machining time required in rough turning a 1 m long, annealed aluminum-alloy round bar, 100 mm in diameter, using a high-speed-steel tool. Estimate the time for a carbide tool.

8.101 A 300-mm long, 100-mm diameter titanium alloy rod is being reduced in diameter to 95 mm by turning on a lathe. The spindle rotates at 400 rpm, and the tool is traveling at an axial speed of 250 mm/min. Calculate the cutting speed, material removal rate, time of cut, power required, and the cutting force.

8.102 Referring to Table 8.9, make a recommendation for a carbide tool for (1) finishing pass in turning a steel shaft at a 0.050 in. depth of cut and (2) high-strength aluminum in which the same tool is to be used for both roughing and finishing passes.

8.103 Calculate the same quantities as in Example 8.6 for high-strength cast iron and at $N = 500$ rpm.

8.104 A 0.5-in. diameter drill is used on a drill press operating at 200 rpm. If the feed is 0.005 in./rev, what is the material removal rate? What is the MRR if the drill diameter is tripled?

8.105 A hole is being drilled in a block of magnesium alloy with a 10 mm drill at a feed of 0.2 mm/rev. The spindle is running at 800 rpm. Calculate the material removal rate and estimate the torque on the drill.

8.106 Show that the distance $l_c$ in slab milling is approximately equal to $\sqrt{Dd}$ for situations where $D \gg d$.

8.107 Calculate the chip depth of cut in Example 8.8.

8.108 In Example 8.8, which of the quantities will be affected when the feed is increased to $f = 0.02$ in./tooth?

8.109 A slab-milling operation is being carried out on a 20 in. long, 6 in. wide high-strength steel block at a feed of 0.01 in./tooth and depth of cut of 0.15 in. The cutter is 3 in. in diameter, has 8 straight cutting teeth, and rotates at 150 rpm. Calculate the material removal rate and the cutting time, and estimate the power required.

8.110 Referring to Fig. 8.65, assume that $D = 200$ mm, $w = 50$ mm, $l = 500$ mm, $d = 2$ mm, $v = 0.01$ mm/s, and $N = 200$ rpm. The cutter has 10 inserts and the workpiece material is 304 stainless steel. Calculate the material removal rate, cutting time, and feed per tooth, and estimate the power required.

8.111 Estimate the time required for face milling an 8 in. long, 2 in. wide brass block with an 8 in. diameter cutter having 10 high-speed-steel teeth.

8.112 A 10 in. long, 1 in. thick plate is being cut on a band saw at 100 ft/min. The saw has 12 teeth per in. If the feed per tooth is 0.003 in., how long will it take to saw the plate along its length?

8.113 A single-thread hob is used to cut 40 teeth on a spur gear. The cutting speed is 100 ft/min and the hob is 4 in. in diameter. Calculate the rotational speed of the spur gear.

8.114 In deriving Eq. (8.20), it is assumed that the friction angle $\beta$ was independent of the shear angle $\phi$. Is this a valid assumption? Explain.

8.115 An orthogonal cutting operation is being carried out under the following conditions: Depth of cut = 0.1 mm, width of cut = 5 mm, chip thickness = 0.2 mm, cutting speed = 2 m/s, rake angle = 10°, cutting force = 500 N, and thrust force = 200 N. Calculate the percentage of the total energy that is dissipated in the shear plane during cutting.

8.116 An orthogonal cutting operation is being carried out under the following conditions: Depth of cut = 0.010 in., width of cut = 0.1 in., cutting ratio = 0.3, cutting speed = 400 ft/min, rake angle = 0°, cutting force = 200 lb, thrust force = 150 lb, workpiece density = 0.26 lb/in$^3$, workpiece specific heat = 0.12 BTU/lb.°F. Assume that (1) the sources of heat are the shear plane and the tool–chip interface, (2) the thermal conductivity of the tool is zero and there is no heat loss to the environment, and (3) the temperature of the chip is uniform throughout. If the temperature rise in the chip is 668°F, calculate the percentage of the energy dissipated in the shear plane that goes into the workpiece.

8.117 With a simple analytical expression, prove the validity of the last sentence in Example 8.2.

8.118 It can be shown that the angle $\psi$ between the shear plane and the direction of maximum grain elongation (see Fig. 8.5a) is given by the expression (see *J. App. Phys.*, v. 18, 1947, 489)

$$\psi = 0.5 \cot^{-1}(\gamma/2),$$

where $\gamma$ is the shear strain, as given by Eq. (8.3). Assume that you are given a piece of the chip obtained from orthogonal cutting of an annealed metal. The rake angle and cutting speed are also given, but you have not seen the setup on which the chip was produced. Outline the procedure that you would follow to estimate the power consumed in producing this chip. Assume that you have access to a fully equipped laboratory and a technical library.

8.119 In Figs. 8.19 and 8.21b, we note that the maximum temperature is about halfway up the face of the tool. We have also described the adverse effects of temperature on various tool materials. Considering the mechanics of cutting operations, describe your thoughts on the technical and economic merits of embedding a small insert, made of materials such as ceramic or carbide, halfway up the face of a tool made of a material with lower resistance to temperature.

8.120 Assume that you are an instructor covering the topics in this chapter, and you are giving a quiz on the quantitative aspects to test the understanding of the students. Prepare several numerical problems and supply the answers.

# 9

# Material-Removal Processes: Abrasive, Chemical, Electrical, and High-Energy Beams

**9.1** Introduction
**9.2** Abrasives
**9.3** Bonded Abrasives (Grinding Wheels)
**9.4** Mechanics of Grinding
**9.5** Grinding Wheel Wear
**9.6** Grinding Operations and Machines
**9.7** Ultrasonic Machining
**9.8** Finishing Operations
**9.9** Deburring
**9.10** Chemical Machining
**9.11** Electrochemical Machining
**9.12** Electrochemical Grinding
**9.13** Electrical-Discharge Machining
**9.14** High-Energy-Beam Machining
**9.15** Water-Jet Machining
**9.16** Abrasive-Jet Machining
**9.17** Process Economics

## 9.1  Introduction

In all the cutting processes described in Chapter 8, the tool is made of a certain material and has a clearly defined geometry. Furthermore, the cutting process is carried out by chip removal, the mechanics of which are reasonably well understood. There are many situations in manufacturing, however, where the workpiece material is either too *hard* or too *brittle,* or its *shape* is difficult to produce with sufficient accuracy by any of the cutting methods described.

One of the best methods for producing such parts is to use **abrasives**. An abrasive is a small, hard particle having sharp edges and an irregular shape, unlike the cutting tools we described earlier. Abrasives are capable of removing small amounts of material from a surface by a cutting process that produces tiny chips.

**Abrasive machining** processes are generally among the last operations performed on manufactured products. These processes, however, are not necessarily confined to fine or small-scale material removal. They are also used for large-scale removal operations and can indeed compete economically with some machining processes, such as milling and turning.

Because they are hard, abrasives are also used in finishing very hard or heat-treated parts; shaping hard nonmetallic materials, such as ceramics and glasses; removing unwanted weld beads; cutting off lengths of bars, structural shapes, masonry, and concrete; and cleaning surfaces with jets of air or water containing abrasive particles.

There are situations, however, where none of these processes is satisfactory or economical for the following reasons:

1.  The hardness and strength of the material is very high, typically above 400 HB.
2.  The workpiece is too flexible or slender to support the cutting or grinding forces, or parts are difficult to clamp in workholding devices.
3.  The shape of the part is complex, such as internal and external profiles, or small-diameter holes.
4.  Surface finish and tolerances better than those obtainable by other processes are required.
5.  Temperature rise or residual stresses in the workpiece are undesirable or unacceptable.

These requirements led to the development of chemical, electrical, and other means of material removal in the 1940s. These methods are called **nontraditional** or **unconventional machining.** When selected and applied properly these processes offer significant economic and technical advantages over the traditional machining methods described in the preceding chapter.

## 9.2  Abrasives

Abrasives commonly used in manufacturing processes are:

*Conventional abrasives*

*   Aluminum oxide ($Al_2O_3$)
*   Silicon carbide (SiC)

*Superabrasives*

*   Cubic boron nitride (CBN)
*   Diamond

These abrasives are considerably harder than conventional cutting-tool materials (see Tables 8.7 and 9.1), the last two of the four listed above being the two hardest materials known, hence the term **superabrasives.** In addition to hardness, an important characteristic is **friability,** that is, the ability of abrasive grains to fracture (breakdown) into smaller pieces. This property gives abrasives *self-sharpening* characteristics, which is important in maintaining the sharpness of the abrasives in their use. High friability indicates low strength or low fracture resistance of the abrasive. Thus a highly friable abrasive grain fragments more rapidly under grinding forces than one with low friability. Aluminum oxide has lower friability than silicon carbide, hence less tendency to fragment.

The **shape** and **size** of the abrasive grain also affect its friability. Blocky grains, for example, which may be analogous to negative-rake angle cutters, are less friable than platelike grains. Also, because the probability of defects in small grains is lower (size effect), they are stronger and less friable than larger grains. The importance of friability in abrasive processes will be described in Section 9.5.

Abrasives found in nature are emery, corundum (alumina), quartz, garnet, and diamond. However, natural abrasives contain unknown amounts of impurities and possess nonuniform properties. Consequently, their performance is inconsistent and unreliable. As a result, aluminum oxides and silicon carbides used in abrasive processes are now made synthetically.

**TABLE 9.1**

**KNOOP HARDNESS FOR VARIOUS MATERIALS AND ABRASIVES**

| | | | |
|---|---|---|---|
| Common glass | 350–500 | Titanium nitride | 2000 |
| Flint, quartz | 800–1100 | Titanium carbide | 1800–3200 |
| Zirconium oxide | 1000 | Silicon carbide | 2100–3000 |
| Hardened steels | 700–1300 | Boron carbide | 2800 |
| Tungsten carbide | 1800–2400 | Cubic boron nitride | 4000–5000 |
| Aluminium oxide | 2000–3000 | Diamond | 7000–8000 |

Synthetic **aluminum oxide,** first made in 1893, is obtained by fusing bauxite, iron filings, and coke. Aluminum oxides are divided into two groups. **Fused** aluminum oxides are categorized as dark (less friable), white (very friable), and monocrystalline. **Unfused** alumina (also known as ceramic aluminum oxides) can be harder than fused alumina, the purest (free of flaws) form of which is **seeded gel.** First introduced in 1987, seeded gel has a particle size on the order of 0.2 $\mu$m, which is much smaller than commonly used abrasive grains. These particles are sintered to form larger sizes. Because of their hardness and relatively high friability, seeded gels maintain their sharpness and are therefore used for difficult-to-grind materials.

**Silicon carbide** (first discovered in 1891) is made with silica sand, petroleum coke, and small amounts of sodium chloride (table salt). Silicon carbides are divided into *black* (less friable) and *green* (more friable), and generally have higher friability than aluminum oxides; hence they have a higher tendency to fracture and remain sharp.

**Superabrasives** are **cubic boron nitride** (first discovered in the 1970s) and **diamond** (first used as an abrasive in 1955), both of which are produced synthetically. Their properties and characteristics are described in Chapters 8 and 11.

**Grain size.**  As used in manufacturing processes, abrasives are generally very small compared to the size of the cutting tools and inserts we described in earlier chapters. Also, abrasives have sharp edges, thus allowing the removal of very small quantities of material from the workpiece surface. Consequently, very fine surface finish and dimensional accuracy can be obtained, as can be seen in Figs. 8.33 and 9.31.

The size of an abrasive **grain,** is identified by a *grit number,* which is a function of sieve size. The smaller the size, the larger the number. For example, grit number 10 is regarded as very coarse, 100 as fine, and 500 as very fine. Sandpaper and emery cloth are also identified in this manner, with the grit number printed on the back of the abrasive paper or cloth.

## 9.3 Bonded Abrasives (Grinding Wheels)

Because each abrasive grain usually removes only a very small amount of material at a time, high rates of material removal can be obtained only if a large number of these grains act together. This is done by using **bonded abrasives,** typically in the form of a **grinding wheel.** A simple grinding wheel is shown schematically in Fig. 9.1. The abrasive grains are held together by a **bonding** material (various types of which are described below) which acts as supporting

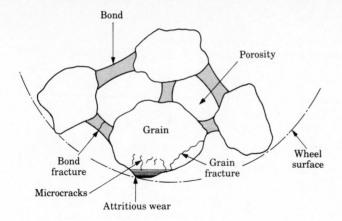

**FIGURE 9.1** Schematic illustration of a physical model of a grinding wheel, showing its structure and wear and fracture patterns.

posts or braces between the grains. Some porosity is essential in bonded wheels to provide clearance for the minute chips being produced and to provide cooling; otherwise they would interfere with the grinding process. (It is thus impossible to utilize a grinding wheel that is fully dense and solid with no porosity.) You can observe porosity by looking at the surface of any grinding wheel. Other features of Fig. 9.1 are described in Sections 9.4 and 9.5.

Some of the more commonly used types of grinding wheels are shown in Fig. 9.2 for conventional abrasives, and in Fig. 9.3 for superabrasives. Note that due to the high cost of the latter, only a small volume of these wheels consists of superabrasives. An estimated 250,000

**FIGURE 9.2** Some common types of grinding wheels made with conventional abrasives. Note that each wheel has a specific grinding face. Grinding on other surfaces is improper and unsafe.

**FIGURE 9.3** Examples of superabrasive wheel configurations. The annular regions (rim) are superabrasive grinding surfaces and the wheel itself (core) is generally made of metal or composites. Note that the basic numbering of wheel type is the same as those shown in Fig. 9.2. The bonding materials for the superabrasives are: (a), (d), and (e) resinoid, metal, or vitrified, (b) metal, (c) vitrified, and (f) resinoid.

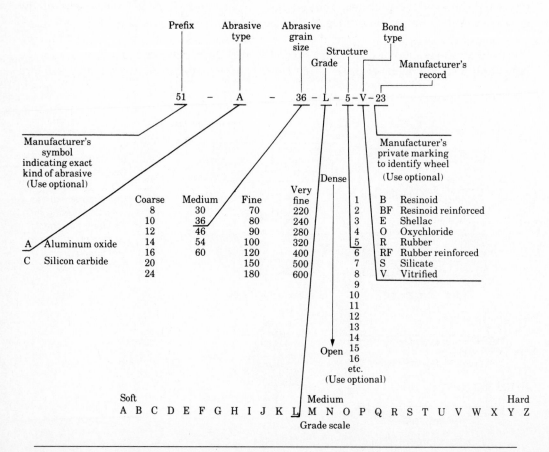

**FIGURE 9.4a** Standard marking system for aluminum-oxide and silicon-carbide bonded abrasives.

different types and sizes of abrasive wheels are made today. The *cost* of grinding wheels depends on the type and size of wheel. Small wheels up to about 25 mm (1 in.) in diameter cost approximately \$2–10 for conventional abrasives, \$30–100 for diamond, and \$50–200 for CBN wheels. For a large wheel of about 500 mm in diameter and 250 mm wide (20 in. by 10 in.), the costs are \$500, \$5000–8000, and almost \$20,000, respectively.

Bonded abrasives are marked with a standardized system of letters and numbers, indicating the type of abrasive, grain size, grade, structure, and bond type. Figure 9.4a shows the marking system for aluminum-oxide and silicon-carbide bonded abrasives and Fig. 9.4b shows the marking system for diamond and cubic boron nitride bonded abrasives.

## 9.3.1   Bond types

The common bond types for bonded abrasives are **vitrified, resinoid, rubber,** and **metal.** These bonds are used for conventional abrasives as well as for superabrasives (except rubber).

**Vitrified.**   Essentially a glass, **vitrified bond** is also called a *ceramic bond,* particularly outside the United States. It is the most common and widely used bond. The raw materials consist of feldspar (a crystalline mineral) and clays. They are mixed with the abrasives, moistened, and molded under pressure into the shape of grinding wheels. These "green" products, which are similar to powder-metallurgy parts, are then fired slowly, up to a temperature of about 1250°C (2300°F), to fuse the glass and develop structural strength. The

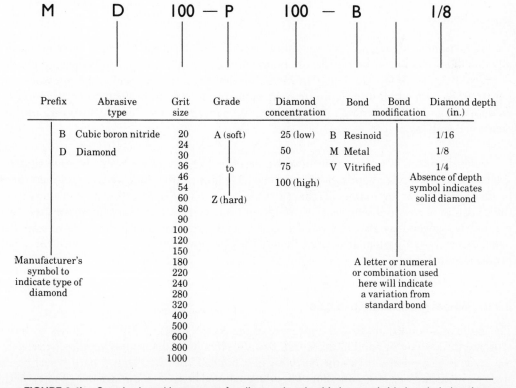

**FIGURE 9.4b**   Standard marking system for diamond and cubic boron nitride bonded abrasives.

wheels are then cooled slowly to avoid thermal cracking, finished to size, inspected for quality and dimensional accuracy, and tested for defects.

Vitrified bonds produce wheels that are strong, stiff, porous, and resistant to oils, acids, and water. Because they are brittle, they lack resistance to mechanical and thermal shock. However, vitrified wheels are also available with steel backing plates or cups for better structural support during their use.

**Resinoid.**    Resinoid bonding materials are *thermosetting resins,* and are available in a wide range of compositions and properties. Because the bond is an organic compound, wheels with **resinoid bonds** are also called **organic wheels**. The basic manufacturing technique consists of mixing the abrasive with liquid or powdered phenolic resins and additives, pressing the mixture into the shape of a grinding wheel, and curing it at temperatures of about 175°C (350°F).

Because the elastic modulus of thermosetting resins is lower than that of glasses, resinoid wheels are more flexible than vitrified wheels. **Reinforced wheels** are widely used, in which one or more layers of fiberglass mats of various mesh sizes provide the reinforcement. Its purpose is to retard the disintegration of the wheel should it break for some reason, rather than to improve its strength. Large-diameter wheels can be supported additionally with one or more internal rings made of round steel bar, which are inserted during molding of the wheel.

**Rubber.**    The most flexible bond used in abrasive wheels is rubber. The manufacturing process consists of mixing crude rubber, sulfur, and the abrasive grains together, rolling the mixture into sheets, cutting out circles, and heating them under pressure to vulcanize the rubber. Thin wheels can be made in this manner and are used like saws for cutting-off operations (cut-off blades).

**Metal bonds.**    Using powder-metallurgy techniques, the abrasive grains, which are usually diamond or cubic boron nitride, are bonded to the periphery of a metal wheel, to depths of 6 mm (0.25 in.) or less (see Fig. 9.3). **Metal bonding** is carried out under high pressure and temperature. The wheel itself (core) may be made of aluminum, bronze, steel, ceramics, or composite materials, depending on wheel requirements such as strength, stiffness, and dimensional stability.

**Other bonds.**    In addition to those we described above, other bonds include silicate, shellac, and oxychloride bonds. However, they have limited uses, and we won't discuss them further here. A new development is the use of polyimide (see Section 10.6) as a substitute for the phenolic in resinoid wheels. It is tough and has resistance to high temperatures. Superabrasive wheels may also be *layered* so that a single abrasive layer is plated or brazed to a metal wheel with a particular desired shape. They are lower in cost and are used for small production quantities.

### 9.3.2  Wheel grade and structure

The **grade** of a bonded abrasive is a measure of the bond's strength. Thus it includes both the *type* and the *amount* of bond in the wheel. Because strength and hardness are directly related, the grade is also referred to as the *hardness* of a bonded abrasive. Thus a hard wheel has a stronger bond and/or a larger amount of bonding material between the grains than a soft wheel.

**FIGURE 9.5** The grinding surface of an abrasive wheel (A46–J8V) showing grains, porosity, wear flats on grains (see also Fig. 9.8), and metal chips from the workpiece adhering to the grains. Note the random distribution and shape of the abrasive grains. Magnification: 50×.

The **structure** is a measure of the *porosity* (spacing between the grains in Fig. 9.1) of the bonded abrasive. Some porosity is essential to provide clearance for the grinding chips; otherwise they would interfere with the grinding process. The structure of bonded abrasives ranges from dense to open.

## 9.4 Mechanics of Grinding

Grinding is basically a *chip-removal* process in which the cutting tool is an individual abrasive grain. The following are major differences between the action of a single grain and a single-point tool:

1. The individual grain has an irregular geometry and is spaced randomly along the periphery of the wheel (Fig. 9.5).
2. The radial positions of the grains vary.
3. The average rake angle of the grains is highly negative: −60° or even lower. Consequently, the shear angles are very low.
4. The cutting speeds are very high, typically 30 m/s (6000 ft/min) (Table 9.2).

**TABLE 9.2**

**TYPICAL RANGE OF SPEEDS AND FEEDS FOR ABRASIVE PROCESSES**

| PROCESS VARIABLE | CONVENTIONAL GRINDING | CREEP-FEED GRINDING | BUFFING | POLISHING |
|---|---|---|---|---|
| Wheel speed (m/min) | 1500–3000 | 1500–3000 | 1800–3600 | 1500–2400 |
| Work speed (m/min) | 10–60 | 0.1–1 | — | — |
| Feed (mm/pass) | 0.01–0.05 | 1–6 | — | — |

**FIGURE 9.6** Grinding chip being produced by a single abrasive grain: (A) chip, (B) workpiece, (C) abrasive grain. Note the large negative rake angle of the grain. The inscribed circle is 0.065 mm (0.0025 in.) in diameter. *Source:* M. E. Merchant.

**FIGURE 9.7** Variables in surface grinding. In actual grinding, the wheel depth of cut $d$ and contact length $\ell$ are much smaller than the wheel diameter $D$. The dimension $t$ is called the grain depth of cut.

An example of chip formation by an abrasive grain is shown in Fig. 9.6. Note the negative rake angle, the low shear angle, and the small size of the chip. A variety of metal chips can be observed in grinding. (Chips are easily collected on a piece of adhesive tape held against the sparks of a grinding wheel.)

The mechanics of grinding and the variables involved can best be studied by analyzing the *surface grinding* operation (Fig. 9.7). A grinding wheel of diameter $D$ is removing a layer of metal at a depth $d$, known as the **wheel depth of cut**. An individual grain on the periphery of the wheel is moving at a tangential velocity $V$ (*up grinding*, as shown in Fig. 9.7), and the workpiece is moving at a velocity $v$. The grain is removing a chip whose *undeformed thickness* (**grain depth of cut**) is $t$ and the undeformed length is $\ell$.

For $v \ll V$, the *undeformed chip length* $\ell$ is approximately

$$\ell \simeq \sqrt{Dd}, \tag{9.1}$$

where $D$ is the diameter of the wheel and $d$ is the depth of cut.

For *external* (cylindrical) grinding (see Section 9.6),

$$\ell = \sqrt{\frac{Dd}{1 + D/D_w}}, \tag{9.2}$$

and for *internal* grinding,

$$\ell = \sqrt{\frac{Dd}{1 - D/D_w}}, \tag{9.3}$$

where $D_w$ is the diameter of the workpiece.

We can derive the relationship between $t$ and other process variables as follows. Let $C$ be the number of cutting points per unit area of wheel surface; $v$ and $V$ are the surface speeds of the workpiece and the wheel, respectively (Fig. 9.7). If we let the width of the workpiece be unity, the number of chips produced per unit time is $VC$, and the volume of material removed per unit time is $vd$.

If we let $r$ be the ratio of the chip width $w$ to the average chip thickness, the volume of a chip with a rectangular cross-sectional area and constant width is

$$\text{Vol}_{\text{chip}} = \frac{wt\ell}{2} = \frac{rt^2\ell}{4}. \tag{9.4}$$

The volume of material removed per unit time, then, is the product of the number of chips produced per unit time and the volume of each chip, or

$$VC\frac{rt^2\ell}{4} = vd,$$

and because $\ell = \sqrt{Dd}$, the undeformed chip thickness in surface grinding is

$$t = \sqrt{\frac{4v}{VCr}\sqrt{\frac{d}{D}}}. \tag{9.5}$$

Experimental observations indicate the value of $C$ to be roughly on the order of 0.1 to 10 per square mm ($10^2$ to $10^3$ per square inch); the finer the grain size of the wheel, the larger this number. The magnitude of $r$ is between 10 and 20 for most grinding operations. If we substitute typical values for a grinding operation into Eqs. (9.1)–(9.5), we obtain very small quantities for $\ell$ and $t$. For example, typical values for $t$ are in the range of 0.3–0.4 $\mu$m (12–160 $\mu$in.).

---

**Example 9.1:** Chip Dimensions

Estimate the undeformed chip length and undeformed chip thickness for a typical surface grinding operation.

SOLUTION.   The formulas for undeformed length and thickness, respectively, are

$$\ell = \sqrt{Dd} \quad \text{and} \quad t = \sqrt{\frac{4v}{VCr}\sqrt{\frac{d}{D}}}.$$

From Table 9.2, we select the following values:

$$v = 0.5 \text{ m/s} \quad \text{and} \quad V = 30 \text{ m/s}.$$

We also assume that

$$d = 0.05 \text{ mm} \quad \text{and} \quad D = 200 \text{ mm},$$

and let

$$C = 2 \text{ per mm}^2 \quad \text{and} \quad r = 15.$$

Then

$$\ell = \sqrt{(200)(0.05)} = 3.2 \text{ mm} = 0.126 \text{ in.}$$

and

$$t = \sqrt{\frac{(4)(0.5)}{(30)(2)(15)}} \sqrt{\frac{0.05}{200}} = 0.006 \text{ mm} = 2.3 \times 10^{-4} \text{ in.}$$

Because of plastic deformation, the actual length of the chip will be shorter and the thickness greater than these values (see Fig. 9.6).

### 9.4.1 Forces

A knowledge of forces is necessary not only in the design of grinding machines, but also in determining the deflections that the workpiece and the machine will undergo, as well as in the design and use of workholding devices. Deflections, in turn, adversely influence dimensional accuracy and are thus critical in precision grinding.

If we assume that the **force** on the grain (see cutting force $F_c$ in Section 8.2.3) is proportional to the cross-sectional area of the undeformed chip, we can show that the **relative grain force** is

$$\text{Relative grain force} \propto \frac{v}{VC} \sqrt{\frac{d}{D}}. \tag{9.6}$$

The actual force is the product of the relative grain force and the strength of the metal being ground.

The **specific energy** consumed in producing a grinding chip consists of three components:

$$u = u_{\text{chip}} + u_{\text{ploughing}} + u_{\text{sliding}} \tag{9.7}$$

where $u_{\text{chip}}$ is the specific energy required for chip formation by plastic deformation and $u_{\text{ploughing}}$ is the specific energy required for ploughing, which is plastic deformation with-

**FIGURE 9.8**  Chip formation and ploughing of the workpiece surface by an abrasive grain.

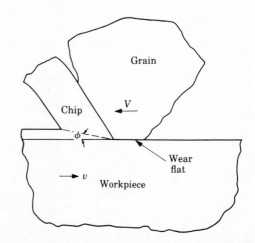

**FIGURE 9.9**  Schematic illustration of chip formation by an abrasive grain. Note the negative rake angle, the small shear angle, and the wear flat on the grain.

**TABLE 9.3**

**APPROXIMATE SPECIFIC ENERGY REQUIREMENTS FOR SURFACE GRINDING**

| WORKPIECE MATERIAL | HARDNESS | SPECIFIC ENERGY | |
|---|---|---|---|
| | | W·s/mm³ | hp·min/in.³ |
| Aluminum | 150 HB | 7–27 | 2.5–10 |
| Cast iron (class 40) | 215 HB | 12–60 | 4.5–22 |
| Low-carbon steel (1020) | 110 HB | 14–68 | 5–25 |
| Titanium alloy | 300 HB | 16–55 | 6–20 |
| Tool steel (T15) | 67 HRC | 18–82 | 6.5–30 |

out chip removal (Fig. 9.8). The last term, $u_{sliding}$, can best be understood by observing the grain in Fig. 9.9.

The grain develops a **wear flat** as a result of the grinding operation (similar to flank wear in cutting tools). The wear flat slides along the surface being ground and, because of friction, requires energy for sliding. The larger the wear flat, the higher the grinding force is.

Typical specific energy requirements in grinding are given in Table 9.3. Note that these energy levels are much higher than those in cutting operations with single-point tools (see Table 8.4). This difference has been attributed to the following factors:

1. **Size effect:** As previously stated, the size of grinding chips is quite small compared to chips from other cutting operations, or by about two orders of magnitude. As described in Section 3.8.3, the smaller the size of a piece of metal, the greater is its strength. Thus grinding involves higher specific energy than cutting. Recent studies have indicated that extremely high dislocation densities occur in the shear zone during chip formation, thus influencing the grinding energies involved.

2. **Wear flat:** Because a wear flat requires frictional energy for sliding, this energy can contribute substantially to the total energy consumed. The size of the wear flat in grinding is much larger than the grinding chip, unlike in metal cutting by a single-point tool where flank wear land is small compared to the size of the chip.

3. **Chip morphology:** Because the average rake angle of a grain is highly negative (see Fig. 9.6), the shear strains are very large. This result indicates that the energy required for plastic deformation to produce a grinding chip is higher than in other cutting processes. Furthermore, ploughing consumes energy without contributing to chip formation.

---

**Example 9.2:   Forces in Surface Grinding**

Assume that you are performing a surface grinding operation on low-carbon steel, with a wheel of diameter $D = 10$ in. and rotating at $N = 4000$ rpm, and a width of cut $w = 1$ in. The depth of cut is $d = 0.002$ in. and the feed rate $v$ of the workpiece is 60 in./min. Calculate the cutting force (force tangential to the wheel) $F_c$ and the thrust force $F_n$ (normal to the workpiece surface).

SOLUTION.   We first determine the material removal rate MRR, as follows:

$$\text{MRR} = dwv = (0.002)(1)(60) = 0.12 \text{ in}^3/\text{min.}$$

The power consumed is given by

$$\text{Power} = (u)(\text{MRR}),$$

where $u$ is the specific energy, as obtained from Table 9.3. For low-carbon steel, let us estimate it to be 15 hp·min/in$^3$. Hence,

$$\text{Power} = (15)(0.12) = 1.8 \text{ hp.}$$

By noting that 1 hp = 33,000 ft·lb/min = 396,000 in.·lb/min, we obtain

$$\text{Power} = (1.8)(396,000) = 712,800 \text{ in.·lb/min.}$$

Since power is defined as

$$\text{Power} = (T)(\omega),$$

where $T$ is the torque and is equal to $T = (F_c)(D/2)$ and $\omega$ is the rotational speed of the wheel in radians per minute, hence $\omega = 2\pi N$.

Thus,

$$712,800 = (F_c)(10/2)(2)(\pi)(4000)$$

and therefore, $F_c = 57$ lb.

The thrust force $F_n$ can be calculated by noting from experimental data in the technical literature that it is about 30 percent higher than the cutting force $F_c$. Consequently,

$$F_n = (1.3)(57) = 74 \text{ lb.}$$

---

## 9.4.2 Temperature

*Temperature rise* in grinding is an important consideration because it can adversely affect the surface properties and cause residual stresses on the workpiece. Furthermore, temperature gradients in the workpiece cause distortions by differential thermal expansion and contraction. When a portion of the heat generated is conducted into the workpiece, it expands the part being ground, thus making it difficult to control dimensional accuracy.

The work expended in grinding is mainly converted into heat. The surface temperature rise $\Delta T$ has been found to be a function of the ratio of the total energy input to the surface area ground. Thus in *surface grinding* if $w$ is the width and $L$ is the length of the surface area ground,

$$\Delta T \propto \frac{uwLd}{wL} \propto ud. \tag{9.8}$$

If we introduce size effect and assume that $u$ varies inversely with the undeformed chip thickness $t$, then the temperature rise is

$$\Delta T \propto \frac{d}{t} \propto d^{3/4} \sqrt{\frac{VC}{v}} \sqrt{D}. \tag{9.9}$$

The *peak temperatures* in chip generation during grinding may be as high as 1650°C (3000°F). However, the time involved in producing a chip is extremely short—on the order of microseconds—hence melting may or may not occur. Because the chips carry away much of the heat generated (as in metal cutting), only a fraction of the heat generated is conducted to the workpiece. Experiments indicate that in grinding, as much as one half the energy is conducted to the workpiece. (This is higher than in metal cutting.) The heat generated by sliding and ploughing is conducted mostly into the workpiece.

**Sparks.** The sparks observed in metal grinding are actually glowing chips. They glow because of the exothermic reaction of the hot chips with oxygen in the atmosphere. Sparks have not been observed with any metal ground in an oxygen-free environment. The color, intensity, and shape of the sparks depend on the composition of the metal being ground.

If the heat generated by exothermic reaction is sufficiently high, the chip may melt and, because of surface tension, acquire a round shape and solidify as a shiny spherical particle. Observation of these particles under scanning electron microscopy has revealed that they are hollow and have a fine dendritic structure, indicating that they were once molten (by exothermic oxidation of hot chips in air) and resolidified rapidly. (It has been suggested that some of the spherical particulars may also be produced by plastic deformation and rolling of chips at the grit–workpiece interface.)

### 9.4.3 Effects of temperature

Temperature rise in grinding can significantly affect surface properties and residual stresses. Temperature gradients distort the part because of differential thermal expansion and contraction. Furthermore, if the heat generated is allowed to be conducted into the workpiece, it expands the part being ground and hence controlling tolerances is difficult.

**Tempering.** Excessive temperature rise caused by grinding can temper and soften the surfaces of steel components, which are often ground in the hardened state. Grinding process parameters must therefore be chosen carefully to avoid excessive temperature rise. The use of grinding fluids can effectively control temperatures.

**Burning.** If the temperature is excessive the surface may burn. Burning produces a bluish color on steels, which indicates oxidation at high temperatures. A burn may not be objectionable in itself. However, the surface layers may undergo metallurgical transformations, with martensite formation in high-carbon steels from reaustenization followed by rapid cooling. This is known as *metallurgical burn,* which also is a serious problem with nickel-base alloys.

High temperatures in grinding may also lead to thermal cracking of the surface of the workpiece, known as **heat checking**. Cracks are usually perpendicular to the grinding direction; however, under severe grinding conditions, parallel cracks may also develop.

**FIGURE 9.10** Residual stresses developed on the workpiece surface in grinding tungsten: (a) effect of wheel speed and (b) effect of grinding fluid. Tensile residual stresses on a surface are detrimental to the fatigue life of ground components. The variables in grinding can be controlled to minimize residual stresses. This is known as low-stress grinding. *Source:* After N. Zlatin, et al., 1963.

**Residual stresses.**  Temperature change and gradients within the workpiece are mainly responsible for residual stresses in grinding. Other contributing factors are the physical interactions of the abrasive grain in chip formation and the sliding of the wear flat along the workpiece surface, causing plastic deformation of the surface. Two examples of residual stresses in grinding are shown in Fig. 9.10, demonstrating the effects of wheel speed and the type of grinding fluid used. The method and direction of the application of grinding fluid can also have a significant effect on residual stresses.

Because of the deleterious effect of tensile residual stresses on fatigue strength, process parameters should be chosen carefully. Residual stresses can usually be lowered (**low-stress** or **gentle grinding**) by using softer grade wheels (free cutting wheels), lower wheel speeds, and higher work speeds.

## 9.5  Grinding Wheel Wear

Grinding wheel wear is an important consideration because it adversely affects the shape and accuracy of ground surfaces, as is the case with cutting tools. Grinding wheels wear by three different mechanisms: attritious grain wear, grain fracture, and bond fracture.

**Attritious wear.**  In **attritious wear,** the cutting edges of a sharp grain become dull by attrition, developing a *wear flat* (see Fig. 9.9) that is similar to flank wear in cutting tools. Wear is caused by the interaction of the grain with the workpiece material, involving both physical and chemical reactions. These reactions are complex and involve diffusion, chemical degradation or decomposition of the grain, fracture at a microscopic scale, plastic deformation, and melting.

Attritious wear is low when the two materials are chemically inert with respect to each other, much like with the use of cutting tools. The more inert the materials, the lower will be the tendency for reaction and adhesion to occur between the grain and the workpiece. For example, because aluminum oxide is relatively inert with respect to iron, its rate of attritious wear when it is used to grind steels is much lower than that for silicon carbide and diamond. On the other hand, silicon carbide can dissolve in iron, and hence it is not suitable for grinding steels. Cubic boron nitride has a higher inertness with respect to steels, and thus it is suitable for use as an abrasive.

The selection of the type of abrasive for low attritious wear is therefore based on the reactivity of the grain and the workpiece and their relative mechanical properties, such as hardness and toughness. The environment and the type of grinding fluid used also have an influence on grain–workpiece material interactions.

**Grain fracture.** Because abrasive grains are brittle, their fracture characteristics in grinding are important. If the wear flat caused by attritious wear is excessive, the grain becomes dull and grinding becomes inefficient and produces undesirably high temperatures. Optimally, the grain should fracture or fragment at a moderate rate, so that new sharp cutting edges are produced continuously during grinding. This is equivalent to breaking a piece of stone or dull chalk into two or more pieces in order to expose new sharp edges. Section 9.2 described *friability* of abrasives, which gives them their self-sharpening characteristics, an important consideration in effective grinding.

The selection of grain type and size for a particular application also depends on the attritious wear rate. A grain–workpiece material combination with high attritious wear and low friability dulls grains and develops a large wear flat. Grinding then becomes inefficient, and surface damage such as burning is likely to occur.

The following combinations are generally recommended:

1. *Aluminum oxide:* with steels, ferrous alloys, and alloy steels.
2. *Silicon carbide:* with cast iron, nonferrous metals, and hard and brittle materials, such as carbides, ceramics, marble, and glass.
3. *Diamond:* with cemented carbide ceramics, and some hardened steels.
4. *Cubic boron nitride:* with steels and cast irons at 50 HRC or above, and high-temperature superalloys.

**Bond fracture.** The strength of the bond (*grade*) is a significant parameter in grinding. If the bond is too strong, dull grains cannot be dislodged so that other sharp grains along the circumference of the grinding wheel begin to contact the workpiece and remove chips; thus the grinding process becomes inefficient. On the other hand, if the bond is too weak, the grains are dislodged easily and the wear rate of the wheel increases. Consequently, maintaining dimensional accuracy becomes difficult. In general, softer bonds are recommended for harder materials and for reducing residual stresses and thermal damage to the workpiece. Hard-grade wheels are used for softer materials and for removing large amounts of material at high rates.

## 9.5.1 Dressing, truing, and shaping of grinding wheels

**Dressing** is the process of conditioning worn grains on the surface of a grinding wheel in order to produce sharp new grains and for truing an out-of-round wheel. Dressing is necessary when excessive attritious wear dulls the wheel—called **glazing** because of the shiny appearance of the wheel surface—or when the wheel becomes loaded. **Loading** occurs when the porosities on the grinding surfaces of the wheel (see Fig. 9.5) become filled or clogged with chips. Loading can occur in grinding soft materials or by improper selection of grinding wheels, such as with low porosity, and improper process parameters. A loaded wheel cuts very inefficiently, generating much frictional heat, causing surface damage and loss of dimensional accuracy.

Dressing is done by various techniques. In one method, a specially shaped diamond point tool or diamond cluster is moved across the width of the grinding face of a rotating wheel and removes a small layer from the wheel surface with each pass. This method can be either dry or wet, depending on whether the wheel is to be used dry or wet, respectively. In another dressing method, a set of star-shaped steel disks is manually pressed against the wheel. Material is removed from the wheel surface by crushing the grains. This method produces a coarse grinding surface on the wheel and is used only for rough grinding operations on bench or pedestal grinders. Abrasive sticks may also be used to dress grinding wheels. New developments in dressing techniques, for metal-bonded diamond wheels, involve the use of electrical-discharge and electrochemical material-removal techniques, which erode away very small layers of the metal bond, thus exposing new diamond cutting edges.

Dressing for form grinding involves **crush dressing**, or *crush forming*, and consists of pressing a metal roll on the surface of the grinding wheel, which is usually a vitrified wheel. The roll, which may be made of high-speed steel, tungsten carbide, or boron carbide, has a machined or ground profile and thus reproduces a replica of this profile on the surface of the grinding wheel being dressed.

Dressing techniques and how frequently the wheel surface is dressed are significant, affecting grinding forces and surface finish. Modern computer-controlled grinders are equipped with automatic dressing features, which dress the wheel as grinding continues. Dressing can also be done to generate a certain shape or form on a grinding wheel for the purpose of grinding profiles on workpieces (see Section 9.6.2).

**Truing** is a dressing operation by which a wheel is restored to its original shape. Thus a round wheel is dressed to make its circumference a true circle (hence the word truing).

Grinding wheels can also be **shaped** to the form to be ground on the workpiece. The grinding face on the Type 1 straight wheel shown in Fig. 9.2a is cylindrical and thus produces a flat surface; however, this surface can be shaped by dressing it to various forms. Although templates have been used for this purpose, new grinders are equipped with computer-controlled shaping features, whereby the diamond dressing tool traverses the wheel face automatically along a certain prescribed path (Fig. 9.11). Note that the axis of the diamond dressing tool remains normal to the wheel face at the point of contact.

Diamond dressing tool

Grinding face

Grinding wheel

**FIGURE 9.11** Shaping the grinding face of a wheel by dressing it with computer control. Note that the diamond dressing tool is normal to the surface at point of contact. *Source:* Okuma Machinery Works Ltd.

### 9.5.2 Grinding ratio

Grinding wheel wear is generally correlated with the amount of material ground by a parameter called the **grinding ratio**, $G$, which is defined as

$$G = \frac{\text{Volume of material removed}}{\text{Volume of wheel wear}}. \tag{9.10}$$

In practice, grinding ratios vary widely, ranging from 2 to 200 and higher, depending on the type of wheel, workpiece material, the grinding fluid, and process parameters (such as depth of cut and speeds of wheel and workpiece). Attempting to obtain a high grinding ratio in practice isn't always desirable because high ratios may indicate grain dulling and possible surface damage. A lower ratio may be quite acceptable when an overall economic analysis justifies it.

**Soft- or hard-acting wheels.** During a grinding operation, a particular wheel may *act soft* (wear is high) or *hard* (wear is low), regardless of its grade. Note, for example, that an ordinary pencil acts soft when you write on rough paper and acts hard when you write on soft paper. This behavior is a function of the force on the grain. The greater the force, the greater the tendency for the grains to fracture or be dislodged from the wheel surface, thus the higher the wheel wear and the lower the grinding ratio. Equation (9.6) shows that the grain force increases with the strength of the workpiece material, work speed, and depth of cut, and decreases with increasing wheel speed and wheel diameter. Thus a wheel acts soft when $v$ and $d$ increase or when $V$ and $D$ decrease.

---

### Example 9.3: Action of a Grinding Wheel

A surface-grinding operation is being carried out with the wheel running at a constant spindle speed. Will the wheel act soft or hard as the wheel wears down over a period of time?

SOLUTION. Referring to Eq. (9.6), we note that the parameters that change by time in this operation are the wheel diameter $D$ (assuming that $d$ remains constant and the wheel is dressed periodically) and $V$. Hence, as $D$ becomes smaller, the relative grain force increases. Therefore the wheel acts softer. Some grinding machines are equipped with variable-speed spindle motors to accommodate these changes, and also to make provision for different diameter wheels.

---

### 9.5.3 Wheel selection and grindability of materials

Selection of a grinding wheel for a certain application greatly influences the quality of surfaces produced, as well as the economics of the operation. The selection involves not only the shape of the wheel with respect to the shape of the part to be produced, but the characteristics of the workpiece material as well.

The **grindability** of materials, as in machinability, is difficult to define precisely. It is a general indication of how easy it is to grind a material and includes considerations such as the quality of the surface produced, surface finish, surface integrity, wheel wear, cycle time, and overall economics. It should also be pointed out that, just as in machinability, grindability of a material can be enhanced greatly by proper selection of process parameters, wheels, and grinding fluids, as well as machine characteristics and fixturing methods.

**TABLE 9.4**

**TYPICAL RECOMMENDATIONS FOR GRINDING WHEELS FOR USE WITH VARIOUS MATERIALS (SEE FIGS. 9.4a AND b)**

| MATERIAL | TYPE OF GRINDING WHEEL |
| --- | --- |
| Aluminum | C46–K6V |
| Brass | C46–K6V |
| Bronze | A54–K6V |
| Cast iron | C60–L6V, A60–M6V |
| Carbides | C60–I9V, D150–R75B |
| Ceramics | D150–N50M |
| Copper | C60–J8V |
| Nickel alloys | B150H 100V |
| Nylon | A36–L8V |
| Steels | A60–M6V |
| Titanium | A60–K8V |
| Tool steels ( > 50 HRC) | B120WB |

Note: These recommendations vary significantly, depending on material composition, the particular grinding operation, and grinding fluids used.

Grinding practices are now well established for a wide variety of metallic and nonmetallic materials, including newly developed materials for aerospace applications. Specific recommendations for selecting wheels and process parameters can be found in various handbooks (see also Table 9.4). As for ceramics, they can now be ground with relative ease by using diamond wheels, as well as carefully selecting process parameters. It has been shown, for example, that with light passes and rigid machines with good damping capacity, it is possible to obtain continuous chips in grinding (see Figs. 9.8 and 9.9), known as **ductile-regime** grinding, and with good surface integrity. However, ceramic chips are typically 1–10 $\mu$m (40–400 $\mu$in.) in size and hence more difficult to remove from grinding fluids than metal chips, thus requiring fine filters and other methods.

## 9.6   Grinding Operations and Machines

Grinding operations are carried out with a variety of wheel–workpiece configurations. The selection of a grinding process for a particular application depends on part shape, part size, ease of fixturing, and the production rate required.

The basic types of grinding operations are surface, cylindrical, internal, and centerless grinding. The relative movement of the wheel may be along the surface of the workpiece (*traverse* grinding, *through feed* grinding, *cross-feeding*), or it may be radially into the workpiece (*plunge* grinding). Surface grinders comprise the largest percentage of grinders in use in industry, followed by bench grinders (usually with two wheels), cylindrical grinders, and tool and cutter grinders. The least used are internal grinders.

Grinding machines are available for various workpiece geometries and sizes. Modern grinding machines may be computer controlled with features such as automatic workpiece

**FIGURE 9.12**   Schematic illustrations of surface-grinding operations. (a) Traverse grinding with a horizontal-spindle surface grinder. (b) Plunge grinding with a horizontal-spindle surface grinder, producing a groove in the workpiece. (c) Vertical-spindle rotary-table grinder (also known as the Blanchard type).

loading and unloading, clamping, cycling, gaging, dressing, and wheel shaping. Grinders can also be equipped with probes and gages for determining the relative position of the wheel and workpiece surfaces as well as tactile sensing features whereby diamond dressing tool breakage, if any, can be detected during the dressing cycle.

### 9.6.1   Surface grinding

**Surface grinding** involves grinding flat surfaces and is one of the most common grinding operations (Fig. 9.12). Typically, the workpiece is secured on a magnetic chuck attached to the work table of a *surface grinder* (Fig. 9.13). Nonmagnetic materials generally are held by vises, special fixtures, vacuum chucks, or double-sided adhesive tapes. A straight wheel is mounted on the *horizontal spindle* of the grinder. Traverse grinding is done as the table reciprocates longitudinally and feeds laterally after each stroke. In *plunge grinding,* the wheel is moved radially into the workpiece, as in grinding a groove (Fig. 9.12b).

The size of a surface grinder is determined by the surface dimensions that can be ground on the machine. In addition to the grinder shown in Fig. 9.13, other types have *vertical spindles* and *rotary tables* (Fig. 9.12c). These configurations allow a number of pieces to be ground in one setup. Steel balls for ball bearings are ground in special setups at high production rates (Fig. 9.14).

**FIGURE 9.13**   Schematic illustration of a horizontal-spindle surface grinder. The majority of grinding operations are done on such machines.

(a)

(b)

Rough-ground balls

**FIGURE 9.14**
(a) Rough grinding of steel balls on a vertical-spindle grinder, guided by a special rotary fixture. (b) Finish grinding of balls in a multiple-groove fixture. The balls are ground to within 0.013 mm (0.0005 in.) of their final size. *Source: American Machinist.*

Grinding wheel

Guide for balls

Finish-ground balls

### 9.6.2   Cylindrical grinding

In **cylindrical grinding** (also called *center-type* grinding; Fig. 9.15), the workpiece's external cylindrical surfaces and shoulders are ground. Typical applications include crankshaft bearings, spindles, pins, bearing rings, and rolls for rolling mills. The rotating cylindrical workpiece reciprocates laterally along its axis, although in grinders used for large and long workpieces the grinding wheel reciprocates. The latter design is called a *roll grinder* and is capable of grinding rolls as large as 1.8 m (72 in.) in diameter.

The workpiece in cylindrical grinding is held between centers or in a chuck, or it is mounted on a faceplate in the headstock of the grinder. For straight cylindrical surfaces, the

(a)

(b)

(c)

**FIGURE 9.15**   Examples of cylindrical-grinding operations. (a) Traverse grinding, (b) plunge grinding, and (c) profile grinding. *Source:* Okuma Machinery Works Ltd.

Grinding wheel

Workpiece

**FIGURE 9.16** Plunge grinding of a workpiece on a cylindrical grinder with the wheel dressed to a stepped shape.

(a)                                    (b)

Grinding wheel

**FIGURE 9.17** Thread grinding by (a) traverse and (b) plunge grinding.

axes of rotation of the wheel and workpiece are parallel. Separate motors drive the wheel and workpiece at different speeds. Long workpieces with two or more diameters are ground on cylindrical grinders. Cylindrical grinding can produce shapes (*form grinding* and *plunge grinding*) in which the wheel is dressed to the form to be ground (Fig. 9.16).

Cylindrical grinders are identified by the maximum diameter and length of the workpiece that can be ground, similar to engine lathes. In *universal grinders,* both the workpiece and the wheel axes can be moved and swiveled around a horizontal plane, thus permitting the grinding of tapers and other shapes. These machines are equipped with computer controls, reducing labor and producing parts accurately and repetitively.

Cylindrical grinders can also be equipped with computer-control features such that *noncylindrical* parts (such as cams) can be ground on rotating workpieces. The workpiece spindle rpm is synchronized such that the distance between the workpiece and wheel axes is varied continuously to produce a particular shape.

**Thread grinding** is done on cylindrical grinders, as well as centerless grinders, with specially dressed wheels matching the shape of the threads (Fig. 9.17). Although costly, threads produced by grinding are the most accurate of any manufacturing process and have a very fine surface finish. The workpiece and wheel movements are synchronized to produce the pitch of the thread, usually in about six passes (see also Section 9.6.4).

### 9.6.3 Internal grinding

In **internal grinding** (Fig. 9.18), a small wheel is used to grind the inside diameter of the part, such as bushings and bearing races. The workpiece is held in a rotating chuck and the wheel

(a) Traverse grinding              (b) Plunge grinding              (c) Profile grinding

Workpiece    Wheel                         Wheel

                                           Work

**FIGURE 9.18** Schematic illustrations of internal grinding operations.

**FIGURE 9.19**   Schematic illustrations of centerless grinding operations.

rotates at 30,000 rpm or higher. Internal profiles can also be ground with profile-dressed wheels that move radially into the workpiece. The headstock of internal grinders can be swiveled on a horizontal plane to grind tapered holes.

### 9.6.4   Centerless grinding

**Centerless grinding** is a high-production process for continuously grinding cylindrical surfaces in which the workpiece is supported not by centers (hence the term centerless) or chucks but by a blade (Figs. 9.19 and 9.20). Typical parts made by centerless grinding are roller bearings, piston pins, engine valves, camshafts, and similar components. This continuous production process requires little operator skill. Parts with diameters as small as 0.1 mm (0.004 in.) can be ground. Centerless grinders are now capable of wheel surface speeds on the order of 10,000 m/min (35,000 ft/min), using cubic boron nitride abrasive wheels.

In *through-feed grinding* the workpiece is supported on a workrest blade and is ground between two wheels. Grinding is done by the larger wheel, while the smaller wheel regulates the axial movement of the workpiece. The *regulating wheel,* which is rubber bonded, is tilted and runs at speeds of only about 1/20 those of the grinding wheel.

**FIGURE 9.20**   A computer-numerical-control centerless grinding machine. The movement of the workpiece is perpendicular to the page. *Source:* Courtesy of Cincinnati Milacron, Inc.

Parts with variable diameters, such as bolts, valve tappets, and distributor shafts, can be ground by centerless grinding. Called *infeed,* or *plunge, grinding* (Fig. 9.19b), the process is similar to plunge or form grinding with cylindrical grinders. Tapered pieces are centerless ground by *end-feed grinding.* High production rate thread grinding can be done with centerless grinders using specially dressed wheels. In *internal centerless grinding,* the workpiece is supported between three rolls and is internally ground. Typical applications are sleeve-shaped parts and rings.

### 9.6.5  Other grinders

A variety of special-purpose grinders are available. *Bench grinders* are used for routine off-hand grinding of tools and small parts. They are usually equipped with two wheels mounted on the two ends of the shaft of an electric motor. One wheel is usually coarse for rough grinding, and the other is fine for finish grinding. *Pedestal,* or *stand, grinders* are placed on the floor and used similarly to bench grinders.

*Universal tool and cutter grinders* are used for grinding single-point or multipoint tools and cutters. They are equipped with special workholding devices for accurate positioning of the tools to be ground. *Tool-post grinders* are self-contained units and are usually attached to the tool post of a lathe. The workpiece is mounted on the headstock and is ground by moving the tool post. These grinders are versatile, but the lathe should be protected from abrasive debris.

*Swing-frame grinders* are used in foundries for grinding large castings. Rough grinding of castings is called **snagging,** and is usually done on floorstand grinders using wheels as large as 0.9 m (36 in.) in diameter. *Portable grinders,* either air or electrically driven, or with a flexible shaft connected to an electric motor or gasoline engine, are available for operations such as grinding off weld beads and *cutting-off* operations using thin abrasive disks, usually on large workpieces.

### 9.6.6  Creep-feed grinding

Grinding has traditionally been associated with small rates of material removal (see Table 9.2) and fine finishing operations. However, grinding can also be used for large-scale metal removal operations similar to milling, broaching, and planing. In **creep-feed grinding,** developed in the late 1950s, the wheel depth of cut $d$ is as much as 6 mm (0.25 in.), and the workpiece speed is low (Fig. 9.21). The wheels are mostly softer grade resin bonded with open structure to keep temperatures low and improve surface finish. Grinders with capabilities for continuously dressing the grinding wheel with a diamond roll are now available. The machines used for creep-feed grinding have special features, such as high power—up to 225 kW (300 hp)—high stiffness (because of the high forces due to the depth of material removed), high damping capacity, variable and well-controlled spindle and work table speeds, and ample capacity for grinding fluids.

Its overall economics and competitive position with other material-removal processes indicate that creep-feed grinding can be economical for specific applications, such as in grinding shaped punches, key seats, twist-drill flutes, the roots of turbine blades (Fig. 9.21c) and various complex superalloy parts. The wheel is dressed to the shape of the workpiece to be produced. Consequently, the workpiece does not have to be previously milled, shaped, or broached. Thus near-net shape castings and forgings are suitable parts for creep-feed grinding. Although generally one pass is sufficient, a second pass may be necessary for improved surface finish.

**FIGURE 9.21**   (a) Schematic illustration of the creep-feed grinding process. Note the large wheel depth of cut. (b) A shaped groove produced on a flat surface in one pass by creep-feed grinding. Groove depth can be on the order of a few mm. (c) An example of creep-feed grinding with a shaped wheel. *Source:* Courtesy of Blohm, Inc., and *Manufacturing Engineering,* Society of Manufacturing Engineers. See also Fig. 9.34.

### 9.6.7   Heavy stock removal by grinding

Grinding processes can be used for large-scale material removal by increasing process parameters. This is a rough grinding operation, with possibly detrimental effects on the workpiece surface, but can be economical in specific applications and compete favorably with machining processes, particularly milling, but also turning and broaching as well. In this operation, surface finish is of secondary importance and the grinding wheel (or belt) is utilized to its maximum for minimum cost per piece. The geometric tolerances in this process are on the same order as those obtained by other machining processes.

### 9.6.8   Grinding chatter

*Chatter* is particularly significant in grinding because it adversely affects surface finish and wheel performance. Vibrations during grinding may be caused by bearings, spindles, and unbalanced wheels, as well as external sources, such as from nearby machinery. The grinding process can itself cause regenerative chatter. The analysis of chatter in grinding is similar to that for machining operations (see Section 8.11). Thus the important variables are stiffness of the machine tool and workholding devices, and damping. Additional factors that are unique to grinding chatter are nonuniformities in the grinding wheel, dressing techniques used, and uneven wheel wear.

   Because these variables produce characteristic *chatter marks* on ground surfaces, careful study of these marks can often lead to the source of the problem. General guidelines have been established to reduce the tendency for chatter in grinding, such as using soft-grade wheels, dressing the wheel frequently, changing, dressing techniques, reducing the material-removal rate, and supporting the workpiece rigidly.

### 9.6.9   Grinding fluids

The functions of grinding fluids are similar to those for cutting fluids (Section 8.7). Although grinding and other abrasive-removal processes can be performed dry, the use of a fluid is important. It prevents temperature rise in the workpiece and improves the part's surface finish

**TABLE 9.5**

**GENERAL RECOMMENDATIONS FOR GRINDING FLUIDS**

| MATERIAL | GRINDING FLUID |
| --- | --- |
| Aluminum | E, EP |
| Copper | CSN, E, MO+FO |
| Magnesium | D, MO |
| Nickel | CSN, EP |
| Refractory metals | EP |
| Steels | CSN, E |
| Titanium | CSN, E |

D: dry; E: emulsion; EP: extreme pressure; CSN: chemicals and synthetics; MO: mineral oil; FO: fatty oil.

and dimensional accuracy. Fluids also improve the efficiency of the operation by reducing wheel loading and wear and lowering power consumption.

Grinding fluids are typically water-base emulsions for general grinding and oils for thread grinding (Table 9.5). They may be applied as a stream (flood) or as mist, which is a mixture of fluid and air. Because of the high surface speeds involved, an airstream or air blanket around the periphery of the wheel usually prevents the fluid from reaching the cutting zone. Special *nozzles* that mate with the shape of the cutting surface of the grinding wheel have been designed in which the grinding fluid is applied effectively under high pressure.

The temperature of water-base grinding fluids can rise significantly during its use as it removes heat from the grinding zone, in which case the workpiece can expand, making it difficult to control dimensional tolerances. The common method employed to maintain even temperature is to use refrigerating systems (chillers) through which the fluid is circulated.

As discussed in Section 8.7.4 on cutting fluids, the biological and ecological aspects, disposal, treatment, and recycling of grinding fluids are among the important considerations in their selection and use. The practices employed must comply with federal, state, and local laws and regulations.

### 9.6.10 Safety

Because grinding wheels are brittle and are operated at high speeds, certain procedures must be carefully followed in their handling, storage, and use. Failure to follow these rules—and the instructions and warnings printed on individual wheel labels—may result in serious injury or death. Grinding wheels should be stored properly and protected from environmental extremes. They should be visually inspected for cracks and damage prior to installing them on grinders.

Damage to a grinding wheel can severely reduce its *bursting speed,* defined as the surface speed at which a freely rotating wheel bursts (also referred to as exploding). The surface speed is, of course, related to the wheel diameter and its rotational speed; hence bursting speed can be defined in terms of bursting rpm for a particular wheel. Bursting speed depends on the type of wheel, its bond, grade, and structure. In diamond and CBN wheels, which are operated at high surface speeds, the core material affects the bursting speed; metal cores, as expected, have the highest speed, on the order of about 250 m/s (800 ft/s).

Wheels should be mounted on spindles of proper size, so that they are neither forced, which may fracture the wheel at its center, nor loose, which can cause unbalance. Flanges should be of appropriate design and dimensions. Wheels should be balanced because otherwise the surface produced will be wavy and the wheel will cause vibrations, possibly leading to wheel fracture. Some machine spindles and flanges provide for balancing wheels. Grinding wheels should be used according to their specifications and maximum operating speeds, and should not be dropped or abused. Wheel guarding, operator protection, and bystander safety are all important.

---

**Example 9.4:    Grinding vs. Hard Turning**

Section 8.8.2 described hard turning, an example of which is the machining of heat-treated steels (usually above 45 HRC) using a single-point polycrystalline cubic-boron-nitride cutting tool. In view of the discussions presented thus far in this chapter, it is evident that grinding and hard turning will be competitive in specific applications. Consequently, there has been considerable debate regarding their respective merits.

Hard turning continues to be increasingly competitive with grinding, and tolerances and surface finish are beginning to approach those for grinding. Turning requires much less energy than grinding (compare Tables 8.4 and 9.3), thermal and other damage to the workpiece surface is less likely to occur, cutting fluids may not be necessary, and the machine tools are less expensive. In addition, finishing the part while still chucked in the lathe eliminates the need for material handling and setting the part in the grinder. However, workholding devices for large and slender workpieces for hard turning can present problems since the cutting forces are higher than in grinding. Furthermore, tool wear and its control can be a significant problem as compared to the automatic dressing of grinding wheels. It is thus evident that the competitive position of hard turning versus grinding must be evaluated individually for each application and in terms of product surface integrity, quality, and overall economics.

---

## 9.7    Ultrasonic Machining

In *ultrasonic machining*, material is removed from a surface by microchipping or erosion with abrasive particles. The tip of the tool (Fig. 9.22a) vibrates at low amplitude [0.05 to

**FIGURE 9.22**    (a) Schematic illustration of the ultrasonic machining process by which material is removed by microchipping and erosion. (b) and (c) Typical examples of holes produced by ultrasonic machining. Note the dimensions of cut and the types of workpiece materials.

0.125 mm (0.002 to 0.005 in.)] and at high frequency (20 kHz). This vibration, in turn, transmits a high velocity to fine abrasive grains between the tool and the surface of the workpiece.

The grains are usually boron carbide, but aluminum oxide and silicon carbide are also used. Grain size ranges from 100 (for roughing) to 1000 (for finishing). The grains are in a water slurry with concentrations ranging from 20 percent to 60 percent by volume. The slurry also carries away the debris from the cutting area.

Ultrasonic machining is best suited for hard, brittle materials, such as ceramics, carbides, glass, precious stones, and hardened steels. The tip of the tool is usually made of low-carbon steel and undergoes wear. It is attached to a transducer through the toolholder. With fine abrasives, tolerances of 0.0125 mm (0.0005 in.) or better can be held in this process. Figures 9.22b and c show two applications of ultrasonic machining.

*Microchipping* in ultrasonic machining is possible because of the high stresses produced by particles striking a surface. The time of contact between the particle and the surface is very short (10 to 100 $\mu$s) and the area of contact is very small.

The time of contact, $t_o$, can be expressed as

$$t_o \simeq \frac{5r}{c_o} \left( \frac{c_o}{v} \right)^{1/5},\tag{9.11}$$

where $r$ is the radius of a spherical particle, $c_o$ is the elastic wave velocity in the workpiece ($c_o = \sqrt{E/\rho}$), and $v$ is the velocity with which the particle strikes the surface. The force $F$ of the particle on the surface is obtained from the rate of change of momentum. That is,

$$F = \frac{d(mv)}{dt},\tag{9.12}$$

where $m$ is the mass of the particle.

The *average force* $F_{ave}$ of a particle striking the surface and rebounding is

$$F_{ave} = \frac{2mv}{t_o}.\tag{9.13}$$

Substitution of numerical values into Eq. (9.13) indicates that even small particles can exert significant forces and, because of the very small contact area, produce very high stresses. In brittle materials, these stresses are sufficiently high to cause microchipping and surface erosion (see also *Abrasive-jet machining*, Section 9.16).

---

**Example 9.5:  Effect of Workpiece Temperature on Impact Force**

Explain what change, if any, takes place in the magnitude of the impact force of a particle in ultrasonic machining as the temperature of the workpiece is increased.

SOLUTION.   The force of a particle is given by Eq. (9.13). For this problem, $m$ and $v$ are constant. For the contact time $t_o$ in Eq. (9.11), we can now write

$$t_o \propto \frac{1}{c_o^{4/5}} \propto \frac{1}{E^{2/5}}.$$

With increasing temperature, the modulus of elasticity $E$ decreases, and thus $t_o$ increases. Therefore the impact force decreases, according to Eq. (9.13).

---

## 9.8  Finishing Operations

In addition to those described thus far, several processes are generally used on workpieces as the final finishing operation. These processes mainly utilize abrasive grains. Commonly used finishing operations are described in this section in the order of improved surface finish produced. Finishing operations can contribute significantly to production time and product cost. Thus they should be specified with due consideration to their costs and benefits.

**Coated abrasives.**   Typical examples of **coated abrasives** are sandpaper and emery cloth. The grains used in coated abrasives are more pointed than those used for grinding wheels. The grains are electrostatically deposited on flexible backing materials, such as paper or cloth (Fig. 9.23), with their long axes perpendicular to the plane of the backing. The matrix (coating) is made of resins.

Coated abrasives are available as sheets, belts, and disks and usually have a much more open structure than the abrasives on grinding wheels. Coated abrasives are used extensively in finishing flat or curved surfaces of metallic and nonmetallic parts, of metallographic specimens, and in woodworking. The precision of surface finish obtained depends primarily on grain size.

Coated abrasives are also used as *belts* for high-rate material removal. **Belt grinding** has become an important production process, in some cases replacing conventional grinding operations such as grinding of camshafts, with 8 to 16 lobes (of approximately elliptic shapes per shaft). Belt speeds are usually in the range of 700–1800 m/min (2500–6000 ft/min). Machines for abrasive-belt operations require proper belt support and rigid construction to minimize vibrations.

**Wire brushing.**   In the **wire brushing** process, the workpiece is held against a circular wire brush that rotates at high speed. The tips of the wire produce longitudinal scratches on the workpiece surface. This process is used to produce a fine surface texture. Some efforts are taking place in developing wire brushing as a light material-removal process.

**Honing.**   **Honing** is an operation used primarily to give holes a fine surface finish. The honing tool (Fig. 9.24) consists of a set of aluminum-oxide or silicon-carbide bonded abra-

FIGURE 9.24  Schematic illustration of a honing tool to improve the surface finish of bored or ground holes.

**FIGURE 9.25** Schematic illustrations of the superfinishing process for a cylindrical part. (a) Cylindrical microhoning, (b) centerless microhoning.

sives, called *stones.* They are mounted on a mandrel that rotates in the hole, applying a radial force with a reciprocating axial motion, thus producing a cross-hatched pattern. The stones can be adjusted radially for different hole sizes. The fineness of surface finish can be controlled by the type and size of abrasive used, the speed of rotation, and the pressure applied. A fluid is used to remove chips and to keep temperatures low. If not done properly, honing can produce holes that are not straight and cylindrical but rather holes that are bell-mouthed, wavy, barrel-shaped, or tapered. Honing is also done on external cylindrical or flat surfaces and to remove sharp edges on cutting tools and inserts (see Fig. 8.42).

In **superfinishing,** the pressure applied is very light and the motion of the stone has a short stroke. The process is controlled so that the grains do not travel along the same path along the surface of the workpiece. Examples of external superfinishing of a round part are shown in Fig. 9.25.

**Lapping.**   **Lapping** is a finishing operation used on flat or cylindrical surfaces. The lap (Fig. 9.26a) is usually made of cast iron, copper, leather, or cloth. The abrasive particles are embedded in the lap, or they may be carried through a slurry. Tolerances on the order of ± .0004 mm (0.000015 in.) can be obtained with the use of fine abrasives—up to size 900. Surface finish can be as smooth as 0.025–0.1 $\mu$m (1–4 $\mu$in.).

**FIGURE 9.26**   (a) Schematic illustration of the lapping process. (b) Production lapping on flat surfaces. (c) Production lapping on cylindrical surfaces.

Production lapping on flat or cylindrical pieces is done on machines such as those shown in Figs. 9.26b and c. Lapping is also done on curved surfaces, such as spherical objects and glass lenses, using specially shaped laps. Running-in of mating gears can be done by lapping. Depending on the hardness of the workpiece, lapping pressures range from 7–140 kPa (1–20 psi).

**Polishing and buffing.**   **Polishing** is a process that produces a smooth, lustrous surface finish. Two basic mechanisms are involved in the polishing process: (1) fine-scale abrasive removal, and (2) softening and smearing of surface layers by frictional heating during polishing. The shiny appearance of polished surfaces results from the smearing action.

Polishing is done with disks or belts made of fabric, leather, or felt and coated with fine powders of aluminum oxide or diamond. Parts with irregular shapes, sharp corners, deep recesses, and sharp projections are difficult to polish.

**Buffing** is similar to polishing, with the exception that very fine abrasives are used on soft disks made of cloth or hide. The abrasive is supplied externally from a stick of abrasive compound. Polished parts may be buffed to obtain an even finer surface finish.

**Electropolishing.**   Mirror-like finishes can be obtained on metal surfaces by **electropolishing,** a process which is the reverse of electroplating (see Section 4.5.1). Because there is no mechanical contact with the workpiece, this process is particularly suitable for polishing irregular shapes. The electrolyte attacks projections and peaks on the workpiece surface at a higher rate than the rest, thus producing a smooth surface. Electropolishing is also used for deburring operations.

**Polishing processes using magnetic fields.**   A recent development in polishing involves the use of magnetic fields to support abrasive slurries in finishing (polishing) ceramic balls and bearing rollers. **Magnetic float polishing** of ceramic balls is illustrated schematically in Fig. 9.27a. A magnetic fluid, containing abrasive grains and extremely fine ferromagnetic particles in a carrier fluid such as water or kerosene, fills the chamber within a guide ring. The ceramic balls are lo-

**FIGURE 9.27**   Schematic illustration of polishing of balls and rollers using magnetic fields. (a) Magnetic float polishing of ceramic balls, and (b) magnetic-field-assisted polishing of rollers. *Source:* R. Komanduri, M. Doc, and M. Fox.

cated between a drive shaft and a float. The abrasive grains, ceramic balls, and the float (which is made of a nonmagnetic material) are all suspended by magnetic forces. The balls are pressed against the rotating drive shaft and are polished by the abrasive action. The forces applied by the abrasive particles on the balls are extremely small and controllable, hence the polishing action is very fine. Polishing times are much lower than those for other polishing methods, hence the process is very economical and the surfaces produced have little or no defects.

**Magnetic-field-assisted polishing** of ceramic rollers is illustrated in Fig. 9.27b. A ceramic or steel roller (the workpiece) is clamped and rotated on a spindle. The magnetic poles are oscillated, introducing a vibratory motion to the magnetic-abrasive conglomerate. This action polishes the cylindrical roller surface. Bearing steels of 63 HRC have been mirror finished in 30 seconds by this process.

# 9.9 Deburring

**Burrs** are thin ridges, usually triangular in shape, that develop along the edges of a workpiece from shearing sheet materials, trimming forgings and castings, and machining. Burrs may interfere with the assembly of parts and can cause jamming of parts, misalignment, and short circuits in electrical components. Furthermore, burrs may reduce the fatigue life of components. Because they are usually sharp, they can be a safety hazard to personnel. The need for deburring may be reduced by adding chamfers to sharp edges on parts. On the other hand, burrs on thin drilled or tapped components, such as tiny parts in watches, can provide extra thickness and, thus, improve the holding torque of screws.

Several **deburring** processes are available: (1) manually with files, (2) mechanically by cutting, (3) wire brushing, (4) flexible abrasive finishing, using rotary nylon brushes with filaments embedded with abrasive grits, (5) abrasive belts, (6) ultrasonics, (7) electropolishing, (8) electrochemical machining, (9) vibratory finishing, (10) shot blasting, (11) abrasive-flow machining, and (12) thermal energy. The last four processes are described below; other processes are covered elsewhere.

**Vibratory** and **barrel finishing** processes are used to improve the surface finish and remove burrs from large numbers of relatively small workpieces. In this batch-type operation, specially shaped *abrasive pellets* or media are placed in a container along with the parts to be deburred. The container is either vibrated or tumbled. The impact of individual abrasives and metal particles removes sharp edges and burrs from the parts. Depending on the application, this is a dry or wet process, and liquid compounds may be added for requirements such as degreasing and corrosion resistance.

In **shot blasting** (also called *grit blasting*), abrasive particles (usually sand) are propelled by a high-velocity jet of air, or by a rotating wheel, onto the surface of the workpiece. Shot blasting is particularly useful in deburring metallic and nonmetallic materials and stripping, cleaning, and removing surface oxides. The surface produced has a matte finish. Small-scale polishing and etching can also be done by this process on bench-type units (*microabrasive blasting*).

In **abrasive-flow machining**, abrasive grains, such as silicon carbide or diamond, are mixed in a putty-like matrix, which is then forced back and forth through the openings and passageways in the workpiece. The movement of the abrasive matrix under pressure erodes away burrs and sharp corners and polishes the part. The process is particularly suitable for workpieces with internal cavities that are inaccessible by other means. Pressures applied range from 0.7 MPa to 22 MPa (100 psi to 3200 psi). External surfaces can also be deburred

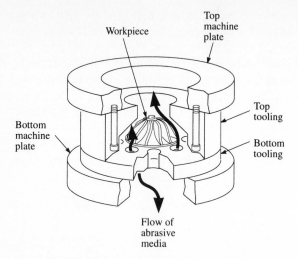

Workpiece

Top machine plate

Bottom machine plate

Top tooling

Bottom tooling

Flow of abrasive media

**FIGURE 9.28** Schematic illustration of abrasive-flow machining to deburr and polish a turbine impeller. The arrows indicate movement of the abrasive media. Note the special fixture, which is usually different for each part design. *Source:* Extrude Hone Corp.

using this process by containing the workpiece within a fixture that directs the abrasive media to the edges and areas to be deburred. An application concerning deburring of a turbine impeller by abrasive-flow machining is illustrated in Fig. 9.28.

The **thermal energy** method of deburring consists of placing the part in a chamber which is then injected with a mixture of natural gas and oxygen. This mixture is ignited, whereby a heat wave is produced with a temperature of 3300°C (6000°F). The burrs heat up instantly and are melted away, while the part temperature reaches only about 150°C (300°F). The process is effective in many applications on noncombustible parts. However, larger burrs or flashes tend to form beads after melting, the process can distort thin and slender parts, and it does not polish or buff the workpiece surfaces as do many of the other deburring processes.

Deburring and flash removal from castings and forgings are now being performed increasingly by *programmable robots* (see Section 13.7), thus eliminating tedious manual labor and resulting in more consistent deburring.

## 9.10  Chemical Machining

We know that certain chemicals attack metals and etch them, thereby removing small amounts of material from the surface. Thus **chemical machining** (CM) was developed, whereby material is removed from a surface by chemical dissolution, using chemical **reagents,** or **etchants,** such as acids and alkaline solutions. Chemical machining is the oldest of the nontraditional machining processes, and has been used for many years for engraving metals and hard stones, and more recently in the production of printed-circuit boards and microprocessor chips. Parts can also be deburred by chemical means.

### 9.10.1  Chemical milling

In **chemical milling,** shallow cavities are produced on plates, sheets, forgings, and extrusions for overall reduction of weight (Fig. 9.29). Chemical milling has been used on a wide variety of metals, with depths of removal to as much as 12 mm (0.5 in.). Selective attack by the

(a)
(b)

4 mm
(before machining)

2 mm
(after machining)

Chemically
machined area

Section

**FIGURE 9.29**   (a) Missile skin-panel section contoured by chemical milling to improve the stiffness-to-weight ratio of the part. (b) Weight reduction of space launch vehicles by chemical milling aluminum-alloy plates. These panels are milled after the plates have first been formed into shape, such as by roll forming or stretch forming. The design of the chemically machined rib patterns can be modified readily at minimal cost. *Source: Advanced Materials and Processes,* p. 43, December 1990. ASM International.

chemical reagent on different areas of the workpiece surfaces is controlled by removable layers of material, called *masking* (Fig. 9.30a), or by partial immersion in the reagent.

Chemical milling is used in the aerospace industry for removing shallow layers of material from large aircraft, missile skin panels, and extruded parts for airframes. Tank capacities for reagents are as large as 3.7 m × 15 m (12 ft × 50 ft). The process is also used to fabricate microelectronic devices. The range of material-removal rates, surface finish, and tolerances obtained by chemical machining and other machining processes are shown in Fig. 9.31 and Table 9.6.

Some surface damage may result from chemical milling because of preferential etching and intergranular attack, which adversely affect surface properties. Chemical milling of welded and brazed structures may produce uneven material removal. Chemical milling of castings may result in uneven surfaces caused by porosity and nonuniformity of structure.

(a)

Agitator

Maskant

Workpiece

Tank

Chemical
reagent

Heating

Cooling
coils

(b)

Edge of maskant

Steps

3rd    2nd    1st

Undercut

Material
removed

Depth

Workpiece

**FIGURE 9.30**   (a) Schematic illustration of the chemical-machining process. Note that no forces or machine tools are involved in this process. (b) Stages in producing a profiled cavity by chemical machining.

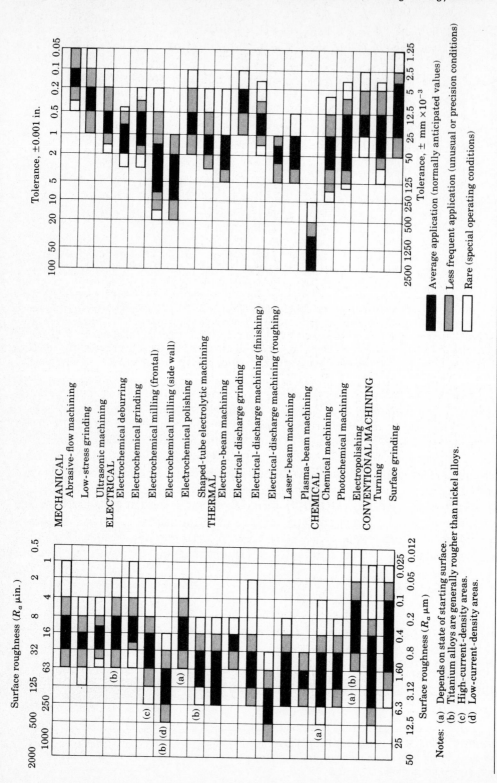

**FIGURE 9.31**   Surface roughness and tolerances obtained in various machining processes. Note the wide range within each process (see also Fig. 8.33). *Source:* Reprinted from *Machining Data Handbook,* 3rd ed. Copyright © 1980, by permission of the Institute of Advanced Manufacturing Sciences.

**TABLE 9.6**

**GENERAL CHARACTERISTICS OF NONTRADITIONAL MACHINING PROCESSES**

| PROCESS | CHARACTERISTICS | PROCESS PARAMETERS AND TYPICAL MATERIAL REMOVAL RATE OR CUTTING SPEED |
|---|---|---|
| Chemical machining (CM) | Shallow removal (up to 12 mm) on large flat or curved surfaces; blanking of thin sheets; low tooling and equipment cost; suitable for low production runs. | 0.025–0.1 mm/min. |
| Electrochemical machining (ECM) | Complex shapes with deep cavities; highest rate of material removal among nontraditional processes; expensive tooling and equipment; high power consumption; medium to high production quantity. | V: 5–25 dc; A: 1.5–8 $A/mm^2$; 2.5–12 mm/min, depending on current density. |
| Electrochemical grinding (ECG) | Cutting off and sharpening hard materials, such as tungsten-carbide tools; also used as a honing process; higher removal rate than grinding | A: 1–3 $A/mm^2$; Typically 1500 $mm^3$/min per 1000 A. |
| Electrical-discharge machining (EDM) | Shaping and cutting complex parts made of hard materials; some surface damage may result; also used as a grinding and cutting process; expensive tooling and equipment. | V: 50–380; A: 0.1–500; Typically 300 $mm^3$/min. |
| Wire EDM | Contour cutting of flat or curved surfaces; expensive equipment. | Varies with material and thickness. |
| Laser-beam machining (LBM) | Cutting and holemaking on thin materials; heat-affected zone; does not require a vacuum; expensive equipment; consumes much energy. | 0.50–7.5 m/min. |
| Electron-beam machining (EBM) | Cutting and holemaking on thin materials; very small holes and slots; heat-affected zone; requires a vacuum; expensive equipment. | 1–2 $mm^3$/min. |
| Water-jet machining (WJM) | Cutting all types of nonmetallic materials to 25 mm and greater in thickness; suitable for contour cutting of flexible materials; no thermal damage; noisy. | Varies considerably with material. |
| Abrasive water-jet machining (AWJM) | Single or multilayer cutting of metallic and nonmetallic materials. | Up to 7.5 m/min. |
| Abrasive-jet machining (AJM) | Cutting, slotting, deburring, deflashing, etching, and cleaning of metallic and nonmetallic materials; manually controlled; tends to round off sharp edges; hazardous. | Varies considerably with material. |

**FIGURE 9.32** An example of photochemical blanking of a 0.1 mm (0.004-in.) thick, 60 mm ($2\frac{3}{8}$ in.) diameter steel sheet with various blanked profiles. Note the fine detail. *Source:* Newcut.

### 9.10.2   Chemical blanking

**Chemical blanking** is similar to blanking of sheet metal in that it is used to produce features which penetrate through the thickness of the material (see Fig. 7.9), with the exception that material is removed by chemical dissolution rather than by shearing. Typical applications for chemical blanking are burr-free etching of printed circuit boards, decorative panels, thin sheet-metal stampings, and the production of complex or small shapes.

### 9.10.3   Photochemical blanking

Also called photoetching, **photochemical blanking** is a modification of chemical milling. Material is removed, usually from flat thin sheet, by photographic techniques. Complex burr-free shapes can be blanked (Fig. 9.32) on metals as thin as 0.0025 mm (0.0001 in.); it is also used for etching.

  Typical applications for photochemical blanking are fine screens, printed-circuit cards, electric-motor laminations, flat springs, and masks for color television. Although skilled labor is required, tooling costs are low, the process can be automated, and it is economical for medium to high production volume.

  Photochemical blanking is capable of forming very small parts where traditional blanking dies (Section 7.3) are difficult to make. Also, the process is effective for blanking fragile workpieces and materials.

  Handling of chemical reagents requires precautions and special safety considerations to protect personnel against both liquid chemicals and volatile chemical exposure. Furthermore, disposal of chemical by-products from this process is a major consideration, although some by-products can be recycled.

## 9.11   Electrochemical Machining

**Electrochemical machining (ECM)** is basically the reverse of electroplating. An electrolyte (Fig. 9.33) acts as current carrier, and the high rate of electrolyte movement in the tool-workpiece gap washes metal ions away from the workpiece (*anode*) before they have a chance to plate onto the tool (*cathode*). Note that the cavity produced is the female mating image of the

Pump for circulating electrolyte

**FIGURE 9.33** Schematic illustration of the electrochemical-machining process. This process is the reverse of electroplating, described in Section 4.5.1.

tool. Modifications of this process are used for turning, facing, slotting, trepanning, and profiling operations in which the electrode becomes the cutting tool.

The shaped tool is generally made of brass, copper, bronze, or stainless steel. The **electrolyte** is a highly conductive inorganic salt solution such as sodium chloride mixed in water or sodium nitrate. It is pumped at a high rate through the passages in the tool. A dc power supply in the range of 5–25 V maintains current densities, which for most applications are 1.5–8 A/mm$^2$ (1000–5000 A/in.$^2$) of active machined surface. Machines having current capacities as high as 40,000 A and as small as 5 A are available. Because the metal removal rate is only a function of ion exchange rate, it is not affected by the strength, hardness, or toughness of the workpiece, which must be electrically conductive.

The material-removal rate, MRR, by electrochemical machining can be calculated from the following equation:

$$MRR = CI\eta, \tag{9.14}$$

where MRR = mm$^3$/min, $I$ = current in amperes, and $\eta$ = current efficiency, which typically ranges from 90 to 100 percent. $C$ is a material constant in mm$^3$/A·min, and for pure metals depends on valence; the higher the valence, the lower its value.

If a cavity of uniform cross-sectional area $A_o$ mm$^2$ is being electrochemically machined, the feed rate $f$ in mm/min would be

$$f = MRR/A_o. \tag{9.15}$$

Note that the feed rate is the speed at which the electrode is penetrating the workpiece. Some examples of the approximate material constant $C$ are given in Table 9.7.

Electrochemical machining is generally used for machining complex cavities in high-strength materials, particularly in the aerospace industry for mass production of turbine blades, jet-engine parts, and nozzles (Fig. 9.34). It is also used for machining forging-die cavities and producing small holes.

The ECM process leaves a burr-free surface; in fact, it can also be used as a deburring process. It does not cause any thermal damage to the part, and the lack of tool forces prevents distortion of the part. Furthermore, there is no tool wear, and the process is capable of

**TABLE 9.7**

**APPROXIMATE VALUES OF THE MATERIAL CONSTANT $C$**

| METAL | VALENCE | $C$(mm³/A·min) |
|---|---|---|
| Aluminum | 3 | 2.0 |
| Copper | 1 | 4.4 |
|  | 2 | 2.2 |
| Iron | 2 | 2.2 |
|  | 3 | 1.5 |
| Nickel | 2 | 2.1 |
|  | 3 | 1.4 |
| Steels | – | 1.5–2.0 |
| Superalloys | – | 1.5–2.1 |
| Titanium | 3 | 2.2 |
|  | 4 | 1.7 |
| Titanium alloys | – | 1.6 |
| Tungsten | 6 | 1.0 |
|  | 8 | 0.7 |

producing complex shapes as well as machining hard materials. However, the mechanical properties of components made by ECM should be compared carefully to those of other material-removal methods. Electrochemical machining systems are now available as numerically controlled machining centers, with the capability for higher production rates, high flexibility, and maintenance of close tolerances.

(a)

140 mm
75 mm
65 mm

Telescoping cover
Insulating layer
Feed
Electrolyte
Forging
Machined workpiece
Copper electrode
Electrode carrier
Ram

(b) 14 holes    86 mm    112 mm    (c)

**FIGURE 9.34** Typical parts made by electrochemical machining. (a) Turbine blade made of a nickel alloy, 360 HB. *Source: Metal Handbook,* 9th ed., vol. 3, 1980, ASM International, Materials Park, OH, p. 849. (b) Thin slots on a 4340-steel roller-bearing cage. (c) Integral airfoils on a compressor disk.

## 9.12 Electrochemical Grinding

**Electrochemical grinding** (ECG) combines electrochemical machining with conventional grinding. The equipment used in electrochemical grinding is similar to a conventional grinder, except that the wheel is a rotating cathode with abrasive particles (Fig. 9.35a). The wheel is metal-bonded with diamond or aluminum-oxide abrasives, and rotates at a surface speed of 1200–2000 m/min (4000–7000 ft/min.) The abrasives serve as insulators between the wheel and the workpiece and mechanically remove electrolytic products from the working area. A flow of electrolyte, usually sodium nitrate, is provided for the electrochemical machining phase of the operation.

The majority of metal removal in ECG is by electrolytic action, and typically less than 5 percent of metal is removed by the abrasive action of the wheel. Therefore, wheel wear is very low, and current densities range from 1–3 A/mm$^2$ (500–2000 A/in$^2$). Finishing cuts are usually made by the grinding action but only to produce a surface with good finish and dimensional accuracy.

The material removal rate, MRR, in electrochemical grinding can be calculated from the following equation:

$$MRR = GI/\rho F \tag{9.16}$$

where MRR = mm$^3$/min, $G$ = equivalent weight in grams, $I$ = current in amperes, $\rho$ = density in g/mm$^3$, and $F$ = Faraday's constant in Coulombs.

The speed of penetration, $V_s$, of the grinding wheel into the workpiece is given by the equation

$$V_s = [G/\rho F][E/gK_p]K, \tag{9.17}$$

where $V_s$ = mm$^3$/min, $E$ = cell voltage in volts, $g$ = wheel–workpiece gap in mm, $K_p$ = coefficient of loss, which is in the range of 1.5 to 3, and $K$ = electrolyte conductivity in $\Omega^{-1}$mm$^{-1}$.

This process is suitable for applications similar to those for milling, grinding, and sawing (Fig. 9.34b). It is not adaptable to cavity-sinking operations, such as die making. The ECG process has been successfully applied to carbides and high-strength alloys. This process offers a distinct advantage over traditional diamond-wheel grinding when processing very hard materials, where wheel wear can be high. ECG machines are now available with numerical controls, thus improving accuracy, repeatability, and increased productivity.

**FIGURE 9.35** (a) Schematic illustration of the electrochemical-grinding process. (b) Thin slot produced on a round nickel-alloy tube by this process.

**Example 9.6:** Machining Time in Electrochemical Machining vs. Drilling

A round hole 12.5 mm in diameter is being produced in a titanium-alloy block by electrochemical machining. Using a current density of 6 A/mm$^2$, estimate the time required for machining a 20 mm deep hole. Assume that the efficiency is 90 percent. Compare this time with that required for ordinary drilling.

SOLUTION.    We note from Eqs. (9.14) and (9.15) that the feed rate can be expressed by the equation

$$f = CI\eta/A_o$$

where, for this case, $C = 1.6$ mm$^3$/A·min from Table 9.7 and $I/A_o = 6$ A/mm$^2$. Hence the feed rate is $f = (1.6)(6)(0.9) = 8.64$ mm/min. Since the hole is 20 mm deep,

$$\text{Machining time} = 20/8.64 = 2.3 \text{ min.}$$

To determine the drilling time, we refer to Table 8.17 and note the data for titanium alloys. Let us select the following values for a 12.5-mm drill: Drill rpm = 300 and feed = 0.15 mm/rev. Thus, the feed rate will be (300 rev/min)(0.15 mm/rev) = 45 mm/min. Since the hole is 20 mm deep,

$$\text{Drilling time} = 20/45 = 0.45 \text{ min,}$$

which is about 20 percent of the time for ECM.

---

**Electrochemical honing** combines the fine abrasive action of honing with electrochemical action. Although the equipment is costly, the process is as much as five times faster than conventional honing and the tool lasts longer by as much as ten times. It is used primarily for finishing internal cylindrical surfaces.

## 9.13  Electrical-Discharge Machining

The principle of **electrical-discharge machining** (EDM), also called *electrodischarge* or *spark-erosion* machining, is based on erosion of metals by spark discharges. We know that when two current-conducting wires are allowed to touch each other, an arc is produced. If we look closely at the point of contact between the two wires, we note that a small portion of the metal has been eroded away, leaving a small crater. Although this phenomenon has been known since the discovery of electricity, it was not until the 1940s that a machining process based on this principle was developed. The EDM process has become one of the most important and widely accepted production technologies in manufacturing industries.

    The EDM system (Fig. 9.36) consists of a shaped tool (*electrode*) and the workpiece, connected to a dc power supply and placed in a **dielectric** (electrically nonconducting) fluid. When the potential difference between the tool and the workpiece is sufficiently high, a transient

Current
Rectifier control    Servo control

Power supply

(+)    (−)

Movable electrode

Tank

Workpiece
Dielectric fluid

**FIGURE 9.36** Schematic illustration of the electrical discharge machining process.

spark discharges through the fluid, removing a very small amount of metal from the workpiece surface. The capacitor discharge is repeated at rates of between 50 kHz and 500 kHz, with voltages usually ranging between 50 V and 380 V, and currents from 0.1 A to 500 A.

The dielectric fluid (1) acts as an insulator until the potential is sufficiently high, (2) acts as a flushing medium and carries away the debris in the gap, and (3) provides a cooling medium. The gap between the tool and the workpiece (called *overcut*) is critical; thus the downward feed of the tool is controlled by a servomechanism, which automatically maintains a constant gap. The most common dielectric fluids are mineral oils, although kerosene and distilled and deionized water may be used in specialized applications. The workpiece is fixtured within the tank containing the dielectric fluid, and its movements are controlled by numerically controlled systems. The machines are equipped with a pump and filtering system for the dielectric fluid.

The EDM process can be used on any material that is an electrical conductor. The melting point and latent heat of melting are important physical properties that determine the volume of metal removed per discharge. As these values increase, the rate of material removal slows. The volume of material removed per discharge is typically in the range of $10^{-6}$ to $10^{-4}$ mm$^3$ ($10^{-10}$ to $10^{-8}$ in$^3$). Since the process doesn't involve mechanical energy, the hardness, strength, and toughness of the workpiece material do not necessarily influence the removal rate. The frequency of discharge or the energy per discharge is usually varied to control the removal rate, as are the voltage and current. The rate and surface roughness increase with increasing current density and decreasing frequency of sparks.

**Electrodes** for EDM are usually made of graphite, although brass, copper, or copper–tungsten alloy may be used. The tools are shaped by forming, casting, powder metallurgy, or machining. Electrodes as small as 0.1 mm (0.005 in.) in diameter have been used. **Tool wear** is an important factor since it affects dimensional accuracy and the shape produced. Tool wear can be minimized by reversing the polarity and using copper tools, a process called **nowear EDM.**

Electrical-discharge machining has numerous applications, such as producing die cavities for large automotive-body components (die sinking machining centers), small-diameter deep holes (see Fig. 9.37c) using tungsten wire as the electrode, narrow slots, turbine blades, and various intricate shapes (Figs. 9.37a and b). Stepped cavities can be produced by controlling the relative movements of the workpiece in relation to the electrode (Fig. 9.38). In another setup, internal cavities are produced by a rotating electrode with a movable tip. The electrode is rotated mechanically during machining (Fig. 9.39).

The material removal rate, MRR, in electrical-discharge machining is basically a function of the current and the melting point of the workpiece material, although other process

(a)    (b)    (c)

Workpiece

Electrode

1.5 mm dia.

8 holes, 0.17 l

1.1 mm wal

**FIGURE 9.37** (a) Examples of cavities produced by the electrical-discharge-machining process, using shaped electrodes. Two round parts (rear) are the set of dies for extruding the aluminum piece shown in front. *Source:* Courtesy of AGIE USA Ltd. (b) A spiral cavity produced by a rotating electrode. *Source: American Machinist.* (c) Holes in a fuel-injection nozzle made by electrical-discharge machining. Material: Heat-treated steel.

variables also have an effect. The following approximate empirical relationships can be used as a guide to estimate the metal removal rate in EDM:

$$\text{MRR} \;=\; 4 \times 10^4 I \, T_w^{-1.23} \tag{9.18}$$

where MRR = mm³/min, $I$ is the current in amperes, and $T_w$ = melting point of the workpiece in °C.

The *wear rate of the electrode*, $W_t$, can be estimated from the following empirical equation:

$$W_t = 11 \times 10^3 I T_t^{-2.38} \tag{9.19}$$

where $W_t$ = mm³/min and $T_t$ = melting point of the electrode material in °C.

The **wear ratio** of the workpiece to electrode, $R$, can be estimated from the following expression:

$$R = 2.25 \, T_r^{-2.3} \tag{9.20}$$

where $T_r$ = ratio of workpiece to electrode melting points. In practice, wear ratios vary over a wide range, between 0.2 and 100. (Note that in abrasive processing, grinding ratios also are in the same range; see Section 9.5.2.)

**FIGURE 9.38** Stepped cavities produced with a square electrode by EDM. The workpiece moves in the two principal horizontal directions, and its motion is synchronized with the downward movement of the electrode to produce various cavities. Also shown is a round electrode capable of producing round or elliptical cavities. *Source:* Courtesy of AGIE USA Ltd.

(a)                                          (b)

**FIGURE 9.39** Schematic illustration of producing an inner cavity by EDM, using a specially designed electrode with a hinged tip, which is rotated along two axes to produce the large cavity. *Source:* Luziesa France.

Metal removal rates usually range from 2 to 400 mm$^3$/min. High rates produce a very rough finish, having a molten and resolidified (recast) structure with poor surface integrity and low fatigue properties. Thus finishing cuts are made at low removal rates, or the recast layer is removed later by finishing operations. New techniques use an oscillating electrode which provides very fine surface finishes, requiring significantly less benchwork to improve lustrous cavities.

**Example 9.7:   Machining Time in Electrical-Discharge Machining vs. Drilling**

Referring to Example 9.6, calculate the machining time for producing the hole by EDM and compare the time with drilling and ECM. Assume that the titanium alloy has a melting point of 1600°C (see Table 3.3) and the current is 100 A. Also calculate the wear rate of the electrode, assuming that the melting point of the electrode is 1100°C.

SOLUTION.   Using Eq. (9.18), we find that

$$\text{MRR} = (4 \times 10^4)(100)(1600^{-1.23}) = 458 \text{ mm}^3/\text{min}.$$

The volume $V$ of the hole is

$$V = \pi \left[(12.5)^2/4\right](20) = 2454 \text{ mm}^3.$$

Thus, the machining time by EDM is $2454/458 = 5.4$ minutes. This is 2.35 times that for ECM and 11.3 times that for drilling. Note, however, that if the current is increased to 300 A, the machining time by EDM will only be 1.8 minutes, which is less than the time for ECM.

The wear rate of the electrode is calculated using Eq. (9.19). Thus,

$$W_t = (11 \times 10^3)(100)(1100^{-2.38}) = 0.064 \text{ mm}^3/\text{min}.$$

### 9.13.1   Electrical-discharge grinding

The grinding wheel in **electrical-discharge grinding** (EDG) is made of graphite or brass and contains no abrasives. Material is removed from the surface of the workpiece by repetitive spark discharges between the rotating wheel and the workpiece.

The material-removal rate, MRR, can be estimated from the following equation:

$$\text{MRR} = KI, \tag{9.21}$$

where MRR = mm³/min, $I$ = current in amperes, and $K$ = workpiece material factor in mm³/A·min; for example, $K$ = 16 for steel and 4 for tungsten carbide.

The EDG process can be combined with electrochemical grinding. The process is then called **electrochemical-discharge grinding** (ECDG). Material is removed by chemical action, with the electrical discharges from the graphite wheel breaking up the oxide film, and is washed away by the electrolyte flow. The process is used primarily for grinding carbide tools and dies but can also be used for fragile parts, such as surgical needles, thin-walled tubes, and honeycomb structures. The ECDG process is faster than EDG, but power consumption is higher.

In *sawing* with EDM, a setup similar to a band or circular saw (but without any teeth) is used with the same electrical circuit as in EDM. Narrow cuts can be made at high rates of metal removal. Because cutting forces are negligible, the process can be used on slender components.

### 9.13.2  Wire EDM

A variation of EDM is **wire EDM** (Fig. 9.40), or *electrical-discharge wire cutting*. In this process, which is similar to contour cutting with a band saw, a slowly moving wire travels along a prescribed path, cutting the workpiece, with the discharge sparks acting like cutting teeth. This process is used to cut plates as thick as 300 mm (12 in.), and for making punches, tools, and dies from hard metals. It can also cut intricate components for the electronics industry.

The wire is usually made of brass, copper, or tungsten and is typically about 0.25 mm (0.01 in.) in diameter, making narrow cuts possible. Zinc- or brass-coated and multi-coated wires are also used. The wire should have sufficient tensile strength and fracture toughness, as well as high electrical conductivity and capacity to flush away the debris produced during cutting. The wire is generally used only once, as it is relatively inexpensive. It travels at a constant velocity in the range of 0.15–9 m/min (6–360 in./min), and a constant gap (kerf) is maintained during the cut. The cutting speed is generally given in terms of the cross-sectional area cut per unit time. Typical examples are: 18,000 mm²/hr (28 in²/hr) for 50-mm (2-in.) thick D2 tool steel, and 45,000 mm²/hr (70 in²/hr) for 150-mm (6-in.) thick aluminum. These removal rates indicate a linear cutting speed of 18,000/50 = 360 mm/hr = 6 mm/min, and 45,000/150 = 300 mm/hr = 5 mm/min, respectively.

The material-removal rate, MRR, in wire EDM can be obtained from the following expression:

$$MRR = V_f h b, \tag{9.22}$$

**FIGURE 9.40** Schematic illustration of the wire EDM process. As much as 50 hours of machining can be performed with one reel of wire, which is then discarded.

where MRR = mm³/min, $V_f$ = feed rate of the wire into the workpiece in mm/min, $h$ = workpiece thickness or height in mm, and the kerf (see Fig. 9.40) is denoted as $b = d_w + 2s$, where $d_w$ = wire diameter in mm and $s$ = gap between wire and workpiece during machining, in mm.

Modern wire EDM machines (*multi-axis EDM wire-cutting machining centers*) are equipped with (1) computer controls to control the cutting path of the wire, (2) automatic self-threading features in case of wire breakage, (3) multiheads for cutting two parts at the same time, (4) features such as controls that prevent wire breakage, and (5) programmed machining strategies. Two-axis computer-controlled machines can produce cylindrical shapes in a manner similar to a turning operation or cylindrical grinding.

## 9.14  High-Energy-Beam Machining

### 9.14.1  Laser-beam machining

In **laser-beam machining (LBM)**, the source of energy is a laser (an acronym for *Light Amplification* by *Stimulated Emission* of *Radiation*), which focuses optical energy on the surface of the workpiece (Fig. 9.41a). The highly focused, high-density energy melts and evaporates portions of the workpiece in a controlled manner. This process, which does not require a vacuum, is used to machine a variety of metallic and nonmetallic materials.

There are several types of lasers used in manufacturing operations: $CO_2$ (pulsed or continuous wave), Nd:YAG (neodymium: yttrium-aluminum-garnet), Nd: glass, ruby, and excimer lasers. It is beyond the scope of this text to describe the principles of these lasers; however, their applications are outlined in Table 9.8.

Important physical parameters in LBM are the *reflectivity* and *thermal conductivity* of the workpiece surface and its specific heat and latent heats of melting and evaporation. The lower these quantities, the more efficient the process.

**FIGURE 9.41**   (a) Schematic illustration of the laser-beam-machining process. (b) and (c) Examples of holes produced in nonmetallic parts by LBM.

**TABLE 9.8**

**GENERAL APPLICATIONS OF LASERS IN MANUFACTURING**

| APPLICATION | LASER TYPE |
|---|---|
| Cutting | |
| Metals | $PCO_2$, $CWCO_2$, Nd:YAG, ruby |
| Plastics | $CWCO_2$ |
| Ceramics | $PCO_2$ |
| Drilling | |
| Metals | $PCO_2$, Nd:YAG, Nd:glass, ruby |
| Plastics | Excimer |
| Marking | |
| Metals | $PCO_2$, Nd:YAG |
| Plastics | Excimer |
| Ceramics | Excimer |
| Surface treatment, metals | $CWCO_2$ |
| Welding, metals | $PCO_2$, $CWCO_2$, Nd:YAG, Nd:glass, ruby |

Note: P = pulsed, CW = continuous wave.

The cutting depth $t$ may be expressed as

$$t \; \alpha \; P/vd, \tag{9.23}$$

where $P$ is the power input, $v$ is the cutting speed, and $d$ is the laser-beam-spot diameter.

The surface produced by LBM is usually rough and has a heat-affected zone which, in critical applications, may have to be removed or heat treated. Kerf width is an important consideration, as it is in other cutting processes such as sawing, wire EDM, and electron-beam machining.

Laser beams may be used in combination with a gas stream, such as oxygen, nitrogen, or argon (*laser-beam torch*), for cutting thin sheet materials. High-pressure inert-gas (nitrogen)–assisted laser cutting is used for stainless steel and aluminum; it leaves an oxide-free edge that can improve weldability. Gas streams also have the important function of blowing away molten and vaporized material from the workpiece surface.

Laser-beam machining is used widely in drilling and cutting metals, nonmetals, and composite materials (Figs. 9.41b and c). The abrasive nature of composite materials and the cleanliness of the operation have made laser-beam machining an attractive alternative to traditional machining methods. Holes as small as 0.005 mm (0.0002 in.), with hole depth-to-diameter ratios of 50 to 1 have been produced in various materials, although a more practical minimum is 0.025 mm (0.001 in.). Extreme caution should be exercised with lasers, as even low-power lasers can cause damage to the retina of the eye if proper precautions are not observed.

Laser-beam machining is being used increasingly in the electronics and automotive industries. Bleeder holes for fuel-pump covers and lubrication holes in transmission hubs, for example, are being drilled with lasers (Fig. 9.42). The part shown, with a hardness of 60 HRC, requires drilling three holes, 1.5 mm in diameter and 9.5 mm deep (0.060 in. and 0.38 in.), and at an angle of 35° to the curved surface. It takes 3.5 s to drill each hole. Significant

**FIGURE 9.42**   Lubrication hole being drilled in an automotive transmission hub, using a 250 W, Nd:YAG laser. *Source:* Courtesy of Conversent Energy and *Mechanical Engineering,* April 1990, p. 41.

cost savings have been achieved by laser-beam machining, a process that is competing with electrical-discharge machining. The cooling holes in the first stage vanes of the Boeing 747 jet engines are also produced by lasers.

Laser beams are also used for (1) *welding* (see Section 12.4), (2) small-scale *heat treating* of metals and ceramics to modify their surface mechanical and tribological properties, and (3) *marking* of parts with letters, numbers, codes, etc. Marking can be done by processes such as with ink; with mechanical devices such as punches, pins, stylus, scroll rolls, or stamping; by etching; and with lasers. Although the equipment is more expensive than for other methods, marking and engraving with lasers has become increasingly common due to its accuracy, reproducibility, flexibility, ease of automation, and on-line application in manufacturing.

The inherent flexibility of the laser cutting process with fiber-optic beam delivery, simple fixturing, low setup times, the availability of multi-kW machines and 2-D and 3-D computer-controlled laser cutting systems are attractive features; thus laser cutting can compete successfully with cutting sheet metal using traditional punching processes described in Chapter 7. There are now efforts to combine the two processes for improved overall efficiency (see example below).

### Example 9.8:   Combining Laser Cutting and Punching of Sheet Metal

As we have seen, laser cutting and punching processes have their respective advantages and limitations regarding both technical and economic aspects. Laser cutting characteristics are generally smaller batches, flexibility of operation, wide range of thicknesses, prototyping capability, materials and composites that may otherwise be cut with difficulty, and complex geometries that can be programmed. Punching characteristics include large lot sizes, relatively simple parts, small range in part thickness, fixed and limited punch geometries even when using turrets, rapid production, and integration with subsequent processing after punching. It is thus evident that the two processes cover different but complementary ranges.

Considerations have been given to the possible benefits of combining the two processes, taking advantage of their individual features to fill the range between the two processes. The purpose is to increase the overall efficiency and productivity of the manufacturing process for parts that are within the capabilities of each of the two processes, somewhat similar to the concept of the machining centers described in Section 8.10. Several factors have to be taken into account in such a combination with respect to the characteristics of each process: processing time; setup times; loading, fixturing, and unloading of parts; programming for cutting; process capabilities of each including dynamic characteristics, vibrations, shock, etc.; ranges of sizes, thicknesses, and shapes to be produced and how they are to be nested; and several individual cost factors.

It is not difficult to visualize parts with certain features that can be produced best by one process, and other features that are best produced by the other process. The combined machine should be designed and constructed in such a manner that the processes and fixturing can be utilized jointly to their full extent but without interference with each other's operational boundaries. During the past few years such special machines have been built and used successfully for specific applications, including rapid-response capabilities for product design changes and varying quantities.

---

### 9.14.2   Electron-beam machining and plasma-arc cutting

The source of energy in **electron-beam machining (EBM)** is high-velocity electrons, which strike the surface of the workpiece and generate heat (Fig. 9.43). Its applications are similar to those of laser-beam machining, except that EBM requires a vacuum. These machines utilize voltages in the range of 50–200 kV to accelerate the electrons to speeds of 50 to 80 percent of the speed of light. The interaction of the electron beam with the workpiece surface produces hazardous x-rays; consequently, the equipment should be used only by highly trained personnel. Electron-beam machining is used for very accurate cutting of a wide variety of metals. Surface finish is better and kerf width is narrower than that for other thermal cutting processes (see also Section 12.4 on electron-beam welding.)

**FIGURE 9.43**  Schematic illustration of the electron-beam-machining process. Unlike LBM, this process requires a vacuum, hence workpiece size is limited.

In **plasma-arc cutting (PAC), plasma beams** (ionized gas) are used for rapid cutting of nonferrous and stainless-steel plates. The temperatures generated are very high (9400°C or 17,000°F in the torch for oxygen as a plasma gas); consequently, the process is fast, kerf width is small, and the surface finish is good. Material-removal rates are much higher than in the EDM and LBM processes, and parts can be machined with good reproducibility. Plasma-arc cutting is highly automated today, using programmable controls.

A more traditional method is **oxyfuel-gas cutting (OFC),** using a torch as in welding. It is particularly useful for cutting steels, cast irons, and cast steels. Cutting occurs mainly by oxidation and burning of the steel, with some melting taking place. Kerf widths are usually in the range of 1.5–10 mm (0.06–0.4 in.).

## 9.15  Water-Jet Machining

We know that when we put our hand across a jet of water or air, we feel a considerable concentrated force acting on it. This force results from the momentum change of the stream—and in fact is the principle on which the operation of water or gas turbines is based. In **water-jet machining (WJM),** also called *hydrodynamic machining* (Fig. 9.44), this force is utilized in cutting and deburring operations.

The water jet acts like a saw and cuts a narrow groove in the material. Although pressures as high as 1400 MPa (200 ksi) can be generated, a pressure level of about 400 MPa (60 ksi) is generally used for efficient operation. Jet-nozzle diameters usually range between 0.05 mm and 1 mm (0.002 in. and 0.040 in.).

A variety of materials can be cut with this technique, including plastics, fabrics, rubber, wood products, paper, leather, insulating materials, brick, and composite materials (Fig. 9.45). Thicknesses range up to 25 mm (1 in.) and higher. Vinyl and foam coverings for some automobile dashboards, for example, are being cut using multiple-axis, robot-guided water-jet machining equipment. Because it is an efficient and clean operation compared with other cutting processes, it is also used in the food processing industry for cutting and slicing food products.

The advantages of this process are that cuts can be started at any location without the need for predrilled holes, no heat is produced, no deflection of the rest of the workpiece takes place

**FIGURE 9.44**
Schematic illustration of water-jet machining.

**FIGURE 9.45** Example of various non-metallic parts cut by a water-jet machine. *Source:* Courtesy of Possis Corporation.

(hence the process is suitable for flexible materials), little wetting of the workpiece takes place, and the burr produced is minimal. It is also an environmentally safe manufacturing process.

**Abrasive water-jet machining.** In **abrasive water-jet machining (AWJM),** the water jet contains abrasive particles such as silicon carbide or aluminum oxide, thus increasing the material-removal rate over water-jet machining. Metallic, nonmetallic, and advanced composite materials of various thicknesses can be cut in single or multilayers, particularly heat-sensitive materials that cannot be machined by processes in which heat is produced. Cutting speeds are as high as 7.5 m/min (25 ft/min) for reinforced plastics, but much lower for metals. With multiple-axis and robotic-control machines, complex three-dimensional parts can be machined to finish dimensions. Nozzle life has been improved by making them from rubies, sapphires, and carbide-base composite materials.

## 9.16 Abrasive-Jet Machining

In **abrasive-jet machining (AJM),** a high-velocity jet of dry air, nitrogen, or carbon dioxide containing abrasive particles is aimed at the workpiece surface under controlled conditions (Fig. 9.46). The impact of the particles develops sufficient concentrated force (see also Section 9.7) to perform operations such as (1) cutting small holes, slots, or intricate patterns in

**FIGURE 9.46** Schematic illustration of the abrasive-jet-machining process.

very hard or brittle metallic and nonmetallic materials, (2) deburring or removing small flash from parts, (3) trimming and beveling, (4) removing oxides and other surface films, and (5) general cleaning of components with irregular surfaces.

The gas supply pressure is on the order of 850 kPa (125 psi) and the abrasive-jet velocity can be as high as 300 m/s (100 ft/s) and is controlled by a valve. The hand-held nozzles are usually made of tungsten carbide or sapphire. The abrasive size is in the range of 10–50 $\mu$m (400–2000 $\mu$in.). Because the flow of the free abrasives tends to round off corners, designs for abrasive-jet machining should avoid sharp corners; also, holes made in metal parts tend to be tapered.

Because of airborne particulates, there is some hazard involved in using this process. This problem can be avoided by using the abrasive water-jet machining process.

## 9.17  Process Economics

We have shown that grinding may be used both as a finishing operation and as a large-scale removal operation (as in creep-feed grinding). The use of grinding as a finishing operation is often necessary because forming and machining processes alone usually cannot produce parts with the desired dimensional accuracy and surface finish. However, because it is an additional operation, grinding contributes significantly to product cost. Creep-feed grinding, on the other hand, has proved to be an economical alternative to machining operations such as milling, even though wheel wear is high.

All finishing operations contribute to product cost. On the basis of the discussion thus far, you can see that as the surface finish improves, more operations are required, and hence the cost increases. Note in Fig. 9.47 how rapidly cost increases as surface finish is improved by processes such as grinding and honing.

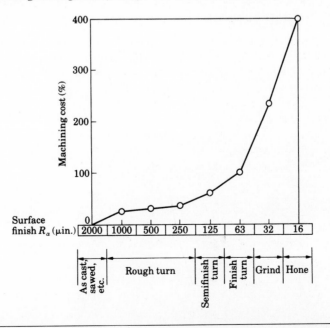

**FIGURE 9.47**   Increase in the cost of machining and finishing a part as a function of the surface finish required.

Much progress has been made in automating the equipment involved in finishing operations, including computer controls. Consequently, labor costs and production times have been reduced, even though such machinery may require significant capital investment. If finishing is likely to be an important factor in manufacturing a particular product, the conceptual and design stages should involve an analysis of the degree of surface finish and dimensional accuracy required.

Furthermore, all processes that precede finishing operations should be analyzed for their capability to produce a more acceptable surface finish and dimensional accuracy. This can be accomplished through proper selection of tools and process parameters and the characteristics of the machine tools involved.

We have also shown the unique applications of nontraditional processes, particularly for difficult-to-machine materials and complex internal and external profiles. The economic production run for a particular process depends on the cost of tooling and equipment, the material-removal rate, operating costs, and the level of operator skill required, as well as secondary and finishing operations that may be necessary. In chemical machining, the costs of reagents, maskants, and disposal, together with the cost of cleaning the parts, are important factors. In electrical-discharge machining, the cost of electrodes and the need to replace them periodically are significant.

The rate of material removal, hence production rate, can vary significantly in these processes. The cost of tooling and equipment also varies significantly, as does the operator skill required. The high capital investment for machines such as electrical and high-energy beam machining should be justified in terms of the production runs and the feasibility of manufacturing the same part by other means, if at all possible.

---

CASE STUDY:    ELECTROCHEMICAL MACHINING OF A BIOMEDICAL IMPLANT

A total knee replacement system consists primarily of a femoral and tibial implant combined with an ultra-high molecular weight polyethylene (UHMWPE) insert, as shown in Fig. 9.48. The polyethylene (see Sections 10.2.2 and 10.5) has superior wear resistance, biocompatibil-

**FIGURE 9.48** Total knee replacement systems showing two metal implants (top and bottom) with an ultra-high molecular weight polyethylene insert (white piece). *Source:* Photo courtesy of Biomet, Inc.

**FIGURE 9.49**    Cross-section of the ECM process.

ity, and low friction against the cobalt-chrome alloy femoral implant. The UHMWPE insert is compression molded or ram extruded (see Section 10.10), and the metal implant is cast and ground on its external surfaces mating with the polyethylene part.

Designers of implants, manufacturing engineers, and clinicians have been particularly concerned with the contact surfaces. As the knee articulates during the normal motion of a person's leg, the polyethylene slides against the metal part, thus becoming a potentially serious wear site. A cavity, typically called the intercondylar box, in the metal implant mates with a protrusion on the polyethylene insert. This geometry is necessary to ensure stability of the knee, i. e., it prevents the femur from sliding forward over the front of the tibia.

In order to produce a smooth surface, grinding of the bearing surfaces of the metal implant (using both hand-held and cam-mounted grinders) is a procedure that has been followed for many years. However, grinding the intercondylar box feature produced marginal repeatability and quality. The interior surfaces of the box feature are extremely difficult to access for grinding, and the cobalt-chrome alloy is difficult to grind. Consequently, and as stated in Section 9.1, nontraditional material-removal processes were an ideal candidate for this operation, particularly electrochemical machining.

As shown in Fig. 9.49, the current procedure consists of placing the metal implant in a fixture and bringing a copper-tungsten electrode of the desired final contour in close proximity to the implant (see also Section 9.11). The electrolyte is a sodium-nitrate and water mixture and is pumped through the tool, filling the gap between the tool and implant. A power source (typically 10 V and 225 A) is applied, causing local electrochemical machining of the high spots on the implant surface and producing a polished surface. Typical machining times are four to six minutes. The electrolyte flow volume can be controlled to maximize surface quality. When the flow rate is too low, defects appear on the machined surface as localized dimples; if the rate is too high, machining times become longer.

*Source:* T. Hershberger and R. Redman, Biomet, Inc., Warsaw, Indiana.

## SUMMARY

- Abrasive machining and nontraditional machining processes are often necessary and economical when workpiece hardness is high, the materials are brittle or flexible, part shapes are complex, and surface finish and tolerance requirements are high. (Section 9.1)

- Conventional abrasives basically consist of aluminum oxide and silicon carbide, and superabrasives consist of cubic boron nitride and diamond. Friability of abrasive grains is an important factor, as are shape and size of the grains, in their performance. (Section 9.2)

- Grinding wheels consist of a combination of abrasive grains and bonding agents. Important characteristics of wheels are abrasive grain and bond types, grade, and hardness. Wheels may also be reinforced to maintain wheel integrity. (Section 9.3)

- The mechanics of grinding processes are studied in order to establish quantitative relationships regarding grinding chip dimensions, grinding forces, temperature rise, residual stresses, and any effects on surface integrity of ground components. (Section 9.4)

- Grinding wheel wear is an important consideration in ground surface quality and product integrity. Dressing and truing of wheels are done by various techniques. Wheel wear is usually monitored in terms of the grinding ratio, defined as the ratio of material removed to the volume of wheel wear. (Section 9.5)

- A variety of abrasive machining processes and machinery are available for surface, external, and internal grinding. Abrasive machining is also used for large-scale material-removal processes, such as creep-feed grinding, thus competing with processes such as milling and turning. The selection of abrasives and process variables, including grinding fluids, is important in order to obtain the desired surface finish and dimensional accuracy. Otherwise, damage to surfaces such as burning, heat checking, harmful residual stresses, and chatter may develop. (9.6)

- Ultrasonic machining removes material by microchipping and is best suited for hard and brittle materials. (Section 9.7)

- Several finishing operations are available to improve surface finish. Because they can contribute significantly to product cost, proper selection and implementation of finishing operations are important. (Section 9.8)

- Deburring may be necessary for certain finished components. Commonly used methods are vibratory and barrel finishing and shot blasting, although thermal energy and other methods are also available. (Section 9.9)

- Machining processes also can involve chemical, electrochemical, and electrical sources of energy. Chemical and electrical methods of machining are particularly suitable for hard materials and complex shapes. They do not produce forces, significant temperatures, or residual stresses. However, the effects of these processes on surface integrity must be understood, as they can damage surfaces considerably, thus reducing fatigue life. (Sections 9.10, 9.11, 9.12, 9.13)

- High-energy-beam machining processes basically utilize laser beams, electron beams, and plasma arc. They have important industrial applications, possess great flexibility of operation, and are economically competitive with various other processes. (Section 9.14)

- Water-jet, abrasive water-jet, and abrasive-jet machining processes can be used for cutting as well as deburring operations. Because they do not utilize hard tooling, they have inherent flexibility of operation. (Sections 9.15, 9.16)
- Abrasive and nontraditional material-removal processes have unique capabilities; however, because they involve different types of machinery and process variables as well as cycle times, the competitive aspects of each process with respect to a particular product must be studied carefully. (Section 9.17)

## SUMMARY OF EQUATIONS

Undeformed chip length, surface grinding, $\ell = \sqrt{Dd}$

external grinding, $\ell = \sqrt{\dfrac{Dd}{1 + D/D_w}}$

internal grinding, $\ell = \sqrt{\dfrac{Dd}{1 - D/D_w}}$

Undeformed chip thickness, surface grinding, $t = \sqrt{\dfrac{4v}{VCr}\sqrt{\dfrac{d}{D}}}$

Relative grain force $\alpha \dfrac{v}{VC}\sqrt{\dfrac{d}{D}}$

Temperature rise, $\Delta T \propto d^{3/4}\sqrt{\dfrac{VC}{v}\sqrt{D}}$

Grinding ratio, $G = \dfrac{\text{Vol. of material removed}}{\text{Vol. of wheel wear}}$

Average force of particle striking a surface in ultrasonic machining, $F_{ave} = \dfrac{2mv}{t_o}$

where $t_o = \dfrac{5r}{c_o}\left(\dfrac{c_o}{v}\right)^{1/5}$

Material-removal rate in electrochemical machining, $\text{MRR} = CI\eta$

Material-removal rate in electrochemical grinding, $\text{MRR} = GI/\rho F$

The speed of penetration in electrochemical grinding, $V_s = [G/dF][E/gK_p]\,K$

Material-removal rate in electrical-discharge machining, $\text{MRR} = 4 \times 10^4 I T_w^{-1.23}$

Wear rate of electrode in electrical-discharge machining, $W_t = 11 \times 10^3 I T_t^{-2.38}$

Material-removal rate in electrical-discharge grinding, $\text{MRR} = KI$

Material-removal rate in wire electrical-discharge machining, $\text{MRR} = V_f h b$,

where $b = d_w + 2s$

Laser-beam cutting, $t = P/vd$

# BIBLIOGRAPHY

## Abrasive processes

Andrew, C., T. D. Howes, and T. R. A. Pearce, *Creep-Feed Grinding.* New York: Industrial Press, 1985.

*ASM Handbook,* Vol. 16: *Machining.* Materials Park, Ohio: ASM International, 1989.

Boothroyd, G., and W. A. Knight, *Fundamentals of Machining and Machine Tools,* 2d ed. New York: Marcel Dekker, 1989.

Borkowski, J., and A. Szymanski, *Uses of Abrasives and Abrasive Tools.* Chichester, Eng.: Ellis Horwood, 1992.

Farago, F. T., *Abrasive Methods Engineering,* Vol. 1, 1976; Vol. 2, 1980. New York: Industrial Press.

Gillespie, L. K., *Deburring Technology for Improved Manufacturing.* Dearborn, Mich.: Society of Manufacturing Engineers, 1981.

King, R. I., and R. S. Hahn, *Handbook of Modern Grinding Technology.* New York: Chapman and Hall/Methuen, 1987.

Krar, S., and E. Ratterman, *Superabrasives: Grinding and Machining with CBN and Diamond.* New York: McGraw-Hill, 1990.

*Machinery Handbook,* revised periodically. New York: Industrial Press.

*Machining Data Handbook,* 3d ed., 2 vols. Cincinnati: Machinability Data Center, 1980.

Malkin, S., *Grinding Technology: Theory and Applications of Machining with Abrasives.* New York: Wiley, 1989.

Maroney, M. L., *A Guide to Metal and Plastic Finishing.* New York: Industrial Press, 1991.

McKee, R. L., *Machining with Abrasives.* New York: Van Nostrand Reinhold, 1982.

Metzger, J. L., *Superabrasive Grinding.* Stoneham, Mass.: Butterworths, 1986.

Salmon, S. C., *Modern Grinding Process Technology.* New York: McGraw-Hill, 1992.

Schneider, A. F., *Mechanical Deburring and Surface Finishing Technology.* New York: Marcel Dekker, 1990.

Shaw, M. C., *Metal Cutting Principles.* New York: Oxford, 1984.

*Tool and Manufacturing Engineers Handbook,* 4th ed., Vol. 1: *Machining.* Dearborn, Mich.: Society of Manufacturing Engineers, 1983.

Webster, J. A., et al., *Abrasive Processes: Theory, Technology, and Practice.* New York: Marcel Dekker, 1996.

## Nontraditional processes

*ASM Handbook,* Vol. 16: *Machining.* Materials Park, Ohio: ASM International, 1989.

Benedict, G. F., *Nontraditional Manufacturing Processes.* New York: Marcel Dekker, 1987.

Crafer, R. C., and P. J. Oakley, *Laser Processing in Manufacturing.* New York: Chapman and Hall, 1993.

Jain, V. K., and P. C. Pandey, *Theory and Practice of Electrochemical Machining.* New York: Wiley, 1993.

*Machinery Handbook,* revised periodically. New York: Industrial Press.

*Machining Data Handbook,* 3d ed., 2 vols. Cincinnati: Machinability Data Center, 1980.

McGeough, J. A., *Advanced Methods of Machining.* London: Chapman and Hall, 1988.

Sommer, C., and S. Sommer, *Wire EDM Handbook.* Houston: Technical Advanced Publishing Co., 1992.

*Tool and Manufacturing Engineers Handbook,* 4th ed., Vol. 1: *Machining.* Dearborn, Mich.: Society of Manufacturing Engineers, 1983.

Weller, E. J., *Nontraditional Machining Processes,* 2d ed. Dearborn, Mich.: Society of Manufacturing Engineers, 1984.

# QUESTIONS

9.1 Why are grinding operations necessary for parts that have been machined by other processes?

9.2 Explain why there are so many different types and sizes of abrasive wheels.

9.3 Explain the reasons for the large difference between the specific energies involved in grinding (Table 9.3) and machining (Table 8.4).

9.4 Describe the advantages of superabrasives over those for conventional abrasives.

9.5 Give examples of applications for the grinding wheels shown in Table 9.2.

9.6 Explain why the same grinding wheel may act soft or hard.

9.7 Describe your understanding of friability of abrasive grains on grinding-wheel performance.

9.8 Explain the factors involved in selecting the appropriate type of abrasive for a particular grinding operation.

9.9 What are the effects of wear flat on the grinding process?

9.10 The grinding ratio G depends on the following: (1) type of grinding wheel, (2) workpiece hardness, (3) wheel depth of cut, (4) wheel and workpiece speeds, and (5) type of grinding fluid. Explain.

9.11 List and explain the precautions you would take when grinding with high precision. Make comments about the machine, process parameters, the grinding wheel, and grinding fluids.

9.12 Describe the methods you would use to determine the number of active cutting points per unit surface area on the periphery of a straight, i.e., Type 1 (see Fig. 9.2a) grinding wheel. What is the significance of this number?

9.13 Describe and explain the difficulties involved in grinding thermoplastic.

9.14 Explain why ultrasonic machining is not suitable for soft, ductile metals.

9.15 It is generally recommended that a soft-grade grinding wheel be used for hardened steels. Explain why.

9.16 Explain the reasons why the processes described in this chapter can adversely affect the fatigue strength of materials.

9.17 Describe the factors that may cause chatter in grinding operations and give reasons.

9.18 Outline the methods that are generally available for deburring parts.

9.19 In which of the processes described in this chapter are physical properties of the workpiece material important? Why?

9.20 Give all possible technical and economic reasons why material-removal processes described in this chapter may be preferred, or even necessary, over those described in Chapter 8.

9.21 What processes would you recommend for die sinking in a die block? Explain.

9.22 The proper grinding surfaces are shown for each type of wheel with an arrow in Fig. 9.2. Explain why grinding on other surfaces of the wheel is improper and unsafe.

9.23 Note that wheel (b) in Fig. 9.3 has serrations along its periphery, somewhat similar to circular metal saws. What does this design offer in grinding with this wheel?

9.24 In Fig. 9.10, we note that wheel speed and grinding fluids can have a major effect on the type and magnitude of residual stresses developed in grinding. Explain the possible reasons for these phenomena.

9.25 Explain the disadvantages of allowing the workpiece temperature to rise too high in grinding operations.

9.26 Comment on your observations regarding the contents of Table 9.4.

9.27 Why has creep-feed grinding become an important process? Explain.

9.28 There has been a trend in industry to increase the spindle speed of grinding wheels. Explain the possible advantages and limitations of such an increase.

9.29 Why is preshaping or premachining of parts sometimes desirable in the nontraditional machining processes described in this chapter?

9.30 Why are finishing operations sometimes necessary? How could they be minimized? Explain, with examples.

9.31 Why has the wire EDM process become so widely accepted in industry?

9.32 Make a list of material-removal processes, described in this chapter, that may be suitable for the following workpiece materials: (1) ceramics, (2) cast iron, (3) thermoplastics, (4) thermosets, (5) diamond, and (6) annealed copper.

9.33 Explain why producing sharp corners and profiles using some of the processes described in this chapter is difficult.

9.34 How do you think specific energy $u$ and grinding ratio $G$ will vary with respect to wheel depth of cut and hardness of the workpiece material? Explain.

9.35 In Example 9.2, it is stated that the thrust force in grinding is about 30 percent higher than the cutting force. Why is it higher?

9.36 Why should we be interested in the magnitude of the thrust force in grinding? Explain.

9.37 Why is the material-removal rate in electrical-discharge machining a function of the melting point of the workpiece material?

9.38 Inspect Table 9.6 and, for each process, list and describe the role of various mechanical, physical, and chemical properties of the workpiece that affect performance.

9.39 Which of the processes listed in Table 9.6 would not be applicable to non-metallic materials? Explain.

9.40 Why does machining cost increase so rapidly as surface finish requirements become finer?

## PROBLEMS

9.41 In a surface-grinding operation, calculate the chip dimensions for the following process variables: $D = 8$ in., $d = 0.001$ in., $v = 100$ ft/min, $V = 6000$ ft/min, $C = 500$ per in$^2$, and $r = 20$.

9.42 If the workpiece strength in grinding is doubled, what should be the percentage decreases in the wheel depth of cut $d$ in order to maintain the same grain force, all other variables being the same?

9.43 Taking a thin Type 1 grinding wheel as an example and referring to texts on stresses in rotating bodies, plot the tangential stress $\sigma_t$ and radial stress $\sigma_r$ as a function of radial distance, i.e., from the whole to the periphery of the wheel. Because the wheel is thin, it can be regarded as a plane-stress problem. How would you determine the maximum combined stress and its location in the wheel?

9.44 Derive a formula for the material-removal rate, MRR, in surface grinding in terms of process parameters. Use the same terminology as in the text.

9.45 Assume that a surface-grinding operation is being carried out under the following conditions: $D = 200$ mm, $d = 0.1$ mm, $v = 0.4$ m/s, and $V = 30$ m/s. These conditions are then changed to the following: $D = 150$ mm, $d = 0.1$ mm, $v = 0.3$ m/s, and $V = 25$ m/s. How different is the temperature rise from the initial condition?

9.46 In a surface-grinding operation, derive an expression for the power dissipated in imparting kinetic energy to the chips. Comment on the magnitude of this energy. Use the same terminology as in the text.

9.47 The shaft of a Type 1 grinding wheel is attached to a flywheel only, which is rotating at a certain initial rpm. With this setup, a surface-grinding operation is being carried out on a long workpiece at a constant workpiece speed $v$. Obtain an expression for estimating the distance ground on the workpiece before the wheel comes to a stop. Ignore wheel wear.

9.48 Calculate the average impact force on a steel plate by a spherical aluminum oxide abrasive particle of grit size 200, dropped from heights (1) 1 ft, (2) 2 ft, and (3) 10 ft.

9.49 A 40 mm deep hole, 20 mm in diameter, is being produced by electrochemical machining. High production rate is more important than machined surface quality. Estimate the maximum current and the time required to perform this operation.

9.50 If the operation in Problem 9.49 were performed on an electrical-discharge machine, what would be the estimated machining time?

9.51 A cutting-off operation is being performed with a laser beam. The workpiece being cut is 1/2 in. thick and 2 in. long. If the kerf is 1/6 in., estimate the time required to perform this operation.

9.52 Refer to Table 3.3 and identify two metals or metal alloys that, when used as workpiece and electrode, respectively, in EDM would give the (1) lowest and (2) highest wear ratios, $W_r$. Calculate these quantities.

9.53 In Section 9.5.2, we have stated that, in practice, grinding ratios typically range from 2 to 200. Based on the information given in Section 9.13, estimate the range of wear ratios in electrical-discharge machining, then compare them with grinding ratios.

9.54 Assume that you are an instructor covering the topics in this chapter, and you are giving a quiz on the quantitative aspects to test the understanding of the students. Prepare several numerical problems and supply the answers.

# 10

## Properties and Processing of Polymers and Reinforced Plastics

**10.1** Introduction

**10.2** The Structure of Polymers

**10.3** Thermoplastics: Behavior and Properties

**10.4** Thermosets: Behavior and Properties

**10.5** Thermoplastics: Applications

**10.6** Thermosets: Applications

**10.7** High-Temperature Polymers, Electrically Conducting Polymers, and Biodegradable Plastics

**10.8** Elastomers (Rubbers): Properties and Applications

**10.9** Reinforced Plastics (Composite Materials)

**10.10** Processing of Plastics

**10.11** Processing of Reinforced Plastics

**10.12** Design Considerations

**10.13** Economics of Processing Plastics

**10.14** Rapid Prototyping

## 10.1  Introduction

Although the word **plastics,** first used around 1909, is a commonly used synonym for **polymers,** plastics are one of numerous polymeric materials and have extremely large molecules (*macromolecules*). Consumer and industrial products made of polymers include food and beverage containers, packaging, signs, housewares, textiles, medical devices, foams, paints, safety shields, toys, appliances, lenses, gears, electronic and electrical products, and automobile bodies and components. Because of their many unique and diverse properties, polymers have increasingly replaced metallic components in applications such as automobiles, civilian and military aircraft, sporting goods, and office equipment. These substitutions reflect the advantages of polymers in terms of high strength-to-weight ratio, design possibilities, wide choice of colors and transparencies, ease of manufacturing, and relatively low cost.

Compared to metals, polymers are generally characterized by low density, low strength and stiffness (Table 10.1), low electrical and thermal conductivity, good resistance to chemicals, and high coefficient of thermal expansion. However, the useful temperature range for most polymers is generally low—up to about 350°C (660°F)—and they are not as dimensionally stable in service, over a period of time, as metals.

The word *plastics* is from the Greek word *plastikos*, meaning it can be molded and shaped. Plastics can be machined, cast, formed, and joined into many shapes with relative ease. Minimal or no additional surface-finishing operations are required, which is an impor-

**TABLE 10.1**

**RANGE OF MECHANICAL PROPERTIES FOR VARIOUS ENGINEERING PLASTICS AT ROOM TEMPERATURE**

| MATERIAL | UTS (MPa) | $E$ (GPa) | ELONGATION (%) | POISSON'S RATIO ($v$) |
|---|---|---|---|---|
| ABS | 28–55 | 1.4–2.8 | 75–5 | — |
| ABS, reinforced | 100 | 7.5 | — | 0.35 |
| Acetal | 55–70 | 1.4–3.5 | 75–25 | — |
| Acetal, reinforced | 135 | 10 | — | 0.35–0.40 |
| Acrylic | 40–75 | 1.4–3.5 | 50–5 | — |
| Cellulosic | 10–48 | 0.4–1.4 | 100–5 | — |
| Epoxy | 35–140 | 3.5–17 | 10–1 | — |
| Epoxy, reinforced | 70–1400 | 21–52 | 4–2 | — |
| Fluorocarbon | 7–48 | 0.7–2 | 300–100 | 0.46–0.48 |
| Nylon | 55–83 | 1.4–2.8 | 200–60 | 0.32–0.40 |
| Nylon, reinforced | 70–210 | 2–10 | 10–1 | — |
| Phenolic | 28–70 | 2.8–21 | 2–0 | — |
| Polycarbonate | 55–70 | 2.5–3 | 125–10 | 0.38 |
| Polycarbonate, reinforced | 110 | 6 | 6–4 | — |
| Polyester | 55 | 2 | 300–5 | 0.38 |
| Polyester, reinforced | 110–160 | 8.3–12 | 3–1 | — |
| Polyethylene | 7–40 | 0.1–1.4 | 1000–15 | 0.46 |
| Polypropylene | 20–35 | 0.7–1.2 | 500–10 | — |
| Polypropylene, reinforced | 40–100 | 3.5–6 | 4–2 | — |
| Polystyrene | 14–83 | 1.4–4 | 60–1 | 0.35 |
| Polyvinyl chloride | 7–55 | 0.014–4 | 450–40 | — |

tant advantage over metals. Plastics are commercially available as sheet, plate, film, rods, and tubing of various cross-sections.

The word *polymer* was first used in 1866. The earliest polymers were made of *natural organic materials* from animal and vegetable products, cellulose being the most common example. With various chemical reactions, cellulose is modified into cellulose acetate, used in making photographic films (celluloid), sheets for packaging, and textile fibers; cellulose is also modified into cellulose nitrate for plastics, explosives, rayon (a cellulose textile fiber), and varnishes. The earliest *synthetic* (man-made) polymer was a phenol-formaldehyde, a thermoset developed in 1906 and called Bakelite (a trade name, after L. H. Baekeland, 1863-1944).

The development of modern plastics technology began in the 1920s, when raw materials necessary for making polymers were extracted from coal and petroleum products. Ethylene was the first example of such raw material use, as it became the building block for polyethylene. Ethylene is the product of the reaction between acetylene and hydrogen, and acetylene is the product of the reaction between coke and methane. Likewise, commercial polymers including polypropylene, polyvinyl chloride, polymethylmethacrylate, polycarbonate, and others are all made in a similar manner; these materials are known as *synthetic organic*

**FIGURE 10.1**   Basic structure of polymer molecules: (a) ethylene molecule; (b) polyethylene, a linear chain of many ethylene molecules; (c) molecular structure of various polymers. These are examples of the basic building blocks for plastics.

*polymers.* An outline of the basic process of making various synthetic polymers is illustrated in Fig. 10.1. Although in polyethylene only carbon and hydrogen atoms are involved, other polymer compounds can be obtained with chlorine, fluorine, sulfur, silicon, nitrogen, and oxygen. As a result, an extremely wide range of polymers with an equally wide range of properties have been developed.

## 10.2   The Structure of Polymers

Many of the polymer properties depend largely on the structure of individual polymer molecules, molecule shape and size, and how they are arranged to form a polymer structure. Polymer molecules are characterized by their extraordinary size, a feature that distinguishes them from other organic chemical compositions. Polymers are *long-chain molecules,* also called *macromolecules* or *giant molecules,* which are formed by polymerization, that is, by linking and cross-linking of different monomers. A **monomer** is the basic building block of polymers. The word **mer,** from the Greek *meros,* meaning part, indicates the smallest repetitive unit, similar to the term unit cell used in connection with crystal structures of metals.

The term *poly-mer* means many mers or units, generally repeated hundreds or thousands of times in a chainlike structure. Most monomers are organic materials in which carbon atoms are joined in *covalent* (electron-sharing) bonds with other atoms, such as hydrogen, oxygen, nitrogen, fluorine, chlorine, silicon, and sulfur. An ethylene molecule, shown in Fig. 10.1a, is a simple monomer consisting of carbon and hydrogen atoms.

### 10.2.1   Polymerization

Monomers in polymers can be linked in repeating units to make longer and larger molecules by a chemical reaction known as the **polymerization** reaction. Polymerization processes are complex and can be described only briefly here. Although there are many variations, two basic polymerization processes are condensation and addition polymerization.

In *condensation polymerization,* polymers are produced by the formation of bonds between two types of reacting mers. One characteristic of this reaction is that reaction byproducts such as water are condensed out, hence the term condensation. This process is also known as *step-growth* or *step-reaction polymerization* because the polymer molecule grows step by step until all of one reactant is consumed.

In *addition polymerization,* also known as *chain-growth* or *chain-reaction polymerization,* bonding takes place without reaction by-products. It is called chain-reaction because of the high rate at which long molecules form simultaneously, usually within a few seconds. This rate is much higher than that for condensation polymerization. In this reaction, an initiator is added to open the double bond between the carbon atoms and begins the linking process by adding many more monomers to a growing chain. For example, ethylene monomers (Fig. 10.1a) link to produce a polymer known as polyethylene (Fig. 10.1b). Other examples of addition-formed polymers are shown in Fig. 10.1c.

**Molecular weight.**   The sum of the molecular weight of the mers in the polymer chain is the **molecular weight** of the polymer. The higher the molecular weight in a given polymer, the greater the chain length. Because polymerization is a random event, the polymer chains produced are not all of equal length, but the chain lengths produced fall into a traditional distribution curve. We determine and express the average molecular weight of a polymer on a statistical basis by averaging. The spread observed in the weight distribution is referred to as the *molecular weight distribution* (MWD). Molecular weight and MWD have a strong influence on the properties of the polymer. For example, tensile and impact strength, resistance to cracking, and viscosity in the molten state all increase with increasing molecular weight (Fig. 10.2). Most commercial polymers have a molecular weight between 10,000 and 10,000,000.

**Degree of polymerization.**   In some cases, it is more convenient to express the size of a polymer chain in terms of the **degree of polymerization (DP)**, defined as the ratio of the molecular weight of the polymer to the molecular weight of the repeat unit. In terms of polymer processing (see Section 10.10), the higher the DP, the greater is the polymer's viscosity, or resistance to flow (Fig. 10.2), thus affecting ease of shaping and overall cost of processing.

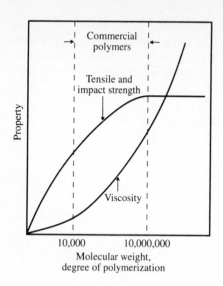

**Bonding.** During polymerization the monomers are linked together in a *covalent bond*, forming a polymer chain. Because of their strength, covalent bonds are also called **primary bonds.** The polymer chains are, in turn, held together by the secondary bonds, such as van der Waals bonds, hydrogen bonds, and ionic bonds. **Secondary bonds** are much weaker than primary bonds, by one to two orders of magnitude. In a given polymer, the increase in strength and viscosity with molecular weight comes in part from the fact that the longer the polymer chain, the greater is the energy needed to overcome the secondary bond strength. For example, ethylene mers having a DP of 1, 6, 35, 140, and 1350 are, respectively, in the form of gas, liquid, grease, wax, and hard plastic at room temperature.

**Linear polymers.** The chainlike polymers shown in Fig. 10.1 are called **linear polymers** because of their linear structure (Fig. 10.3a). A linear molecule is not straight in shape. In addition to those shown, other linear polymers are polyamides (nylon 6,6) and polyvinyl fluoride. Generally, a polymer consists of more than one type of structure. Thus a linear polymer may contain some branched and cross-linked chains. As a result of branching and cross-linking, the polymer's properties can change.

---

**Example 10.1:** Degree of polymerization in polyvinyl chloride (PVC)

Determine the molecular weight of a polyvinyl chloride mer. If a PVC polymer has an average molecular weight of 50,000, what is the degree of polymerization?

SOLUTION. From Fig. 10.1c, we note that each PVC mer has 3 hydrogen atoms, 2 carbon atoms, and 1 chlorine atom. Since the atomic number of each element is 1, 12, and 35.5, respectively, we have the weight of a PVC mer as $(3)(1) + (2)(12) + (1)(35.5) = 62.5$. Hence the degree of polymerization is $50,000/62.5 = 800$.

---

**Branched polymers.** The properties of a polymer depend not only on the type of monomers, but also on their arrangement in the molecular structure. In **branched polymers**

(a) Linear

(b) Branched

(c) Cross-linked

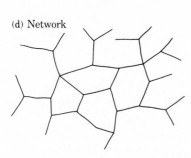

(d) Network

**FIGURE 10.3** Schematic illustration of polymer chains. (a) Linear structure. Thermoplastics such as acrylics, nylons, polyethylene, and polyvinyl chloride have linear structures. (b) Branched structure, such as in polyethylene. (c) Cross-linked structure. Many rubbers or elastomers have this structure. Vulcanization of rubber produces this structure. (d) Network structure, which is basically highly cross-linked. Examples are thermosetting plastics, such as epoxies and phenolics.

(Fig. 10.3b), side-branch chains are attached to the main chain during the synthesis of the polymer. Branching interferes with the relative movement of the molecular chains. As a result, resistance to deformation and stress-crack resistance are affected. Also, the density of branched polymers is lower than that of linear-chain polymers as branches interfere with the packing efficiency of polymer chains. The behavior of branched polymers can be compared to that of linear-chain polymers by making an analogy with a pile of tree branches (branched polymers) and a bundle of straight logs (linear). You will note that it is more difficult to move a branch within the pile of branches than to move a log in its bundle. The three-dimensional entanglements of branches make movements more difficult, a phenomenon akin to increased strength.

**Cross-linked polymers.** Generally three-dimensional in structure, **cross-linked polymers** have adjacent chains linked by covalent bonds (Fig. 10.3c). Polymers with cross-linked chain structure are called **thermosets,** or *thermosetting* plastics, such as epoxies, phenolics, and silicones. Cross-linking has a major influence on the properties of polymers (generally imparting hardness, strength, stiffness, brittleness, and better dimensional stability; see Fig. 10.4), as well as in the *vulcanization* of rubber. **Network polymers** consist of spatial (three-dimensional) networks of three active covalent bonds (Fig. 10.3d). A highly cross-linked polymer is also considered a network polymer. Thermoplastic polymers that have already been formed or shaped can be cross linked to obtain greater strength by subjecting them to high-energy radiation, such as ultraviolet light, x-rays, or electron beams. However, excessive radiation can cause degradation of the polymer.

**Copolymers and terpolymers.** If the repeating units in a polymer chain are all of the same type, we call the molecule a *homopolymer.* However, as with solid-solution metal alloys, two or three different types of monomers can be combined to impart certain special properties and characteristics to the polymer, such as improving both the strength and toughness as well as the formability of the polymer. *Copolymers* contain two types of polymers, such as styrene-butadiene, used widely for automobile tires. *Terpolymers* contain three types, such as ABS (acrylonitrile-butadiene-styrene) used for helmets, telephones, and refrigerator liners.

### 10.2.2 Crystallinity

Polymers such as polymethylmethacrylate, polycarbonate, and polystyrene are generally *amorphous;* that is, the polymer chains exist without long-range order (see also amorphous alloys in Section 3.11.8). The amorphous arrangement of polymer chains is often described as a bowl of spaghetti, or worms in a bucket, all intertwined with each other. However, in some polymers it is possible to impart some crystallinity and thereby modify their characteristics. This may be done either during the synthesis of the polymer or by deformation during its subsequent processing.

The crystalline regions in polymers are called **crystallites** (Fig. 10.5). These crystals are formed when the long molecules arrange themselves in an orderly manner, similar to folding a fire hose in a cabinet or facial tissues in a box. Thus we can regard a partially crystalline (semi-crystalline) polymer as a two-phase material, one phase being crystalline and the other amorphous.

By controlling the rate of solidification during cooling and chain structure, it is possible to impart different degrees of crystallinity to polymers, although never completely 100 percent. Crystallinity ranges from an almost complete crystal, up to about 95 percent by volume in the case of polyethylene, to slightly crystallized but mostly amorphous polymers.

**Effects of crystallinity.** The mechanical and physical properties of polymers are greatly influenced by the *degree of crystallinity.* As it increases, polymers become stiffer, harder, less ductile, more dense, less rubbery, and more resistant to solvents and heat (Fig. 10.4). The in-

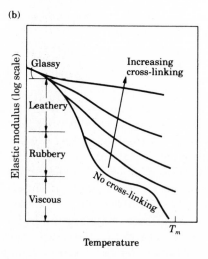

**FIGURE 10.4**  Behavior of polymers as a function of temperature and (a) degree of crystallinity and (b) cross-linking. The combined elastic and viscous behavior of polymers is known as viscoelasticity.

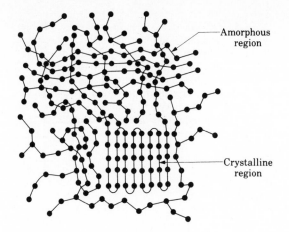

Amorphous region

Crystalline region

**FIGURE 10.5** Amorphous and crystalline regions in a polymer. The crystalline region (crystallite) has an orderly arrangement of molecules. The higher the crystallinity, the harder, stiffer, and less ductile is the polymer.

crease in density with increasing crystallinity is caused by crystallization shrinkage and a more efficient packing of the molecules in the crystal lattice. For example, the highly crystalline form of polyethylene, known as high-density polyethylene (HDPE), has a specific gravity in the range of 0.941 to 0.970 (80 to 95 percent crystalline) and is stronger, stiffer, tougher, and less ductile than low-density polyethylene (LDPE), which is about 60 to 70 percent crystalline and has a specific gravity of about 0.910 to 0.925.

Optical properties are also affected by the degree of crystallinity. The reflection of light from the boundaries between the crystalline and amorphous regions in the polymer causes opaqueness. Furthermore, because the index of refraction is proportional to density, the greater the density difference between the amorphous and crystalline phases, the greater is the opaqueness of the polymer. Polymers that are completely amorphous can be transparent, such as polycarbonate and acrylics.

### 10.2.3 Glass-transition temperature

Amorphous polymers do not have a specific melting point, but they undergo a distinct change in their mechanical behavior across a narrow range of temperature. At low temperatures they are hard, rigid, brittle, and glassy and at high temperatures are rubbery or leathery. The temperature at which this transition occurs is called the **glass-transition temperature**, $T_g$, and is also called the *glass point* or *glass temperature*. The term glass is included in this definition because glasses, which are amorphous solids, behave in the same manner (as you can see by holding a glass rod over a flame and observing its behavior).

Although most amorphous polymers exhibit this behavior, there are some exceptions, such as polycarbonate, which is not rigid or brittle below its glass-transition temperature. Polycarbonate is tough at ambient temperature and is thus used for safety helmets and shields.

To determine $T_g$, we measure the specific volume of the polymer and plot it against temperature to find the sharp change in the slope of the curve (Fig. 10.6). However, in the case of highly cross-linked polymers, the slope of the curve changes gradually near $T_g$, making it difficult to determine $T_g$ for these polymers. The glass-transition temperature varies with different polymers (Table 10.2). For example, room temperature is above $T_g$ for some polymers

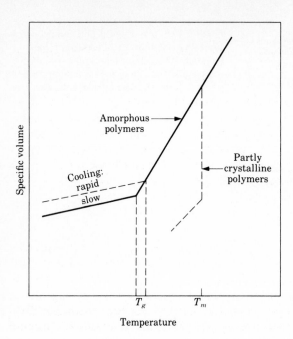

**FIGURE 10.6**   Specific volume of polymers as a function of temperature. Amorphous polymers, such as acrylic and polycarbonate, have a glass-transition temperature, $T_g$, but do not have a specific melting point, $T_m$. Partly crystalline polymers, such as polyethylene and nylons, contract sharply at their melting points during cooling.

and below it for others. Unlike amorphous polymers, partly crystalline polymers have a distinct melting point, $T_m$ (Fig. 10.6; see also Table 10.2). Because of the structural changes (first-order changes) occurring, the specific volume of the polymer drops suddenly as its temperature is reduced.

**TABLE 10.2**

**GLASS-TRANSITION AND MELTING TEMPERATURES OF SOME POLYMERS**

| MATERIAL | $T_g$ (°C) | $T_m$ (°C) |
|---|---|---|
| Nylon 6,6 | 57 | 265 |
| Polycarbonate | 150 | 265 |
| Polyester | 73 | 265 |
| Polyethylene | | |
|    High density | −90 | 137 |
|    Low density | −110 | 115 |
| Polymethylmethacrylate | 105 | — |
| Polypropylene | −14 | 176 |
| Polystyrene | 100 | 239 |
| Polytetrafluoroethylene | −90 | 327 |
| Polyvinyl chloride | 87 | 212 |
| Rubber | −73 | — |

### 10.2.4   Blends

To improve the brittle behavior of amorphous polymers below their glass-transition temperature, we can blend them (mix them with another polymer), usually with small quantities of an **elastomer.** These tiny particles are dispersed throughout the amorphous polymer, enhancing its toughness and impact strength by improving its resistance to crack propagation. These are known as *rubber modified polymers.* More recent trends in blending involve several components, or **polyblends,** that utilize the favorable properties of different polymers. Some advances have been made in *miscible blends* (mixing without separation of two phases), a process similar to alloying of metals, enabling polymer blends to become more ductile. Polymer blends account for about 20 percent of all polymer production.

### 10.2.5   Additives

In order to impart certain specific properties, polymers are usually compounded with **additives.** These additives modify and improve certain characteristics of polymers, such as their stiffness, strength, color, weatherability, flammability, arc resistance for electrical applications, and ease of subsequent processing.

**Fillers** used are generally wood flour (fine sawdust), silica flour (fine silica powder), clay, powdered mica, and short fibers of cellulose, glass, and asbestos. Because of their low cost, fillers are important in reducing the overall cost of polymers. Depending on their type, fillers improve the strength, hardness, toughness, abrasion resistance, dimensional stability, and/or stiffness of plastics. These properties are greatest at various percentages of different types of polymer–filler combinations. As in reinforced plastics, a filler's effectiveness depends on the nature of the bond between the filler material and the polymer chains.

**Plasticizers** are added to some polymers to impart flexibility and softness by lowering their glass-transition temperature. Plasticizers are low molecular weight solvents with high boiling points (nonvolatile). They reduce the strength of the secondary bonds between the long-chain molecules, thus making the polymer soft and flexible. The most common use of plasticizers is in polyvinyl chloride (PVC), which remains flexible during its many uses. Other applications of plasticizers are in thin sheet, film, tubing, shower curtains, and clothing materials.

Most polymers are adversely affected by ultraviolet radiation (sunlight) and oxygen, which weaken and break the primary bonds, resulting in the scission (splitting) of the long-chain molecules. The polymer then degrades and becomes brittle and stiff. On the other hand, degradation may be beneficial, as in the disposal of plastic objects by subjecting them to environmental attack.

A typical example of protection against ultraviolet radiation is the compounding of certain plastics and rubber with *carbon black* (soot). The carbon black absorbs a high percentage of the ultraviolet radiation. Protection against degradation by oxidation, particularly at elevated temperatures, is done by adding antioxidants to the polymer. Various coatings are another means of protecting polymers.

The wide variety of colors available in plastics is obtained by adding **colorants.** These materials are either organic (dyes) or inorganic (pigments). The selection of a colorant depends on service temperature and exposure to light. Pigments are dispersed particles and generally have greater resistance than dyes to temperature and light.

If the temperature is sufficiently high, most polymers will ignite. The *flammability* (ability to support combustion) of polymers varies considerably, depending on their composition (such as the chlorine and fluorine content). Polymethylmethacrylate, for example, continues to burn when ignited, whereas polycarbonate self-extinguishes. The flammability of polymers can be reduced either by making them from less flammable raw materials, or by the addition of **flame retardants,** such as compounds of chlorine, bromine, and phosphorus.

**Lubricants** may be added to polymers to reduce friction during their subsequent processing into useful products and to prevent parts from sticking to the molds. Lubrication is also important in preventing thin polymer films from sticking to each other.

## 10.3   Thermoplastics: Behavior and Properties

We noted earlier that the bonds between adjacent long-chain molecules (secondary bonds) are much weaker than the covalent bonds (primary bonds) within each molecule. Also, it is the strength of the secondary bonds that determines the overall strength of the polymer. Linear and branched polymers have weak secondary bonds.

For certain polymers, as the temperature is raised above the glass-transition temperature $T_g$ or melting point $T_m$, we find that it becomes easier to form or mold them into desired shapes. The increased temperature weakens the secondary bonds via thermal vibration of the long molecules, and the adjacent chains can thus move more easily under the shaping forces. If we then cool the polymer, it returns to its original hardness and strength. In other words, the process is reversible. Polymers that exhibit this behavior are known as **thermoplastics.** Typical examples are acrylics, cellulosics, nylons, polyethylenes, and polyvinyl chloride.

In addition to the structure and composition of thermoplastics, their behavior depends on numerous variables. Among the most important are temperature and rate of deformation. Below the glass-transition temperature, $T_g$, most polymers are glassy (described as rigid, brittle, or hard), and they behave like an elastic solid. If the load exceeds a certain critical value the polymer fractures, just as a piece of glass does at room temperature. In the glassy region, the relationship between stress and strain is linear, or

$$\sigma = E\varepsilon. \tag{10.1}$$

If we test the polymer in torsion, we have

$$\tau = G\gamma. \tag{10.2}$$

The glassy behavior can be represented by a spring having a stiffness equivalent to the modulus of elasticity of the polymer (Fig. 10.7a). Note that the strain is completely recovered when the load is removed at time $t_1$.

When the applied stress is increased, the polymer fractures, just as a piece of glass does at ambient temperature. The mechanical properties of several polymers listed in Table 10.1 indicate that thermoplastics are about two orders of magnitude less stiff than metals. Their ultimate tensile strength is about one order of magnitude lower than that of metals (see Table 2.1).

**FIGURE 10.7** Various deformation modes for polymers: (a) elastic; (b) viscous; (c) viscoelastic (Maxwell model); and (d) viscoelastic (Voigt or Kelvin model).

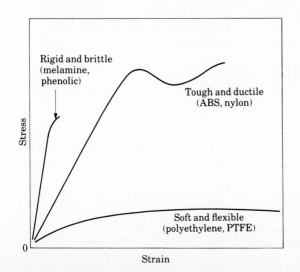

**FIGURE 10.8** General terminology describing the behavior of three types of plastics. PTFE (polytetrafluoroethylene) is Teflon, a trade name. *Source:* R.L.E. Brown.

**FIGURE 10.9**   Effect of temperature on the stress-strain curve for cellulose acetate, a thermoplastic. Note the large drop in strength and increase in ductility with a relatively small increase in temperature. *Source:* After T. S. Carswell and H. K. Nason.

Typical stress-strain curves for some thermoplastics and thermosets at room temperature are shown in Fig. 10.8. Note that these plastics exhibit different behaviors, which we may describe as rigid, soft, brittle, flexible, and so on. Plastics, like metals, undergo fatigue and creep phenomena.

**Effects of temperature and rate of deformation.**   The typical effects of temperature on the strength and elastic modulus of thermoplastics are similar to those for metals. Thus with increasing temperature, the strength and modulus of elasticity decrease, and toughness increases (Fig. 10.9). The effect of temperature on impact strength is shown in Fig. 10.10. Note the large difference in the impact behavior of various polymers.

**FIGURE 10.10**   Effect of temperature on the impact strength of various plastics. Small changes in temperature can have a significant effect on impact strength. *Source:* P. C. Powell.

**FIGURE 10.11** Parameters used to describe viscosity; see Eq. (10.3).

If we raise the temperature of a thermoplastic polymer above its $T_g$, it first becomes leathery, and then rubbery with increasing temperature (Fig. 10.4). Finally, at higher temperatures, e.g., above $T_m$ for crystalline thermoplastics, it becomes a viscous fluid, with viscosity decreasing with increasing temperature.

This is known as *viscoelastic* behavior and can be shown by the spring and dashpot models in Fig. 10.7c and d, known as the *Maxwell* and *Kelvin* (or *Voigt*) models, respectively. When a constant load is applied, the polymer first stretches at a high strain rate, and then continues to elongate over a period of time because of its viscous behavior. Note in these models that the elastic portion of the elongation is reversible (elastic recovery), but the viscous portion is not.

The viscous behavior is expressed by

$$\tau = \eta(dv/dy) = \eta\dot{\gamma} \tag{10.3}$$

where $\eta$ is the *viscosity,* and $dv/dy$ is the *shear strain rate* $\dot{\gamma}$, as shown in Fig. 10.11. When the shear stress $\tau$ is directly proportional to the shear strain rate, the behavior is known as *Newtonian.* Note from Eq. (10.3) that the viscous behavior is similar to the strain rate sensitivity of metals, described in Section 2.2.7, and given by the expression

$$\sigma = C\dot{\epsilon}^m \tag{10.4}$$

where we note that for Newtonian behavior, $m = 1$. Thermoplastics have high $m$ values, indicating that they can undergo large uniform deformations in tension before fracture. Note how, unlike ordinary metals, the necked region elongates considerably (Fig. 10.12). This behavior can easily be demonstrated by stretching a piece of the plastic (polypropylene) holder for 6-pack beverage cans. This characteristic, which is the same as in superplastic metals, enables thermoplastics to be formed into complex shapes, such as bottles for soft drinks, cookie and meat trays, and lighted signs, as described in Section 10.10. Note also that the effect of increasing the rate of load application is to increase the strength of the polymer.

Between $T_g$ and $T_m$, thermoplastics also exhibit leathery and rubbery behavior, depending on their structure and degree of crystallinity, as shown in Fig. 10.4. A term combining the strains caused by elastic behavior ($e_e$) and viscous flow ($e_v$) is called the *viscoelastic modulus* $E_r$, and is expressed as

$$E_r = \frac{\sigma}{(e_e + e_v)}. \tag{10.5}$$

**FIGURE 10.12** (a) Load-elongation curve for polycarbonate, a thermoplastic. *Source:* R. P. Kambour and R. E. Robertson. (b) High-density polyethylene tensile-test specimen, showing uniform elongation (the long, narrow region in the specimen).

This modulus, shown as the ordinate in Fig. 10.4, essentially represents a time-dependent elastic modulus.

The viscosity $\eta$ of polymers—a measure of the resistance of their molecules in sliding along each other—depends on temperature and on their structure, molecular weight, and pressure. The effect of temperature can be represented by

$$\eta = \eta_0 e^{E/kT} \qquad (10.6)$$

where $\eta_0$ is a material constant, $E$ is the activation energy (energy required to initiate a reaction), $k$ is Boltzmann's constant (thermal energy constant, or $13.8 \times 10^{-24}$ J/K), and $T$ is the temperature (K). Thus as the temperature of the polymer is increased, $\eta$ decreases because of the higher mobility of the molecules. Note that viscosity may be regarded as the inverse of fluidity of molten metals in casting. Increasing the molecular weight (hence increasing chain length) increases $\eta$ because of the greater number of secondary bonds present. As the molecular weight distribution widens, there are more shorter chains and $\eta$ decreases. The effect of increasing pressure is to increase viscosity because of reduced *free volume* or *free space,* defined as the volume in excess of the true volume of the crystal in the crystalline regions of the polymer.

Based on experimental observations that, at the glass-transition temperature $T_g$, polymers have a viscosity $\eta$ of about $10^{12}$ Pa·s, an empirical relationship between viscosity and temperature has been developed for linear thermoplastics:

$$\log \eta = 12 - \frac{17.5 \, \Delta T}{52 + \Delta T}, \tag{10.7}$$

where $\Delta T = T - T_g$, in K or °C. Thus we can estimate the viscosity of the polymer at any temperature.

---

**Example 10.2: Lowering the viscosity of a polymer**

In processing a batch of polycarbonate at 170°C to make a part, we find that its viscosity is twice that desired. We want to determine the temperature at which this polymer should be processed.

SOLUTION.   From Table 10.2, we find that $T_g$ for polycarbonate is 150°C. We find its viscosity at 170°C by using Eq. (10.7):

$$\log \eta = 12 - \frac{17.5(20)}{52 + 20} = 7.14.$$

Hence $\eta = 13.8$ MPa·s. Because this magnitude is twice what we want, the new viscosity should be 6.9 MPa·s. Therefore

$$\log (6.9 \times 10^6) = 12 - \frac{17.5(\Delta T)}{52 + \Delta T},$$

and $\Delta T = 21.7$, or the new temperature is $150 + 21.7 = 171.7$°C. Note that we rounded the numbers for temperature and that viscosity is very sensitive to temperature.

---

**Creep and stress relaxation.**   We defined the terms creep and stress relaxation in Section 2.8. Because of their viscoelastic behavior, thermoplastics are particularly susceptible to these phenomena. Referring back to the dashpot models in Fig. 10.7b, c, and d, we note that under a constant load, the polymer undergoes further strain—thus it creeps. The recovered strain depends on the stiffness of the spring, hence the modulus of elasticity.

Stress relaxation in polymers occurs over a period of time, according to

$$\tau = \tau_0 e^{-t/\lambda}, \tag{10.8}$$

where $\tau_0$ is the shear stress at time zero, $\tau$ is the stress at time $t$, and $\lambda$ is the *relaxation time.* This expression may also be given in terms of the normal stresses $\sigma$ and $\sigma_0$. The relaxation time is defined as

$$\lambda = \eta / G, \tag{10.9}$$

where $G$ is the shear modulus of the polymer and $\gamma = \tau/G$. Thus the relaxation time is a function of the material and its temperature. As both $\eta$ and $G$ depend on temperature to varying degrees, the relaxation time also depends on temperature.

---

**Example 10.3:   Stress relaxation in a thermoplastic member under tension**

A piece of thermoplastic is stretched between two rigid supports at a stress level of 5 MPa. After 30 days, the stress level has decayed to half the original level. How long will it take for the stress level to reach one-tenth of the original value?

SOLUTION.   Substituting these data into Eq. (10.8) as modified for normal stress, we find that

$$2.5 = 5e^{-30/\lambda}.$$

Hence

$$\ln\left(\frac{2.5}{5}\right) = \frac{-30}{\lambda}, \quad \text{or} \quad \lambda = 43.3 \text{ days}.$$

Therefore

$$\ln\left(\frac{0.5}{5}\right) = \frac{-t}{43.3}, \quad \text{or} \quad t = 99.7 \text{ days}.$$

---

**Orientation.**   When thermoplastics are deformed, say by stretching, the long-chain molecules align in the general direction of elongation. This process is called **orientation** and, just as in metals, the polymer becomes anisotropic. The specimen becomes stronger and stiffer in the elongated (stretched) direction than in its transverse direction. This is an important technique to enhance the strength and toughness of polymers. However, orientation weakens the polymer in the transverse direction.

**Crazing.**   Some thermoplastics, such as polystyrene and polymethylmethacrylate, develop localized, wedge-shaped, narrow regions of highly deformed material when subjected to tensile stresses or to bending. This phenomenon is called **crazing.** Although they may appear to be like cracks, crazes are spongy material, typically containing about 50 percent voids. With increasing tensile load on the specimen, these voids coalesce to form a crack and eventually lead to fracture of the polymer. Crazing has been observed both in transparent glassy polymers and in other types.

The environment and the presence of solvents, lubricants, and water vapor enhance the formation of crazes (environmental stress cracking and solvent crazing). Residual stresses in the material contribute to crazing and cracking of the polymer. Radiation and ultraviolet radiation can adversely affect the crazing behavior in certain polymers.

A related phenomenon is **stress whitening.** When subjected to tensile stresses, such as by folding or bending, the plastic becomes lighter in color. This phenomenon is usually attributed to the formation of microvoids in the material. As a result, the material becomes less translucent (transmits less light), or more opaque. You can easily demonstrate this result by bending plastic components commonly found in household products and toys.

**Water absorption.**   An important limitation of some polymers, such as nylons, is their ability to absorb water (*hygroscopy*). Water acts as a plasticizing agent; that is, it makes the polymer more plastic. Thus in a sense, it lubricates the chains in the amorphous region. Typically, with increasing moisture absorption, the glass-transition temperature, the yield stress, and the elastic modulus of the polymer are lowered severely. Dimensional changes also occur because of water absorption, such as in a humid environment.

**Thermal properties.**   Compared to metals, plastics are generally characterized by low thermal and electrical conductivity, low specific gravity (ranging from 0.90 to 2.2), and relatively high coefficient of thermal expansion, which is about an order of magnitude higher.

# 10.4   Thermosets: Behavior and Properties

When the long-chain molecules in a polymer are cross-linked in a three-dimensional arrangement, the structure in effect becomes one giant molecule with strong covalent bonds. As we previously stated, these polymers are called thermosetting polymers, or thermosets, because during polymerization the network is completed and the shape of the part is permanently set. This *curing* (*cross-linking*) reaction, unlike that of thermoplastics, is irreversible. We can liken the response of a thermosetting plastic to temperature to baking a cake or boiling an egg. Once the cake is baked and cooled, or the egg boiled and cooled, reheating it will not change its shape.

Some thermosets, such as epoxy, polyester, and urethane, cure at room temperature. Although curing takes place at ambient temperature, the heat of the exothermic reaction cures the plastic. Thermosetting polymers do not have a sharply defined glass-transition temperature. The polymerization process for thermosets generally takes place in two stages. The first is at the chemical plant where the molecules are partially polymerized into linear chains. The second stage is at the parts-producing plant where cross-linking is completed under heat and pressure during the molding and shaping of the part.

Because of the nature of their bonds, the strength and hardness of the thermosets, unlike thermoplastics, are not affected by temperature or rate of deformation. A typical thermoset is phenolic, which is a product of the reaction between phenol and formaldehyde. Common products made from this polymer are the handles and knobs on cooking pots and pans and components of light switches and outlets. Thermosetting plastics generally possess better mechanical, thermal, and chemical properties; electrical resistance; and dimensional stability than do thermoplastics. However, if the temperature is increased sufficiently, the thermosetting polymer begins to burn up, degrade, and char.

# 10.5   Thermoplastics: Applications

This section outlines the general characteristics and typical applications of major thermoplastics, particularly as they relate to manufacturing plastic products and their service life. General recommendations for various plastics applications are given in Table 10.3.

**Acetals** (from acetic and alcohol) have good strength, stiffness, and resistance to creep, abrasion, moisture, heat, and chemicals. Typical applications are mechanical parts and components where high performance is required over a long period: bearings, cams, gears,

**TABLE 10.3**

**GENERAL RECOMMENDATIONS FOR PLASTIC PRODUCTS**

| DESIGN REQUIREMENT | APPLICATIONS | PLASTICS |
|---|---|---|
| Mechanical strength | Gears, cams, rollers, valves, fan blades, impellers, pistons | Acetal, nylon, phenolic, polycarbonate |
| Functional and decorative | Handles, knobs, camera and battery cases, trim moldings, pipe fittings | ABS, acrylic, cellulosic, phenolic, polyethylene, polypropylene, polystyrene, polyvinyl chloride |
| Housings and hollow shapes | Power tools, pumps, housings, sport helmets, telephone cases | ABS, cellulosic, phenolic, polycarbonate, polyethylene, polypropylene, polystyrene |
| Functional and transparent | Lenses, goggles, safety glazing, signs, food-processing equipment, laboratory hardware | Acrylic, polycarbonate, polystyrene, polysulfone |
| Wear resistance | Gears, wear strips and liners, bearings, bushings, roller-skate wheels | Acetal, nylon, phenolic, polyimide, polyurethane, ultrahigh-molecular-weight polyethylene |

bushings, rollers, impellers, wear surfaces, pipes, valves, shower heads, and housings. Common trade name: Delrin.

**Acrylics (polymethylmethacrylate, PMMA)** possess moderate strength, good optical properties, and weather resistance. They are transparent but can be made opaque, are generally resistant to chemicals, and have good electrical resistance. Typical applications are lenses, lighted signs, displays, window glazing, skylights, bubble tops, automotive lenses, windshields, lighting fixtures, and furniture. Common trade names: Plexiglas, Lucite.

**Acrylonitrile-butadiene-styrene (ABS)** is dimensionally stable and rigid and has good impact, abrasion, and chemical resistance; strength and toughness; low-temperature properties; and electrical resistance. Typical applications are pipes, fittings, chrome-plated plumbing supplies, helmets, tool handles, automotive components, boat hulls, telephones, luggage, housing, appliances, refrigerator liners, and decorative panels.

**Cellulosics** have a wide range of mechanical properties, depending on composition. They can be made rigid, strong, and tough. However, they weather poorly and are affected by heat and chemicals. Typical applications are tool handles, pens, knobs, frames for eyeglasses, safety goggles, machine guards, helmets, tubing and pipes, lighting fixtures, rigid containers, steering wheels, packaging film, signs, billiard balls, toys, and decorative parts.

**Fluorocarbons** possess good resistance to temperature, chemicals, weather, and electricity. They also have unique nonadhesive properties and low friction. Typical applications are linings for chemical process equipment, nonstick coatings for cookware, electrical insulation for high-temperature wire and cable, gaskets, low-friction surfaces, bearings, and seals. Common trade name: Teflon.

**Polyamides** (from the words poly, amine, and carboxyl acid) are available in two main types: nylons and aramids. **Nylons** (a coined word) have good mechanical properties and abrasion resistance. They are self-lubricating and resistant to most chemicals. All nylons are hygroscopic (absorb water). Moisture absorption reduces mechanical properties and increases part dimensions. Typical applications are gears, bearings, bushings, rollers, fasteners, zippers, electrical parts, combs, tubing, wear-resistant surfaces, guides, and surgical equipment. **Aramids** (aromatic polyamides) have very high tensile strength and stiffness. Typical applications are fibers for reinforced plastics (composite materials), bulletproof vests, cables, and radial tires. Common trade name: Kevlar.

**Polycarbonates** are versatile and have good mechanical and electrical properties. They also have high impact resistance, and can be made resistant to chemicals. Typical applications are safety helmets, optical lenses, bullet-resistant window glazing, signs, bottles, food-processing equipment, windshields, load-bearing electrical components, electrical insulators, medical apparatus, business machine components, guards for machinery, and parts requiring dimensional stability. Common trade name: Lexan.

**Polyesters** (thermoplastic; see also Section 10.6) have good mechanical, electrical, and chemical properties; good abrasion resistance; and low friction. Typical applications are gears, cams, rollers, load-bearing members, pumps, and electromechanical components. Common trade names: Dacron, Mylar, Kodel.

**Polyethylenes** possess good electrical and chemical properties. Their mechanical properties depend on composition and structure. Three major classes are low density (LDPE), high density (HDPE), and ultrahigh molecular weight (UHMWPE). Typical applications for LDPE and HDPE are housewares, bottles, garbage cans, ducts, bumpers, luggage, toys, tubing, bottles, and packaging material; UHMWPE are used in parts requiring high-impact toughness and abrasive wear resistance.

**Polyimides** have the structure of a thermoplastic but the nonmelting characteristic of a thermoset. See Section 10.6.

**Polypropylenes** have good mechanical, electrical, and chemical properties and good resistance to tearing. Typical applications are automotive trim and components, medical devices, appliance parts, wire insulation, TV cabinets, pipes, fittings, drinking cups, dairy-product and juice containers, luggage, ropes, and weather stripping.

**Polystyrenes** are inexpensive, have generally average properties, and are somewhat brittle. Typical applications are disposable containers; packaging; trays for meats, cookies, and candy; foam insulation; appliances; automotive and radio/TV components; housewares; and toys and furniture parts (as a wood substitute).

**Polysulfones** have excellent resistance to heat, water, and steam; are highly resistant to some chemicals; but are attacked by organic solvents. Typical applications are steam irons, coffeemakers, hot water containers, medical equipment that requires sterilization, power-tool and appliance housings, aircraft cabin interiors, and electrical insulators.

**Polyvinyl chloride (PVC)** has a wide range of properties, is inexpensive and water-resistant, and can be made rigid or flexible. It is not suitable for applications requiring strength and heat resistance. Rigid PVC is tough and hard. Rigid PVC is used for signs and in the construction industry, such as pipes and conduits. Flexible PVC is used in wire and cable coatings, low-pressure flexible tubing and hose, footwear, imitation leather, upholstery, records, gaskets, seals, trim, film, sheet, and coatings. Common trade names: Saran, Tygon.

# 10.6 Thermosets: Applications

This section outlines the general characteristics and typical applications of major thermosetting plastics.

**Alkyds** (from alkyl, meaning alcohol, and acid) possess good electrical insulating properties, impact resistance, and dimensional stability and have low water absorption. Typical applications are electrical and electronic components.

**Aminos** (urea and melamine) have properties that depend on composition. Generally, aminos are hard and rigid and are resistant to abrasion, creep, and electrical arcing. Typical applications are small appliance housings, countertops, toilet seats, handles, and distributor caps. Urea is used for electrical and electronic components, melamine for dinnerware.

**Epoxies** have excellent mechanical and electrical properties, dimensional stability, strong adhesive properties, and good resistance to heat and chemicals. Typical applications are electrical components requiring mechanical strength and high insulation, tools and dies, and adhesives. Fiber-reinforced epoxies have excellent mechanical properties and are used in pressure vessels, rocket motor casings, tanks, and similar structural components.

**Phenolics,** although brittle, are rigid, dimensionally stable, and have high resistance to heat, water, electricity, and chemicals. Typical applications are knobs, handles, laminated panels, telephones, bond material to hold abrasive grains together in grinding wheels, and electrical components, such as wiring devices, connectors, and insulators.

**Polyesters** (thermosetting; see also Section 10.5) have good mechanical, chemical, and electrical properties. Polyesters are generally reinforced with glass or other fibers. Typical applications are boats, luggage, chairs, automotive bodies, swimming pools, and material for impregnating cloth and paper. They also are available as casting resins.

**Polyimides** possess good mechanical, physical, and electrical properties at elevated temperatures. They also have creep resistance and low friction and wear characteristics. Polyimides have the nonmelting characteristics of a thermoset but the structure of a thermoplastic. Typical applications are pump components (bearings, seals, valve seats, retainer rings, and piston rings), electrical connectors for high-temperature use, aerospace parts, high-strength impact-resistant structures, sports equipment, and safety vests.

**Silicones** have properties that depend on composition. Generally, they weather well, possess excellent electrical properties over a wide range of humidity and temperature, and resist chemicals and heat (see also Section 10.8). Typical applications are electrical components requiring strength at elevated temperatures, oven gaskets, heat seals, and waterproof materials.

---

### Example 10.4: Materials for refrigerator door liner

In the selection of candidate materials for a door liner for refrigerators—where eggs, butter, salad dressings, and small bottles are stored—the following factors should be considered:

- *Mechanical requirements:* Strength, toughness (to withstand impact, door slamming, racking), stiffness, resilience, and resistance to scratching and wear at operating temperatures.

- *Physical requirements:* Dimensional stability and electrical insulation.

- *Chemical requirements:* Staining, odor, chemical reactions with food and beverages, and resistance to cleaning fluids.

- *Appearance:* Color, stability of color, surface finish, texture, and feel.

- *Manufacturing properties:* Methods of manufacturing and assembly, effects of processing on material properties and behavior over a period of time, compatibility with other components in the door, and cost of materials and manufacturing.

An extensive study, considering all the factors involved, identified two candidate materials for door liners: ABS (acrylonitrile-butadiene-styrene) and HIPS (high-impact polystyrene). One aspect of the study involved the effect of vegetable oils, such as salad dressing stored in the door shelf, on the strength of these plastics. Experiments showed that the presence of vegetable oils significantly reduced the load-bearing capacity of HIPS. It was found that HIPS becomes brittle in the presence of oils (solvent stress cracking), whereas ABS is not affected to any significant extent.

---

## 10.7 High-Temperature Polymers, Electrically Conducting Polymers, and Biodegradable Plastics

We have thus far described the structure, behavior and fundamental properties of polymers. As in all other materials, important developments are taking place in polymers regarding their capabilities and in response to environmentally conscious design and manufacturing. Among these advances are high-temperature polymers, electrically conducting polymers, and biodegradable plastics. This section describes the nature and extent of these developments to date.

### 10.7.1 High-temperature polymers

During the past several years, there have been important advances in polymers and polymer blends for high-temperature applications, particularly in the aerospace industry. Properties of polymers such as high strength-to-weight ratio and fatigue strength are being further enhanced with high-temperature resistance. This resistance, as in metals, may be short term at relatively high temperatures or long term at lower temperatures (see, for example, Section 3.11.6 for a similar consideration regarding titanium alloys).

Short-term exposure generally requires *ablative* materials to dissipate the heat, such as phenolic-silicone copolymers, which have been used for rocket and missile components at temperatures of thousands of degrees. Long-term exposure for polymers is presently confined to temperatures on the order of 260°C (500°F). High-temperature thermoplastic polymers include fluorine-containing thermoplastics, polyketones, and polyimides. High-temperature thermosetting polymers include phenolic resins, epoxy resins, silicone-based thermosetting polymers, and phenolic-fiberglass systems. Several other polymers with complex structures are being developed for high-temperature applications.

### 10.7.2. Electrically conducting polymers

Most polymers are electrical insulators and hence are used as insulators for electrical equipment (light switches, outlets, circuit breakers, etc.), and as packaging material for electronic components. However, the electrical conductivity of some polymers can be increased by doping, i.e., introducing certain impurities in the polymer, such as metal powder, salts, and iodides. The conductivity of polymers also increases with moisture absorption, and the electronic properties of polymers can be altered by irradiation as well.

One of the earliest conducting polymers developed was polyacetylene. Applications of electrically conducting polymers include microelectronic devices, electrical power equipment, and conducting adhesives for surface-mount technologies (see Section 12.13). An early application was in rechargeable batteries. The cathode is made of a conducting polymer and the anode is lithium (a soft, silver-white element that is the lightest metal known) or lithium alloy. Developments are now taking place in making battery cells in which both electrodes are made of conducting polymers; one such cell has been constructed with a capacity of 3.5 V.

### 10.7.3    Biodegradable plastics

One-third of plastic production is in the disposable products sector, such as bottles, packaging, garbage bags, etc. With the growing use of plastics and increasing concern over environmental issues regarding disposal of plastic products and limited landfills, major efforts are underway to develop biodegradable plastics. Although earlier efforts were not successful in producing true biodegradable plastics, three different bioplastics are currently available: starch-based bioplastics, lactic-based bioplastics, and bioplastics made from the fermentation of sugar. These three types of material degrade over various periods of time (from a few months to a few years) and have different degradability characteristics. These plastics are designed to degrade completely when exposed to microorganisms in soil or water, without producing toxic by-products.

The starch-based system is the farthest along in terms of production capacity. In this system, starch granules are processed into a powder, which is heated and becomes a sticky liquid. The liquid is then cooled, formed into pellets, and processed in conventional plastic-processing equipment, described in Section 10.10. Various additives are blended with the starch to impart special characteristics to the bioplastic materials.

In the lactic-based system, fermenting corn or other feedstocks produces lactic acid, which is then polymerized to form a polyester resin. In the third system, organic acids are added to a sugar feedstock. With the use of a specially developed process, the resulting reaction produces a highly crystalline and very stiff polymer which, after further processing, behaves in a manner similar to polymers developed from petroleum.

Because the development of bioplastics is relatively new, their long-range performance, both during their useful life cycle as products and in landfills, has not been fully assessed. There is also concern that emphasis on biodegradability will divert attention from the issue of recyclability of plastics and the efforts for conservation of materials and energy.

## 10.8    Elastomers (Rubbers): Properties and Applications

Elastomers comprise a large family of amorphous polymers having a low glass-transition temperature. They have the characteristic ability to undergo large elastic deformations without rupture; they are soft and have a low elastic modulus. The term elastomer is derived from the words *elastic* and *mer*. The structure of these polymers is highly kinked (tightly twisted or curled). They stretch but then return to their original shape after the load is removed (Fig. 10.13). They can be cross-linked. The best example is the elevated-temperature **vulcanization** of rubber with sulfur, discovered by Charles Goodyear in 1839 and named for Vulcan, the Roman god of fire. Once the elastomer is cross-linked, it cannot be reshaped. For example, an automobile tire, which is one giant molecule, cannot be softened and reshaped.

The terms rubber and elastomer are often used interchangeably. Generally, an **elastomer** is defined as being capable of recovering substantially in shape and size after the load has been removed. **Rubber** is defined as being capable of recovering from large deformations quickly.

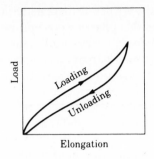

**FIGURE 10.13** Typical load-elongation curve for rubbers. The clockwise loop, indicating loading and unloading paths, is the hysteresis loss. Hysteresis gives rubbers the capacity to dissipate energy, damp vibration, and absorb shock loading, as in automobile tires and vibration dampeners placed under machinery.

The hardness of elastomers, which is measured with a durometer, increases with increasing cross-linking of the molecular chains. A variety of additives can be blended with elastomers to impart specific properties, as with plastics. Elastomers have a wide range of applications, such as high-friction and nonskid surfaces, protection against corrosion and abrasion, electrical insulation, and shock and vibration insulation. Examples include tires, hoses, weather stripping, footwear, linings, gaskets, seals, printing rolls, and flooring.

A characteristic of elastomers is their hysteresis loss in stretching or compression (Fig. 10.13). The clockwise loop indicates energy loss, whereby mechanical energy is converted into heat. This property is desirable for absorbing vibrational energy (damping) and sound deadening.

**Natural rubber.** The base for natural rubber is **latex,** a milklike sap obtained from the inner bark of a tropical tree. It has good resistance to abrasion and fatigue and high frictional properties, but it has low resistance to oil, heat, ozone, and sunlight. Typical applications are tires, seals, shoe heels, couplings, and engine mounts.

**Synthetic rubbers.** Further developed than natural rubbers are the synthetic rubbers. Examples are synthetic natural rubber, butyl, styrene butadiene, polybutadiene, and ethylene propylene. Compared to natural rubbers, they have improved resistance to heat, gasoline, and chemicals and higher useful temperature range. Examples of synthetic rubbers that are resistant to oil are neoprene, nitrile, urethane, and silicone. Typical applications of synthetic rubbers are tires, shock absorbers, seals, and belts.

Silicones (see also Section 10.6) have the highest useful temperature range, up to 315°C (600°F), but their other properties—such as strength and resistance to wear and oils—are generally inferior to other elastomers. Typical applications are seals, gaskets, thermal insulation, high-temperature electrical switches, and electronic apparatus.

Polyurethane has very good overall properties of high strength, stiffness, and hardness and exceptional resistance to abrasion, cutting, and tearing. Typical applications are seals, gaskets, cushioning, diaphragms for rubber forming of sheet metals, and auto body parts.

## 10.9 **Reinforced Plastics (Composite Materials)**

Among the major developments in materials are **composite materials.** In fact, composites are now one of the most important classes of *engineered materials,* as they offer several outstanding properties as compared to conventional materials. We define these materials as a combination of two or more chemically distinct and insoluble phases whose properties and structural performance are superior to those of the constituents acting independently. We showed that plastics, for example, possess mechanical properties (particularly strength, stiffness, and creep resistance) that are generally inferior to those of metals and alloys. These properties can be improved by embedding reinforcements of various types, such as glass or graphite fibers, to produce **reinforced plastics.** As shown in Table 10.1, reinforcements improve the strength, stiffness, and creep resistance of plastics—and their strength-to-weight and stiffness-to-weight ratios.

Reinforced plastics have found increasingly wider applications in aircraft, space vehicles, offshore structures, piping, electronics, automobiles, boats, ladders, and sporting goods. Metals and ceramics also can be embedded with fibers or particles to improve their properties.

The oldest example of composites is the addition of straw to clay for making mud huts and bricks for structural use, dating back to 4000 B.C. In that application, the straws are the reinforcing fibers and the clay is the matrix. Another example of a composite material is the reinforcing of masonry and concrete with iron rods, begun in the 1800s. In fact, concrete itself is a composite material, consisting of cement, sand, and gravel. In reinforced concrete, steel rods impart the necessary tensile strength to the composite, since concrete is brittle and generally has little or no useful tensile strength.

### 10.9.1    Structure of reinforced plastics

Reinforced plastics consist of **fibers** (the discontinuous or dispersed phase) in a plastic **matrix** (the continuous phase). Commonly used fibers are glass, graphite, aramids, and boron (Table 10.4). These fibers are strong and stiff and have high specific strength (strength-to-weight ratio) and specific modulus (stiffness-to-weight ratio), as shown in Fig. 10.14.

**TABLE 10.4**

**PROPERTIES OF REINFORCING FIBERS**

| TYPE | TENSILE STRENGTH (MPa) | ELASTIC MODULUS (GPa) | DENSITY (kg/m$^3$) | RELATIVE COST |
|---|---|---|---|---|
| Boron | 3500 | 380 | 2600 | Highest |
| Carbon | | | | |
|   High strength | 3000 | 275 | 1900 | Low |
|   High modulus | 2000 | 415 | 1900 | Low |
| Glass | | | | |
|   E type | 3500 | 73 | 2480 | Lowest |
|   S type | 4600 | 85 | 2540 | Lowest |
| Kevlar | | | | |
|   29 | 2800 | 62 | 1440 | High |
|   49 | 2800 | 117 | 1440 | High |

Note: These properties vary significantly depending on the material and method of preparation.

**FIGURE 10.14** Specific tensile strength (tensile strength-to-density ratio) and specific tensile modulus (modulus of elasticity-to-density ratio) for various fibers used in reinforced plastics. Note the wide range of specific strengths and stiffnesses available.

However, they are generally brittle and abrasive and lack toughness. Thus fibers, by themselves, have little structural value. The plastic matrix is less strong and less stiff but tougher than the fibers. Thus reinforced plastics combine the advantages of each of the two constituents (Table 10.5). When more than one type of fiber is used in a reinforced plastic, the composite is called a *hybrid,* which generally has even better properties.

In addition to high specific strength and specific modulus, reinforced plastic structures have improved fatigue resistance, greater toughness, and higher creep resistance than unreinforced plastics. These structures are relatively easy to design, fabricate, and repair.

The percentage of fibers (by volume) in reinforced plastics usually ranges between 10 percent and 60 percent. Practically, the percentage of fiber in a matrix is limited by the average distance between adjacent fibers or particles. The highest practical fiber content is 65 percent; higher percentages generally result in diminished structural properties.

**TABLE 10.5**

**TYPES AND GENERAL CHARACTERISTICS OF REINFORCED PLASTICS AND COMPOSITE MATERIALS**

| MATERIAL | CHARACTERISTICS |
| --- | --- |
| Fibers | |
| Glass | High strength, low stiffness, high density; lowest cost; E (calcium aluminoborosilicate and S (magnesia-aluminosilicate) types commonly used. |
| Graphite | Available as high-modulus or high-strength; low cost; less dense than glass. |
| Boron | High strength and stiffness; highest density; highest cost; has tungsten filament at its center. |
| Aramids (Kevlar) | Highest strength-to-weight ratio of all fibers; high cost. |
| Other fibers | Nylon, silicon carbide, silicon nitride, aluminum oxide, boron carbide, boron nitride, tantalum carbide, steel, tungsten, molybdenum. |
| Matrix materials | |
| Thermosets | Epoxy and polyester, with the former most commonly used; others are phenolics, fluorocarbons, polyethersulfone, silicon, and polyimides. |
| Thermoplastics | Polyetheretherketone; tougher than thermosets but lower resistance to temperature. |
| Metals | Aluminum, aluminum-lithium, magnesium, and titanium; fibers are graphite, aluminum oxide, silicon carbide, and boron. |
| Ceramics | Silicon carbide, silicon nitride, aluminum oxide, and mullite; fibers are various ceramics. |

## 10.9.2    Reinforcing fibers

**Glass.**    Glass fibers are the most widely used and least expensive of all fibers. The composite material is called *glass-fiber reinforced plastic* (GFRP) and may contain between 30 percent and 60 percent glass fibers by volume. Glass fibers are made by drawing molten glass through small openings in a platinum die. There are two principal types of glass fibers: (1) the E type, a calcium aluminoborosilicate glass, which is used most; and (2) the S type, a magnesia-aluminosilicate glass, which has higher strength and stiffness and is more expensive.

**Graphite.**    Graphite fibers (Fig. 10.15a), although more expensive than glass fibers, have a combination of low density, high strength, and high stiffness. The product is called *carbon-fiber reinforced plastic* (CFRP). All graphite fibers are made by pyrolysis of organic **precursors,** commonly polyacrylonitrile (PAN) because of its lower cost. Rayon and pitch (the residue from catalytic crackers in petroleum refining) can also be used as precursors. **Pyrolysis** is the term for inducing chemical changes by heat, such as burning a length of yarn, which becomes carbon and black in color. The temperatures for carbonizing range up to about 1500°C (2730°F) and for graphitizing to 3000°C (5400°F).

The difference between carbon and graphite, although the words are often used interchangeably, depends on the temperature of pyrolysis and the purity of the material. Carbon fibers are generally 80 to 95 percent carbon, and graphite fibers are usually more than 99 percent carbon.

**Aramids.**    Aramids are among the toughest fibers and have very high specific strength (Fig. 10.14). A common aramid is marketed under the trade name *Kevlar.* They can undergo some plastic deformation before fracture and thus have higher toughness than brittle fibers. However, aramids absorb moisture, which reduces their properties and complicates their application, as hygrothermal stresses must be considered.

**Boron.**    Boron fibers consist of boron deposited (by chemical vapor-deposition techniques) on tungsten fibers (Fig. 10.15b), although boron can also be deposited on carbon

**FIGURE 10.15**   (a) Cross-section of a tennis racket, showing graphite and aramid (Kevlar) reinforcing fibers. *Source:* J. Dvorak, Mercury Marine Corporation, and F. Garrett, Wilson Sporting Goods Co. (b) Cross-section of boron-fiber reinforced composite material.

fibers. These fibers have favorable properties, such as high strength and stiffness in tension and compression and resistance to high temperatures. However, because of the use of tungsten, they have high density and are expensive, thus increasing the cost and weight of the reinforced plastic component.

**Other fibers.** Other fibers that are being used are nylon, silicon carbide, silicon nitride, aluminum oxide, sapphire, steel, tungsten, molybdenum, boron carbide, boron nitride, and tantalum carbide. **Whiskers** are also used as reinforcing fibers. They are tiny needlelike single crystals that grow to 1 $\mu$m to 10 $\mu$m (40 $\mu$in. to 400 $\mu$in.) in diameter and have aspect ratios (defined as the ratio of fiber length to diameter) ranging from 100 to 15,000. Because of their small size, either they are free of imperfections or the imperfections they contain do not significantly affect their strength, which approaches the theoretical strength of the material. Recently, a high-performance polyethylene fiber was introduced. Called *Spectra* fiber (a trade name), it has ultra-high molecular weight and high molecular-chain orientation. The fiber has better abrasion resistance and flexural-fatigue resistance than the aramid fiber (see above) and at a similar cost. In addition, because of its lower density (970 kg/m$^3$) it has higher specific strength and specific stiffness than the aramid fiber. However, a low melting point and poor adhesion characteristics to other polymers are its major limitations to application.

### 10.9.3 Fiber size and length

The mean diameter of fibers used in reinforced plastics is usually less than 0.01 mm (0.0004 in.). The fibers are very strong and rigid in tension. The reason is that the molecules in the fibers are oriented in the longitudinal direction, and their cross-sections are so small that the probability is low that any defects exist in the fiber. Glass fibers, for example, can have tensile strengths as high as 4600 MPa (650 ksi), whereas the strength of glass in bulk form is much lower. Thus glass fibers are stronger than steel.

Fibers are classified as *short* or *long* fibers, also called *discontinuous* or *continuous* fibers, respectively. Short fibers generally have an aspect ratio between 20 and 60, and long fibers between 200 and 500. The short and long fiber designations are, in general, based on the following observations: In a given fiber, if the mechanical properties improve as a result of increasing the fiber length, then it is denoted as a short fiber. When no additional improvement in properties occurs, it is denoted as a long fiber. In addition to the discrete fibers that we have described, reinforcements in composites may be in the form of continuous *roving* (slightly twisted strand of fibers), *woven* fabric (similar to cloth), *yarn* (twisted strand), and *mats* of various combinations. Reinforcing elements may also be in the form of particles and flakes.

### 10.9.4 Matrix materials

The matrix in reinforced plastics has three functions:

1. Support and transfer the stresses to the fibers, which carry most of the load.
2. Protect the fibers against physical damage and the environment.
3. Reduce propagation of cracks in the composite by virtue of the ductility and toughness of the plastic matrix.

Matrix materials are usually epoxy, polyester, phenolic, fluorocarbon, polyethersulfone, and silicon. The most commonly used are epoxies (80 percent of all reinforced plastics) and polyesters, which are less expensive than epoxies. Polyimides, which resist exposure to temperatures in excess of 300°C (575°F), are being developed for use with graphite fibers. Some thermoplastics, such as polyetheretherketone, are also being developed as matrix materials. They generally have higher toughness than thermosets, but their resistance to temperature is lower, being limited to 100−200°C (200−400°F).

### 10.9.5   Properties of reinforced plastics

The properties of reinforced plastics depend on the kind, shape, and orientation of the reinforcing material, the length of the fibers, and the volume fraction (percentage) of the reinforcing material. Short fibers are less effective than long fibers (Fig. 10.16), and their properties are strongly influenced by time and temperature. Long fibers transmit the load through the matrix better and thus are commonly used in critical applications, particularly at elevated temperatures. Fiber reinforcement affects many other properties of composites (Fig. 10.17).

A critical factor in reinforced plastics is the strength of the bond between the fiber and the polymer matrix, since the load is transmitted through the fiber-matrix interface. Weak

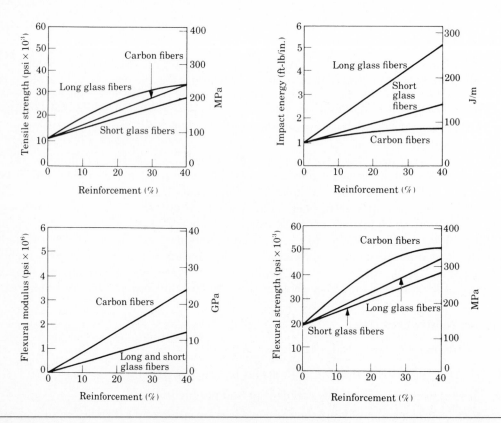

**FIGURE 10.16**   Effect of the amount of reinforcing fibers and fiber length on the mechanical properties of reinforced nylon. Note the significant improvement with increasing percentage of fiber reinforcement. *Source:* Courtesy of Wilson Fiberfill International.

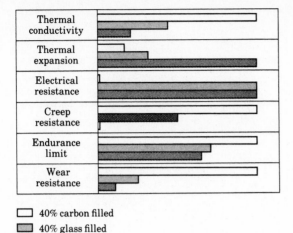

Thermal
conductivity

Thermal
expansion

Electrical
resistance

Creep
resistance

Endurance
limit

Wear
resistance

☐ 40% carbon filled
▨ 40% glass filled
▪ Unfilled

**FIGURE 10.17** The effect of type of fibers on various properties of fiber-reinforced nylon 6,6. *Source:* NASA.

bonding causes **fiber pullout** and **delamination** of the structure, particularly under adverse environmental conditions. Poor bonding in composites is analogous to a brick structure with poor bonding between the bricks and the mortar.

Adhesion at the interface can be improved by special surface treatments, such as coatings and coupling agents. Glass fibers, for example, are treated with a chemical called *silane* for improved wetting and bonding between the fiber and the matrix. You can appreciate the importance of proper bonding by looking at Figs. 10.18a and b, which show the fracture surfaces of reinforced plastics.

Generally, the greatest stiffness and strength in reinforced plastics is obtained when the fibers are aligned in the direction of the tension force. This composite, of course, is highly anisotropic (Fig. 10.19). As a result, other properties of the composite, such as stiffness, creep resistance, thermal and electrical conductivity, and thermal expansion, are also anisotropic. The transverse properties of such a unidirectionally reinforced structure are much lower than the longitudinal properties. Note, for example, how easily you can split fiber-reinforced packaging tape, yet how strong it is when you pull on it (tension).

For a specific service condition, we can give a reinforced plastic part an optimal configuration. For example, if the reinforced plastic part is to be subjected to forces in different

(a)   (b)

**FIGURE 10.18** (a) Fracture surface of glass-fiber reinforced epoxy composite. The fibers are 10 $\mu$m (400 $\mu$in.) in diameter and have random orientation. (b) Fracture surface of a graphite-fiber reinforced epoxy composite. The fibers, 9-11 $\mu$m in diameter, are in bundles and are all aligned in the same direction. *Source:* L. J. Broutman.

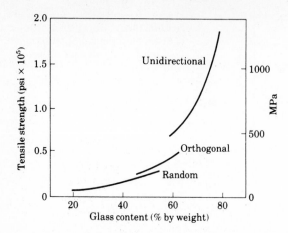

**FIGURE 10.19**   The tensile strength of glass-reinforced polyester as a function of fiber content and fiber direction in the matrix. *Source:* R. M. Ogorkiewicz, *The Engineering Properties of Plastics.* Oxford: Oxford University Press, 1977.

directions (such as thin-walled, pressurized vessels), the fibers are criss-crossed in the matrix. Reinforced plastics may also be made with various other materials and shapes of the polymer matrix in order to impart specific properties (such as permeability and dimensional stability), make processing easier, and reduce costs.

## Strength and elastic modulus of reinforced plastics

The strength of a reinforced plastic with longitudinal fibers can be determined in terms of the strength of the fibers and matrix, respectively, and the volume fraction of fibers in the composite. In the following equations, $c$ refers to the composite, $f$ to the fiber, and $m$ to the matrix. The total load $P_c$ on the composite is shared by the fiber ($P_f$) and the matrix ($P_m$) thus,

$$P_c = P_f + P_m, \tag{10.10}$$

which we can write as

$$\sigma_c A_c = \sigma_f A_f + \sigma_m A_m, \tag{10.11}$$

where $A_c$, $A_f$, and $A_m$ are the cross-sectional areas of the composite, fiber, and matrix, respectively, and thus $A_c = A_f + A_m$. Let's now denote $x$ as the area fraction of the fibers in the composite. (Note that $x$ also represents the volume fraction, because the fibers are uniformly longitudinal in the matrix.) We may now rewrite Eq. (10.11) as follows:

$$\sigma_c = x\sigma_f + (1 - x)\sigma_m. \tag{10.12}$$

We can now calculate the fraction of the total load carried by the fibers. First, we note that, in the composite under a tension load, the strains sustained by the fibers and the matrix are the same (that is, $e_c = e_f = e_m$), and then recall from Section 2.2 that

$$e = \frac{\sigma}{E} = \frac{P}{AE}.$$

Consequently,

$$\frac{P_f}{P_m} = \frac{A_f E_f}{A_m E_m}. \tag{10.13}$$

As we know the relevant quantities for a specific situation, and using Eq. (10.10), we can determine the fraction $P_f/P_c$. Then, using the foregoing relationships, we can also calculate the elastic modulus $E_c$ of the composite by replacing $\sigma$ in Eq. (10.12) with $E$. Thus,

$$E_c = x E_f + (1 - x) E_m. \tag{10.14}$$

---

**Example 10.5:  Properties of a graphite-epoxy reinforced plastic**

Let us assume that a graphite-epoxy reinforced plastic with longitudinal fibers contains 20 percent graphite fibers with a strength of 2500 MPa and an elastic modulus of 300 GPa. The strength of the epoxy matrix is 120 MPa, with an elastic modulus of 100 GPa. Calculate the elastic modulus of the composite and the fraction of the load supported by the fibers.

SOLUTION.   The data given are $x = 0.2$, $E_f = 300$ GPa, $E_m = 100$ GPa, $\sigma_f = 2500$ MPa, and $\sigma_m = 120$ MPa. Using Eq. (10.14), we find that

$$E_c = 0.2(300) + (1 - 0.2)100 = 60 + 80 = 140 \text{ GPa}.$$

We obtain the load fraction $P_f/P_m$ from Eq. (10.13):

$$\frac{P_f}{P_m} = \frac{0.2(300)}{0.8(100)} = \frac{60}{80} = 0.75.$$

Since

$$P_c = P_f + P_m \quad \text{and} \quad P_m = \frac{P_f}{0.75},$$

we find that

$$P_c = P_f + \frac{P_f}{0.75} = 2.33 \, P_f, \quad \text{or} \quad P_f = 0.43 P_c.$$

Thus, the fibers support 43 percent of the load, even though they occupy only 20 percent of the cross-sectional area (hence volume) of the composite.

---

### 10.9.6  Applications

The first application of reinforced plastics (in 1907) was for an acid-resistant tank, made of a phenolic resin with asbestos fibers. Formica, commonly used for counter tops, was developed in the 1920s. Epoxies were first used as a matrix in the 1930s and beginning in the 1940s, boats were made with fiberglass and reinforced plastics were used for aircraft, electrical equipment, and sporting goods. Major developments in composites began in the 1970s, and these materials are now called *advanced composites*. Glass- or carbon-fiber reinforced hybrid plastics are now being developed for high-temperature applications, with continuous use ranging up to about 300°C (550°F).

Reinforced plastics are typically used in military and commercial aircraft and rocket components, helicopter blades, automotive bodies, leaf springs, drive shafts, pipes, ladders, pressure vessels, sporting goods, sports and military helmets, boat hulls, and various other structures. Applications of reinforced plastics include components in DC-10, L-1011, and Boeing 727, 757, 767, and 777 commercial aircraft. The new Boeing 777 is made of about 9 percent composites by total weight, which is triple the composite content of existing Boeing transport aircraft. The floor beams and panels and most of the vertical and horizontal tail are made of composite materials. By virtue of the resulting weight savings, reinforced plastics have reduced fuel consumption by about 2 percent.

Substituting aluminum in large commercial aircraft with graphite-epoxy reinforced plastics could reduce both weight and production costs by 30 percent, with improved fatigue and corrosion resistance. The structure of the Lear Fan 2100 passenger aircraft is almost totally made of graphite-epoxy reinforced plastic. Nearly 90 percent of the structure of the lightweight Voyager craft, which circled the earth without refueling, was made of carbon-reinforced plastic. The contoured frame of the Stealth bomber is made of composites consisting of carbon and glass fibers, epoxy-resin matrices, high-temperature polyimides, and other materials. Boron-fiber reinforced composites are used in military aircraft, golf-club shafts, tennis rackets, fishing rods, and sailboard.

The processing of reinforced plastics presents significant challenges; several innovative techniques have been developed to manufacture both large and small parts by molding, forming, cutting and assembly. Careful inspection and testing of reinforced plastics is essential in critical applications, in order to ensure that good bonding between the reinforcing fiber and the matrix has been obtained throughout. In some instances, the cost of inspection can be as high as one-quarter of the total cost of the composite product.

---

**Example 10.6:    Composite blades for helicopters**

Metal blades on helicopters are now being replaced with blades made of composite materials, principally S-glass fibers in an epoxy matrix. These blades have high stiffness, strength, resilience, and temperature and fatigue resistance. They also have high impact strength and consequently, as compared to metal blades, composites can withstand a major ballistic impact without catastrophic failure and the helicopter can return safely to its base. Furthermore, repairs of damaged blades can be made on the field whereas failure of metal blades requires more extensive maintenance. S glass has a higher fatigue life than E glass. It has also been shown that glass-reinforced composite blades have a better cost/performance basis and are superior to aramid- or carbon-reinforced composites.

---

# 10.10    Processing of Plastics

The processing of plastics involves operations similar to those used to form and shape metals, as described in the preceding chapters. Plastics can be molded, cast, and formed, as well as machined and joined, into many shapes with relative ease and with little or no additional

**TABLE 10.6**

**CHARACTERISTICS OF PROCESSING PLASTICS AND REINFORCED PLASTICS**

| PROCESS | CHARACTERISTICS |
|---------|-----------------|
| Extrusion | Long, uniform, solid or hollow complex cross-sections; high production rates; low tooling costs; wide tolerances. |
| Injection molding | Complex shapes of various sizes, eliminating assembly; high production rates; costly tooling; good dimensional accuracy. |
| Structural foam molding | Large parts with high stiffness-to-weight ratio; less expensive tooling than in injection molding; low production rates. |
| Blow molding | Hollow thin-walled parts of various sizes; high production rates and low cost for making containers. |
| Rotational molding | Large hollow shapes of relatively simple design; low tooling cost; low production rates. |
| Thermoforming | Shallow or relatively deep cavities; low tooling costs; medium production rates. |
| Compression molding | Parts similar to impression-die forging; relatively inexpensive tooling; medium production rates. |
| Transfer molding | More complex parts than compression molding and higher production rates; some scrap loss; medium tooling cost. |
| Casting | Simple or intricate shapes made with flexible molds; low production rates. |
| Processing of composite materials | Long cycle times; tolerances and tooling cost depend on process. |

operations required (Table 10.6). Plastics melt or cure at relatively low temperatures (see Table 10.2) and so, unlike metals, are easy to handle and require less energy to process. However, because the properties of plastic parts and components are influenced greatly by the method of manufacture and processing conditions, the proper control of these conditions is important for part quality.

Plastics are usually shipped to manufacturing plants as pellets or powders and are melted just before the shaping process. Plastics are also available as sheet, plate, rod, and tubing, which may be formed into a variety of products. Liquid plastics are used especially in making reinforced plastic parts. An important development is stereo-lithography, which is a means of making prototypes rapidly by curing a photopolymer into a specific shape.

### 10.10.1  Extrusion

In **extrusion,** raw materials in the form of thermoplastic pellets, granules, or powder are placed into a hopper and fed into the extruder barrel (Fig. 10.20). The barrel is equipped with a *screw* that blends and conveys the pellets down the barrel. The internal friction from the mechanical action of the screw, along with heaters around the extruder's barrel, heats the pellets and liquefies them. The screw action also builds up pressure in the barrel.

**FIGURE 10.20**   Schematic illustration of a typical extruder. *Source: Encyclopedia of Polymer Science and Engineering,* 2d ed. Copyright © 1985. Reprinted by permission of John Wiley & Sons, Inc.

Screws have three distinct sections: (1) a feed section that conveys the material from the hopper area into the central region of the barrel; (2) a melt, or transition, section where the heat generated from shearing of the plastic causes melting to begin; and (3) a pumping section where additional shearing and melting occurs, with pressure buildup at the die. The lengths of these sections can be changed to accommodate the melting characteristics of different plastics.

The molten plastic is forced through a die—similar to extruding metals. The extruded product is then cooled, either by air or by passing it through a water-filled channel. Controlling the rate and uniformity of cooling is important to minimize product shrinkage and distortion. The extruded product can also be drawn (sized) by a puller after it has cooled. The extruded product is then coiled or cut into desired lengths. Complex shapes with constant cross-section can be extruded with relatively inexpensive tooling. This process is also used to extrude elastomers.

Because this operation is continuous, long products with cross-sections of solid rods, channels, pipe, window frames, and architectural components, as well as sheet, are extruded through dies of various geometries. Plastic-coated wire, cable, or strips for electrical applications are also extruded and coated by this process. The wire is fed into the die opening at a controlled rate with the extruded plastic.

**Pellets,** which are used for other plastics-processing methods described in this chapter, are also made by extrusion. Here the extruded product is a small-diameter rod and is chopped into short lengths, or pellets, as it is extruded. With some modifications, extruders can also be used as simple melters for other shaping processes, such as injection molding and blow molding.

The basic *process parameters* are extruder screw speed (typically one rev/s), barrel-wall temperatures, die design, and cooling rates and drawing speeds. These parameters should be controlled carefully in order to extrude products having uniform cross-sections and dimensional accuracy. There are temperature and velocity gradients present in the barrel, both in

**FIGURE 10.21**  Die geometry (coat-hanger die) for extruding sheet. *Source: Encyclopedia of Polymer Science and Engineering,* 2d ed. Copyright © 1985. Reprinted by permission of John Wiley & Sons, Inc.

radial and longitudinal directions. The pressure is zero at the hopper and builds up toward the die exit. The material flows in a viscous manner between the stationary barrel wall and the rotating screw (see also Fig. 10.11); consequently, the flow of the material and its pressure distribution in the extruder is a complex phenomenon and includes a number of variables.

Because the material being extruded is still soft as it leaves the die and the pressure is relieved, the cross-section of the extruded product is different from the shape of the die opening. This is known as **die swell** (see Fig. 10.40b); thus, the diameter of a round extruded part, for example, is larger than the die diameter, the difference depending on the type of polymer. Considerable experience is required to design proper dies for extruding complex cross-sections of desired shape and dimensions.

Extruders are generally rated by the diameter of the barrel and by the length-to-diameter (L/D) ratio of the barrel. Typical commercial units are 25-200 mm (1-8 in.) in diameter, with L/D ratios of 5 to 30. Production-size extrusion equipment costs $30,000-$80,000, with an additional $30,000 cost for equipment for downstream cooling and winding of the extruded product. Thus large production runs are generally required to justify such an expenditure.

**Sheet and film extrusion.**    Polymer sheet and film can be produced using a flat extrusion die, such as that shown in Fig. 10.21. The polymer is extruded by forcing it through a specially designed die, after which the extruded sheet is taken up first on water-cooled rolls and then by a pair of rubber-covered pull-off rolls.

Thin polymer films and common plastic bags are made from a tube produced by an extruder (Fig. 10.22). In this process, a thin-walled tube is extruded vertically and expanded into a balloon shape by blowing air through the center of the extrusion die until the desired film thickness is reached. The balloon is usually cooled by air from a cooling ring around it, which can also act as a barrier to further expansion of the balloon. **Blown film** is sold as wrapping film (after slitting the cooled bubble) or as bags (where the bubble is pinched and cut off). Film is also produced by shaving solid round billets of plastics, especially polytetrafluoroethylene (PTFE), by skiving, that is, shaving with specially designed knives.

Pinch rolls

Wind-up

Guide rolls

Blown
tube

Mandrel

Die

Extruder

**FIGURE 10.22** Schematic illustration of production of thin film and plastic bags from tube produced by an extruder and then blown by air. *Source:* D. C. Miles and J. H. Briston, *Polymer Technology*, 1979. Reproduced by permission of Chemical Publishing Co., Inc.

---

### Example 10.7:   Blown film

Assume that a typical plastic shopping bag, made by blown film, has a lateral (width) dimension of 400 mm. (1) What should be the extrusion die diameter? (2) These bags are relatively strong, so how is this strength achieved?

SOLUTION.

1.   The perimeter of the bag is $(2)(400) = 800$ mm. As the original cross-section of the film was round, we find that the blown diameter should be $\pi D = 800$, or $D = 255$ mm. Recall that in this process a tube is expanded 1.5-2.5 times the extrusion die diameter. Taking the maximum value of 2.5, we calculate the die diameter as $255/2.5 = 100$ mm.

2.   Note in Fig. 10.22 that after being extruded, the bubble is being pulled upward by the pinch rolls. Thus, in addition to diametral stretching and the attendant molecular orientation, the film is stretched and oriented in the longitudinal direction. The biaxial orientation of the polymer molecules significantly improves the strength and toughness of the blown film.

---

### 10.10.2   Injection molding

**Injection molding** is essentially the same process as hot-chamber die casting. The pellets or granules are fed into a heated cylinder, where they are melted. The melt is then forced into a split-die chamber (Fig. 10.23a), either by a hydraulic plunger or by the rotating screw system of an extruder. Newer equipment is of the *reciprocating screw* type (Fig. 10.23b). As the pressure builds up at the mold entrance, the rotating screw begins to move backward under pressure to a predetermined distance, thus controlling the volume of material to be injected. The

(a)

(b)

Rotating and reciprocating
screw

**FIGURE 10.23** Injection molding with (a) plunger and (b) reciprocating rotating screw. Telephone receivers, plumbing fittings, tool handles, and housings are made by injection molding.

screw stops rotating and is pushed forward hydraulically, forcing the molten plastic into the mold cavity. Injection-molding pressures usually range from 70 MPa to 200 MPa (10,000 psi to 30,000 psi), which is on the same order of magnitude as in cold-chamber die casting.

Typical injection-molded products are cups, containers, housings, tool handles, knobs, electrical and communication components (such as telephone receivers), toys, and plumbing fittings. Although for thermoplastics the molds are relatively cool, thermosets are molded in heated molds where *polymerization* and *cross-linking* take place. In either case, after the part is sufficiently cooled (for thermoplastics) or set or cured (for thermosets), the molds are opened and the part is ejected. The molds are then closed and the process is repeated automatically. Elastomers also are injection molded by these processes.

Because the material is molten when injected into the mold, complex shapes and good dimensional accuracy can be achieved, as in die casting. However, as in metal casting, the molded part shrinks during cooling (note also that plastics have a higher thermal expansion than metals). Shrinkage in plastics molding is compensated for by making the molds a little larger. Linear shrinkage for plastics typically ranges between 0.005 and 0.025 mm/mm or in./in. Thus, the volumetric shrinkage is typically in the range of 1.5 to 7 percent (see also Table 5.1 and Section 11.10). Molds with moving and unscrewing mandrels are not unusual, and allow the molding of parts with multiple cavities and internal and external threads.

Injection molds have several components, depending on part design, such as runners (as in metal-casting dies), cores, cavities, cooling channels, inserts, knockout pins, and ejectors. There are three basic types of molds: (1) *cold-runner two-plate* mold, which is the basic and simplest mold design, (2) *cold-runner three-plate* mold, in which the runner system is separated from the part when the mold opens, and (3) *hot-runner* mold, also called *runnerless* mold, in which the molten plastic is kept hot in a heated runner plate. In cold-runner molds, the solidified plastic in the channels that connect the mold cavity to the end of the barrel must be removed, usually by trimming. This scrap can be chopped and recycled. In hot-runner molds, which are more expensive, there are no gates, runners, or sprues attached to the part. Cycle times are shorter because only the injection-molded part must be cooled and ejected.

**FIGURE 10.24**   Products made by injection molding. Metallic components are embedded in these parts during molding. *Source:* Plainfield Molding, Inc.

Metallic components, such as screws, pins, and strips, can also be placed in the mold cavity to become an integral part of the injection-molded product (**insert molding,** Fig. 10.24). The most common examples are electrical components. *Multi-component* injection molding (also called *co-injection* or *sandwich* molding) allows the forming of parts with a combination of colors and shapes. Examples are multicolor molding of rear-light covers for automobiles and ball joints made of different materials. Also, printed film can be placed in the mold cavity; thus parts need not be decorated or labeled after molding.

Injection molding is a high-rate production process, with good dimensional control. Typical cycle times range from 5 to 60 seconds, but can be several minutes for thermosetting materials. The molds, generally made of tool steels or beryllium-copper, may have multiple cavities so that more than one part can be made in one cycle. Proper mold design and control of material flow in the die cavities are important factors in the quality of the product. Other factors affecting quality are injection pressure, temperature, and condition of the resin. Computer models have been developed to study material flow in dies and thus improve die design and establish appropriate process parameters. Modern machines are equipped with microprocessors and microcomputers in a control panel and monitor all aspects of the operation.

Injection-molding machines are usually horizontal and the clamping force on the dies is supplied generally by hydraulic means, although electrical types are now available. Electrically driven models weigh less and are quieter than hydraulic machines. Vertical machines are used for making small, close-tolerance parts, and for insert molding. Injection-molding machines are rated according to the capacity of the mold and the clamping force. Although in most machines this force generally ranges from 0.9 to 2.2 MN (100 to 250 tons), the largest machine in operation has a capacity of 45 MN (5000 tons), and can produce parts weighing 25 kg (55 lb). However, parts typically weigh 100-600 g (3-20 oz). Because of the high cost of dies, ranging from $20,000 to $200,000, high-volume production is required to justify such an expenditure.

**Example 10.8:** Injection molding of gears

A 250-ton injection molding machine is to be used to make 4.5-in. diameter spur gears, 0.5 in. thick. The gears have fine teeth profile. How many gears can be injection molded in one set of molds? Does the thickness of the gears influence the answer?

SOLUTION. Because of the fine detail involved, the pressures required in the mold cavity will probably be on the order of 100 MPa (15 ksi). The cross-sectional (projected) area of the gear is $\pi(4.5)^2/4 = 15.9$ in$^2$. If we assume that the parting plane of the two halves of the mold is in the middle of the gear, the force required is $(15.9)(15,000) = 238,500$ lb. The capacity of the machine is 250 tons, so we have $(250)(2000) = 500,000$ lb of clamping force available. Therefore, the mold can accommodate two cavities, thus producing two gears per cycle. Because it does not influence the cross-sectional area of the gear, the thickness of the gear does not directly influence the pressures involved, and the answer is the same.

## Reaction-injection molding

In the **reaction-injection molding** (RIM) process, a mixture of two or more reactive fluids is forced into the mold cavity (Fig. 10.25). Chemical reactions take place rapidly in the mold and the polymer solidifies, producing a thermoset part. Major applications are automotive bumpers and fenders, thermal insulation for refrigerators and freezers, and stiffeners for structural components. Various reinforcing fibers, such as glass or graphite, may also be used to improve the product's strength and stiffness. Mold costs are low.

## Structural foam molding

The *structural foam molding* process is used to make plastic products that have a solid skin and a cellular inner structure. Typical products are furniture components, TV cabinets, business-machine housings, and storage-battery cases. Although there are several foam molding processes, they are basically similar to injection molding or extrusion. Both thermoplastics and

**FIGURE 10.25** Schematic illustration of the reaction-injection molding process. *Source:* Modern Plastics Encyclopedia.

thermosets can be used for foam molding, but thermosets are in the liquid processing form, similar to polymers for reaction-injection molding.

In *injection foam molding,* thermoplastics are mixed with a blowing agent (usually an inert gas such as nitrogen), which expands the material. The core of the part is cellular and the skin is rigid. The thickness of the skin can be as much as 2 mm (0.08 in.), and part densities are as low as 40 percent of the density of the solid plastic. Thus parts have a high stiffness-to-weight ratio and can weigh as much as 55 kg (120 lb).

### 10.10.3   Blow molding

**Blow molding** is a modified extrusion and injection molding process. In *extrusion blow molding,* a tube is extruded (usually turned so that it is vertical), clamped into a mold with a

**FIGURE 10.26**   Schematic illustrations of (a) the blow-molding process for making plastic beverage bottles and (b) a three-station injection blow-molding machine. *Source: Encyclopedia of Polymer Science and Engineering,* 2d ed. Copyright © 1985. Reprinted by permission of John Wiley & Sons, Inc.

cavity much larger than the tube diameter, and then blown outward to fill the mold cavity (Fig. 10.26a). Blowing is usually done with an air blast, at a pressure of 350-700 kPa (50-100 psi). In some operations the extrusion is continuous, and the molds move with the tubing. The molds close around the tubing, close off both ends (thereby breaking the tube into sections), and then move away as air is injected into the tubular piece. The part is then cooled and ejected. Corrugated pipe and tubing are made by continuous blow molding, whereby the pipe or tube is extruded horizontally and blown into moving molds. (See also section 10.10.1).

In *injection blow molding,* a short tubular piece (**parison**) is first injection molded (Fig. 10.26b). The dies then open and the parison is transferred to a blow-molding die. Hot air is injected into the parison, which then expands and fills the mold cavity. Typical products are plastic beverage bottles and hollow containers.

*Multilayer blow molding* involves the use of coextruded tubes or parisons, thus allowing the use of multilayer structures. Typical examples of multilayer structures are plastic packaging for food and beverages with characteristics such as odor and permeation barrier, taste and aroma protection, scuff resistance, printing capability, and ability to be filled with hot fluids. Other applications are in the cosmetics and pharmaceutical industries.

### 10.10.4 Rotational molding

Most thermoplastics and some thermosets can be formed into large hollow parts by **rotational molding.** The thin-walled metal mold is made of two pieces (split female mold) and is designed to be rotated about two perpendicular axes (Fig. 10.27). A premeasured quantity of powdered plastic material is placed inside a warm mold. The powder is obtained from a polymerization process that precipitates a powder from a liquid. The mold is then heated, usually in a large oven, while it is rotated about the two axes. This action tumbles the powder against the mold where heating fuses the powder without melting it. In some parts, a chemical cross-linking agent is added to the powder, and cross-linking occurs after the part

**FIGURE 10.27** The rotational molding (roto-molding or roto-casting) process. Trash cans, buckets, and plastic footballs can be made by this process.

is formed in the mold by continued heating. Typical parts made by rotational molding are tanks of various sizes, trash cans, boat hulls, buckets, housings, toys, carrying cases, and footballs. Various metallic or plastic inserts may also be molded into the parts made by this process.

Liquid polymers, called **plastisols** (vinyl plastisols being the most common), can also be used in a process called *slush molding*. The mold is simultaneously heated and rotated. The particles of plastic material are forced against the inside walls of the heated mold by the tumbling action. Upon contact, the material melts and coats the walls of the mold. The part is cooled while still rotating and is then removed by opening the mold.

Rotational molding can produce parts with complex hollow shapes, with wall thicknesses as small as 0.4 mm (0.016 in.). Parts as large as 1.8 m $\times$ 1.8 m $\times$ 3.6 m (6 ft $\times$ 6 ft $\times$ 12 ft) have been formed. The outer surface finish of the part is a replica of the surface finish of the mold walls. Cycle times are longer than in other processes, but equipment costs are low. Quality control considerations usually involve proper weight of powder placed in the mold, proper rotation of the mold, and the temperature-time relationship during the oven cycle.

### 10.10.5  Thermoforming

**Thermoforming** is a series of processes for forming thermoplastic sheet or film over a mold with the application of heat and pressure differentials (Fig. 10.28). In this process, a sheet is heated in an oven to the sag (softening) point—but not to the melting point. The sheet is then removed from the oven, placed over a mold, and pulled against the mold through the application of a vacuum. Since the mold is usually at room temperature, the shape of the plastic is set upon contacting the mold. Because of the low strength of the materials formed, the pressure differential caused by the vacuum is usually sufficient for forming, although air pressure or mechanical means are also applied for some parts.

Typical parts made this way are advertising signs, refrigerator liners, packaging, appliance housings, and panels for shower stalls. Parts with openings or holes cannot be formed because the pressure differential cannot be maintained during forming. Because thermoforming is a drawing and stretching operation, much like sheet-metal forming, the material should exhibit high uniform elongation—otherwise, it will neck and fail. Thermoplastics

| Straight vacuum forming | Drape vacuum forming | Force above sheet | Plug and ring forming |

a. Heater       d. Mold
b. Clamp        e. Vacuum line
c. Plastic sheet

**FIGURE 10.28**  Various thermoforming processes for thermoplastic sheet. These processes are commonly used in making advertising signs, cookie and candy trays, panels for shower stalls, and packaging.

have a high capacity for uniform elongation by virtue of their high strain-rate sensitivity exponent, $m$. The sheets used in thermoforming are made by sheet extrusion.

Molds for thermoforming are usually made of aluminum since high strength is not a requirement. The holes in the molds are usually less than 0.5 mm (0.02 in.) in order not to leave any marks on the formed sheets. Tooling is inexpensive. Quality considerations include tears, nonuniform wall thickness, improperly filled molds, and poor part definition (surface details).

### 10.10.6 Compression molding

In **compression molding,** a preshaped charge of material, a premeasured volume of powder, or a viscous mixture of liquid resin and filler material is placed directly in a heated mold cavity. Forming is done under pressure with a plug or the upper half of the die (Fig. 10.29). Compression molding results in flash formation; it is removed by trimming or some other

**FIGURE 10.29** Types of compression molding, a process similar to forging: (a) positive, (b) semipositive, and (c) flash. The flash in part (c) has to be trimmed off. (d) Die design for making a compression-molded part with undercuts. Such designs are also used in other molding and shaping operations.

means. Typical parts made are dishes, handles, container caps, fittings, electrical and electronic components, washing-machine agitators, and housings. Fiber-reinforced parts with long chopped fibers are exclusively formed by this process.

Compression molding is used mainly with thermosetting plastics, with the original material in a partially polymerized state. Cross-linking is completed in the heated die, with curing times ranging from 0.5 to 5 min, depending on the material and part geometry and its thickness. The thicker the material, the longer it will take to cure. Elastomers also are shaped by compression molding.

Because of their relative simplicity, die costs in compression molding are generally lower than in injection molding. Three types of compression molds are available: (1) *flash-type* for shallow or flat parts, (2) *positive* for high density, and (3) *semipositive* for quality production. Undercuts in parts are not recommended; however, dies can be designed to open sideways (Fig. 10.29d) to allow removal of the part. In general, the part complexity is less than with injection molding and dimensional control is better.

### 10.10.7  Transfer molding

**Transfer molding** represents a further development of compression molding. The uncured thermosetting material is placed in a heated transfer pot or chamber (Fig. 10.30). After the material is heated, it is injected into heated, closed molds. Depending on the type of machine used, a ram, plunger, or rotating screw-feeder forces the material to flow through the narrow channels into the mold cavity. This flow generates considerable heat, which raises the temperature of the

**FIGURE 10.30**  Sequence of operations in transfer molding for thermosetting plastics. This process is particularly suitable for intricate parts with varying wall thickness.

material and homogenizes it. Curing takes place by cross-linking. Because the resin is molten as it enters the molds, the complexity of the part and dimensional control approach those for injection molding.

Typical parts made by transfer molding are electrical and electronic components and rubber and silicone parts. The process is particularly suitable for intricate shapes having varying wall thicknesses. Molds tend to be more expensive than those for compression molding, and material is wasted in the channels of the mold during filling. (See also resin transfer molding, Section 10.11.1.)

### 10.10.8  Casting

Some thermoplastics, such as nylons and acrylics, and thermosetting plastics, such as epoxies, phenolics, polyurethanes, and polyester, can be cast in rigid or flexible molds into a variety of shapes (Fig. 10.31a). Typical parts **cast** are large gears, bearings, wheels, thick sheets, and components requiring resistance to abrasive wear.

In casting thermoplastics, a mixture of monomer, catalyst, and various additives is heated and poured into the mold. The part is formed after polymerization takes place at ambient pressure. Intricate shapes can be formed with flexible molds, which are then peeled off. Centrifugal casting is also used with plastics, including reinforced plastics with short fibers. Thermosets are cast in a similar manner. Typical parts produced are similar to those made by thermoplastic castings. Degassing may be necessary for product integrity.

**Potting and encapsulation.**    A variation of casting that is important to the electrical and electronics industry is potting and encapsulation. This involves casting the plastic around an electrical component, thus embedding it in the plastic. **Potting** (Fig. 10.31b) is done in a

**FIGURE 10.31**    (a) Schematic illustration of (a) casting, (b) potting, and (c) encapsulation of plastics.

housing or case, which is an integral part of the product. In **encapsulation** (Fig. 10.31c), the component is covered with a layer of the solidified plastic. In both applications the plastic serves as a dielectric (nonconductor). Structural members, such as hooks and studs, may also be partly encapsulated.

**Foam molding/casting.**   Products such as styrofoam cups and food containers, insulating blocks, and shaped packaging materials (such as for cameras, appliances, and electronics), are made by **foam molding.** The material is made of expandable polystyrene in which polystyrene *beads* (obtained by polymerization of styrene monomer), containing a blowing agent, are placed in a mold and exposed to heat, usually by steam. As a result, the beads expand to as much as 50 times their original size and take the shape of the mold. The amount of expansion can be controlled through temperature and time. A common method of molding is using pre-expanded beads, in which beads are expanded by steam (although hot air, hot water, or an oven can also be used) in an open-top chamber. They are then placed in a storage bin and allowed to stabilize for a period of 3 to 12 hours. The beads can then be molded into shapes as described above.

Polystyrene beads are available in three sizes: small for cups, medium for molded shapes, and large for molding of insulating blocks (which can then be cut to size). Thus, the bead size chosen depends on the minimum wall thickness of the product. Beads can also be colored prior to expansion, or are available as integrally colored.

Polyurethane *foam* processing, for products such as cushions and insulating blocks, involves several processes but basically they consist of mixing two or more chemical components. The reaction forms a cellular structure, which solidifies in a mold. Various low-pressure and high-pressure machines, with computer controls for proper mixing, are available.

### 10.10.9   Cold forming and solid-phase forming

Processes that have been used in cold working of metals, such as rolling, deep drawing, extrusion, closed-die forging, coining, and rubber forming, can also be used to form many thermoplastics at room temperature (**cold forming**). Typical materials formed are polypropylene, polycarbonate, ABS, and rigid PVC. The important considerations are that (a) the material be sufficiently ductile at room temperature (hence polystyrenes, acrylics, and thermosets cannot be formed); and (b) the material's deformation must be nonrecoverable (to minimize springback and creep).

The advantages of cold forming of plastics over other methods of shaping are:

*   Strength, toughness, and uniform elongation are increased.
*   Plastics with high molecular weight can be used to make parts with superior properties.
*   Forming speeds are not affected by part thickness since there is no heating or cooling involved. Typical cycle times are shorter than molding processes.

**Solid-phase forming** is carried out at a temperature about $10-20°C$ $(20-40°F)$ below the melting temperature of the plastic, if a crystalline polymer, and formed while still in a solid state. The advantages over cold forming are that forming forces and springback are lower. These processes are not as widely used as hot-processing methods and are generally restricted to special applications.

### 10.10.10 Processing elastomers

Elastomers can be formed by a variety of processes used for shaping thermoplastics. Thermoplastic elastomers are commonly shaped by extrusion and injection molding, extrusion being the most economical and the fastest process. In terms of its processing characteristics, a thermoplastic elastomer is a polymer; in terms of its function and performance, it is a rubber. These polymers can also be formed by blow molding and thermoforming. Thermoplastic polyurethane can be shaped by all conventional methods. It can also be blended with thermoplastic rubbers, polyvinyl chloride compounds, ABS, and nylon. Dryness of the materials is important. For extrusion, the temperatures are in the range of 170°C to 230°C (340°F to 440°F), and for molding up to 60°C (140°F). Typical extruded products are tubing, hoses, moldings, and inner tubes. Injection-molded products cover a broad range of applications such as components for automobiles and appliances.

Rubber and some thermoplastic sheets are formed by the **calendering** process (Fig. 10.32), wherein a warm mass of the compound is fed through a series of rolls (*masticated*) and is then stripped off in the form of a sheet. The rubber may also be formed over both surfaces of a fabric liner. Discrete rubber products, such as gloves, are made by dipping a form repeatedly into a liquid compound that adheres to the form. It is then vulcanized, usually in steam, and stripped from the form.

## 10.11 Processing of Reinforced Plastics

As described earlier, reinforced plastics are among the most important materials and can be engineered to meet specific design requirements such as high strength-to-weight and stiffness-to-weight ratios and creep resistance. Because of their unique structure, reinforced plastics require special methods to shape them into useful products (Fig. 10.33).

The care required and the several steps involved in manufacturing reinforced plastics make processing costs substantial, and generally are not competitive with processing traditional materials and shapes. This situation has necessitated the careful assessment and integration of the design and manufacturing processes (concurrent engineering) in order to minimize costs while maintaining product integrity and production rate. An important environmental concern in reinforced plastics is the dust generated during processing, such as airborne carbon fibers which are known to remain in the work area long after fabrication of parts has been completed.

Reinforced plastics can usually be fabricated by the methods described in this chapter, with some provision for the presence of more than one type of material in the composite. The

**FIGURE 10.33** Reinforced plastic components for a Honda motorcycle. Parts shown are front and rear forks, rear swingarm, wheel, and brake disks.

reinforcement may be chopped fibers, woven fabric or mat, roving or yarn (slightly twisted fiber), or continuous lengths of fiber. In order to obtain good bonding between the reinforcing fibers and the polymer matrix as well as protect the fibers during subsequent processing, fibers are surface treated by impregnation (*sizing*). Short fibers are commonly added to thermoplastics for injection molding, milled fibers can be used for reaction-injection molding, and longer chopped fibers are primarily used in compression molding of reinforced plastics.

When the impregnation is done as a separate step, the resulting partially cured sheets are called (1) **prepregs,** (2) **bulk-molding compound (BMC),** or (3) **sheet-molding compound (SMC),** depending on their form. They should be stored at a sufficiently low temperature to delay curing. Alternatively, the resin and the fibers can be mixed together at the time they are placed in the mold. More recently, a **thick molding compound** (**TMC,** a trademark) has been developed that combines the characteristics of BMC (lower cost) and SMC (higher strength). Thick molding compounds are usually injection molded using chopped fibers of various lengths. One application is for electrical components because of the high dielectric strength of TMC.

Commercial reinforced plastics are available as sheet-molding compounds. Continuous strands of reinforcing fiber are chopped into short fibers (Fig. 10.34) and deposited over a layer of resin paste, usually a polyester mixture, carried on a polymer film such as polyethylene. A second layer of resin paste is deposited on top, and the sheet is pressed between rollers.

**FIGURE 10.34** Manufacturing process for producing reinforced plastic sheets. The sheet is still viscous at this stage and can later be shaped into various products. *Source:* T.-W. Chou, R. L. McCullough, and R. B. Pipes.

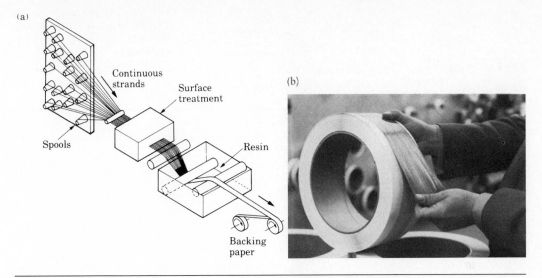

**FIGURE 10.35**  (a) Manufacturing process for polymer-matrix composite. *Source:* T.-W. Chou, R. L. Mc-Cullough, and R. B. Pipes. (b) Boron-epoxy prepreg tape. *Source:* Textron Systems.

The product is gathered into rolls, or placed into containers in layers, and stored until it undergoes a maturation period, reaching the desired molding viscosity. The maturing process is under controlled temperature and humidity, and usually takes one day. The matured SMC, which has a leatherlike feel and is tack-free, has a shelf life of about 30 days and must be processed within this period.

A typical procedure for making reinforced plastic prepregs is shown in Fig. 10.35a. The continuous fibers are aligned and subjected to surface treatment to enhance adhesion to the polymer matrix. They are then coated by dipping them in a resin bath and made into a sheet or tape (Fig. 10.35b). Individual pieces of the sheet are then assembled into laminated structures, such as the horizontal stabilizer for the F-14 fighter aircraft. Special computer-controlled tape-laying machines have been developed for this purpose. Typical products are flat or corrugated architectural paneling, panels for construction and electric insulation, and structural components of aircraft.

**Example 10.9:**  Tennis rackets made of composite materials

In order to impart certain desirable characteristics, such as light weight and stiffness, composite-material tennis rackets are being manufactured with graphite, fiberglass, boron, ceramic (silicon carbide), and Kevlar as the reinforcing fibers. Rackets have a foam core; some have unidirectional and others have braided reinforcement. Rackets with boron fibers have the highest stiffness, followed by graphite (carbon), glass, and Kevlar. The racket with the lowest stiffness has 80 percent fiberglass, whereas the stiffest has 95 percent graphite and 5 percent boron. Thus it has the highest percentage of inexpensive reinforcing fiber and the smallest percentage of the most expensive fiber.

(a)

(b)

**FIGURE 10.36** (a) Vacuum-bag forming. (b) Pressure-bag forming. *Source:* T. H. Meister.

### 10.11.1 Molding

In compression molding, the material is placed between two molds and pressure is applied. Depending on the material, the molds may be either at room temperature or heated to accelerate hardening. The material may be in bulk form (bulk-molding compound), which is a viscous, sticky mixture of polymers, fibers, and additives. It is generally shaped into a log which is cut into the desired mass. Fiber lengths generally range from 3 to 50 mm (0.125 to 2 in.), although longer fibers (75 mm or 3 in.) may also be used.

Sheet-molding compounds can also be used in molding. Sheet-molding compound is similar to BMC, except that the resin-fiber mixture is laid between plastic sheets to make a sandwich that can be easily handled. The sheets are removed when the SMC is placed in the mold.

In **vacuum-bag molding** (Fig. 10.36), prepregs are laid in a mold to form the desired shape. In this case, the pressure required to form the shape and create good bonding is obtained by covering the lay-up with a plastic bag and creating a vacuum. If additional heat and pressure are desired, the entire assembly is put in an autoclave. Care should be exercised to maintain fiber orientation, if specific fiber orientations are desired. In chopped-fibers materials, no specific orientation is intended.

In order to prevent the resin from sticking to the vacuum bag and to facilitate removal of any excess, several sheets of various materials (*release cloth, bleeder cloth*) are placed on top of the prepreg sheets. The molds can be made of metal, usually aluminum, but more often are made from the same resin (with reinforcement) as the material to be cured. This eliminates any problem with differential thermal expansion between the mold and the part.

**Contact molding** processes use a single male or female mold (Fig. 10.37) made of materials such as reinforced plastics, wood, or plaster. Contact molding is used in making products with high surface area-to-thickness ratios, such as swimming pools, boats, tub and shower units, and housings. This is a "wet" method, in which the reinforcement is impregnated with the resin at the time of molding. The simplest method is called *hand lay-up*. The

**FIGURE 10.37**  Manual methods of processing reinforced plastics: (a) hand lay-up and (b) spray-up. These methods are also called *open-mold* processing.

materials are placed and formed in the mold by hand (Fig. 10.37a), and the squeezing action expels any trapped air and compacts the part.

Molding may also be done by spraying (*spray-up*; Fig. 10.37b). Although spraying can be automated, these processes are relatively slow and labor costs are high. However, they are simple and the tooling is inexpensive. Only the mold-side surface of the part is smooth and the choice of materials is limited. Many types of boats are made by this process.

**Resin transfer molding** is based on transfer molding (Section 10.10.7) whereby a resin, mixed with a catalyst, is forced by a piston-type positive displacement pump into the mold cavity filled with fiber reinforcement. The process is a viable alternative to hand lay-up, spray-up, and compression molding for low or intermediate volume production. Most of the exterior body of the 1992 Dodge Viper RT/10 (hood, front fenders, rear quarter panels, decklid, and roof support) is made from resin-transfer-molded components. They are 40 percent lighter than similar metal parts.

### 10.11.2  Filament winding, pultrusion, and pulforming

**Filament winding** is a process whereby the resin and fibers are combined at the time of curing. Axisymmetric parts, such as pipes and storage tanks, as well as asymmetric parts are produced on a rotating mandrel. The reinforcing filament, tape, or roving is wrapped continuously around the form. The reinforcements are impregnated by passing them through a polymer bath (Fig. 10.38a). The process can be modified by wrapping the mandrel with prepreg material.

The products made by filament winding are very strong because of their highly reinforced structure. Filament winding has also been used for strengthening cylindrical or spherical pressure vessels (Fig. 10.38b) made of materials such as aluminum and titanium. The presence of a metal inner lining makes the part impermeable. Filament winding can be used directly over solid-rocket propellant forms. Seven-axis computer-controlled machines have been developed for making asymmetric parts that automatically dispense several unidirectional prepregs. Typical asymmetric parts made are aircraft engine ducts, fuselages, propellers, blades, and struts.

**FIGURE 10.38** (a) Schematic illustration of the filament-winding process. (b) Fiberglass being wound over aluminum liners for slide-raft inflation vessels for the Boeing 767 aircraft. *Source:* Brunswick Corporation, Defense Division.

**Pultrusion.** Long shapes with various constant profiles, such as rods, profiles, or tubing (similar to drawn metal products), are made by the **pultrusion** process. Typical products are golf clubs, drive shafts, and structural members such as ladders, walkways, and handrails. In this process, developed in the early 1950s, the continuous reinforcement (roving or fabric) is pulled through a thermosetting polymer bath, and then through a long heated steel die (Fig. 10.39). The product is cured during its travel through the die and cut into desired lengths. The most common material used in pultrusion is polyester with glass reinforcements.

**Pulforming.** Continuously reinforced products other than constant cross-sectional profiles are made by **pulforming**. After being pulled through the polymer bath, the composite is clamped between the two halves of a die and cured into a finished product. The dies recirculate and shape the products successively. Common examples are glass-fiber reinforced hammer handles and curved automotive leaf springs.

### 10.11.3    Quality considerations

The major quality considerations for the processes described in this section involve internal voids and gaps between successive layers of material. Volatile gases developing during processing must be allowed to escape from the lay-up through the vacuum bag to avoid porosity due to trapped gases within the lay-up. Microcracks also may develop due to improper

**FIGURE 10.39** Schematic illustration of the pultrusion process.

curing or during transportation and handling. These defects can be detected using ultrasonic scanning and other techniques (see Section 4.8.1).

## 10.12 Design Considerations

Design considerations for forming and shaping plastics are somewhat similar to those for processing of metals. However, the mechanical and physical properties of plastics should be carefully considered during design, material, and process selection.

Selection of an appropriate material from an extensive list requires consideration of service requirements and possible long-range effects on properties and behavior, such as dimensional stability and wear as well as their ultimate disposal after completion of their life cycle. Some of these issues are described in Section 1.10.

Compared to metals, plastics have lower strength and stiffness, although the strength-to-weight and stiffness-to-weight ratios for reinforced plastics are higher than for many metals. Thus section sizes should be selected accordingly, with a view to maintaining a sufficiently high section modulus for improved stiffness. Reinforcement with fibers and particles can also be highly effective in achieving this objective, as can designing sections with a high ratio of moment of inertia to cross-sectional area.

One of the major design advantages of reinforced plastics is the directional nature of the strength of the material (see, for example, Fig. 10.19). Forces applied to the material are transferred by the resin matrix to the fibers, which are much stronger and stiffer than the matrix. When fibers are all oriented in one direction, the resulting material is exceptionally strong in the fiber direction. This property is often utilized in designing reinforced plastic structures. For strength in two principal directions, the unidirectional materials are often laid at different angles to each other. If strength in the third (thickness) direction is desired, a different type of material is used to form a sandwich structure.

**FIGURE 10.40** Examples of design modifications to eliminate or minimize distortion of plastic parts. (a) Suggested design changes to minimize distortion. *Source:* F. Strasser. (b) Die design (exaggerated) for extrusion of square sections. Without this design, product cross-sections swell because of the recovery of the material, which is known as *die swell*. (c) Design change in a rib to minimize pull-in caused by shrinkage during cooling. (d) Stiffening of bottom of thin plastic containers by doming, which is similar to the process used to make the bottoms of aluminum beverage cans.

Physical properties, especially high coefficient of thermal expansion, and hence contraction, are important. Improper part design or assembly can lead to warping and shrinking (Fig. 10.40a). Plastics can easily be molded around metallic parts and inserts. However, their compatibility with metals when so assembled is an important consideration.

The overall part geometry often determines the particular forming or molding process. Even after a particular process is selected, the design of the part and die should be such that it will not cause problems concerning shape generation (Fig. 10.40b), dimensional control, and surface finish. As in casting metals and alloys, material flow in the mold cavities should be controlled properly. The effects of molecular orientation during processing should also be considered, especially in processes such as extrusion, thermoforming, and blow molding.

Large variations in cross-section sizes (Fig. 10.40c) and abrupt changes in geometry should be avoided for better product quality and increased mold life. Furthermore, contraction in large cross-sections tends to cause porosity in plastic parts. Conversely, because of a lack of stiffness, removing thin sections from molds after shaping may be difficult. The low elastic modulus of plastics further requires that shapes be selected properly for improved stiffness of the component (Fig. 10.40d), particularly when saving materials is important. These considerations are similar to those in designing metal castings and forgings.

The properties of the final product depend on the original material and its processing history. Cold working of polymers improves their strength and toughness. On the other hand, because of the nonuniformity of deformation (even in simple rolling), residual stresses develop in polymers, as they do in metals. These stresses can also be generated by thermal cycling of the part. Also, as in metals, the magnitude and direction of residual stresses, however produced, are important factors. These stresses can relax over a period of time and cause distortion of the part during its service life.

## 10.13  Economics of Processing Plastics

We have described a number of processes that are used to form and shape plastics and composite materials. As in all other processes, design and manufacturing decisions are ultimately based on performance and cost, including the costs of equipment, tooling, and production. The final selection of a process depends greatly on production volume. High equipment and tooling costs can be acceptable if the production run is large, as is the case in casting and forging.

Various types of equipment are used in plastics forming and shaping processes. The most expensive are injection-molding machines, with costs being directly proportional to clamping force. A machine with a 2000-kN (225-ton) clamping force costs about $100,000, and one with a 20,000-kN (2250-ton) clamping force costs about $450,000.

The optimum number of cavities in the die for making the product in one cycle is an important consideration, as in die casting. For small parts, a number of cavities can be made in a die, with runners to each cavity. If the part is large, then only one cavity may be accommodated. As the number of cavities increases, so does the cost of the die. Larger dies may be considered for larger numbers of cavities, increasing die cost even further. On the other hand, more parts will be produced per machine cycle, thus increasing the production rate. Hence a detailed analysis has to be made to determine the optimum number of cavities, die size, and machine capacity. Similar considerations apply to other plastics processing methods. Tables 10.7 and 10.8 are general guides to economical processing of plastics and composite

**TABLE 10.7**

**COMPARATIVE COSTS AND PRODUCTION VOLUMES FOR PROCESSING OF PLASTICS**

| | EQUIPMENT CAPITAL COST | PRODUC-TION RATE | TOOLING COST | TYPICAL PRODUCTION VOLUME, NUMBER OF PARTS (10 – $10^7$) | | | | | | |
|---|---|---|---|---|---|---|---|---|---|---|
| | | | | 10 | $10^2$ | $10^3$ | $10^4$ | $10^5$ | $10^6$ | $10^7$ |
| Machining | Med | Med | Low | ⊢————⊣ | | | | | | |
| Compression molding | High | Med | High | | ⊢——————————⊣ | | | | | |
| Transfer molding | High | Med | High | | ⊢——————————⊣ | | | | | |
| Injection molding | High | High | High | | | | ⊢————————⊣ | | | |
| Extrusion | Med | High | Low | * | | | | | | |
| Rotational molding | Low | Low | Low | | ⊢——⊣ | | | | | |
| Blow molding | Med | Med | Med | | ⊢——————————————⊣ | | | | | |
| Thermoforming | Low | Low | Low | | ⊢———⊣ | | | | | |
| Casting | Low | Very low | Low | ⊢———⊣ | | | | | | |
| Forging | High | Low | Med | ⊢———⊣ | | | | | | |
| Foam molding | High | Med | Med | | | | ⊢————————⊣ | | | |

Source: After R.L.E. Brown, *Design and Manufacture of Plastic Parts.* Copyright © 1980 by John Wiley & Sons, Inc. Reprinted by permission of John Wiley & Sons, Inc.
*Continuous process.

materials. Note, for example, the high capital costs for molding plastics and the wide range of production rates. For composite materials, equipment and tooling costs for most molding operations are generally high. Production rates and economic production quantities vary widely.

**TABLE 10.8**

**ECONOMIC PRODUCTION QUANTITIES FOR VARIOUS MOLDING METHODS***

| MOLDING METHOD | RELATIVE INVESTMENT REQUIRED | | RELATIVE PRODUCTION RATE | ECONOMIC PRODUCTION QUANTITY |
|---|---|---|---|---|
| | EQUIPMENT | TOOLING | | |
| Hand lay-up | VL | L | L | VL |
| Spray-up | L | L | L | L |
| Casting | M | L | L | L |
| Vacuum-bag molding | M | L | VL | VL |
| Compression-molded BMC | H | VH | H | H |
| SMC and preform | H | VH | H | H |
| Pressure-bag molding | H | H | L | L |
| Centrifugal casting | H | H | M | M |
| Filament winding | H | H | L | L |
| Pultrusion | H | H | H | H |
| Rotational molding | H | H | L | M |
| Injection molding | VH | VH | VH | VH |

*Source*: J.G. Bralla (ed.), *Handbook of Product Design for Manufacturing.* Copyright © 1986, McGraw-Hill Publishing Company. Reproduced by permission of the McGraw-Hill Companies.
*VL, very low; L, low; M, medium; H, high; VH, very high.

## 10.14    Rapid Prototyping

Making a prototype—that is, a first full-scale model of a product (see Fig. 1.2)—has traditionally involved actual manufacturing processes using a variety of tooling and machines and usually taking weeks or months. An important advance in manufacturing is **rapid prototyping,** also called *desktop manufacturing* or *free-form fabrication,* a process by which a solid physical model of a part is made directly from a three-dimensional CAD drawing. Developed in the mid-1980s, rapid prototyping entails several different consolidation techniques: resin curing, deposition, solidification, and sintering. In this section, we first describe one of the most successful methods, called **stereolithography,** because it involves curing of a resin with a laser beam to make the part.

The importance and economic impact of rapid prototyping can be appreciated by noting that (1) the conceptual design is viewed in its entirety and at different angles on the monitor through a three-dimensional CAD system, as described in Section 14.4; (2) a prototype from various nonmetallic and metallic materials is manufactured and studied thoroughly from functional, technical, and esthetic aspects; and (3) prototyping is accomplished in a much shorter time and at lower cost than by traditional methods.

### 10.14.1    Stereolithography

The **stereolithography** process is based on the principle of curing (hardening) a photopolymer into a specific shape (Fig. 10.41). A vat, containing a mechanism whereby a platform can be lowered and raised vertically, is filled with a photocurable liquid acrylate polymer. The liquid is a mixture of acrylic monomers, oligomers (polymer intermediates), and a photoinitiator. When the platform is at its highest position, the layer of liquid above it is shallow. A *laser,* generating an ultraviolet beam, is then focused along a selected surface area of the photopolymer at surface *a* and moved in the $x-y$ direction. The beam cures that portion of the photopolymer, say, a ring-shaped portion, producing a solid body. The platform is then lowered enough to cover the cured polymer with another layer of liquid polymer, and the sequence is repeated. In Fig. 10.41, the process is repeated until level *b* is reached. Hence, thus far we have a cylindrical part with a constant wall thickness. Note that the platform is now lowered by a vertical distance *ab.*

**FIGURE 10.41**
Schematic illustration of the stereolithography process. *Source:* Courtesy of A. Sidney Alpert, 3D Systems, 26081 Ave. Hall, Valencia, CA 91355.

At level $b$, the $x-y$ movements of the beam are wider so that we now have a flange-shaped piece that is being produced over the previously formed part. After the proper thickness of the liquid has been cured, the process is repeated, producing another cylindrical section between levels $b$ and $c$. Note that the surrounding liquid polymer is still fluid because it has not been exposed to the ultraviolet beam, and that the part has been produced from the bottom up in individual "slices." The unused portion of the liquid polymer can be used again to make another part or another prototype. Note that the word stereolithography, as used to describe this process, comes from the fact that the movements are three-dimensional and the process is similar to lithography, in which the image to be printed on a flat surface is ink-receptive and the blank areas are ink-repellent. After completion, the part is removed from the platform, blotted, cleaned ultrasonically and with an alcohol bath, and subjected to a final curing cycle.

By controlling the movements of the beam and the platform through a servo-control system, a variety of parts can be formed by this process (Fig. 10.42). Total cycle times range from a few hours to a day. Progress is being made toward improvements in (1) accuracy and dimensional stability of the prototypes produced, (2) less expensive liquid modeling materials, (3) CAD interfaces to transfer geometrical data to model-making systems, and (4) strength so that the prototypes made by this process can be truly considered prototypes and models in the traditional sense. Sloping surfaces tend to be relatively rough because of the manner in which layer-by-layer curing takes place.

Depending on capacity, the cost of these machines is in the range from \$100,000 to \$500,000, while the cost of the liquid polymer is on the order of \$300 per gallon. Maximum part size is 0.5 m × 0.5 m × 0.6 m (20 in. × 20 in. × 24 in.). Automotive, aerospace, electronics, and medical industries are among those using stereolithography as a rapid and inexpensive method of producing prototypes, and thus reducing product development cycle times. One major application is in the area of making molds and dies for casting and injection molding.

(a)   (b)

(c)   (d)

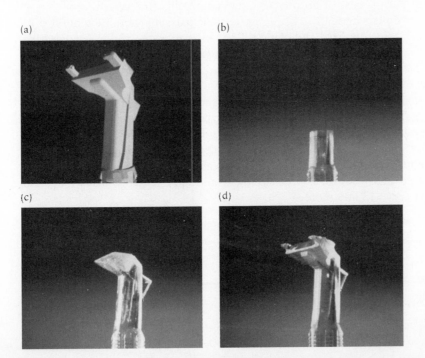

**FIGURE 10.42**
An application of stereolithography. (a) CAD model of a razor, as viewed on the monitor screen. (b)-(d) The physical model (prototype) being built, one layer at a time from the bottom up, using CAD data. *Source:* Courtesy of The Gillette Company.

### 10.14.2   Other methods of rapid prototyping

Several other methods of rapid prototyping are in various stages of development as described below. All use a CAD system interfaced with the machine.

**Optical fabrication.**   Similar to stereolithography, this system uses a visible light argon-ion laser, and the part is built on a stationary platform. Point-by-point solidification of the liquid takes place at the intersection of two laser beams perpendicular to each other. In another method, solidification takes place by holographic interference.

**Fused deposition modeling.**   In this process a thermoplastic filament, similar to wire, is fed through a heated extruding head. The filament, made of materials such as wax or nylon, melts at a temperature just above its solidification state prior to its deposition on a platform to produce the part. The prototype made requires no further curing.

**Solid base curing.**   In this process the part is formed by multiple layers. The photopolymer liquid in each layer is covered with a photomask and cured in a few seconds by a strong ultraviolet lamp. The unexposed liquid polymer is then removed and the voids are filled with molten wax to support the next layer. This sequence is repeated until the entire part is formed.

Additional methods of rapid prototyping that are based on techniques other than those described in this section include three-dimensional printing, ballistic particle manufacturing, photochemical machining, and laminated object manufacturing. These processes are described in Section 11.7.

---

**Example 10.10:   Rapid prototyping in the automotive industry**

It has been estimated that the cost of prototyping new automobile components can run to hundreds of millions of dollars annually. Furthermore, some parts may require a number of design iterations before being finalized and could take at least a year to make. Among various large manufacturing organizations, Ford Motor Company has been one of the early users of rapid prototyping technology. In a specific example, Ford engineers used stereolithography to make a plastic prototype of a new rocker arm, which took about 24 hours to make. In contrast, a conventionally made prototype would have taken about three months to produce and would cost about $30,000 more. The plastic prototype enabled the engineers to make design modifications within a matter of hours and finalize the design.

In another example, the engineers made a model of an experimental manifold with stereolithography. The model was then used in tests to increase air flow and optimize engine power. It was estimated that traditional prototyping would have taken 83 days longer to make and would have cost $12,000 more.

*Source: Manufacturing Engineering, June 1991.*

---

CASE STUDY:   MANUFACTURE OF A TOOTHBRUSH

The toothbrush is a common household item which has evolved to a nearly all-polymer design (the bristles are held in place with metal staples) as plastics have become more available and inexpensive.

## Material Selection

Appropriate materials are defined as those that meet product functional requirements and manufacturing process requirements. In defining product requirements, one should consider loads on the product, environment, expected life, and appearance.

The functional requirements for the toothbrush handle and bristles should be familiar to you. The toothbrush handle and bristle combination must be stiff enough to effectively deliver the brushing strokes from the user to the teeth. Lightweight materials are desirable since the user must support the entire weight of the toothbrush for up to three minutes. Both the handle and bristles must survive a broad range of temperature and humidity conditions. The toothbrush handle must be impact resistant, i.e., not brittle, since it is not uncommon to drop a toothbrush on a hard surface. The bristles must resist wear and be chemically inert (to toothpaste). Typical life for a toothbrush is between three and six months. As for appearance, several handle designs are currently available: clear, colored (translucent), transparent colored, and transparent colored with design stamp. The handle surface finish should be smooth. Neither the bristles nor handle should discolor due to exposure to water or sunlight.

Because the quantity of toothbrushes produced is high, the material should be recyclable (to minimize scrap). The handle shape is slightly curved, with no intricate details. The handle material must be compatible with manufacturing processes that minimize scrap, have a high production-volume capacity, and produce noncomplex shapes. The uniform circular cross-section of the bristle suggests that the bristle material must be compatible with extrusion processes. The product's functional requirements and manufacturing process requirements for a toothbrush are summarized in Table 10.9.

One helpful method for selecting a material objectively is to construct a table that lists critical properties for the materials under consideration. Critical material properties would be those required for product function as well as material cost and availability. For the toothbrush handle, we can narrow down the materials choices to thermoplastic materials, given

**TABLE 10.9**

**FUNCTIONAL AND MANUFACTURING REQUIREMENTS FOR A TOOTHBRUSH**

| CHARACTERISTIC | REQUIREMENT |
| --- | --- |
| Stiffness | Moderate |
| Temperature | $-25°C - > 40\ °C$ |
| Humidity | Up to 100% relative humidity |
| Environment | Chemically inert, sunlight resistant |
| Impact resistance | Tough |
| Appearance | Transparent or color tinted |
| Wear (bristles) | Durable |
| Life | 3-6 months |
| Surface | Smooth, moderate detail |
| Recyclable | Yes |
| Geometry (handle) | Noncomplex, thick cross-section |
| Geometry (bristle) | Cylindrical cross-section |

the requirement for recyclability. Among thermoplastics, the selection can be further narrowed by requiring that the material be amorphous (for transparency). A summary of material properties for various transparent thermoplastics, as obtained from technical literature, is shown in Table 10.10. Glass-transition temperature $T_g$ is included as a critical property because this temperature must be higher than the highest environmental temperature encountered. The data in Table 10.10 indicate that acrylic and cellulosic materials would be appropriate choices (see also Table 10.3). Cellulosic materials are easy to process, exhibit good transparency and surface gloss, and are available in many colors. These materials meet the functional requirements at the lowest cost and are readily available.

Following a procedure similar to materials selection for the handle, the material choices for the bristles can be narrowed down to those thermoplastics that exhibit good wear resistance: acetal, nylon, polyurethane, and ultrahigh-molecular-weight polyethylene. The material should be available as a fiber. A comparison of these materials (Table 10.11) indicates that nylon should be selected. It is a low-cost material that meets the functional requirements and is available as a fiber.

## Manufacturing process selection

In selecting a manufacturing process, consider Tables 10.7 and 10.8, which list process characteristics and material limitations for various processes. From these, we can see that the following part qualities must be considered: (1) part shape, (2) part wall thickness, (3) part detail and tolerances, (4) material, and (5) production volume and rate.

For the toothbrush handle, both cellulosics and acrylic parts can be produced by compression molding, injection molding, extrusion, thermoforming, and calendering. Among these processes, the method that can best produce the handle shape is injection molding. Compression molding is also a possibility; however, as discussed in this chapter, this process is mainly used with thermosets. The one advantage of compression molding for this application is the relatively lower cost (compared to injection molding) of the molds. On the other hand, since the production volume of toothbrushes is high, the fixed cost associated with molds can be easily offset by the increase in production capacity associated with injection molding.

Once the process is selected, we would like to know the process settings. In injection molding, the main process variables are cycle time, injection pressure, and temperature. In specifying injection-molding equipment, clamping force and shot size are also important. Typical injection pressures are in the range of 5–20 ksi. Specific guidelines for acrylic and cellulosics fall within these ranges as well. In the toothbrush handle, the area where bristles are stapled consists of several precisely aligned rows of cylindrical holes. To achieve such detail and to ensure adequate resin flow to the area, high pressures on the order of 15-20 ksi should be used. (Note that this is consistent with Example 10.8, in which 15 ksi is recommended for injection molding of gears to achieve part detail.)

Process temperatures are based on the polymer melt temperature and can be obtained from material suppliers. As described in the text, part cooling time is determined by the ratio of the volume to the surface area. A particular challenge with toothbrush handles is the handle thickness (see Fig. 10.43). Most injection-molded parts have much thinner cross-sections than a toothbrush (typical thicknesses are less than 0.125 in.). High injection pressure to ensure complete filling and long cooling times (up to 1 minute) are required.

**TABLE 10.10**

**TRANSPARENT THERMOPLASTICS FOR HANDLE**

| | $T_g$ (°C) | MAX SERVICE TEMPERATURE (°C) | ELASTIC MODULUS (GPa) | STRENGTH (MPa) | CHEMICALLY INERT | IMPACT RESISTANCE, IZOD (ft.lb/in) | COST $/kg |
|---|---|---|---|---|---|---|---|
| Acrylic | | 60–110 | 1.4–3.5 | High 40–75 | Yes | Moderate 0.8–2.3 | 1.50 |
| Polycarbonate | 150 | 120 | 2.5–3.0 | High 55–70 | Yes | High 12–18 | 2.75 |
| Polystyrene | 100 | 60–80 | 1.4–4 | 14–83 | No | No, brittle 0.2–0.4 | 0.80 |
| Polysulfone | | 150–260 | 2.5 | 69 | Somewhat | Good, ductile to −100 °C 1.3 | 7.90 |
| Cellulosics | | 50–90 | 0.4–1.4 | 10–48 | Moderate | Tough | 2.20 |

**TABLE 10.11**

**PLASTICS FOR BRISTLES**

| | STRENGTH (MPa) | ELASTIC MODULUS (GPa) | WEAR RESISTANT | CHEMICALLY INERT | $T_g$ (°C) | COST $/kg | AVAILABILITY | APPLICATIONS |
|---|---|---|---|---|---|---|---|---|
| Acetal | 55–70 | 1.4–3.5 | Yes | Yes, moisture resistant | 27–40.5 | 2.45 | Inj. mold, extrusion, blow molding, | High performance components |
| Nylon | 55–83 | 1.4–2.8 | Yes | Yes, some types moisture resistant | 57 | 4.30 | Inj. molding, extrusion | Bristles, sutures |
| Polyurethane (elastomer) | 34 | 0.035 | Yes | Yes, good sunlight | | 5.15 | Usually cast | Seals, gaskets |
| UHMWPE* | 21.4–42.8 | 0.69–1.03 | Yes | Yes | 120 | Very expensive | Extrusion, compression molding | Refrig. parts, sterilizable housewares |

*Ultra-high molecular weight polyethylene

(a) Handle, top view    Holes for bristles
0.05" diameter

Ejector pins leave
circular imprint on
part

0.38"

0.36"

0.90"

0.47"

4.0"

6.25"

(b) Side view    0.150" deep

0.215"    Parting line

Gate inlet, rough
area on handle
0.10" × 0.10", approximately

(c) Head    Staple:  0.055" wide
0.080" high
0.010" thick
Bristle:  1.10" in length
0.005" in diameter

26 bristles

Metal "staple" to fix
bristles into handle

**FIGURE 10.43**  Model for children's toothbrush with animal design on the handle (not shown). The figure also shows some features of the injection mold.

Equipment requirements can be determined by considering the injection pressure and part projected area. From inspection, the parting line occurs longitudinally along the handle at the midplane (Fig. 10.43). For each toothbrush, the clamping force required is = (20,000 psi) (projected area). The amount of material required per toothbrush handle is = (projected area)(thickness). If the machine shot size and total clamping force are known, then the number of toothbrushes per cycle is defined as total clamping force/clamping force per toothbrush handle, or number of toothbrushes = machine shot size/volume per toothbrush handle.

Nylon bristles are extruded by a supplier and purchased in spools. The constant circular cross-section and high volume of material supplied in spools indicate that the bristles have been manufactured by extrusion. Bristles are stapled into the toothbrush handle with corrosion-resistant staples (Ni-Cr plated steel or stainless steel). They are then trimmed to the desired shape and dimensions.

*Source:* Susan C. Mantell, University of Minnesota

# SUMMARY

- Polymers are an important class of materials because they possess a very wide range of mechanical, physical, chemical, and optical properties. Compared to metals, plastics are generally characterized by lower density, strength, elastic modulus, and thermal and electrical conductivity, and a higher coefficient of thermal expansion. (Section 10.1)

- Plastics are composed of polymer molecules and various additives. The smallest repetitive unit in a polymer chain is called a mer. Monomers are linked by polymerization processes to form larger molecules. The properties of the polymer depend on the molecular structure (linear, branched, cross-linked, and network), the degree of crystallinity, and additives (with various functions, such as improving strength, flame retardation, and lubrication, as well as imparting flexibility, color, and stability against ultraviolet radiation and oxygen). The glass-transition temperature separates the brittle and ductile regions of polymers. (Section 10.2)

- Two major classes of polymers are thermoplastics and thermosets. Thermoplastics become soft and are easy to form at elevated temperatures. Their mechanical behavior can be characterized by various spring and dashpot models, and includes phenomena such as creep and stress relaxation, crazing, and water absorption. (Section 10.3)

- Thermosets, which are obtained by cross-linking polymer chains, do not become soft to any significant extent with increasing temperature. (Section 10.4)

- Thermoplastics and thermosets have a very wide range of consumer and industrial applications with a variety of characteristics. (Sections 10.5, 10.6)

- Among important developments in polymers is biodegradable plastics; several formulations have been developed. (Section 10.7)

- Elastomers have the characteristic ability to undergo large elastic deformations and return to their original shapes when unloaded. Consequently, they have important applications as tires, seals, footwear, hoses, belts, and shock absorbers. (Section 10.8)

- Reinforced plastics are an important class of materials that have superior mechanical properties and are lightweight. The reinforcing fibers are usually glass, graphite, aramids, and boron; epoxies commonly serve as a matrix material. The properties of reinforced plastic depend on its composition and the orientation of fibers. Reinforced plastics are being developed rapidly, with a wide range of important applications. (Section 10.9)

- Plastics can be formed and shaped by a variety of processes, such as extrusion, various molding operations, casting, and thermoforming. Thermosets are generally molded or cast, and thermoplastics are shaped by these processes as well as by thermoforming and other techniques commonly used in metalworking operations. (Section 10.10)

- Reinforced plastics are processed into structural components using liquid plastics, prepregs, and bulk-and sheet-molding compounds. Fabricating techniques include various molding methods, filament winding, and pultrusion. Important factors in fabricating reinforced-plastic components are the type and orientation of the fibers, and the strength of the bond between fibers and matrix and between layers of materials. Inspection techniques are available to determine the integrity of these products. (Section 10.11)

- The design of plastic parts includes considerations of their low strength and stiffness, as well as physical properties such as high thermal expansion and generally low resistance to temperature. (Section 10.12)

- Because of the variety of low-cost materials and processing techniques available, the economics of processing plastics and reinforced plastics is an important consideration, particularly when compared to metal components, and includes die costs, cycle times, and production volume as important parameters. (Section 10.13)
- Rapid prototyping has become an increasingly important technology because of its inherent flexibility and the much shorter times required for making prototypes. A major example is stereolithography, which involves curing of a photopolymer into desired shapes using a laser beam and computer-aided design and controls. (Section 10.14)

## SUMMARY OF EQUATIONS

Linearly elastic behavior in tension, $\sigma = E\varepsilon$

Linearly elastic behavior in shear, $\tau = G\gamma$

Viscous behavior, $\tau = \eta\,(dv/dy) = \eta\dot{\gamma}$

Strain-rate sensitivity, $\sigma = C\dot{\varepsilon}^m$

Viscoelastic modulus, $E_r = \sigma/(e_e + e_v)$

Viscosity, $\eta = \eta_o e^{E/kT}$

$\log \eta = 12 - (17.5\Delta T)/(52 + \Delta T)$

Stress relaxation, $\tau = \tau_o e^{-t/\lambda}$

Relaxation time, $\lambda = \eta/G$

Strength of composite, $\sigma_c = x\sigma_f + (1 - x)\sigma_m$

Elastic modulus of composite, $E_c = xE_f + (1 - x)E_m$

## BIBLIOGRAPHY

Agarwal, B.D., and L.J. Broutman, *Analysis and Performance of Fiber Composites*, 2d ed. New York: Wiley, 1990.

*ASM Engineered Materials Handbook, Desk Edition.* Materials Park, Ohio: ASM International, 1995.

Barlow, F., *Rubber Compounding: Principles, Materials, and Techniques*, 2d ed. New York: Marcel Dekker, 1993.

Bhowmick, A.K., and H.L. Stephens, *Handbook of Elastomers.* New York: Marcel Dekker, 1988.

———, *Rubber Products Manufacturing Technology.* New York: Marcel Dekker, 1994.

Bicerano, J., *Prediction of Polymer Properties.* New York: Marcel Dekker, 1995.

Bittence, J.C., (ed.), *Engineering Plastics and Composites.* Materials Park, Ohio: ASM International, 1990.

Burns, M., *Automated Fabrication: Improving Productivity in Manufacturing.* Englewood Cliffs, New Jersey: PTR Prentice-Hall, 1993.

Chabot, J.F., Jr., *The Development of Plastics Processing Machinery and Methods.* New York: Wiley, 1992.

Chanda, M., and S.K. Roy, *Plastics Technology Handbook,* 2d ed. New York: Marcel Dekker, 1992.

Charrier, J.-M., *Polymeric Materials and Processing: I Plastics, Elastomers, and Composites.* New York: Hanser, 1991.

Cheremisinoff, N.P., *Handbook of Ceramics and Composites,* 3 vols. New York: Marcel Dekker, 1991.

——, *Polymer Mixing and Extrusion Technology.* New York: Marcel Dekker, 1987.

——, *Product Design and Testing of Polymeric Materials.* New York: Marcel Dekker, 1990.

Daniel, I.M., and O. Ishai, *Engineering Mechanics of Composite Materials.* New York: Oxford, 1994.

Dumitriu, S., *Polymeric Biomaterials.* New York: Marcel Dekker, 1994.

Easterling, K., *Tomorrow's Materials,* 2d ed. London: The Institute of Metals, 1990.

*Engineered Materials Handbook,* Vol. 1: *Composites,* 1987; Vol. 2: *Engineering Plastics,* 1988. Metals Park, Ohio: ASM International.

*Engineering Plastics and Composites,* 2d ed. Materials Park, Ohio: ASM International, 1993.

Florian, J., *Practical Thermoforming: Principles and Applications.* New York: Marcel Dekker, 1987.

Gibson, R.F., *Principles of Composite Material Mechanics.* New York: McGraw-Hill, 1994.

Harper, C., *Handbook of Plastics, Elastomers, and Composites,* 2d ed. New York: McGraw-Hill, 1992.

Hensen, F. (ed.), *Plastic Extrusion Technology.* New York: Oxford, 1988.

Isayev, A.l., *Injection and Compression Molding Fundamentals.* New York: Marcel Dekker, 1987.

Jacobs, P., *Stereolithography and other RP&M Technologies: From Rapid Prototyping to Rapid Tooling.* New York: American Society of Mechanical Engineers, 1995.

Jang, B.Z., *Advanced Polymer Composites: Principles and Applications.* Materials Park, Ohio: ASM International, 1994.

Kelly, A. (ed.), *Concise Encyclopedia of Composite Materials.* New York: Pergamon, 1989.

Kricheldorf, H.R., *Handbook of Polymer Synthesis,* 2 vols. New York: Marcel Dekker, 1991.

Kroschwitz, J.I. (ed.), *Concise Encyclopedia of Polymer Science and Engineering.* New York: Wiley, 1990.

Levy, S., and J.E. Carley, *Plastics Extrusion Technology Handbook.* New York: Industrial Press, 1989.

Levy, S., and J.H. DuBois, *Plastics Product Design Engineering Handbook,* 2d ed. New York: Van Nostrand Reinhold, 1985.

Lutz, J.T., Jr., *Thermoplastic Polymer Additives: Theory and Practice.* New York: Marcel Dekker, 1989.

Macosko, C.W., *Fundamentals of Reaction-Injection Molding.* New York: Oxford University Publishers, 1989.

Mallick, P.K., *Fiber-Reinforced Composites: Materials, Manufacturing, and Design,* 2d ed. New York: Marcel Dekker, 1993.

Malloy, R.A., *Plastic Part Design for Injection Molding, An Introduction.* Cincinnati, Ohio: Hanser-Gardner, 1994.

Margolis, J.M., *Engineering Thermoplastics: Properties and Applications.* New York: Marcel Dekker, 1985.

McCrum, N.G., C.P. Buckley, and C.B. Bucknall, *Principles of Polymer Engineering.* Oxford: Oxford University Press, 1988.

Meyer, R.W., *Handbook of Pultrusion Technology.* New York: Methuen, 1985.

Miller, E., *Introduction to Plastics and Composites: Mechanical Properties and Engineering Applications.* New York: Marcel Dekker, 1995.

Miller, E. (ed.), *Plastics Product Design Handbook. Part A: Materials and Components,* 1981; *Part B: Processes and Design for Processes.* 1983. New York: Marcel Dekker.

*Modern Plastics Encyclopedia.* New York: McGraw-Hill, annual.

Morena, J.J., *Advanced Composite Mold Making.* New York: Van Nostrand Reinhold, 1988.

Morton-Jones, D.H., *Polymer Processing.* London: Chapman and Hall, 1989.

Muccio, E.A., *Plastic Part Technology.* Materials Park, Ohio: ASM International, 1991.

——, *Plastics Processing Technology.* Materials Park, Ohio: ASM International, 1994.

Mustafa, N., *Plastics Waste Management: Disposal, Recycling, and Reuse.* New York: Marcel Dekker, 1993.

Nielsen, L.E., and R.F. Landel, *Mechanical Properties of Polymers and Composites,* 2d ed. New York: Marcel Dekker, 1994.

*Polymers for Engineering Applications.* Metals Park, Ohio: ASM International, 1987.

Pye, R.G.W., *Injection Mold Design,* 4th ed. New York: Wiley, 1989.

Rosato, D.V., D.P. DiMattia, and D.V. Rosato, *Designing with Plastics and Composites: A Handbook.* New York: Van Nostrand Reinhold, 1991.

Rosato, D., and D. Rosato (eds.), *Injection Molding Handbook,* 2d ed. New York: Chapman and Hall, 1994.

Rubin, I.I. (ed.), *Handbook of Plastic Materials and Technology.* New York: Wiley, 1990.

Schwartz, M., *Composite Materials Handbook,* 2d ed. New York: McGraw-Hill, 1992.

Seymour, R.B., *Engineering Polymer Sourcebook.* New York: McGraw-Hill, 1990.

———, *Reinforced Plastics: Properties and Applications.* Materials Park, Ohio: ASM International, 1991.

Shastri, R., *Plastics Product Design.* New York: Marcel Dekker, 1996.

Skotheim, T.A., *Handbook of Conducting Polymers,* 2 vols. New York: Marcel Dekker, 1986.

Strong, A.B., *Fundamentals of Composites Manufacturing: Materials, Methods, and Applications.* Dearborn, Mich.: Society of Manufacturing Engineers, 1989.

———,*Plastics: Materials and Processing.* Englewood Cliffs, New Jersey: Prentice-Hall, 1996.

Sweeney, F.M., *Reaction-Injection Molding Machinery and Processes.* New York: Marcel Dekker, 1987.

Tres, P.A., *Designing Plastic Parts for Assembly.* Cincinnati, Ohio: Hanser-Gardner, 1994.

Ulrich, H., *Introduction to Industrial Polymers.* New York: Hanser, 1993.

Vollrath, L., and H.G. Haldenwanger, *Plastics in Automotive Engineering: Materials, Components, Systems.* Cincinnati, Ohio: Hanser-Gardner, 1994.

Wang, S., *Deformation and Fracture of High-Temperature Thermoplastic Composites.* New York: Marcel Dekker, 1996.

Wendle, B., *What Every Engineer Should Know About Developing Plastics Products.* New York: Marcel Dekker, 1991.

Winchaeli, W., *Extrusion Dies for Plastics and Rubber,* 2d ed., New York: Hanser, 1992.

Wood, L., *Rapid Automated Prototyping: An Introduction.* New York: Industrial Press, 1993.

Young, R.J., and P. Lovell, *Introduction to Polymers.* London: Chapman and Hall, 1991.

Zachariades, A.E., and R.S. Porter, *High-Modulus Polymers: Approaches to Design and Development.* New York: Marcel Dekker, 1995.

## QUESTIONS

10.1 Summarize the most important mechanical and physical properties of plastics.

10.2 What are the major differences between the properties of plastics and metals?

10.3 What properties are influenced by the degree of polymerization?

10.4 Give applications for which flammability of plastics would be a major concern.

10.5 What properties do elastomers have that thermoplastics in general do not have?

10.6 Is it possible for a material to have a hysteresis behavior that is the opposite of that shown in Fig. 10.13 whereby the arrows are counterclockwise? Explain.

10.7 Observe the behavior of the tensile-test specimen shown in Fig. 10.12, and state whether the material has a high or low $m$ value (see Section 2.2.7). Explain why.

10.8 Why would we want to synthesize a polymer with a high degree of crystallinity?

10.9 Add more to the applications column in Table 10.3.

10.10 Discuss the significance of the glass-transition temperature, $T_g$, in engineering applications.

10.11 Why does cross-linking improve the strength of polymers?

10.12 Describe the methods by which optical properties of polymers can be altered.

10.13 Explain the reasons why elastomers were developed. Are there any substitutes for elastomers? Explain.

10.14 Give several examples of plastic products or components where creep and stress relaxation are important considerations.

10.15 Describe your opinions regarding recycling of plastics versus developing plastics that are biodegradable.

10.16 Explain how you would go about determining the hardness of plastics described in this chapter.

10.17 Distinguish between composites and alloys.

10.18 Describe the functions of the matrix and the reinforcing fibers. What fundamental differences are there in the characteristics of the two materials?

10.19 What products have you personally seen that are made of reinforced plastics? How can you tell that they are reinforced?

10.20 Identify metals and alloys that have strengths comparable to those of reinforced plastics.

10.21 Compare the advantages and disadvantages of metal-matrix composites, reinforced plastics, and ceramic-matrix composites.

10.22 You studied the many advantages of composite materials in this chapter. What limitations or disadvantages do these materials have? What suggestions would you make to overcome these limitations?

10.23 A hybrid composite is defined as one containing two or more different types of reinforcing fibers. What advantages would such a composite have over others?

10.24 Why are fibers capable of supporting a major portion of the load in composite materials?

10.25 What design considerations are involved in replacing a metal beverage container with one made of plastic?

10.26 Assume that you are manufacturing a product in which all the gears are made of metal. A salesman visits you and asks you to consider replacing some of these metal gears with plastic ones. Make a list of the questions that you would raise before making a decision.

10.27 Make a list of products or parts that are not currently made of plastics and explain the possible reasons why they are not.

10.28 Review the three curves in Fig. 10.8 and name applications for each type of behavior. Explain your choices.

10.29 Repeat Question 10.28 for the curves in Fig. 10.10.

10.30 Do you think honeycomb structures could be used in passenger cars? If so, where? Explain.

10.31 Other than those described in this chapter, what materials can you think of that can be regarded as composite materials?

10.32 What applications for composite materials can you think of in which high thermal conductivity would be desirable?

10.33 As with other materials, the mechanical properties of these composites are obtained by preparing appropriate specimens and testing them. Explain what problems you might encounter in preparing specimens for and testing these materials.

10.34 Design and describe a test method to determine the mechanical properties of reinforced plastics in their thickness direction.

10.35 As described in this chapter, reinforced plastics can be adversely affected by environment, such as moisture, chemicals, and temperature variations. Design and describe test methods to determine the mechanical properties of composite materials under these conditions.

10.36 Make a survey of a variety of sports equipment and identify the components made of composite materials. Explain the reasons for and advantages of using composites for these specific applications.

10.37 We have described several material combinations and structures in this chapter. In relative terms, identify those that would be suitable for applications involving one of the following: (1) very low temperatures, (2) very high temperatures, (3) vibrations, (4) high humidity.

10.38 Would a composite material with a strong and stiff matrix and soft and flexible reinforcement have any practical uses? Explain.

10.39 Make a list of products for which the use of composite materials could be advantageous because of their anisotropic properties.

10.40 Name applications in which both specific strength and specific stiffness (see Fig. 10.14) are important.

10.41 Explain how you would go about determining the hardness of the reinforced plastics and composite materials described in this chapter. Are hardness measurements on these types of materials meaningful? Does the size of the indentation make a difference in your answer? Explain.

10.42 Describe the advantages of applying traditional metalworking techniques to forming plastics.

10.42 Describe the advantages of cold forming of plastics over other processing methods.

10.43 Explain the reasons why some forming processes are more suitable for certain plastics than for others.

10.44 Would you use thermosetting plastics for injection molding? Explain.

10.45 By inspecting plastic containers, such as for baby powder, you can see that the lettering on them is raised rather than sunk. Can you offer an explanation as to why they are molded in that way?

10.46 Give examples of several parts suitable for insert molding. How would you manufacture these parts if insert molding were not available?

10.47 With specific examples, discuss the design issues involved in products made of plastics versus reinforced plastics.

10.48 What manufacturing considerations are involved in making a metal beverage container versus a plastic one?

10.49 Inspect several electrical components, such as light switches, outlets, circuit breakers, etc., and describe the process(es) used in making them.

10.50 Inspect several similar products that are made of metals and plastics, such as a metal bucket and a plastic bucket of similar shape and size. Comment on their respective thicknesses and explain the reasons for their differences, if any.

10.51 Make a list of processing methods used for reinforced plastics. Identify each with the following fiber orientation and arrangement capabilities: (1) uniaxial, (2) cross-ply, (3) in-plane random, and (4) three-dimensional random.

10.52 Some plastic products have lids with integral hinges; that is, no other material or part is used at the junction of the two parts. Identify such products and describe a method for making them.

10.53 Explain why operations such as blow molding and film-bag making are done vertically, and why buildings housing equipment for these operations have ceilings 10-15 m (35-50 ft) high.

## PROBLEMS

10.54 Calculate the areas under the stress-strain curve (toughness) for the material in Fig. 10.9, plot them as a function of temperature, and describe your observations.

10.55 Note in Fig. 10.9 that, as expected, the elastic modulus of the polymer decreases as temperature increases. Using the stress-strain curves in the figure, make a plot of the modulus of elasticity versus temperature.

10.56 Calculate the percentage increase in mechanical properties of reinforced nylon from the data shown in Fig. 10.16.

10.57 A rectangular cantilever beam 100 mm high, 25 mm wide, and 1 m long is subjected to a concentrated force of 100 N at its end. Select two different unreinforced and reinforced materials each from Table 10.1 and calculate the maximum deflection of the beam. Then select aluminum and steel, and for the same beam dimensions, calculate the maximum deflection and compare the results.

10.58 In Sections 10.5 and 10.6, we listed several plastics and their applications. Rearrange this information, respectively, by making a table of products and the type of plastics that can be used to make these products.

10.59 Determine the dimensions of a tubular steel drive shaft for a typical automobile. If you now replace this shaft with ones of unreinforced and reinforced plastic, respectively, what should be its new dimensions to transmit the same torque? Choose the materials from Table 10.1 and assume a Poisson's ratio of 0.4.

10.60  Calculate the average increase in the properties of plastics in Table 10.1 as a result of their reinforcement and describe your observations.

10.61  In Example 10.5, what would be the percentage of the load supported by the fibers if their strength is 1250 MPa and the matrix strength is 240 MPa?

10.62  Estimate the die clamping force required for injection molding the identical 6 in. diameter disks in one die. Include the runners of appropriate length and diameter.

10.63  A two-liter plastic beverage bottle is made from a parison with the same diameter as the threaded neck of the bottle, and has a length of 5 in. Assuming uniform deformation during blow molding, estimate the wall thickness of the tubular portion of the parison and identify the processes that could be used in making it.

# 11

# Properties and Processing of Metal Powders, Ceramics, Glasses, and Composites

**11.1** Introduction
**11.2** Production of Metal Powders
**11.3** Compaction of Metal Powders
**11.4** Sintering
**11.5** Secondary and Finishing Operations
**11.6** Design Considerations for Powder Metallurgy
**11.7** Selective Laser Sintering (Rapid Prototyping)
**11.8** Economics of Powder Metallurgy
**11.9** Ceramics: Structure, Properties, and Applications
**11.10** Processing of Ceramics
**11.11** Glasses: Structure, Properties, and Applications
**11.12** Processing of Glasses
**11.13** Design Considerations for Ceramics and Glasses
**11.14** Graphite and Diamond
**11.15** Composites: Structure, Properties, and Applications
**11.16** Processing Metal-Matrix and Ceramic-Matrix Composites

## 11.1  Introduction

In the manufacturing processes described in the preceding chapters, the raw materials used are either in a molten state or in solid form. This chapter describes how metal parts can be made by compacting metal powders in tool-steel or carbide dies and sintering them (heating without melting). This process is called **powder metallurgy (P/M).** One of its first uses was in the early 1900s to make the tungsten filaments for incandescent light bulbs. The availability of a wide range of powder compositions, the capability to produce parts to net dimensions (*net-shape forming*), and the economics of the overall operation make this process attractive for many applications.

Typical products made by powder metallurgy techniques are gears, cams, bushings, cutting tools, porous products such as filters and oil-impregnated bearings, and automotive components, such as piston rings, valve guides, connecting rods, and hydraulic pistons (Fig. 11.1 and Table 11.1). Advances in technology now permit structural parts of aircraft—such as landing gears, engine-mount supports, engine disks, impellers, and engine nacelle frames—to be made by P/M.

**FIGURE 11.1**  Examples of parts made by powder-metallurgy processes. (a) Left, steel chain-drive sprocket; right, race and cam for automatic transmissions. These are high-volume conventional parts. *Source:* Reprinted with permission of the Metal Powder Industries Federation, 105 College Road East, Princeton, NJ. (b) Titanium-alloy nacelle frame for F-14 fighter aircraft made of four pieces and welded (at arrows) by the electron-beam process. This is an example of specialty low-volume aerospace parts. *Source:* Colt-Crucible.

A unique advantage of powder metallurgy is that, unlike any other metalworking process, material density in P/M is a controllable variable. For example, even with the same material composition, it is possible to produce porous, low-density stainless-steel filters, as well as nearly full-density structural parts. In general, P/M processing has great diversity in terms of applications, materials, part sizes and shapes, and their properties, as we shall see throughout this chapter. Although specific materials and processes in P/M are important for

**TABLE 11.1**

**TYPICAL APPLICATIONS FOR METAL POWDERS**

| APPLICATION | METALS | USES |
|---|---|---|
| Abrasives | Fe, Sn, Zn | Cleaning, abrasive wheels |
| Aerospace | Al, Be, Nb | Jet engines, heat shields |
| Automotive | Cu, Fe, W | Valve inserts, bushings, gears |
| Electrical/electronic | Ag, Au, Mo | Contacts, diode heat sinks |
| Heat treating | Mo, Pt, W | Furnace elements, thermocouples |
| Joining | Cu, Fe, Sn | Solders, electrodes |
| Lubrication | Cu, Fe, Zn | Greases, abradable seals |
| Magnetic | Co, Fe, Ni | Relays, magnets |
| Manufacturing | Cu, Mn, W | Dies, tools, bearings |
| Medical/dental | Ag, Au, W | Implants, amalgams |
| Metallurgical | Al, Ce, Si | Metal recovery, alloying |
| Nuclear | Be, Ni, W | Shielding, filters, reflectors |
| Office equipment | Al, Fe, Ti | Electrostatic copiers, cams |

Source: R.M. German.

**TABLE 11.2**

**CHARACTERISTICS OF POWDER-METALLURGY PROCESSING**

ADVANTAGES

- Availability of a wide range of compositions to obtain special mechanical and physical properties, such as stiffness, damping characteristics, hardness, density, toughness, and electrical and magnetic properties. Some of the high alloyed new superalloys can be manufactured into parts only by P/M processing.
- A net- or near-net-shape technique for making parts from high-melting-point refractory metals, which would be difficult or uneconomical to make by other methods.
- High production rates on relatively complex parts, with automated equipment requiring little labor.
- Good dimensional control and, in many instances, elimination of machining and finishing operations, thus eliminating scrap and waste and saving energy.
- Capability for impregnation and infiltration for special applications.

LIMITATIONS

- Size of parts, complexity of part shapes, and press capacity.
- High cost of powder metals compared to other raw materials.
- High cost of tooling and equipment for small production runs.
- Mechanical properties, such as strength and ductility, that are generally lower than those obtained by forging. However, the properties of full-density P/M parts made by HIP or additional forging can be better than those made by other processes.

their own niches, from an overall viewpoint the P/M industry is strongly dominated by parts production from ferrous- or copper-base materials by the most common press-and-sinter method.

Pure metals, alloys, or mixtures of metallic and nonmetallic materials can be used in powder metallurgy. The most commonly used metals are iron and steel, followed by aluminum, copper, nickel, tungsten, stainless steel, and tin. For parts made of brass, bronze, steels, and stainless steels, prealloyed powders are used, where each powder particle itself is an alloy.

Powder metallurgy has become competitive with processes such as casting, forging, and machining, particularly for relatively complex parts made of high-strength and hard alloys (Table 11.2). Nearly 70 percent of the P/M part production is for automotive applications. Parts made by this process have good dimensional accuracy, and their sizes range from tiny balls for ball-point pens to parts weighing about 50 kg (100 lb), although most parts weigh less than 2.5 kg (5 lb). A typical family car now contains on the average 11 kg (25 lb) of precision metal parts made by powder metallurgy, an amount that is expected to reach 22 kg (50 lb) per car by the year 2000.

This chapter also describes the structure, properties, and processing of ceramics, glasses, and composites. These topics are introduced and discussed in Sections 11.9 through 11.16.

## 11.2   Production of Metal Powders

Basically, the powder metallurgy process consists of the following operations:

- Powder production
- Blending
- Compaction
- Sintering
- Finishing operations

For improved quality and dimensional accuracy, or for special applications, additional processing such as coining, sizing, machining, infiltration, and resintering may be carried out.

### 11.2.1   Methods of powder production

There are several methods of producing metal powder. Metal sources are generally bulk metals and alloys, ores, salts, and other compounds. Most metal powders can be produced by more than one method, the choice depending on the requirements of the end product. Particle sizes range from 0.1 $\mu$m to 1000 $\mu$m (4 $\mu$in. to 0.04 $\mu$in.).

The shape, size distribution, porosity, chemical purity, and bulk and surface characteristics of the particles depend on the particular process used (Fig. 11.2). These characteristics

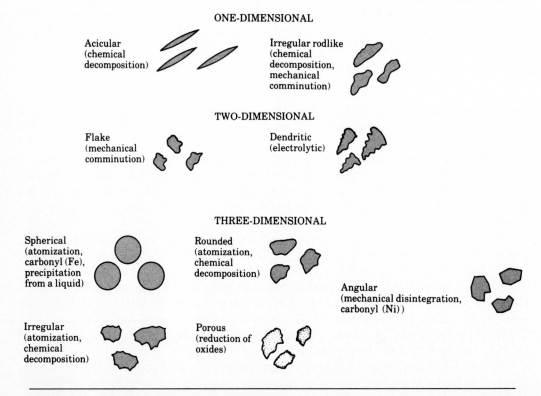

**FIGURE 11.2**   Particle shapes in metal powders and the processes by which they are produced. Iron powders are produced by many of these processes.

**FIGURE 11.3** Methods of metal powder production by atomization: (a) melt atomization; and (b) atomization with a rotating consumable electrode.

are important because they significantly affect flow characteristics during compaction and reactivity in subsequent sintering operations.

**Atomization.** Automization produces a liquid-metal stream by injecting molten metal through a small orifice (Fig. 11.3a). The stream is broken up by jets of inert gas, air, or water. The size of particles formed depends on the temperature of the metal, the rate of flow, nozzle size, and jet characteristics. Melt atomization methods are widely used for production of powders for P/M. In one variation of this method, a consumable electrode is rotated rapidly in a helium-filled chamber (Fig. 11.3b). The centrifugal force breaks up the molten tip of the electrode, producing metal particles.

**Reduction.** Reduction of metal oxides (removal of oxygen) uses gases, such as hydrogen and carbon monoxide, as reducing agents. Thus very fine metallic oxides are reduced to the metallic state. The powders produced by this method are spongy and porous and have uniformly sized spherical or angular shapes.

**Electrolytic deposition.** Electrolytic deposition utilizes either aqueous solutions or fused salts. The powders produced are among the purest.

**Carbonyls.** Metal **carbonyls**—such as iron carbonyl, $Fe(CO)_5$, and nickel carbonyl, $Ni(CO)_4$—are formed by letting iron or nickel react with carbon monoxide. The reaction products are then decomposed to iron and nickel, producing small, dense, and uniform spherical particles of high purity.

**Comminution.** Mechanical **comminution** (*pulverization*) involves crushing, milling in a *ball mill,* or grinding brittle or less ductile metals into small particles. A ball mill is a machine with a rotating hollow cylinder partly filled with steel or white cast iron balls. With brittle materials, the powder particles have angular shapes, whereas with ductile metals they are flaky and are not particularly suitable for powder metallurgy applications.

**Mechanical alloying.** In **mechanical alloying**, developed in the 1960s, powders of two or more pure metals are mixed in a ball mill. Under the impact of the hard balls, the powders repeatedly fracture and weld together by diffusion, forming alloy powders.

**Other methods.** Other less commonly used methods are *precipitation* from a chemical solution, production of fine metal chips by *machining,* and *vapor condensation.* New developments include techniques based on high-temperature extractive metallurgical processes. Metal powders are being produced using high-temperature processing techniques based on the reaction of volatile halides with liquid metals and based on controlled reduction and reduction/carburization of solid oxides.

### 11.2.2   Particle size, distribution, and shape

Particle *size* is measured usually by screening, that is, passing the metal powder through screens (sieves) of various mesh sizes. Screen analysis is achieved using a vertical stack of screens with increasing mesh size as the powder flows downward through the screens. The larger the mesh size, the smaller the opening in the screen. For example, a mesh size of 30 has an opening of 600 $\mu$m, size 100 has 150 $\mu$m, and size 400 has 38 $\mu$m. (This method is similar to numbering abrasive grains; the larger the number, the smaller the size of the abrasive particle; see Section 9.2).

In addition to screen analysis, several other methods are also used for particle size analysis, particularly for powders finer than 45 $\mu$m:

1.  *Sedimentation,* which involves measuring the rate at which particles settle in a fluid.

2.  *Microscopic* analysis, including the use of transmission and scanning electron microscopy.

3.  *Light scattering* from a laser that illuminates a sample consisting of particles suspended in a liquid medium. The particles cause the light to be scattered, which is then focused on a detector that digitizes the signals and computes the particle size distribution.

4.  *Optical* means, such as particles blocking a beam of light, which is then sensed by a photocell.

5.  Suspending particles in a liquid and detecting particle size and distribution by *electrical sensors.*

The *size distribution* of particles is an important consideration since it affects the processing characteristics of the powder. The distribution of particle size is given in terms of a *frequency distribution* plot, as shown in Fig. 11.4 (see Section 4.9.1 for details). In this plot,

**FIGURE 11.4**   A plot of the weight of particles as a function of particle size. The most populous size is termed the *mode.* In this case, it is between 75 $\mu$m and 90 $\mu$m. *Source:* Reprinted with permission from Randall M. German, *Powder Metallurgy Science.* Princeton, NJ: Metal Powder Industries Federation, 1984.

**FIGURE 11.5** Cumulative particle size distribution as a function of weight. *Source:* Reprinted with permission from Randall M. German, *Powder Metallurgy Science.* Princeton, NJ: Metal Powder Industries Federation, 1984.

note that the highest percentage of the particles (by weight) have a size in the range of 75 $\mu$m to 90 $\mu$m. This maximum is called the *mode size*. The same particle size data are also plotted in the form of a *cumulative distribution* (Fig. 11.5). Thus, for example, about 75 percent of the weight of the particles have a size finer than 100 $\mu$m. As expected from the distribution plot in Fig. 11.5, the cumulative weight reaches 100 percent at a particle size of about 200 $\mu$m.

The *shape* of particles has a major influence on their processing characteristics. The shape is usually described in terms of aspect ratio or shape index. *Aspect ratio* is the ratio of the largest dimension to the smallest dimension of the particle. This ratio ranges from unity for a spherical particle, to about 10 for flakelike or needlelike particles. *Shape index* or **shape factor (SF)** is a measure of the surface area to the volume of the particle with reference to a spherical particle of equivalent diameter. Thus the shape factor for a flake is higher than that for a sphere.

## Example 11.1: Particle shape-factor determination

Determine the shape factor for (1) a spherical particle, (2) a cubic particle, and (3) a cylinder with a length-to-diameter ratio of 2.

SOLUTION.　(1) In general, the surface area $A$ to volume $V$ ratio can be expressed as

$$A/V = k/D_{eq},$$

where $k$ is the shape factor and $D_{eq}$ is the diameter of a sphere having the same volume as the particle or object being considered. In the case of a sphere, its diameter $D = D_{eq}$, $A = \pi D^2$, and $V = \pi D^3/6$. Thus,

$$k = (A/V)D_{eq} = \left[(\pi D^2)/(\pi D^3/6)\right]D = 6.$$

(2) The surface area of a cube of length $L$ is $6L^2$ and its volume is $L^3$. Hence,

$$A/V = 6L^2/L^3 = 6/L.$$

Its equivalent diameter is

$$D_{eq} = (6V/\pi)^{1/3} = (6L^3/\pi)^{1/3} = 1.24\,L.$$

Thus, the shape factor is

$$k = (A/V)\,D_{eq} = (6/L)(1.24L) = 7.44.$$

(3) The surface area of a cylinder with length-to-diameter ratio of 2 is

$$A = (2\pi\,D^2/4) + \pi D\,L = (\pi D^2/2) + 2\pi D^2 = 2.5\,\pi D^2.$$

Its volume is

$$V = (\pi D^2/4)(L) = (\pi D^2/4)(2D) = \pi D^3/2.$$

Its equivalent diameter is

$$D_{eq} = (6V/\pi)^{1/3} = (6\pi D^3/2\pi)^{1/3} = 1.442D.$$

Hence its shape factor is

$$k = (A/V)D_{eq} = \left[(2.5\pi D^2)/(\pi D^3/2)\right](1.442D) = 7.21.$$

### 11.2.3  Blending metal powders

**Blending** (mixing) powders is the second step in powder metallurgy processing and is carried out for the following purposes:

- Because the powders made by various processes may have different sizes and shapes, they must be mixed to obtain uniformity. The ideal mix is one in which all the particles of each material are distributed uniformly.
- Powders of different metals and other materials may be mixed in order to impart special physical and mechanical properties and characteristics to the P/M product.
- Lubricants may be mixed with the powders to improve their flow characteristics. The results are reduced friction between the metal particles, improved flow of the powder metals into the dies, and longer die life. Lubricants typically are stearic acid or zinc stearate, in proportions of 0.25 to 2 percent by weight.

Powder mixing must be carried out under controlled conditions to avoid contamination or deterioration. Deterioration is caused by excessive mixing, which may alter the shape of the particles and work-harden them, thus making the subsequent compacting operation more difficult. Powders can be mixed in air, in inert atmospheres (to avoid oxidation), or in liquids, which act as lubricants and make the mix more uniform. Several types of blending equipment are available. These operations are being increasingly controlled by microprocessors to improve and maintain quality.

**Hazards.** Because of their high surface-area-to-volume ratio, metal powders are explosive, particularly aluminum, magnesium, titanium, zirconium, and thorium. Great care must be exercised both during blending and during storage and handling. Precautions include grounding equipment; preventing sparks (by using nonsparking tools and avoiding friction as a source of heat); and avoiding dust clouds, open flames, and chemical reactions.

---

**Example 11.2:** Density of metal powder–lubricant mix

We have stated that zinc stearate is a lubricant that commonly is mixed with metal powders prior to compaction, in proportions up to 2 percent by weight. Calculate the theoretical and apparent densities of an iron powder–zinc stearate mix, assuming that (1) 1000 g of iron powder is mixed with 20 g of lubricant, (2) the density of the lubricant is 1.10 g/cm$^3$, (3) the theoretical density of the iron powder is 7.86 g/cm$^3$ (from Table 3.3), and (4) the apparent density of the iron powder is 2.75 g/cm$^3$ (from Fig. 11.7a).

SOLUTION. The volume of the mixture is

$$V = \left(\frac{1000}{7.86}\right) + \left(\frac{20}{1.10}\right) = 145.41 \text{ cm}^3.$$

The combined weight of the mix is 1020 g, so its theoretical density is $1020/145.41 = 7.01$ g/cm$^3$. The apparent density of the iron powder is 2.75 g/cm$^3$, so it is $(2.75/7.86)100 = 35\%$ of the theoretical density. Assuming a similar percentage for the mix, we can estimate the apparent density of the mix as $(0.35)(7.01) = 2.45$ g/cm$^3$. (Note: Although g and g/cm$^3$ are not SI units, they are commonly used in the powder metallurgy industry.)

---

## 11.3 Compaction of Metal Powders

**Compaction** is the step in which the blended powders are pressed into shapes in dies (Figs. 11.6a and 11.6b), using presses that are either hydraulically or mechanically actuated. The purposes of compaction are to obtain the required shape, density, and particle-to-particle contact and to make the part strong enough to be processed further. The as-pressed powder is known as a **green compact**. The powder must flow easily to properly feed into the die cavity. Pressing is generally carried out at room temperature, although it can be done at elevated temperatures.

In powder metallurgy, one refers to *density* at three different stages: (1) as loose powder, (2) as a green compact, and (3) after sintering. The particle shape, average size, and size distribution dictate the packing density of loose powder. The tennis-ball analogy applies in this case. Spherical powder with a wide size distribution gives a high packing density. But a compact made from such powder has poor green strength, and, hence, such powder is unsuitable for P/M parts to be made by the (most common) die-pressing method. For P/M

**FIGURE 11.6**   (a) Compaction of metal powder to form a bushing. The pressed powder part is called green compact. (b) Typical tool and die set for compacting a spur gear. *Source:* Reprinted with the permission of the Metal Powder Industries Federation, 105 College Road East, Princeton, NJ.

parts production, powders with some irregularity of shape (for example, water-atomized ferrous powders) are preferred even though the fill density of powder in the die is somewhat lower. For reproducibility of part dimensions, the fill density of the powder should be consistent from one powder lot to the next. (Spherical powder is preferred for hot isostatic pressing, a special technique described in Section 11.3.3.)

The density after compaction (*green density*) depends primarily on the compaction pressure, the metal powder composition, and the hardness of the powder (Fig. 11.7). High pressure and softer powder gives a higher green density. Also, pure iron powder will compact to a higher density than alloyed steels. The green density and its uniformity within a compact improve with addition of a small quantity of admixed (blend of) lubricant.

To understand the effect of particle shape, consider two powder grades with the same chemical composition and hardness, but one grade having spherical particle shape. This grade will have a higher apparent (or fill) density, but after compaction at a higher pressure, compacts from both grades will have similar green densities. When comparing two similar powders pressed under some standard conditions, the powder that gives a higher green density is said to have a higher *compressibility*. (See also sintered density in Section 11.4.)

The higher the density, the higher will be the strength and elastic modulus of the part (Fig. 11.7b). The reason is that the higher the density, the higher will be the amount of solid

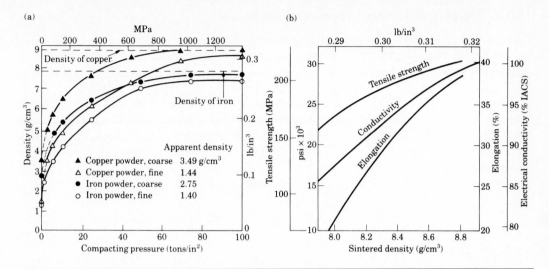

**FIGURE 11.7**  (a) Density of copper and iron powder compacts as a function of compacting pressure. Density greatly influences the mechanical and physical properties of P/M parts. *Source:* F. V. Lenel, *Powder Metallurgy: Principles and Applications.* Princeton, NJ: Metal Powder Industries Federation, 1980. Reprinted by permission of Metal Powder Industries Federation, Princeton, NJ. (b) Effect of density on tensile strength, elongation, and electrical conductivity of copper powder. IACS means International Annealed Copper Standard for electrical conductivity.

metal in the same volume—hence the greater its resistance to external forces. Because of friction between the metal particles in the powder, and the friction between the punches and the die walls, the density can vary considerably within the part. This variation can be minimized by proper punch and die design and friction control. For example, it may be necessary to use multiple punches, with separate movements in order to ensure that the density is more uniform throughout the part (Fig. 11.8).

**FIGURE 11.8**  Density variation in compacting metal powders in different dies: (a) and (c) single-action press; (b) and (d) double-action press. Note in (d) the greater uniformity of density in pressing with two punches with separate movements compared with (c). Generally, uniformity of density is the goal; although there are situations in which density variation, hence property variation, within a part may be desirable. (e) Pressure contours in compacted copper powder in a single-action press. *Source:* P. Duwez and L. Zwell.

### 11.3.1 Pressure distribution in powder compaction

Figure 11.8e shows that, in single-action pressing, the pressure decays rapidly toward the bottom of the compact. We can determine the pressure distribution along the length of the compact by using the slab method of analysis of deformation processes described in Section 6.2.2. As we did in Fig. 6.4, we first describe the operation in terms of its coordinate system, as shown in Fig. 11.9: $D$ is the diameter of the compact, $L$ is its length, and $p_0$ is the pressure applied by the punch. We take an element $dx$ thick and place on it all the relevant stresses, namely, the compacting pressure $p_x$, die-wall pressure $\sigma_x$, and frictional stress $\mu\sigma_x$. Note that the frictional stresses act upward on the element, because the punch movement is downward.

Balancing the vertical forces acting on this element, we have

$$\left(\frac{\pi D^2}{4}\right)p_x - \left(\frac{\pi D^2}{4}\right)(p_x + dp_x) - (\pi D)(\mu\sigma_x)dx = 0, \tag{11.1}$$

which we can simplify to

$$D\,dp_x + 4\mu\sigma_x dx = 0.$$

We have one equation but two unknowns ($p_x$ and $\sigma_x$). Let's now introduce a factor $k$:

$$\sigma_x = kp_x,$$

which is a measure of the interparticle friction during compaction. Thus if there is no friction between the particles, $k = 1$, the powder behaves like a fluid, and we have $\sigma_x = p_x$, signifying a state of hydrostatic pressure. We now have the expression

**FIGURE 11.9** Coordinate system and stresses acting on an element in compaction of powders. The pressure is assumed to be uniform across the cross-section (see also Fig. 6.4).

$$dp_x + \frac{4\mu k p_x dx}{D} = 0, \text{ or } \frac{dp_x}{p_x} = \frac{-4\mu k \, dx}{D}.$$

This expression is similar to that given in Section 6.2.2 for upsetting. In the same manner as there, we integrate this expression, noting that the boundary condition in this case is $p_x = p_0$ when $x = 0$

$$p_x = p_0 e^{-4\mu kx/D}. \tag{11.2}$$

Thus the pressure within the compact decays as the coefficient of friction, the parameter $k$, and the length-to-diameter ratio increase.

---

**Example 11.3:** Pressure decay in compaction

Assume that a powder mix has $k = 0.5$ and $\mu = 0.3$. At what depth will the pressure in a straight cylindrical compact 10 mm in diameter become (a) zero and (b) one-half the pressure at the punch?

SOLUTION.   For case (1) from Eq. 11.2, $p_x = 0$. Consequently, we have the expression

$$0 = p_0 e^{-(4)(0.3)(0.5)x/10}, \text{ or } e^{-0.06x} = 0.$$

The value of $x$ must approach $\infty$ for the pressure to decay to 0.
    For case (2), we have $p_x/p_o = 0.5$. Therefore

$$e^{-0.06x} = 0.5, \text{ or } x = 11.55 \text{ mm}.$$

In a practical case, a pressure drop of 50 percent is quite severe, as the compact density will then be unacceptably low. This example shows that, under the conditions assumed, uniaxial compaction of even a cylinder of length/diameter ratio of about 1.2 will be unsatisfactory.

---

### 11.3.2  Equipment

The pressure required for pressing metal powders ranges from 70 MPa (10 ksi) for aluminum to 800 MPa (120 ksi) for high-density iron parts (Table 11.3). The compacting pressure required depends on the characteristics and shape of the particles, method of blending, and lubrication.

Press capacities are on the order of 1.8–2.7 MN (200–300 tons), although presses with much higher capacities are used for special applications. Most applications require less than 100 tons. For small tonnage, crank or eccentric type mechanical presses are used; for higher capacities, toggle or knucklejoint presses are employed (see Fig. 6.30). Hydraulic presses with capacities as high as 45 MN (5000 tons) can be used for large parts.

The selection of the press depends on part size and configuration, density requirements, and production rate. The higher the pressing speed, the greater the tendency for the press to trap air in the die cavity. Therefore, good die design, including provision of vents, is important so that trapped air does not hamper proper compaction.

**TABLE 11.3**

**COMPACTING PRESSURES FOR VARIOUS METAL POWDERS**

| METAL | PRESSURE (MPa) |
|---|---|
| Aluminum | 70–275 |
| Brass | 400–700 |
| Bronze | 200–275 |
| Iron | 350–800 |
| Tantalum | 70–140 |
| Tungsten | 70–140 |
| **OTHER MATERIALS** | |
| Aluminum oxide | 110–140 |
| Carbon | 140–165 |
| Cemented carbides | 140–400 |
| Ferrites | 110–165 |

### 11.3.3 Isostatic pressing

Compaction can also be carried out or improved by a number of additional processes, such as isostatic pressing, rolling, and forging. Because the density of die-compacted powders can vary significantly, powders are subjected to *hydrostatic pressure* in order to achieve more uniform compaction. This process is similar to pressing your cupped hands together when making snow balls.

In **cold isostatic pressing (CIP)**, the metal powder is placed in a flexible rubber mold made of neoprene rubber, urethane, polyvinyl chloride, or other elastomers (Fig. 11.10). The assembly is then pressurized hydrostatically in a chamber, usually with water. The most com-

Plug

Solid core

Powder

Flexible wall

Stopper

Fluid

Chamber

**FIGURE 11.10** Schematic diagram of cold isostatic pressing as applied to forming a tube. The powder is enclosed in a flexible container around a solid core rod. Pressure is applied isostatically to the assembly inside a high-pressure chamber. *Source:* Reprinted with permission from Randall M. German, *Powder Metallurgy Science.* Princeton, NJ: Metal Powder Industries Federation, 1984.

**FIGURE 11.11** Capabilities of part size and shape complexity according to various P/M operations. P/F means powder forging. *Source:* Metal Powder Industries Federation.

mon pressure is 400 MPa (60 ksi), although pressures of up to 1000 MPa (150 ksi) have been used. The applications of CIP and other compacting methods, in terms of size and complexity of part, are shown in Fig. 11.11.

In **hot isostatic pressing (HIP)**, the container is usually made of a high-melting-point sheet metal, and the pressurizing medium is inert gas or vitreous (glasslike) fluid (Fig 11.12). Common conditions for HIP are 100 MPa (15 ksi) at 1100°C (2000°F), although the trend is toward higher pressures and temperatures. The main advantage of HIP is its ability to produce compacts with essentially 100 percent density, good metallurgical bonding among the particles, and good mechanical properties.

The HIP process is relatively expensive and is used mainly in making superalloy components for the aerospace industry. It is routinely used as a final densification step for tungsten-carbide cutting tools and P/M tool steels. The process is also used to close internal porosity and improve properties in superalloy and titanium-alloy castings for the aerospace industry.

**FIGURE 11.12** Schematic illustration of hot isostatic pressing. The pressure and temperature variation versus time are shown in the diagram. *Source:* Reprinted with permission from Randall M. German, *Powder Metallurgy Science*. Princeton, NJ: Metal Powder Industries Federation, 1984.

The main advantage of isostatic pressing is that, because of uniformity of pressure from all directions and the absence of die-wall friction, it produces compacts of practically uniform grain structure and density, irrespective of shape. Parts with high length-to-diameter ratios have been produced, with very uniform density, strength, toughness, and good surface details.

### 11.3.4    Other compacting and shaping processes

**Forging.**    An important development in P/M is the use of unsintered or sintered alloy powder preforms, which are subsequently hot forged in heated, confined dies to the desired final shapes. The process is usually referred to as *powder-metallurgy forging, P/F* (see also Fig. 11.11), and when the preform is sintered, it is usually referred to as sinter forging. Ferrous and nonferrous powders can be processed in this manner, resulting in full-density products (on the order of 99.9 percent of the theoretical density of the material). The deformation of the preform may consist of the following two modes:

1. *Upsetting,* in which the material flows laterally outward, as shown in Fig. 6.1. The preform is subjected to compressive and shear stresses during deformation; any interparticle residual oxide films are broken up, as a result of which the part has better toughness, ductility, and fatigue strength.

2. *Repressing,* in which the material flow is mainly in the direction of punch movement with little lateral flow; as a result, the mechanical properties are not as high as those in upsetting.

The powder-metallurgy forged products have good surface finish, dimensional tolerances, little or no flash formation, and uniform fine grain size. The superior properties obtained make this technology particularly suitable for applications such as highly stressed components for automotive (e.g., connecting rods), jet-engine, military, and off-the-road equipment.

**Rolling.**    In *powder rolling,* also called *roll compaction,* the powder is fed to the roll gap in a two-high rolling mill (see Fig. 6.44a), and is compacted into a continuous strip at speeds of up to 0.5 m/s (100 ft/min). The process can be carried out at room or elevated temperatures. Sheet metal for electrical and electronic components and for coins can be made by powder rolling.

**Extrusion.**    Powders may be compacted by *hot extrusion;* the powder is encased in a metal container and extruded. Superalloy powders, for example, are hot extruded for improved properties. Preforms made from the extrusions may be reheated and forged in a closed die to their final shape.

**Injection molding.**    In *injection molding,* also called *metal injection molding (MIM),* very fine metal powders (generally $< 45 \ \mu m$ and often $< 10 \ \mu m$ ) are blended with a polymer or a wax-based binder. The mixture then undergoes a process similar to injection molding of plastics (see Section 10.10.2). The molded greens are placed in a low-temperature oven to burn off the plastic, or the binder is partially removed by solvent extraction, and then the greens are sintered in a furnace. (The term *powder injection molding, PIM,* is used with a broader reference to molding of both metal and ceramic powders.)

The major advantage of injection molding over conventional compaction is that relatively complex shapes, with wall thicknesses as small as 0.5 mm (0.02 in.), can be molded and removed easily from the dies. The process is particularly well suited for small complex parts required in large quantities; most commercially produced parts weigh from a fraction of a gram to about 50 g each. Typical parts made are components for guns, surgical instruments, automobiles, and watches.

**Pressureless compaction.** In **pressureless compaction,** the die is filled with metal powder by gravity, and the powder is sintered directly in the die. Because of the resulting low density, pressureless compaction is used principally for porous parts, such as filters.

**Ceramic molds.** Molds for shaping metal powders are made by the technique used in investment casting. After the ceramic mold is made, it is filled with metal powder and placed in a steel container. The space between the mold and the container is filled with particulate material. The container is then evacuated, sealed, and subjected to hot isostatic pressing. Titanium-alloy compressor rotors for missile engines have been made by this process.

### 11.3.5  Punch and die materials

The selection of punch and die materials for P/M depends on the abrasiveness of the powder metal and the number of parts to be made. Most common die materials are air- or oil-hardening tool steels, such as D2 or D3, with a hardness range of 60–64 HRC. Because of their greater hardness and wear resistance, tungsten-carbide dies are used for more severe applications. Punches are generally made of similar materials.

Close control of die and punch dimensions is essential for proper compaction and die life. Too large a clearance between the punch and the die will allow the metal powder to enter the gap, interfere with the operation, and result in eccentric parts. Diametral clearances are generally less than 25 $\mu$m (0.001 in.). Die and punch surfaces must be lapped or polished—and in the direction of tool movements for improved die life and overall performance.

# 11.4  Sintering

**Sintering** is the process whereby compressed metal powder is heated in a controlled-atmosphere furnance to a temperature below its melting point, but sufficiently high to allow bonding (fusion) of the individual particles. Prior to sintering, the compact is brittle and its strength, known as **green strength,** is low. The nature and strength of the bond between the particles, and hence of the sintered compact, depend on the mechanisms of diffusion, plastic flow, evaporation of volatile materials in the compact, recrystallization, grain growth, and pore shrinkage.

The sintered density of a part depends mainly on its green density and on sintering conditions in terms of temperature, time, and furnace atmosphere. The sintered density increases with the increasing values of these parameters and usually with a more deoxidizing type of furnace atmosphere. For structural P/M parts, a higher sintered density is very desirable, as it leads to better mechanical properties. At the same time, in conventional P/M parts production, for better dimensional precision it is preferable to minimize increase in density during sintering. These two features can be achieved by using a powder with a high compressibility, i.e., powder that will give a high green density and keep

**TABLE 11.4**

**SINTERING TEMPERATURE AND TIME FOR VARIOUS METALS**

| MATERIAL | TEMPERATURE (°C) | TIME (min) |
| --- | --- | --- |
| Copper, brass, and bronze | 760–900 | 10–45 |
| Iron and iron-graphite | 1000–1150 | 8–45 |
| Nickel | 1000–1150 | 30–45 |
| Stainless steels | 1100–1290 | 30–60 |
| Alnico alloys (for permanent magnets) | 1200–1300 | 120–150 |
| Ferrites | 1200–1500 | 10–600 |
| Tungsten carbide | 1430–1500 | 20–30 |
| Molybdenum | 2050 | 120 |
| Tungsten | 2350 | 480 |
| Tantalum | 2400 | 480 |

the sintering temperature moderate (not high). Such powder also has another important benefit in that larger parts can be produced with a specific press tonnage. Therefore, a considerable amount of development work by powder producers is aimed at producing powder grades with higher compressibility.

Sintering temperatures (Table 11.4) are generally within 70 to 90 percent of the melting point of the metal or alloy. Sintering times range from a minimum of about 10 minutes for iron and copper alloys to as much as 8 hours for tungsten and tantalum. Continuous sintering furnaces are used for most production today. These furnaces have three chambers: (1) a burn-off chamber to volatilize the lubricants in the green compact in order to improve bond strength and prevent cracking; (2) a high-temperature chamber for sintering; and (3) a cooling chamber.

Proper control of the furnace atmosphere is essential for successful sintering and to obtain optimum properties. An oxygen-free atmosphere is essential to control the carburization and decarburization of iron and iron-base compacts and to prevent oxidation of powders. Oxide inclusions have a detrimental effect on mechanical properties. For the same volume of inclusions, the smaller inclusions have a larger effect because there are more of them per unit volume of the part. A vacuum is generally used for sintering refractory metal alloys and stainless steels. The gases most commonly used for sintering a variety of other metals are hydrogen, dissociated or burned ammonia, partially combusted hydrocarbon gases, and nitrogen.

Sintering mechanisms are complex and depend on the composition of metal particles as well as processing parameters (Fig. 11.13). As temperature increases, two adjacent particles begin to form a bond by **diffusion** (*solid-state bonding*). As a result, the strength, density, ductility, and thermal and electrical conductivities of the compact increase (Fig. 11.14). At the same time, however, the compact shrinks; hence allowances should be made for shrinkage, as in casting (see Section 5.11.1).

If two adjacent particles are of different metals, *alloying* can take place at the interface of the two particles. One of the particles may have a lower melting point than the other. In that case, one particle may melt and, because of surface tension, surround the particle that has not melted (*liquid-phase sintering*). An example is cobalt in tungsten-carbide tools and dies. Stronger and denser parts can be obtained in this way.

(a)

(b)

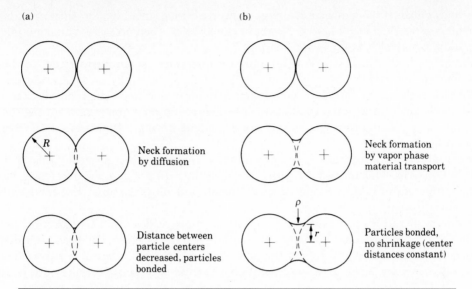

Neck formation
by diffusion

Distance between
particle centers
decreased, particles
bonded

Neck formation
by vapor phase
material transport

Particles bonded,
no shrinkage (center
distances constant)

**FIGURE 11.13**  Schematic illustration of two mechanisms for sintering metal powders: (a) solid-state material transport; and (b) liquid-phase material transport. $R$ = particle radius, $r$ = neck radius, and $\rho$ = neck profile radius.

Depending on temperature, time, and processing history, different structures and porosities can be obtained in a sintered compact. However, porosity cannot be completely eliminated

**FIGURE 11.14**  Effect of sintering temperature and time on elongation and dimensional change during sintering of type 316L stainless steel. *Source:* ASM International.

because voids remain after compaction and gases evolve during sintering. Porosities can consist of either a network of interconnected pores or closed holes. Their presence is an important consideration in making P/M filters and bearings.

Another method, which is still at an experimental stage, is **spark sintering.** In this process, loose metal powders are placed in a graphite mold, heated by an electric current, subjected to a high-energy discharge, and compacted, all in one step. The rapid discharge strips contaminants or any oxide coating, such as that found on aluminum, from the surfaces of the particles, and thus encourages good bonding during compaction at elevated temperatures.

Typical mechanical properties for several sintered P/M alloys are given in Table 11.5. Note the effect of heat treating on the properties of metals. To evaluate the differences between the properties of P/M, wrought, and cast metals and alloys, compare this table with those in other chapters.

The effects of various processes on the mechanical properties of a titanium alloy are shown in Table 11.6. Note that hot isostatically pressed (HIP) titanium has properties that are similar to those for cast and forged titanium. It should be remembered, however, that forged components are likely to require additional machining processes (unless precision

**TABLE 11.5**

**MECHANICAL PROPERTIES OF SELECTED P/M MATERIALS**

| DESIGNATION | MPIF TYPE | CONDITION | ULTIMATE TENSILE STRENGTH (MPa) | YIELD STRENGTH (MPa) | HARDNESS | ELONGATION IN 25 mm (%) | ELASTIC MODULUS (GPa) |
|---|---|---|---|---|---|---|---|
| *FERROUS* | | | | | | | |
| FC-0208 | N | AS | 225 | 205 | 45 HRB | < 0.5 | 70 |
| | | HT | 295 | — | 95 HRB | < 0.5 | 70 |
| | R | AS | 415 | 330 | 70 HRB | 1 | 110 |
| | | HT | 550 | — | 35 HRC | < 0.5 | 110 |
| | S | AS | 550 | 395 | 80 HRB | 1.5 | 130 |
| | | HT | 690 | 655 | 40 HRC | < 0.5 | 130 |
| FN-0405 | S | AS | 425 | 240 | 72 HRB | 4.5 | 145 |
| | | HT | 1060 | 880 | 39 HRC | 1 | 145 |
| | T | AS | 510 | 295 | 80 HRB | 6 | 160 |
| | | HT | 1240 | 1060 | 44 HRC | 1.5 | 160 |
| *ALUMINUM* | | | | | | | |
| 601 AB, pressed bar | | AS | 110 | 48 | 60 HRH | 6 | — |
| | | HT | 252 | 241 | 75 HRH | 2 | — |
| *BRASS* | | | | | | | |
| CZP-0220 | T | — | 165 | 76 | 55 HRH | 13 | — |
| | U | — | 193 | 89 | 68 HRH | 19 | — |
| | W | — | 221 | 103 | 75 HRH | 23 | — |
| *TITANIUM* | | | | | | | |
| Ti-6Al-4V | | HIP | 917 | 827 | — | 13 | — |
| *SUPERALLOYS* | | | | | | | |
| Stellite 19 | | — | 1035 | — | 49 HRC | < 1 | — |

MPIF: Metal Powder Industries Federation. AS: as sintered, HT: heat treated, HIP: hot isostatically pressed.

**TABLE 11.6**

**MECHANICAL PROPERTY COMPARISON FOR Ti-6Al-4V**

| PROCESS (*) | DENSITY (%) | YIELD STRENGTH (MPA) | ULTIMATE STRENGTH (MPa) | ELONGATION (%) | REDUCTION OF AREA (%) |
|---|---|---|---|---|---|
| Cast | 100 | 840 | 930 | 7 | 15 |
| Cast and forged | 100 | 875 | 965 | 14 | 40 |
| Blended elemental (P + S) | 98 | 786 | 875 | 8 | 14 |
| Blended elemental (HIP) | > 99 | 805 | 875 | 9 | 17 |
| Prealloyed (HIP) | 100 | 880 | 975 | 14 | 26 |

(*) P + S = pressed and sintered, HIP = hot isostatically pressed.
Source: R.M. German.

forged to net shape), which the P/M component may not. Thus powder metallurgy can be a competitive alternative.

---

**Example 11.4:** Shrinkage in sintering

In solid-state bonding during sintering of a powder-metal green compact, the linear shrinkage is 4 percent. If the desired sintered density is 95 percent of the theoretical density of the metal, what should be the density of the green compact? Ignore the small changes in mass during sintering.

SOLUTION.   We define linear shrinkage as $\Delta L / L_o$, where $L_o$ is the original length. We can then express the volume shrinkage during sintering as

$$V_{\text{sint}} = V_{\text{green}}\left(1 - \frac{\Delta L}{L_o}\right)^3. \tag{11.3}$$

The volume of the green compact must be larger than that of the sintered part. However, the mass does not change during sintering, so we can rewrite this expression in terms of the density $\rho$ as

$$\rho_{\text{green}} = \rho_{\text{sint}}\left(1 - \frac{\Delta L}{L_o}\right)^3. \tag{11.4}$$

Thus,

$$\rho_{\text{green}} = 0.95(1 - 0.04)^3 = 0.84, \text{ or } 84 \text{ percent.}$$

---

### 11.4.1  Spray deposition

Spray deposition is a shape generation process in which the basic components are an atomizer, a spray chamber with inert atmosphere, and a mold for producing preforms. Although there are several variations, the best known is the *Osprey* process (Fig. 11.15). After the

**FIGURE 11.15**  Spray casting (*Osprey* process) in which molten metal is sprayed over a rotating mandrel to produce seamless tubing and pipe. *Source:* J. Szekely, *Scientific American,* July 1987, © George V. Kelvin/Scientific American.

metal is atomized, it is deposited on a cooled preform mold, usually made of copper or ceramic, where it solidifies. The metal particles weld together, developing a density that is normally above 99 percent of the solid metal density. The mold may be of various shapes, such as billets, tubes, disks, and cylinders. Spray-deposited preforms may be subjected to additional shaping and consolidation processes, such as forging, rolling, and extrusion. The grain size is fine and the mechanical properties are comparable to those for wrought products of the same alloy.

## 11.5   Secondary and Finishing Operations

In order to improve further the properties of sintered P/M products or to give them special characteristics, several additional operations may be carried out after sintering. **Repressing** (*coining and sizing*) are additional compacting operations, performed under high pressure in presses. The purposes of these operations are to impact dimensional accuracy to the sintered part and to improve its strength and surface finish by additional densification.

The inherent porosity of powder metallurgy components can be utilized by **impregnating** them with a fluid. A typical application is to impregnate the sintered part with oil, which is usually done by immersing the part in heated oil. Bearings and bushings that are internally lubricated, with up to 30 percent oil by volume, are made by this method. Internally lubricated components have a continuous supply of lubricant during their service lives. Universal joints are now being made with grease-impregnated P/M techniques, no longer requiring grease fittings.

**Infiltration** is a process whereby a slug of lower melting-point metal is placed against the sintered part, and the assembly is heated to a temperature sufficient to melt the slug. The molten metal infiltrates the pores by capillary action, resulting in a relatively pore-free part with good density and strength. The most common application is the infiltration of iron-base compacts with copper. The advantages are that hardness and tensile strength are improved and the pores are filled, thus preventing moisture penetration, which could cause corrosion.

Infiltration may also be done with lead whereby, because of the low shear strength of lead, the infiltrated part has lower frictional characteristics than the uninfiltrated one. Some bearing materials are formed in this way.

Powder-metal parts may be subjected to other operations, including:

- *Heat treating* (quench and temper, steam treat) for improved strength, hardness, and wear resistance.

- *Machining* (turning, milling, drilling, tapping, grinding) for producing undercuts and slots, improved surface finish and dimensional accuracy, threaded holes and other surface features.

- *Finishing* (deburring, burnishing, mechanical finishing, plating, coating) for improved surface characteristics, corrosion and fatigue resistance, and appearance.

---

**Example 11.5:   Production of tungsten carbide for tools and dies**

Tungsten carbide is an important tool and die material mostly because of its hardness, strength, and wear resistance over a wide range of temperatures. Powder metallurgy techniques are used in making these carbides. First, powders of tungsten and carbon are blended together in a ball mill or rotating mixer. The mixture (basically 94 percent tungsten and 6 percent carbon, by weight) is heated to approximately 1500°C (2800°F) in a vacuum-induction furnace. As a result, the tungsten is carburized, forming tungsten carbide in fine powder form. A binding agent (usually cobalt) is then added to the tungsten carbide (with an organic fluid such as hexane), and the mixture is ball milled to produce a uniform and homogeneous mix, a process that can take several hours and even days.

The mixture is then dried and consolidated, usually by cold compaction, at pressures in the range of 200 MPa (30,000 psi). It is then sintered in a hydrogen-atmosphere or vacuum furnace at a temperature of 1350°C to 1600°C (2500°F to 2900°F), depending on its composition. At this temperature, the cobalt is in a liquid phase and acts as a binder for the carbide particles. (Powders may also be hot pressed at the sintering temperature, using graphite dies.) During sintering, the tungsten carbide undergoes a linear shrinkage of about 16 percent, corresponding to a volume shrinkage of about 40 percent. Thus, control of size and shape is important to produce tools with accurate dimensions. A combination of other carbides such as titanium carbide and tantalum carbide can also be produced, using mixtures made by the method described.

---

# 11.6   Design Considerations for Powder Metallurgy

Because of the unique properties of metal powders, their flow characteristics in the die, and the brittleness of green compacts, there are certain design principles that should be followed (Fig. 11.16):

- The shape of the compact must be kept as simple and uniform as possible. Sharp changes in contour, thin sections, variations in thickness, and high length-to-diameter ratios should be avoided.

**FIGURE 11.16** Examples of P/M parts, showing poor and good designs. Note that sharp radii and reentry corners should be avoided and that threads and transverse holes have to be produced separately by additional machining operations.

- Provision must be made for ejection of the green compact from the die without damaging the compact. Thus holes or recesses should be parallel to the axis of punch travel. Chamfers should also be provided.

- As with most other processes, P/M parts should be made with the widest tolerances, consistent with their intended applications, in order to increase tool and die life and reduce production costs.

Tolerances of sintered P/M parts are usually on the order of $\pm 0.05-0.1$ mm ($\pm 0.002-0.004$ in.). Tolerances improve significantly with additional operations such as sizing, machining, and grinding (see Figs. 9.31 and 15.5).

## 11.7 Selective Laser Sintering (Rapid Prototyping)

There is a growing trend in making prototypes of parts and components (**rapid prototyping**) rapidly and economically, without the need for complex and expensive equipment (see also Section 10.14). There are several such processes, one of which is **selective laser sintering** of metal powders.

### 11.7.1 Selective laser sintering

This is a process based on sintering of metallic or nonmetallic powders selectively into an individual object. The basic elements in this process are shown in Fig. 11.17. The bottom of the processing chamber is equipped with two cylinders: (1) the powder-feed cylinder, which is raised incrementally to supply powder to the part-build cylinder through a roller mechanism, and (2) the part-build cylinder, which is lowered incrementally and is where the sintered part is formed. A thin layer of powder is first deposited in the part-build cylinder. A laser beam, guided by a process control computer and based on a 3-D CAD program of the part to be produced, is then focused on that layer, tracing and sintering a particular cross-section into a solid mass. The powder in other areas remains loose, and supports the sintered portion. Another layer of powder is then deposited and the cycle is repeated. Successive cycles are repeated until the entire three-dimensional part is produced. The loose particles are then shaken off and the part is recovered. The part does not require further curing, unless it is ceramic.

A variety of materials can be used in this process, including polymers (ABS, PVC, nylon, polyester, polystyrene, epoxy), wax, metals, and ceramics with appropriate binders. Its applications are similar to those for stereolithography. The cost of these machines is in the range of $200,000 to $450,000.

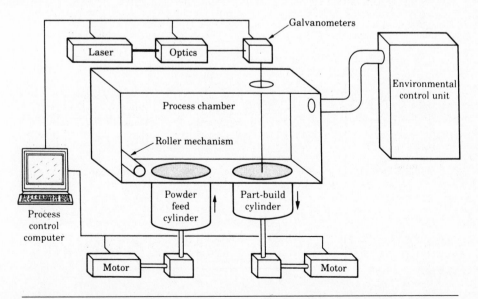

**FIGURE 11.17** Schematic illustration of the selective laser sintering process. *Source:* C. Deckard and P.F. McClure.

## 11.7.2 Other methods of rapid prototyping

Several other methods of rapid prototyping, with a price range between $50,000 and $500,000, are in various stages of development as described below. All use a CAD system interface with the machine.

**Three-dimensional printing.**   This process deposits powdered materials in thin layers and selectively binds the powder by ink-jet printing of an inorganic binder material (such as colloidal silica), without the use of a laser (Fig. 11.18). A piston supporting the powder bed is lowered incrementally, and with each step a layer is deposited and joined by the binder. Powder materials used are aluminum oxide, silicon carbide, silica, and zirconia.

**Ballistic particle manufacturing.**   In this process streams of material, such as plastics, ceramics, metals, and wax, are ejected using an ink-jet mechanism through a small orifice at a surface (target). The mechanism uses a piezoelectric pump, which operates when an electric charge is applied, generating a shock wave that propels 50-$\mu$m droplets of wax at a rate of 10,000 per second. The operation is repeated in a manner similar to other processes described above to form the part, with layers of wax deposited on top of each other. The ink-jet head is guided by a three-axis robot.

**Photochemical machining.**   This process is similar to stereolithography but uses two laser beams intersecting each other to form the part. One beam moves in the $x$–$y$ plane and the other in the $x$–$z$ plane; thus part production is more versatile and no longer done in layers as in other processes.

**Laminated object manufacturing.**   This process is based on the principle of layers of sheet, such as metal foil, paper, or plastic, bonded together in a stack. Beginning with the top

Spread powder          Print layer          Piston movement

Intermediate stage          Last layer printed          Finished part

**FIGURE 11.18**   Schematic illustration of the three-dimensional printing process.
*Source:* E. Sachs and M. Cima.

sheet, a properly adjusted laser cuts each sheet successively, one at a time, to a particular profile. The unused portions are discarded and the individual pieces are joined or bonded together. Sheet thickness may be as small as 0.05–0.12 mm (0.002–0.005 in.).

See Section 10.14 for additional methods of rapid prototyping that are based on techniques other than those described in this section. These include stereolithography, optical fabrication, fused deposition modeling, and solid base curing.

## 11.8  Economics of Powder Metallurgy

Because P/M can produce parts at or near net shape, thus eliminating many secondary manufacturing and assembly operations, it has become increasingly competitive with casting, forging, and machining. However, because of the high initial cost of punches, dies, and equipment for P/M processing, production volume must be high enough to warrant this expenditure. Although there are exceptions, the process is generally economical for quantities above 10,000 pieces.

The near-net-shape capability of P/M reduces or eliminates scrap. Weight comparisons of aircraft components produced by forging and P/M processes are shown in Table 11.7. Note that these P/M parts are subjected to material-removal processes; thus the final parts weigh less than those made by either of the two processes.

A large proportion of conventional (die-pressed and sintered) parts are used without any machining. Some require simple finishing operations such as grinding on a diameter or for thickness control. Even in very complex parts, finish machining of P/M parts involves relatively simple operations such as drilling and tapping of some holes and grinding of some surfaces. In some cases, a single P/M part may replace an assembly of several parts made by other manufacturing processes (see the Case Study at the end of this chapter). Thus, the near-net-shape-making capability of P/M is an important factor because of its cost effectiveness in numerous applications.

The major cost elements in P/M parts production are costs for metal powders, compacting dies, and equipment for compacting and sintering. The unit cost of metal powders is generally much higher than that for molten metal for casting or wrought bar stock for machining. The least expensive powder metal is iron, followed by lead, aluminum, zinc, copper and its alloys, chromium, stainless steel, tin, molybdenum, tungsten, cobalt, niobium, zirconium, and tantalum. The cost also depends on the method of powder production. The high powder cost is offset by the high material utilization of the near-net P/M methods.

**TABLE 11.7**

**FORGED AND P/M TITANIUM PARTS AND POTENTIAL COST SAVINGS**

| PART | WEIGHT (kg) | | | POTENTIAL COST SAVING (%) |
| --- | --- | --- | --- | --- |
| | FORGED BILLET | P/M | FINAL PART | |
| F-14 Fuselage brace | 2.8 | 1.1 | 0.8 | 50 |
| F-18 Engine mount support | 7.7 | 2.5 | 0.5 | 20 |
| F-18 Arrestor hook support fitting | 79.4 | 25.0 | 12.9 | 25 |
| F-14 Nacelle frame | 143 | 82 | 24.2 | 50 |

**TABLE 11.8**

**COMPETITIVE FEATURES OF P/M AND OTHER MANUFACTURING PROCESSES**

| PROCESS | ADVANTAGES OVER P/M | LIMITATIONS OVER P/M |
|---------|---------------------|----------------------|
| Casting | Generally low mold and setup cost; very wide range of part sizes and shapes produced | Not feasible for some high-temperature alloys; waste of material in processing |
| Forging, hot | High production rate of a wide range of sizes and shapes; high mechanical properties through control of grain flow | Waste of material in processing; die wear; poor dimensional control |
| Extrusion, hot | High production rate of long parts; complex cross-sections | Constant cross-sectional shape; die wear; poor dimensional control |
| Machining | Wide range of sizes and shapes; all materials; short lead time; good dimensional control; simple tooling | Waste of material in processing; low productivity; possible adverse effects on surface integrity |

Because of the high tooling (die) and equipment costs, P/M usually requires a high production volume, typically upwards of 50,000 parts/year, for it to be cost effective. Today's equipment is highly automated and P/M production is ideal for many automotive parts made in millions per year. Especially for such large quantities, the labor cost per part tends to be low. Although some larger parts are produced, a vast majority of conventional P/M parts weigh from a few grams to about a kilogram each.

Cost effectiveness of many specialized P/M methods results from special considerations that are different from those discussed above for conventional die-press and sinter methods. For example, powder-metallurgy forging (P/F) is used mainly for critical applications in which full density and accompanying superior fatigue resistance are essential. Automotive connecting rods are made in millions every year by P/M forging. Metal injection molding is cost effective for many small but highly complex parts required in large quantities. Examples of parts made by MIM are many components for air bags in cars, surgical instruments and orthodontic devices, hinges for eyeglasses, hand-gun parts, and small gears for electric toothbrushes.

At the other end of the spectrum are relatively large aerospace parts produced in small quantities, but for which there are critical property requirements or special metallurgical considerations. For example, some nickel-base superalloys are so highly alloyed that they are prone to segregation during casting and can only be processed by P/M methods. The superalloy powders may first be consolidated by hot extrusion and then hot forged in dies maintained at a high temperature. All beryllium processing is also based on powder metallurgy, using methods such as CIP-and-sinter, HIP, and hot pressing and machining. Many of these materials are difficult to process and may require much finish machining even when the parts are made by P/M methods. This is in sharp contrast to the numerous ferrous press-and-sinter P/M parts that require practically no finish machining. The competitive features of P/M versus other manufacturing processes are shown in Table 11.8.

# 11.9  Ceramics: Structure, Properties, and Applications

Ceramics are compounds of metallic and nonmetallic elements. The term *ceramics* refers both to the material and to the ceramic product itself. In Greek the word *keramos* means potter's clay and *keramikos* means clay products. Because of the large number of possible combinations of elements, a great variety of ceramics is now available for widely different consumer and industrial applications.

The earliest use of ceramics was in pottery and bricks, dating back to before 4000 B.C. Ceramics have been used for many years in automotive spark plugs as an electrical insulator and for high-temperature strength. They are becoming increasingly important in heat engines and various other applications (Table 11.9), as well as tool and die materials. More recent applications of ceramics are in automotive components, such as exhaust-port liners, coated pistons, and cylinder liners, with the desirable properties of strength and corrosion resistance at high operating temperatures.

### 11.9.1 Structure and types

The structure of ceramic crystals is among the most complex of all materials, containing various elements of different sizes. The bonding between these atoms is generally covalent

**TABLE 11.9**

**TYPES AND GENERAL CHARACTERISTICS OF CERAMICS**

| TYPE | GENERAL CHARACTERISTICS |
|---|---|
| **Oxide Ceramics** | |
| Alumina | High hardness, moderate strength; most widely used ceramic; cutting tools, abrasives, electrical and thermal insulation. |
| Zirconia | High strength and toughness; thermal expansion close to cast iron; suitable for heat engine components. |
| **Carbides** | |
| Tungsten carbide | Hardness, strength, and wear resistance depend on cobalt binder content; commonly used for dies and cutting tools. |
| Titanium carbide | Not as tough as tungsten carbide; has nickel and molybdenum as the binder; used as cutting tools. |
| Silicon carbide | High-temperature strength and wear resistance; used for heat engines and as abrasives. |
| **Nitrides** | |
| Cubic boron nitride | Second-hardest substance known, after diamond; used as abrasives and cutting tools. |
| Titanium nitride | Gold in color; used as coatings because of low frictional characteristics. |
| Silicon nitride | High resistance to creep and thermal shock; used in heat engines. |
| Sialon | Consists of silicon nitrides and other oxides and carbides; used as cutting tools. |
| Cermets | Consist of oxides, carbides, and nitrides; used in high-temperature applications. |
| Silica | High temperature resistance; quartz exhibits piezoelectric effect; silicates containing various oxides are used in high-temperature nonstructural applications. |
| Glasses | Contain at least 50 percent silica; amorphous structures; several types available with a range of mechanical and physical properties. |
| Glass ceramics | Have a high crystalline component to their structure; good thermal-shock resistance and strong. |
| Graphite | Crystalline form of carbon; high electrical and thermal conductivity; good thermal-shock resistance. |
| Diamond | Hardest substance known; available as single crystal or polycrystalline form; used as cutting tools and abrasives and as dies for fine wire drawing. |

Silicon ions

Oxygen ions

Aluminum ions

OH ions

**FIGURE 11.19**  The crystal structure of kaolinite, commonly known as clay.

(electron sharing, hence strong bonds) and ionic (primary bonding between oppositely charged ions, thus strong bonds). The properties of ceramics are significantly higher than those for metals, particularly their hardness and thermal and electrical resistance.

Ceramics are available as a single crystal or in polycrystalline form, consisting of many grains. Grain size has a major influence on the strength and properties of ceramics. The finer the grain size, the higher are the strength and toughness—hence the term *fine ceramics.*

## Raw materials

Among the oldest raw materials for ceramics is *clay,* a fine-grained sheetlike structure, the most common example being *kaolinite* (from Kao-ling, a hill in China). It is a white clay, consisting of silicate of aluminum with alternating weakly bonded layers of silicon and aluminum ions (Fig 11.19). When added to kaolinite, water attaches itself to the layers (adsorption), makes them slippery, and gives wet clay its well-known softness and plastic properties (*hydroplasticity*) that make it formable.

Other major raw materials for ceramics that are found in nature are *flint* (rock of very fine grained silica, $SiO_2$) and *feldspar* (a group of crystalline minerals consisting of aluminum silicates, potassium, calcium, or sodium). In their natural state, these raw materials generally contain impurities of various kinds, which have to be removed prior to further processing of the materials into useful products with reliable performance. Highly refined raw materials produce ceramics with improved properties.

## Oxide ceramics

**Alumina.**   Also called *corundum* or *emery, alumina* (aluminum oxide, $Al_2O_3$) is the most widely used *oxide ceramic,* either in pure form or as a raw material to be mixed with other oxides. It has high hardness and moderate strength. Although alumina exists in nature, it contains unknown amounts of impurities and possesses nonuniform properties. As a result, its behavior is unreliable. Aluminum oxide, as well as silicon carbide and many other ceramics, are now almost totally manufactured synthetically so that we can control their quality.

First made in 1893, synthetic aluminum oxide is obtained by the fusion of molten bauxite (an aluminum oxide ore that is the principal source of aluminum), iron filings, and coke in electric furnaces. It is then crushed and graded by size by passing the particles through standard screens. Parts made of aluminum oxide are cold pressed and sintered (*white ceramics*). Their properties are improved by minor additions of other ceramics, such as titanium oxide and titanium carbide. Structures containing various alumina and other oxides are known as *mullite* and *spinel* and are used as refractory materials for high-temperature applications. The mechanical and physical properties of alumina are particularly suitable for applications such as electrical and thermal insulation and as cutting tools and abrasives.

**Zirconia and partially stabilized zirconia.** *Zirconia* (zirconium oxide, $ZrO_2$, white in color) has good toughness, resistance to thermal shock, wear, and corrosion, low thermal conductivity, and low friction coefficient. A more recent development is *partially stabilized zirconia* (PSZ), which has high strength and toughness and better reliability in performance than zirconia. It is obtained by doping the zirconia with oxides of calcium, yttrium, or magnesium. This process forms a material with fine particles of tetragonal zirconia in a cubic lattice. Typical applications include dies for hot extrusion of metals and zirconia beads for grinding and dispersion media for aerospace coatings, for automotive primers and topcoats, and for fine glossy print on flexible food packaging.

Another important characteristic of PSZ is the fact that its coefficient of thermal expansion is only about 20 percent lower than that of cast iron, and its thermal conductivity is about one-third that of other ceramics. Consequently, it is very suitable for heat-engine components, such as cylinder liners and valve bushings, to keep the cast-iron engine assembly intact. New developments to further improve the properties of PSZ include *transformation-toughened zirconia* (TTZ) which has higher toughness because of dispersed tough phases in the ceramic matrix.

## Other ceramics

**Carbides.** Typical examples of *carbides* are those of tungsten (WC) and titanium (TiC), used as cutting tools and die materials, and silicon carbide (SiC), used as abrasives, as in grinding wheels. *Tungsten carbide* consists of tungsten-carbide particles with cobalt as a binder. The amount of binder has a major influence on the material's properties. Toughness increases with cobalt content, whereas hardness, strength, and wear resistance decrease. *Titanium carbide* has nickel and molybdenum as the binder and is not as tough as tungsten carbide.

*Silicon carbide* (SiC) has good wear, thermal shock, and corrosion resistance. It has a low friction coefficient and retains strength at elevated temperatures. It is suitable for high-temperature components in heat engines and is also used as an abrasive. First produced in 1891, synthetic silicon carbide is made from silica sand, coke, and small amounts of sodium chloride and sawdust. The process is similar to making synthetic aluminum oxide.

**Nitrides.** Another important class of ceramics are the *nitrides*, particularly cubic boron nitride (CBN), titanium nitride (TiN), and silicon nitride ($Si_3N_4$).

*Cubic boron nitride,* the second hardest known substance, after diamond, has special applications, such as abrasives in grinding wheels and as cutting tools. It does not exist in nature and was first made synthetically in the 1970s, with techniques similar to those used in making synthetic diamond.

*Titanium nitride* is used widely as coatings on cutting tools. It improves tool life by virtue of its low frictional characteristics.

*Silicon nitride* has high resistance to creep at elevated temperatures, low thermal expansion, and high thermal conductivity, and hence resists thermal shock. It is suitable for high-temperature structural applications, such as in automotive engine and gas-turbine components.

**Sialon.**   *Sialon* consists of silicon nitride, with various additions of aluminum oxide, yttrium oxide, and titanium carbide (see Section 8.6.8). The word sialon is derived from silicon, aluminum, oxygen, and nitrogen. It has higher strength and thermal-shock resistance than silicon nitride and thus far is used primarily as a cutting-tool material.

**Cermets.**   *Cermets* are combinations of ceramics bonded with a metallic phase. Introduced in the 1960s, they combine the high-temperature oxidation resistance of ceramics and the toughness, thermal-shock resistance, and ductility of metals. An application of cermets is cutting tools, a typical composition being 70 percent aluminum oxide and 30 percent titanium carbide. Other cermets contain various oxides, carbides, and nitrides. They have been developed for high-temperature applications such as nozzles for jet engines and aircraft brakes. Cermets can be regarded as composite materials and can be used in various combinations of ceramics and metals bonded by powder-metallurgy techniques.

## Silica

Abundant in nature, *silica* is a polymorphic material; that is, it can have different crystal structures. The cubic structure is found in refractory bricks used for high-temperature furnace applications. Most glasses contain more than 50 percent silica. The most common form of silica is *quartz,* which is a hard, abrasive hexagonal crystal. It is used extensively as oscillating crystals of fixed frequency in communications applications, since it exhibits the piezoelectric effect.

*Silicates* are products of the reaction of silica with oxides of aluminum, magnesium, calcium, potassium, sodium, and iron. Examples are clay, asbestos, mica, and silicate glasses. *Lithium aluminum silicate* has very low thermal expansion and thermal conductivity and good thermal-shock resistance. However, it has very low strength and fatigue life. Thus it is suitable only for nonstructural applications, such as catalytic converters, regenerators, and heat-exchanger components.

## Nanophase ceramics

A recent development is *nanophase ceramics,* which consist of atomic clusters containing a few thousand atoms. Nanophase ceramics exhibit ductility at significantly lower temperatures than conventional ceramics, and are stronger and easier to fabricate, with fewer flaws. Possible applications in the automotive industry, for example, include valves, rocker arms, turbocharger rotors, and cylinder liners.

## 11.9.2   Properties and applications of ceramics

Compared to metals, ceramics have the following relative characteristics: brittle, high strength and hardness at elevated temperatures, high elastic modulus, low toughness, low density, low thermal expansion, and low thermal and electrical conductivity. However, because of the wide variety of ceramic material composition and grain size, the mechanical and physical properties of ceramics vary significantly. For example, the electrical conductivity of ceramics can be modified from poor to good, which is the principle behind semiconductors. Because of their sensitivity to flaws, defects, and cracks (surface or internal), the presence of different types and levels of impurities, and different methods of manufacturing, ceramics can have a wide range of properties.

## Mechanical properties

The mechanical properties of several engineering ceramics are presented in Table 11.10. Note that their strength in tension (transverse rupture strength) is approximately one order of magnitude lower than their compressive strength. The reason is their sensitivity to cracks, impurities, and porosity. Such defects lead to the initiation and propagation of cracks under tensile stresses, severely reducing tensile strength. Thus reproducibility and reliability (acceptable performance over a specified period of time) is an important aspect in the service life of ceramic components. Tensile strength of polycrystalline ceramic parts increases with decreasing grain size.

Tensile strength is empirically related to porosity as follows:

$$UTS \simeq UTS_o e^{-nP}, \tag{11.5}$$

where $P$ is the volume fraction of pores in the solid, $UTS_o$ is the tensile strength at zero porosity, and the exponent $n$ ranges between 4 and 7.

The modulus of elasticity is likewise affected by porosity, as given by

$$E \simeq E_o(1 - 1.9P + 0.9P^2), \tag{11.6}$$

where $E_o$ is the modulus at zero porosity. Equation (11.5) is valid up to 50 percent porosity. Common earthenware has a porosity ranging between 10 percent and 15 percent, whereas hard porcelain is about 3 percent.

**TABLE 11.10**

**PROPERTIES OF VARIOUS CERAMICS AT ROOM TEMPERATURE**

| MATERIAL | SYMBOL | TRANSVERSE RUPTURE STRENGTH (MPa) | COMPRESSIVE STRENGTH (MPa) | ELASTIC MODULUS (GPa) | HARD-NESS (HK) | POISSON'S RATIO $(v)$ | DENSITY $(kg/m^3)$ |
|---|---|---|---|---|---|---|---|
| Aluminum oxide | $Al_2O_3$ | 140–240 | 1000–2900 | 310–410 | 2000–3000 | 0.26 | 4000–4500 |
| Cubic boron nitride | CBN | 725 | 7000 | 850 | 4000–5000 | — | 3480 |
| Diamond | — | 1400 | 7000 | 830–1000 | 7000–8000 | — | 3500 |
| Silica, fused | $SiO_2$ | — | 1300 | 70 | 550 | 0.25 | — |
| Silicon carbide | SiC | 100–750 | 700–3500 | 240–480 | 2100–3000 | 0.14 | 3100 |
| Silicon nitride | $Si_3N_4$ | 480–600 | — | 300–310 | 2000–2500 | 0.24 | 3300 |
| Titanium carbide | TiC | 1400–1900 | 3100–3850 | 310–410 | 1800–3200 | — | 5500–5800 |
| Tungsten carbide | WC | 1030–2600 | 4100–5900 | 520–700 | 1800–2400 | — | 10,000–15,000 |
| Partially stabilized zirconia | PSZ | 620 | — | 200 | 1100 | 0.30 | 5800 |

Note: These properties vary widely depending on the condition of the material.

Although there are exceptions and unlike most metals and thermoplastics, ceramics generally lack impact toughness and thermal-shock resistance because of their inherent lack of ductility. Once initiated, a crack propagates rapidly. In addition to undergoing fatigue failure under cyclic loading, ceramics (and particularly glasses) exhibit a phenomenon called **static fatigue**. When subjected to a static tensile load over a period of time, these materials may suddenly fail. This phenomenon occurs in environments where water vapor is present. Static fatigue, which does not occur in a vacuum or dry air, has been attributed to a mechanism similar to stress-corrosion cracking of metals.

Ceramic components that are to be subjected to tensile stresses may be prestressed, much like prestressed concrete. Prestressing shaped ceramic components subjects them to compressive stresses. Methods used include (1) heat treatment and chemical tempering, (2) laser treatment of surfaces, (3) coating with ceramics with different thermal expansion, and (4) surface-finishing operations, such as grinding, in which compressive residual stresses are induced on the surfaces.

Significant advances are being made in improving the toughness and other properties of ceramics. Among these are proper selection and processing of raw materials, control of purity and structure, and use of reinforcements, with particular emphasis during design on advanced methods of stress analysis in ceramic components.

### Physical properties

Most ceramics have relatively low specific gravity, ranging from about 3 to 5.8 for oxide ceramics, compared to 7.86 for iron. They have very high melting or decomposition temperatures. Thermal conductivity of ceramics varies by as much as three orders of magnitude, depending on their composition, whereas metals vary by one order. Thermal conductivity of ceramics, as well as other materials, decreases with increasing temperature and porosity because air is a poor thermal conductor.

The thermal conductivity $k$ is related to porosity by

$$k = k_o(1 - P), \tag{11.7}$$

where $k_o$ is the thermal conductivity at zero porosity.

The thermal expansion characteristics of ceramics are shown in Fig. 11.20. Thermal expansion and thermal conductivity induce thermal stresses that can lead to thermal shock or thermal fatigue. The tendency for *thermal cracking* (called *spalling* when a piece or a layer from the surface breaks off) is lower with low thermal expansion and high thermal conductivity. For example, fused silica has high thermal shock resistance because of its virtually zero thermal expansion.

A familiar example that illustrates the importance of low thermal expansion is the heat-resistant ceramics for cookware and stove tops. They can sustain high thermal gradients, from hot to cold, and vice versa. Moreover, the relative thermal expansion of ceramics and metals is an important reason for the use of ceramic components in heat engines. The fact that the thermal conductivity of partially stabilized zirconia components is close to that of the cast iron in engine blocks is an additional advantage in the use of PSZ in heat engines.

An additional characteristic is the *anisotrophy of thermal expansion* exhibited by oxide ceramics, whereby thermal expansion varies in different directions of the ceramic. This behavior causes thermal stresses that can lead to cracking of the ceramic component.

**FIGURE 11.20** Effect of temperature on thermal expansion for several ceramics, metals, and plastics. Note that the expansion for cast iron and for partially stabilized zirconia (PSZ) are within about 20 percent. This makes the two materials compatible for internal-combustion engine applications.

The optical properties of ceramics can be controlled by various formulations and control of structure, imparting different degrees of transparency and colors. Single-crystal sapphire, for example, is completely transparent, zirconia is white, and fine-grained polycrystalline aluminum oxide is a translucent gray. Porosity influences the optical properties of ceramics, much like trapped air in ice cubes, which makes the ice less transparent and gives it a white appearance.

Although basically ceramics are resistors, it has been shown that they can be made electrically conducting by alloying them with certain elements, thus making the ceramic act like a semiconductor or even a superconductor.

**Example 11.6:** Effect of porosity on properties

If a fully dense ceramic has the properties of $UTS_o = 100$ MPa, $E_o = 400$ GPa, and $k_o = 0.5$ W/m · K, what are these properties at 10% porosity? Let $n = 5$ and $P = 0.1$.

SOLUTION. Using Eqs. 11.5–11.7, we have

$$\text{UTS} = 100e^{-(5)(0.1)} = 61 \text{ MPa},$$
$$E = 400\big[1 - (1.9)(0.1) + (0.9)(0.1)^2\big] = 328 \text{ GPa},$$

and

$$k = 0.5(1 - 0.1) = 0.45 \text{ W/m·K}.$$

## Applications

As shown in Table 11.9, ceramics have numerous consumer and industrial applications. Several types of ceramics are used in the electrical and electronics industry because of their high electrical resistivity, dielectric strength (voltage required for electrical breakdown per unit thickness), and magnetic properties suitable for applications such as magnets for speakers. An example is *porcelain*, which is a white ceramic composed of kaolin, quartz, and feldspar. Certain ceramics also have good piezoelectric properties.

The capability of ceramics to maintain their strength and stiffness at elevated temperatures (Figs. 11.21 and 11.22) makes them very attractive for high-temperature applications. Their high resistance to wear makes them suitable for applications such as cylinder liners, bushings, seals, and bearings. The higher operating temperatures made possible by the use of ceramic components means more efficient fuel burning and reduced emissions. Currently, internal-combustion engines are only about 30 percent efficient, but with the use of ceramic components the operating performance can be improved by at least 30 percent. Ceramics being used successfully, especially in gasoline and diesel-engine components and as rotors,

**FIGURE 11.21** Effect of temperature on the strength of various engineering ceramics. Note that much of the strength is maintained at high temperatures.

**FIGURE 11.22** Effect of temperature on the modulus of elasticity for several ceramics. *Source:* D.W. Richerson, *Modern Ceramic Engineering.* New York: Marcel Dekker, Inc., 1982.

are silicon nitride, silicon carbide, and partially stabilized zirconia. Coating metal with ceramics is another application, which may be done to reduce wear, prevent corrosion, and provide a thermal barrier.

Other attractive properties of ceramics are their low density and high elastic modulus. Thus engine weight can be reduced and, in other applications, the inertial forces generated by moving parts are lower. High-speed components for machine tools, for example, are candidates for ceramics. Furthermore, the higher elastic modulus of ceramics makes them attractive for improving the stiffness, while reducing the weight, of machines. Silicon-nitride ceramics are also used as ball bearings and rollers. Because of their strength and inertness, ceramics are used as biomaterials to replace joints in the human body, as prosthetic devices, and for dental work.

## 11.10   Processing of Ceramics

Several techniques have been developed for processing ceramics into useful products (Table 11.11). Generally, the procedure involves the following steps: crushing or grinding the raw materials into very fine particles, mixing them with additives to impart certain desirable characteristics, and shaping, drying, and firing the material.

The first step in processing ceramics is *crushing* (also called comminution or milling) of the raw materials. Crushing is generally done in a ball mill (Fig. 11.23), either dry or wet. Wet crushing is more effective because it keeps the particles together and prevents the suspension of fine particles in air. The ground particles are then mixed with *additives,* the functions of which are one or more of the following:

**TABLE 11.11**

**CHARACTERISTICS OF CERAMICS PROCESSING**

| PROCESS | ADVANTAGES | LIMITATIONS |
|---|---|---|
| Slip casting | Large parts, complex shapes; low equipment cost. | Low production rate; limited dimensional accuracy. |
| Extrusion | Hollow shapes and small diameters; high production rate. | Parts have constant cross-section; limited thickness. |
| Dry pressing | Close tolerances; high production rate with automation. | Density variation in parts with high length-to-diameter ratios; dies require high abrasive-wear resistance; equipment can be costly. |
| Wet pressing | Complex shapes; high production rate. | Part size limited; limited dimensional accuracy; tooling costs can be high. |
| Hot pressing | Strong, high-density parts. | Protective atmospheres required; die life can be short. |
| Isostatic pressing | Uniform density distribution. | Equipment can be costly. |
| Jiggering | High production rate with automation; low tooling cost. | Limited to axisymmetric parts; limited dimensional accuracy. |
| Injection molding | Complex shapes; high production rate. | Tooling can be costly. |

1. Binder for the ceramic particles.
2. Lubricant for mold release and to reduce internal friction between particles during molding.
3. Wetting agent to improve mixing.
4. Plasticizer to make the mix more plastic and formable.
5. Deflocculent to make the ceramic–water suspension. Deflocculation changes the electrical charges on the particles of clay so that they repel instead of attract each other. Water is added to make the mixture more pourable and less viscous. Typical deflocculants are $Na_2CO_3$ and $Na_2SiO_3$ in amounts of less than 1 percent.
6. Various agents to control foaming and sintering.

The three basic shaping processes for ceramics are casting, plastic forming, and pressing.

(a)  (b)  (c)

**FIGURE 11.23** Methods of crushing ceramics to obtain very fine particles: (a) roll crushing, (b) ball mill, and (c) hammer milling.

(a)     (b)     (c)     (d)     (e)

**FIGURE 11.24** Sequence of operations in slip casting a ceramic part. After the slip has been poured, the part is dried and fired in an oven to give it strength and hardness. *Source:* F.H. Norton, *Elements of Ceramics* (Fig. 8.4, p. 97). Copyright © 1974, Addison-Wesley Publishing Company, Inc. Reprinted by permission of Addison Wesley Longman, Inc.

### 11.10.1  Casting

The most common casting process is *slip casting,* also called *drain casting* (Fig. 11.24). A *slip* is a suspension of ceramic particles in a liquid, generally water. In this process, the slip is poured into a porous mold made of plaster of paris. The slip must have sufficient fluidity and low viscosity to flow easily into the mold, much like the fluidity of molten metals. After the mold has absorbed some of the water from the outer layers of the suspension, it is inverted and the remaining suspension is poured out (for making hollow objects, as in slush casting of metals). The top of the part is then trimmed, the mold is opened, and the part is removed.

Large and complex parts, such as plumbing ware, art objects, and dinnerware, can be made by slip casting. Although dimensional control is limited and the production rate is low, mold and equipment costs are low. In some applications, components of the product (such as handles for cups and pitchers) are made separately and then joined, using the slip as an adhesive.

Thin sheets of ceramics, less than 1.5-mm (0.06-in.) thick, can be made by a casting technique called the *doctor-blade process.* The slip is cast over a moving plastic belt and its thickness is controlled by a blade. Other processes include *rolling* the slip between pairs of rolls and casting the slip over a paper tape, which is then burned off during firing.

For solid ceramic parts, the slip is supplied continuously into the mold to replenish the absorbed water; the suspension is not drained from the mold. At this stage the part is a soft solid or semirigid. The higher the concentration of solids in the slip, the less water has to be removed. The part, called *green,* as in powder metallurgy, is then fired.

### 11.10.2  Plastic forming

*Plastic forming* (also called *soft, wet,* or *hydroplastic forming*) can be done by various methods, such as extrusion, injection molding, or molding and jiggering (as done on a potter's wheel). Plastic forming tends to orient the layered structure of clays along the direction of material flow. This leads to anisotropic behavior of the material, both in subsequent processing and in the final properties of the ceramic product.

In *extrusion,* the clay mixture, containing 20–30 percent water, is forced through a die opening by screw-type equipment. The cross-section of the extruded product is constant, but there are limitations to wall thickness for hollow extrusions. Tooling costs are low and production rates are high. The extruded products may be subjected to additional shaping operations.

### 11.10.3  Pressing

**Dry pressing.**   Similar to powder-metal compaction, *dry pressing* is used for relatively simple shapes. Typical parts are whiteware, refractories, and abrasive products. The process has the same high production rates and close control of tolerances as in P/M. The moisture content of the mixture is generally below 4 percent but may be as high as 12 percent. Organic and inorganic binders, such as stearic acid, wax, starch, and polyvinyl alcohol, are usually added to the mixture and also act as lubricants. The pressure for pressing is between 35 MPa and 200 MPa (5 ksi and 30 ksi). Modern presses used for dry pressing are highly automated. Dies, usually made of carbides or hardened steel, must have high wear resistance to withstand the abrasive ceramic particles, and can be expensive.

Density can vary greatly in dry-pressed ceramics (Fig. 11.25) because of friction between particles and at the mold walls, as in P/M compaction. Density variations cause warping during firing. Warping is particularly severe for parts having high length-to-diameter ratios; the recommended maximum ratio is 2:1. Several methods may be used to minimize density variations. Design of tooling is important. Vibratory pressing and impact forming are used, particularly for nuclear-reactor fuel elements. Isostatic pressing also reduces density variations.

**Wet pressing.**   In *wet pressing* the part is formed in a mold while under high pressure in a hydraulic or mechanical press. This process is generally used to make intricate shapes. Moisture content usually ranges from 10 percent to 15 percent. Production rates are high but part size is limited, dimensional control is difficult because of shrinkage during drying, and tooling costs can be high.

**FIGURE 11.25**   Density variation in pressed compacts in a single-action press. The variation increases with increasing $L/D$ ratio. See also Fig. 11.8e. *Source:* Adapted from W.D. Kingery, et al., *Ceramic Fabrication Processes,* MIT Press, Cambridge, Mass., 1963.

**FIGURE 11.26** Extruding and jiggering operations. *Source:* R.F. Stoops.

**Isostatic pressing.** Used extensively in powder metallurgy, as you have seen, *isostatic pressing* is also used for ceramics in order to obtain uniform density distribution throughout the part. Automotive spark-plug insulators are made by this method. Silicon-nitride vanes for high-temperature use are also made by hot isostatic pressing.

**Jiggering.** A combination of processes is used to make ceramic plates. Clay slugs are first extruded, then formed into a *bat* over a plaster mold, and finally jiggered on a rotating mold (Fig. 11.26). *Jiggering* is a motion in which the clay bat is formed with templates or rollers. The part is then dried and fired. The process is limited to axisymmetric parts and has limited dimensional accuracy, but the operation can be automated.

**Injection molding.** We have previously described the advantages of *injection molding* of plastics and powder metals. This process is now being used extensively for precision forming of ceramics for high-technology applications, as in rocket-engine components. The raw material is mixed with a binder, such as a thermoplastic polymer (polypropylene, low-density polyethylene, ethylene vinyl acetate, or wax). The binder is usually removed by pyrolysis, and the part is sintered by firing. This process can produce thin sections, typically less than 10–15 mm (0.4–0.6 in.), using most engineering ceramics such as alumina, zirconia, silicon nitride, and silicon carbide. Thicker sections require careful control of the materials used and processing parameters to avoid internal voids and cracks, such as those due to shrinkage.

**Hot pressing.** In *hot pressing,* also called *pressure sintering,* pressure and temperature are applied simultaneously. This method reduces porosity, making the part denser and stronger. *Hot isostatic pressing* may also be used in this operation, particularly to improve the quality of high-technology ceramics. Because of the presence of both pressure and temperature, die life in hot pressing can be short. Protective atmospheres are usually employed, and graphite is a commonly used punch and die material.

### 11.10.4 Drying and firing

After the ceramic has been shaped by any of the methods described, the next step is to dry and fire the part to give it the proper strength. *Drying* is a critical stage because of the

**FIGURE 11.27**  Shrinkage of wet clay caused by removal of water during drying. Shrinkage may be as much as 20 percent by volume. *Source:* F.H. Norton, *Elements of Ceramics* (Fig. 9.5, p. 117). Copyright © 1974, Addison-Wesley Publishing Company, Inc. Reprinted by permission of Addison Wesley Longman, Inc.

tendency for the part to warp or crack from variations in moisture content and thickness within the part and the complexity of its shape. Control of atmospheric humidity and temperature is important in order to reduce warping and cracking. Loss of moisture results in shrinkage of the part by as much as 15–20 percent of the original moist size (Fig. 11.27). In a humid environment the evaporation rate is low, and consequently the moisture gradient across the thickness of the part is lower than that in a dry environment. The low moisture gradient, in turn, prevents a large, uneven gradient in shrinkage from the surface to the interior during drying.

*Firing* (sintering) involves heating the part to an elevated temperature in a controlled environment, similar to sintering in powder metallurgy. Some shrinkage occurs during firing. Firing gives the ceramic part its strength and hardness. The improvement in properties results from (1) development of a strong bond between the complex oxide particles in the ceramic and (2) reduced porosity.

Nanophase ceramics (see Section 11.9.1) can be sintered at lower temperatures than conventional ceramics. They are easier to fabricate and can be compacted at room temperature to high densities, hot pressed to theoretical density, and formed into net-shape parts without binders or sintering aids.

## 11.10.5  Finishing operations

After firing, additional operations may be performed to give the part its final shape, remove surface flaws, and improve surface finish and tolerances. The processes used can be grinding, lapping, and ultrasonic, chemical, and electrical-discharge machining. The choice of the process is important in view of the brittle nature of most ceramics and the additional costs involved in these processes. The effect of the finishing operation on properties of the product must also be considered. Because of notch sensitivity, the finer the finish, the higher the part's strength. To improve appearance and strength, and to make them impermeable, ceramic products are often coated with a *glaze* material, which forms a glassy coating after firing.

**Example 11.7:**  Dimensional changes during shaping of ceramic components

A solid cylindrical ceramic part is to be made whose final length must be $L = 20$ mm. It has been established that for this material, linear shrinkages during drying and firing are 7 percent and 6 percent, respectively, based on the dried dimension $L_d$. Calculate (1) the initial length $L_o$ of the part and (2) the dried porosity $P_d$ if the porosity of the fired part, $P_f$, is 3 percent.

SOLUTION.    (1) On the basis of the information given and remembering that firing is preceded by drying, we can write

$$(L_d - L)/L_d = 0.06$$

or

$$L = (1 - 0.06)L_d;$$

hence

$$L_d = 20/0.94 = 21.28 \text{ mm}$$

and

$$L_o = (1 + 0.07)L_d = (1.07)(21.28) = 22.77 \text{ mm.}$$

(2) Since the final porosity is 3 percent, the actual volume $V_a$ of the ceramic material is

$$V_a = (1 - 0.03)V_f = 0.97V_f,$$

where $V_f$ is the fired volume of the part. Since the linear shrinkage during firing is 6 percent, we can determine the dried volume $V_d$ of the part as

$$V_d = \frac{V_f}{(1 - 0.06)^3} = 1.2V_f.$$

Hence

$$\frac{V_a}{V_d} = \frac{0.97}{1.2} = 0.81 = 81 \text{ percent.}$$

Therefore, the porosity $P_d$ of the dried part is 19 percent.

## 11.11    Glasses: Structure, Properties, and Applications

*Glass* is an amorphous solid with the structure of a liquid. In other words, it has been *supercooled*, that is, cooled at a rate too high for crystals to form. Generally, we define glass as an inorganic product of fusion that has cooled to a rigid condition without crystallizing. Glass has no distinct melting or freezing point; thus its behavior is similar to amorphous polymers.

Glass beads were produced in about 2000 B.C., followed by glass blowing in about 200 B.C. Silica was used for all glass products until the late 1600s. Rapid developments in glasses began in the early 1900s. Presently there are some 750 different types of commercially available glasses. The uses of glass range from window glass, bottles, and cookware to glasses with special mechanical, electrical, high-temperature, chemical, corrosion, and optical characteristics. Special glasses are used in fiber optics for communication by light with little loss in signal power and in glass fibers with very high strength for reinforced plastics.

All glasses contain at least 50 percent silica, which is known as a *glass former*. The composition and properties of glasses, except strength, can be modified greatly by the addition of oxides of aluminum, sodium, calcium, barium, boron, magnesium, titanium, lithium, lead, and potassium. Depending on their function, these oxides are known as *intermediates*, or *modifiers*. Glasses are generally resistant to chemical attack, and are ranked by their resistance to acid, alkali, or water corrosion.

### 11.11.1   Types of glasses

Almost all commercial glasses are categorized by type (Tables 11.12 and 11.13):

1.  Soda–lime glass (the most common)
2.  Lead–alkali glass
3.  Borosilicate glass
4.  Aluminosilicate glass
5.  96 percent silica glass
6.  Fused silica

Glasses are also classified as colored, opaque (white and translucent), multiform (variety of shapes), optical, photochromatic (darkens when exposed to light, as in sunglasses), pho-

**TABLE 11.12**

**PROPERTIES OF VARIOUS GLASSES**

|  | SODA–LIME GLASS | LEAD GLASS | BOROSILICATE GLASS | 96 PERCENT SILICA | FUSED SILICA |
|---|---|---|---|---|---|
| Density | High | Highest | Medium | Low | Lowest |
| Strength | Low | Low | Moderate | High | Highest |
| Resistance to thermal shock | Low | Low | Good | Better | Best |
| Electrical resistivity | Moderate | Best | Good | Good | Good |
| Hot workability | Good | Best | Fair | Poor | Poorest |
| Heat treatability | Good | Good | Poor | None | None |
| Chemical resistance | Poor | Fair | Good | Better | Best |
| Impact-abrasion resistance | Fair | Poor | Good | Good | Best |
| Ultraviolet-light transmission | Poor | Poor | Fair | Good | Good |
| Relative cost | Lowest | Low | Medium | High | Highest |

**TABLE 11.13**

**COMPOSITION AND USES OF SOME GLASSES**

| TYPE | MAJOR COMPONENT (%) | | | | | | | | TYPICAL USES |
|------|------------------|-----------|-----|-------------------|-----------|-----|-----|-------|-------------|
| | $SiO_2$ | $Al_2O_3$ | CaO | $Na_2O$ | $B_2O_3$ | MgO | PbO | OTHER | |
| Fused silica | 99.5 + | | | | | | | | High-frequency electrical insulation |
| 96% silica (Vycor) | 96.3 | 0.4 | | < 0.2 | 2.9 | | | $K_2O$ < 2 | Chemical ware, home appliances, sun lamps |
| Borosilicate (Pyrex) | 81 | 2 | | 4 | 13 | | | $K_2O$ 0.4 | Oven ware, laboratory glasses, industrial glass piping, gauge glasses |
| Soda-lime (Plate glass) | 71–73 | 1 | 10–12 | 12–14 | | 1–4 | | | Windows, containers, ash trays, glass blocks, electric bulbs |
| Fiber (E-glass) | 54 | 14 | 16–22 | 1.5 | 10 | 4 | | | Fiber reinforcement |
| Lead glass | 67 | | | 6 | 6 | | 17 | $K_2O$ 10 | Table ware, optical lenses, crystal glassware, neon tubes, capacitors |
| Glass ceramics | 40–70 | 10–35 | | | 10–30 | | | $TiO_2$–7–15 | Cookware, heat exchangers |
| Aluminosilicate | 57 | 20.5 | 5.5 | 1 | 4 | 12 | | | Resistors, stove top ware |

tosensitive (changing from clear to opal), fibrous (drawn into long fibers, as in fiberglass), and foam or cellular glass (containing bubbles, thus a good thermal insulator).

Glasses are referred to as hard and soft, usually in the sense of a thermal property rather than mechanical, as in hardness. Thus a soft glass softens at a lower temperature than does a hard glass. Soda–lime and lead–alkali glasses are considered soft and the rest as hard.

### 11.11.2 Mechanical properties

For all practical purposes, we regard the behavior of glass, as for most ceramics, as perfectly elastic and brittle. The modulus of elasticity range for most commercial glasses is 55–90 GPa (8–13 million psi), and their Poisson's ratio 0.16–0.28. Hardness of glasses, as a measure of resistance to scratching, range from 5 to 7 on the Mohs scale, equivalent to a range of approximately 350–500 HK.

Glass in bulk form has a strength of less than 140 MPa (20 ksi). The relatively low strength of bulk glass is attributed to the presence of small flaws and microcracks on its surface, some or all of which may be introduced during normal handling of the glass by inadvertent abrading. These defects reduce the strength of glass by two to three orders of magnitude, compared to its ideal (defect free) strength. Glasses can be strengthened by thermal or chemical treatments to obtain high strength and toughness.

The strength of glass can theoretically reach as high as 35 GPa (5 million psi). When molten glass is freshly drawn into fibers (*fiberglass*), its tensile strength ranges from 0.2 GPa to 7 GPa (30 ksi to 1000 ksi), with an average value of about 2 GPa (300 ksi). Thus, glass fibers are stronger than steel and are used to reinforce plastics in applications such as boats, automobile bodies, furniture, and sports equipment.

The strength of glass is usually measured by bending it. The surface of the glass is first thoroughly abraded (roughened) to ensure that the test gives a reliable strength level in actual service under adverse conditions. The phenomenon of static fatigue observed in ceramics is also exhibited by glasses. If a glass item must withstand a load for 1000 hours or longer, the maximum stress that can be applied to it is approximately one-third the maximum stress that the same item can withstand during the first second of loading.

### 11.11.3   Physical properties

Glasses have low thermal conductivity and high electrical resistivity and dielectric strength. Their thermal expansion coefficient is lower than those for metals and plastics, and may even approach zero. Titanium silicate glass (a clear synthetic high-silica glass), for example, has a near-zero coefficient of expansion. Fused silica, a clear, synthetic amorphous silicon dioxide of very high purity, also has a near-zero coefficient of expansion (see Fig. 11.20). Optical properties of glasses, such as reflection, absorption, transmission, and refraction, can be modified by varying their composition and treatment.

### 11.11.4   Glass ceramics

Although glasses are amorphous, *glass ceramics* (such as Pyroceram, a trade name) have a high crystalline component to their microstructure. Glass ceramics contain large proportions of several oxides, and thus their properties are a combination of those for glass and ceramics. Most glass ceramics are stronger than glass. These products are first shaped and then heat treated, with *devitrification* (recrystallization) of the glass occurring. Unlike most glasses, which are clear, glass ceramics are generally white or gray in color.

The hardness of glass ceramics ranges approximately from 520 HK to 650 HK. They have a near-zero coefficient of thermal expansion; hence they have good thermal shock resistance and are strong because of the absence of porosity usually found in conventional ceramics. The properties of glass ceramics can be improved by modifying their composition and by heat-treatment techniques. First developed in 1957, glass ceramics are suitable for cookware, heat exchangers for gas-turbine engines, radomes (housings for radar antenna), and electrical and electronics applications.

---

**Example 11.8:**   Ovenware

Common household ovenware carries the following instruction: For oven and microwave, no stove top or broiler. Explain the reasons for this type of instruction for this product.

SOLUTION.   This product is made of borosilicate, which has the property of low thermal expansion; hence thermal stresses are low when the product is subjected to thermal gradients. Under extreme temperature conditions, however, thermal stresses can be high enough to fracture this material. The contact temperatures on a stovetop, or the radiation temperatures in a broiler, are very high compared to those in an oven. Furthermore, this product is subjected to a rapid temperature rise on its exposed surface on a stovetop or in a broiler. Thus temperature gradients are much more severe compared to a preheated oven, where heat is transmitted mainly through convection at temperatures generally below 230°C (450°F).

---

## 11.12 Processing of Glasses

Glass products can generally be categorized as:

- Flat sheet or plate, ranging in thickness from about 0.8 mm to 10 mm (0.03 in. to 0.4 in.), such as window glass, glass doors, and table tops.
- Rods and tubing used for chemicals, neon lights, and decorative artifacts.
- Discrete products, such as bottles, vases, headlights, and television tubes.
- Glass fibers to reinforce composite materials and for fiber optics.

All glass forming and shaping processes begin with molten glass, which has the appearance of red-hot viscous syrup, supplied from a melting furnace or tank.

Traditionally, flat sheet glass has been made by drawing or rolling from the molten state, and more recently by the floating method, all of which are continuous processes. In the *drawing* process, the molten glass passes through a pair of rolls, similar to an old-fashioned clothes wringer. The solidifying glass is squeezed between these rolls, forming a sheet, which is then moved forward over a set of smaller rolls. In the *rolling* process, the molten glass is squeezed between rollers, forming a sheet. The surfaces of the glass can be embossed with a pattern by shaping the roller surfaces accordingly. Glass sheet produced by drawing and rolling has a rough surface appearance. In making plate glass, both surfaces have to be ground parallel and polished.

In the *float* method, developed in the 1950s (Fig. 11.28), molten glass from the furnace is fed into a bath in which the glass, under controlled atmosphere, floats on a bath of molten tin. The glass then moves over rollers into another chamber (called a *lehr*) and solidifies. *Float glass* has a smooth (*fire polished*) surface and needs no further grinding or polishing.

Glass tubing is manufactured by the process shown in Fig. 11.29. Molten glass is wrapped around a rotating hollow cylindrical or cone-shaped mandrel and is drawn out by a set of

**FIGURE 11.29** Manufacturing process for glass tubing. Air is blown through the mandrel to keep the tube from collapsing. *Source:* Corning Glass Works.

rolls. Air is blown through the mandrel to keep the glass tube from collapsing. These machines may be horizontal, vertical, or slanted downward. Glass rods are made in a similar manner, but air is not blown through the mandrel; thus the drawn product becomes a solid rod.

Continuous fibers are drawn through multiple (200–400) orifices in heated platinum plates, at speeds as high as 500 m/s (1700 ft/s). Fibers as small as 2 $\mu$m (80 $\mu$in.) in diameter can be produced by this method. In order to protect their surfaces, the fibers are subsequently coated with chemicals. Short glass fibers, used as thermal insulating material (*glass wool*) or for acoustic insulation, are made by a *centrifugal spraying process* in which molten glass is fed into a rotating head.

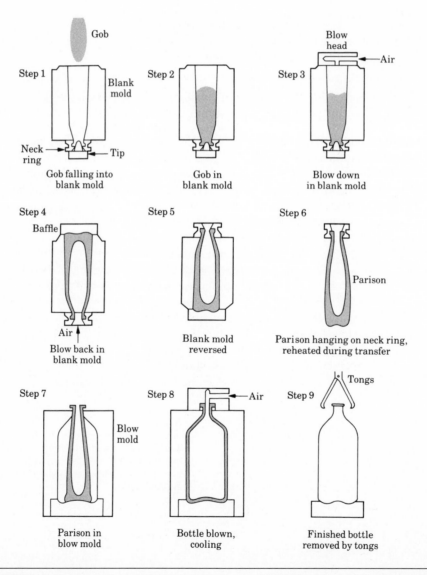

**FIGURE 11.30** Stages in manufacturing an ordinary glass bottle. *Source:* F.H. Norton, *Elements of Ceramics* (Fig. 14.4, p. 187). Copyright © 1974, Addison-Wesley Publishing Company, Inc. Reprinted by permission of Addison Wesley Longman, Inc.

Step 1   Step 2   Step 3   Step 4

Empty mold   Loaded mold   Glass pressed   Finished piece

**FIGURE 11.31** Manufacturing a glass item by pressing in a mold. *Source:* Corning Glass Works.

## 11.12.1 Manufacturing discrete glass products

Several processes are used in making discrete glass objects. These processes include blowing, pressing, centrifugal casting, and sagging.

**Blowing.** The *blowing* process is used to make hollow thin-walled glass items, such as bottles and flasks, and is similar to blow molding of thermoplastics. The steps involved in the production of an ordinary glass bottle by the blowing process are shown in Fig. 11.30. Blown air expands a hollow gob of heated glass against the walls of the mold. The molds are usually coated with a parting agent, such as oil or emulsion, to prevent the part from sticking to the mold.

The surface finish of products made by the blowing process is acceptable for most applications. Although it is difficult to control the wall thickness of the product, the process is used for high rates of production. Light bulbs are made in automatic blowing machines, at a rate of over 1000 bulbs per minute.

**Pressing.** In *pressing,* a gob of molten glass is placed in a mold and is pressed into shape with the use of a plunger. The mold may be made in one piece (Fig. 11.31), or it may be a split mold (Fig. 11.32). After pressing, the solidifying glass acquires the shape of the mold-plunger cavity. Because of the confined environment, the product has greater dimensional accuracy than can be obtained with blowing. However, pressing cannot be used on thin-walled items, or for parts such as bottles from which the plunger cannot be retracted.

**Centrifugal casting.** Also known as *spinning* in the glass industry, the *centrifugal casting* process is similar to that for metals. The centrifugal force pushes the molten glass against the mold wall where it solidifies. Typical products are TV picture tubes and missile nose cones.

Step 1   Step 2   Step 3   Step 4

Empty mold   Loaded mold   Glass pressed   Finished product

**FIGURE 11.32** Pressing glass in a split mold. *Source:* E.B. Shand, *Glass Engineering Handbook.* New York: McGraw-Hill, 1958. Reproduced by permission of the McGraw-Hill Companies.

**Sagging.**   Shallow dish-shaped or lightly embossed glass parts can be made by the *sagging* process. A sheet of glass is placed over the mold and is heated. The glass sags by its own weight and takes the shape of the mold. The process is similar to thermoforming with thermoplastics but without pressure or a vacuum. Typical applications are dishes, sunglass lenses, mirrors for telescopes, and lighting panels.

### 11.12.2   Techniques for treating glass

As produced by the methods we described, glass can be strengthened by thermal tempering, chemical tempering, and laminating. Glass products may also undergo annealing and other finishing operations.

In **thermal tempering** (also called *physical tempering* or *chill tempering*), the surfaces of the hot glass are cooled rapidly (Fig. 11.33). As a result, the surfaces shrink, and because the bulk is still hot, tensile stresses develop on the surfaces. As the bulk of the glass begins to cool, it contracts. The solidified surfaces are now forced to contract, thus developing residual compressive surface stresses and interior tensile stresses. Compressive surface stresses improve the strength of the glass, as they do in other materials. Note that the higher the coefficient of thermal expansion of the glass and the lower its thermal conductivity, the higher the level of residual stresses developed and hence the stronger the glass becomes. Thermal tempering takes a relatively short time (minutes) and can be applied to most glasses. Because of the large amount of energy stored from residual stresses, **tempered glass** shatters into a large number of pieces when broken.

In **chemical tempering,** the glass is heated in a bath of molten $KNO_3$, $K_2SO_4$, or $NaNO_3$, depending on the type of glass. Ion exchange takes place, with larger atoms replacing the smaller atoms on the surface of the glass. As a result, residual compressive stresses are developed on the surface. This condition is similar to that created by forcing a wedge between two bricks in a wall. The time required for chemical tempering is longer (about one hour) than for thermal tempering. Chemical tempering may be performed at various temperatures. At low temperatures, part distortion is minimal and complex shapes can be treated. At elevated temperatures, there may be some distortion of the part, but the product can be used at higher temperatures without loss of strength.

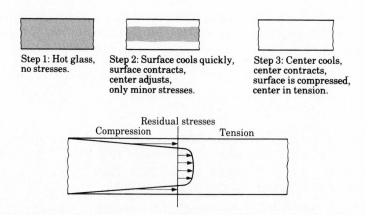

Step 1: Hot glass, no stresses.

Step 2: Surface cools quickly, surface contracts, center adjusts, only minor stresses.

Step 3: Center cools, center contracts, surface is compressed, center in tension.

Residual stresses
Compression     Tension

**FIGURE 11.33**   Residual stresses in tempered glass plate and stages involved in inducing compressive surface residual stresses for improved strength.

Another strengthening method is to make **laminated glass**, which is two pieces of flat glass with a thin sheet of tough plastic between them. The process is also called **laminate strengthening**. When laminated glass is broken, its pieces are held together by the plastic sheet. You probably have seen this phenomenon in a shattered automobile windshield.

As in metal products, residual stresses can develop in glass products if they are not cooled slowly enough. In order to ensure that the product is free from these stresses, it is *annealed* by a process similar to stress-relief annealing of metals. The glass is heated to a certain temperature and cooled gradually. Depending on the size, thickness, and type of glass, annealing times may range from a few minutes, to as long as 10 months, as in the case of a 600-mm (24-in.) mirror for a telescope.

In addition to annealing, glass products may be subjected to further operations, such as cutting, drilling, grinding, and polishing. Sharp edges and corners can be smoothened by grinding, as in glass tops for desks and shelves, or by holding a torch against the edges (*fire polishing*), which rounds them by localized softening and surface tension.

## 11.13 Design Considerations for Ceramics and Glasses

Ceramic and glass products require careful selection of composition, processing methods, finishing operations, and methods of assembly into other components. Limitations such as general lack of tensile strength, sensitivity to defects, and low impact toughness are important. These limitations have to be balanced against desirable characteristics, such as their hardness, scratch resistance, compressive strength at room and elevated temperatures, and diverse physical properties.

Control of processing parameters and the quality and level of impurities in the raw materials used are important. As in all design decisions, there are priorities, limitations, and various factors that should be considered, including the number of parts needed and the costs of tooling, equipment, and labor.

Dimensional changes and warping and cracking possibilities during processing are significant factors in selecting methods for shaping these materials. When a ceramic or glass component is part of a larger assembly, compatibility with other components is another important consideration. Particularly important are thermal expansion (as in seals) and the type of loading. The potential consequences of part failure are always a significant factor in designing ceramic products.

## 11.14 Graphite and Diamond

**Graphite** is a crystalline form of carbon having a *layered structure* of basal planes or sheets of close-packed carbon atoms. Consequently, graphite is weak when sheared along the layers. This characteristic, in turn, gives graphite its low frictional properties as a solid lubricant. However, frictional properties are low only in an environment of air or moisture; graphite is abrasive and a poor lubricant in vacuum.

A new development is the production of soccer-ball-shaped carbon molecules, called **buckyballs** and named after Buckminster Fuller (1895–1983), inventor of the geodesic dome. Also called **fullerenes,** these chemically inert spherical molecules are produced from soot and act much like solid lubricant particles. Fullerenes become superconductors when mixed with metals.

Although brittle, graphite has high electrical and thermal conductivity and resistance to thermal shock and high temperature (although it begins to oxidize at 500°C (930°F)). It is therefore an important material for applications such as electrodes, heating elements, brushes for motors, high-temperature fixtures and furnace parts, mold materials such as crucibles for melting and casting of metals, and seals (because of low friction and wear). Unlike other materials, the strength and stiffness of graphite increases with temperature.

An important use of graphite is as fibers in composite materials and reinforced plastics. A characteristic of graphite is its resistance to chemicals; hence it is used as filters for corrosive fluids. Also, its low thermal neutron absorption cross-section and high scattering cross-section make graphite suitable for nuclear applications. Ordinary pencil "lead" is a graphite and clay mixture.

Graphite is generally graded in decreasing order of grain size: industrial, fine grain, and micrograin. As in ceramics, the mechanical properties of graphite improve with decreasing grain size. Micrograin graphite can be impregnated with copper and is used as electrodes for electrical discharge machining and for furnace fixtures. Amorphous graphite is known as lampblack (black soot) and is used as a pigment. Graphite is usually processed by molding or forming, oven baking, and then machining to the final shape. It is available commercially in square, rectangular, or round shapes of various sizes.

The second principal form of carbon is **diamond**, which has a covalently bonded structure. It is the hardest substance known (7000–8000 HK). This characteristic makes diamond an important cutting-tool material, as a single crystal or in polycrystalline form, as an abrasive in grinding wheels for grinding hard materials, and for dressing of grinding wheels (sharpening of abrasive grains). Diamond is also used as a die material for drawing thin wire, of less than 0.06 mm (0.0025 in.) in diameter. Diamond is brittle and it begins to decompose in air at about 700°C (1300°F). In nonoxidizing environments it resists higher temperatures.

**Synthetic,** or **industrial diamond**, was first made in 1955 and is used extensively for industrial applications. One method of manufacturing it is to subject graphite to a hydrostatic pressure of 14 GPa (2 million psi) and a temperature of 3000°C (5400°F). Synthetic diamond is identical to natural diamond for industrial applications, and has superior properties because of lack of impurities. It is available in various sizes and shapes, the most common abrasive grain size being 0.01 mm (0.004 in.) in diameter.

Gem-quality synthetic diamond is now being made with electrical conductivity 50 times higher than that for natural diamond and 10 times more resistant to laser damage. Potential applications are for heat sinks for computers, telecommunications, integrated-circuit industries, and windows for high-power lasers.

## 11.15   Composites: Structure, Properties, and Applications

New developments in composite materials are continually taking place, with a wide range and form of polymeric, metallic, and ceramic materials being used both as fibers and as matrix materials. Research and development activities in this area are concerned with improving strength, toughness, stiffness, resistance to high-temperatures, and reliability in service.

### 11.15.1   Metal-matrix composites

The advantage of a metal matrix over a polymer matrix is its higher resistance to elevated temperatures and higher ductility and toughness. The limitations are higher density and greater difficulty in processing components. Matrix materials in these composites are usually aluminum, aluminum-lithium, magnesium, and titanium, although other metals are also being investigated. Fiber materials are graphite, aluminum oxide, silicon carbide, and boron, with beryllium and tungsten as other possibilities.

Because of their high specific stiffness, light weight, and high thermal conductivity, boron fibers in aluminum matrix have been used for structural tubular supports in the space shuttle orbiter. Other applications are bicycle frames and sporting goods. Studies of techniques for optimum bonding of fibers to the metal matrix are in progress. Metal-matrix composite applications are presently in gas turbines, electrical components, and various structural components (Table 11.14).

### 11.15.2   Ceramic-matrix composites

Ceramic-matrix composites are another important recent development in engineered materials. As we described earlier, ceramics are strong and stiff, resist high temperatures, but generally lack toughness. New matrix materials that retain their strength to 1700°C (3100°F) are silicon carbide, silicon nitride, aluminum oxide, and mullite (a compound of aluminum, silicon, and oxygen). Also under development are carbon–carbon matrix composites that retain much of their strength up to 2500°C (4500°F), although they lack oxidation resistance at high temperatures. Present applications for ceramic-matrix composites are in jet and automotive engines, deep-sea mining equipment, pressure vessels, and various structural components.

### 11.15.3   Other composites

Composites may also consist of coatings of various kinds on base metals or substrates. Examples are plating of aluminum and other metals over plastics for decorative purposes and

**TABLE 11.14**

**METAL-MATRIX COMPOSITE MATERIALS AND APPLICATIONS**

| FIBER | MATRIX | APPLICATIONS |
|---|---|---|
| Graphite | Aluminum | Satellite, missile, and helicopter structures |
| | Magnesium | Space and satellite structures |
| | Lead | Storage-battery plates |
| | Copper | Electrical contacts and bearings |
| Boron | Aluminum | Compressor blades and structural supports |
| | Magnesium | Antenna structures |
| | Titanium | Jet-engine fan blades |
| Alumina | Aluminum | Superconductor restraints in fusion power reactors |
| | Lead | Storage-battery plates |
| | Magnesium | Helicopter transmission structures |
| Silicon carbide | Aluminum, titanium | High-temperature structures |
| | Superalloy (cobalt-base) | High-temperature engine components |
| Molybdenum, tungsten | Superalloy | High-temperature engine components |

enamels, dating to before 1000 B.C., or similar vitreous (glasslike) coatings on metal surfaces for various functional or ornamental purposes.

Composites are made into cutting tools and dies, such as cemented carbides, usually tungsten carbide and titanium carbide, with cobalt and nickel, respectively, as a binder. Other composites are grinding wheels made of aluminum oxide, silicon carbide, diamond, or cubic boron nitride abrasive particles, held together with various organic, inorganic, or metallic binders. Another category of composites is cermets.

A composite developed relatively recently is granite particles in an epoxy matrix. It has high strength, good vibration damping capacity (better than gray cast iron), and good frictional characteristics. It is used as machine-tool beds for some precision grinders.

## 11.16   Processing Metal-Matrix and Ceramic-Matrix Composites

This section outlines the basic manufacturing processes for these composites. In all these processes, important considerations involve the nature and strength of the interface between the matrix and the fibers, as well as uniform distribution of the fibers and the residual stresses developed during processing.

### 11.16.1   Processing metal-matrix composites (MMC)

There are three methods of manufacturing these composites into near-net-shape parts: liquid-phase processes, solid-phase processes, and two-phase (liquid/solid) processes.

*Liquid-phase processing* basically consists of casting the liquid matrix and the solid reinforcement, using either conventional casting processes or pressure infiltration casting. In the latter process, pressurized gas is used to force the liquid matrix metal into a preform (usually as sheet or wire) made of the reinforcing fibers.

*Solid-phase processes* consist basically of powder-metallurgy techniques, including cold and hot isostatic pressing. Proper mixing for homogeneous distribution of the fibers is important. An example of this technique, employed in tungsten-carbide tool and die manufacturing with cobalt as the matrix material, is given in Example 11.5. In making complex MMC parts with whisker or fiber reinforcement, die geometry and control of process variables are very important in ensuring proper distribution and orientation of the fibers within the part. MMC parts made by powder-metallurgy processes are generally heat treated for optimum properties.

The techniques used for *two-phase processing* consist of rheocasting, described in Section 5.9.6, and spray atomization and deposition. In the latter processes, the reinforcing fibers are mixed with a matrix that contains both liquid and solid phases.

### 11.16.2   Processing ceramic-matrix composites (CMC)

There are several processes used in making these composites, as briefly described below. In addition to these processes, new techniques such as melt infiltration, controlled oxidation, and hot-press sintering, which are still largely in the experimental stage, are being developed for improving the properties and performance of these composites.

*Slurry infiltration,* the most common process, involves the preparation of a fiber preform that is hot pressed, and then impregnated with a slurry that contains the matrix powder, a carrier liquid, and an organic binder. A further improvement is *reaction bonding* or *reaction sintering* of the slurry. High strength, toughness, and uniform structure is obtained by the slurry infiltration process, but the product has limited high-temperature properties due to the low melting temperature of the matrix materials used.

*Chemical synthesis* processes involve the sol-gel and polymer precursor techniques. In the *sol-gel* process, a sol (a colloidal fluid with the liquid as the continuous phase) containing fibers is converted to a gel, which is then subjected to heat treatment to produce a ceramic-matrix composite. The *polymer precursor* method is analogous to the process used in making ceramic fibers.

In *chemical vapor infiltration* a porous fiber preform is infiltrated with the matrix phase using the chemical vapor deposition technique (see Section 4.5.1). The product has very good high-temperature properties, but the process is costly and time-consuming.

---

**Example 11.9:    Metal-matrix composite brake rotors and cylinder liners**

Brake rotors are being made of composites consisting of an aluminum matrix with 20 percent silicon carbide (SiC) particles. The particles are stirred into molten aluminum alloys and cast into ingots. The ingots are then remelted and cast into shapes, such as brake rotors and drums, using processes such as green sand, bonded sand, investment, permanent mold, and squeeze casting (see Chapter 5). The MMC brake rotors weigh about one-half of gray cast iron rotors, conduct heat three times faster, combine the stiffness and wear-resistance of ceramics, and reduce noise and vibration.

To improve the wear-and heat-resistance of cast-iron cylinder liners in aluminum engine blocks, aluminum-matrix liners are being developed. The MMC layer consists of 12 percent aluminum oxide ($Al_2O_3$) fiber and 9 percent graphite fiber, in a thickness that ranges from 1.5 to 2.5 mm (0.06 to 0.1 in.).

---

**CASE STUDY:    PRODUCTION OF AUTOMOTIVE-ENGINE MAIN BEARING CAPS BY POWDER METALLURGY**

Main bearing caps are essential structural elements in internal-combustion engines (Fig. 11.34). The crankshaft has to be held securely in place but must turn freely. This is accomplished by bolting the main bearing caps (MBC) of the engine block (Fig. 11.35) and installing thin bearing shells. In this way, the main bearing caps both guide and retain the crankshaft, withstanding the combustion and inertia loads of the engine.

The majority of MBCs are made from gray cast iron (see Section 5.6), which is cast into a sand mold to produce a "loaf." Extensive machining includes broaching, drilling, spot facing, and milling before the loaf is cut into individual caps (similar to slicing a loaf of bread). The yield of finished MBCs from the casting is typically 60 percent by weight. Grey cast iron is a very versatile material exhibiting excellent castability with good machinability, due to its low hardness and flake graphite in the microstructure, compressive strength, and sound dampening properties. However, it has low fatigue strength and is very brittle, which limits its use for highly stressed applications.

In high-power engines, gray cast iron may not have sufficient strength, which could result in fatigue cracking. To date, the material of choice for these engines has been ductile (nodular) cast iron. Ductile iron has much higher strength and ductility, but is more

**FIGURE 11.34** Typical piston engine crankshaft arrangement.

Piston

Connecting rod

Crankshaft

Main bearing cap

Engine block

Oil pan

expensive, difficult to machine, and requires close control to produce consistent casting quality. Both cast irons are prone to traditional casting defects, which include nonmetallic inclusions, burned-in sand, blowholes, and variation in graphite size and form.

Powder-metallurgy MBCs (Fig. 11.36) were used for the first time in 1993 in General Motors 3100 and 3800 V6 engines. P/M was chosen to replace the cast iron because of its design flexibility and functional advantages, and, because of its net-shape capability, it eliminated major investments in machining lines required for finishing the cast parts to final shape and dimensions. It is estimated that MBCs will consume more than 5,000 tons of steel powder annually.

GM wanted to improve the sound quality of its vehicles by reducing engine noise. It was determined that the main bearing caps play a role in noise generation; the addition of bolts into the sides of the MBC increases the stiffness of the crankcase and significantly improves sound quality. The use of side bolts involved a substantial redesign of the MBCs and would require major investments in machine tools for the machining lines. This capital expense was avoided by adopting P/M as the manufacturing process.

There are four principal requirements of an MBC material: (1) strength, principally fatigue strength, and ductility, (2) consistent and robust quality, (3) integrity, i.e., free of defects and (4) machinability. In order to be cost competitive, it was essential to develop a low-

**FIGURE 11.35** Typical main bearing cap (MBC).

Area under bolt head

Bolt hole

Indentification

Bearing notch

Installation profile

Joint faces

Snap width

**FIGURE 11.36**  Main bearing powder-metal caps for 3.8 and 3.1 liter GM engines. *Source:* Courtesy of Zenith Sintered Products, Inc., Milwaukee, Wisconsin.

cost/high-strength P/M material that also exhibited good ductility, machinability, and ease of manufacture. A high-strength, low-cost P/M alloy steel was developed for this application involving liquid-phase sintering (see Section 11.4) and using an optimized low-alloy combination in a carbon–steel matrix, designated as Zenith Material 833 (ZM833).

The P/M process molds individual MBCs into net shape, except for the new side bolt holes, which are drilled and tapped. Critical to the success of the P/M approach was the molding of the long bolt holes, which would be costly and difficult to machine. This molding step requires "upright" compaction, i.e., vertical bolt holes, which presents significant powder compaction challenges due to the irregular shape conferred by the bearing arch. The advantages, however, include pressing in the bearing notch and the top face features, which provide identification and orientation for block line assembly. Upright compaction also permits use of a much smaller compacting press than for "flat" pressing, i.e., horizontal bolt holes, with associated cost benefits. In addition, the 3800 engine rear MBC has several features that would be impossible to compact in the flat orientation; thus, the upright compaction challenge was inevitable. Principal features include recessed bolt hole bosses, rectangular oil drain hole, vertical side grooves, a double arch to accept an oil seal, and a raised arch feature for added strength.

A shape feature to provide smooth installation of the MBC into the block channel is desirable to avoid crooked entry from cocking to one side. The P/M compaction process allows a sophisticated installation design to be pressed that consists of a location radius followed by a shallow lead-in ramp angle. This has proven to be superior to the previously machined simple chamfer in the cast iron MBCs, as evidenced by the lower installation loads and a smoother pulldown action when the bolts are tightened.

Having achieved the desired fatigue strength, the ZM833 was used to produce actual MBCs that were subjected to severe engine-endurance testing in both engines. An initial test involved running several engines for hundreds of hours under the highest stress regime available. The P/M parts passed the test. The next phase involved designing and building a fatigue test machine that accepts an engine block and simulates the combustion stresses of a powered engine. The testing was so severe that the cast iron engine block generally broke before the P/M cap cracked, and often several engine blocks had to be sacrificed to obtain a fatigue test point.

GM engineers then decided to construct a high-strength steel block to enable extended stress tests to be run cost effectively to generate S–N fatigue curves (see Fig. 2.28) for the P/M material and establish the functional safety margin. The P/M was proven to be a robust design by repeated successful fatigue testing, simulating 1.8 times the most severe running conditions predicted for the engine.

Material ductility is a functional benefit since it is possible to induce cracks in a very brittle material during press-fit installation. The ZM833 P/M material had the following properties: (1) density of 6.6 g/cm$^3$; (2) tensile strength of 450 MPa (65,000 psi); (3) hardness of 70 HRB; (4) modulus of elasticity of 107 GPa ($15.5 \times 10^6$ psi)—the P/M material developed for this application has a lower modulus than cast iron, resulting in significantly lower initial tensile stresses in highly stressed areas of the MBC; (5) tensile elongation of 3 percent to 4 percent, which compares with gray iron at 0.5 percent and ductile iron at typically 3 percent; and (6) the fatigue endurance limit of the P/M materials was determined by flexural bending fatigue tests; it was found that the P/M alloy endurance limit was intermediate between the gray and ductile irons, being double the gray iron level.

Other experiments were also carried out to ensure the quality level of the P/M material, including the effect of various powder mixes, density, time and temperature, critical defect size, and inclusions. Also studied was surface damage from handling; even the most severely notched MBC passed the engine fatigue testing successfully.

Because the P/M MBC is essentially a net-shape product, the machining requirement is limited to side bolt drill and tapping performance (see Section 8.8.4). The results showed that the drilling performance of the P/M alloy was intermediate between gray iron (best) and ductile iron (worst). For tap life, the drill bit was replaced by a tap on the CNC machine. Testing indicated that a life of over 3000 tapped holes was achievable.

(Adapted from "Powder Metallurgy Main Bearing Caps for Automotive Engines," by T.M. Cadle, M.A. Jarrett, and W.L. Miller, *Int. Jr. of Powder Metallurgy*, Vol. 30, No. 3, 1994, pp. 275–282.)

## SUMMARY

- The powder-metallurgy process is capable of producing relatively complex parts economically, in net-shape form to close tolerances, from a wide variety of metal and alloy powders. (Section 11.1)

- The steps in powder metallurgy processing can be summarized as follows:

  1. Powder production (Section 11.2)

  2. Blending of powders (Section 11.2)

3. Compaction (Section 11.3)

4. Sintering (Section 11.4)

5. Additional processing, as appropriate, to improve dimensional accuracy, surface finish, mechanical or physical properties, or appearance (Section 11.5)

- Design considerations pertaining to powder metallurgy include the shape of the compact, the ejection of the green compact from the die, and the acceptable tolerances of the application. From an economic standpoint, the powder-metallurgy process is suitable for medium- to high-volume production runs and for relatively small parts, and has competitive advantages over other methods of production, such as casting, forging, and machining. (Sections 11.6, 11.8)

- Among newer developments in P/M techniques is selective laser sintering as a method of rapid prototyping. (Section 11.7)

- A variety of ceramic materials are available. Ceramics have the following relative characteristics: brittleness, high strength and hardness at elevated temperatures, high elastic modulus, low toughness, low density, low thermal expansion, and low thermal and electrical conductivity. (Section 11.9)

- The first step in processing ceramics is crushing. The three basic shaping processes for ceramics are casting, plastic forming, and pressing. After the ceramic has been shaped, it is dried and fired to give it the proper strength. Finishing operations, such as machining and grinding, may be performed to give the part its final shape, or the part may be subjected to surface treatments to improve tribological properties. (Section 11.10)

- Almost all commercial glasses are categorized as one of six types that indicate their composition. In general, the behavior of glass can be thought of as perfectly elastic and brittle. Glass in bulk form has relatively low strength, but glasses can be strengthened by thermal or chemical treatments to obtain high strength and toughness. (Section 11.11)

- Continuous methods of glass processing are drawing, rolling, and floating. Discrete glass products can be manufactured by blowing, pressing, centrifugal casting, and sagging. After initial processing, glasses can be strengthened by thermal or chemical tempering, or laminating. They may also be annealed to relieve residual stresses. (Section 11.12)

- Design considerations for ceramics and glasses are guided by factors such as general lack of tensile strength and toughness, and sensitivity to external and internal defects. Warping and cracking are important considerations, as are the methods employed for production and assembly. (Section 11.13)

- Graphite, fullerenes, and diamond are forms of carbon that display unusual combinations of properties. These materials have unique and emerging applications in engineering and manufacturing. (Section 11.14)

- As engineering materials, metal-matrix and ceramic-matrix composites possess useful combinations of properties and have increasingly broad and important applications. Metal-matrix composites are processed via liquid-phase, solid-phase, and two-phase processes. Ceramic-matrix composites can be processed via slurry infiltration, chemical synthesis, and chemical vapor infiltration. (Sections 11.15, 11.16)

## SUMMARY OF EQUATIONS

Shape factor of particle, $k = (A/V)D_{eq}$

Pressure distribution in compaction, $p_x = p_o e^{-4\mu k x/D}$

Volume, $V_{sint} = V_{green}(1 - \Delta L/L_o)^3$

Density, $\rho_{green} = \rho_{sint}(1 - \Delta L/L_o)^3$

Tensile strength, $UTS = UTS_o e^{-nP}$

Elastic modulus, $E = E_o(1 - 1.9P + 0.9P^2)$

Thermal conductivity, $k = k_o(1 - P)$

## BIBLIOGRAPHY

### Powder metallurgy

Allen, T., *Particle Size Measurement,* 4th ed. London: Chapman and Hall, 1990.
*ASM Handbook,* Vol. 7: *Powder Metallurgy.* Materials Park, Ohio: ASM International, 1984.
Bradbury, S. (ed.), *Powder Metallurgy Equipment Manual.* Princeton, New Jersey: Powder Metallurgy Equipment Association, 1986.
Burns, M., *Automated Fabrication: Improving Productivity in Manufacturing.* Englewood Cliffs, New Jersey: PTR Prentice-Hall, 1993.
Cumberland, D.J., and R.J. Crawford, *The Packing of Particles.* Amsterdam: Elsevier, 1987.
German, R.M., *Particle Packing Characteristics.* Princeton, New Jersey: Metal Powder Industries Federation, 1989.
———, *Powder Injection Molding.* Princeton, New Jersey: Metal Powder Industries Federation, 1990.
———, *Powder Metallurgy Science,* 2d ed. Princeton, New Jersey: Metal Powder Industries Federation, 1994.
Gessinger, G.H., *Powder Metallurgy of Superalloys.* London: Butterworth, 1984.
Hanes, E.D., D.A. Seifert, and C.R. Watts, *Hot Isostatic Pressing.* New York: Springer-Verlag, 1980.
Hausner, H.H., and M.K. Mal, *Handbook of Powder Metallurgy,* 2d ed. New York: Chemical Publishing Co., 1982.
Jenkins, I., and J.V. Wood (eds.), *Powder Metallurgy: An Overview.* London: The Institute of Metals, 1991.
Kuhn, H.A., and B.L. Ferguson, *Powder Forging.* Princeton, New Jersey: Metal Powder Industries Federation, 1990.
Kulkarni, K.M. (ed.), *Powder Metallurgy for Full-Density Products.* Princeton, New Jersey: Metal Powder Industries Federation, 1987.
Lawley, A., *Atomization: The Production of Metal Powders.* Princeton, New Jersey: Metal Powder Industries Federation, 1992.
Lenel, F.V., *Powder Metallurgy: Principles and Applications.* New York: American Powder Metallurgy Institute, 1980.
Lowell, S., and J.E. Shields, *Introduction to Powder Surface Area and Porosity.* New York: Wiley, 1984.
*Powder Metallurgy Design Guidebook.* New York: American Powder Metallurgy Institute, revised periodically.

Rhodes, M.J. (ed.), *Principles of Powder Technology.* New York: Wiley, 1990.

Uskokovic, P., H. Palmour III, and R.M. Spriggs (eds.), *Science of Sintering.* New York: Plenum, 1989.

Wood, L., *Rapid Prototyping: An Introduction.* New York: Industrial Press, 1993.

## Ceramics, glasses, and composites

*ASM Engineered Materials Handbook, Desk Edition.* Materials Park, Ohio: ASM International, 1995.

Belitskus, D.L., *Fiber and Whisker Reinforced Ceramics for Structural Applications.* New York: Marcel Dekker, 1993.

Buchanan, R.C., *Ceramic Materials for Electronics: Processing, Properties, and Applications,* 2d ed. New York: Marcel Dekker, 1991.

Cheremisinoff, N.P., *Handbook of Ceramics and Composites,* 3 vols. New York: Marcel Dekker, 1991.

Clyne, T.W., and P.J. Withers, *An Introduction to Metal Matrix Composites.* New York: Cambridge University Press, 1993.

*Concise Encyclopedia of Advanced Ceramics.* Cambridge, Mass.: The MIT Press, 1991.

Cranmer, D., *Mechanical Testing Methodology for Ceramic Design and Reliability.* New York: Marcel Dekker, 1996.

*Engineered Materials Handbook,* Vol. 4: *Ceramics and Glasses.* Materials Park, Ohio: ASM International, 1991.

Ford, R.W., *Ceramics Drying.* Elmsford, New York: Pergamon, 1986.

Hlavak, J., *The Technology of Glass and Ceramics.* Amsterdam: Elsevier, 1983.

Hoa, S.V., *Computer-Aided Design for Composite Structures.* New York: Marcel Dekker, 1996.

Jahanmir, S., *Friction and Wear of Ceramics.* New York: Marcel Dekker, 1994.

Kelly, A. (ed.), *Concise Encyclopedia of Composite Materials.* New York: Pergamon, 1989.

Lu, H.Y., *Introduction to Ceramic Science.* New York: Marcel Dekker, 1996.

McCauley, R.A., *Corrosion of Ceramics.* New York: Marcel Dekker, 1995.

Ochiai, S., *Mechanical Properties of Metallic Composites.* New York: Marcel Dekker, 1994.

Rahaman, M.N., *Ceramic Processing and Sintering.* New York: Marcel Dekker, 1996.

Reed, J.S., *Introduction to the Principles of Ceramic Processing.* New York: Wiley, 1988.

Richerson, D.W., *Modern Ceramic Engineering: Properties, Processing, and Use in Design,* 2d ed. New York: Marcel Dekker, 1992.

Schwartz, M.M., *Ceramic Joining.* Materials Park, Ohio: ASM International, 1990.

——— *Composite Materials Handbook,* 2d ed. New York: McGraw-Hill, 1992.

——— (ed.), *Handbook of Structural Ceramics.* New York: McGraw-Hill, 1992.

Yosomiya, R., *Adhesion and Bonding in Composites.* New York: Marcel Dekker, 1990.

## QUESTIONS

### Powder metallurgy

11.1 Explain why metal powders are blended.

11.2 Is green strength important in powder metals processing? Explain.

11.3 Explain why density differences may be desirable within some P/M parts. Give specific examples.

11.4 Give the reasons why injection molding of metal powders is becoming an important process.

11.5 Describe what happens during sintering.

11.6 What is mechanical alloying, and what are its advantages over conventional alloying of metals?

11.7 How are design considerations for powder-metallurgy parts different from casting and forging?

11.8 Describe the effects of different shapes and sizes of metal powders in P/M processing, including comments on the shape factor (SF) of the particles.

11.9 Are there any shapes or design details that are not suitable for fabrication by P/M? Explain.

11.10 Comment on the shapes of the curves and their relative positions shown in Fig. 11.7.

11.11 Should green compacts be brought up to the sintering temperature slowly or rapidly? Explain the advantages and limitations of each.

11.12 Explain the effects of using fine powders and coarse powders, respectively, in making P/M parts.

11.13 Are the requirements for punch and die materials in powder metallurgy different from those for forging and extrusion? Explain.

11.14 Describe the relative advantages and limitations of cold and hot isostatic pressing, respectively.

11.15 Why do mechanical and physical properties depend on the density of P/M parts? Explain with appropriate diagrams.

11.16 What type of press is required to compact parts by the set of punches shown in Fig. 11.8d?

11.17 Explain the difference between impregnation and infiltration. Give some typical applications.

11.18 Tool steels are now being made by P/M techniques. Explain the advantages of this method over traditional methods such as casting and subsequent metalworking techniques.

11.19 One method of producing superconducting wire and strip is by compacting powders of these materials, placing them in a tube, and drawing them through dies, as well as rolling them. Describe your thoughts on the steps and the problems involved in such production.

## Ceramics and others

11.20 Compare the major differences between ceramics and metals and plastics.

11.21 Explain why ceramics are weaker in tension than in compression.

11.22 Explain why the mechanical and physical properties of ceramics decrease with increasing porosity.

11.23 What engineering applications could benefit from the fact that ceramics maintain their modulus of elasticity at elevated temperatures?

11.24 Explain why the mechanical property data in Table 11.10 have such a broad range. What is the significance of this in engineering practice?

11.25 List the factors that you would take into account when replacing a metal component with a ceramic component.

11.26 How are ceramics made tougher?

11.27 Describe applications in which static fatigue can be important.

11.28 Explain the difficulties involved in making large ceramic components.

11.29 Explain why ceramics are effective cutting-tool materials. Would ceramics also be suitable as die materials for metal forming? Explain.

11.30 Describe applications in which a ceramic material with a zero coefficient of thermal expansion is desirable.

11.31 Give reasons for the development of ceramic-matrix components. Name some possible applications.

11.32 Name the factors that are important in drying ceramic components, and explain why they are important.

11.33 It has been stated that the higher the coefficient of thermal expansion of the glass and the lower its thermal conductivity, the higher the level of residual stresses developed during processing. Explain why.

11.34 What types of finishing operations are used on ceramics? Why are they done?

11.35 What should be the requirements for the metal balls in a ball mill?

11.36 Which property of glasses allows them to be expanded and shaped into bottles by blowing?

11.37 What properties should plastic sheet have when used in laminated glass? Why?

11.38 Consider some ceramic products that you are familiar with and outline a sequence of processes to manufacture them.

11.39 What is the difference between physical and chemical tempering of glass?

11.40 What do you think is the purpose of the operation shown in Fig. 11.24d?

11.41 As you have seen, injection molding is a process that is used for plastics and powder metals, as well as ceramics. Why is this so?

11.42 Are there any similarities between the strengthening mechanisms for glass and those for other metallic and nonmetallic materials? Explain.

11.43 Describe and explain the differences in the manner in which each of the following flat surfaces would fracture when struck with a large piece of rock: (1) ordinary window glass, (2) tempered glass, and (3) laminated glass.

11.44 Describe similarities and differences between the processes described in this chapter and in Chapters 5 through 10.

## PROBLEMS

11.45 Estimate the number of particles in a 500 g sample of iron powder, if the particle size is 100 $\mu$m.

11.46 Assume that the surface of a copper particle is covered with an oxide layer 0.1 $\mu$m in thickness. What is the volume occupied by this layer if the copper particle itself is 50 $\mu$m in diameter? What would be the role of this oxide layer in subsequent processing of the powder metal?

11.47 Determine the shape factor (SF) of (1) a flakelike particle with a surface-area-to-thickness ratio of 15 by 15 by 1, and (2) ellipsoid with an axial ratio of 5 by 2 by 1.

11.48 We have seen in Section 3.3.2 and Example 3.1 that the energy in brittle fracture is dissipated as surface energy. We have also noted that the comminution process of powder preparation generally involves brittle fracture. What are the relative energies involved in making spherical powders of diameters 1, 10, and 100 $\mu$m, respectively?

11.49 Referring to Fig. 11.7a, what should be the volume of loose, fine iron powder in order to make a solid cylindrical compact 20 mm in diameter and 10 mm high?

11.50 In Fig. 11.8e we note that the pressure is not uniform across the diameter of the compact. What is the reason for it?

11.51 Plot the family of pressure ratio $p_x/p_o$ curves as a function of $x$ for the following ranges of process parameters: $\mu = 0$ to 1, $k = 0$ to 1, $D = 5$mm to 50 mm.

11.52 Derive an expression similar to Eq. (11.2) for compaction in a square die with dimensions $a$ by $a$.

11.53 In the Example 11.7, calculate (1) the porosity of the dried part if the porosity of the fired part is to be 9 percent, and (2) the initial length $L_o$ of the part if the linear shrinkages during drying and firing are 8 and 7 percent, respectively.

11.54 Plot the UTS, $E$, and $k$ values for ceramics as a function of porosity $P$, and describe and explain the trends that you observe in their behavior.

# 12

# Joining and Fastening Processes

**12.1** Introduction
**12.2** Arc-Welding Processes: Consumable Electrode
**12.3** Arc-Welding Processes: Nonconsumable Electrode
**12.4** Laser-Beam and Electron-Beam Welding
**12.5** The Welded Joint; Weld Design
**12.6** Cold Welding
**12.7** Ultrasonic Welding
**12.8** Friction Welding
**12.9** Resistance Welding
**12.10** Explosion Welding
**12.11** Diffusion Bonding (Welding)
**12.12** Brazing and Soldering
**12.13** Adhesive Bonding
**12.14** Mechanical Fastening
**12.15** Joining Plastics

## 12.1 Introduction

**Joining** is an all-inclusive term covering numerous processes that are important and necessary aspects of manufacturing operations for the following reasons:

- The product is impossible to manufacture as a single piece.

- The product, such as a cooking pot with a handle, is easier and more economical to manufacture as individual components, which are then assembled.

- Products may have to be taken apart for repair or maintenance during their service lives.

- Different properties may be desirable for functional purposes of the product. For example, surfaces subjected to friction and wear or corrosion and environmental attack generally require characteristics different from those of the component's bulk.

- Transporting the product in individual components and assembling them at home or at the customer's plant may be easier and less costly.

Although there are different ways of categorizing the wide variety of joining and fastening processes, this book will follow the sequence shown in the chapter heading above. The choice of a particular process depends on several factors, such as its application and environment, joint design, the materials involved, and the thickness, size, and shape of the components to be joined. Other considerations are location of the joint, number of parts involved, operator skill required, and equipment and labor costs (Table 12.1).

**Fusion welding** is defined as melting together and coalescing materials by means of heat, usually supplied by chemical or electrical means; filler metals may or may not be used. This

**TABLE 12.1**

**COMPARISON OF VARIOUS JOINING METHODS**

| METHOD | Strength | Design Variability | Small Parts | Large Parts | Tolerances | Reliability | Ease of Maintenance | Visual Inspection | Cost |
|---|---|---|---|---|---|---|---|---|---|
| Arc welding | 1 | 2 | 3 | 1 | 3 | 1 | 2 | 2 | 2 |
| Resistance welding | 1 | 2 | 1 | 1 | 3 | 3 | 3 | 3 | 1 |
| Brazing | 1 | 1 | 1 | 1 | 3 | 1 | 3 | 2 | 3 |
| Bolts and nuts | 1 | 2 | 3 | 1 | 2 | 1 | 1 | 1 | 3 |
| Riveting | 1 | 2 | 3 | 1 | 1 | 1 | 3 | 1 | 2 |
| Fasteners | 2 | 3 | 3 | 1 | 2 | 2 | 2 | 1 | 3 |
| Seaming, crimping | 2 | 2 | 1 | 3 | 3 | 1 | 3 | 1 | 1 |
| Adhesive bonding | 3 | 1 | 1 | 2 | 3 | 2 | 3 | 3 | 2 |

Note: 1, very good; 2, good; 3, poor.

is a major category of welding and consists of consumable or nonconsumable electrode arc welding, and high-energy-beam welding processes. The welded joint undergoes important metallurgical and physical changes that, in turn, have a major effect on the properties and performance of the welded component or structure. Because they are not used in a modern industrial setting and are used basically for maintenance and repair work, the oxyfuel gas welding and thermit welding processes are not described in this chapter.

In **solid-state welding,** joining takes place without fusion; thus, there is no liquid (molten) phase in the joint. The basic categories are cold, ultrasonic, friction, resistance, and explosion welding and diffusion bonding. The latter, combined with superplastic forming, has become an important manufacturing process for complex shapes. **Brazing** and **soldering** use filler metals and involve lower temperatures than welding; the heat required is supplied externally.

**Adhesive bonding** has been developed into an important technology because of its various advantages and unique applications requiring strength, sealing, insulating, vibration damping, and resistance to corrosion between dissimilar metals. Included in this category are electrically conducting adhesives for surface-mount technologies. **Mechanical fastening** processes involve traditional methods of using various fasteners, bolts, nuts, rivets, etc. **Joining plastics** can be accomplished by adhesive bonding, fusion by various external or internal heat sources, and mechanical fastening.

The individual groups of joining processes outlined above have, as in all other manufacturing processes described throughout this text, several characteristics such as joint strength, design, size of parts to be joined, reliability, ease of maintenance, and cost to be considered. These characteristics are outlined, in relative terms, in Table 12.1 as a guide to process selection.

**TABLE 12.2**

**GENERAL CHARACTERISTICS OF JOINING PROCESSES**

| JOINING PROCESS | OPERATION | ADVANTAGE | SKILL LEVEL REQUIRED | WELDING POSITION | CURRENT TYPE | DISTOR-TION* | COST OF EQUIP-MENT |
|---|---|---|---|---|---|---|---|
| Shielded metal-arc | Manual | Portable and flexible | High | All | ac, dc | 1 to 2 | Low |
| Submerged arc | Automatic | High deposition | Low to medium | Flat and horizontal | ac, dc | 1 to 2 | Medium |
| Gas metal-arc | Semiautomatic or automatic | Most metals | Low to high | All | dc | 2 to 3 | Medium to high |
| Gas tung-sten-arc | Manual or automatic | Most metals | Low to high | All | ac, dc | 2 to 3 | Medium |
| Flux-cored arc | Semiautomatic or automatic | High deposition | Low to high | All | dc | 1 to 3 | Medium |
| Oxyfuel | Manual | Portable and flexible | High | All | — | 2 to 4 | Low |
| Electron-beam, Laser-beam | Semiautomatic or automatic | Most metals | Medium to high | All | — | 3 to 5 | High |

*1, highest; 5, lowest.

## 12.2   Arc-Welding Processes: Consumable Electrode

In **arc welding,** developed in the mid-1800s, the heat required is obtained through electrical energy. Using either a *consumable* or *nonconsumable electrode* (rod or wire), an arc is produced between the tip of the electrode and the workpiece to be welded, using ac or dc power supplies. This arc produces temperatures in the range of 30,000°C (54,000°F), which is much higher than those developed in oxyfuel gas welding. Arc welding includes various welding processes (Table 12.2), which are described below.

### 12.2.1   Shielded metal arc welding

**Shielded metal arc welding (SMAW)** is one of the oldest, simplest, and most versatile joining processes. Currently, about 50 percent of all industrial and maintenance welding is performed by this process. The electric arc is generated by touching the tip of a coated electrode against the workpiece and then withdrawing it quickly to a distance sufficient to maintain the arc (Fig. 12.1). The electrodes are in the shape of thin, long sticks (see Section 12.2.7), so this process is also known as *stick welding.*

The heat generated melts a portion of the tip of the electrode, its coating, and the base metal in the immediate area of the arc. A weld forms after the molten metal—a mixture of the base metal (workpiece), electrode metal, and substances from the coating on the electrode—solidifies in the weld area. The electrode coating deoxidizes and provides a shielding gas in the weld area to protect it from oxygen in the environment.

A bare section at the end of the electrode is clamped to one terminal of the power source, while the other terminal is connected to the workpiece being welded (Fig. 12.2). The current

**FIGURE 12.1** Schematic illustration of the shielded metal arc-welding process. About 50 percent of all large-scale industrial welding operations use this process.

usually ranges between 50 A and 300 A, with power requirements generally less than 10 kW. Too low a current causes incomplete fusion, and too high a current can damage the electrode coating and reduce its effectiveness. The current may be dc or ac. For sheet-metal welding, dc is preferred because of the steady arc.

The *polarity* of the dc current, i.e., direction of current flow, can be important, depending on factors such as type of electrode, the metals to be welded, and the depth of the heated zone. In *straight polarity,* the workpiece is positive and the electrode negative; it is preferred for sheet metals because of shallow penetration and for joints with very wide gaps. In *reverse polarity,* where the electrode is positive, deeper weld penetration is possible. In the ac method, the arc pulsates rapidly; it is suitable for welding thick sections and using large-diameter electrodes at maximum currents.

The SMAW process has the advantage of being relatively simple and versatile, requiring a relatively small variety of electrodes. The equipment consists of a power supply, power cables, and electrode holder. The use of safety equipment, similar to that used with oxyfuel gas welding, is essential.

This process is commonly used in general construction, shipbuilding, and pipelines, as well as for maintenance work, since the equipment is portable and can be easily maintained. It is especially useful for work in remote areas where portable fuel-powered generators can be used as the power supply. The SMAW process is best suited for workpiece thicknesses of

**FIGURE 12.2** Schematic illustration of the shielded metal arc-welding operation, also known as stick welding because the electrode is in the shape of a stick.

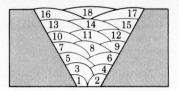

**FIGURE 12.3** A weld zone showing the build-up sequence of individual weld beads in deep welds.

3–19 mm (0.12–0.75 in.), although this range can be easily extended using multiple-pass techniques (Fig. 12.3), and highly skilled welders. This process requires that slag be cleaned after each weld bead. Unless removed completely, the solidified slag can cause severe corrosion of the weld area and lead to failure of the weld. Slag should also be completely removed, such as by wire brushing, before another weld is applied for multiple-pass welding. Thus labor costs are high, as are material costs.

### 12.2.2 Submerged arc welding

In **submerged arc welding (SAW),** the weld arc is shielded by granular flux, consisting of lime, silica, manganese oxide, calcium fluoride, and other elements. The flux is fed into the weld zone by gravity flow through a nozzle (Fig. 12.4). The thick layer of flux completely covers the molten metal and prevents spatter and sparks—and the intense ultraviolet radiation and fumes of the SMAW process. The flux also acts as a thermal insulator, allowing deep penetration of heat into the workpiece. The operator must wear gloves, but other than tinted safety glasses, face shields generally are unnecessary.

The consumable electrode is a coil of bare round wire 1.5–10 mm ($\frac{1}{16}-\frac{3}{8}$ in.) in diameter, and is fed automatically through a tube (*welding gun*). Electric currents usually range between 300 A and 2000 A. The power supplies are usually connected to standard single- or three-phase power lines with a primary rating up to 440 V.

Because the flux is fed by gravity, the SAW process is somewhat limited to welds in a flat or horizontal position with a backup piece. Circular welds can be made on pipes, provided that they are rotated during welding. As Fig. 12.4 shows, the unfused flux can be recovered, treated, and reused.

**FIGURE 12.4** Schematic illustration of the submerged arc-welding process and equipment. Unfused flux is recovered and reused.

Developed in the 1940s, the SAW process can be automated for greater economy. This process is used to weld a variety of carbon and alloy steel and stainless steel sheet or plate, at speeds as high as 5 m/min (16 ft/min). The quality of the weld is very high, with good toughness, ductility, and uniformity of properties. The SAW process provides very high welding productivity, depositing 4–10 times the amount of weld metal per hour as the SMAW process. Typical applications include thick plate welding for shipbuilding and pressure vessels.

### 12.2.3   Gas metal arc welding

In **gas metal arc welding (GMAW)**, the weld area is shielded by an external source, such as argon, helium, carbon dioxide, or various other gas mixtures (Fig. 12.5). The consumable bare wire is fed automatically through a nozzle into the weld arc (Fig. 12.6). In addition to the use of inert shielding gases, deoxidizers are usually present in the electrode metal itself, in order to prevent oxidation of the molten weld puddle. Multiple weld layers can be deposited at the joint.

Metal can be transferred three ways in the GMAW process: spray, globular, and short circuiting. In *spray transfer,* small droplets of molten metal from the electrode are transferred to the weld area at rates of several hundred droplets per second. The transfer is spatter-free and very stable. High dc current and voltages and large-diameter electrodes are used, with argon or argon-rich gas mixtures used as the shielding gas. The average current required in this process can be reduced by *pulsed arcs,* which are high-amplitude pulses superimposed over a low, steady current, and the process can be used in all welding positions.

In *globular transfer,* carbon dioxide-rich gases are utilized, and globules propelled by the forces of the electric arc transfer the metal, resulting in considerable spatter. High welding currents are used, with greater weld penetration and welding speed than in spray transfer. Heavier sections are commonly joined by this method.

**FIGURE 12.5** Gas metal arc-welding process, formerly known as MIG (for metal inert gas).

**FIGURE 12.6** Basic equipment used in gas metal arc-welding operations.

In *short circuiting*, the metal is transferred in individual droplets, at rates of more than 50 per second, as the electrode tip touches the molten weld pool and short circuits. Low currents and voltages are utilized, with carbon dioxide-rich gases and electrodes made of small-diameter wire. The power required is about 2 kW. The temperatures involved are relatively low. Thus this method is suitable only for thin sheets and sections (less than 6 mm; 0.25 in.); otherwise, incomplete fusion may occur. This process is very easy to use and may be the most popular for welding ferrous metals in thin sections. However, pulsed-arc systems are gaining wide usage for thin ferrous and nonferrous metals.

The GMAW process was developed in the 1950s and was formerly called *metal inert gas* (MIG) welding. It is suitable for welding a variety of ferrous and nonferrous metals and is used extensively in the metal-fabrication industry. Because of the relatively simple nature of the process, training operators is easy. This process is rapid, versatile, and economical; welding productivity is double that of the SMAW process. The cost of equipment usually ranges from $1000 to $3000. The GMAW process can easily be automated and lends itself readily to robotics and flexible manufacturing systems (Chapters 13 and 14).

### 12.2.4  Flux-cored arc welding

The **flux-cored arc welding** (FCAW) process (Fig. 12.7) is similar to gas metal arc welding, with the exception that the electrode is tubular in shape and is filled with flux (hence the term flux-cored). Cored electrodes produce a more stable arc, improve weld contour, and improve the mechanical properties of the weld metal. The flux in these electrodes is much more flexible than the brittle coating used on SMAW electrodes. Thus the tubular electrode can be provided in long coiled lengths. The electrodes are usually 0.5–4 mm (0.020–0.15 in.) in diameter. The power required is about 20 kW.

Self-shielded cored electrodes are also available. These electrodes do not require external gas shielding because they contain emissive fluxes that shield the weld area against the surrounding atmosphere. Advances in manufacturing of electrodes for FCAW, as well as in the chemistry of the flux, have made this process the fastest growing in welding. Small-diameter electrodes have made welding of thinner materials not only possible but often desirable. Also, small-diameter electrodes make it relatively easy to weld parts out of position, and the flux chemistry enables welding of many base metals.

The flux-cored arc-welding process combines the versatility of SMAW with the continuous and automatic electrode-feeding feature of GMAW. It is economical and is used for weld-

**FIGURE 12.7**   Schematic illustration of the flux-cored arc-welding process. This operation is similar to gas metal arc welding, shown in Fig. 12.5.

ing a variety of joints, mainly with steels, stainless steels, and nickel alloys. The higher weld metal deposition rate of the FCAW process, over that of GMAW, has led to its use in joining sections of all thicknesses. Recent development of tubular electrodes with very small diameters has extended the use of this process to smaller workpiece section sizes.

A major advantage of FCAW is the ease with which specific weld metal chemistries can be developed. By adding alloys to the flux core, virtually any alloy composition can be developed. This process is easy to automate and is readily adaptable to flexible manufacturing systems and robotics.

## 12.2.5   Electrogas welding

**Electrogas welding (EGW)** is used primarily for welding the edges of sections vertically in one pass with the pieces placed edge to edge (butt). It is classified as a machine-welding process because it requires special equipment (Fig. 12.8). The weld metal is deposited into a weld cavity between the two pieces to be joined. The space is enclosed by two water-cooled copper dams (shoes) to prevent the molten slag from running off. Mechanical drives move the shoes upward. Circumferential welds such as on pipes are also possible, with the workpiece rotating.

Single or multiple electrodes are fed through a conduit and a continuous arc is maintained, using flux-cored electrodes at up to 750 A, or solid electrodes at 400 A. Power requirements are about 20 kW. Shielding is by inert gas, such as carbon dioxide, argon, or

Electrode conduit    Welding wire    Drive rolls

Gas

Oscillator

Welding gun

Water

Welding wire

Gas

Gas box

Gas

Water out

Water out

Movable shoe

Water in

Water in

Fixed shoe    Primary shielding gas    Supplementary shielding gas

**FIGURE 12.8**  Schematic illustration of the electrogas-welding process. *Source:* Courtesy of the American Welding Society.

helium, depending on the type of material being welded. The gas may be provided from an external source, or it may be produced from a flux-cored electrode, or both.

The equipment for electrogas welding is reliable, and training operators is relatively simple. Weld thickness ranges from 12 mm to 75 mm (0.5 in. to 3 in.) on steels, titanium, and aluminum alloys. Typical applications are in the construction of bridges, pressure vessels, thick-walled and large-diameter pipes, storage tanks, and ships. The cost of machines typically ranges from $15,000 to $25,000, although portable machines with less power cost as little as $5000.

### 12.2.6  Electroslag welding

Developed in the 1950s, **electroslag welding (ESW)** and its applications are similar to electrogas welding (Fig. 12.9). The main difference is that the arc is started between the electrode tip and the bottom of the part to be welded. Flux is added and melted by the heat of the arc. After the molten slag reaches the tip of the electrode, the arc is extinguished. Energy is supplied continuously through the electrical resistance of the molten slag. Thus because the arc is extinguished, ESW is not strictly an arc welding process. Single or multiple solid as well as flux-cored electrodes may be used. The guide may be nonconsumable (conventional method) or consumable.

Electroslag welding is capable of welding plates with thicknesses ranging from 50 mm to more than 900 mm (2 in. to more than 36 in.). Welding is done in one pass. The current required is about 600 A at 40–50 V, although higher currents are used for thick plates. Travel speed of the weld is 12–36 mm/min (0.5–1.5 in./min). The weld quality is good and the process is used for heavy structural steel sections, such as heavy machinery and nuclear-reactor vessels.

### 12.2.7  Electrodes

**Electrodes** for the consumable arc-welding processes described are classified according to the strength of the deposited weld metal, current (ac or ± dc), and the type of coating.

**FIGURE 12.9** Equipment used for electroslag-welding operations. *Source:* Courtesy of the American Welding Society.

Electrodes are identified by numbers and letters (Table 12.3) or by color code, particularly if they are too small to imprint with identification. Typical coated electrode dimensions are 150–460 mm (6–18 in.) long and 1.5–8 mm ($\frac{1}{16}-\frac{5}{16}$ in.) in diameter. The thinner the sections to be welded and the lower the current required, the smaller the diameter of the electrode should be.

The specifications for electrodes and filler metals, including tolerances, quality control procedures, and processes, are stated by the American Welding Society (AWS) and the American National Standards Institute (ANSI), and in the Aerospace Materials Specifications (AMS) by the Society of Automotive Engineers (SAE). Among others, the specifications require that the wire diameter not vary more than 0.05 mm (0.002 in.) from nominal size, and that the coatings be concentric with the wire. Electrodes are sold by weight and are available in a wide variety of sizes and specifications. Selection and recommendations for electrodes for a particular metal and its application can be found in supplier literature and various handbooks and references listed at the end of this chapter.

**Electrode coatings.** Electrodes are *coated* with claylike materials that include silicate binders and powdered materials such as oxides, carbonates, fluorides, metal alloys, and cellulose (cotton cellulose and wood flour). The coating, which is brittle and has complex interactions during welding, has the following basic functions:

- Stabilize the arc.

- Generate gases to act as a shield against the surrounding atmosphere. The gases produced are carbon dioxide and water vapor, and carbon monoxide and hydrogen in small amounts.

- Control the rate at which the electrode melts.

**TABLE 12.3**

---

**DESIGNATIONS FOR MILD-STEEL COATED ELECTRODES**

---

The prefix "E" designates arc-welding electrode.

The first two digits of four-digit numbers and the first three digits of five-digit-numbers indicate minimum tensile strength:

| | |
|---|---|
| E60XX | 60,000 psi minimum tensile strength |
| E70XX | 70,000 psi minimum tensile strength |
| E110XX | 110,000 psi minimum tensile strength |

The next-to-last digit indicates position:

| | |
|---|---|
| EXX1X | All positions |
| EXX2X | Flat position and horizontal fillets |

The last two digits together indicates the type of covering and the current to be used.

The suffix (Example: EXXXX-A1) indicates the approximate alloy in the weld deposit:

| | |
|---|---|
| —A1 | 0.5% Mo |
| —B1 | 0.5% Cr, 0.5% Mo |
| —B2 | 1.25% Cr, 0.5% Mo |
| —B3 | 2.25% Cr, 1% Mo |
| —B4 | 2% Cr, 0.5% Mo |
| —B5 | 0.5% Cr, 1% Mo |
| —C1 | 2.5% Ni |
| —C2 | 3.25% Ni |
| —C3 | 1% Ni, 0.35% Mo, 0.15% Cr |
| —D1 and D2 | 0.25–0.45% Mo, 1.75% Mn |
| —G | 0.5% min. Ni, 0.3% min. Cr, 0.2% min. Mo, 0.1% min. V, 1% min. Mn (only one element required) |

---

• Act as a flux to protect the weld against formation of oxides, nitrides, and other inclusions, and with the resulting slag, protect the molten weld pool.

• Add alloying elements to the weld zone to enhance the properties of the weld, including deoxidizers to prevent the weld from becoming brittle.

The deposited electrode coating or slag must be removed after each pass in order to ensure a good weld. A manual or power wire brush can be used for this purpose. Bare electrodes and wire, made of stainless steels and aluminum alloys, are also available. They are used as filler metals in various welding operations.

## 12.3  Arc-Welding Processes: Nonconsumable Electrode

Unlike the arc-welding processes that use consumable electrodes, which were described in Section 12.2, nonconsumable-electrode processes typically use a *tungsten electrode*. As one pole of the arc, it generates the heat required for welding. A shielding gas is supplied from an external source. Described below are the advantages, limitations, and typical applications of these processes.

**FIGURE 12.10**  Gas tungsten arc welding process, formerly known as TIG (for tungsten inert gas).

### 12.3.1  Gas tungsten arc welding

In **gas tungsten arc welding (GTAW)**, formerly known as TIG welding (for tungsten inert gas), the filler metal is supplied from a *filler wire* (Fig. 12.10). Because the tungsten electrode is not consumed in this operation, a constant and stable arc gap is maintained at a constant current level. The filler metals are similar to the metals to be welded, and flux is not used. The shielding gas is usually argon or helium, or a mixture of the two. Welding with GTAW may be done without filler metals, as in welding close-fit joints.

The power supply (Fig. 12.11) is either dc at 200 A, or ac at 500 A, depending on the metals to be welded. In general, ac is preferred for aluminum and magnesium because the cleaning action of ac removes oxides and improves weld quality. Thorium or zirconium may be used in the tungsten electrodes to improve their electron emission characteristics. Power requirements range from 8 kW to 20 kW. Contamination of the tungsten electrode by the molten metal can be a significant problem, particularly in critical applications, as it can cause discontinuities in the weld. Thus contact of the electrode with the molten metal pool should be avoided.

The GTAW process is used for a wide variety of metals and applications, particularly aluminum, magnesium, titanium, and refractory metals. It is especially suitable for thin metals. The cost of the inert gas makes this process more expensive than SMAW, but it provides welds with very high quality and surface finish. It is used in a variety of critical applications with a wide range of workpiece thicknesses and shapes.

**FIGURE 12.11**  Equipment for gas tungsten arc-welding operations.

**FIGURE 12.12**
Two types of plasma arc-welding processes: (a) transferred and (b) nontransferred. Deep and narrow welds are made by this process at high welding speeds.

### 12.3.2   Plasma arc welding

In **plasma arc welding** (**PAW**), developed in the 1960s, a concentrated plasma arc is produced and aimed at the weld area. The arc is stable and reaches temperatures as high as 33,000°C (60,000°F). A *plasma* is ionized hot gas, composed of nearly equal numbers of electrons and ions. The plasma is initiated between the tungsten electrode and the orifice, using a low-current pilot arc. Unlike other processes, the plasma arc is concentrated because it is forced through a relatively small orifice. Operating currents are usually below 100 A, but they can be higher for special applications. When a filler metal is used, it is fed into the arc, as in GTAW. Arc and weld-zone shielding is supplied through an outer shielding ring by gases such as argon, helium, or mixtures.

There are two methods of plasma-arc welding. In the *transferred-arc* method (Fig. 12.12a), the workpiece being welded is part of the electrical circuit. The arc thus transfers from the electrode to the workpiece—hence the term transferred. In the *non-transferred* method (Fig. 12.12b), the arc is between the electrode and the nozzle, and the heat is carried to the workpiece by the plasma gas. The thermal transfer mechanism is similar to that for oxyfuel flame.

Compared to other arc-welding processes, plasma arc welding has greater energy concentration (hence deeper and narrower welds can be made), better arc stability, less thermal distortion, and higher welding speeds, such as 120–1000 mm/min (5–40 in./min). A variety of metals can be welded, with part thicknesses generally less than 6 mm (0.25 in.). The high heat concentration can penetrate completely through the joint (*keyhole* technique), with thicknesses as much as 20 mm (0.75 in.) for some titanium and aluminum alloys. In the keyhole technique, the force of the plasma arc displaces the molten metal and produces a hole at the leading edge of the weld pool. Plasma arc welding, rather than the GTAW process, is often used for butt and lap joints because of higher energy concentration, better arc stability, and higher welding speeds.

## 12.4   Laser-Beam and Electron-Beam Welding

In Section 9.14, we described the characteristics of high-energy beams and their unique applications in machining operations. Joining with high-energy beams, principally laser-beam and electron-beam welding, are among increasingly important applications in modern manufacturing because of their high quality and technical and economic advantages.

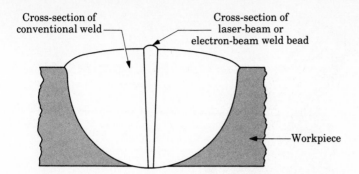

Cross-section of conventional weld

Cross-section of laser-beam or electron-beam weld bead

Workpiece

**FIGURE 12.13** Comparison of the size of weld beads in conventional (tungsten arc) and laser-beam and electron-beam welding.

### 12.4.1 Laser-beam welding

**Laser-beam welding (LBW)** utilizes a high-power laser beam as the source of heat (see Fig. 9.41 and Table 9.8) to produce a fusion weld. Because the beam can be focused to a very small area, it has high-energy density and, therefore, has deep penetrating capability. The beam can be directed, shaped, and focused precisely on the workpiece. Consequently, this process is particularly suitable for welding deep and narrow joints (Fig. 12.13), with depth-to-width ratios typically ranging from 4 to 10. The laser beam may be pulsed (milliseconds) for applications such as spot welding of thin materials, with power levels up to 100 kW. Continuous multi-kW laser systems are used for deep welds on thick sections.

Laser-beam welding produces welds of good quality, with minimum shrinkage and distortion. Laser welds have good strength and are generally ductile and free of porosity. As we described in Example 7.9, an important application of laser welding is the forming of butt-welded sheet metal, particularly for automotive body panels, of two different thicknesses. The LBW process can be automated and used successfully on a variety of materials with thicknesses of up to 25 mm (1 in.), and is particularly effective on thin workpieces. Typical metals and alloys welded include aluminum, titanium, ferrous metals, copper, superalloys, and refractory metals. Welding speeds range from 2.5 m/min (8 ft/min) to as high as 80 m/min (250 ft/min) for thin metals.

Because of the nature of the process, welding can be done in otherwise inaccessible locations. Safety is a particularly important consideration in laser-beam welding because of the extreme hazards to the eye as well as the skin. Solid-state (YAG) lasers are particularly dangerous. In the automotive industry, welding of transmission components is the most widespread application. Among numerous other applications is the welding of thin parts for electronic components.

The major advantages of LBW over EBW are:

- The beam can be transmitted through air, so a vacuum is not required.
- Because laser beams can be shaped, manipulated, and focused optically (using fiber optics), the process can easily be automated.
- The beams do not generate x-rays as compared to EBW.
- The quality of the weld is better, with less tendency for incomplete fusion, spatter and porosity, and less distortion.

**Example 12.1:  Laser welding of razor blades**

Figure 12.14 shows a closeup of the Gillette Sensor razor cartridge. Each of the two narrow, high-strength blades has 13 pinpoint welds, 11 of which can be seen (as darker spots, about

**FIGURE** 12.14

0.5 mm in diameter) on each blade in the photograph. The welds are made with a Nd: YAG laser equipped with fiber-optic delivery, thus providing very flexible beam manipulation to the exact locations along the length of the blade. With a set of these machines, production is at a rate of 3 million welds per hour, with accurate and consistent weld quality.

*Source:* Courtesy of Lumonics Corporation, Industrial Products Division.

### 12.4.2  Electron-beam welding

In **electron-beam welding** (EBW), heat is generated by high-velocity narrow-beam electrons. The kinetic energy of the electrons is converted into heat as they strike the workpiece. This process requires special equipment to focus the beam on the workpiece in a vacuum. The higher the vacuum, the more the beam penetrates and the greater the depth-to-width ratio. Almost any metal can be welded by EBW, with workpiece thicknesses ranging from foil to plate. The intense energy is also capable of producing holes in the workpiece. Generally, no shielding gas, flux, or filler metal is required. Capacities of electron beam guns range to 100 kW.

Developed in the 1960s, EBW has the capability to make high-quality welds that are almost parallel sided, are deep and narrow, and have small heat-affected zones (see Section 12.5). Depth-to-width ratios range between 10 and 30 (Fig. 12.13). Using servo controls, parameters can be controlled accurately, with welding speeds as high as 12 m/min (40 ft/min).

Almost any metal can be butt or lap welded with this process, with thicknesses to as much as 150 mm (6 in.). Distortion and shrinkage in the weld area are minimal. Weld quality is good, with very high purity, and typical applications include welding of aircraft, missile, nuclear, and electronic components, as well as gears and shafts in the automotive industry. Electron-beam welding equipment generates x-rays, and hence proper monitoring and periodic maintenance are important.

## 12.5 The Welded Joint; Weld Design

In this section we describe the following aspects of welding processes:

- The nature, properties, and quality of the welded joint
- Weldability of metals
- Testing welds
- Weld design and process selection

The mechanical properties of a welded joint depend on many factors. For example, the rate of heat application and the thermal properties of metals are important, in that they control the magnitude and distribution of temperature in a joint during welding. The microstructure and grain size of the welded joint depends on the magnitude of heat applied and temperature rise, the degree of prior cold work of the metals, and the rate of cooling after the weld is made.

Weld quality depends on factors such as the geometry of the weld bead and the presence of cracks, residual stresses, inclusions, and oxide films. Their control is essential to reliable welds that have acceptable mechanical properties.

A typical fusion weld joint is shown in Fig. 12.15, where three distinct zones can be identified:

1. The *base metal,* that is, the metal to be joined.
2. The *heat-affected zone* (HAZ).
3. The *fusion zone,* that is, the region that has melted during welding.

The metallurgy and properties of the second and third zones depend strongly on the metals joined, the welding process, filler metals used, if any, and process variables. A joint produced without a filler metal is called *autogenous,* and the weld zone is composed of the resolidified **base metal.** A joint made with a filler metal has a central zone called the **weld metal** and is composed of a mixture of the base and filler metals.

**FIGURE 12.15** Characteristics of a typical fusion weld zone in oxyfuel gas and arc welding.

(a)

(b)

**FIGURE 12.16**   Grain structure in (a) a deep and (b) a shallow weld. Note that the grains in the solidified weld metal are perpendicular to their interface with the base metal. In a good weld, the solidification line at the center in the deep weld shown in (a) has grain migration, thus developing uniform strength in the weld bead.

**Solidification of the weld metal.**   After applying heat and introducing filler metal, if any, into the weld area, the molten weld joint is allowed to cool to ambient temperature. The *solidification* process is similar to that of casting and begins with the formation of columnar (dendritic) grains. These grains are relatively long and form parallel to the heat flow (see Fig. 5.9). Because metals are much better heat conductors than the surrounding air, the grains lie parallel to the plane of the two plates or sheets being welded (Fig. 12.16a). The grains in a shallow weld are shown in Figs. 12.16b and 12.17. Grain structure and size depend on the specific alloy, the welding process, and the filler metal used.

The weld metal is basically a *cast* structure and, because it has cooled slowly, it generally has coarse grains. Consequently, this structure has generally low strength, toughness, and ductility. However, the proper selection of filler-metal composition or heat treatments following welding can improve the joint's mechanical properties. The results depend on the particular alloy, its composition, and the thermal cycling to which the joint is subjected. Cooling rates may, for example, be controlled and reduced by preheating the general weld area prior to welding. Preheating is particularly important for metals with high thermal conductivity, such as aluminum and copper; otherwise, the heat during welding rapidly dissipates.

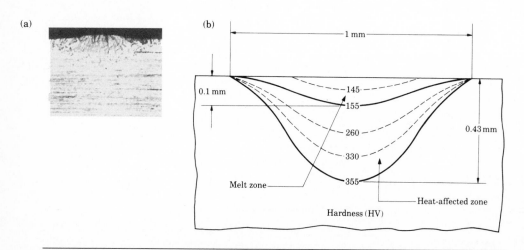

**FIGURE 12.17**   (a) Weld bead on a cold-rolled nickel strip produced by a laser beam. (b) Micro-hardness profile across the weld bead. Note the softer condition of the weld bead compared to the base metal. *Source:* IIT Research Institute.

**FIGURE 12.18** Schematic illustration of various regions in a fusion weld zone and the corresponding phase diagram for 0.30 percent carbon steel. *Source:* Courtesy of the American Welding Society.

**Heat-affected zone.** The **heat-affected zone (HAZ)** is within the base metal itself. It has a microstructure different from that of the base metal before welding, because it has been subjected to elevated temperatures for a period of time during welding. The portions of the base metal that are far enough away from the heat source do not undergo any structural changes during welding.

The properties and microstructure of the HAZ depend on (1) the rate of heat input and cooling; and (2) the temperature to which this zone was raised. The HAZ and the corresponding phase diagram for 0.3 percent carbon steel are shown in Fig. 12.18. In addition to metallurgical factors (such as original grain size, grain orientation, and degree of prior cold work), the specific heat and thermal conductivity of the metals influence the HAZ's size and characteristics.

The strength and hardness of the heat-affected zone depend partly on how the original strength and hardness of the particular alloy was developed prior to welding. As described in Chapters 3 and 5, they may have been developed by cold working, solid-solution strengthening, precipitation hardening, or by various heat treatments. Of these strengthening methods, the simplest to analyze is base metal that has been cold worked, say, by cold rolling or forging.

The heat applied during welding *recrystallizes* the elongated grains (preferred orientation) of the cold-worked base metal. Grains that are away from the weld metal will recrystallize into fine equiaxed grains. However, grains close to the weld metal, having been subjected to elevated temperatures for a longer period of time, will grow. This growth will result in a region that is softer and has less strength. Such a joint will be weakest in its heat-affected zone. The grain structure of such a weld—exposed to corrosion by chemical reaction—is shown in Fig. 12.19. The center vertical line is where the two workpieces meet.

The effects of heat during welding on the HAZ for joints made with dissimilar metals, and for alloys strengthened by other methods, are complex and beyond the scope of this text. Details can be found in the more specialized texts listed in the bibliography at the end of this chapter.

## 12.5.1  Weld quality

Because of a history of thermal cycling and attendant microstructural changes, a welded joint may develop certain **discontinuities.** Welding discontinuities can also be caused by inadequate or careless application of established welding technologies or substandard operator training. The major discontinuities that affect weld quality are described below.

### Porosity

**Porosity** in welds is caused by trapped gases released during melting of the weld area and trapped during solidification, chemical reactions during welding, or contaminants. Most welded joints contain some porosity, which is generally spherical in shape or in the form of elongated pockets (see also Section 5.4.6). The distribution of porosity in the weld zone may be random, or it may be concentrated in a certain region.

Porosity in welds can be reduced by the following methods:

- Proper selection of electrodes and filler metals.
- Improving welding techniques, such as preheating the weld area or increasing the rate of heat input.
- Proper cleaning and preventing contaminants from entering the weld zone.
- Slowing the welding speed to allow time for gas to escape.

### Slag inclusions

**Slag inclusions** are compounds such as oxides, fluxes, and electrode-coating materials that are trapped in the weld zone. If shielding gases are not effective during welding, contamination from the environment may also contribute to such inclusions. Welding conditions are important, and with proper techniques the molten slag will float to the surface of the molten weld metal and not be entrapped. Slag inclusions may be prevented by:

- Cleaning the weld-bead surface before the next layer is deposited by using a hand or power wire brush.
- Providing adequate shielding gas.
- Redesigning the joint to permit sufficient space for proper manipulation of the puddle of molten weld metal.

(a)

Weld

Base Metal

Incomplete fusion in fillet welds. *B* is often
termed 'bridging'

**FIGURE 12.20** Examples of various discontinuities in fusion welds.

(b)

Weld

Incomplete fusion from oxide or dross at
the center of a joint, especially in aluminum

(c)

Weld

Incomplete fusion in a groove weld

### Incomplete fusion and penetration

**Incomplete fusion** (or lack of fusion) produces poor weld beads, such as those shown in Fig. 12.20. A better weld can be obtained by:

- Raising the temperature of the base metal.
- Cleaning the weld area prior to welding.
- Changing the joint design and type of electrode.
- Providing adequate shielding gas.

**Incomplete penetration** occurs when the depth of the welded joint is insufficient. Penetration can be improved by:

- Increasing the heat input.
- Lowering travel speed during welding.
- Changing the joint design.
- Ensuring that the surfaces to be joined fit properly.

### Weld profile

**Weld profile** is important not only because of its effects on the strength and appearance of the weld, but also because it can indicate incomplete fusion or the presence of slag inclusions in multiple-layer welds. **Underfilling** results when the joint is not filled with the proper amount of weld metal (Fig. 12.21). **Undercutting** results from melting away the base metal and subsequently generating a groove in the shape of a sharp recess or notch. Unless it is not deep or sharp, an undercut can act as a stress raiser and reduce the fatigue strength of the joint—and may lead to premature failure. **Overlap** is a surface discontinuity generally caused by poor welding practice and selection of the wrong materials. A proper weld is shown in Fig. 12.21c.

**FIGURE 12.21**   Examples of various discontinuities in fusion welds.

## Cracks

**Cracks** may occur in various locations and directions in the weld area. The types of cracks are typically longitudinal, transverse, crater, underbead, and toe cracks (Fig. 12.22). These cracks generally result from a combination of the following factors:

**FIGURE 12.22**   Types of cracks in welded joints caused by thermal stresses that develop during solidification and contraction of the weld bead and the welded structure. (a) Crater cracks. (b) Various types of cracks in butt and T joints.

**FIGURE 12.23**   Crack in a weld bead. The two components were not allowed to contract after the weld was completed. *Source:* S.L. Meiley, Packer Engineering Associates, Inc.

- Temperature gradients that cause thermal stresses in the weld zone.
- Variations in the composition of the weld zone that cause different contractions.
- Embrittlement of grain boundaries by segregation of elements, such as sulfur, to the grain boundaries as the solid–liquid boundary moves when the weld metal begins to solidify.
- Hydrogen embrittlement.
- Inability of the weld metal to contract during cooling (Fig. 12.23)—a situation similar to *hot tears* that develop in castings and related to excessive restraint of the workpiece.

Cracks are classified as hot or cold cracks. *Hot cracks* occur while the joint is still at elevated temperatures. *Cold cracks* develop after the weld metal has solidified. Some crack-prevention measures are:

1. Change the joint design to minimize stresses from shrinkage during cooling.
2. Change welding-process parameters, procedures, and sequence.
3. Preheat components being welded.
4. Avoid rapid cooling of the components after welding.

### Lamellar tears

In describing the anisotropy of plastically deformed metals (see Section 3.5), we stated that because of the alignment of nonmetallic impurities and inclusions (stringers), the workpiece is weaker when tested in its thickness direction. This condition is particularly evident in rolled plates and structural shapes. In welding such components, *lamellar tears* may develop because of shrinkage of the restrained members in the structure during cooling. Such tears can be avoided by providing for shrinkage of the members or by changing the joint design to make the weld bead penetrate the weaker member more deeply.

### Surface damage

During welding, some of the metal may spatter and be deposited as small droplets on adjacent surfaces. In arc-welding processes, the electrode may inadvertently contact the parts being welded at places not in the weld zone (*arc strikes*). Such surface discontinuities may be objectionable for reasons of appearance or subsequent use of the welded part. If severe, these discontinuities may adversely affect the properties of the welded structure, particularly for notch-sensitive metals. Using proper welding techniques and procedures is important in avoiding surface damage.

## Residual stresses

Because of localized heating and cooling during welding, expansion and contraction of the weld area causes **residual stresses** in the workpiece (see also Section 2.10). Residual stresses can cause:

- Distortion, warping, and buckling of the welded parts (Fig. 12.24).
- Stress–corrosion cracking.
- Further distortion if a portion of the welded structure is subsequently removed, say, by machining or sawing.
- Reduced fatigue life.

The type and distribution of residual stresses in welds is best described by reference to Fig. 12.25a. When two plates are being welded, a long narrow region is subjected to elevated temperatures, whereas the plates as a whole are essentially at ambient temperature. As the weld is completed and time elapses, the heat from the weld area dissipates laterally to the plates as the weld area cools. The plates thus begin to expand longitudinally while the welded length begins to contract.

These two opposing effects cause residual stresses that are typically distributed as shown in Fig. 12.25b. Note that the magnitude of compressive residual stresses in the plates diminishes to zero at a point away from the weld area. Because no external forces are acting on the welded plates, the tensile and compressive forces represented by these residual stresses must balance each other.

**FIGURE 12.24** Distortion of parts after welding: (a) butt joints and (b) fillet welds. Distortion is caused by differential thermal expansion and contraction of different parts of the welded assembly. Warping can be reduced or eliminated by proper weld design and parts fixturing prior to welding.

**FIGURE 12.25**
Residual stresses developed in a straight butt joint. *Source:* Courtesy of the American Welding Society.

In complex welded structures, residual stress distributions are three dimensional and difficult to analyze. The preceding example involves two plates that are not restrained from movement. In other words, the plates are not an integral part of a larger structure. If they are restrained, reaction stresses will be generated because the plates are not free to expand or contract. This situation arises particularly in structures with high stiffness.

**Stress relieving of welds**

The problems caused by residual stresses, such as distortion, buckling, or cracking, can be reduced by *preheating* the base metal or the parts to be welded. Preheating reduces distortion by reducing the cooling rate and the level of thermal stresses (by reducing the elastic modulus). This technique also reduces shrinkage and possible cracking of the joint. The workpieces may be heated in a furnace or electrically or inductively, and for thin sections, by radiant lamps or a hot-air blast. For optimum results, preheating temperatures and cooling rates must be controlled carefully in order to maintain acceptable strength and toughness in the welded structure.

Residual stresses can be reduced by **stress relieving** the welded structure. The temperature and time required for stress relieving depend on the type of material and magnitude of the residual stresses developed. Other methods of stress relieving include peening, hammering, or surface rolling the weld bead area (see Section 4.5.1). These processes induce compressive residual stresses, thus reducing or eliminating tensile residual stresses in the weld. For multilayer welds, the first and last layers should not be peened in order to protect them against possible peening damage.

Residual stresses can also be relieved, or reduced, by plastically deforming the structure by a small amount. This technique can be used in some welded structures, such as pressure vessels, by pressurizing the vessels internally (*proof-stressing*). In order to reduce the possibility of sudden fracture under high internal pressure, the weld must be made properly and be free from notches and discontinuities, which could act as points of stress concentration. In addition to stress relieving, welds may also be heat treated by various techniques in order to modify their properties (see Section 5.10). These techniques include annealing, normalizing, or quenching and tempering of steels, and solution treatment and aging of various alloys.

### 12.5.2  Weldability

We may define **weldability** of a metal as its capacity to be welded into a specific structure that has certain properties and characteristics and that will satisfactorily meet its service requirements. Weldability involves a large number of variables, making generalizations difficult. As you have seen, the material characteristics—such as alloying elements, impurities, inclusions, grain structure, and processing history—of the base metal and filler metal are important.

Because of the melting, solidification, and microstructural changes involved, a thorough knowledge of the phase diagram and the response of the metal or alloy to elevated temperatures over a period of time is essential. Also influencing weldability are the mechanical and physical properties of strength, toughness, ductility, notch sensitivity, elastic modulus, specific heat, melting point, thermal expansion, surface-tension characteristics of the molten metal, and corrosion.

Preparation of surfaces for welding is important, as are the nature and properties of surface oxide films and adsorbed gases. The welding process employed significantly affects the temperatures developed and their distribution in the weld zone. Other factors are shielding gases, fluxes, moisture content of the coatings on electrodes, welding speed, welding position, cooling rate, preheating, and post-welding techniques (such as stress relieving and heat treating).

The following list states the general weldability of specific metals, which can vary if special welding techniques are used.

- *Plain-carbon steels:* Excellent for low-carbon steels; fair to good for medium-carbon steels; poor for high-carbon steels.
- *Low-alloy steels:* Similar to that for medium-carbon steels.
- *High-alloy steels:* Generally good under well-controlled conditions.
- *Stainless steels:* Weldable by various processes.
- *Aluminum alloys:* Weldable at a high rate of heat input. Alloys containing zinc or copper generally are considered unweldable.
- *Copper alloys:* Similar to that of aluminum alloys.
- *Magnesium alloys:* Weldable with the use of protective shielding gas and fluxes.
- *Nickel alloys:* Similar to that of stainless steels.
- *Titanium alloys:* Weldable with the proper use of shielding gases.
- *Tantalum:* Similar to that of titanium.
- *Tungsten:* Weldable under well-controlled conditions.
- *Molybdenum:* Similar to that of tungsten.
- *Niobium (columbium):* Weldable by various processes.

### 12.5.3  Testing welded joints

As in all manufacturing processes, the *quality* of a welded joint is established by testing. Several standardized tests and test procedures have been established and are available from organizations such as the American Society for Testing and Materials (ASTM), American Welding Society (AWS), American Society of Mechanical Engineers (ASME), American Society of Civil Engineers (ASCE), and federal agencies. Welded joints may be tested either destructively or nondestructively (see Section 4.8). Each technique has certain capabilities, sensitivity, limitations, reliability, and need for special equipment and operator skill.

**FIGURE 12.26** Types of specimens for tension-shear testing of welds.

Longitudinal    Transverse

## Destructive techniques

Five methods of destructively testing welded joints are commonly used. A brief review of each method is provided below.

   **Tension test.**   Longitudinal and transverse tension tests are performed on specimens removed from actual welded joints and from the weld metal area. Stress–strain curves are obtained by the procedures described in Section 2.2. These curves indicate the yield strength ($Y$), ultimate tensile strength (UTS), and ductility of the welded joint in different locations and directions. Ductility is measured in terms of percentage elongation and percentage reduction of area.

   **Tension-shear test.**   The specimens in the tension-shear test (Fig. 12.26) are specially prepared to simulate actual welded joints and procedures. The specimens are subjected to tension, and the shear strength of the weld metal and the location of fracture are determined.

   **Bend test.**   Several bend tests have been developed to determine the ductility and strength of welded joints. In one test, the welded specimen is bent around a fixture (*wrap-around* bend test; Fig. 12.27a). In another test, the specimens are tested in three-point transverse bending (Fig. 12.27b). These tests help establish the relative ductility and strength of welded joints.

**FIGURE 12.27**
(a) Wrap-around bend test method.
(b) Three-point bending of welded specimens (see also Fig. 2.23).

(a)

Clamp

Roller

(b)

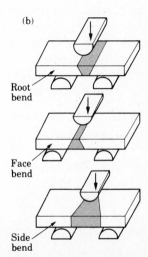

Root bend

Face bend

Side bend

**FIGURE 12.28**   (a) Tension-shear test for spot welds. (b) Cross-tension test. (c) Twist test. (d) Peel test.

**Fracture toughness test.**   Fracture toughness tests commonly utilize the impact testing techniques described in Section 2.9. Charpy V-notch specimens are prepared and tested for toughness. Other toughness tests include the drop-weight test in which the energy is supplied by a falling weight.

**Corrosion and creep tests.**   In addition to mechanical tests, welded joints may be tested for corrosion and creep resistance. Because of the difference in the composition and microstructure of the materials in the weld zone, preferential corrosion may take place in it (see Fig. 12.19). Creep tests are important in determining the behavior of welded joints at elevated temperatures. Testing weld hardness may also be used to indicate weld strength and microstructural changes in the weld zone.

**Testing of spot welds.**   Spot welded joints may be tested for weld-nugget strength using the (1) *tension-shear,* (2) *cross-tension,* (3) *twist,* and (4) *peel tests* (Fig. 12.28). Because they are easy to perform and inexpensive, tension-shear tests are commonly used in fabricating facilities. The cross-tension and twist tests are capable of revealing flaws, cracks, and porosity in the weld area. The peel test is commonly used for thin sheets. After bending and peeling the joint, the shape and size of the torn-out weld nugget is observed.

## Nondestructive techniques

Welded structures often have to be tested nondestructively, particularly for critical applications where weld failure can be catastrophic, such as in pressure vessels, load-bearing structural members, and power plants. Nondestructive testing techniques for welded joints usually consist of visual, radiographic, magnetic-particle, liquid-penetrant, and ultrasonic testing methods. The details of these tests are described in Section 4.8.1.

### 12.5.4  Weld design and process selection

In addition to the material characteristics described thus far, selection of a joint and a welding process involves the following considerations:

- Configuration of the parts or structure to be welded and their thickness and size.
- The methods used to manufacture component parts.
- Service requirements, such as the type of loading and stresses generated.
- Location, accessibility, and ease of welding.
- Effects of distortion and discoloration.
- Appearance.
- Costs involved in edge preparation, welding, and post-processing of the weld, including machining and finishing operations.

As in all manufacturing processes, the optimum choice is the one that meets all design and service requirements at minimum cost. Some examples of weld characteristics are shown in Fig. 12.29, emphasizing the need for careful consideration of some of the factors just identified. General design guidelines may be summarized as follows:

1. Product design should minimize the number of welds, as welding can be costly unless automated. The weld location should be selected to avoid excessive stresses or stress concentrations in the welded structure and for appearance.

**FIGURE 12.29**  Design guidelines for welding. *Source:* J.G. Bralla (ed.), *Handbook of Product Design for Manufacturing.* Copyright © 1986, McGraw-Hill Publishing Company. Reproduced by permission of the McGraw-Hill Companies.

| Basic arc and gas weld symbols | | | | | | | |
|---|---|---|---|---|---|---|---|
| Bead | Fillet | Plug or slot | Groove | | | | |
| | | | Square | V | Bevel | U | J |
| ⌒ | ◺ | ⏖ | ‖ | ∨ | ⋁ | ⋃ | ⋃ |

| Basic resistance weld symbols | | | |
|---|---|---|---|
| Spot | Projection | Seam | Flash or upset |
| ✳ | ⤬ | ⋙ | │ |

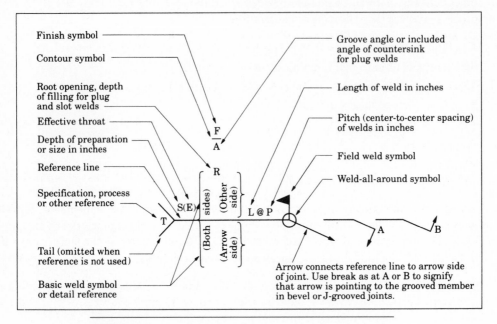

Finish symbol

Contour symbol

Root opening, depth of filling for plug and slot welds

Effective throat

Depth of preparation or size in inches

Reference line

Specification, process or other reference

Tail (omitted when reference is not used)

Basic weld symbol or detail reference

Groove angle or included angle of countersink for plug welds

Length of weld in inches

Pitch (center-to-center spacing) of welds in inches

Field weld symbol

Weld-all-around symbol

Arrow connects reference line to arrow side of joint. Use break as at A or B to signify that arrow is pointing to the grooved member in bevel or J-grooved joints.

**FIGURE 12.30** Standard identification and symbols for welds.

2. Parts should fit properly before welding. The method used to produce edges (sawing, machining, shearing, and flame cutting) can affect weld quality.

3. Some designs can avoid the need for edge preparation.

4. Weld-bead size should be kept to a minimum to conserve weld metal. Weld location should be selected so as not to interfere with further processing of the part or with its intended use and appearance.

Standardized symbols used in engineering drawings to describe the type of weld and its characteristics are shown in Fig. 12.30. These symbols identify the type of weld, groove design, weld size and length, welding process, sequence of operations, and various other information. Further details about these symbols can be found in the references cited at the end of this chapter.

**Example 12.2:** Weld design selection

Three different types of weld design are shown in Fig. 12.31. In 12.31a, the two vertical joints can be welded externally or internally. Full-length external welding takes considerable

(a) Continuous weld  Intermittent welds

(b) M  3M  Welds

(c) Weld  Bnetal

Single V groove  Double V groove

**FIGURE 12.31**

time and requires more weld material than the alternative design, which consists of intermittent internal welds. Moreover, in the alternative method, the appearance of the structure is improved and distortion is reduced. In 12.31b, although both designs require the same amount of weld material and welding time, it can be shown that the design on the right can carry three times the moment $M$ of the one on the left.

In 12.31c, the weld on the left requires about twice the amount of weld material than the design on the right. Note also that because more material must be machined, the design on the left will require more time for edge preparation and more base metal will be wasted.

## 12.6 Cold Welding

In **cold welding (CW)**, pressure is applied to the workpieces, either through dies or rolls. Because of the *plastic deformation* involved, it is necessary that at least one, but preferably both, of the mating parts be ductile. The interface is usually cleaned by wire brushing prior to welding.

However, in joining two dissimilar metals that are mutually soluble, brittle intermetallic compounds may form (see Section 5.2), resulting in a weak and brittle joint. An example is the bonding of aluminum and steel, where a brittle intermetallic compound is formed at the interface. The best bond strength and ductility is obtained with two similar materials. Cold welding can be used to join small workpieces made of soft, ductile metals. Applications include electrical connections and welding wire stock.

**Roll bonding**  The pressure required for cold welding can be applied through a pair of rolls (Fig. 12.32). Hence the process is called *roll bonding*. Developed in the 1960s, roll bonding is used for manufacturing certain U.S. coins. The process can be carried out at elevated temperatures (hot roll bonding). Typical examples are cladding pure aluminum over aluminum-alloy sheet and stainless steel over mild steel for corrosion resistance.

Cladding metal

Base metal

Rolls

**FIGURE 12.32** Schematic illustration of the roll-bonding, or cladding, process.

**Example 12.3:    Roll bonding of the U.S. quarter**

The technique used for manufacturing composite U.S. quarters is roll bonding of two outer layers of 75 percent copper–25 percent nickel (cupronickel), each 1.2 mm (0.048 in.) thick, with an inner layer of pure copper 5.1 mm (0.20 in.) thick. To obtain good bond strength, the faying surfaces are chemically cleaned and wire brushed. The strips are first rolled to a thickness of 2.29 mm (0.090 in.). A second rolling operation reduces the final thickness to 1.36 mm (0.0535 in.). The strips thus undergo a total reduction in thickness of 82 percent. Since volume constancy is maintained in plastic deformation, there is a major increase in the surface area between the layers, thus generating clean interfacial surfaces. This extension in surface area under the high pressure of the rolls, combined with the solid solubility of nickel in copper (see Section 5.2.1), produces a strong bond.

## 12.7    Ultrasonic Welding

In **ultrasonic welding (USW)**, the faying surfaces of the two members are subjected to a static normal force and oscillating shearing (tangential) stresses. The shearing stresses are applied by the tip of a *transducer* (Fig. 12.33a), similar to that used for ultrasonic machining (see Fig. 9.22a). The frequency of oscillation generally ranges from 10 kHz to 75 kHz, although both lower and higher frequencies than these can be employed. The energy required increases with the thickness and hardness of the materials being joined. Proper coupling between the transducer and the tip (called *sonotrode,* from the word *sonic,* as contrasted to electrode) is important for efficient operation.

The shearing stresses cause plastic deformation at the workpiece interfaces, breaking up oxide films and contaminants and thus allowing good contact and producing a strong solid-state bond. Temperatures generated in the weld zone are usually in the range of one-third to one-half the melting point (absolute scale) of the metals joined. Therefore no melting and fusion take place. In certain situations, however, temperatures can be sufficiently high to cause metallurgical changes in the weld zone. The mechanism of joining plastics by ultrasonic welding (see Section 12.15.1) is different from that for metals, and melting takes place at the interface, because of their much lower melting temperatures (see Table 10.2).

The ultrasonic welding process is reliable and versatile. It can be used with a wide variety of metallic and nonmetallic materials, including dissimilar metals (bimetallic strips). It is used extensively in joining plastics and in the automotive and consumer electronics industries for lap welding of sheet, foil, and thin wire and in packaging with foils. The welding tip can be replaced with rotating disks (Fig. 12.33b) for seam welding structures, similar to those shown in Fig. 12.38 for resistance seam welding, one component of which is a sheet or foil. Moderate skill is required to operate the equipment.

**FIGURE 12.33** (a) Components of an ultrasonic welding machine for lap welds. The lateral vibrations of the tool tip cause plastic deformation and bonding at the interface of the workpieces. (b) Ultrasonic seam welding using a roller.

## 12.8 Friction Welding

In the joining processes that have been described so far, the energy required for welding such as chemical, electrical, and ultrasonic, is supplied externally. In **friction welding (FRW)**, the required heat for welding is, as the name implies, generated through *friction* at the interface of the two members being joined. Thus the source of energy is mechanical. You can demonstrate the significant rise in temperature from friction by rubbing your hands together fast or sliding down a rope rapidly.

In friction welding, one of the members remains stationary while the other is placed in a chuck or collet and rotated at a high constant speed. The two members to be joined are then brought into contact under an axial force (Fig. 12.34). After sufficient contact is established,

**FIGURE 12.34** Sequence of operations in the friction-welding process. (a) Left part is rotated at high speed. (b) Right part is brought into contact under an axial force. (c) Axial force is increased; flash begins to form. (d) Left part stops rotating. Weld is completed. Flash can be removed by machining or grinding.

(a) High pressure
    or low speed

(b) Low pressure
    or high speed

(c) Optimum

**FIGURE 12.35** Shape of fusion zone in friction welding, as a function of force applied and rotational speed.

the rotating member is brought to a quick stop, so that the weld is not destroyed by shearing, while the axial force is increased. The rotating member must be clamped securely to the chuck or collet to resist both torque and axial forces without slipping.

The pressure at the interface and the resulting friction produce sufficient heat for a strong joint to take place. The weld zone is usually confined to a narrow region, depending on (1) the amount of heat generated, (2) the thermal conductivity of the materials, and (3) the materials' mechanical properties at elevated temperatures. The shape of the welded joint depends on the rotational speed and the axial pressure applied (Fig. 12.35). These factors must be controlled to obtain a uniformly strong joint. Oxides and other contaminants at the interface are removed by the radially outward movement of the hot metal at the interface (Fig. 12.34d).

Developed in the 1940s, friction welding can be used to join a wide variety of materials, provided that one of the components has some rotational symmetry. Solid, as well as tubular parts, can be joined by this method, with good joint strength. Solid steel bars up to 100 mm (4 in.) in diameter and pipes up to 250 mm (10 in.) outside diameter have been welded successfully by this process. The surface speed of the rotating member may be as high as 900 m/min (3000 ft/min). Because of the combined heat and pressure, the interface in FRW develops a flash by plastic deformation (upsetting) of the heated zone. This flash, if objectionable, can easily be removed by machining or grinding.

**Inertia friction welding.**   **Inertia friction welding** is a modification of FRW although the term has been used interchangeably with friction welding. The energy required for frictional heating in inertia friction welding is supplied through the kinetic energy of a flywheel. The flywheel is accelerated to the proper speed, the two members are brought into contact, and an axial force is applied. As friction at the interface slows the flywheel, the axial force is increased. The weld is completed when the flywheel comes to a stop. The timing of this sequence is important. If the timing is not properly controlled, weld quality will be poor. The rotating mass of inertia friction welding machines can be adjusted for applications requiring different levels of energy, which depend on workpiece size and properties. In one application of inertia friction welding, a 10-mm (0.4-in.) diameter shaft is welded to automotive turbocharger impellers at a rate of one every 15 seconds.

**Linear friction welding.**   A recent development in friction welding involves subjecting the interface of the two parts to be joined to a *linear* reciprocating motion, as opposed to the rotary motion required of at least one of the components described thus far. In **linear friction welding,** the parts do not have to be circular or tubular in cross-section. The process is capable of welding square or rectangular components, as well as round parts, made of metals or plastics. In this process, one part is moved across the face of the other part using a balanced reciprocating mechanism. In one application, a rectangular titanium-alloy part was friction welded at a linear frequency of 25 Hz, amplitude of $\pm 2$ mm (0.08 in.), and a pressure of 100 MPa (15,000 psi) acting on a 240 mm$^2$ (0.38 in$^2$) interface. Various other metal parts have been welded successfully, with rectangular cross-sections as large as 50 mm $\times$ 20 mm (2 in. $\times$ 0.8 in.).

## 12.9 Resistance Welding

**Resistance welding (RW)** covers a number of processes in which the heat required for welding is produced by means of the *electrical resistance* between the two members to be joined. These processes have major advantages, such as not requiring consumable electrodes, shielding gases, or flux.

The heat generated in resistance welding is given by the general expression

$$H = I^2Rt, \tag{12.1}$$

where

$H$ = heat generated, in joules (watt-seconds);

$I$ = current, in amperes;

$R$ = resistance, in ohms; and

$t$ = time of current flow, in seconds.

The equation above is often modified to represent the actual heat energy available in the weld by including a factor $K$ which represents the energy losses through radiation and conduction. Thus the equation becomes $H = I^2RtK$, where the value of $K$ is less than unity.

The total resistance in these processes, such as in the resistance spot welding shown in Fig. 12.36, is the sum of the following:

1. Resistance of the electrodes.
2. Electrode–workpiece contact resistances.

(a) Pressure applied / Current on / Current off, pressure on / Pressure released

Electrodes / Weld nugget / Lap joint

(b) Electrode tip / Electrode / Indentation / Sheet separation / Weld nugget / Heat-affected zone / Electrode

**FIGURE 12.36**
(a) Sequence in the resistance spot-welding process.
(b) Cross-section of a spot weld, showing weld nugget and light indentation by the electrode on sheet surfaces. This is one of the most common processes used in sheet-metal fabrication and automotive-body assembly.

3. Resistances of the individual parts to be welded.

4. Workpiece–workpiece contact resistances (*faying surfaces*).

The actual temperature rise at the joint depends on the specific heat and thermal conductivity of the metals to be joined. Thus because they have high thermal conductivity, metals such as aluminum and copper require high heat concentrations. Similar as well as dissimilar metals can be joined by resistance welding. The magnitude of the current in resistance welding operations may be as high as 100,000 A, although the voltage is typically only 0.5–10 V.

Developed in the early 1900s, resistance welding processes require specialized machinery. Many are now operated by programmable computer control. The machinery is generally not portable, and the process is more suitable for use in manufacturing plants and machine shops. Operator skill required is minimal, particularly with modern machinery. Safety precautions are similar to those for other welding operations.

There are five basic methods of resistance welding: spot, seam, projection, flash, and upset welding. Lap joints are used in the first three processes, and butt joints in the last two.

### 12.9.1 Resistance spot welding

In **resistance spot welding (RSW),** the tips of two opposing solid cylindrical electrodes contact the lap joint of two sheet metals, and resistance heating produces a spot weld (Fig. 12.36a). In order to obtain a good bond in the **weld nugget,** pressure is also applied until the current is turned off. Accurate control and timing of the electric current and pressure are essential in resistance welding. The strength of the bond depends on surface roughness and the cleanliness of the mating surfaces. Thus oil, paint, or thick oxide layers should be removed before welding. The presence of uniform, thin oxide layers and other contaminants is not critical.

The weld nugget (Fig. 12.36b) is generally 6–10 mm (0.25–0.375 in.) in diameter. The surface of the weld spot has a slightly discolored indentation. Currents range from 3000 A to 40,000 A, depending on the materials being welded and their thickness.

Spot welding is the simplest and most commonly used resistance welding process. Welding may be performed by means of single (most common) or multiple electrodes, and the required pressure is supplied through mechanical or pneumatic means. The *rocker-arm*-type spot-welding machines are normally used for smaller parts, with *press-type* machines being used for larger workpieces. The shape and surface condition of the electrode tip and accessibility are important factors in spot welding. A variety of electrode shapes are used to spot weld areas that are difficult to reach (Fig. 12.37).

(a) (b)

**FIGURE 12.37** Types of special electrodes designed for easy access in spot-welding operations for complex shapes.

Spot welding is widely used for fabricating sheet metal. Examples range from attaching handles to stainless-steel cookware, to rapid spot welding of automobile bodies using multiple electrodes. An automobile may have as many as 10,000 spot welds. Modern equipment used for spot welding is computer controlled for optimum timing of current and pressure, and the spot-welding guns are manipulated by programmable robots.

---

**Example 12.4:  Heat generated in resistance spot welding**

Assume that two 1 mm (0.04 in.) thick steel sheets are being spot welded at a current of 5000 A and current flow time $t = 0.1$ s, and using electrodes 5 mm (0.2 in.) in diameter. Estimate the heat generated and its distribution in the weld zone.

SOLUTION.   Let's assume that the effective resistance in this operation is 200 $\mu\Omega$. Then according to Eq. (12.1),

$$\text{Heat} = (5000)^2(0.0002)(0.1) = 500 \text{ J.}$$

From the information given, we estimate the weld-nugget volume to be 30 mm³ (0.0018 in³). If we assume the density for steel to be 8000 kg/m³ (0.008 g/mm³), the weld nugget has a mass of 0.24 g. Since the heat required to melt 1 g of steel is about 1400 J, the heat required to melt the weld nugget is $(1400)(0.24) = 336$ J. Consequently, the remaining heat (164 J) is dissipated into the metal surrounding the nugget.

---

### 12.9.2    Resistance seam welding

**Resistance seam welding (RSEW)** is a modification of resistance spot welding, wherein the electrodes are replaced by rotating wheels or rollers (Fig. 12.38). With continuous ac power supply, the electrically conducting rollers produce continuous spot welds whenever the current reaches a sufficiently high level in the ac cycle. These are actually overlapping spot welds and

**FIGURE 12.38**   (a) Seam-welding process, with rolls acting as electrodes. (b) Overlapping spots in a seam weld. (c) Roll spot welds.

(a)

(b)

**FIGURE 12.39**
Methods of high-frequency butt welding of tubes.

Contacts

Vee

Apex

Current
Squeeze roll

Tube travel

High-frequency coil

Apex

Current

Squeeze roll

produce a joint that is liquid tight and gas tight (Fig. 12.38b). With intermittent application of current to the rollers, a series of spot welds at various intervals can be made along the length of the seam (Fig. 12.38c). This procedure is called *roll spot welding*. The RSEW process is used to make the longitudinal (side) seam of cans for household products, mufflers, gasoline tanks, and other containers. The typical welding speed is 1.5 m/min (60 in./min) for thin sheet.

### 12.9.3   High-frequency resistance welding

**High-frequency resistance welding (HFRW)** is similar to seam welding, except that high-frequency current (up to 450 kHz) is employed. A typical application is making butt-welded tubing. In one method, the current is conducted through two sliding contacts (Fig. 12.39a) to the edges of roll-formed tubes. The heated edges are then pressed together by passing the tube through a pair of squeeze rolls.

In another method, the roll-formed tube is subjected to high-frequency induction heating (Fig. 12.39b). Structural sections such as I-beams can also be fabricated by HFRW. Spiral pipe and tubing, and finned tubes (for heat exchangers), may be made by these techniques.

### 12.9.4   Resistance projection welding

In **resistance projection welding (RPW)**, high electrical resistance at the joint is developed by embossing one or more projections (dimples) on one of the surfaces to be welded (Fig. 12.40). High localized temperatures are generated at the projections, which are in con-

(a)

(b)

Flat electrodes

Sheet

Workpiece

Projections

Force

Product

Weld nuggets

Force

**FIGURE 12.40**   Schematic illustration of resistance projection welding: (a) before and (b) after. The projections are produced by embossing operations, as described in Section 7.12.2.

**FIGURE 12.41**   Projection welding of (a) nuts or threaded bosses and (b) studs. The projections are produced by forging or machining operations. *Source:* Courtesy of the American Welding Society.

tact with the flat mating part. The electrodes—made of copper-base alloys and water cooled to keep their temperature low—are large and flat. Weld nuggets, similar to those in spot welding, are formed as the electrodes exert pressure to compress the projections. The projections may be round or oval for design or strength purposes.

Spot-welding equipment can be used for RPW by modifying the electrodes. Although embossing workpieces is an added expense, this process produces a number of welds in one stroke, extends electrode life, and is capable of welding metals of different thicknesses. Nuts and bolts are also welded to sheet and plate by this process (Fig. 12.41), with projections that are produced by machining or forging. Joining a network of wires such as metal baskets, grills, oven racks, and shopping carts is also considered resistance projection welding because of the small contact area between crossing wires (grids).

### 12.9.5   Flash welding

In **flash welding (FW),** also called *flash butt welding,* heat is generated from the arc as the ends of the two members begin to make contact, developing an electrical resistance at the joint (Fig. 12.42a). Because of the arc's presence, this process is also classified as arc welding. After the proper temperature is reached and the interface begins to soften, an axial force is applied at a controlled rate, and a weld is formed by plastic deformation (upsetting) of the joint. Some metal is expelled from the joint as a shower of sparks during the flashing process.

Poor                Good

**FIGURE 12.42**   (a) Flash-welding process for end-to-end welding of solid rods or tubular parts. (b) Design guidelines for flash welding.

Because impurities and contaminants are squeezed out during this operation, the quality of the weld is good, although a significant amount of material may be burned off during the welding process. The joint may later be machined to improve its appearance. The machines for FW are usually automated and large, with a variety of power supplies ranging from 10 to 1500 kVA.

The flash-welding process is suitable for end-to-end or edge-to-edge joining of similar or dissimilar metals 1–75 mm (0.05–3 in.) in diameter and sheet and bars 0.2–25 mm (0.01–1 in.) thick. Thinner sections have a tendency to buckle under the axial force applied during welding. Rings made by forming processes, such as those shown in Fig. 7.26b and c, can also be flash butt welded. This process is also used to repair broken band-saw blades, with fixtures that are attached to the band-saw frame.

This process can be automated for reproducible welding operations. Typical FW applications are joining pipe and tubular shapes for metal furniture and windows. It is also used for welding the ends of coils of sheet or wire in continuously operating rolling mills and wire-drawing equipment. Once the appropriate process parameters are established, the required operator skill is minimal. Design guidelines for mating surfaces in flash welding are shown in Fig. 12.42b. Note the importance of uniform cross-sections at the joint.

### 12.9.6   Stud arc welding

**Stud arc welding (SW)** is similar to flash welding. The stud, such as a small part or a threaded rod or hanger, serves as one of the electrodes while being joined to another member, which is usually a flat plate (Fig. 12.43). In order to concentrate the heat generated, prevent oxidation, and retain the molten metal in the weld zone, a disposable ceramic ring (*ferrule*) is placed around the joint. The equipment for stud welding can be automated, with various controls for arcing and applying pressure. Portable stud welding equipment is also available.

In **capacitor-discharge stud welding,** a direct-current arc is produced from a capacitor bank. No ferrule or flux is required because the welding time is very short, on the order of 1 to 6 ms. The choice between this process and stud arc welding depends on factors such as the type of metals to be joined, workpiece thickness, stud diameter, and shape of the joint.

**Stud welding** is a more general term for joining a metal stud to a workpiece, and can be done by various joining processes such as resistance, friction, or any other suitable means described in this chapter. Shielding gases may or may not be used.

**FIGURE 12.43** Sequence of operations in stud arc welding, which is used for welding bars, threaded rods, and various fasteners on metal plates.

### 12.9.7  Percussion welding

The resistance-welding processes already described require a transformer to meet power requirements. However, the electrical energy for welding may be stored in a capacitor. **Percussion welding (PEW)** utilizes this technique, in which the power is discharged in a very short time (1–10 ms), developing high localized heat at the joint. This process is useful where heating of the components adjacent to the joint is to be avoided, such as in electronic components.

---

**Example 12.5:    Resistance welding vs. laser-beam welding in the can-making industry**

The cylindrical bodies of cans for food and household products have been resistance seam welded (lap joint along the height of the can) for many years. Beginning in about 1987, the laser-beam welding technology was introduced to the can-making industry whereby the joints are made by lasers, with the same productivity but offering the following advantages over resistance welding. (1) As opposed to lap joints used in resistance welding, laser welding utilizes butt joints; thus some material is saved which, multiplied by the billions of cans made, becomes a very significant savings. (2) Because laser welds produce a very narrow heat-affected zone (see Fig 12.13), the unprinted area on the can surface (printing margin) is greatly reduced, thus improving its appearance and customer acceptance. (3) The resistance lap-welded joint can be subject to corrosion due to the contents of the can (tomato juice, for example), thus affecting taste and any potential liability risk. A butt joint made by laser avoids this problem.

*Source:* G. F. Benedict.

---

## 12.10  Explosion Welding

In **explosion welding (EXW)**, pressure is applied by detonating a layer of explosive that has been placed over one of the members being joined, called the *flyer plate* (Fig. 12.44). The contact pressures developed are extremely high, and the plate's kinetic energy striking the

**FIGURE 12.44**  Schematic illustration of the explosion-welding process: (a) constant interface clearance gap and (b) angular interface clearance gap.

(a)    (b)

**FIGURE 12.45** Cross-sections of explosion-welded joints: (a) titanium (top) on low-carbon steel (bottom) and (b) incoloy 800 (iron–nickel-base alloy) on low-carbon steel. The wavy interfaces shown improve the shear strength of the joint. Some combinations of metals, such as tantalum and vanadium, produce a much less wavy interface. If the two metals have little metallurgical compatibility, an interlayer may be added that has compatibility with both metals. *Source:* Courtesy of DuPont Company.

mating member produces a wavy interface. This impact mechanically interlocks the two surfaces (Fig. 12.45). Cold pressure welding by plastic deformation also takes place. The flyer plate is placed at an angle, and any oxide films present at the interface are broken up and propelled from the interface. As a result, bond strength in explosion welding is very high.

The explosive may be flexible plastic sheet, cord, granulated, or liquid, which is cast or pressed onto the flyer plate. Detonation speeds are usually 2400–3600 m/s (8000–12,000 ft/s), depending on the type of explosive, thickness of the explosive layer, and its packing density. There is a minimum detonation speed for welding to occur in this process. Detonation is carried out using a standard commercial blasting cap.

This process was developed in the 1940s. It is particularly suitable for cladding plates and slab with dissimilar metals. Plates as large as 6 m × 2 m (20 ft × 7 ft) have been explosively clad. These plates may then be rolled into thinner sections. Tube and pipe are often joined to the holes in head plates of boilers and heat exchangers by placing the explosive inside the tube; explosion expands the tube. Explosion welding is inherently dangerous and requires safe handling by well-trained and experienced personnel.

## 12.11   Diffusion Bonding (Welding)

**Diffusion bonding,** or **diffusion welding (DFW),** is a solid-state joining process in which the strength of the joint results primarily from diffusion (movement of atoms across the interface) and, to a lesser extent, small plastic deformation of the faying surfaces. This process requires temperatures of about $0.5T_m$ (where $T_m$ is the melting point of the metal on the absolute scale) in order to have a sufficiently high diffusion rate between the parts being joined. The bonded interface in DFW has essentially the same physical and mechanical properties as the base metal. Its strength depends on pressure, temperature, time of contact, and the cleanliness of the faying surfaces. These requirements can be lowered by using filler metal at the interfaces.

Although developed in the 1970s as a modern welding technology, the principal of diffusion bonding actually dates back centuries, when goldsmiths bonded gold over copper. Called *filled gold*, a thin layer of gold foil is obtained by hammering; the gold is then placed over copper, and a weight is placed on top of it. The assembly is then placed in a furnace and left until a good bond is obtained.

In diffusion bonding, pressure may be applied by dead weights or a press, as well as by using differential gas pressure or utilizing relative thermal expansion of the parts to be joined. The parts are usually heated in a furnace or by electrical resistance. High-pressure autoclaves are also used for bonding complex parts.

Diffusion bonding is generally most suitable for dissimilar metal pairs. However, it is also used for reactive metals, such as titanium, beryllium, zirconium, and refractory metal alloys, and for composite materials. Because diffusion involves migration of the atoms across the joint, this process is slower than other welding processes. Although DFW is used for fabricating complex parts in low quantity for the aerospace, nuclear, and electronics industries, it has been automated to make it suitable and economical for moderate-volume production.

### 12.11.1 Diffusion bonding/superplastic forming

An important development is the ability to fabricate sheet-metal structures by combining *diffusion bonding* with *superplastic forming* (see Section 7.10). Typical structures in which flat sheets are diffusion bonded and formed are shown in Fig. 12.46. After diffusion bonding of

**FIGURE 12.46** (a) Sequence of operations in diffusion bonding and superplastic forming of a structure with three flat sheets. *Source:* After D. Stephen and S. J. Swadling. (b) and (c) Typical structures fabricated. *Source:* Rockwell International Corp.

selected locations of the sheets, the unbonded regions are expanded into a mold by air pressure. These structures are thin and have high stiffness-to-weight ratios. Hence they are particularly important in aircraft and aerospace applications.

This process improves productivity by eliminating mechanical fasteners and produces parts with good dimensional accuracy and low residual stresses. The technology is now well advanced for titanium structures (typically Ti-6Al-4V alloy) for aerospace applications. Structures made of 7475-T6 aluminum alloy and other alloys are also being developed using this technique.

## 12.12    Brazing and Soldering

This section describes two joining processes that involve lower temperatures than those required for welding. In both **brazing** and **soldering**, filler metals are placed in or supplied to the joint. They are then melted using an external source of heat and, upon solidification, a solid joint is produced. Brazing temperatures are higher than those for soldering, and the brazed joint has higher strength. Thus brazing and soldering are arbitrarily distinguished by temperature.

### 12.12.1    Brazing

**Brazing** is a joining process in which a filler metal is placed at or between the faying surfaces to be joined, and the temperature is raised to melt the filler metal but not the workpieces (Fig. 12.47a). The molten metal fills the closely fitting space by *capillary* action. Upon cooling and solidification of the filler metal, a strong joint is obtained. Brazing was first used as far back as 3000–2000 B.C. Actually, there are two types of brazing processes: (1) that which we have already described, and (2) **braze welding** (Fig. 12.47b), in which the filler metal is deposited at the joint with a technique similar to oxyfuel gas welding.

Filler metals used for brazing melt above 450°C (840°F). The temperatures employed in brazing are below the melting point (solidus temperature) of the metals to be joined. Thus this process is unlike liquid-state welding processes in which the workpieces must melt in the weld area for fusion to occur. Problems associated with heat-affected zones, warping, and residual stresses are therefore reduced in brazing.

The strength of the brazed joint depends on (1) joint design and (2) the adhesion at the interfaces of the workpiece and filler metal. Consequently, the surfaces to be brazed should be chemically or mechanically cleaned to ensure full capillary action; hence the use of a flux is important.

**FIGURE 12.47**    (a) Brazing and (b) braze welding operations.

**TABLE 12.4**

**TYPICAL FILLER METALS FOR BRAZING VARIOUS METALS AND ALLOYS**

| BASE METAL | FILLER METAL | BRAZING TEMPERATURE (°C) |
|---|---|---|
| Aluminum and its alloys | Aluminum–silicon | 570–620 |
| Magnesium alloys | Magnesium–aluminum | 580–625 |
| Copper and its alloys | Copper–phosphorus | 700–925 |
| Ferrous and nonferrous alloys (except aluminum and magnesium) | Silver and copper alloys, copper–phosphorus | 620–1150 |
| Iron-, nickel-, and cobalt-base alloys | Gold | 900–1100 |
| Stainless steels, nickel- and cobalt-base alloys | Nickel–silver | 925–1200 |

## Filler metals

Several **filler metals** (*braze metals*) are available and have a range of brazing temperatures (Table 12.4). They come in a variety of shapes, such as wire, rings, shims, and filings. The choice of filler metal and its composition are important in order to avoid embrittlement of the joint (by grain boundary penetration of liquid metal), formation of brittle intermetallic compounds at the joint, and galvanic corrosion in the joint. Note that filler metals for brazing, unlike other welding operations, generally have significantly different compositions than the metals to be joined.

Because of diffusion between the filler metal and the base metal, mechanical and metallurgical properties of joints can change in subsequent processing or during the service life of brazed components. For example, when titanium is brazed with pure tin filler metal, it is possible for the tin to completely diffuse into the titanium base metal by subsequent aging or heat treatment. When that happens, the joint no longer exists.

## Fluxes

The use of a **flux** is essential in brazing in order to prevent oxidation and to remove oxide films from workpiece surfaces. Brazing fluxes are generally made of borax, boric acid, borates, fluorides, and chlorides. Wetting agents may also be added to improve both the wetting characteristics of the molten filler metal and capillary action.

Surfaces to be brazed must be clean and free from rust, oil, and other contaminants. Clean surfaces are essential to obtain the proper wetting and spreading characteristics of the molten filler metal in the joint, and maximum bond strength. Sand blasting may also be used to improve surface finish of faying surfaces. Because they are corrosive, fluxes should be removed after brazing (usually by washing with hot water).

## Brazing methods

The heating methods used also identify the various brazing processes as described below.

**Torch brazing.**  The heat source in *torch brazing* (TB) is oxyfuel gas with a carburizing flame. Brazing is performed by first heating the joint with the torch, then depositing the brazing

(a)

Filler-metal wire

(b)

Filler metal

Filler

**FIGURE 12.48**
An example of furnace brazing: (a) before and (b) after. Note that the filler metal is a shaped wire.

rod or wire in the joint. Suitable part thicknesses are usually in the range of 0.25–6 mm (0.01–0.25 in.). More than one torch may be used in this process. Although it can be automated as a production process, torch brazing is difficult to control and requires skilled labor. This process can also be used for repair work.

**Furnace brazing.**   As the name suggests, *furnace brazing* (FB) is carried out in a furnace. The parts are precleaned and preloaded with brazing metal in appropriate configurations before being placed in the furnace (Fig. 12.48). Furnaces may be batch type for complex shapes or continuous type for high production runs, especially for small parts with simple joint designs. Vacuum furnaces or neutral atmospheres are used for metals that react with the environment. This process is similar to using shielding gas in welding operations. Skilled labor is not required, and complex shapes can be brazed since the whole assembly is heated uniformly in the furnace.

**Induction brazing.**   The source of heat in *induction brazing* (IB) is induction heating by high-frequency ac current. Parts are preloaded with filler metal and are placed near the induction coils for rapid heating. Unless a protective atmosphere is utilized, fluxes are generally used. Part thicknesses are usually less than 3 mm (0.125 in.). Induction brazing is particularly suitable for brazing parts continuously.

**Resistance brazing.**   In *resistance brazing* (RB), the source of heat is through electrical resistance of the components to be brazed. Electrodes are utilized for this purpose, as in resistance welding. Parts are either preloaded with filler metal, or it is supplied externally during brazing. Parts that are commonly brazed by this process have thicknesses of 0.1–12 mm (0.004–0.5 in.). As in induction brazing, the process is rapid, heating zones can be confined to very small areas, and the process can be automated to produce uniform quality.

**Dip brazing.**   *Dip brazing* (DB) is carried out by dipping the assemblies to be brazed into either a molten filler-metal bath or a molten salt bath (at a temperature just above the melting point of the filler metal), which serves as the heat source. All workpiece surfaces are thus coated with the filler metal. Consequently, dip brazing in metal baths is used only for small parts, such as sheet, wire, and fittings, usually of less than 5 mm (0.2 in.) in thickness or diameter. Molten salt baths, which also act as fluxes, are used for complex assemblies of various thicknesses. Depending on the size of the parts and the bath, as many as 1000 joints can be made at one time by dip brazing.

**Infrared brazing.**   The heat source in *infrared brazing* (IRB) is a high-intensity quartz lamp. This process is particularly suitable for brazing very thin components, usually less than

**FIGURE 12.49**
Joint designs commonly used in brazing operations.

1 mm (0.04 in.) thick, including honeycomb structures. The radiant energy is focused on the joint, and the process can be carried out in a vacuum.

**Diffusion brazing.** *Diffusion brazing* (DFB) is carried out in a furnace where—with proper control of temperature and time—the filler metal diffuses into the faying surfaces of the components to be joined. The brazing time required may range from 30 minutes to 24 hours. Diffusion brazing is used for strong lap or butt joints and for difficult joining operations. Because the rate of diffusion at the interface does not depend on the thickness of the components, part thicknesses may range from foil to as much as 50 mm (2 in.).

**Braze welding.** The joint in *braze welding* is prepared as in fusion welding. Using an oxyacetylene torch with an oxidizing flame, filler metal is deposited at the joint (see Fig. 12.47b), rather than by capillary action as in brazing. Thus considerably more filler metal is used, compared to brazing. However, temperatures in braze welding are generally lower than in fusion welding, and part distortion is minimal. The use of a flux is essential in this process. The principal use of braze welding is to maintain and repair parts, such as ferrous castings and steel components.

Examples of typical brazed joints are shown in Fig. 12.49. In general, dissimilar metals can be assembled with good joint strength, including carbide drill bits (*masonry drill*) or carbide inserts on steel shanks (Fig 8.40d). The shear strength of brazed joints can reach 800 MPa (120 ksi) using brazing alloys containing silver (*silver solder*). Intricate, light-weight shapes can be joined rapidly and with little distortion. Brazing can be automated and used for mass production.

As in all joining processes, joint design is important in brazing. Some design guidelines are given in Fig. 12.50. Note that strong joints require greater contact area for brazing than for welding. The typical joint clearance in brazing ranges from 0.025 mm to 0.2 mm (0.001 in. to 0.008 in.). The clearances must fit within a very small tolerance range because larger clearances reduce the strength of the brazed joint. A variety of special fixtures may be used during brazing to hold the parts together, some with provision for thermal expansion and contraction.

| Good | Poor | Comments |
|------|------|----------|
| | | Too little joint area in shear |
| | | Improved design when fatigue loading is a factor to be considered |
| | | Insufficient bonding |

**FIGURE 12.50** Examples of good and poor design for brazing.

## 12.12.2 Soldering

In **soldering,** the filler metal, called *solder,* melts below 450°C (840°F). As in brazing, the solder fills the joint by capillary action between closely fitting or closely placed components. Heat sources for soldering are usually soldering irons, torches, or ovens. Soldering with copper–gold and tin–lead alloys was first practiced as long ago as 4000–3000 B.C.

There are several soldering methods, which are similar to brazing methods:

- Torch soldering (TS)
- Furnace soldering (FS)
- Iron soldering (INS), using a soldering iron
- Induction soldering (IS)
- Resistance soldering (RS)
- Dip soldering (DS)
- Infrared soldering (IRS)
- Wave soldering (WS), used for automated soldering of printed circuit boards
- Ultrasonic soldering, in which a transducer subjects the molten solder to ultrasonic cavitation, which removes the oxide films from the surfaces to be joined. The need for a flux is thus eliminated.

Solders are usually tin–lead alloys in various proportions. For better joint strength and special applications, other solder compositions that can be used are tin–zinc, lead–silver, cadmium–silver, and zinc–aluminum alloys (Table 12.5). Because of the toxicity of lead and its adverse effects on the environment, *lead-free solders* are being developed and are now in wider use. Typical compositions are 96.5% Sn–3.5% Ag and 42% Sn–58% Bi.

**TABLE 12.5**

**TYPES OF SOLDERS AND THEIR APPLICATIONS**

| | |
|---|---|
| Tin–lead | General purpose |
| Tin–zinc | Aluminum |
| Lead–silver | Strength at higher than room temperature |
| Cadmium–silver | Strength at high temperatures |
| Zinc–aluminum | Aluminum; corrosion resistance |
| Tin–silver | Electronics |
| Tin–bismuth | Electronics |

In soldering, fluxes are used as in welding and brazing and for the same purposes. Fluxes are generally of two types:

1.  Inorganic acids or salts, such as zinc ammonium chloride solutions, which clean the surface rapidly. After soldering, the flux residues should be removed by washing thoroughly with water to avoid corrosion.

2.  Noncorrosive resin-based fluxes, used in electrical applications.

Soldering is used extensively in the electronics industry. Unlike brazing, soldering temperatures are relatively low. Thus a soldered joint has very limited use at elevated temperatures. Moreover, because solders do not generally have much strength, they are not used for load-bearing structural members. Because of the small faying surfaces, butt joints are rarely made with solders. In other situations, joint strength is improved by mechanical interlocking of the joint.

Copper and precious metals, such as silver and gold, are easy to solder. Aluminum and stainless steels are difficult to solder because of their strong, thin oxide film (see Section 4.2). However, these and other metals can be soldered using special fluxes that modify surfaces.

Soldering can be used to join various metals and thicknesses. Although manual operations require skill and are time consuming, soldering speeds can be high with automated equipment. Design guidelines for soldering are similar to those for brazing. Some frequently used joint designs are shown in Fig. 12.51. Note again the importance of large contact surfaces to develop sufficient joint strength in soldered products.

(a) Flanged T  (b) Flush lap  (c) Flanged corner

(d) Line contact  (e) Flat lock seam  (f) Flanged bottom

**FIGURE 12.51** Joint designs commonly used for soldering.

# 12.13   Adhesive Bonding

Numerous components and products can be joined and assembled using an *adhesive,* rather than by any of the joining methods described thus far. **Adhesive bonding** has been a common method of joining and assembly for applications such as labeling, packaging, bookbinding, home furnishings, and footware. Plywood, developed in 1905, is a typical example of adhesive bonding of several layers of wood with glue. Adhesive bonding has been gaining increased acceptance in manufacturing ever since its first use on a large scale in assembling load-bearing components in aircraft during World War II (1939–1945).

## 12.13.1   Adhesives

Many types of adhesives are available—and continue to be developed—which provide adequate joint strength, including fatigue strength. The three basic types of adhesives are:

1. *Natural adhesives,* such as starch, dextrin (a gummy substance obtained from starch), soya flour, and animal products.
2. *Inorganic adhesives,* such as sodium silicate and magnesium oxychloride.
3. *Synthetic organic adhesives,* which may be thermoplastics (used for nonstructural and some structural bonding) or thermosetting polymers (used primarily for structural bonding).

Because of their strength, synthetic organic adhesives are the most important in manufacturing processes, particularly for load-bearing applications. They are classified as:

1. *Chemically reactive;* such as polyurethanes, silicones, epoxies, cyanoacrylates, modified acrylics, phenolics, polyimides, and anaerobics.
2. *Pressure sensitive;* examples are natural rubber, styrene-butadiene rubber, butyl rubber, nitrile rubber, and polyacrylates.
3. *Hot melt;* which are thermoplastics such as ethylene-vinyl acetate copolymers, polyolefins, polyamides, polyester, and thermoplastic elastomers.
4. *Reactive hot melt;* which have a thermoset portion based on the urethane chemistry, and with improved properties.
5. *Evaporative* or *diffusion;* including vinyls, acrylics, phenolics, polyurethanes, synthetic rubbers, and natural rubbers.
6. *Film* and *tape;* such as nylon-epoxies, elastomer-epoxies, nitrile-phenolics, vinyl-phenolics, and polyimides.
7. *Delayed tack;* examples are styrene-butadiene copolymers, polyvinyl acetates, polystyrenes, and polyamides.
8. *Electrically* and *thermally conductive;* including epoxies, polyurethanes, silicones, and polyimides. Electrical conductivity is obtained by addition of fillers such as silver (used most commonly), copper, aluminum, and gold. Fillers that improve the electrical conductivity of adhesives generally also improve their thermal conductivity.

Adhesive systems may also be classified based on their specific chemistries:

1. *Epoxy-based systems.* They have high strength and high-temperature properties, to as high as 200°C (400°F), with typical applications such as automotive brake linings and bonding agents for sand molds for casting.

2. *Acrylics.* Suitable for applications with substrates that are not clean.

3. *Anaerobic systems.* Curing of these adhesives is by oxygen deprivation and the bond is usually hard and brittle. Curing times can be reduced by external heat or ultraviolet (UV) radiation.

4. *Cyanoacrylate.* The bond lines are thin and the bond sets within 5 s to 40 s.

5. *Urethanes.* They have high toughness and flexibility at room temperature, and are widely used as sealants.

6. *Silicones.* Highly resistant to moisture and solvents, they have high impact and peel strength; however, curing times are typically in the range of 1–5 days.

Many of these adhesives can be combined to optimize their properties. Examples are epoxy-silicon, nitrile-phenolic, and epoxy-phenolic.

The least expensive of adhesives are epoxies and phenolics, followed by polyurethanes, acrylics, silicones, and cyanoacrylates. Adhesives for high-temperature applications in a range up to about 260°C (500°F), such as polyimides and polybenzimidazoles, are generally the most expensive.

Adhesives are available in various forms, such as liquids, pastes, solutions, emulsions, powder, tape, and film. When applied, adhesives generally are about 0.1 mm (0.004 in.) thick. Depending on the particular application, an adhesive must have one or more of the following properties (Table 12.6):

- Strength (shear and peel)
- Toughness

**TABLE 12.6**

**TYPICAL PROPERTIES AND CHARACTERISTICS OF CHEMICALLY REACTIVE STRUCTURAL ADHESIVES**

| | EPOXY | POLYURETHANE | MODIFIED ACRYLIC | CYANOACRYLATE | ANAEROBIC |
|---|---|---|---|---|---|
| Impact resistance | Poor | Excellent | Good | Poor | Fair |
| Tension–shear strength, MPa ($10^3$ psi) | 15.4 (2.2) | 15.4 (2.2) | 25.9 (3.7) | 18.9 (2.7) | 17.5 (2.5) |
| Peel strength, N/m (lb/in.) | < 525 (3) | 14,000 (80) | 5250 (30) | < 525 (3) | 1750 (10) |
| Substrates bonded | Most | Most smooth, nonporous | Most smooth, nonporous | Most nonporous metals or plastics | Metals, glass, thermosets |
| Service temperature range, °C (°F) | −55 to 120 (−70 to 250) | −160 to 80 (−250 to 175) | −70 to 120 (−100 to 250) | −55 to 80 (−70 to 175) | −55 to 150 (−70 to 300) |
| Heat cure or mixing required | Yes | Yes | No | No | No |
| Solvent resistance | Excellent | Good | Good | Good | Excellent |
| Moisture resistance | Excellent | Fair | Good | Poor | Good |
| Gap limitation, mm (in.) | None | None | 0.75 (0.03) | 0.25 (0.01) | 0.60 (0.025) |
| Odor | Mild | Mild | Strong | Moderate | Mild |
| Toxicity | Moderate | Moderate | Moderate | Low | Low |
| Flammability | Low | Low | High | Low | Low |

Source: *Advanced Materials & Processes,* July 1990, ASM International.

**FIGURE 12.52**  Characteristic behavior of (a) brittle and (b) tough adhesives in a peeling test. This test is similar to peeling adhesive tape from a solid surface. Adhesive joints should not be subjected to this type of loading in service.

- Resistance to various fluids and chemicals
- Resistance to environmental degradation, including heat and moisture
- Capability to wet the surfaces to be bonded

Adhesive joints are designed to withstand shear, compressive, and tensile forces, but they should not be subjected to peeling forces (Fig. 12.52). Note, for example, how easily you can peel adhesive tape from a surface. During peeling the behavior of an adhesive may be brittle or it may be ductile and tough, requiring large forces to peel it.

### 12.13.2   Surface preparation and application

Surface preparation is very important in adhesive bonding. Joint strength depends greatly on the absence of dirt, dust, oil, and various other contaminants. It is virtually impossible to apply an adhesive tape over a dusty or oily surface. Contaminants also affect the wetting ability of the adhesive, and they can prevent spreading of the adhesive evenly over the interface. Thick, weak, or loose oxide films on workpieces are detrimental to adhesive bonding. On the other hand, a porous or thin, strong oxide film may be desirable, particularly one with some surface roughness to improve adhesion. Various compounds and primers are available to modify surfaces to increase adhesive bond strength. Liquid adhesives may be applied by brushes, sprayers, and rollers.

### 12.13.3   Process capabilities

A wide variety of similar and dissimilar metallic and nonmetallic materials and components with different shapes, sizes, and thicknesses can be bonded to each other by adhesives. Adhesive bonding can be combined with mechanical fastening methods (see Section 12.14) to further improve the strength of the bond.

An important consideration in the use of adhesives in production is curing time, which can range from a few seconds at high temperatures to many hours at room temperature, particularly for thermosetting adhesives. Thus production rates may be low. Furthermore, adhesive bonds for structural applications are rarely suitable for service above 250°C (500°F). Joint design and bonding methods require care and skill. Special equipment, such as fixtures, presses, tooling, and autoclaves and ovens for curing, is usually needed.

Nondestructive inspection of the quality and strength of adhesively bonded components can be difficult. Some of the techniques described in Section 4.8.1, such as acoustic impact (tapping), holography, infrared detection, and ultrasonic testing, are effective nondestructive testing methods.

Major industries that use adhesive bonding extensively are aerospace, automotive, appliances, and building products. Applications include attaching rear-view mirrors to windshields, automotive brake-lining assemblies, laminated windshield glass, appliances, helicopter blades, honeycomb structures, and aircraft bodies and control surfaces.

The major advantages of adhesive bonding are:

1. It provides a bond at the interface either for structural strength or for nonstructural applications such as sealing, insulating, preventing electrochemical corrosion between dissimilar metals, and reducing vibration and noise through internal damping at the joints.

2. It distributes the load at an interface, thus eliminating localized stresses that result from joining the components with welds or mechanical fasteners, such as bolts and screws. Moreover, structural integrity of the sections is maintained, since no holes are required, and the appearance of the components is generally improved.

3. Very thin and fragile components can be bonded with adhesives without contributing significantly to weight.

4. Porous materials and materials of very different properties and sizes can be joined.

5. Because it is usually carried out at a temperature between room temperature and about 200°C (400°F), there is no significant distortion of the components or change in their original properties. This is particularly important for materials that are heat sensitive.

The major limitations of adhesive bonding are:

1. Limited service temperatures.
2. Possibly long bonding time.
3. Need for great care in surface preparation.
4. Difficulty in testing bonded joints nondestructively, particularly for large structures.
5. Limited reliability of adhesively bonded structures during their service life.

The cost of adhesive bonding depends on the particular operation. However, in many applications the overall economics of the process makes adhesive bonding an attractive joining process, sometimes the only one that is feasible or practical. The cost of equipment varies greatly, depending on the size and type of operation.

### 12.13.4 Design for adhesive bonding

Designs for adhesive bonding should ensure that joints are subjected to compressive, tensile, and shear forces, but not peeling or cleavage (see Fig. 12.52). Several joint designs for adhesive bonding are shown in Figs. 12.53 and 12.54. They vary considerably in strength; hence selection of the appropriate design is important and should include considerations such as type of loading and the environment.

Butt joints require large bonding surfaces. Simple lap joints tend to distort under tension because of the force couple at the joint (see Fig. 12.28a). The coefficients of expansion of the components to be bonded should preferably be close in order to avoid internal stress during adhesive bonding. Also, situations where thermal cycling can cause differential movement across the joint should be avoided.

(a) Poor

Adhesive

(b) Good

(c) Very good

(d) Combination joints

Adhesive

Rivet

Adhesive

Spot weld

**FIGURE 12.53**
Various joint designs in adhesive bonding. Note that good design requires large contact areas between the members to be joined.

(a)

Simple

Beveled

Radiused

(b)

Simple

Beveled

Radiused

(c)

Single taper

Double taper

Increased thickness

(d)

Single

Double

Beveled

**FIGURE 12.54** Various configurations for adhesively bonded joints: (a) single lap, (b) double lap, (c) scarf, and (d) strap.

### 12.13.5   Electrically conducting adhesives

Although the majority of adhesive bonding use is for mechanical strength, a relatively recent advance is the development and application of electrically conducting adhesives to replace lead-base solder alloys, particularly in the electronics industry. They also require curing or setting temperatures that are lower than that for soldering. In these adhesives the polymer is the matrix, containing conducting metals (fillers) in forms such as flakes and particles (see also Section 10.7.2 on conducting polymers). There is a minimum volume concentration of fillers in order to make the adhesive electrically conducting, typically in the range of 40–70 percent. They are available as film or as paste.

The size, shape, and distribution of the metallic particles, as well as how heat and pressure are applied and how the individual conducting particles contact each other, can be controlled so as to impart isotropic or anisotropic electrical conductivity to the adhesive. Typical fillers are silver, carbon, nickel, copper, and gold. New developments in fillers include the use of polymeric particles, such as polystyrene, coated with thin films of silver or gold. Matrix materials are generally epoxies, although thermoplastics are also used. Applications of electrically conducting adhesives include their use in calculators, remote controls, and control panels, as well as high-density uses in electronic assemblies, liquid-crystal displays, electronic games, and pocket TVs.

## 12.14   Mechanical Fastening

Two or more components may have to be joined or fastened in such a way that they can be taken apart sometime during the product's service life. Numerous objects, including mechanical pencils, caps and lids on containers, mechanical watches, engines, and bicycles, have components that are fastened mechanically. **Mechanical fastening** may be preferred for the following reasons:

- Ease of manufacturing
- Ease of assembly and transportation
- Ease of parts replacement, maintenance, or repair
- Ease in creating designs that require movable joints, such as hinges, sliding mechanisms for drawers and doors, and adjustable components and fixtures
- Lower overall cost of manufacturing the product

The most common method of mechanical fastening is by using bolts, nuts, screws, pins, and a variety of other **fasteners.** These processes are also known as mechanical assembly. Mechanical fastening generally requires that the components have *holes* through which fasteners are inserted. These joints may be subjected to both shear and tensile stresses and should be designed to resist these forces.

### 12.14.1   Hole preparation

**Hole preparation** is an important aspect of mechanical fastening. A hole in a solid body can be produced by punching, drilling, chemical and electrical means, and high-energy beams, depending on the type of material, its properties, and thickness. Recall from Chapters 5, 6, and 11 that holes may also be produced as an integral part of the product during casting, forging, extrusion, and powder metallurgy. For improved accuracy and surface finish, many of these holemaking operations may be followed by finishing operations, such as shaving, deburring, reaming, and honing.

Because of their fundamental differences, each holemaking operation produces a hole with different surface finish and properties and dimensional characteristics. The most significant influence of a hole in a solid body is its tendency to reduce the component's fatigue life (stress concentration). Fatigue life can best be improved by inducing compressive residual stresses on the cylindrical surface of the hole. These stresses are usually developed by pushing a round rod (*drift pin*) through the hole and expanding it by a very small amount. This process plastically deforms the surface layers of the hole, in a manner similar to shot peening or roller burnishing (Section 4.5.1).

### 12.14.2 Threaded fasteners

Bolts, screws, and nuts are among the most commonly used *threaded fasteners*. Texts on machine design describe numerous standards and specifications, including thread dimensions, tolerances, pitch, strength, and the quality of materials used to make these fasteners.

Bolts and screws may be secured with nuts (machine screws), or they may be self-tapping, whereby the screw either cuts or forms the thread into the part to be fastened. This method is particularly effective and economical in plastic products. In this way, fastening does not require a tapped hole and the need for—and cost of—a nut.

If the joint is to be subjected to vibration, such as in aircraft and various types of engines and high-speed machinery, several specially designed nuts and lock washers are available. They increase the frictional resistance in the torsional direction, thus preventing vibrational loosening of the fasteners.

### 12.14.3 Rivets

The most common method of permanent or semipermanent mechanical joining is by *riveting* (Fig. 12.55). Hundreds of thousands of rivets may be used in the construction and assembly of large commercial aircraft. Holes for the rivets are required in the components to be joined. Installing a rivet consists of placing the rivet in the hole and deforming the end of its shank by upsetting (heading). Riveting may be done either at room temperature or hot, using special tools or explosives in the rivet cavity. Some design guidelines for riveting are illustrated in Fig. 12.56.

(a)            (b)            (c)            (d)

**FIGURE 12.55**  Examples of rivets: (a) solid, (b) tubular, (c) split (or bifurcated), and (d) compression.

**FIGURE 12.56** Design guidelines for riveting. *Source:* J. G. Bralla.

(a)

(b)

(c)

(d)

Poor          Good

### 12.14.4 Design for mechanical fastening

The design of mechanical joints requires consideration of the type of loading, such as shear and tension, to which the structure will be subjected and the size and spacing of holes.

Compatibility of the fastener material with that of the components to be joined is important. Incompatibility may lead to galvanic corrosion, also known as *crevice corrosion*. In a system where, for example, a steel bolt or rivet is used to fasten copper sheets, the bolt is anodic and the copper plate cathodic, thus leading to rapid corrosion and loss of joint strength. Aluminum or zinc fasteners on copper products react similarly.

Other general design guidelines for mechanical joining include the following (see also Section 13.10):

1. Using fewer but larger fasteners is generally less costly than using a large number of smaller ones.
2. Part assembly should be accomplished with a minimum number of fasteners.
3. The fit between parts to be joined should be as loose as possible to reduce costs and facilitate the assembly process.
4. Fasteners of standard size should be used whenever possible.
5. Holes should not be too close to edges or corners to avoid tearing the material when it is subjected to external forces.

(a) Standard loop

(b) Flat clinch

(c) Nonmetal / Metal channel

(d)

**FIGURE 12.57** Examples of metal stitching.

(a)  (b)  (c)  (d)

**FIGURE 12.58** Stages in forming a double-lock seam.

### 12.14.5  Other methods of fastening

Many types of fasteners are used in numerous joining and assembly applications. The most common types are described below.

**Metal stitching or stapling.**  The process of *metal stitching* or *stapling* (Fig. 12.57) is much like that of ordinary stapling of papers. This operation is fast and particularly suitable for joining thin metallic and nonmetallic materials, and it does not require holes in the components. A common example is the stapling of cardboard containers for appliances and other consumer products.

**Seaming.**  Seaming is based on the simple principle of folding two thing pieces of material together. Seaming is much like joining two pieces of paper, in the absence of a paper clip, by folding them at the corner. The most common examples of seaming are the tops of beverage cans and containers for food and household products (Fig. 12.58).

In seaming, the materials should be capable of undergoing bending and folding at very small radii (see Section 7.4.1). Otherwise, they will crack and the seams will not be airtight or watertight. The performance of seams may be improved with adhesives, coatings, seals, or by soldering.

**Crimping.**  The **crimping** process is a method of joining without using fasteners. It can be done with beads or dimples (Fig. 12.59), which can be produced by shrinking or swaging operations. Crimping can be used on both tubular and flat parts, provided that the materials are thin and ductile enough to withstand the large localized deformations. Caps on glass bottles are attached by crimping, as are connectors for electrical wiring.

**Snap-in fasteners.**  Various spring and snap-in fasteners are shown in Fig. 12.60. Such fasteners are widely used in automotive bodies and household appliances. They are economical and permit easy and rapid component assembly.

**FIGURE 12.59** Two examples of mechanical joining by crimping.

**FIGURE 12.60** Examples of spring and snap-in fasteners to facilitate assembly.

**Shrink** and **press fits.** Components may also be assembled by shrink fitting and press fitting. **Shrink fitting** is based on the principle of differential thermal expansion and contraction of two components. Typical applications are the assembly of die components and mounting gears and cams on shafts. In **press fitting,** one component is forced over another, resulting in high joint strength.

## 12.15 Joining Plastics

Thermoplastics and thermosets can be joined by many of the methods already described for joining metals and nonmetallic materials, including:

- Fusion welding with heat from external sources
- Ultrasonic welding
- Friction welding
- Adhesive bonding
- Bonding with solvents
- Mechanical fastening, particularly with self-tapping screws. Mechanical fastening is particularly effective with most plastics because of their inherent toughness and resilience.

### 12.15.1   Joining thermoplastics

As we described in Chapter 10, thermoplastics soften and melt as temperature increases; consequently, they can be joined by means whereby heat is generated at the interface, either through external means or internally. The heat softens the thermoplastic at the interface to a viscous or molten state and, with the application of pressure to ensure a good bond, allows *fusion* to take place. Filler materials of the same type of polymer may also be used.

*External* heat sources may consist of the following, depending on the compatibility of the polymers to be joined:

1.  Hot air or gases, or infrared radiation using high-intensity quartz heat lamps.

2.  Heated tools and dies, known as *hot-plate welding,* that contact and heat the surfaces to be joined by interdiffusion of molecular chains. This process is commonly used to butt-weld pipe and tubing, i.e., end to end.

3.  Radio-frequency or dielectric heating, particularly useful for thin films.

4.  Electrical-resistance wire or braids, or carbon-based tapes, sheet, and ropes, which are placed at the interface and pass electrical current, known as *resistive-implant welding.* The elements at the interface may be subjected to a radio-frequency field (*induction welding*). The elements at the interface must be compatible with the use of the joined product since they are left in the weld zone.

5.  Lasers using defocused beams at low power to prevent degradation of the polymer.

*Internal* heat sources are developed through the following means:

1.  *Ultrasonic welding* (see Section 12.7) which is the most commonly used process for thermoplastics, particularly amorphous polymers such as ABS and high-impact polystyrene.

2.  *Friction welding* (see Section 12.8), also called *spin welding* for polymers; it includes linear friction welding, also called *vibration welding,* which is particularly useful for joining polymers with a high degree of crystallinity (see Section 10.2.2) such as acetal, polyethylene, nylons, and polypropylene.

3.  *Orbital welding,* which is similar to friction welding except that the rotary motion of one part is in an orbital path (see also Fig. 6.16).

Coextruded multiple food wrappings consist of different types of films, which are bonded by heat during *extrusion.* Each film has a different function, such as to keep out moisture, keep out oxygen, and facilitate heat sealing during packaging. Some wrappings have as many as seven layers, all bonded together during production of the film.

Oxidation can be a problem in joining some polymers, such as polyethylene, causing degradation. In these cases an inert shielding gas such as nitrogen is used to prevent oxidation. Because of low thermal conductivity of thermoplastics, the heat source may burn or char the surfaces of the components if applied at a high rate, possibly causing difficulties in obtaining sufficiently deep fusion.

*Adhesive bonding* of plastics is best illustrated in joining sections of polyvinyl chloride (PVC) pipe, used extensively in home plumbing systems, and ABS pipe, used in drain, waste, and vent systems. The adhesive is applied to the connecting sleeve and pipe surfaces, using a primer to improve adhesion (much like using primers in painting), and the pieces are pushed together. Adhesive bonding of polyethylene, polypropylene, and polytetrafluoroethylene (Teflon) can be difficult because they have low surface energy and adhesives do not stick readily to their surfaces, which usually have to be treated chemically to improve bonding. The use of adhesive primers and double-sided adhesive tapes are also effective.

### 12.15.2  Joining thermosets

Because they do not soften or melt with temperature, thermosetting plastics, such as epoxy and phenolics, are usually joined by using (1) threaded or other molded-in inserts, (2) mechanical fasteners, and (3) **solvent bonding.**

The basic process of thermoset bonding with solvents involves:

- Roughening the surfaces with an abrasive.
- Wiping the surfaces with a solvent.
- Pressing the surfaces together and holding until sufficient bond strength is developed.

---

## CASE STUDY:  WELD CRACKING

### Description of the welded component

Figure 12.61 shows a die mold subassembly, which consists of a base plate 305 mm × 152 mm × 25 mm (12 in. × 6 in. × 1 in.) welded along its full length to two vertical risers 305 mm × 152 mm × 38 mm (12 in. × 6 in. × 1.5 in.). Both the base plate and risers are made of low-carbon steel (ANSI-SAE 1025). The high deposition rates of the submerged arc welding process allowed each weld to be made in a single pass. The joint preparation consisted of a double-bevel groove along the sides of the base plate. Welding was accomplished in the flat position, and the assembly was not preheated prior to welding.

After welding, the slag was removed to reveal the weld. The weld surface was noted to be convex, and the operator reported seeing no cracks. The finish-machining operation consisted of removing the weld cap in order to conform to the thickness of the base plate. After the machining process, a crack along the entire weld length was evident.

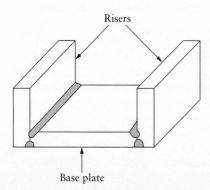

Risers

Base plate

**FIGURE 12.61**  Die mold subassembly. Cracks were noted in the submerged arc welds that connect the base plate to the risers.

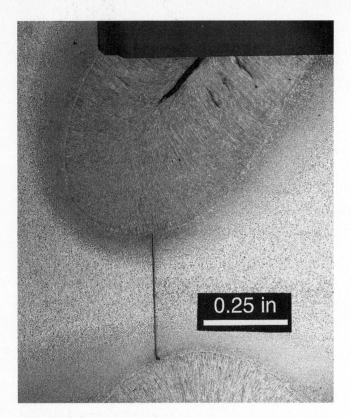

**FIGURE 12.62**  Cracking in the upper weld is noted to follow the solidification grain structure of the weld.

### Diagnosis of the cracking problem

The cracks in the die mold subassembly could have formed either during the solidification of the weld or as a result of the machining process. Weld metal contracts as it solidifies; the still-molten interfaces between dendrites pull apart and solidify smoothly along the contours of the dendrites. This phenomenon is referred to as *solidification cracking.*

Alternately, stress from the machining process, combined with the residual tensile stresses present in the weldment after cooling, could produce cracking. This type of cracking generally results in a jagged or dimpled fracture surface. To determine which explanation is correct, a cross-section of the weld was removed from the subassembly, ground flat, and then polished. The weld structure was revealed by etching with a solution of 2 percent nitric acid in ethanol, common for use with low-alloy steels.

Figure 12.62 shows a major crack down the centerline of the weld; smaller cracks follow the solidification grain structure out from the weld center. This type of crack morphology suggests solidification cracking, which can be verified by breaking open the crack and examining its surface under a scanning electron microscope. The crack surface (Fig. 12.63) reveals smooth dendritic contours and confirms that the cracks occurred during solidification.

### Methods to eliminate solidification cracking

Weld-metal composition and shrinkage can affect the susceptibility to solidification cracking. The weld-metal composition can be altered by varying the electrode and flux combinations used. Shrinkage of the weld metal can be affected by the joint design and welding parameters.

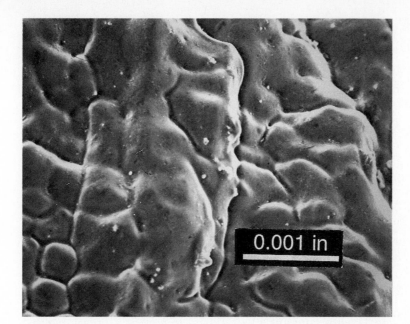

**FIGURE 12.63** A scanning electron microscope image reveals a smooth fracture surface, which outlines the dendritic solidification structure. This feature is indicative of solidification cracking.

As a first attempt to solve the cracking problem, the die-mold maker consulted its welding consumable supplier for advice. An electrode-flux combination was selected that would increase the manganese and silicon content of the weld deposit. These elements are known to decrease the susceptibility to solidification cracking. This new electrode-flux combination, however, failed to solve the cracking problem.

Next, the die-mold producer checked several welding handbooks for further information. A significant amount of research into weld cracking, especially as related to the welding process and joint geometry, has been documented. Welds produced by the submerged arc process may crack internally if they are deeper than they are wide. A width-to-depth ratio of 1.25 is noted to minimize internal shrinkage stresses. Further, a flat to slightly convex bead surface is preferred.

The weld surface was noted to be convex by the operator prior to machining. Measurement of the weld width and depth, however, revealed a weld bead that was 11 mm wide and 12 mm deep. For a depth of 12 mm, a weld-bead width of 15 mm is required to meet the optimum width-to-depth ratio of 1.25. This observation prompted a review of the weld joint design.

### Review of the joint design

*AWS D1. 1: Structural Welding Code—Steel* provides a range of prequalified joint designs for use with the arc-welding processes. The term prequalified implies that satisfactory service performance of a particular joint design has been documented. The code provides a further description of welding procedure specifications and procedure qualification records, both of which are integral parts of a quality control system for welded assembly.

Figure 12.64 shows a prequalified geometry for the corner joint on the mold subassembly. This is a partial-penetration weld; as revealed in Fig. 12.63, the top and bottom welds do not meet. The code details minimum weld size, weld position, root opening and root face dimensions, and groove angles for a given plate thickness.

**FIGURE 12.64** Partial penetration joint design for use with submerged arc welding on plate thicknesses greater than 3/4 in. The new joint design is consistent with *ANSI/AWS D1. 1: Structural Welding Code—Steel.*

Current joint design parameters were checked against the geometry provided in the code. Weld size, weld position, root opening and root face dimensions were consistent with the code. The code required a groove angle of 60°; the current design, however, called for only 30°. The change from a 30°- to 60°-degree groove angle would provide for a greater weld width. This increased width would in turn provide a greater width-to-depth ratio of the weld bead, easily approaching the optimal ratio of 1.25. Steels with a thickness of 25–50 mm (1–2 in.) require a minimum preheat and interpass temperature of 40°C (100°F) to prevent hydrogen-induced cracking, a phenomenon peculiar to steels. If low hydrogen practice is employed, however, welding can occur at ambient temperatures above −12°C (10°F). Low hydrogen practice involves the elimination of all sources of hydrogen, including moisture, oil, machining and cleaning fluids. For submerged arc welding, low hydrogen practice includes storing the flux in a heated hopper to drive off moisture, as well as degreasing and drying the plates prior to welding.

The die-mold producer had consistently employed low hydrogen practice during production. Therefore, hydrogen was not initially considered a potential source of the cracking. Avoiding the time-consuming preheating process increases productivity, saves energy, and is well worth the trade-off in terms of stringent safeguards against hydrogen.

## General approach to welding problems

A complete description of the welded assembly, including both the processing steps and information about the type of material being fabricated, is vital in order to diagnose and correct a welding problem efficiently. Small details, such as the shape of the weld or whether preheating was employed, can offer clues that aid in determining a solution. The first step in this case study was to identify the type and cause of cracking. In turn, that information shortened the list of possible corrective measures.

Small manufacturers generally do not employ welding engineers and will rely heavily on their welding staff and welding consumable suppliers for advice concerning such topics as electrode selection and joint design. General welding handbooks are a prime, but often untapped, source of information about welding procedures and common problems or defects. The American Welding Society publishes a wide range of codes, procedures, and recommended practices that cover both arc- and non-arc-welding processes.

*Source*: R. J. Bowers, Edison Welding Institute, Columbus, Ohio.

# SUMMARY

- Almost all products are an assemblage of many components; thus, joining and fastening processes are an important and often necessary aspect of manufacturing, as well as servicing or transportation. (Section 12.1)

- One major category of joining processes is fusion welding, in which the two pieces to be joined melt together and coalesce by means of heat, usually supplied by chemical or electrical means. Filler metals may or may not be used. Two modern methods are arc welding with consumable or nonconsumable electrodes. These processes are, in turn, divided into various categories. Their selection depends on factors such as workpiece material, shape complexity, size, and thickness, and type of joint. (Sections 12.2, 12.3)

- Laser-beam and electron-beam welding are major categories of joining using high-energy beams. They produce small and high-quality weld zones and thus have important and unique applications in manufacturing. (Section 12.4)

- Because the joint undergoes important metallurgical and physical changes, the nature, properties, and quality of the weld joint are important factors to consider. Weldability of metals, weld design, and process selection are other factors that should be considered. (Section 12.5)

- In solid-state welding, joining takes place without fusion and pressure is applied either mechanically, using tools, electrodes, or rollers, or by explosives. There is no liquid (molten) phase in the joint. Surface preparation and cleanliness can be important. The less common examples are cold welding, friction welding (in which heat is generated internally), and explosion welding. (Sections 12.6, 12.8, 12.10)

- Ultrasonic welding and resistance welding are two major examples of solid-state welding with a wide variety of applications, particularly for foil and sheet metals. Although there is no significant heat generated in ultrasonic welding, in resistance welding heat is generated by electrical-resistance heating between electrodes or rollers. (Sections 12.7, 12.9)

- Diffusion bonding is another example of solid-state joining. Combined with superplastic forming, various complex sheet-metal structures with high strength-to-weight and stiffness-to-weight ratios can be fabricated. (Section 12.11)

- Brazing and soldering utilize filler metal at the interfaces to be joined. These processes require lower temperatures than those in welding and are capable of joining dissimilar metals of intricate shapes and thicknesses. (Section 12.12)

- Adhesive bonding has gained increased acceptance in major industries, such as aerospace, automotive, and building products. Adhesive-bonded joints have favorable characteristics such as strength, sealing, insulating, vibration damping, and resistance to corrosion between dissimilar metals. An important development is electrically conducting adhesives for surface-mount technologies. (Section 12.13)

- Mechanical fastening is among the oldest and most common techniques. A wide variety of shapes and sizes of fasteners and fastening techniques have been developed for numerous permanent or semipermanent applications. (Section 12.14)

- With increasing use of polymers in a wide range of consumer and industrial applications, joining of thermoplastics and thermosetting resins has undergone many advances. Various techniques can be used, including fusion welding, mechanical fastening, and adhesive bonding. (Section 12.15)

# BIBLIOGRAPHY

## General

Arata, Y. (ed.), *Plasma, Electron, and Laser Beam Technology*. Metals Park, Ohio: American Society for Metals, 1986.

*ASM Handbook,* Vol. 6: *Welding, Brazing and Soldering*. Materials Park, Ohio: ASM International, 1993.

Bowditch, W.A., and K.E. Bowditch, *Welding Technology Fundamentals*. South Holland, Ill.: Goodheart-Wilcox, 1991.

Cary, H.B., *Modern Welding Technology*, 3d ed. Englewood Cliffs, New Jersey: Prentice-Hall, 1994.

Davies, A.C., *The Science and Practice of Welding*, 10th ed., 2 vols. New York: Cambridge University Press, 1993.

Easterling, K.E., *Introduction to the Physical Metallurgy of Welding*. Stoneham, Mass.: Butterworth, 1983.

*Electronic Materials Handbook*, Vol. 1: *Packaging*. Materials Park, Ohio: ASM International, 1989.

Galyen, J., G. Sear., and C. Tuttle, *Welding: Fundamentals and Procedures*. New York: Wiley, 1985.

Granjon, H., *Fundamentals of Welding Metallurgy*. Cambridge, England: Woodhead Publishing, 1991.

Houldcroft, P.T., *Welding and Cutting: A Guide to Fusion Welding and Associated Cutting Processes*. New York: Industrial Press, 1988.

Koellhoffer, L., *Welding Processes and Practices*. Englewood Cliffs, New Jersey: Prentice-Hall, 1988.

Kou, S., *Welding Metallurgy*. New York: Wiley, 1987.

Lancaster, J.F., *Metallurgy of Welding*. London: Chapman and Hall, 1993.

Messler, R.W., Jr., *Joining of Advanced Materials*. Stoneham, Mass.: Butterworth-Heinemann, 1993.

Mohler, R., *Practical Welding Technology*. New York: Industrial Press, 1983.

North, T.H., *Advanced Joining Technologies*. London: Chapman and Hall, 1990.

Powell, J., $CO_2$ *Laser Cutting*. New York: Springer, 1992.

*Resistance Welding Manual*, 4th ed. Philadelphia: Resistance Welder Manufacturers' Association, 1989.

Steen, W.M., *Laser Material Processing*. New York: Springer, 1991.

Stout, R.D., *Weldability of Steels*. New York: Welding Research Council, 1987.

*Tool and Manufacturing Engineers Handbook*, Vol 4, *Quality Control and Assembly*. Dearborn, Mich.: Society of Manufacturing Engineers, 1986.

*Welding Handbook*. 7th ed., 5 vols. Miami: American Welding Society, 1976.

*Welding Handbook,* 8th ed., 3 vols. Miami: American Welding Society, 1987.

## Brazing and soldering

*ASM Handbook,* Vol 6: *Welding, Brazing and Soldering*. Materials Park, Ohio: ASM International, 1993.

*Brazing Handbook*. Miami: American Welding Society, 1991.

Frear, D.R., W.B. Jones, and K.R. Kinsman (eds.), *Solder Mechanics: A State of the Art Assessment*. Warrendale, Penn.: The Minerals, Metals & Materials Society, 1991.

Humpston, G., and D.M. Jacobson, *Principles of Soldering and Brazing*. Materials Park, Ohio: ASM International, 1993.

Lambert, L., *Soldering for Electronic Assemblies*. New York: Marcel Dekker, 1988.

Manko, H.H., *Solders and Soldering-Materials, Design, Production, and Analysis for Reliable Bonding*. New York: McGraw-Hill, 1992.

Pecht, M.G., *Soldering Processes and Equipment*. New York: Wiley, 1993.

Rahn, A., *The Basics of Soldering*. New York: Wiley Interscience, 1993.

Schwartz, M.M., *Brazing*. Metals Park, Ohio: ASM International, 1987.

## Adhesive bonding

*ASM Engineered Materials Handbook,* Vol. 3: *Adhesives and Sealants.* Materials Park, Ohio: ASM International, 1993.

Baghdachi, J., *Adhesive Bonding Technology.* New York: Marcel Dekker, 1996.

Haviland, G.S., *Machinery Adhesives for Locking, Retaining, and Sealing.* New York: Marcel Dekker, 1986.

Landrock, A.H., *Adhesives Technology Handbook.* Park Ridge, New Jersey: Noyes, 1985.

Lee, L.-H., *Adhesive Bonding.* New York: Plenum, 1991.

Parmley, R.O. (ed.), *Standard Handbook of Fastening and Joining,* 2d ed. New York: McGraw-Hill, 1989.

Satas, D., *Handbook of Pressure-Sensitive Adhesive Bonding.* New York: Van Nostrand Reinhold, 1989.

Schneberger, G.L. (ed.), *Adhesives in Manufacturing.* New York: Marcel Dekker, 1983.

Shields, J., *Adhesives Handbook,* 3d ed. London: Butterworth, 1984.

Skeist, I., *Handbook of Adhesives.* New York: Van Nostrand Reinhold, 1990.

*Tool and Manufacturing Engineers Handbook,* Vol. 4, *Quality Control and Assembly.* Dearborn, Mich.: Society of Manufacturing Engineers, 1986.

## Mechanical fastening

Blake, A., *What Every Engineer Should Know About Threaded Fasteners.* New York: Marcel Dekker, 1983.

Lincoln, B., K.J. Gomes, and J.F. Braden, *Mechanical Fastening of Plastics.* New York: Marcel Dekker, 1984.

Parmley, R.O. (ed.), *Standard Handbook of Fastening and Joining,* 2d ed. New York: McGraw-Hill, 1989.

*Tool and Manufacturing Engineers Handbook,* Vol. 4, *Quality Control and Assembly.* Dearborn, Mich.: Society of Manufacturing Engineers, 1986.

## QUESTIONS

12.1 Explain why so many different welding processes have been developed.

12.2 Why is the quality of submerged arc welding very good?

12.3 What are the similarities and differences between consumable and nonconsumable electrodes?

12.4 What is the effect of the thermal conductivity of the workpiece on kerf width in cutting using the processes described in this chapter?

12.5 Comment on your observations regarding Fig. 12.3.

12.6 Discuss the need and role of fixtures in holding workpieces in the welding operations described in this chapter.

12.7 Describe the factors that influence the size of the two weld beads in Fig. 12.13.

12.8 What determines whether a certain welding process can be used for workpieces in horizontal, vertical, or upside-down positions, or all? Explain, giving appropriate examples.

12.9 Explain the factors involved in electrode selection in arc-welding processes.

12.10 Explain why the electroslag-welding process is suitable for thick plates and heavy structural sections.

12.11 What are the similarities and differences between consumable- and nonconsumable-electrode arc-welding processes.

12.12 In Table 12.2, there is a column on the distortion of welded components, ranging from lowest to highest. Explain why the degree of distortion varies among different welding processes.

12.13 In Fig. 12.16a and b, explain why the grains grow in the directions shown.

12.14 Prepare a table of the processes described in this chapter and give the range of welding speeds as a function of workpiece material and thicknesses.

12.15 Explain what is meant by solid-state welding.

12.16 Describe your observations concerning Figs. 12.18 and 12.19.

12.17 What advantages does friction welding have over other methods described in this chapter?

12.18 Why is diffusion bonding, when combined with superplastic forming of sheet metals, an attractive fabrication process? Does it have any disadvantages?

12.19 Can roll bonding be applied to a variety of part configurations? Explain.

12.20 Comment on your observations concerning Fig. 12.28.

12.21 Describe your observations concerning the welding design guidelines illustrated in Fig. 12.29.

12.22 Discuss the factors that influence the strength of a (1) diffusion-bonded and (2) cold-welded component.

12.23 Describe the difficulties you might encounter in applying explosion welding in a factory environment.

12.24 Inspect the edges of a U.S. quarter and comment on your observations. Is the cross-section, i.e., thickness of individual layers, symmetrical? Explain.

12.25 What advantages do resistance-welding processes have over others described in this chapter?

12.26 What does the strength of a weld nugget in resistance spot welding depend on?

12.27 Explain the significance of the magnitude of the pressure applied through the electrodes during resistance-welding operations.

12.28 Design a machine that can perform friction welding of two cylindrical pieces, as well as remove the flash from the welded joint (see Fig. 12.35).

12.29 How would you modify your design in Question 12.28 if one of the pieces to be welded is noncircular?

12.30 Inspect Fig. 12.35 and explain why those particular fusion-zone shapes are developed as a function of pressure and speed. Comment on the influence of material properties.

12.31 Which applications could be suitable for the roll spot-welding process shown in Fig. 12.38c? Give specific examples.

12.32 Could the process shown in Fig. 12.39 also be applicable for part shapes other than tubular? Explain and give specific examples.

12.33 Could the projection-welded parts shown in Fig. 12.41 be made by any of the processes described in other parts of this text? Explain.

12.34 In resistance projection welding, describe the factors that influence flattening of the interface after welding.

12.35 What factors influence the shape of the upset joint in flash welding, as shown in Fig. 12.42a?

12.36 Discuss your observations concerning the flash-welding design guidelines illustrated in Fig. 12.42b.

12.37 Explain how you would fabricate the structures shown in Fig. 12.46 with methods other than diffusion bonding and superplastic forming.

12.38 Make a survey of metal containers for household products and foods and beverages. Identify those that have utilized any of the processes described in this chapter. Describe your observations.

12.39 Describe designs that cannot be joined by friction welding processes.

12.40 Explain why some joints may have to be preheated prior to welding.

12.41 What are the similarities and differences between casting of metals (see Chapter 5) and fusion welds?

12.42 Explain the role of the excessive restraint (stiffness) of various components to be welded on weld defects.

12.43 Discuss the weldability of several metals and explain why some metals are easier to weld than others. Cast iron is generally difficult to weld. Why?

12.44 Must the filler metal be of the same composition as the base metal to be welded? Explain.

12.45 Describe the factors that contribute to the difference in properties across a welded joint.

12.46 How does the weldability of steel change as its carbon content increases? Why?

12.47 Are there common factors among the weldability, castability, formability, and machinability of metals? Explain, with appropriate examples.

12.48 Assume that you are asked to inspect a weld for a critical application. Describe the procedure.

12.49 If you find a flaw in a welded joint during inspection, how would you go about determining whether or not this flaw is important?

12.50 Do you think it is acceptable to differentiate brazing and soldering arbitrarily by temperature of application? Comment.

12.51 Why is surface preparation important in adhesive bonding?

12.52 Why have mechanical joining methods been developed? Give several specific examples of their applications.

12.53 Explain why hole preparation may be important in mechanical joining.

12.54 What precautions should be taken in mechanical joining of dissimilar metals?

12.55 What difficulties are involved in joining plastics? Why?

12.56 Comment on your observations concerning the joints shown in Figs. 12.50 and 12.51.

12.57 How different is adhesive bonding from other joining methods? What limitations does it have?

12.58 Soldering is generally applied to thinner components. Why?

12.59 Explain why adhesively bonded joints tend to be weak in peeling.

12.60 Inspect various household products and describe how they are joined and assembled. Explain why those particular processes were used.

12.61 Name several products that have been assembled by (1) seaming, (2) stitching, and (3) soldering.

12.62 Suggest methods of attaching a round bar (made of thermosetting plastic) perpendicular to a flat metal plate.

12.63 In Fig. 12.56, why are the designs on the left labeled poor and the ones on the right good?

12.64 Describe the tooling and equipment that are necessary to perform the double-lock seaming operation shown in Fig. 12.58, starting with flat sheet. (See Chapter 7.)

12.65 What joining methods would be suitable to assemble a thermoplastic cover over a metal frame? Assume that the cover is removed periodically.

12.66 Do the same as Question 12.65, but for a cover made of (1) a thermoset, (2) metal, and (3) ceramic. Describe the factors involved in your selection of methods.

12.67 Do you think the strength of an adhesively bonded structure is as high as that obtained by diffusion bonding? Explain.

12.68 Comment on workpiece size limitations, if any, for each of the processes described in this chapter.

12.69 Describe part shapes that cannot be joined by the processes described in this chapter. Gives specific examples.

12.70 Make a comprehensive outline of joint designs (relating to the processes described in this chapter). Give specific examples of engineering applications for each type of joint.

12.71 Give several applications for fasteners in various products, and explain why other joining methods have not been used.

12.72 Comment on workpiece shape limitations, if any, for each of the processes described in this chapter.

12.73 List and explain the rules that must be followed to avoid cracks in welded joints, such as hot tearing, hydrogen-induced cracking, lamellar tearing, etc.

# PROBLEMS

12.74 Two 1.2 mm thick, flat copper sheets are being spot welded using a current of 6000 A and a current flow time of $t = 0.18$ s. The electrodes are 5 mm in diameter. Estimate the heat generated in the weld zone.

12.75 Calculate the temperature rise in the problem above, assuming that the heat generated is confined to the volume of material directly between the two electrodes and that the temperature is distributed uniformly.

12.76 Plot the hardness in Fig. 12.17b as a function of the distance from the top surface, and discuss the steps that you would follow during your inspection. Would the size of the part or structure you are inspecting have an influence on your methodology? Explain.

12.77 In Fig. 12.25, assume that most of the top portion of the top piece is cut horizontally with a sharp saw. Thus, the residual stresses will be disturbed and, as described in Section 2.10, the part will undergo shape change. For this case, how do you think the part will distort? Explain.

12.78 Review the two weld designs in Fig. 12.31 in Example 12.2, and, based on the topics covered in courses on the strength of materials, show that the de-

sign on the right is capable of supporting a larger moment as shown.

12.79 The figure below shows a metal sheave that consists of two matching pieces of hot-rolled low-carbon steel sheets. These two pieces can be joined either by spot welding or by V-groove welding. Discuss the advantages and limitations of each process for this application.

(a)

(b)  0.135″  (c)

$2\frac{7}{16}''$

Spot weld  V-groove weld

# 13

## Automation of Manufacturing Processes

**13.1** Introduction
**13.2** Automation
**13.3** Numerical Control
**13.4** Programming for Numerical Control
**13.5** Adaptive Control
**13.6** Material Handling and Movement
**13.7** Industrial Robots
**13.8** Sensor Technology
**13.9** Flexible Fixturing
**13.10** Assembly, Disassembly, and Service

## 13.1 Introduction

Until about the 1950s, most manufacturing operations were carried out on traditional machinery, such as lathes, milling machines, and presses, which lacked flexibility and required considerable skilled labor. Each time a different product was manufactured, the machinery had to be retooled and the movement of materials had to be rearranged. The development of new products and parts with complex shapes required numerous trial-and-error attempts by the operator to set the proper processing parameters on the machine. Furthermore, because of human involvement, making parts that were exactly alike was difficult.

These circumstances meant that processing methods were generally inefficient and that labor costs were a significant portion of overall production costs. The need to reduce the labor share of product cost gradually became apparent, as did the need to improve the efficiency and flexibility of manufacturing operations. These needs were particularly significant in terms of increased competition, both domestic and global.

**Productivity** also became a major concern. Defined as the optimum use of all resources—materials, energy, capital, labor, and technology—or as output per employee per hour, productivity basically measures operating efficiency. How could productivity be improved? Mechanization of machinery and operations had, by and large, reached its peak by the 1940s. **Mechanization** runs a process or operation with the use of various mechanical, hydraulic, pneumatic, or electrical devices. In mechanized systems, the operator still directly controls the process, and must check each step of the machine's performance.

The next step in improving the efficiency of manufacturing operations was **automation,** from the Greek word *automatos,* meaning self-acting. The word automation was coined in the mid-1940s by the U.S. automobile industry to indicate automatic handling of parts between production machines and the continuous processing of parts on the machines. During the past four decades, major advances and breakthroughs in the types and extent of automation have occurred. These important developments were made possible largely through rapid advances in the capacity and sophistication of **computers** and **control systems.**

This chapter will describe the technical and economic aspects of automation in manufacturing processes and operations.

## 13.2   Automation

**Automation** can generally be defined as the process of following a predetermined sequence of operations with little or no human labor, using specialized equipment and devices that perform and control manufacturing processes. As described in Section 13.8 and Chapter 14, automation, in its full sense, is achieved through the use of a variety of devices, sensors, actuators, techniques, and equipment that are capable of observing the manufacturing process, making decisions concerning the changes that should be made in the operation, and controlling all aspects of it. Automation is and will continue to be an *evolutionary*, rather than a revolutionary, concept.

Although metalworking processes were developed as early as 4000 B.C., it was not until the beginning of the Industrial Revolution in the 1750s that automation began to be used in the production of goods. The major breakthrough in automation began with numerical control (NC) of machine tools in the 1950s. Since this historic development, rapid progress has been made in automating many aspects of manufacturing (see Table 1.3). This progress involves the introduction of computers into automation, computerized numerical control (CNC), adaptive control (AC), industrial robots, and computer-integrated manufacturing (CIM) systems, including computer-aided design, engineering and manufacturing (CAD/CAE/CAM).

### 13.2.1   Goals and applications of automation

Automation has several primary goals:

- Integrate various aspects of manufacturing operations so as to improve product quality and uniformity, minimize cycle times and effort, and thus reduce labor costs.

- Improve productivity by reducing manufacturing costs through better control of production. Parts are loaded, fed, and unloaded on machines more efficiently. Machines are used more effectively and production is organized more efficiently.

- Improve quality by employing more repeatable processes.

- Reduce human involvement, boredom, and possibilities of human error.

- Reduce workpiece damage caused by manual handling of parts.

- Raise the level of safety for personnel, especially under hazardous working conditions.

- Economize on floor space in the manufacturing plant by arranging machines, material movement, and related equipment more efficiently.

## Automation and production quantity

Production quantity is crucial in determining the type of machinery and equipment—and the level of automation—required to produce parts economically. Before we proceed further, let's define some basic production terms. *Total production quantity* is defined as the total number of parts to be made. This quantity can be produced in individual batches of various *lot sizes*. Lot size greatly influences the economics of production, as described in Chapter 14. *Production rate* is defined as the number of parts produced per unit of time, such as per day, month, or year.

Small quantities per year (Fig. 13.1) can be manufactured in *job shops* using various standard general-purpose machine tools (*stand-alone* machines) or machining centers. These operations have high part variety, meaning that different parts can be produced in a short time without extensive changes in production operations and tooling. However, machinery in job shops generally requires skilled labor to operate; production quantity and rate are low; and hence cost per part can be high (Fig. 13.2). Production of parts involving a large labor component is known as *labor intensive*.

*Piece-part* production usually involves very small quantities and is suitable for job shops. The majority of piece-part production is in lot sizes of 50 or less. *Small-batch* production quantities usually range from 10 to 100, and general-purpose machines and machining centers with various computer controls are used. *Batch* production usually involves lot sizes between 100 and 5000 and utilizes machinery similar to that used for small-batch production, but with specially designed fixtures for higher production rates.

*Mass production* involves quantities of 100,000 and over, and requires special-purpose machinery, called *dedicated* machines, and automated equipment for transferring materials and parts. Although the machinery, equipment, and specialized tooling are expensive, the labor skills required and labor costs are relatively low because of the high level of automation.

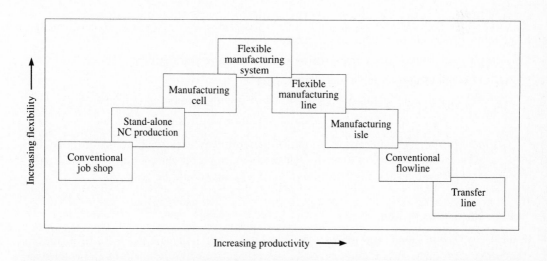

**FIGURE 13.1** Flexibility and productivity of various manufacturing systems. Note the overlap between the systems; this is because of various levels of automation and computer control that are possible in each group. See also Chapter 14 for details. *Source:* U. Rembold, et. al., *Computer Integrated Manufacturing and Engineering.* Addison-Wesley, 1993.

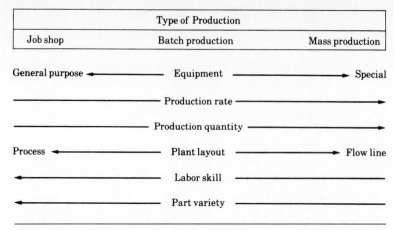

**FIGURE 13.2** General characteristics of three types of production methods.

However, these production systems are organized for a specific product and, consequently, lack flexibility. Most manufacturing facilities operate with a variety of machines in combination and with various levels of automation and computer control.

### Applications of automation

Automation can be applied to manufacturing all types of goods, from raw materials to finished products, and in all types of production from job shops to large manufacturing facilities. The decision to automate a new or existing production facility requires the following additional considerations:

- Type of product manufactured
- Quantity and rate of production required
- The particular phase of manufacturing operation to be automated
- Level of skill in the available work force
- Reliability and maintenance problems associated with automated systems
- Economics, i.e., labor cost versus costs of machines and various equipment

It should be noted that because automation generally involves a high initial cost of equipment, as well as a knowledge of the principles of operation and maintenance required, *selective automation,* rather than total automation, of a facility could be desirable.

### 13.2.2 Hard automation

In **hard,** or **fixed-position, automation,** the production machines are designed to produce a standardized product, such as engine blocks, valves, gears, or spindles. Although product size and processing parameters (such as speed, feed, and depth of cut) can be changed, these machines are specialized. They lack flexibility and cannot be modified to any significant extent to accommodate products that have different shapes and dimensions (see Group Technology,

Section 14.8). Because these machines are expensive to design and construct, their economic use requires mass production of parts in very large quantities.

Machines used in hard-automation applications are usually built on the *building block,* or *modular,* principle. They are generally called *transfer machines,* and consist of the following two major components: powerhead production units and transfer mechanisms.

**Powerhead production units.** Consisting of a frame or bed, electric driving motors, gearboxes, and tool spindles, *powerhead production units* are self-contained. The components are commercially available in various standard sizes and capacities. They can be easily regrouped for producing a different part and thus have certain adaptability and flexibility. Transfer machines consist of two or more powerhead units, which can be arranged on the shop floor in linear, circular, or U patterns. The weight and shape of parts influence the arrangement selected. The arrangement is also important for continuity of operation in the event of tool failure or machine breakdown in one or more of the units. Buffer storage features are incorporated into these machines to permit continued operation.

**Transfer mechanisms and transfer lines.** Transfer mechanisms are used to move the workpiece from one station to another in the machine, or from one machine to another, to enable various operations to be performed on the part. Workpieces are transferred by several methods: (1) rails along which the parts, usually placed on pallets, are pushed or pulled by various mechanisms (Fig. 13.3a); (2) rotary indexing tables (Fig. 13.3b); and (3) overhead conveyors. The transfer of parts from station to station is usually controlled by sensors and other devices. Tools on transfer machines can be changed easily using toolholders with quick-change features. These machines may be equipped with various automatic gaging and inspection systems. These systems are utilized between operations to ensure that the dimensions of a part produced in one station are within acceptable tolerances before that part is transferred to the next station. As described in Section 13.10, transfer machines are also used extensively in automated assembly.

**FIGURE 13.3** Two types of transfer mechanisms: (a) straight and (b) circular patterns.

**Figure 13.4**   A large transfer line for producing cylinder heads. *Source:* Ford Motor Company and *American Machinist,* July 1985.

The **transfer lines** or **flow lines** for a very large system for producing cylinder heads for engine blocks, consisting of a number of transfer machines, are shown in Fig. 13.4. This system is capable of producing 100 cylinder heads per hour. Note the various machining operations performed: milling, drilling, reaming, boring, tapping, and honing, as well as washing and gaging. Such systems are very expensive. For example, it has been estimated that the cost of such a system for high-volume production of engine blocks ranges from $60 million to $130 million.

### 13.2.3   Soft automation

It has been stated that hard automation generally involves mass-production machines that lack flexibility. In **soft** (or **flexible** or **programmable**) **automation,** greater flexibility is achieved through computer control of the machine and its various functions, using programs described in detail in Sections 13.3 and 13.4. Soft automation is an important development, because the machine can be easily and readily reprogrammed to produce a part that has a different shape or dimensions than the one just produced. Because of this capability, soft automation can produce parts with complex shapes. Advances in flexible automation, with extensive use of modern computers, have led to the development of **flexible manufacturing systems** (see Section 14.10) with high levels of efficiency and productivity.

### 13.2.4   Programmable controllers

The control of manufacturing processes in a proper sequence, involving groups of machines and various material-handling equipment and accessories, has traditionally been done by timers, switches, relays, counters, and similar hard-wired devices based on mechanical, electromechanical, and pneumatic principles. Beginning in 1968, **programmable logic controllers (PLCs;** also called PCs) were introduced to replace these devices. A programmable logic controller has been defined by the National Electrical Manufacturers Association (NEMA) as "a digitally operating electronic apparatus which uses a programmable memory for the internal storage of instructions for implementing specific functions such as logic, sequencing, timing, counting, and arithmetic to control, through digital or analog input/output modules, various types of machines or processes." The digital computer, which is used to control the functions of a programmable controller, is considered to be within this scope.

Because relay control panels can now be eliminated and PLCs can be reprogrammed and take less space, programmable controllers have become widely adopted in manufacturing systems and operations. Their basic functions are on-off, motion, sequential operations, and feedback control. In addition to their use in manufacturing process control, they are also used in system control with high-speed digital processing and communication capabilities. These controllers perform reliably in industrial environments and improve the overall efficiency of the operation. PLCs are becoming less popular in new installations, however, due to advances in numerical-control machines, but they still represent a very large installation base.

## 13.3 Numerical Control

**Numerical control** (NC) is a method of controlling the movements of machine components by directly inserting coded instructions in the form of numerical data (numbers and letters) into the system. The system automatically interprets these data and converts them to output signals. These signals, in turn, control various machine components, such as turning spindles on and off, changing tools, moving the workpiece or the tools along specific paths, and turning cutting fluids on and off.

The importance of numerical control can be illustrated by the following example. Assume that holes have to be drilled on a part in the positions shown in Fig. 13.5. In the traditional manual method of machining this part, the operator positions the drill with respect to the workpiece, using as reference points any of the three methods shown. The operator then proceeds to drill these holes. Let's assume that 100 parts, having exactly the same shape and

**FIGURE 13.5** Positions of drilled holes in a workpiece. Three methods of measurements are shown; (a) absolute dimensioning, referenced from one point at the lower left of the part; (b) incremental dimensioning, made sequentially from one hole to another; and (c) mixed dimensioning, a combination of both methods.

dimensional accuracy, have to be drilled. Obviously, this operation is going to be tedious because the operator has to go through the same motions again and again. Moreover, the probability is high that, for various reasons, some of the parts machined will be different from others. Let's further assume that during this production run, the order for these parts is changed, so that 10 of the parts now require holes in different positions. The machinist now has to reposition the work table, which will be time consuming and subject to error. Such operations can be performed easily by numerical-control machines that are capable of producing parts repeatedly and accurately and of handling different parts by simply loading different part programs.

In numerical control, data concerning all aspects of the machining operation, such as locations, speeds, feeds, and cutting fluids, are stored on magnetic tape, cassette, floppy or hard disks, or paper or plastic (Mylar, a thermoplastic polyester). The concept of NC is that specific information can be relayed from these storage devices to the machine tool's control panel. On the basis of input information, relays and other devices (called *hard-wired* controls) can be actuated to obtain a desired machine setup. Complex operations, such as turning a part having various contours or die sinking in a milling machine, can be carried out.

### 13.3.1  Computer numerical control

In the next step of numerical-control development, the control hardware mounted on the NC machine was converted to local computer control with software. Two types of computerized systems were developed: direct numerical control and computer numerical control.

In **direct numerical control (DNC)**, as originally conceived and developed in the 1960s, several machines are directly controlled step by step by a central mainframe computer. In this system, the operator has access to the central computer through a remote terminal. Thus, handling of tapes and the need for computers on each machine are eliminated. With DNC, the status of all machines in a manufacturing facility can be monitored and assessed from the central computer. However, the crucial disadvantage of DNC is that if the computer goes down, all the machines become inoperative.

A more recent definition of DNC (now meaning **distributed numerical control**) includes the use of a central computer serving as the control system over a number of individual computer numerical control machines with onboard microcomputers. This system provides large memory and computational capabilities, thus offering flexibility while overcoming the previous disadvantage of DNC.

**Computer numerical control (CNC)** is a system in which a microcomputer or microprocessor is an integral part of the control of a machine or equipment (onboard computer). The part program may be prepared at a remote site by the programmer, and may incorporate information obtained from drafting software packages and machining simulations to ensure that the part program is bug-free. However, the machine operator can now easily and manually program onboard computers. The operator can modify the programs directly, and prepare and store programs for different parts. Because of the availability of small computers with large memory, microprocessors, and program-editing capabilities, CNC systems are widely used today. The importance of the availability of low-cost, programmable controllers in the successful implementation of CNC in manufacturing plants should be recognized.

The advantages of CNC over conventional NC systems are:

- Increased flexibility. The machine can produce a certain part, followed by other parts with different shapes, at reduced cost.

**FIGURE 13.6** Schematic illustration of the major components of a numerical control machine tool. *Source: Flexible Automation 87/88, The International CNC Reference Book.*

- Greater accuracy. Computers have a higher sampling rate and faster speed.
- More versatility. Editing and debugging programs, reprogramming, and plotting and printing part shape are simpler.

### 13.3.2  Principles of NC machines

The basic elements and operation of a typical NC machine are outlined in Fig. 13.6. The functional elements in numerical control and the components involved are:

1. *Data input.* Numerical information is read and stored in the tape reader or in computer memory.
2. *Data processing.* The programs are read into the machine control unit for processing.
3. *Data output.* This information is translated into commands, typically pulsed commands to the servomotor (Fig. 13.7). The servomotor moves the table on which the workpiece is placed to specific positions, through linear or rotary movements, by means of stepping motors, leadscrews, and other devices.

### Types of control circuits

An NC machine can be controlled through two types of circuits: open-loop and closed-loop. In the **open-loop** system (Fig. 13.7a), the signals are given to the servomotor by the processor, but the movements and final destinations of the worktable are not checked for accuracy. The **closed-loop** system (Fig. 13.7b) is equipped with various transducers, sensors, and counters that measure the position of the table accurately. Through *feedback* control, the position of the work table is compared against the signal. Table movements terminate when the proper coordinates are reached. The closed-loop system is more complicated and more expensive than the open-loop system.

*Position measurement* in NC machines can be accomplished through direct or indirect methods. In direct measuring systems, a sensing device reads a graduated scale on the machine

**FIGURE 13.7**   Schematic illustration of the components of (a) open-loop and (b) closed-loop control systems for a numerical-control machine. DAC means digital-to-analog converter.

table or slide for linear movement (Fig. 13.8a). This system is the most accurate because the scale is built into the machine, and backlash (the play between two adjacent mating gear teeth) in the mechanisms is not significant. In indirect measuring systems, rotary encoders or resolvers (Figs. 13.8b and c) convert rotary movement to translation movement. In this system, backlash can significantly affect measurement accuracy. Position feedback mechanisms use various sensors, based mainly on magnetic and photoelectric principles.

### 13.3.3   Types of control systems

There are two basic types of control systems in numerical control: point-to-point and contouring. In the **point-to-point** system, also called **positioning,** each axis of the machine is driven separately by leadscrews and, depending on the type of operation, at different velocities.

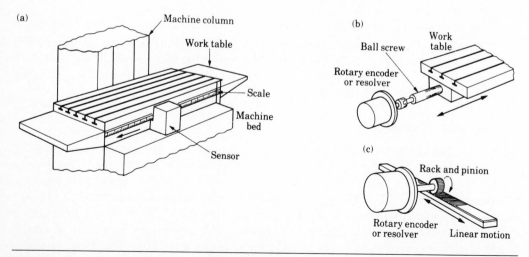

**FIGURE 13.8**   (a) Direct measurement of linear displacement of a machine-tool work table. (b) and (c) Indirect measurement methods.

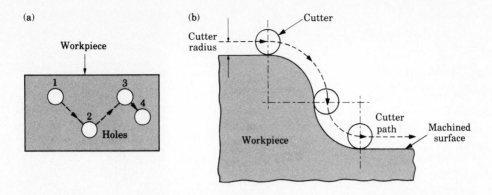

**FIGURE 13.9** Movement of tools in numerical-control machining. (a) Point to point, in which the drill bit drills a hole at position 1, is retracted and moved to position 2, and so on. (b) Continuous path by a milling cutter. Note that the cutter path is compensated for by the cutter radius. This path can also be compensated for to include cutter wear.

The machine moves initially at maximum velocity in order to reduce nonproductive time, but decelerates as the tool reaches its numerically defined position. Thus in an operation such as drilling or punching, the positioning and cutting take place sequentially (Fig.13.9a). After the hole is drilled or punched, the tool retracts, moves rapidly to another position, and repeats the operation. The path followed from one position to another is important in only one respect: The time required should be minimized for efficiency. Point-to-point systems are used mainly in drilling, punching, and straight (one-axis) milling operations.

In the **contouring** system, also known as the **continuous path** system, both positioning and cutting operations take place along controlled paths but at different velocities. Because the tool cuts as it travels along a prescribed path (Fig. 13.9b), accurate control and synchronization of velocities and movements are important. The contouring system is used on lathes, milling machines, grinders, welding machinery, and machining centers.

Movement along the path, or **interpolation,** occurs incrementally, by one of several basic methods (Fig. 13.10). Examples of actual paths in drilling, boring, and milling operations are shown in Fig.13.11. In all interpolations, the path controlled is that of the center of rotation

**FIGURE 13.10** Types of interpolation: (a) linear, (b) continuous path approximated by incremental straight lines, and (c) circular.

(a)

Point-to-point

Drilling and
boring

Workpiece

Point-to-point and
straight line

Milling

(b)

2-axis contouring with
switchable plane

2-axis contour
milling

3-axis contouring
continuous path

3-axis contour
milling

**FIGURE 13.11**    (a) Schematic illustration of drilling, boring, and milling with various paths. As described in Section 8.9, the relationship of milling cutters and the workpiece and the direction of cut are important considerations. (b) Machining a sculptured surface on a 5-axis numerical control machine tool. *Source:* Courtesy of The Ingersoll Milling Machine Company.

of the tool. Compensation for different tools, different diameter tools, or tool wear during machining can be made in the NC program.

In **linear** interpolation, the tool moves along a straight line from start to end, in two or three axes. Theoretically, all types of profiles can be produced by this method by decreasing the increments between the points. However, a large amount of data must be processed in order to do so. In **circular** interpolation, the inputs required for the path are the coordinates of the end points, the coordinates of the center of the circle and its radius, and the direction of the tool along the arc. In **parabolic** and **cubic** interpolation, the path is approximated by curves using higher order mathematical equations. This method is effective in five-axis machines and is particularly useful in die-sinking operations for automotive bodies. These interpolations are also used for industrial robot movements, which are discussed in Section 13.7.

### 13.3.4  Accuracy in numerical control

Positioning accuracy in NC machines is defined by how accurately the machine can be positioned to a certain coordinate system. An NC machine usually has a positioning accuracy of at least $\pm$ 3 $\mu$m (0.0001 in.). **Repeatability,** defined as the closeness of agreement of repeated position movements under the same operating conditions of the machine, is usually around $\pm$ 8 $\mu$m (0.0003 in.). **Resolution,** defined as the smallest increment of motion of the machine components, is usually about 2.5 $\mu$m (0.0001 in.).

The *stiffness* of the machine tool and *backlash* in its gear drives and leadscrews are important for accuracy. Backlash can be eliminated with special backlash take-up circuits, whereby the tool always approaches a particular position on the workpiece from the same direction, as should be done in traditional machining operations. Rapid response to command signals requires that friction and inertia be minimized—for example, by reducing the mass of moving components of the machine.

### 13.3.5 Advantages and limitations

Numerical control has the following advantages over conventional methods of machine control:

- Flexibility of operation is improved, as is the ability to produce complex shapes with good dimensional accuracy, repeatability, reduced scrap loss, and high production rates, productivity, and product quality.

- Tooling costs are reduced, since templates and other fixtures are not required.

- Machine adjustments are easy to make with microcomputers and digital readouts.

- More operations can be performed with each setup, and less lead time for setup and machining is required compared to conventional methods. Design changes are facilitated and inventory is reduced.

- Programs can be prepared rapidly and can be recalled at any time utilizing microprocessors. Less paperwork is involved.

- Faster prototype production is possible.

- Required operator skill is less than that for a qualified machinist, and the operator has more time to attend to other tasks in the work area.

The major limitations of NC are the relatively high initial cost of the equipment, the need and cost for programming and computer time, and special maintenance requiring trained personnel. Because NC machines are complex systems, breakdowns can be very costly, so preventive maintenance is essential. However, these limitations are often easily outweighed by the overall economic advantages of NC.

## 13.4 Programming for Numerical Control

A program for numerical control consists of a sequence of directions that causes an NC machine to carry out a certain operation; machining is the most commonly used process. **Programming for NC** may be performed by an internal programming department, done on the shop floor, or purchased from an outside source. Also, programming may be done manually or with computer assistance.

**Manual part programming** consists of first calculating the dimensional relationships of the tool, workpiece, and work table—based on the engineering drawings of the part—and the manufacturing operations to be performed and their sequence. A program sheet is then prepared, which consists of the necessary information to carry out the operation, such as cutting tools, spindle speeds, feeds, depth of cut, cutting fluids, power, and tool or workpiece relative positions and movements. On the basis of this information, the part program is prepared. Manual programming can be done by someone knowledgeable about the particular process and able to understand, read, and change part programs. Because they are familiar with machine tools and process capabilities, skilled machinists can, with some training in programming, do manual programming. However, the work is tedious, time consuming, and uneconomical—and is used mostly in simple point-to-point applications.

**Computer-aided part programming** involves special symbolic programming languages that determine the coordinate points of corners, edges, and surfaces of the part.

The **programming language** is the means of communicating with the computer and involves the use of symbolic characters. The programmer describes the component to be processed in this language, and the computer converts it to commands for the NC machine. Several languages having various features and applications are commercially available. The first language that used English-like statements was developed in the late 1950s and is called **APT** (for Automatically Programmed Tools). This language, in its various expanded forms, is still the most widely used for both point-to-point and continuous-path programming. (See also various sections in Chapter 14.)

Computer-aided part programming has the following significant advantages over manual part programming:

- Use of symbolic language; several programs have been developed.
- Reduced programming time. Computer-aided programming is capable of accommodating a large amount of data concerning machine characteristics and process variables, such as power, speeds, feed, tool shape, compensation for tool shape changes, tool wear, deflections, and coolant use.
- Reduced possibility of human error, which can occur in manual programming.
- Ability to view a machining sequence on the screen for debugging.
- Capability of simple changeover of machining sequence or from machine to machine.
- Lower cost because less time is required for programming.

The use of programming languages or *compilers* not only results in greater part quality for the reasons mentioned above, but also allows for more rapid development of machining instructions. In addition, simulations can be run on remote computer terminals to ensure that the program functions as intended. This method prevents unnecessary occupation of expensive machinery for debugging procedures.

Selection of a particular NC programming language depends on the following factors:

- Level of expertise of the personnel in the manufacturing facility
- Complexity of the part
- Type of equipment and computers available
- Time and costs involved in programming

---

**Example 13.1:   NC Programming for Machining**

Figure 13.12 shows a 2 in. square and 0.75 in. thick workpiece that is machined around its circumference with an end mill to a depth of 0.125 in. Before the NC program is run, the origin of the axes is set at the lower left corner of the part (near D) and the dimensions are given in the absolute dimensioning method (see Fig. 13.5a). Also, machining tolerances are not taken into consideration in this example.

The cutter has a diameter of 0.250 in. and starts at a position of $X = -1.500$ in., $Y = 0$, and $Z = 1.500$ in.

The approach position of the cutter is shown in Fig. 13.12. The NC program and its codes are given in Table 13.1. After rapid travel to within $-1.500 + 1.200 = 0.300$ in. of the part, the cutter moves along the following paths: (1) It travels in the $+X$ direction in linear interpolation, (2) machines the radius at A by counterclockwise circular interpolation,

**Figure 13.12**

(3) travels in the $+Y$ direction, (4) machines the radius at B, (5) travels in the $-X$ direction, (6) machines the radius at C, (7) travels in the $-Y$ direction, (8) machines the radius at D, and (9) travels in the $-X$ direction to its starting position. Note in the program that the sums of all $X$ and $Y$ movements, respectively, are zero, indicating that the tool has returned to its

## TABLE 13.1

### NC MACHINING PROGRAM

| N | G (M) | X | Y | Z | OPERATION |
|---|---|---|---|---|---|
| N00 | G91 | | | | Set incremental programming mode. |
| N01 | M03 | | | | Start cutter spindle. |
| N02 | G00 | 0 | 0 | −1625 | Move cutter to depth of 0.125″. |
| N03 | G00 | 1200 | 0 | 0 | Rapid move to 0.300″ away from part. |
| N04 | G01 | 2025 | 0 | 0 | Cut until cutter reaches start of radius A. |
| N05 | G03 | 125 | 125 | | Cut radius A using relative endpoints. |
| N06 | G01 | 0 | 1250 | 0 | Cut until cutter reaches start of radius B. |
| N07 | G03 | −125 | 125 | | Cut radius B using relative endpoints. |
| N08 | G01 | −1450 | 0 | 0 | Cut until cutter reaches start of radius C. |
| N09 | G03 | −125 | −125 | | Cut radius C using relative endpoints. |
| N10 | G01 | 0 | −1250 | 0 | Cut until cutter reaches start of radius D. |
| N11 | G03 | 125 | −125 | | Cut radius D using relative endpoints. |
| N12 | G00 | 0 | 0 | 1625 | Rapid rise of cutter to original z-position. |
| N13 | G00 | −1775 | 0 | 0 | Rapid movement of cutter to original x-position. |
| N14 | M30 | | | | End of program. |

Source: EMCO.

starting position. Codes: G91—incremental value programming; M03—cutter spindle on, clockwise rotation; G00—rapid traverse; G01—linear interpolation; G03—circular interpolation, counterclockwise; M30—program end.

## 13.5   Adaptive Control

In **adaptive control (AC)**, the operating parameters in an operation automatically adapt themselves to conform to new circumstances, such as changes in the dynamics of the particular process and any disturbances that may arise. The sense in which these phenomena are adaptive is that the control strategy adapts to changes in the operating conditions. Likewise, a feedback-controlled braking system in a car would use measurements of speed in achieving a controlled braking profile. On the other hand, an *adaptive controller* would check load conditions, adapt an appropriate desired braking profile, and then use feedback to implement it.

One can contrast dynamic feedback control with adaptive control as follows: (1) Dynamic feedback control has a fixed controller mechanism that adapts or adjusts controller signals in response to measured changes in system behavior. A constant-gain control is a special case of dynamic feedback control, gain being defined as the ratio of output to input in an amplifier. (2) Adaptive control adjusts not only the controller signals, but also the controller mechanism.

*Gain scheduling* is perhaps the simplest form of what now is called adaptive control. In gain scheduling, a different gain for the feedback is selected depending on the measured operating conditions. A different gain is assigned to each region of the system's operating space. With advanced adaptive controllers, the gain may vary continuously with changes in operating conditions.

Several adaptive-control systems are now commercially available for applications such as ship steering, chemical-reactor control, rolling mills, and medical technology. In manufacturing engineering, specifically, the purposes of adaptive control are to:

- Optimize production rate.
- Optimize product quality.
- Minimize cost.

**Principles and applications of adaptive control.** The basic functions common to adaptive control systems are to:

1. Determine the operating conditions of the process, including measures of performance. This is typically achieved by using sensors that measure process parameters such as force, torque, vibration, and temperature.

2. Configure the process control in response to the operating conditions. Large changes in the operating conditions may provoke a decision to make a major switch in control strategy. More modest alterations may be the modification of process parameters, such as changing the speed of operation or the feed in machining.

3. Continue to monitor the process, making further changes in the controller when and as needed.

Systems that place a constraint on a process variable (such as forces, torque, or temperature) are called **adaptive control constraint (ACC)** systems. Thus, if the thrust force, cutting force, and, hence, the torque increase excessively—e.g., because of a hard region in a cast workpiece—the adaptive control system changes the speed and/or feed to lower the cutting forces to acceptable levels. Without adaptive control or direct operator intervention as in traditional machining operations, high forces may cause tools to chip or break and the workpiece to deflect excessively, thus losing dimensional accuracy. Systems that optimize an operation are called **adaptive control optimization (ACO)** systems. Optimization may involve maximizing the material-removal rate between tool changes or improving surface finish. Most systems are currently based on ACC because development and implementation of ACO is complex.

For adaptive control to be effective in manufacturing operations, quantitative relationships must be known and stored in the computer software as mathematical models. For example, if the tool wear rate in a machining operation is excessive, the computer must be able to know how much of a change (and whether increase or decrease) in speed and/or feed is necessary to reduce the wear rate to an acceptable level. The system should also be able to compensate for dimensional changes in the workpiece caused by tool wear and temperature rise.

# 13.6  Material Handling and Movement

During a typical manufacturing operation, raw materials and parts are moved from storage to machines, from machine to machine, from inspection to assembly and inventory, and finally to shipment. These materials must be moved either manually or by some mechanical means, and time is required to transport them from one location to another. Thus we may define **material handling** as the functions and systems associated with the transportation, storage, and control of materials and parts in the total manufacturing cycle of a product. The total time required for manufacturing depends on part size and shape and the set of operations required. Idle time and the time required for transporting materials can constitute the majority of the time consumed.

**Plant layout** is an important aspect of the flow of materials and components throughout the manufacturing cycle. As noted in the discussion of flexible manufacturing systems in Chapter 14, the arrangement of production machinery and material-handling equipment should be orderly and efficient. The time and distances required for moving raw materials and parts should be minimized, and storage areas and service centers should be organized accordingly. For parts requiring multiple operations, equipment should be grouped around the operator or the industrial robot (see Cellular Manufacturing, Section 14.9).

## 13.6.1  Methods of material handling

Several factors must be considered in choosing a suitable material-handling method for a particular manufacturing operation:

- Shape, weight, and characteristics of parts
- Types and distances of movement and the position and orientation of parts during movement and at their final destination
- Conditions of the path along which parts are to be transported
- Degree of automation, amount of control desired, and integration with other systems and equipment

- Operator skill required
- Economic considerations

For small-batch manufacturing operations, raw materials and parts can be handled and transported by hand, though this method is generally costly. Moreover, because it involves human beings, this practice can be unpredictable and unreliable and can be unsafe to the operator, depending on the weight and shape of the parts to be moved and environmental factors, such as heat and smoke in foundries and forging plants. In automated manufacturing plants, computer-controlled material and parts flow is rapidly being implemented. Such changes have improved repeatability and lowered labor costs.

### 13.6.2 Equipment

Various types of equipment can be used to move materials, such as conveyors, rollers, self-powered monorails, carts, forklift trucks, and various mechanical, electrical, magnetic, pneumatic, and hydraulic devices and manipulators. **Manipulators** are designed to be controlled directly by the operator, or they are automated for repeated operations, such as loading and unloading parts from machine tools, presses, and furnaces. Manipulators are capable of gripping and moving heavy parts and orienting them as required between manufacturing and assembly operations. Machines are often used in a sequence where workpieces are transferred directly from machine to machine. Machines capable of conveying parts without need for additional material-handling schemes are called *integral transfer devices.*

Industrial robots, specially designed pallets, and **automated guided vehicles (AGVs)** are used extensively in flexible manufacturing systems to move parts and orient them as required (Fig. 13.13). Thus, flexible material handling and movement with real-time control has become an integral part of modern manufacturing.

Automated guided vehicles, which are the latest development in material movement in plants, operate automatically along pathways with in-floor wiring or tapes for optical scanning and without any operator intervention. This transport system has great flexibility and is capable of random delivery to different workstations. It optimizes the movement of materials and parts in cases of congestion around workstations, machine breakdown (downtime), or failure of part of the system. The movements of AGVs are planned so that they interface with **automated storage/retrieval systems (AS/RS)**, which utilizes warehouse spaces efficiently and reduces labor costs.

Coding systems have been developed to locate and identify parts throughout the manufacturing system and to transfer them to their appropriate stations. *Bar codes* are the most widely used and the least costly of these systems. They are printed on labels, magnetic stripes, and RF (radio frequency) tags.

## 13.7    Industrial Robots

The word robot was coined in 1920 by the Czech author K. Capek in his play *R.U.R.* (Rossum's Universal Robots), and is derived from the word *robota,* meaning worker. An **industrial robot** has been defined as a reprogrammable multifunctional manipulator, designed to move materials, parts, tools, or other specialized devices by means of variable pro-

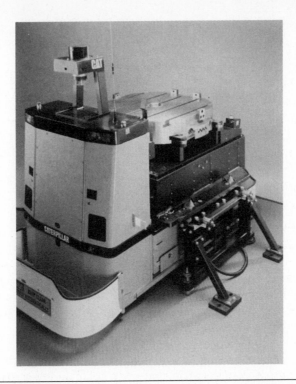

**FIGURE 13.13** A self-guided vehicle (Caterpillar Model SGC-M) carrying a machining pallet. The vehicle is aligned next to a stand on the floor. Instead of following a wire or stripe path on the factory floor, this vehicle calculates its own path and automatically corrects for any deviations. *Source:* Permission to reprint granted by Caterpillar, Inc.

grammed motions and to perform a variety of other tasks. In a broader context, the term robot also includes manipulators that are activated directly by an operator.

Introduced in the early 1960s, the first industrial robots were used in hazardous operations, such as handling toxic and radioactive materials and loading and unloading hot workpieces from furnaces and handling them in foundries. Some rule-of-thumb applications for robots are the three D's (dull, dirty, and dangerous, including demeaning but necessary tasks), and the three H's (hot, heavy, and hazardous). From their early uses for worker protection and safety in manufacturing plants, industrial robots have been further developed and have become important components in manufacturing processes and systems. They have helped improve productivity, increase product quality, and reduce labor costs. Computer-controlled robots were commercialized in the early 1970s; the first robot controlled by a minicomputer appeared in 1974.

### 13.7.1 Components

The basic components of an industrial robot are manipulators, end effectors, power supply, and control system.

**Manipulator.**  Also called *arm and wrist,* the *manipulator* is a mechanical unit that provides motions (trajectories) similar to that of a human arm and hand. The end of the wrist can reach a point in space with a specific orientation. There are three degrees of freedom each in linear and rotational movements, respectively. Manipulation is carried out using mechanical devices, such as linkages, gears, and various joints.

**End effector.**   The end of the wrist in a robot is equipped with an end effector, also called *end-of-arm tooling*. Depending on the type of operation, conventional end effectors are equipped with:

- Grippers, hooks, scoops, electromagnets, vacuum cups, and adhesive fingers, for materials handling
- Spray guns for painting
- Attachments for spot and arc welding and arc cutting
- Power tools, such as drills, nut drivers, and burs
- Measuring instruments, such as dial indicators

**Power supply.**   Each motion of the manipulator, in linear and rotational axes, is controlled and regulated by independent actuators, using electric, pneumatic, or hydraulic power supplies. Each source of energy and the motors involved have their own characteristics, advantages, and limitations.

**Control system.**   Also known as the *controller*, the *control system* is the communications and information processing system that gives commands for the movements of the robot. It is the brain of the robot and stores data to initiate and terminate the movements of the manipulator. It interfaces with computers and other equipment, such as manufacturing cells or assembly systems. *Feedback devices*, such as transducers, are an important part of the control system.

Robots with a fixed set of motions have open-loop control. In this system, commands are given and the robot arm goes through its motions, but—unlike feedback in closed-loop systems—accuracy of the movements is not monitored. The system does not have self-correcting capability.

As in numerical control machines, the types of control in industrial robots are point-to-point and continuous path. Depending on the particular task, the positioning repeatability required may be as small as 0.050 mm (0.002 in.), as in assembly operations for electronic printed circuitry. Specialized robots can achieve such accuracy, although most robots are unable to do so.

### 13.7.2 Classification

Robots may be classified by basic type: (1) cartesian or rectilinear, (2) cylindrical, (3) spherical or polar, and (4) articulated, or revolute, jointed, or anthropomorphic (Fig. 13.14). Robots may be attached permanently to the floor of a manufacturing plant, may move along overhead rails (*gantry robot*), or may be equipped with wheels to move along the factory floor (*mobile robot*). However, a broader classification of robots currently in use is most helpful for our purposes here. This classification is used below to describe various types of robots.

**FIGURE 13.14** Types of industrial robots: (a) cartesian (rectilinear), (b) cylindrical, (c) spherical (polar), (d) articulated (revolute, jointed, or anthropomorphic).

**Fixed- and variable-sequence robots.** The *fixed-sequence robot,* also called a *pick-and-place robot,* is programmed for a specific sequence of operations. Its movements are from point to point, and the cycle is repeated continuously. These robots are simple and relatively inexpensive. The *variable-sequence robot* can be programmed for a specific sequence of operations, but also can be reprogrammed to perform another sequence.

**Playback robot.** An operator leads, or walks, the *playback robot* and its end effector through the desired path. In other words, the operator teaches the robot by showing it what to do. The robot memorizes and records the path and sequence of motions and can repeat them continuously without any further action or guidance by the operator. Another method, the *teach pendant,* utilizes hand-held button boxes that are connected to the control panel and used to control and guide the robot and its tooling through the work to be performed. These movements are then registered in the memory of the controller and automatically reenacted by the robot.

**Numerically controlled robot.** The *numerically controlled robot* is programmed and operated much like a numerically controlled machine. The robot is servocontrolled by digital data, and its sequence of movements can be changed with relative ease. As in NC machines, there are two basic types of controls: point to point and continuous path. Point-to-point robots are easy to program and have a higher load-carrying capacity and *work envelope* (the maximum extent or reach of the robot hand or working tool in all directions, also called *working envelope;* Fig. 13.15). Continuous-path robots have greater accuracy than point-to-point robots, but have lower load-carrying capacity. Some advanced robots have a complex system of path control, enabling high-speed movements with great accuracy.

**Intelligent (sensory) robot.** The *intelligent,* or *sensory, robot* is capable of performing some of the functions and tasks carried out by human beings. The intelligent robot is equipped with a variety of sensors with visual (computer vision) and tactile (touching) capabilities (see Section 13.8). Much like humans, the robot observes and evaluates the immediate environment and proximity to other objects, machinery, etc., by perception and pattern recognition. It then makes appropriate decisions for the next movement and proceeds with it. Because its operation is so complex, powerful computers are required to control this type of robot.

(a) Rectangular

(b) Cylindrical

(c) Spherical

Work envelopes

**FIGURE 13.15**   Work envelopes for three types of robots. The choice depends on the particular application.

### 13.7.3   Applications and selection of robots

Major applications of industrial robots include the following:

- Material handling, loading, unloading, and transferring workpieces in manufacturing operations

- Spot welding automobile and truck bodies, producing welds of good quality. Robots perform other, similar operations, such as arc welding, arc cutting and riveting.

- Machining operations, such as deburring, grinding, and polishing, with appropriate tools attached to their end effectors

- Applying adhesives and sealants, as in the automobile frame

- Spray painting—particularly of complex shapes—and cleaning operations

- Automated assembly (see also Section 13.10)

- Inspection and gaging in various stages of manufacture

**Selection of robots.**   Factors that influence the selection of robots in manufacturing plants are: (1) load-carrying capacity, (2) speed of movement, (3) reliability, (4) repeatability, (5) arm configuration, (6) degrees of freedom, (7) control system, (8) program memory, and (9) work envelope.

**Economics.**   In addition to technical factors, cost and benefit considerations are significant aspects of robot selection and use. The increasing availability and reliability, and reduced costs of sophisticated intelligent robots are having a major economic impact on manufacturing operations and are gradually replacing human labor. Whereas hourly wages are steadily rising, particularly in industrialized nations, the cost of robot operation per hour has increased more slowly.

**Robot safety.**   Depending on the size of the robot's work envelope, its speed, and its proximity to humans, safety in a robot environment is an important consideration. This is particularly so for programmers and maintenance personnel who interact directly with robots. In addition, the movements of the robot with respect to other machinery require a high level of reliability in order to avoid collisions and serious damage, and its material-handling activities require proper securing of raw materials and parts in the robot gripper at various stages along the production line.

## 13.8   Sensor Technology

A sensor is a device that produces a signal for purposes of detecting or measuring a property, such as position, force, torque, pressure, temperature, humidity, speed, acceleration, and vibration. Traditionally, sensors, actuators, and switches have been used to set limits on the performance of machines. Familiar examples are stops on machine tools to restrict work table movements, pressure and temperature gages with automatic shut-off features, and governors on engines to prevent excessive speed of operation. Sensor technology has become an important component of manufacturing processes and systems and is essential to data acquisition, monitoring, communication, and computer control of machines and systems.

Because they convert one quantity to another, sensors are also often referred to as *transducers,* meaning to transfer. *Analog* sensors produce a signal, such as voltage, that is proportional to the measured quantity. *Digital* sensors have numeric or digital outputs that can be directly transferred to computers. *Analog-to-digital converters (ADC)* are available to interface analog sensors with computers.

### 13.8.1   Classification

Sensors that are of interest in manufacturing may be classified generally as:

- *Mechanical*—for measuring quantities such as position, shape, velocity, force, torque, pressure, vibration, strain, and mass.
- *Electrical*—for measuring voltage, current, charge, and conductivity.
- *Magnetic*—for measuring magnetic field, flux, and permeability.
- *Thermal*—for measuring temperature, flux, conductivity, and specific heat.
- Others—such as acoustic, ultrasonic, chemical, optical, radiation, laser, and fiber optic.

Sensors are being developed with capabilities that resemble those of human beings, as in the two principal methods of sensing: tactile and visual.

**Tactile sensing.**   *Tactile sensing* is the continuous sensing of variable contact forces, commonly by an array of sensors. Such a system is capable of performing within an arbitrary three-dimensional space. Fragile parts, such as glass bottles and electronic devices, can be handled by robots with *compliant,* or *smart,* end effectors. These effectors can sense the force applied to the object being handled, using piezoelectric devices, strain gages, magnetic induction, ultrasonics, and optical systems of fiber optics and light-emitting diodes. Tactile sensors capable of measuring and controlling gripping forces and moments in three axes are available commercially.

The force sensed is monitored and controlled through closed-loop feedback devices. Compliant grippers with force feedback and sensory perception can be complicated, require powerful computers, and hence are costly. Anthropomorphic end effectors are being designed to simulate the human hand and fingers, with the capability of sensing touch, force, movement, and pattern. The ideal tactile sensor must also sense slip, a capability of the human fingers and hand that we tend to take for granted, which is so important in the use of robots.

**Visual sensing (machine vision or computer vision).**   In *visual sensing*, cameras optically sense the presence and shape of the object (Fig. 13.16). There are two basic systems of machine vision: linear array and matrix array. In *linear array*, only one dimension is sensed, such as the presence of an object or some feature on its surface. *Matrix arrays* sense up to three dimensions and are, for example, capable of detecting a properly inserted component in a printed circuit or a properly made solder joint. When used in automated inspection systems (see Section 4.8.3) they can also detect cracks and flaws.

Figure 13.16   Examples of machine-vision application. (a) On-line inspection of parts. (b) Identification of parts with various shapes, inspection, and rejection of defective parts. (c) Use of cameras to provide positional input to a robot relative to the workpiece. (d) Painting parts with different shapes with input from a camera. The system's memory allows the robot to correctly identify the particular shape to be painted and to proceed with the correct movements of a paint spray attached to the end effector. *Source: Manufacturing Engineering,* November 1982.

In visual sensing, a microprocessor processes the image, usually in less than one second. The image is measured and the measurements are digitized (**image recognition**). Machine vision is capable of on-line identification and inspection of parts—and rejection of defective parts. Several applications of machine vision in manufacturing are shown in Fig. 13.16. With visual sensing capabilities, end effectors are able to pick up parts and grip them in proper orientation and location. However, picking parts from a bin has proven to be a difficult task because of the random orientation of parts in close proximity to each other.

### 13.8.2  Sensor fusion

Although there is no clear definition of the term **sensor fusion,** it is generally understood as the integration of multiple sensors in a manner in which the individual data from each of the sensors (force, vibration, temperature, dimensions, etc.) are combined to provide a higher level of information. It has been suggested that a common example of sensor fusion is someone drinking from a cup of hot tea or coffee. Although we take such an everyday event for granted, it can readily be seen that this process involves data input from the person's eyes, hands, lips, and tongue. There is real-time monitoring, through various senses, of relative movement and positions, as well as temperature. Thus, if the fluid is too hot, the hand movement of the cup toward the lip is controlled accordingly.

An important aspect in sensor fusion is sensor validation, whereby the failure of a sensor is detected so that the control system has high reliability. Thus, redundant data from different sensors are essential. It can be seen that receiving, integrating and processing of all data from various sensors can be a complex problem. Artificial neural networks (see Section 14.13) are now being used for sensor fusion.

Among new developments are **smart sensors,** which have the capability of performing a logic function and two-way communication, and making decisions and taking appropriate actions. The necessary input and knowledge required to make a decision can be built into a smart sensor, whereby a *chip* with sensors can be programmed to turn a machine tool off when, for example, a cutting tool fails. Likewise, a smart sensor can stop a mobile robot or a robot arm from accidentally coming in contact with an object or people by sensing quantities such as distance, heat, and noise.

## 13.9  Flexible Fixturing

Previous chapters have described several workholding devices, such as chucks, collets, mandrels, and various fixtures, many of which are operated manually. Other devices are designed and operated at various levels of mechanization and automation, such as power chucks driven by mechanical, hydraulic, or electrical means. Workholding devices have certain ranges of capacity: collets can accommodate bars within a certain range of dimensions; four-jaw chucks can accommodate square or prismatic workpieces having certain dimensions; and other devices and fixtures are made for specific workpiece shapes and dimensions.

In manufacturing operations, the words clamp, jig, and fixture are often used interchangeably and sometimes in pairs, as in jigs and fixtures. **Clamps** are simple multifunctional devices, whereas **fixtures** are generally designed for specific purposes, with on and off features and shapes that are usually replicas of workpieces. **Jigs** have various reference surfaces and points for accurate alignment of parts and tools and are widely used in mass production.

The emergence of flexible manufacturing systems, described in Chapter 14, has necessitated the use of workholding devices and fixtures that have a certain built-in flexibility. Generally called **flexible fixturing**, these devices are capable of quickly accommodating a range of part shapes and dimensions, without the necessity of changing or extensively adjusting the fixtures, or requiring operator intervention, both of which would adversely affect productivity.

**Design considerations for flexible fixturing.**    Proper design of flexible workholding devices and fixtures is essential to the operation of advanced manufacturing systems. These devices should position the workpiece automatically and accurately and maintain its location precisely and with sufficient clamping force during the manufacturing operation. Fixtures should accommodate parts repeatedly in the same position and must have sufficient stiffness to resist, without excessive deflection, the forces developed. Fixtures and clamps should have low profiles so as to avoid colliding with tools and dies. Collision avoidance is also an important factor in programming tool paths.

## 13.10    Assembly, Disassembly, and Service

The individual parts and components made by various manufacturing processes are *assembled* into finished products by various methods. Some products are simple, having only two or three components to assemble, which can be done with relative ease, such as an ordinary pencil with an eraser, a frying pan with a wooden handle, or a beverage can. Most products, however, consist of many parts and their assembly requires considerable care and planning.

Traditionally, assembly has involved much manual work, contributing significantly to cost. The total assembly operation is usually broken into individual assembly operations, with an operator assigned to carry out each step. Assembly costs are typically 25 to 50 percent of the total cost of manufacturing, with the percentage of workers involved in assembly operations ranging from 20 to 60 percent. In the electronics industries, some 40 to 60 percent of total wages are paid to assembly workers.

The choice of an assembly method and system depends on the required production rate and total quantities, product market life, labor availability, and most important, cost. **Automated assembly** can effectively minimize overall product cost.

**Random assembly and selective assembly.**    We have seen that parts are manufactured with certain tolerances. Taking roller bearings as an example, we know that although they have the same nominal dimensions, some rollers in a lot are smaller than others by a small amount. Likewise, some bearing races are smaller than others in the lot.

There are two methods of assembly for such high-volume products: random assembly and selective assembly. In *random assembly,* parts are put together by selecting them randomly from the lots. In *selective assembly,* the rollers and races are segregated by groups of sizes, from smallest to largest. The parts are then selected to mate properly. Thus, the smallest diameter rollers are mated with inner races having the largest outside diameter and with outer races having the smallest inside diameters.

### 13.10.1    Assembly methods and systems

There are three basic methods of assembly: manual, high-speed automatic, and robotic. These methods can be used individually or in combination, as is the case in most applications in prac-

**FIGURE 13.17** Stages in design-for-assembly analysis. *Source:* Reprinted from *Product Design for Assembly,* 1989, by G. Boothroyd and P. Dewhurst, courtesy of Marcel Dekker Inc.

tice. An analysis of the design should first be made with regard to an appropriate and economical method of assembly (Fig. 13.17).

**Manual assembly** uses relatively simple tools and is economical for small lots. Because of the dexterity of the human hand and fingers, and their capability for feedback through various senses, workers can manually assemble even complex parts without much difficulty. (However, there are potential problems of cumulative trauma disorders.) In spite of the use of sophisticated mechanisms, robots, and computer controls, which are described below, aligning and placing a simple square peg in a square hole involving small clearances can be difficult in automated assembly. Yet the human hand is capable of doing it with relative ease.

**High-speed automated assembly** utilizes *transfer mechanisms* designed specially for assembly. In **robot assembly,** one or two general-purpose robots operate at a single workstation, or the robots operate at a multistation assembly system.

Whether human, automatic, or robotic labor performs individual assembly tasks, a system is needed to organize the various stages of assembly of products with multiple parts. There are three basic types of *assembly systems:* synchronous, nonsynchronous and continuous. In **synchronous systems,** also called *indexing,* individual parts and components are supplied and assembled at a constant rate at fixed individual stations. The rate of movement is based on the station that takes the longest time to complete its portion of the assembly. This system is used primarily for high-volume, high-speed assembly of small products. Transfer systems move the partial assemblies from workstation to workstation by various mechanical means.

Two typical transfer systems are *rotary* indexing and *in-line* indexing. These systems can operate in either fully automatic or semiautomatic mode. However, a breakdown of one station will shut down the whole assembly operation. The part feeders supply the individual parts to be assembled and place them on other components, which are secured on work carriers or fixtures. The feeders move the individual parts by vibratory or other means through delivery chutes and ensure their proper orientation by various ingenious means. Properly orienting parts and avoiding jamming are essential in all automated assembly operations.

In **nonsynchronous systems,** each station operates independently, and any imbalance is accommodated in storage (a buffer) between stations. The station continues working until the next buffer is full or the previous buffer is empty. Furthermore, if one station becomes inoperative for some reason, the assembly line continues to operate until all the parts in the buffer have been used. Nonsynchronous systems are suitable for large assemblies with many parts to be assembled. For types of assembly in which the times required for individual assembly operations vary widely, the output is controlled by the slowest station.

In **continuous systems,** the product is assembled while moving at a constant speed on pallets or similar work carriers. The parts to be assembled are brought to the product by various workheads, and their movements are synchronized with the constant movement of the product. Typical applications of this system are in bottling and packaging plants, although the method has been used in mass-production lines for automobiles and appliances.

Although assembly systems are set up for a certain product line, they can be modified for increased flexibility in order to assemble product lines that have a variety of models. Such **flexible assembly systems (FAS)** utilize computer controls, interchangeable and programmable workheads and feeding devices, coded pallets, and automated guiding devices. The General Motors plant for the Saturn subcompact automobile is designed with a flexible assembly system, whereby the system is capable of assembling up to a dozen different transmission and engine combinations, as well as power steering and air conditioning systems.

### 13.10.2  Guidelines for design for assembly, disassembly, and service

Although the functions of a product and its design for manufacturing have been matters of considerable interest for some time, only recently has **design for assembly (DFA)** attracted special attention, particularly design for automated assembly, necessitated by the need to reduce labor costs.

Certain guidelines have been established as an aid in the design of parts for ease of assembly. The general guidelines for manual assembly may be summarized as follows:

1.  Reduce the number and different types of parts in a product so that fewer steps and fixtures are required, thus lowering assembly costs.

2.  Parts to be assembled should have a high degree of symmetry, or they should be highly asymmetrical to avoid errors by the worker.

3.  Parts should be designed so that they cannot be installed incorrectly, or so they do not require locating, aligning, or adjusting.

4.  Designs should allow parts to be assembled without obstructions or lack of direct line of sight.

Design guidelines for *high-speed automated assembly* include, in addition to some of those for manual assembly, the fact that parts have to be handled not manually, but automatically using various devices. Humans can easily pick parts from bulk, such as from a nearby bin, but automatic handling requires that parts be separated from bulk and conveyed by hoppers or vibratory feeders in the proper orientation for assembly. Consequently, the guidelines involve the following:

1.  Part designs should consider factors such as size, shape, weight, flexibility, tangling with other parts, abrasiveness, etc.

2.  Parts should be designed so that they can be inserted from a single direction, preferably vertically to take advantage of gravity, as assembly from two or more directions can present problems.

3.  Products should be designed, or existing products redesigned, so that there are no physical obstructions to the free movement of parts during assembly (see Fig. 1.3); sharp external and internal corners should be replaced with chamfers, tapers, or radii.

*Robotic assembly* design guidelines have similar rules to those for manual and high-speed automated assembly. In addition,

1.  Parts should be designed so that they can be gripped and manipulated by the same gripper of the robot, thus avoiding the need for different grippers; also, parts should be delivered to the gripper in the proper orientation.
2.  Assembly requiring threaded fasteners, such as bolts, nuts, and screws, may be difficult for robots, whereas robots can handle snap fits, rivets, welds, and adhesives faster. One exception is the use of self-threading screws for sheet metal, plastics, and wooden parts.

**Design for disassembly.**    The manner and ease with which a product may be taken apart for maintenance or replacement of its parts is another important consideration in product design. Consider, for example, the difficulties one has in removing certain components from under the hood of some automobiles, such as spark plugs or oil filters. Similar difficulties exist in the disassembly of several other products. Consequently, *design for disassembly* has become an important factor in the overall design process.

Although there is no established set of guidelines, the general approach takes into consideration factors similar to those outlined for design for assembly. Analysis of computer or physical models of products and their components with regard to disassembly should indicate potential problems, such as obstructions, size of passageways, lack of line of sight, and difficulty of firmly grasping and guiding objects.

**Design for service.**    The latest trend in design for assembly and disassembly includes taking into account the ease with which a product can be repaired. Called *design for service,* this approach is based on the concept that the elements most likely to need servicing should be at the outer layers of the product. This way, individual parts are easier to reach and service, without having to remove various other parts.

## SUMMARY

- There are several levels of automation, from simple automation of machines to untended manufacturing cells to, ultimately, the factory of the future. Automation has been implemented successfully in manufacturing processes, material handling, inspection, assembly, and packaging. Production quantity and rate are important factors in determining the economical levels of automation. (Sections 13.1, 13.2)
- True automation began with the numerical control of machines, by which are achieved flexibility of operation, lower cost, and ease of making different parts with lower operator skill. (Sections 13.3, 13.4)
- Manufacturing operations are further optimized, both in quality and cost, by adaptive control techniques that continuously monitor the operation and make necessary adjustments in process parameters. (Section 13.5)

- Great advances have been made in material handling, particularly with the implementation of industrial robots and automated guided vehicles. (Sections 13.6, 13.7)

- The role of sensors is crucial in the implementation of these technologies, and a wide variety of sensors based on various principles have been developed and installed. (Section 13.8)

- Other advances include flexible fixturing and automated assembly techniques that reduce the need for worker intervention and that reduce manufacturing costs. The efficient and economic implementation of these techniques requires that design for assembly, disassembly, and servicing be recognized as an important factor in the total design and manufacturing process. (Sections 13.9, 13.10)

# BIBLIOGRAPHY

Asfahl, C.R., *Robots and Manufacturing Automation,* 2d ed. New York: Wiley, 1992.

Astrom, K.J., and B. Wittenmark, *Adaptive Control.* Reading, Mass.: Addison-Wesley, 1989.

Boothroyd, G., *Assembly Automation and Product Design.* New York: Marcel Dekker, 1991.

Boothroyd, G., P. Dewhurst, and W. Knight, *Product Design for Manufacture and Assembly.* New York: Marcel Dekker, 1994.

Chang, C.-W., and M.A. Melkanoff, *NC Machine Programming and Software Design.* Englewood Cliffs, N.J.: Prentice-Hall, 1989.

Chow, W., *Assembly Line Design: Methodology and Applications.* New York: Marcel Dekker, 1990.

Galbiati, L.J., *Machine Vision and Digital Image Processing Fundamentals.* Englewood Cliffs, N.J.: Prentice-Hall, 1990.

Gibbs, D., and T. Crandell, *CNC: An Introduction to Machining and Part Programming.* New York: Industrial Press, 1991.

Groover, M.P., *Automation, Production Systems, and Computer-Integrated Manufacturing.* Englewood Cliffs, N.J.: Prentice-Hall, 1987.

Hang, C.C., T.H. Lee, and W.K. Ho, *Adaptive Control.* Research Triangle Park, NC: Instrument Society of America, 1993.

Lee, M.H., *Intelligent Robotics.* New York: Halsted Press, 1989.

Lynch, M., *Computer Numerical Control for Machining.* New York: McGraw-Hill, 1992.

Miller, R.K., *Industrial Robot Handbook.* New York: Van Nostrand Reinhold, 1989.

Murphy, S.D., *In-Process Measurement and Control.* New York: Marcel Dekker, 1990.

Nanfara, F., T. Uccello, and D. Murphy, *The CNC Workbook: An Introduction to Computer Numerical Controls.* Reading, Mass.: Addison-Wesley, 1995.

*NC/CIM Guidebook,* published annually by *Modern Machine Shop,* Cincinnati, Ohio.

Polywka, J., and S. Gabriel, *Programming of Computer Numerically Controlled Machines.* New York: Industrial Press, 1992.

Rembold, U. (ed.), *Robot Technology and Applications.* New York: Marcel Dekker, 1990.

Sandler, B.-Z, *Robotics: Designing the Mechanisms for Automated Machinery.* Englewood Cliffs, N.J.: Prentice-Hall, 1991.

Seames, W., *Computer Numerical Control: Concepts and Programming,* 2d ed. Albany, N.Y.: Delmar, 1990.

Soloman, S., *Sensors and Control Systems in Manufacturing.* New York: McGraw-Hill, 1994.

Thyer, G., *Computer Numerical Control of Machine Tools,* 2d ed. New York: Industrial Press, 1993.

*Tool and Manufacturing Engineers Handbook*, 4th ed., Vol. 4: *Assembly, Testing, and Quality Control.* Dearborn, Mich.: Society of Manufacturing Engineers, 1986.

Valentino, J.V., and J. Goldenberg, *Introduction to Computer Numerical Control*. Englewood Cliffs, N.J.: Regents/Prentice-Hall, 1993.

Williams, D.J., *Manufacturing Systems*, 2d ed. New York: Chapman & Hall, 1994.

Zuech, N., *Applying Machine Vision*. New York: Wiley, 1988.

## QUESTIONS

13.1 Describe the differences between mechanization and automation. Give several specific examples for each.

13.2 Why is automation generally regarded as evolutionary rather than revolutionary?

13.3 Are there activities in manufacturing operations that cannot be automated? Explain.

13.4 Explain the difference between hard and soft automation. Why are they so called?

13.5 Describe the principle of numerical control of machines. What factors led to the need and development of numerical control? Name typical applications.

13.6 Explain the differences between direct numerical control and computer numerical control. What are their relative advantages?

13.7 Describe open-loop and closed-loop control circuits.

13.8 What are the advantages of computer-aided NC programming?

13.9 Describe the principle and purposes of adaptive control. Give some examples of present applications and others that you think can be implemented.

13.10 What factors have led to the development of automated guided vehicles? Do they have any disadvantages? Explain.

13.11 Describe the features of an industrial robot. Why are these features necessary?

13.12 Discuss the principles of various types of sensors and give two applications for each type.

13.13 Describe the concept of design for assembly. Why has it become an important factor in manufacturing?

13.14 Is it possible to have partial automation in assembly? Explain.

13.15 What are the relative advantages and limitations of the two arrangements for power heads shown in Fig. 13.3?

13.16 Discuss methods of on-line gaging of workpiece diameters in turning operations.

13.17 Is drilling the only application for the point-to-point system shown in Fig. 13.9a? Explain.

13.18 What determines the number of robots in an automated assembly line?

13.19 Why should the level of automation in a manufacturing facility depend on production quantity and production rate?

13.20 Explain why sensors have become crucial in the development of automated manufacturing systems.

13.21 Why is there a need for flexible fixturing for holding workpieces? Are there any disadvantages? Explain.

13.22 This text describes a wide variety of casting, forming, shaping, machining and grinding, and joining and assembly processes. Which of these are most suitable for transfer lines?

13.23 Explain the difference between an automated guided vehicle and a self-guided vehicle.

## PROBLEMS

13.24 List and discuss the factors that should be considered in choosing a suitable material-handling system.

13.25 Giving specific examples, discuss your observations concerning Fig. 13.2.

13.26  Describe possible applications for industrial robots not discussed in this chapter.

13.27  Describe situations in which the shape and size of the work envelope of a robot (see Fig. 13.15) can be critical.

13.28  It has been commonly acknowledged that, at their early stages of development and implementation, the usefulness and cost effectiveness of industrial robots were overestimated. What reasons can you think of to explain this situation?

13.29  Describe the type of manufacturing operations (see Fig. 13.1) that are likely to make the best use of a machining center. Comment on the influence of product quantity and part variety. (See also Section 8.10.)

13.30  Give a specific example where (1) an open-loop and (2) a closed-loop control system would be desirable.

13.31  Describe situations in manufacturing in which you would not want to apply numerical control.

13.32  Describe situations in which each of the three positioning methods shown in Fig. 13.5 would be desirable.

13.33  Describe applications of machine vision for specific products, similar to the examples shown in Fig. 13.16.

13.34  Give an example of a forming operation that is suitable for adaptive control.

13.35  Design two different systems of mechanical grippers for widely different applications.

13.36  Give some applications for the systems shown in Fig. 13.16a and c.

13.37  Add others to the examples shown in Fig. 1.3 in Chapter 1.

13.38  Give examples of products that are suitable for the types of production shown in Fig. 13.2.

13.39  Give examples of situations in which tactile sensors would not be suitable. Explain why.

13.40  Give examples in which machine vision cannot be applied properly and reliably. Explain why.

13.41  Choose one machine each from Chapters 5 through 12, and design a system in which sensor fusion can be used effectively. How would you convince a prospective customer of the merits of such a system?

13.42  Do the same as for Problem 13.40, but for a flexible fixturing system.

13.43  Think of a product and design a transfer line for it. Specify the types and number of machines required.

13.44  Considering the wide variety of parts made, prepare design guidelines for flexible fixturing. Make simple sketches illustrating the principle for each type of fixturing, describing its range of applications and limitations.

13.45  Describe your thoughts on the usefulness and applications of modular fixturing, consisting of various individual clamps, pins, supports, and attachments mounted on a base plate.

13.46  Inspect several household products and describe the way they have been assembled. Comment on any changes you would make so that assembly, disassembly, and servicing are simpler and faster.

13.47  In Sections 10.14 and 11.7 we described several methods of rapid prototyping. In Section 13.3 we described computer numerical control with its capabilities and flexibility as applied, for example, to machining. Note that in spite of its name, rapid prototyping still requires time to produce the part and NC machining requires time to program it, although machining the part may not take much time. Taking a relatively simple part as an example, describe your thoughts regarding the competitive aspects of rapid prototyping versus NC machining of the part, including the mechanical properties of the part required in service.

# 14

# Computer-Integrated Manufacturing Systems

**14.1** Introduction
**14.2** Manufacturing Systems
**14.3** Computer-Integrated Manufacturing
**14.4** Computer-Aided Design and Engineering
**14.5** Computer-Aided Manufacturing
**14.6** Computer-Aided Process Planning
**14.7** Computer Simulation of Manufacturing Processes and Systems
**14.8** Group Technology
**14.9** Cellular Manufacturing
**14.10** Flexible Manufacturing Systems
**14.11** Just-In-Time Production
**14.12** Manufacturing Communications Networks
**14.13** Artificial Intelligence
**14.14** Factory of the Future

## 14.1  Introduction

This chapter focuses on the *integration* of manufacturing activities. Integration means that manufacturing processes, operations, and their management are treated as a *system,* allowing complete control of the manufacturing facility. In this way, productivity, product quality, and reliability can be optimized and costs minimized.

Beginning in the 1960s, certain trends developed that had a major impact on manufacturing.

- Global competition increased rapidly.
- Market conditions fluctuated widely.
- Customers demanded high-quality, low-cost products and on-time delivery.
- Product variety increased substantially.
- Product life cycles became shorter.

These trends led to recognition of the need for an integrated approach to manufacturing. In **computer-integrated manufacturing,** the traditionally separate functions of research and development, design, production, assembly, inspection, and quality control are all linked. Consequently, integration requires that quantitative relationships among product design, materials, manufacturing process and equipment capabilities, and related activities be well understood. Thus, changes in material requirements, product types, and market demand can be accommodated. The goal is to find the optimum method of producing high-quality, low-cost products efficiently.

Machines, tooling, and manufacturing operations must have a certain built-in *flexibility* to respond to change and ensure *on-time delivery* of products to the customer. In industry, failure to deliver on time can upset production plans and schedules and, consequently, have a significant economic impact. In a highly competitive global environment, failure of on-time product delivery can cost a company its *competitive edge* because the customer will simply change suppliers.

This chapter concludes with a discussion and overview of the *factory of the future*. The prediction is that manufacturing operations in the future will be performed with little human labor; hence these facilities are also being called *untended factories*. Because of its potential impact on the work force, the social implications of this trend are also discussed.

## 14.2   Manufacturing Systems

Because manufacturing entails a large number of interdependent activities consisting of distinct entities such as materials, tools, machines, power, and human beings, it should properly be regarded as a *system*. It is, in fact, a complex system because it is comprised of many diverse physical and human elements, some of which are difficult to predict and control, such as raw-material prices, market changes, and human behavior and performance.

Ideally, we should be able to represent a system by *mathematical* and *physical models* that show us the nature and extent of interdependence. In a manufacturing system, a change or disturbance anywhere in the system requires that it adjust itself in order to continue functioning efficiently. For example, if the supply of a particular raw material is reduced and hence its cost is increased because of, say, geopolitical reasons, alternative materials must be selected. This selection should be made only after careful consideration of the effect this change may have on product quality, production rate, and costs.

Similarly, if demand for a product is such that its shape, size, or capacity fluctuates randomly and rapidly, the system must be able to produce the modified product on short lead time and without the need for major capital investment in machinery and tooling. Although the labor share of product cost has been decreasing steadily over the years (in some cases it is now less than 10 percent of total product cost), the system must also be capable of absorbing some of the cost if worker pay escalates.

## 14.3   Computer-Integrated Manufacturing

The various levels of automation in manufacturing operations described in Chapter 13 can be extended further by including information-processing functions utilizing an extensive network of interactive computers. The result is **computer-integrated manufacturing (CIM)**, which is a broad term describing the computerized integration of *all* aspects of design, planning, manufacturing, distribution, and management. Computer-integrated manufacturing is a *methodology* and a *goal*, rather than an assemblage of equipment and computers. The technology for implementing CIM is well understood and available. However, because CIM ideally involves the total operation of a company, it must be comprehensive and have an extensive database. Consequently, if implemented all at once, CIM can be prohibitively expensive, particularly for small- and medium-sized companies.

Computer-integrated manufacturing systems consist of subsystems that are integrated into a whole. These subsystems consist of business planning and support, product design, manufacturing process planning, process control, shop floor monitoring systems, and process automation (Fig. 14.1). The subsystems are designed, developed, and applied in such a manner that the output of one subsystem serves as the input of another subsystem. Organizationally, these subsystems are usually divided into business planning and business execution functions, respectively. **Planning** functions include activities such as forecasting, scheduling, material-requirements planning, invoicing, and accounting. **Execution** functions include production and process control, material handling, testing, and inspection.

The effectiveness of CIM depends greatly on the presence of a large-scale, *integrated communications system* involving computers, machines, and their controls. Major problems

**FIGURE 14.1** A schematic illustration of a computer-integrated manufacturing system. *Source:* U. Rembold et al., *Computer-Integrated Manufacturing and Engineering.* Addison-Wesley, 1993.

have arisen in factory communication because of the difficulty of interfacing different types of computers purchased from different vendors by the company at various times. As described in Section 14.12, the trend is strongly toward standardization to make communications equipment compatible. The benefits of CIM includes its:

- Responsiveness to shorter product life cycles, changing market demand, and global competition.

- Emphasis on product quality and its uniformity through better process control.

- Better use of materials, machinery, and personnel, and reduction of work-in-progress (WIP) inventory, thus improving productivity and lowering product cost.

- Better control of production, scheduling, and management of the total manufacturing operation, resulting in lower product cost.

**Database.**   An efficient computer-integrated manufacturing system requires a single database, which is shared by the entire manufacturing organization. A **database** consists of up-to-date, detailed, and accurate data relating to products, designs, machines, processes, materials, production, finances, purchasing, sales, marketing, and inventory. This vast array of data is stored in computer memory and recalled or modified as necessary, either by individuals in the organization or by the CIM system itself in controlling various aspects of design and production.

Databases are built by individuals and various sensors in the machinery and equipment used in production. Data from the latter are collected automatically by a **data acquisition system (DAS),** and may include the number of parts being produced per unit of time, their dimensional accuracy, surface finish, weight, and so on, at specified rates of sampling. The components of a DAS include microprocessors, transducers, and analog-to-digital converters (ADC). Data acquisition systems are also capable of analyzing data and transferring them to other computers for purposes such as statistical analysis, data presentation, and forecasting product demand.

Several attributes are important in the use and implementation of databases. They should be timely, accurate, easily accessible, easily shared, and user friendly. In the event that something goes wrong with the data, correct data should be recovered and restored. Because it is used for many purposes and by many people, the database must be flexible and responsive to the needs of different users. Companies must protect data against tampering or unauthorized use. CIM systems can be accessed by designers, manufacturing engineers, process planners, financial officers, and the management of the company by using appropriate access codes.

## 14.4   Computer-Aided Design and Engineering

**Computer-aided design (CAD)** involves the use of computers to create design drawings and product models (Fig. 14.2). Computer-aided design is usually associated with *interactive computer graphics,* known as a *CAD system.* Computer-aided design systems are powerful tools and are used in mechanical design and geometric modeling of products and components. **Computer-aided engineering (CAE)** refers to a system of computerized analytical tools that share a common database, permitting several applications such as finite-element analysis of stresses, strains, deflections, and temperature distribution in structures and load-bearing

| (a) | (b) | (c) | (d) | (e) | (f) |
|---|---|---|---|---|---|
| 2-D lateral model | 2½-D profile body | 2½-D rotating body | 3-D wire-frame model | 3-D surface model | 3-D volume model |

**FIGURE 14.2**  Types of modeling for CAD.

members and the generation, storage, and retrieval of NC data. These systems are also used extensively in the design of integrated circuits and other electronic devices.

Because of the availability of a wide variety of CAD systems with different characteristics supplied by different vendors, proper communication and exchange of data between these systems have become a significant problem (see also Section 14.12). The need for a single neutral format for better compatibility is presently filled largely by the **Initial Graphics Exchange Specification (IGES)**. Vendors need only provide translators for their own systems to preprocess the data into the neutral format and postprocess from the neutral format into their system. IGES is used for translation in two directions (into and out of a system) and is also used widely for translation of 3-D line and surface data.

A more recent development is a solid-model-based standard called **Product Data Exchange Specification (PDES)**, which is based on IGES. Although IGES is adequate for most requirements, PDES has less memory size, requires less time for execution, and is less error prone. Because of the existence of various standards in other countries as well, it is expected that in the near future these standards will culminate in an international standard, called the **Standard for the Exchange of Product Model Data (STEP)**.

### Elements of CAD systems

The design process in a CAD system consists of four stages, as described below.

**Geometric modeling.**  In *geometric modeling,* a physical object or any of its parts is described mathematically or analytically. The designer first constructs a geometric model by giving commands that create or modify lines, surfaces, solids, dimensions, and text that together are an accurate and complete two- or three-dimensional representation of the object. The models can be presented in three different ways. In **line** representation (**wire-frame;** Fig. 14.2d), all edges are visible as solid lines. This image can be ambiguous, particularly for complex shapes. However, various colors are generally used for different parts of the object, thus making the object easier to visualize. In the **surface** model, all visible surfaces are shown in the model, and in the **solid** model, all surfaces are shown, but the data describe the interior volume (Fig. 14.2e and f).

The three types of wire-frame representations are 2-D, $2\frac{1}{2}$-D, and 3-D. The 2-D image shows the profile of the object or part. The $2\frac{1}{2}$-D image can be obtained by translational

**FIGURE 14.3** (a) Boundary representation of solids, showing the enclosing surfaces of the solid model and the generated solid model. (b) A solid model represented as compositions of solid primitives. (c) Three representations of the same part by CAD. *Source:* P. Ranky.

sweep—that is, moving the 2-D object along the z axis. For round objects, a 2 1/2-D model can be generated by simply rotating a 2-D model around its axis.

Solid models can be constructed from "swept volumes" (Fig. 14.2b and c) or by the techniques shown in Fig. 14.3. In *boundary representation* (B-rep), surfaces are combined together to develop a solid model (Fig. 14.3a). In *constructive solid geometry (CSG),* simple shapes such as spheres, cubes, blocks, cylinders, and cones (called *primitives of solids*) are combined to develop a solid model (Fig. 14.3b). Programs are available whereby the user selects any combination of these primitives and their sizes, and combines them into the desired solid model. Although solid models have advantages, such as ease of design analysis and preparation for manufacturing the part, they require more memory and processing time than the wire-frame and surface models shown in Fig. 14.2.

**Design analysis and optimization.** After the design's geometric features have been determined, the design is subjected to an engineering analysis. This phase may consist of analyzing stresses, strains, deflections, vibrations, heat transfer, temperature distribution, or tolerance. Various sophisticated software packages capable of computing these quantities accurately and rapidly are available.

**Design review and evaluation.** An important design stage is review and evaluation to check for any interference between various components in order to avoid difficulties during assembly or use of the part, and whether moving members, such as linkages, are going to operate as intended. Software is now available with animation capabilities to identify potential problems with moving members and other dynamic situations. During the design review and evaluation stage, the part is precisely dimensioned and toleranced, as required for manufacturing it.

**Documentation and drafting.** After the preceding stages have been completed, the design is reproduced by automated drafting machines for documentation and reference. Detail and working drawings are also developed and printed. The CAD system is capable of developing and drafting sectional views of the part, scaling the drawings, and performing transformations to present various views of the part. In CAD systems, much of the design process was formerly carried out on workstations connected to a mainframe computer, but the trend has changed rapidly to powerful, high-performance, and much less expensive stand-alone desktop 32-bit Unix workstations; 64-bit processors with even better performance are also becoming available.

## 14.5 Computer-Aided Manufacturing

**Computer-aided manufacturing (CAM)** is the use of computers and computer technology to assist in all phases of manufacturing a product, including process and production planning, machining, scheduling, management, and quality control. Computer-aided manufacturing encompasses many of the technologies described in Chapter 13 and in this chapter. Because of the benefits of doing so, computer-aided design and computer-aided manufacturing are often combined into CAD/CAM systems. This combination allows the transfer of information from design into the planning for manufacture of a product without the need to reenter the part geometry manually. The database developed in CAD is stored and processed further by CAM into the necessary data and instructions for operating and controlling production machinery, material-handling equipment, and automated testing and inspection for product quality.

An important feature of CAD/CAM in machining operations is its ability to describe tool paths in operations such as NC turning, milling, and drilling. The programmer provides instructions (programs) to the CAD/CAM system, which automatically determines and optimizes the tool path. The engineer or technician can display and visually check the tool path for possible tool collisions with clamps, fixtures, or other interferences. The tool path can be modified at any time to accommodate other part shapes to be machined. CAD/CAM systems are also capable of coding and classifying parts into groups that have similar shapes (see Section 14.8), using alphanumeric coding.

The emergence of CAD/CAM has had a major impact on manufacturing by standardizing product development and reducing design effort, tryout, and prototype work, thus resulting in significantly reduced costs and improved productivity. The two-engine Boeing 777 passenger airplane, for example, is designed completely by computer ("paperless design"), with 2000 workstations linked to eight computers. The airplane is constructed directly from the CAD/CAM software developed (an enhanced CATIA system) and no prototypes or mockups are built, unlike for previous models.

---

**Example 14.1:** Application of CAD/CAM to make a mold

The metal mold used for injection molding of plastic blender jars is made on a CNC milling machine using a ball-nosed end mill, as illustrated in Fig. 14.4. The jar is about 230 mm (9 in.) high and has a maximum diameter of 127 mm (5 in.). First, a 3-D wireframe model of

**FIGURE 14.4**

the mold is made, as seen in the photograph on the left, and viewed and inspected for various geometric features. Next, an offset is added to each surface to account for the nose radius of the end mill, thus determining the cutter path, that is, the center of the machine spindle. The NC programming software then executes this program on the CNC milling machine, producing the die cavity with proper dimensions and accuracy. Electrical-discharge machining can be used to make this mold; however, it was found that EDM is about twice as expensive as machining the mold and has less dimensional accuracy.

*Source:* Model Threads Inc., and *Manufacturing Engineering,* March 1991.

## 14.6   Computer-Aided Process Planning

In order for a manufacturing operation to be efficient, all of its diverse activities must be planned. **Process planning** is concerned with selecting methods of production, tooling, fixtures and machinery, sequence of operations, and assembly. It is traditionally done by process planners. The sequence of processes and operations to be performed, machines to be used, standard time for each operation, and similar information are documented on a *routing sheet* (Fig. 14.5). This task is highly labor intensive and time consuming and relies heavily on the experience of the process planner.

**Computer-aided process planning (CAPP)** accomplishes this complex task by viewing the total operation as an integrated system, so that the individual operations and steps involved in making each part are coordinated with others and are performed efficiently and reliably. Computer-aided process planning is an important adjunct to CAD and CAM. Although CAPP requires extensive software and good coordination with CAD/CAM, as well as other aspects of integrated manufacturing systems discussed in the rest of this chapter, it is a powerful tool for efficiently planning and scheduling manufacturing operations. It is particularly effective in small-volume, high-variety parts production involving machining, forming, and assembly operations.

ROUTING SHEET

CUSTOMER'S NAME: Midwest Valve Co.    PART NAME: Valve body

QUANTITY: 15    PART NO.: 302

| Operation no. | Description of operation | Machine |
|---|---|---|
| 10 | Inspect forging, check hardness | Rockwell tester |
| 20 | Rough machine flanges | Lathe No. 5 |
| 30 | Finish machine flanges | Lathe No. 5 |
| 40 | Bore and counter bore hole | Boring mill No. 1 |
| 50 | Turn internal grooves | Boring mill No. 1 |
| 60 | Drill and tap holes | Drill press No. 2 |
| 70 | Grind flange end faces | Grinder No. 2 |
| 80 | Grind bore | Internal grinder No. 1 |
| 90 | Clean | Vapor degreaser |
| 100 | Inspect | Ultrasonic tester |

**FIGURE 14.5**   An example of a simple routing sheet. These *operation sheets* may include additional information on materials, tooling, estimated times for each operation, processing parameters such as cutting speeds, feeds, etc., and other information. The routing sheet travels with the parts from operation to operation.

### 14.6.1  Elements of CAPP systems

There are two types of computer-aided process planning systems: variant and generative process planning.

In the **variant system,** also called a *derivative system,* the computer files contain a standard process plan for the part to be manufactured. The search for a standard plan is made in the database by a code number for the part based on its shape and manufacturing characteristics (see Group Technology, Section 14.8). The standard plan is then retrieved, displayed for review, and printed as a routing sheet. The process plan includes information such as types of tools and machines to be used, the sequence of manufacturing operations to be performed, speeds, feeds, time required for each sequence, and so on. Minor modifications of an existing process plan, which are usually necessary, can also be made. If the standard plan for a particular part is not in computer files, one that is close to it with a similar code number and an existing routing sheet is retrieved. If a routing sheet does not exist, one is made for the new part and stored in computer memory.

In the **generative system,** a process plan is automatically generated on the basis of the same logical procedures that would be followed by a traditional process planner in making that particular part. Such a system is complex because it must contain comprehensive and detailed knowledge of part shapes, dimensions, process capabilities, selection of manufacturing methods, machinery, and tools required, and the sequence of operations to be performed. (These capabilities of computers, known as **expert systems,** are discussed in Section 14.13.) The generative system is thus capable of creating a new plan instead of using and modifying existing plans, as in the variant system. Although presently less common than the variant system, the generative system has advantages such as (1) flexibility and consistency for process planning for new parts and (2) higher overall planning quality because of the capability of the decision logic in the system to optimize the planning and utilize up-to-date manufacturing technology.

Process planning capabilities of computers can be integrated with the planning and control of the production systems. These activities are a subsystem of computer-integrated manufacturing, described in Section 14.3. Several functions, such as *capacity planning* for plants to meet production schedules, control of *inventory, purchasing,* and production *scheduling,* can be performed.

### 14.6.2  Advantages of CAPP systems

The advantages of CAPP systems over traditional process planning methods include standardization of process plans, thus improving the productivity of process planners, reducing lead times, reducing planning costs, and improving the consistency of product quality and reliability. Process plans can be prepared for parts having similar shapes and features, and they can be retrieved easily to produce new parts. Process plans can be modified to suit specific needs.

Routing sheets can be prepared more quickly. Compared to traditional handwritten routing sheets, computer printouts are neater and much more legible. Other functions, such as cost estimating and work standards, can be incorporated into CAPP.

### 14.6.3  Material-requirements planning and manufacturing resource planning

Computer-based systems for managing inventories and delivery schedules of raw materials and tools are called **material-requirements planning (MRP).** This activity, sometimes regarded as a method of *inventory control,* involves keeping complete records of inventories of materials, supplies, parts in various stages of production (work in progress), orders, purchasing, and scheduling. Several files of data are usually involved in a master production schedule. These pertain to the raw materials required (bill of materials), product structure levels (individual items that comprise a product, such as components, subassemblies, and assemblies), and scheduling.

A further development is **manufacturing resource planning (MRP-II),** which controls all aspects of manufacturing planning through feedback. Although the system is complex, MRP-II is capable of final production scheduling, monitoring actual results in terms of performance and output, and comparing those results against the master production schedule.

## 14.7  Computer Simulation of Manufacturing Processes and Systems

With increasing sophistication of computer hardware and software, computer simulation of manufacturing processes and systems has grown rapidly. Process simulation takes two basic forms. The first is a model of a specific operation, intended to determine the viability of a process or to optimize or improve its performance. The second one models multiple processes and their interactions, and helps process planners and plant designers in the layout of machinery and facilities.

Individual processes have been modeled using various mathematical schemes. Finite element analysis has been increasingly performed using software packages that are commercially available and inexpensive. Typical problems addressed are process viability (such as formability of sheet metal in a certain die) and process optimization (such as material flow in forging in a given die to identify potential defects, or mold design in casting to eliminate hot spots, promote uniform cooling, and minimize defects).

Simulation of an entire manufacturing system involving multiple processes and equipment helps plant engineers in the organization of machinery and identification of critical machinery elements. In addition, such models can assist manufacturing engineers with scheduling and routing. Commercially available software packages are often used for such simulations, but dedicated software programs written for a particular company are not unusual.

## 14.8   Group Technology

Group technology (GT) is a manufacturing concept that seeks to take advantage of the *design* and *processing similarities* among parts. This concept was first developed in Europe around the turn of the century. It involved categorizing parts and recording them manually in card files or catalogs; designs were retrieved manually, as needed. This concept began to evolve further in the 1950s, and the term *group technology* was first used in 1959. However, not until the use of interactive minicomputers became widespread in the 1970s did the use of GT grow significantly.

In large manufacturing facilities, tens of thousands of different parts are made and placed in inventory. Although many of these parts have certain similarities in shape and method of manufacture, each part is traditionally viewed as a separate entity and produced in batches, involving considerable duplication and redundancy. The characteristic of similar parts (Fig. 14.6) suggests that benefits can be obtained by *classifying* and *coding* these parts into families. Results of surveys in manufacturing plants have repeatedly shown the prevalence of similar parts. One survey found that 90 percent of 3000 parts made by a company fell into five major families of parts.

Such surveys consist of breaking each product into its components and identifying similar parts. For example, a pump can be broken into the basic components of motor, housing, shaft, seals, and flanges. In spite of the variety of pumps manufactured, each of these components is basically the same in terms of design and manufacturing methods. Consequently,

**FIGURE 14.6**   (a) Grouping parts according to their geometric similarities. (b) Grouping parts according to their manufacturing similarities. *Source:* Organization for Indusrial Research

all shafts can be placed in one family, unless there are major differences in shapes, materials, and production methods.

This approach becomes more attractive in view of consumer demand for an ever greater variety of products in smaller quantities. Under these conditions, maintaining high efficiency in batch operations is difficult. Overall manufacturing efficiency is hurt because nearly 75 percent of manufacturing today is batch production. Questions have been raised, for example, about why a particular part should have so many different sizes of fasteners.

The traditional product flow in a batch manufacturing operation is shown in Fig. 14.7a. Note that machines of the same type are arranged in groups, that is, groups of lathes, of

**FIGURE 14.7** (a) Functional layout of maching tools in a traditional plant. Arrows indocate the flow of materials and parts in various stages of completion. (b) Group-technology (cellular) layout. Legend: L = lathe, M = milling maching, D = drilling, G = grinding, A = assembly. *Source:* M. P. Groover.

milling machines, of drill presses, and of grinders. In such a layout (**functional layout**), there is usually considerable random movement, as shown by the arrows that indicate movement of materials and parts. Such an arrangement is not efficient because it wastes time and effort. The machines in *cellular manufacturing* are arranged in a more efficient product flow line (**group layout**; Fig. 14.7b). This is described in further detail in Section 14.9.

### 14.8.1  Advantages of group technology

The advantages of group technology include standardization of part design and minimization of design duplication. New part designs can be developed using similar previous designs, saving a significant amount of time and effort. The product designer can quickly determine from the computer files whether a similar part already exists.

Data reflecting the experience of the designer and manufacturing process planner are stored in the database. A new and less experienced engineer can quickly benefit from that experience by retrieving any of the previous designs and process plans.

Costs can be estimated more easily, and relevant statistics on materials, processes, number of parts produced, and other factors can be more easily obtained. Process plans are standardized and scheduled more efficiently, orders are grouped for more efficient production, and machine utilization is improved. Setup times are reduced, and parts are produced more efficiently with better and more consistent product quality. Similar tools, fixtures, and machinery are shared in the production of a family of parts. Programming for NC is more automatic.

With the implementation of CAD/CAM, cellular manufacturing, and CIM, group technology is capable of improving productivity and reducing costs in batch production that approach those of mass production. Potential savings in each of the various design and manufacturing phases can range from 5 to 75 percent.

### 14.8.2  Classification and coding of parts

Parts are identified and grouped into families by classification and coding (C/C) systems. Classification and coding of parts is a complex and critical first step in GT. It may be done according to the part's design attributes or manufacturing attributes (see Fig. 14.6).

**Design attributes** pertain to similarities in geometric features and consist of the following:

- External and internal shapes and dimensions
- Aspect ratio (length-to-width or length-to-diameter)
- Tolerances specified
- Surface finish specified
- Part function

**Manufacturing attributes** pertain to similarities in the methods and sequence of manufacturing operations performed on the part. As we have seen, selection of a manufacturing process or processes depends on many factors, among which are the shape, dimensions, and other geometric features of the part. Consequently, manufacturing and design attributes are interrelated. Manufacturing attributes consist of the following:

- Primary processes
- Secondary and finishing processes
- Tolerances and surface finish

- Sequence of operations performed
- Tools, dies, fixtures, and machinery
- Production quantity and production rate

Parts may also be classified by studying the production flow of parts during their manufacturing cycle (called **production flow analysis,** or **PFA**). Recall from Section 14.6 that routing sheets clearly show process plans and operations to be performed. One drawback to PFA is that a particular routing sheet does not necessarily indicate that the total operation is optimized. In fact, depending on the experience of the process planner, routing sheets for manufacturing the same part can be quite different. The benefits of computer-aided process planning in avoiding such problems have already been noted.

### 14.8.3  Coding systems

Coding can be based on a company's own system or one of several classification and coding systems that are available commercially. Because of widely varying product lines and organizational needs, none of the C/C systems has been universally adopted. Whether the system was developed in-house or was purchased, it should be compatible with the company's other systems, such as NC machinery and CAPP systems.

The code structure for part families consists of numbers, letters, or a combination of both. Each specific part or component of a product is assigned a code, which may pertain to design attributes only (generally less than 12 digits) or manufacturing attributes only, although most advanced systems include both, using as many as 30 digits.

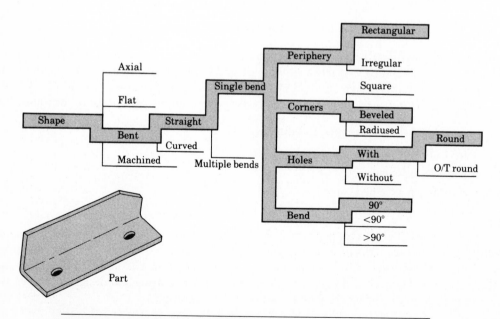

**FIGURE 14.8**  Decision-tree classification for a sheet-metal bracket. *Source:* G. W. Millar, *Manufacturing Engineering,* November 1985.

The three basic levels of coding vary in degree of complexity. They are described below.

**Hierarchical coding.**   In this code, also called *monocode,* the interpretation of each succeeding digit depends on the value of the preceding digit. Each symbol amplifies the information contained in the preceding digit, so a digit in the code cannot be interpreted alone. The advantage of this system is that a short code can contain a large amount of information. However, this method is difficult to apply in a computerized system.

**Polycoding.**   Each digit in this code, also known as *chain type,* has its own interpretation, which does not depend on the preceding digit. This structure tends to be relatively long, but allows identification of specific part attributes and is well suited to computer implementation.

**Decision-tree coding.**   This system (Fig. 14.8), also called *hybrid code,* is the most advanced and combines both design and manufacturing attributes.

Three major industrial coding systems are the Opitz system, the MultiClass system, and the KK-3 system. The **Opitz system,** developed in the 1960s in Germany by H. Opitz (1905–1977), was the first comprehensive coding system presented. The basic code consists of nine digits (12345 6789) representing design and manufacturing data (Fig. 14.9). Four

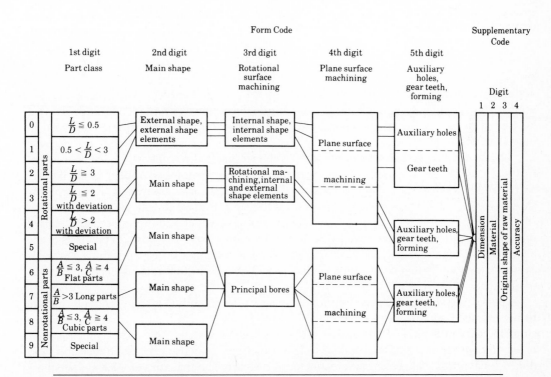

**FIGURE 14.9**  Classification and coding system according to Opitz, consisting of 5 digits and a supplementary code of 4 digits.

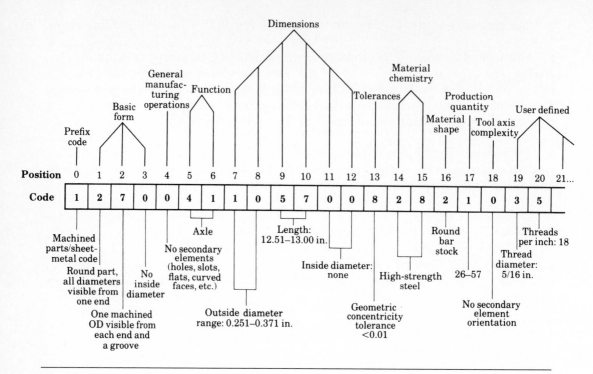

**FIGURE 14.10**   Typical MultiClass code for a machined part. *Source:* Organization for Industrial Research.

additional codes (ABCD) may be used to identify the type and sequence of production operations. This system has two drawbacks: (1) it is possible to have different codes for parts that have similar manufacturing attributes, and (2) a number of parts with different shapes can have the same code.

The **MultiClass system** was originally developed under the name MICLASS (for Metal Institute Classification System) by the Netherlands Organization for Applied Scientific Research and marketed in the United States by The Organization for Industrial Research; it is shown in Fig. 14.10. This system was developed to help automate and standardize several design, production, and management functions. MultiClass, which uses up to 30 digits, is used interactively with a computer, which asks the user a number of questions. On the basis of the answers, the computer automatically assigns a code number to the part. The software is available in modules that can be linked, costing $50,000–$500,000 each, depending on their capabilities.

The **KK-3 system** is a general-purpose classification and coding system for parts that are to be machined and/or ground. Developed by the Japan Society for the Promotion of Machine Industry in the late 1970s, it uses a 21-digit decimal system. This code is much greater in length than the two previous coding systems described, and also classifies dimensions and dimensional ratios such as length to diameter of the part. The structure of a KK-3 system for rotational components is shown in Fig. 14.11.

| Digit | Items | | (Rotational component) |
|---|---|---|---|
| 1 | Parts name | | General classification |
| 2 | | | Detail classsification |
| 3 | Materials | | General classification |
| 4 | | | Detail classification |
| 5 | Major dimensions | | Length |
| 6 | | | Diameter |
| 7 | | | Primary shapes and ratio of major dimensions |
| 8 | Shape details and kinds of processes | External surface | External surface and outer primary shape |
| 9 | | | Concentric screw threaded parts |
| 10 | | | Functional cut-off parts |
| 11 | | | Extraordinary shaped parts |
| 12 | | | Forming |
| 13 | | | Cylindrical surface |
| 14 | | Internal surface | Internal primary shape |
| 15 | | | Internal curved surface |
| 16 | | | Internal flat surface and cylindrical surface |
| 17 | | End surface | |
| 18 | | Nonconcentric holes | Regularly located holes |
| 19 | | | Special holes |
| 20 | | Noncutting process | |
| 21 | Accuracy | | |

**FIGURE 14.11** The structure of a KK-3 system for rotational components. *Source:* Japan Society for the Promotion of Machine Industry.

## 14.9 Cellular Manufacturing

The concept of group technology can be implemented effectively in **cellular manufacturing,** consisting of one or more manufacturing cells. A **manufacturing cell** is a small unit of one to several workstations within a manufacturing system. The workstations usually contain one (*single machine cell*) or more machines (*group machine cell*), each performing a different operation on the part. The machines can be modified, retooled, and regrouped for different product lines within the same family of parts. Manufacturing cells are particularly effective in producing families of parts that have relatively constant demand.

Cellular manufacturing thus far has been utilized primarily in machining and sheet-metal forming operations. The machine tools commonly used in manufacturing cells are lathes, machining centers, milling machines, drills, and grinders. For sheet forming, equipment consists of shearing, punching, bending, and other forming machines. This equipment may be special-purpose machines or CNC machines.

Cellular manufacturing has some degree of automatic control for:

- Loading and unloading raw materials and workpieces at workstations.
- Changing tools at workstations.
- Transferring workpieces and tools between workstations.
- Scheduling and controlling the total operation in the cell.

Central to these activities is a material-handling system to transfer materials and parts among workstations. In attended (manned) machining cells, materials can be moved and transferred manually by the operator, unless the parts are too heavy or the movements are hazardous, or by an industrial robot located centrally in the cell. Automated inspection and testing equipment can also be part of the cell.

The important feature of cellular manufacturing is the economics of reduced work in progress and product quality problems that are detected right away, thus improving productivity. Furthermore, because of the variety of machines and processes involved, the operator becomes multifunctional and is not subjected to the tedium involved with working on the same machine all the time. The result is increased productivity, an important benefit of attended cells.

### 14.9.1  Manufacturing cell design

Because of the unique features of manufacturing cells, their design and implementation in traditional plants require reorganization of the plant and rearrangement of existing product flow lines. The machines may be arranged along a line, in a U-shape, in an L-shape, or in a loop. For a group-machine cell in which materials are handled by the operator, the U-shaped arrangement is convenient and efficient because the operator can easily reach the various machines. For mechanized material handling, the linear arrangement and loop layout are more efficient. Selecting the best machine and material-handling equipment arrangement also depends on factors such as production rate and type of product and its shape, size, and weight.

### 14.9.2  Flexible manufacturing cells

In the introduction to this chapter and in Chapter 13, it was stressed that, in view of rapid changes in market demand and the need for a variety of products in smaller quantities, flexibility of manufacturing operations is highly desirable. Manufacturing cells can be made flexible by using CNC machines and machining centers, with industrial robots or other mechanized systems for handling materials. An example of **flexible manufacturing cells** (**FMC**) for machining operations is shown in Fig. 14.12.

**FIGURE 14.12**
Schematic view of a flexible manufacturing cell, showing two maching tools, an automated part inspection system, and a central robot serving these machines. *Source:* P. K. Wright and D.A. Bourne, *Manufacturing Intelligence* (Fig. 2.4, p. 29), © 1988, Addison-Wesley Publishing Company, Inc. Reprinted by permission of Addison Wesley Longman, Inc.

Flexible manufacturing cells are usually untended, and their design and operation are more exacting than for other cells. The selection of machines and robots, including the types and capacities of end effectors and their control systems, is important to the proper functioning of the FMC. Significant changes in demand for part families should be considered during design to ensure that the equipment involved has the proper flexibility and capacity.

As in other flexible manufacturing systems (described in Section 14.10), the cost of flexible cells is high. Cellular manufacturing generally requires more machine tools, thus increasing the cost and decreasing utilization; however, this is balanced by increased speed of manufacture, flexibility, and controllability. Proper maintenance of tools and machinery is essential, as are two or three shift operations of the cells for economic reasons.

## 14.10 Flexible Manufacturing Systems

A **flexible manufacturing system (FMS)** integrates all the major elements of manufacturing that have been described into a highly automated system. First utilized in the late 1960s, FMS consists of a number of manufacturing cells, each containing an industrial robot serving several CNC machines and an automated material-handling system, all interfaced with a central computer. Different computer instructions for the manufacturing process can be downloaded for each successive part passing through the workstation. This system is highly automated and is capable of optimizing each step of the total manufacturing operation.

Flexible manufacturing systems represent the highest level of efficiency, sophistication, and productivity that has been achieved in manufacturing plants (see Fig. 1.9 in Chapter 1). The flexibility of FMS is such that it can handle a *variety of part configurations* and produce them *in any order*. Hence FMS can be regarded as a system that combines the benefits of two systems: (1) the highly productive but inflexible transfer lines (see Section 13.2.2); and (2) job-shop production, which can produce large product variety on stand-alone machines but is inefficient. Table 14.1 shows the relative characteristics of transfer lines and FMS. Note

**TABLE 14.1**

**COMPARISON OF THE CHARACTERISTICS OF TRANSFER LINES AND FLEXIBLE MANUFACTURING SYSTEMS**

| CHARACTERISTIC | TRANSFER LINE | FMS |
|---|---|---|
| Types of parts made | Generally few | Infinite |
| Lot size | > 100 | 1–50 |
| Part changing time | 1/2 to 8 hr | 1 min |
| Tool change | Manual | Automatic |
| Adaptive control | Difficult | Available |
| Inventory | High | Low |
| Production during breakdown | None | Partial |
| Efficiency | 60–70% | 85% |
| Justification for capital expenditure | Simple | Difficult |

that in FMS the time required for changeover to a different part is very short. The quick response to product and market-demand variations is a major attribute of FMS. A variety of FMS technologies are available from machine-tool manufacturers.

Compared to conventional systems, the benefits of FMS are:

- Parts can be produced randomly and in batch sizes as small as one and at lower unit cost.
- Direct labor and inventories are reduced, with major savings over conventional systems.
- Machine utilization is as high as 90 percent, thus improving productivity.
- Shorter lead times are needed for product changes.
- Production is more reliable because the system is self-correcting and product quality is uniform.
- Work-in-progress inventories are reduced.

### 14.10.1   Elements of FMS

The basic elements of a flexible manufacturing system are workstations, automated handling and transport of materials and parts, and control systems. The types of machines in workstations depend on the type of production. For machining operations, they usually consist of a variety of three- to five-axis machining centers, CNC lathes, milling machines, drill presses, and grinders. Also included are various other pieces of equipment for automated inspection (including coordinate-measuring machines), assembly, and cleaning. Other types of operations suitable for FMS include sheet-metal forming, punching and shearing, and forging, which include heating furnaces, forging machines, trimming presses, heat-treating facilities, and cleaning equipment. The workstations in FMS are arranged to yield the greatest efficiency in production, with an orderly flow of materials, parts, and products through the system.

Because of FMS flexibility, the material-handling, storage, and retrieval systems are very important. Material handling is controlled by a central computer and performed by automated guided vehicles, conveyors, and various transfer mechanisms. The system is capable of transporting raw materials, blanks, and parts in various stages of completion to any machine, in random order, at any time. Prismatic parts are usually moved in specially designed *pallets*. Parts with rotational symmetry, such as those for turning operations, are usually moved by mechanical devices and robots.

### 14.10.2   Scheduling

Because FMS involves a major capital investment, efficient machine utilization is essential; that is, machines should not stand idle. Consequently, proper scheduling and process planning are crucial. Scheduling for FMS is *dynamic,* unlike that in job shops where a relatively rigid schedule is followed to perform a set of operations during a certain period of time. The scheduling system for FMS clearly specifies the types of operations to be performed on each part and identifies the machines or manufacturing cells to be used. Dynamic scheduling is capable of responding to quick changes in product type and hence is responsive to real-time decisions. Because of the flexibility in FMS, no setup time is wasted in switching manufacturing operations since the system is capable of performing different operations in different orders on different machines. However, the characteristics, performance, and reliability of each unit in the system must be checked to ensure that parts moving from workstation to workstation are of acceptable quality and dimensional accuracy.

### 14.10.3 Economic justification of FMS

FMS installations are very capital intensive, typically starting at well over $1 million. Consequently, a thorough cost/benefit analysis must be conducted before a final decision is made. This analysis should include factors such as costs of capital, energy, materials, and labor, expected markets for the products to be manufactured, and anticipated fluctuations in market demand and product type. An additional factor is the time and effort required for installing and debugging the system. Typically, an FMS system can take two to five years to install and at least six months to debug. Although FMS requires few, if any, machine operators, the personnel involved with the total operation must be trained and highly skilled. These personnel include manufacturing engineers, computer programmers, and maintenance engineers.

The most effective FMS applications have been in medium-volume batch production. High-volume, low-variety parts production is best obtained from transfer machines (dedicated equipment). Low-volume, high-variety parts production can best be done on standard machinery, with or without numerical control, or by machining centers. When a variety of parts is to be produced, FMS is suitable for production volumes of 15,000–35,000 aggregate parts per year. For individual parts of the same configuration, production may reach 100,000 units per year.

## 14.11 Just-In-Time Production

The **just-in-time (JIT) production** concept, although it originated in the United States decades ago, was first implemented in Japan to eliminate waste of materials, machines, capital, manpower, and inventory throughout the manufacturing system. The JIT concept has the following goals:

* Receive supplies just in time to be used.
* Produce parts just in time to be made into subassemblies.
* Produce subassemblies just in time to be assembled into finished products.
* Produce and deliver finished products just in time to be sold.

In traditional manufacturing, parts are made in batches, placed in inventory, and used whenever necessary. This approach is known as a **push system,** meaning that parts are made according to a schedule and are in inventory to be used if and when they are needed. Just-in-time is a **pull system,** meaning that parts are produced to order, and production is matched with demand for final assembly of products. There are no stockpiles, with the ideal production quantity being one (**zero inventory, stockless production, demand scheduling**). Moreover, parts are inspected by the worker as they are manufactured and are used within a short period of time. In this way, the worker maintains continuous production control, immediately identifying defective parts and reducing process variation to produce quality products. The worker takes pride in product quality. Unnecessary motions in stockpiling and retrieving parts from storage are eliminated.

The JIT concept includes all-important timely delivery of supplies and parts from outside sources as well as from other sections or divisions of a company, thus significantly reducing in-plant inventory. As a result, major reductions in storage facilities have already taken place and storage space has been reclaimed for other purposes; in fact, building large warehouses for parts is now a thing of the past. Suppliers are expected to deliver pre-inspected goods as they are needed for production, often on a daily basis. This requires reliable suppliers, close cooperation and trust between the company and its vendors, and a reliable system of transportation. Also important is the reduction of the number of suppliers for smoother operation.

## Kanban

The basic concept of JIT was first demonstrated on a large scale in 1953 at the Toyota Motor Company, under the name *kanban,* meaning visible record. These records usually consist of two types of cards (kanbans). One is the *production card,* which authorizes the production of one container or cart of identical, specified parts at a workstation. The other card is the *conveyance* or *move card,* which authorizes the transfer of one container or cart of parts from that workstation to the workstation where the parts will be used. The cards contain information on the type of part, place of issue, part number, and the number of items in the container or cart. Recently, these cards have been replaced by bar-coded plastic tags and other devices. The number of containers in circulation at any time is completely controlled and can be scheduled as desired for maximum efficiency.

## Advantages of JIT

The advantages of JIT are:

- Low inventory carrying costs
- Fast detection of defects in production or in delivery of supplies and hence low scrap loss
- Reduced inspection and rework of parts
- High-quality parts at low cost

Although there can be significant variations, implementation of just-in-time production has resulted in reductions of 20 to 40 percent in product cost; 60 to 80 percent in inventory; up to 90 percent in rejection rates; 90 percent in lead times; 50 percent in scrap, rework, and warranty costs; and increases of 30 to 50 percent in direct labor productivity and 60 percent in indirect labor productivity.

---

**Example 14.2:   An application of JIT in the U.S. automotive industry**

In one application, automotive seats are made in the supplier's plant just two hours before they are needed at the assembly plant, which is 75 miles away. The seats are unloaded at the assembly plant and quickly transferred in proper sequence of style and color, each seat marked for a car moving down an assembly line. Each seat arrives on the assembly line just in time for installation. Because the system works efficiently, there have been trends by the big three U.S. automakers to get suppliers of various components to relocate plants closer to assembly lines. However, this is a costly and capital-intensive undertaking on the part of suppliers. An alternative is to have a dependable delivery system of transportation by trucks and rail for reliable and timely delivery of supplies.

## 14.12 Manufacturing Communications Networks

Thus far we have described the use of computers in all phases of product planning, manufacture, assembly, and inspection. We have seen that, because of its many benefits, total integration of diverse manufacturing activities is a major goal of modern manufacturing. In order to maintain a high level of coordination and efficiency of operation in integrated manufacturing, an extensive, high-speed, and interactive **communications network** is required.

A significant advance in communications technology is the **local-area network (LAN)**. In this hardware and software system, logically related groups of machines and equipment communicate with each other. Recall the importance and efficiency of grouping related machines and activities by such means as group technology and manufacturing cells using robots and various other material-handling equipment. A LAN links these groups to each other, bringing different phases of manufacturing into a unified operation. A local-area network can be quite large and complex, linking hundreds or even thousands of machines and devices in several buildings. Various network layouts (Fig. 14.13) of copper cables or fiber optics are used over distances ranging from a few meters to as much as 30 km (20 mi). For larger distances, **wide-area networks (WAN)** are used.

**FIGURE 14.13** Three basic types of topology for a local-area network (LAN). (a) The *star* topology is suitable for situations that are not subject to frequent configuration changes. All messages pass through a central station. Telephone systems in office buildings usually have this type of topology. (b) In the *ring* topology, all individual user stations are connected in a continuous ring. The message is forwarded from one station to the next until it reaches its assigned destination. Although the wiring is relatively simple, the failure of one station shuts down the entire network. (c) In the *bus* topology, all stations have independent access to the bus. This system is reliable and is easier to service. Because its arrangement is similar to the layout of machines in a factory, its installation is relatively easy and can be rearranged when machines are rearranged.

Different types of networks can be linked or integrated through "gateways" and "bridges." Access control to the network is important; otherwise, collisions can occur when several workstations transmit simultaneously. Thus, continuous scanning of the transmitting medium is essential. In the 1970s, a carrier sense multiple-access network with collision detection (CSMA/CD) was developed and implemented in Ethernet. Used by a majority of workstations and minicomputers, Ethernet has now become the industry standard. Other access control methods are *token ring* and *token bus,* in which a token is passed from device to device, and the device that has the token is allowed to transmit while other devices receive only.

### Communications standard

Typically, a manufacturing cell is built with machines and equipment purchased from one vendor, another cell with machines from another vendor, and yet another from a third vendor. Thus a variety of programmable devices are involved, driven by several computers and microprocessors purchased at various times from different vendors and having various capacities and levels of sophistication. Each cell's computers have their own specifications and proprietary standards and cannot communicate beyond the cell with others unless equipped with custom-built interfaces. This situation has created "islands of automation," and in some cases up to 50 percent of the cost of automation has been related to difficulties in communication between manufacturing cells and other parts of the organization.

The existence of automated cells that could only function independently from each other—and without a common base for information transfer—led to the need for a **communications standard** to improve communications and the efficiency of computer-integrated manufacturing. The first step toward standardization began in 1980. After considerable effort, and on the basis of existing national and international standards, a set of communications standards known as *manufacturing automation protocol (MAP)* was developed. Its capabilities and effectiveness were demonstrated in 1984 by the successful interconnection of devices from a number of vendors. As a result, the importance of a worldwide communications standard is now recognized and vendors are designing their products in compliance with these standards.

The International Organization for Standardization (ISO)/Open System Interconnect (OSI) reference model has now been accepted worldwide. The ISO/OSI model has a hierarchical structure in which communication between two users is divided into seven layers (Fig. 14.14). Each layer has a special task, such as mechanical and electronic means of data transmission, error detection and correction, correct message transmission, controlling the dialog between users, translating messages into a common syntax, and ensuring that the data transferred is understood. The operation of this system is complex. Basically, each message or data from user A is transmitted to user B sequentially through successive layers. Additional messages are added to the original message as it travels from layer 7 to layer 1. The complete message (packet) is then transmitted through the communications medium to user B, through layer 1 to layer 7. The transmission takes place through coaxial cable, fiber-optic cable, microwave, and similar devices.

Communication protocols are now being extended to office automation as well, with the development of the **technical and office protocol (TOP)** based on the ISO/OSI reference model. In this way, total communication (MAP/TOP) is being established between the factory floor and offices at all levels of an organization.

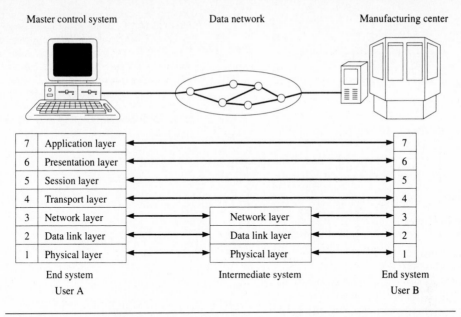

**FIGURE 14.14** The ISO/OSI reference model for open communication. *Source:* U. Rembold et al., *Computer Integrated Manufacturing and Engineering.* Addison-Wesley, 1993.

# 14.13 Artificial Intelligence

**Artificial intelligence (AI)** is a part of computer science concerned with systems that exhibit the characteristics usually associated with intelligence in human behavior, such as learning, reasoning, problem solving, understanding language, and so on. The goal of AI is to simulate human behavior on the computer. The art of bringing relevant principles and tools of AI to difficult application problems is known as **knowledge engineering.**

Artificial intelligence is likely to have a profound effect on design, automation, and the overall economics of manufacturing operations, due in large part to the advances in memory expansion—that is, VLSI chip design—and decreasing costs. Artificial intelligence packages costing on the order of a few thousand dollars have been developed, many of which can be run on personal computers. Thus, AI has become accessible to office desks and shop floors.

### Elements of artificial intelligence

In general, artificial-intelligence applications in manufacturing encompass the following activities: expert systems, natural language, and machine (computer) vision. Each of these elements is described below.

**Expert systems.** An **expert system (ES),** also called a *knowledge-based system,* can be defined as an intelligent computer program that has the capability to solve difficult real-life

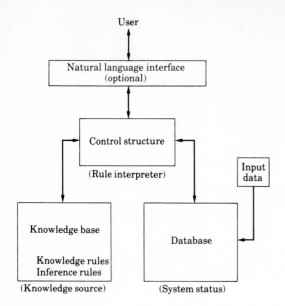

**FIGURE 14.15** Basic structure of an expert system. The knowledge base consists of knowledge rules (general information about the problem) and the inference rules (the way conclusions are reached). The results may be communicated to the user through the natural language interface. *Source:* K. W. Goff, *Mechanical Engineering,* October 1985.

problems using a *knowledge base* and *inference* procedures (Fig. 14.15). The goal of an expert system is to perform an intellectually demanding task as well as human experts would. The knowledge required to perform this task is called the *domain* of the expert system. Expert systems utilize a knowledge base containing facts, data, definitions, and assumptions. They also have the capacity for a *heuristic* approach—that is, making good judgments by discovery and revelation, and making good guesses, just as an expert would. The knowledge base is expressed in computer codes, usually in the form of *if-then* rules, with a series of questions. The mechanism by which to use these rules to solve problems is called an *inference engine.* Expert systems can also communicate with other computer-software packages.

To construct expert systems for solving the complex design and manufacturing problems encountered, one needs both a large amount of knowledge and some mechanisms for manipulating that knowledge to create solutions. Because of the difficulty involved in accurately modeling the many years of experience of an expert or a team of experts, and the complex inductive reasoning and decision-making capabilities of humans (including the capacity to learn from mistakes), developing knowledge-based systems requires considerable time and effort.

Expert systems work on a real-time basis, and their short reaction times provide rapid responses to problems. The most common programming languages are C++, LISP, and PRO-LOG; other languages can also be used. A more recent development is expert-system software **shells** or *environments,* also called *framework systems.* These software packages are essentially expert-system outlines that allow a person to write specific applications to suit special needs. Writing these programs requires considerable experience and time.

Several expert systems have been developed and used since the early 1970s, utilizing computers with various capacities, for specialized applications such as:

- Problem diagnosis in various types of machines and equipment and determination of corrective actions
- Missile guidance systems

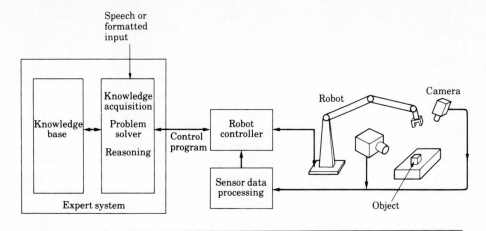

**FIGURE 14.16** Expert system as applied to an industrial robot guided by machine vision.

- Modeling and simulation of production facilities
- Computer-aided design, process planning, and production scheduling
- Management of a company's manufacturing strategy

**Natural-language procession.**   Traditionally, obtaining information from a database in computer memory has required utilization of computer programs that translate natural language to machine language. Natural-language interfaces with database systems are now in various stages of development. These systems allow a user to obtain information by entering English-language commands in the form of simple, typed queries.

Software shells are available and are used in applications such as scheduling material flow in manufacturing and analyzing information in databases. Significant progress is being made on computers that will have speech synthesis and recognition (*voice recognition*) capabilities, thus eliminating the need to type commands on keyboards.

**Machine vision.**   The basic features of machine vision are described in Section 13.8. Computers and software for artificial intelligence can be combined with cameras and other optical sensors. These machines can then perform operations such as inspecting, identifying, and sorting parts and guiding robots (intelligent robots; Fig. 14.16), which would otherwise require human intervention.

### 14.13.1  Artificial neural networks

Although computers are much faster than the human brain, humans are much better at recognizing features (faces, voices, etc.), even under noisy conditions, assessing situations quickly, and adjusting to new and dynamic conditions. These advantages are due to the ability of humans to use several senses in parallel (data fusion) and in real time. The artificial intelligence branch of **artificial neural networks (ANN)** attempts to gain some of these capabilities through computer imitation of the way data is processed by the human brain.

The human brain has about 100 billion linked neurons (cells that are the fundamental functional units of nervous tissue) and more than a thousand times that many connections. Each neuron performs only a simple task; it receives a set of input signals. When the combined input is above a certain critical level, the neuron transfers its own electrochemical signal to adjacent neurons at the output. It is now believed that human learning is accomplished by changes in the strength of the connections among neurons.

A fully developed feed-forward network is the most common type of ANN, and is built according to this principle from several layers of processing elements (neurons). The elements in the first (input) layer are fed with input data, such as forces, velocities, voltages, etc. Each element sums all the inputs, with one input per element in the input layer but many inputs per element in the next layer. Each element transfers the data, according to a transfer function, to all the elements in the next layer. However, each element receives a different signal due to different connection weights between the elements.

The last layer is the output layer. The output of each element in this layer is compared to the desired or actual output. The difference between the desired output and the calculated one (the error) is fed back to the network so that the weights of the connections are adjusted in a way that minimizes this error. After repeating this procedure several times, the network is "trained" and can be used with input data not shown previously to the system. Other kinds of ANNs are: associative memories, self-organizing ANN, and adaptive-resonance ANN. The common feature of such neural networks is that they must be trained with concrete exemplars. It is, therefore, very difficult mathematically to formulate input-output relations and to predict ANN behavior with untrained inputs.

Artificial neural networks are being used today in applications such as noise reduction in telephones, speech recognition, and process control. They have been used for predicting the surface finish of a workpiece obtained by end milling with input parameters such as forces, acoustic emission, and spindle acceleration. Although still very controversial, it is believed by many that true artificial intelligence will evolve only through advances in ANN.

### 14.13.2  Fuzzy logic

Another element of artificial intelligence that is gaining in popularity and applications in control systems and pattern recognition is **fuzzy logic** or fuzzy models. Introduced in 1965 and based on the observation that people make decisions based on imprecise and nonnumerical information, fuzzy models or sets are mathematical means of representing vagueness and imprecise information, hence the term fuzzy. These models have the capability of recognizing, representing, manipulating, interpreting, and utilizing data and information that are vague and lack certainty. Linguistic examples are: a few, almost all, more or less, very important, good, poor, acceptable, etc. These methods deal with reasoning and decision making at a higher level than do neural networks.

Fuzzy technologies and devices have been developed and applied successfully in areas such as robotics and motion control, image processing and machine vision, machine learning, evaluation of design alternatives, decision making, and the design of intelligent systems. Examples are (1) automatic transmission, engine and cruise control, and anti-lock braking systems of automobiles; (2) a washing machine that automatically adjusts the washing cycle for load size, fabric type, and amount of dirt; (3) a voice-controlled helicopter that obeys commands such as forward, up, left, right, hover, and land; (4) decision making concerning the type of computer equipment to be acquired for education and research in a college department; (5) material selection involving multicriteria such as high hardness, a minimum machinability rating, good corrosion resistance, and low cost for a particular engineering application; (6) a vacuum cleaner with a motor-suction strategy based on dust quantity and floor type; and (7) a copy machine in which the drum voltage is adjusted based on picture density, temperature, and humidity.

## 14.14  Factory of the Future

On the basis of the advances made to date in all aspects of manufacturing technology and computer integration, we may envisage the **factory of the future** as a fully automated facility in which human beings would not be directly involved in production on the shop floor (hence the term untended factories). All manufacturing, material handling, assembly, and inspection would be done by computer-controlled machinery and equipment.

Similarly, activities such as processing incoming orders, production planning and scheduling, cost accounting, and various decision-making processes (usually performed by management) would also be done automatically by computers. The role of human beings would be confined to activities such as supervising, maintaining (especially preventive maintenance), and upgrading machines and equipment; shipping and receiving supplies and finished products; providing security for the plant facilities; programming, upgrading, and monitoring computer programs; and monitoring, maintaining, and upgrading hardware.

Virtually untended manufacturing cells already make products such as engine blocks, axles, and housings for clutches and air compressors. For large-scale, flexible manufacturing systems, however, highly trained and skilled personnel will always be needed to plan, maintain, and oversee operations.

The reliability of machines, control systems, and power supplies is crucial to full factory automation. A local or general breakdown in machinery, computers, power, or communications networks will, without rapid human intervention, cripple production. The computer-integrated factory of the future should be capable of automatically rerouting materials and production flows to other machines and to the control of other computers in case of such emergencies.

An important consideration is the nature and extent of the fully automated factory's impact on employment, with all its social and political ramifications. Although projections indicate that there will be a decline in the number of machine tool operators and tool and die workers, there will be major increases in service industries such as computer service technicians and maintenance electricians. Thus, the generally low skilled, direct labor force engaged in traditional manufacturing will shift to an indirect labor force, with special training or retraining required in areas such as computer programming, information processing, CAD/CAM, and other high-technology tasks that are essential parts of CIM. Development of more user-friendly computer software is making retraining of the work force much easier.

Opinions about the impact of untended factories diverge widely. Consequently, predicting the nature of future manufacturing strategies with any certainty is difficult. Although economic considerations and trade-offs are crucial, companies that do not install computer-integrated operations are in jeopardy. It is now recognized that, in a highly competitive global marketplace, rapid adaptability is essential to the survival of a manufacturing organization. The term **world class** is being used more and more frequently to describe the quality and level of manufacturing activities, signifying that products made must meet high international standards and be acceptable worldwide. It has also been pointed out that world class is not so much a fixed target for a manufacturing company or a country to reach, but that it is a *moving* target and that it acquires a higher and higher level as time passes.

## SUMMARY

- The computerized integration of manufacturing in all its system aspects, including design, planning, production, distribution, and management, falls under the broad descriptive term CIM, or computer-integrated manufacturing. (Sections 14.1, 14.2, 14.3)

- CAD (computer-aided design) involves the use of computers to create design drawings and product models. CAE (computer-aided engineering) refers to a system of automated analytical tools that share a common database and that can evaluate such parameters as stresses, volumes, heat transfer, and tolerances. (Section 14.4)

- CAM (computer-aided manufacturing) is the use of computers and computer technology to assist in various phases of manufacturing a product, including process and production planning, machining, scheduling, management and quality control. CAM is often combined with computer-aided design techniques (CAD/CAM) to transfer information from the design stage to the planning stage without reentering part geometry into the computer system. (Section 14.5)

- Under the rubric of CIM, a great number of trends are contributing significantly to improved productivity, quality, and cost control, including:

  1. Computer-assisted process planning (CAPP) (Section 14.6)
  2. Computer simulation of processes and systems (Section 14.7)
  3. Group technology (GT) (Section 14.8)
  4. Cellular manufacturing (Section 14.9)
  5. Flexible manufacturing systems (FMS) (Section 14.10)
  6. Just-in-time manufacturing (JIT) (Section 14.11)

- Communications networks (local area and wide area) are required to implement CIM strategies. A worldwide communications standard has been developed so that regardless of which vendor produced a particular piece of equipment, the various pieces can exchange data without difficulty. (Section 14.12)

- Artificial intelligence is likely to open new opportunities and have a profound effect on the design of products, automation and overall economics of manufacturing operations. (Section 14.13)

- A fully automated factory of the future now seems quite possible, given the computer-aided and computer-integrated manufacturing techniques described in this chapter and Chapter 13. However, both partially automated and fully automated factories will have a significant impact on employment issues. (Section 14.14)

# BIBLIOGRAPHY

Amirouche, F.M.L., *Computer-Aided Design and Manufacturing*. Englewood Cliffs, New Jersey: Prentice-Hall, 1993.

Anderson, D.M., *Design for Manufacturability, Optimizing Cost, Quality, and Time-to-Market*. Lafayette, Calif.: CIM Press, 1991.

*Artificial Intelligence Handbook*. Vol. 1: *Principles;* Vol. 2: *Applications*. Research Triangle Park, N.C.: Instrument Society of America, 1989.

Askin, R.G., and C.R. Standridge, *Modeling and Analysis of Manufacturing Systems*. New York: Wiley, 1993.

Badiru, A.B., *Expert Systems Applications in Engineering and Manufacturing*. Englewood Cliffs, N.J.: Prentice-Hall, 1992.

Bedworth, D.D., M.R. Henderson, and P.M. Wolf, *Computer-Integrated Design and Manufacturing*. New York: McGraw-Hill, 1991.

Biekert, R., *CIM Technology: Fundamentals and Applications*. South Holland, Ill.: Goodheart-Wilcox, 1993.

Black, J.T., *The Design of the Factory With a Future*. New York: McGraw-Hill, 1991.

Boucher, T.O., *Introduction to Computer Control of Manufacturing Systems*. London: Chapman & Hall, 1995.

Chang, T.-C., *Expert Process Planning for Manufacturing*. Reading, Mass.: Addison-Wesley, 1990.

———, R.A. Wysk, and H.P. Wang, *Computer Aided Manufacturing*. Englewood Cliffs, N.J.: Prentice-Hall, 1991.

Chorafas, D., *Expert Systems in Manufacturing*. New York: Van Nostrand Reinhold, 1992.

Corbett, J., M. Dooner, J. Meleka, and C. Pym, *Design for Manufacture: Strategies, Principles and Techniques*. Reading, Mass.: Addison-Wesley, 1991.

Coticchia, M, et al., *CAD/CAM/CAE Systems: Justification, Implementation, Productivity Measurement*, 2d ed. New York: Marcel Dekker, 1993.

Foston, A.L., C.L. Smith, and T. Au, *Fundamentals of Computer-Integrated Manufacturing*. Englewood Cliffs, N.J.: Prentice-Hall, 1991.

Galbiati, L.J., *Machine Vision and Digital Image Processing Fundamentals*. Englewood Cliffs, N.J.: Prentice-Hall, 1990.

Goldratt, E.M., *Theory of Constraints*. Croton-on-Hudson, New York: North River Press, 1990.

Groover, M.P., *Automation, Production Systems, and Computer-Integrated Manufacturing*. Englewood Cliffs, N.J.: Prentice-Hall, 1987.

Hernandez, A., *Just-In-Time Manufacturing: A Practical Approach*. Englewood Cliffs, N.J.: Prentice-Hall, 1989.

Holt, D.R., *Integrating Manufacturing Engineering Systems*. New York: McGraw-Hill, 1992.

Kaewert, J.W., and J.M. Frost, *Developing Expert Systems for Manufacturing: A Case Study Approach*. New York: McGraw-Hill, 1990.

Koenig, D.T., *Manufacturing Engineering: Principles for Optimization*, 2d ed. Bristol, PA: Taylor & Francis, 1994.

Kusiak, A., *Intelligent Manufacturing Systems*. Englewood Cliffs, N.J.: Prentice-Hall, 1990.

———, *Intelligent Design and Manufacturing*. New York: Wiley, 1992.

———(ed.), *Concurrent Engineering: Automation, Tools, and Techniques*. New York: Wiley, 1993.

Luggen, W.W., *Flexible Manufacturing Cells and Systems*. Englewood Cliffs, N.J.: Prentice-Hall, 1991.

Maleki, R., *Flexible Manufacturing Systems: The Technology and Management*. Englewood Cliffs, N.J.: Prentice-Hall, 1991.

Maus, R., and J. Keyes (eds.), *Handbook of Expert Systems in Manufacturing*. New York: McGraw-Hill, 1991.

McMahon, C., and J. Browne, *CAD/CAM - From Principles to Practice,* Wokingham, England: Addison-Wesley, 1993.

Mitchell, F.H., Jr., *CIM Systems: An Introduction to Computer-Integrated Manufacturing.* Englewood Cliffs, N.J.: Prentice-Hall, 1991.

*NC/CIM Guidebook,* published annually by *Modern Machine Shop,* Cincinnati, Ohio.

Nee, A.Y.C., K. Whybrew, and A.S. Kumar, *Advanced Fixture Design for FMS.* New York: Springer, 1994.

O'Sullivan, D., *Manufacturing Systems Redesign: Creating the Integrated Manufacturing Environment.* Englewood Cliffs, N.J.: Prentice-Hall, 1995.

Pimentel, J.R., *Communications Networks for Manufacturing.* Englewood Cliffs, N.J.: Prentice-Hall, 1990.

Popovic, D., and V. Bhatkar, *Methods and Tools for Applied Artificial Intelligence.* New York: Marcel Dekker, 1994.

Prasad, B., *Concurrent Engineering Fundamentals: Integrated Product and Process Organization,* Vol. 1. Englewood Cliffs, N.J.: Prentice-Hall, 1996.

Ranky, P.G., *Computer-Integrated Manufacturing: An Introduction with Case Studies.* Englewood Cliffs, N.J.: Prentice-Hall, 1986.

———, *Computer-Integrated Manufacturing.* Englewood Cliffs, N.J.: Prentice-Hall, 1989.

Rehg, J.A., *Computer-Integrated Manufacturing.* Englewood Cliffs, N.J.: Prentice-Hall, 1994.

Rembold, U., B.O. Nnaji, and A. Storr, *Computer Integrated Manufacturing and Engineering.* Wokingham, England: Addison-Wesley, 1993.

Rodd, M.G., and D. Farzin, *Communication Systems for Industrial Automation.* Englewood Cliffs, N.J.: Prentice-Hall, 1990.

Scholz-Reiter, B., *CIM Interfaces: Concepts, Standards and Problems of Interfaces in Computer Integrated Manufacturing.* London: Chapman & Hall, 1992.

Shah, J.J., and M. Mantyla, *Parametric and Feature-Based CAD/CAM: Concepts, Techniques, and Applications.* New York: Wiley, 1995.

Shunta, J.P., *Achieving World-Class Manufacturing Through Process Control.* Englewood Cliffs, N.J.: Prentice-Hall, 1995.

Singh, N., and D. Rajamani, *Cellular Manufacturing Systems: Design, Planning and Control.* London: Chapman & Hall, 1995.

Soloman, S., *Sensors and Control Systems in Manufacturing.* New York: McGraw-Hill, 1994.

Talavage, J., and R.G. Hannam, *Flexible Manufacturing Systems in Practice: Applications, Design, and Simulation.* New York: Marcel Dekker, 1988.

Taylor, D., *Computer-Aided Design.* Reading, Mass.: Addison-Wesley, 1991.

*Tool and Manufacturing Engineers Handbook,* 4th ed., Vol. 5: *Manufacturing Management,* Dearborn, Mich.: Society of Manufacturing Engineers, 1988.

*Tool and Manufacturing Engineers Handbook,* 4th ed., Vol. 7: *Continuous Improvement.* Dearborn, Mich.: Society of Manufacturing Engineers, 1993.

Ulsoy, A.G., and W.R. DeVries, *Microcomputer Applications in Manufacturing.* New York: Wiley, 1988.

Vajpayee, S.K., *Principles of Computer-Integrated Manufacturing.* Englewood Cliffs, N.J.: Prentice-Hall, 1995.

Valliere, D., *Computer-Aided Design in Manufacturing.* Englewood Cliffs, N.J.: Prentice-Hall, 1990.

Wang, H.P., and J.K. Li, *Computer-Aided Process Planning.* Amsterdam: Elsevier, 1991.

Williams, D.J., *Manufacturing Systems,* 2d ed. New York: Chapman & Hall, 1994.

Wright, P.K., and D.A. Bourne, *Manufacturing Intelligence.* Reading, Mass.: Addison-Wesley, 1988.

Wu, J.-K., *Neural Networks and Simulation Methods.* New York: Marcel Dekker, 1994.

Zeid, I., *CAD/CAM Theory and Practice.* New York: McGraw-Hill, 1991.

Zuech, N., *Applying Machine Vision.* New York: Wiley, 1988.

# QUESTIONS

14.1 In what ways have computers had an impact on manufacturing science, engineering, and technology?

14.2 What advantages are there in viewing manufacturing as a system? What are the components of a manufacturing system?

14.3 Discuss the benefits and limitations of computer-integrated manufacturing operations.

14.4 Explain how a CAD system operates.

14.5 What are the advantages of CAD systems over traditional methods of design? Are there any limitations?

14.6 Describe the purposes of process planning. How are computers used in such planning?

14.7 Explain the features of two types of CAPP systems.

14.8 Describe the features of a route sheet. Why is it necessary?

14.9 What is group technology? Why was it developed? Explain its advantages.

14.10 Why were manufacturing cells developed?

14.11 Describe the principle of flexible manufacturing systems. Why do they require major capital investment?

14.12 Why is a flexible manufacturing system capable of producing a wide range of lot sizes?

14.13 What are the benefits of just-in-time production? Why is it called a pull system?

14.14 Elaborate on how the Kanban system is related to JIT.

14.15 What are the advantages of a communications standard?

14.16 Describe the elements of artificial intelligence. Why is machine vision a part of it?

14.17 Explain why humans will still be needed in the factory of the future.

14.18 How would you go about describing the principle of computer-aided manufacturing to an older worker in a manufacturing facility who is not familiar with computers?

14.19 Explain the logic behind the arrangements shown in Fig. 14.7b.

14.20 Describe your observations concerning the flexible manufacturing system shown in Fig. 1.9 in Chapter 1.

14.21 What should be the characteristics of the guidance system for automated guided vehicles?

14.22 Would machining centers be suitable for just-in-time-type production? Explain.

14.23 Is there a minimum to the number of machines in a manufacturing cell? Explain.

# PROBLEMS

14.24 Give examples of primitives of solids other than those shown in Fig. 14.3.

14.25 Review various manufactured parts discussed in this text and group them in a manner similar to those shown in Fig. 14.6.

14.26 Think of a product and make a decision-tree chart similar to that shown in Fig. 14.8.

14.27 Think of a commonly used product and design a manufacturing cell for making it, describing the features of the machines and equipment involved.

14.28 Give an example of (1) push and (2) pull systems to describe the fundamental difference between the two methods.

14.29 Give a specific example each where the (1) variant system and the (2) generative system of CAPP, respectively, are desirable.

14.30 This chapter stated that artificial neural networks are particularly useful where the problems are ill-defined and the data are fuzzy. Give examples in manufacturing where this is the case.

14.31 Give examples in manufacturing engineering in which artificial intelligence could be particularly effective.

14.32 Describe your opinions concerning voice recognition capabilities of future machines and controls.

14.33 Discuss your thoughts on how the machines, AGVs, and computer facilities in a flexible manufacturing system should be arranged (physical layout) in a manufacturing plant.

14.34 Surveys have indicated that 95 percent of all parts made in the United States are produced in lots of 50 or less. Comment on this observation and describe your thoughts regarding the implementation of the technologies outlined in Chapters 13 and 14.

14.35 Assume that you are asked to rewrite Section 14.14 on the factory of the future. Outline your thoughts regarding this topic and present them in a brief paper.

14.36 Assume that you own a manufacturing company and you are aware that you have not taken full advantage of the technological advances in manufacturing described in this chapter, but would like to do so and have the necessary capital. Describe how you would go about analyzing your company's needs and how you would plan to implement these advances.

# 15

# Competitive Aspects of Manufacturing

**15.1** Introduction
**15.2** Selection of Materials
**15.3** Product Design and Quantity of Materials
**15.4** Substitution of Materials
**15.5** Selection of Manufacturing Processes
**15.6** Process Capabilities
**15.7** Manufacturing Costs
**15.8** Value Engineering

## 15.1 Introduction

Manufacturing high-quality products at the lowest possible cost requires an understanding of the often complex relationships among many factors. We have seen that product design, selection of materials, and manufacturing processes are interrelated. Designs are often modified to improve product performance, take advantage of the characteristics of new materials, and make manufacturing, assembly, and disassembly easier, thus reducing overall costs.

Because of the wide variety of materials and processes available today, the task of producing a high-quality product by selecting the best materials and processes while minimizing costs has become a major challenge. Meeting this challenge requires not only a thorough knowledge of the characteristics of materials and processes, but also innovative and creative approaches to design and manufacturing technology. The cost of a product often determines its marketability and customer satisfaction. Consequently, each production step must be analyzed with a view toward ensuring an optimal economic outcome (Fig. 15.1).

Although the economics of various manufacturing processes have been emphasized throughout this text, this chapter takes a broader view and summarizes the important overall manufacturing cost factors. We will look at cost-reduction methods, particularly value engineering, which is a powerful tool that can be used to evaluate the cost of each manufacturing step with respect to its share of the value added to the product.

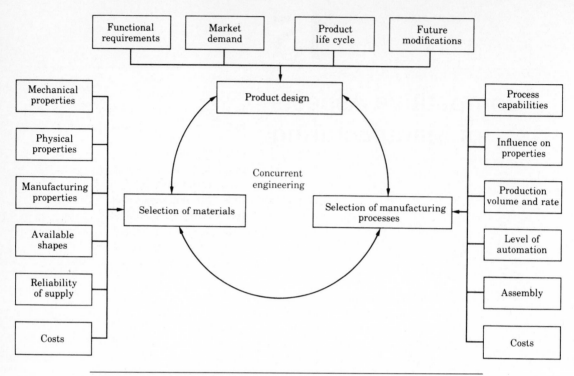

**FIGURE 15.1**    Interrelationships among various factors influencing product cost.

## 15.2    Selection of Materials

In selecting materials for a product, we must have a clear understanding of the functional requirements for each of its individual components or subassemblies. The general criteria for selecting materials were described in Chapter 1. This section will discuss further details of selecting materials.

### Mechanical, physical, and chemical properties

Mechanical properties are strength, stiffness, toughness, ductility, hardness, and resistance to fatigue, creep, and impact. Physical properties include density, melting point, specific heat, thermal and electrical conductivity, thermal expansion, and magnetic properties. Chemical properties with which we are primarily concerned are oxidation and corrosion resistance. These properties and their relevance to product design and manufacturing were described in previous chapters.

Selection of materials is now easier and faster with the availability of computer software programs, which are in a sense computerized handbooks, thus providing greater accessibility. However, to facilitate the selection of materials, as well as other parameters described below, expert-system software (*smart databases*) has been developed. With proper input of product design and functional requirements, these systems are capable of selecting appropriate materials for a particular application just as an expert or a team of experts would.

## Shapes of commercially available materials

When selecting materials, we need to know the shapes and sizes of materials that are commercially available to avoid additional processing, unless it is economically justified. Materials are available in various forms: cast, extruded, forged, bar, plate, sheet, foil, rod, tube, pipe, wire, or powder. Purchasing materials in shapes that generally require the least additional processing means that we also have to consider characteristics such as surface quality, tolerances, and straightness. The better and more consistent these characteristics are, the less additional processing is required.

For example, if you want to produce a simple shaft with good dimensional accuracy and surface finish, you should purchase round bar that is then turned and centerless ground to the dimensions needed. Unless your production facilities are capable of making round bars economically, it is cheaper to purchase them. On the other hand, if you have to make a stepped shaft with different diameters along its length, you should purchase a round bar (with a diameter equal to the largest diameter of the stepped shaft) and turn it on a lathe or process it by some other means to reduce the diameter. For stock that has wide tolerances or is warped or out of round, you must order larger sizes to ensure proper finished product size.

Each manufacturing step produces stock or parts that have their own shape, surface finish, and tolerance characteristics. Hot-rolled or hot-drawn products, for example, have coarser surface finishes and wider tolerances than cold-rolled or cold-drawn products. The wall thickness of welded tubing is more uniform than that of seamless tubing. Extrusions have smaller cross-sectional tolerances than those made by roll forming. Turned round bars have a coarser surface finish than ground bars. Information on these characteristics can generally be found in catalogs and various publications furnished by suppliers of these materials.

## Manufacturing properties

Manufacturing properties are typically castability, formability, machinability, weldability, and hardenability by heat treatment. Because materials have to be formed, shaped, machined, ground, fabricated, and heat treated into individual components with specific shapes and dimensions, these properties are crucial to proper selection of materials. Recall that the quality of a raw material can greatly influence its manufacturing properties. A blank cut from a rod or bar with a longitudinal seam will develop cracks during upsetting and heading operations. Bars with internal defects and inclusions will crack during seamless-tube production. Porous cast workpieces will produce a poor surface finish when machined. Blanks that are heat treated nonuniformly or bars that are not stress relieved will distort during subsequent operations.

Similarly, incoming stock that has variations in composition and microstructure cannot be heat treated or machined consistently. Sheet-metal stock with variations in its cold-worked conditions will spring back differently during bending and other forming operations. If prelubricated sheet-metal blanks are supplied with a lubricant with nonuniform thickness, their formability, surface finish, and overall quality may be adversely affected.

## Reliability of supply

Geopolitical factors can affect the supply of strategic materials. Other factors such as strikes, shortages, and reluctance of suppliers to produce materials of a particular shape, quality, or quantity also affect the reliability of supply. Even though availability of materials may not be

**TABLE 15.1**

**APPROXIMATE AMOUNT OF SCRAP PRODUCED IN VARIOUS MANUFACTURING PROCESSES**

| PROCESS | SCRAP (%) | PROCESS | SCRAP (%) |
|---|---|---|---|
| Machining | 10–60 | Cold or hot extrusion; forging | 15 |
| Hot closed-die forging | 20–25 | Permanent-mold casting | 10 |
| Sheet-metal forming | 10–25 | Powder metallurgy | 5 |

a problem for a country as a whole, it can be a problem for certain industries, such as the location of a manufacturing plant. The reliability of supply is particularly important in just-in-time manufacturing.

## Costs of materials and processing

Because of its processing history, the unit cost (cost per unit weight or volume) of a material depends not only on the material itself, but also on its shape, size, and condition. Because more operations are involved in its production, the unit cost of thin wire, for example, is higher than that of a round bar made of the same material. The unit cost of metal foil is higher than that of metal plate. Similarly, powder metals are more expensive than bulk metals. Also, the cost of materials generally decreases as the purchase quantity increases, much like packaged food products you purchase at supermarkets: the larger the quantity per package, the lower the cost per unit weight.

The cost of a particular material is subject to fluctuations that may be caused by factors as simple as supply and demand or as complex as geopolitics. If a product is no longer cost-competitive in the marketplace, alternative and less costly materials should be selected. For example, the copper shortage around 1915 caused the U.S. government to mint pennies from zinc-plated steel. Similarly, when the price of copper increased substantially during the 1960s, electrical wiring for residential use was, for a time, made from aluminum.

If scrap is produced during manufacturing, as in sheet-metal fabricating, forging, and machining (Table 15.1), the value of the scrap is deducted from the material's cost to obtain net material cost. The value of the scrap depends on the type of metal and the demand for it. The value is usually 10 to 15 percent of the original cost of the material.

---

**Example 15.1:** Effect of workpiece hardness on cost in drilling

Gear blanks, forged from 8617 alloy steel and having a hardness range of 149–156 HB, required drilling a hole 75-mm (3-in.) in diameter in the hub. The blanks were drilled with a standard helix drill. However, after only 10 pieces the drill began to gall, temperature increase was excessive, the drill became dull, and the holes had a rough surface finish. To improve machinability and reduce galling, the hardness of the gear blanks was increased to the range of 217–241 HB by heating to 815°C (1515°F) and quenching in oil. When the blanks were drilled at this hardness, galling was reduced, the surface finish was improved, drill life increased to 50 pieces, and the cost of drilling was reduced by 80 percent.

*Source:* ASM International.

## 15.3 Product Design and Quantity of Materials

With high production rates and less labor, the cost of materials becomes a significant portion of product cost. Although the cost of materials cannot be reduced below a certain level, efforts can be made to reduce the amount of material involved in components that are to be mass produced. Since the overall shape of the part is usually optimized during the design and prototype stages, further reductions in the amount of material used can be obtained only by reducing the part's thickness. To do so requires selection of materials having high strength-to-weight or stiffness-to-weight ratios. Note also that higher ratios can be obtained by improved design and better selection of cross-sections, such as high moment of inertia as in I-beams.

New techniques such as finite-element analysis, minimum-weight design, design optimization, and computer-aided design and manufacturing have greatly facilitated design analysis and optimization, modeling, and product simulation. Implementing new design techniques and minimizing the amount of materials used can, however, present significant problems in manufacturing. Producing parts with thin cross-sections can be difficult, such as welding, casting, forging and sheet-stretching operations on thin sections. For example, (1) in forging, thin parts require high forces (see Section 6.2), (2) in welding, they tend to distort, (3) in casting, thin sections present difficulties in mold-cavity filling and maintaining flatness, dimensional accuracy, good surface finish, and (4) in sheet-forming operations, formability may be reduced.

Conversely, using parts with thick sections can slow production in processes such as casting and injection molding because of the time required for cooling and removing the part from the mold. Also, abrupt changes in part cross-section, such as from thick to thin, contribute to various problems and may lead to defective parts as well as reducing die and mold life.

## 15.4 Substitution of Materials

There is hardly a product on the market today for which substitution of materials has not played a major role in helping companies maintain their competitive positions. Automobile and aircraft manufacturing are examples of major industries in which substitution of materials is an important ongoing activity. A similar trend is evident in sporting goods and other consumer products. Although new products continually appear on the market, the majority of both design and manufacturing activities are concerned with improving existing products. Significant product improvements have resulted from substitution of materials, as well as implementation of new or improved processing techniques, better control of processing parameters, and plant automation.

There are several reasons for substituting materials in existing products:

- Reduction in costs of materials and processing.
- Ease of manufacturing and assembly, installation, and conversion to automated parts assembly.
- Improvements in performance, such as reduction in weight and better resistance to wear, fatigue, corrosion, and other characteristics.
- Increase in stiffness-to-weight and strength-to-weight ratios.
- Ease of maintenance and repair.
- Unreliable domestic and overseas supply of materials.
- Legislation and regulations prohibiting the use of certain materials in products because of toxicity or environmental considerations.

**Substitution of materials in the automobile industry.**    The automobile is a good example of effective substitution of materials in order to achieve one or more of the objectives just listed. Several examples are:

- All or part of the metal body replaced by plastic or reinforced plastic parts.

- Metal bumpers, fuel tanks, housings, covers, gears, pumps, clamps, and many other similar components replaced by plastic substitutes.

- Metal engine components replaced by ceramic and reinforced-plastic parts.

- Composite-material drive shafts used in place of all-metal drive shafts.

- Various other substitutions, such as cast-iron to cast-aluminum engine blocks, forged to cast crankshafts, and forged to cast, powder-metallurgy, or composite-material connecting rods.

The automobile industry is a major consumer of both metallic and nonmetallic materials. Consequently, there is constant competition among suppliers, particularly in the steel, aluminum, and plastics industries. Automotive industry engineers and management are con-

**FIGURE 15.2**    Advanced materials used on the Lockheed C-5A transport aircraft (FRP = fiber-reinforced plastic).

stantly investigating the advantages and limitations of these principal materials with regard to their applications, costs versus benefits, and recycling and other environmental considerations.

**Substitution of materials in the aircraft industry.**   In the aircraft and aerospace industries, conventional aluminum alloys (2000 and 7000 series) are being replaced by aluminum-lithium alloys and titanium alloys because of their higher strength-to-weight ratios. Forged parts are being replaced with powder-metallurgy parts that are manufactured with better control of impurities and microstructure, which enhance the overall economics of the substitutions. Furthermore, the P/M parts require less machining and produce less scrap of expensive materials. Advanced composite materials and honeycomb structures are replacing traditional aluminum airframe components (Fig. 15.2), and metal-matrix composites are replacing some of the aluminum and titanium structural components.

## 15.5   Selection of Manufacturing Processes

This section describes the importance of proper manufacturing process and machinery selection, and how the selection process relates to characteristics of materials, tolerances and surface finishes obtained, and cost. Most manufacturing processes have now been automated and are being computer controlled to optimize the operations. Computerization is also effectively increasing product reliability and quality and reducing labor costs.

The choice of a manufacturing process is dictated by various considerations, such as:

- Characteristics and properties of the workpiece material
- Shape, size, and thickness of the part
- Tolerances and surface-finish requirements
- Functional requirements for the expected service life of the component
- Production volume (quantity)
- Level of automation needed to meet production volume and rate requirements
- Costs involved in individual and combined aspects of the manufacturing operation

We have seen that some materials can be processed at room temperature, but others require elevated temperatures, so furnaces and appropriate tooling and equipment are required. Some materials are soft and ductile, whereas others are hard, brittle, and abrasive, requiring special processing techniques and tool and die materials. Materials have different manufacturing characteristics, such as castability, forgeability, workability, machinability, and weldability. Few materials have the same favorable characteristics in all categories. For example, a material that is castable or forgeable may later present problems in machining, grinding, or finishing operations that may be required to produce a product with acceptable surface finish, dimensional accuracy, and quality.

---

**Example 15.2:**   **Process selection in making a part**

Assume that you are asked to make the part shown in Fig. 15.3. You should first determine the part's function, the types of loads and environment to which it is to be subjected, the tolerances

**FIGURE 15.3**

and surface finish required, and so on. For the sake of discussion, let's assume that the part is round, 125 mm (5 in.) long and that the large and small diameters are 38 mm and 25 mm (1.5 in. and 1.0 in.), respectively. Let's further assume that, for functional requirements such as stiffness, hardness, and resistance to elevated temperatures, this part should be made of metal.

Which manufacturing process would you choose, and how would you organize the production facilities to manufacture a cost-competitive, high-quality product? Remember that parts should be produced at or near their final shape (net- or near-net-shape manufacturing). This approach eliminates much secondary processing, such as machining, grinding, and other finishing operations, reduces total manufacturing time, and thus lowers the cost.

This part is relatively simple and could be suitably manufactured by different methods. You can readily see that the part can be made by (a) casting or powder metallurgy, (b) upsetting, (c) extrusion, (d) machining, or (e) joining two pieces together. For net-shape processing, the two suitable processes are casting and powder metallurgy, each process having its own characteristics and various tooling and labor costs. This part can also be made by cold, warm, or hot forming. One method is by upsetting a 25 mm (1 in.) diameter round bar in a suitable die to form the larger end. The partial extrusion of a 38 mm (1.5 in.) diameter bar to reduce the diameter to 25 mm is another possibility. Note that each of these processes shapes the material without producing any waste.

This part can also be made by machining a 38 mm diameter bar stock to obtain the 25 mm diameter section. However, machining will take much longer than forming, and inevitably some material will be wasted as metal chips. On the other hand, unlike net-shape processes that require special dies, machining does not require special tooling and the operation can easily be carried out on a lathe. Note also that you can make this part in two pieces and then join them by welding, brazing, or adhesive bonding.

Each process has its own characteristics, production rate, surface finish, dimensional accuracy, and final product properties. Note that all these processes require raw materials in different shapes, ranging from metal powders to round bars. Because of the various operations involved in producing the materials, costs depend not only on the type of material (ingot, powder, drawn rod, extrusion), but also on its size and shape. Per unit weight, square bars are more expensive than round bars, and cold-rolled plate and sheet are more expensive than hot-rolled plate and sheet. Also, hot-rolled bars per unit weight are much less expensive than powders of the same metal.

Along with technical requirements, process selection also depends on factors such as the required production quantity and rate, as described in the next section. Summarizing our discussion, we can say that if only a few parts are needed, machining this part is the economical method. However, as the production quantity increases, producing this part by a heading operation or by cold extrusion would be the proper choice. Method (e) would be the most appropriate choice if the top and bottom portions of this part were made of different metals.

## 15.6  Process Capabilities

We have seen that all manufacturing processes have certain advantages and limitations. Casting and injection molding, for example, can generally produce more complex shapes than can forging and powder metallurgy. The reason is that the molten metal or plastic is capable of filling complex mold cavities. On the other hand, forgings generally can be made into complex shapes by additional machining and finishing operations, and have toughnesses that are generally superior to castings and powder-metallurgy products.

Recall that the shape of a product may be such that it has to be fabricated from several parts, then joined with fasteners or by brazing, welding, or adhesive-bonding techniques. The reverse may be true for other products, whereby manufacturing them in one piece might be more economical because of significant assembly-operation costs. Other factors that have to be considered in process selection are the minimum section size or dimensions that can be satisfactorily produced (Fig. 15.4). For example, very thin sections can be obtained by cold rolling, but processes like sand casting or forging prohibit forming such thin sections.

**FIGURE 15.4**  Process capabilities for minimum part dimensions. *Source:* J.A. Schey, *Introduction to Manufacturing Processes,* 2d ed. New York: McGraw-Hill, 1987. Reprinted with permission of the McGraw-Hill Companies.

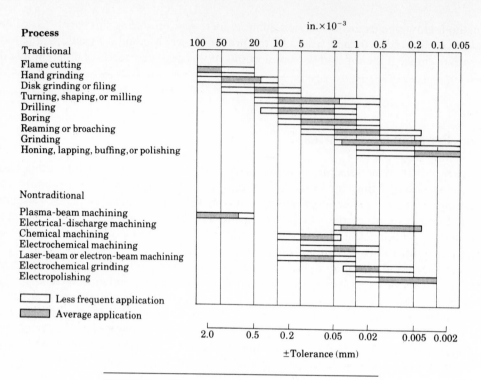

**FIGURE 15.5**  Tolerance capability of various processes.

**Tolerances and surface finish.**    Tolerances and surface finishes produced are important aspects of manufacturing. They have a particular bearing on subsequent assembly operations, proper operation of various machines and instruments, and consistency of overall appearance. The range of surface finishes and tolerances obtained in manufacturing processes are given in various chapters and are summarized in Fig. 15.5. We have seen that to obtain finer surface finishes and closer tolerances, additional finishing operations, better control of processing parameters, and the use of higher quality equipment may be required.

The closer the tolerance required, the higher the cost of manufacturing (Fig. 15.6); and, the finer the surface finish, the longer the manufacturing time, thereby increasing the cost

**FIGURE 15.6**  Relationship between relative cost and tolerance.

**FIGURE 15.7** Relative production time as a function of surface finish produced by various manufacturing methods. *Source: American Machinist.* See also Fig. 9.47.

(Fig. 15.7). In machining aircraft structural members made of titanium alloys, for example, as much as 60 percent of the cost of machining the part may be consumed in the final machining pass in order to hold proper tolerances and surface finish. Unless specifically required with proper technical and economic justification, parts should be made with as coarse a surface finish and with as wide a tolerance as are functionally and aesthetically acceptable. Here the importance of interaction and continuous communication between the product designer and the manufacturing engineer is obvious.

**Production volume.**   Depending on the type of product, the production volume or quantity (lot size) can vary greatly. For example, paper clips, bolts, washers, spark plugs, bearings, and ballpoint pens are produced in very large quantities. On the other hand, jet engines for large commercial aircraft, diesel engines for locomotives, and propellers for cruise ships are manufactured in limited quantities. Production quantity plays a significant role in process and equipment selection. An entire manufacturing discipline, known as *economic order quantity,* is now devoted to determining mathematically the optimum production quantity.

**Production rate.**   A significant factor in manufacturing process selection is the production rate, defined as the number of pieces to be produced per unit time, such as per hour, month, or year. Processes such as powder metallurgy, die casting, deep drawing, and roll forming are high-production-rate operations. On the other hand, sand casting, conventional and electrochemical machining, spinning, superplastic forming, adhesive and diffusion bonding, and processing of reinforced plastics are relatively slow operations. These rates can be increased by automation or the use of multiple machines. However, a slower production rate does not necessarily mean that the manufacturing process is inherently uneconomical.

**Lead time.**   The choice of a process is greatly influenced by the time required to start production, or lead time. Processes such as forging, extrusion, die casting, roll forming, and sheet-metal forming may require extensive and expensive dies and tooling. In contrast, most machining and grinding processes have an inherent flexibility and utilize tooling that can be adapted to many requirements in a relatively short time. Recall also our discussion of machining centers, flexible manufacturing cells, and flexible manufacturing systems, all of which are capable of responding quickly and effectively to product changes, both in type and quantity.

## 15.7    Manufacturing Costs

We have indicated repeatedly that economics is one of the most important aspects of manufacturing operations. Designing and manufacturing a product according to certain specifications and to meet service requirements are only one aspect of production. In order for the product to be successfully marketed, its cost must be competitive with similar products in the global marketplace.

The total cost of a product consists of items such as the costs for materials, tooling, and fixed, direct, and indirect labor costs. There are presently several methods of *cost accounting* used by manufacturing organizations. Because the methodologies of various accounting procedures can be complex and even controversial, and their selection dependent on the particular company and its type of operations and product lines, a description of these methods is beyond the scope of this book. Furthermore, because of the technical and operational factors involved, calculating the individual cost factors is difficult, time consuming, and not always accurate or reliable.

Recent trends in costing systems (*cost justification*) include consideration of factors such as the intangible benefits of quality improvements and inventory reduction, as well as factors regarding full life-cycle costs, machine usage, purchasing or leasing of machinery, and the risks involved in implementing automation and new manufacturing technologies. Because of the significant costs that can be incurred, a more recent consideration is *return on quality (ROQ)*. Thus, an incremental improvement in quality with respect to the additional costs involved must be studied carefully.

The significant costs that are directly attributable to *product liability* defense claims also have been a matter of concern and discussion among various parties involved, particularly manufacturers and members of the legal profession. For example, it is claimed that liability suits against car manufacturers in the United States add about $500 to the price of an automobile. In another example, it is claimed that 20 percent of the price of a step ladder is attributed to product liability costs.

### Material costs

Material costs were described in Section 15.2. We have also presented some cost data throughout this text.

### Tooling costs

Tooling costs are the costs involved in making the tools, dies, molds, patterns, and special jigs and fixtures necessary for manufacturing a product or component. The tooling cost is greatly influenced by the production process selected. For example, if a part is to be made by casting, the tooling cost for die casting is higher than that for sand casting. Similarly, the tooling cost in machining or grinding is much lower than that for powder metallurgy, forging, or extrusion. In machining operations, carbide tools are more expensive than high-speed-steel tools, but tool life is longer.

If a part is to be manufactured by spinning, the tooling cost for conventional spinning is much lower than that for power spinning. Tooling for rubber-forming processes is less expensive than that for male-and-female die sets used for drawing and stamping of sheet metals. High tooling costs, on the other hand, can be justified for high-volume production of a single item. The expected life of tools and dies—and their obsolescence because of product changes—are also important considerations.

## Fixed costs

Fixed costs include the costs of power, fuel, taxes on real estate, rent, insurance, and capital, including depreciation and interest. The company must pay these costs whether or not it makes a particular product. Thus, fixed costs are not sensitive to production volume. *Capital costs* represent the capital investment in land, buildings, machinery, and equipment and represent major expenses for most manufacturing operations (Table 15.2). Note the wide range of prices in each category and that some machines cost a million dollars or more. Interestingly, the cost of traditional machine tools, per unit weight, is historically roughly equivalent to the cost of steak per pound. This cost can easily double for machinery having computer controls.

**TABLE 15.2**

**APPROXIMATE RANGES OF MACHINERY BASE PRICES**

| TYPE OF MACHINERY | PRICE RANGE ($000) | TYPE OF MACHINERY | PRICE RANGE ($000) |
|---|---|---|---|
| Broaching | 10–300 | Machining center | 50–1000 |
| Drilling | 10–100 | Mechanical press | 20–250 |
| Electrical discharge | 30–150 | Milling | 10–250 |
| Electromagnetic and electrohydraulic | 50–150 | Robots | 20–200 |
| Gear shaping | 100–200 | Roll forming | 5–100 |
| Grinding | | Rubber forming | 50–500 |
|   Cylindrical | 40–150 | Stretch forming | 400– > 1000 |
|   Surface | 20–100 | Transfer machines | 100– > 1000 |
| Headers | 100–150 | Welding | |
| Injection molding | 30–150 |   Electron beam | 200–1000 |
| Boring | |   Spot | 10–50 |
|   Jig | 50–150 |   Ultrasonic | 50–200 |
|   Horizontal boring mill | 100–400 | | |
| Flexible manufacturing system | > 1000 | | |
| Lathe | 10–100 | | |
|   Single- and multi-spindle automatic | 30–250 | | |
|   Vertical turret | 100–400 | | |

Note: Prices vary considerably, depending on size, capacity, options, and level of automation and computer controls.

Let's assume that a company has decided to begin manufacturing a variety of valves. A new plant has to be built, or an old plant remodeled, with all the necessary machinery, support equipment, and facilities. In order to cast and machine the valve bodies, melting furnaces, casting equipment, machine tools, quality control equipment, and related equipment and machinery of various sizes and types must be purchased. If valve designs are to be changed often, or if product lines vary greatly, the manufacturing equipment must have sufficient flexibility to accommodate these requirements. Machining centers and flexible manufacturing systems are especially suitable for this purpose. The equipment and machinery just listed are capital cost items and, as you can appreciate, require major investment.

In view of generally high equipment costs, particularly those involving transfer lines and flexible manufacturing systems, high production rates are often required to justify large expenditures and hold product unit cost at a competitive level. Lower unit costs can be achieved by continuous production, involving around-the-clock operation, so long as demand warrants. Proper equipment maintenance is essential to ensure high productivity. Any breakdown of machinery (*downtime*) can be very expensive, ranging from a few hundred to thousands of dollars per hour.

## Labor costs

Labor costs are generally divided into direct and indirect costs. The *direct labor cost* is for the labor directly involved in manufacturing the part (*productive labor*). This cost includes all labor from the time materials are first handled to the time the product is finished. This time is generally referred to as *floor-to-floor time*. For example, a machine operator picks up a round bar from a bin, machines it to the shape of a threaded rod, and returns it to another bin. The direct labor cost is calculated by multiplying the labor rate (hourly wage, including benefits) by the time that the worker spends producing the part.

The time required for producing a part depends not only on required part size, shape, and dimensional accuracy, but also on the workpiece material. In Chapter 8, we stated that the cutting speeds for high-temperature alloys are lower than those for aluminum or plain-carbon steels. Consequently, the cost of machining certain aerospace materials is much higher than that for the more common alloys.

*Indirect labor costs* are those that are involved in servicing the total manufacturing operation. This cost is composed of activities such as supervision, repair, maintenance, quality control, engineering, research, and sales, and also generally includes the cost of office staff. Because these costs do not contribute directly to the production of finished parts or are not chargeable to a specific product, they are referred to as *overhead,* or the *burden rate,* which is charged proportionally to all products. The personnel involved are categorized as *nonproductive labor.*

## Manufacturing costs and production volume

One of the most significant factors in manufacturing costs is production volume. Large production volumes require high production rates. High production rates require the use of mass-production techniques, involving special machinery and proportionately less labor, and plants operating on two or three shifts per day. On the other hand, small production volumes usually mean larger direct labor involvement. Figure 15.8 shows a downward trend in the unit-cost curves as production volume increases. This result reflects proportionately lower

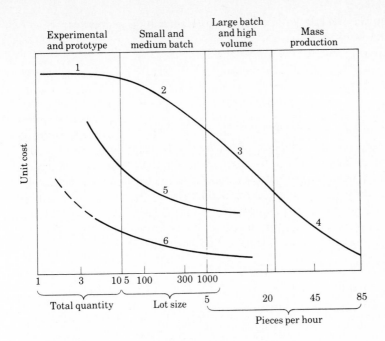

**FIGURE 15.8** Unit cost of product as a function of production method and quantity.
1 = toolroom machinery;
2 = general-purpose machine tools;
3 = special-purpose machines;
4 = automatic transfer lines;
5 = NC and CNC machine tools;
6 = computer-integrated manufacturing systems.
*Source:* Cincinnati Milacron, Inc.

tooling and capital costs per piece. Once the tooling is made, a greater number of produced parts will lower the cost of tooling per part. At low production volumes, material costs are low relative to direct labor costs, whereas at high volumes, the opposite is true.

As described in Section 13.2, *small-batch production* is usually done on general-purpose machines, such as lathes, milling machines, and hydraulic presses. The equipment is versatile, and parts with different shapes and sizes can be produced by appropriate changes in the tooling. However, direct labor costs are high because these machines are usually operated by skilled labor. For larger quantities (*medium-batch production*), these general-purpose machines can be equipped with various jigs and fixtures, or they can be computer controlled. To reduce labor costs further, machining centers and flexible manufacturing systems have been developed. For quantities of 100,000 and higher, the machines involved are designed for specific purposes (*dedicated machines*) and perform a variety of specific operations with very little labor involved.

## Cost reduction

The costs described above are interrelated, with *relative costs* depending on many factors. Consequently, the unit cost of the product can vary widely. For example, some parts may be made from expensive materials but require very little processing, such as minting gold coins. Thus, the cost of materials relative to direct labor costs is high. On the other hand, some products may require several production steps to process relatively inexpensive materials, such as carbon steels. An electric motor, for example, is made of relatively inexpensive materials, yet many different manufacturing operations are involved in making the housing, rotor, bearings, brushes, and other components of the motor. In such cases, assembly costs can become a significant portion of the overall cost.

An approximate but typical breakdown of costs in manufacturing is as follows:

Design: 5%

Material: 50%

Direct labor: 15%

Overhead: 30%

Although labor has accounted for as much as 15 percent of the production cost, today it can be as low as 5 percent, depending on the type of product and level of automation. Note above the small contribution of design. However, design, in its comprehensive sense of design for manufacture and assembly, including concurrent engineering, generally has the largest influence on the *quality* and *success* of a product in the marketplace. The *rule of ten* is an effective way to appreciate the role of defects in a product and the costs involved in rectifying them at different stages, from product concept to its use in the customer's facilities. The relative cost for rectifying a defect or fault at each stage is summarized below:

Eliminate design defect in product concept stage: 0.1%

Eliminate defect during planning stage: 1%

Eliminate defect during during production: 10%

Eliminate defect in customer's facilities during its use: 100%

Thus, identification and rectification of defects must be done as early as possible.

Cost reductions can be achieved by careful analysis of *all* the costs incurred in each step in manufacturing a product, with methods that are described in detail in some of the references in the bibliography at the end of this chapter. Opportunities for cost reduction have been emphasized in various chapters throughout this text. Among these are simplifying part design and the number of subassemblies required, specifying broader tolerances and surface finish, using less expensive materials, investigating alternative methods of manufacturing and assembly, and using more efficient machines and equipment.

The introduction of automation and up-to-date technology in a manufacturing facility is an obvious means of reducing costs. These changes must, however, be considered with care and after thorough **cost/benefit analyses,** with reliable data input and due consideration of technical as well as human factors. Implementation of high technology, which can be very expensive, should be made only after full assessment of the more obvious cost factors (**return on investment, ROI**).

Undoubtedly you have noted that, over time, the costs of products such as calculators, computers, and digital watches have decreased, whereas the costs of other products such as automobiles, aircraft, houses, and books have gone up. These differences result from the changes in various costs over time, including labor, machinery, materials, market demand, domestic and international competition, and worldwide economic trends (demand, exchange rates, and tariffs), and the inevitable impact of computers on all aspects of product design and manufacturing.

---

### Example 15.3:    Design changes and cost reduction

The original design for an aluminum bell crank assembly is shown in Fig. 15.9. This part was made by die casting. However, quality was inconsistent and there were excessive dimensional

**FIGURE 15.9**

Old design               New design

variations and porosity in critical regions and problems in finish machining it. Consequently, scrap was significant. Alternative design and material were considered (right), consisting of a fabricated stamped-steel crank with a separate round hub, which was pressed into a hole. This design had the same load-deflection characteristics, but its cost was about 32 percent less than the die-cast bell crank.

*Source: Metals Handbook,* 9th ed., Vol. 3, 1980, ASM International, Materials Park, Ohio, p. 849.

## 15.8 Value Engineering

Cost reduction is possible in several areas of activity in manufacturing. As stated in Chapter 1, manufacturing adds value to materials as they become discrete products and are marketed. Because this value is added in individual steps during creation of the product, utilization of value engineering (value analysis, value control, and value management) is important.

*Value engineering* is a system that evaluates each step in design, materials, processes, and operations to manufacture a product that performs all its intended functions at the lowest possible cost. A monetary value is established for each of two product attributes: (1) use value, reflecting the functions of the product, and (2) esteem or prestige value, reflecting the attractiveness of the product that makes its ownership desirable. The value of a product is then defined as the ratio of product function and performance to the cost of the product. Thus, the goal of value engineering is to obtain maximum performance per unit cost.

Value analysis generally consists of five phases:

1. Information phase to gather facts and determine costs.
2. Analysis phase to define functions and identify problem areas and opportunities.
3. Creativity phase to seek ideas that respond to problems and opportunities, without judging the values of these ideas.

4.  Evaluation phase to select ideas to be developed and the costs involved.

5.  Implementation phase to present facts, costs, and values to management; develop a plan; and motivate positive action to obtain a commitment of necessary resources to accomplish the task.

Value engineering is an important and all-encompassing interdisciplinary activity. It is usually coordinated by a value engineer and conducted jointly by designers, engineers, quality control, purchasing, and marketing personnel, and managers. In order for value engineering to be effective, it requires the full support of the company's top management. Implementation of value engineering in manufacturing companies has resulted in significant benefits such as cost savings, reduced lead times, better product quality and performance, reduced product weight and size, and shorter manufacturing times.

To properly assess the value of each step in manufacturing a product, several groups of questions must be asked:

## I.  Product design

-  Can the design be simplified without adversely affecting its functions? Have all alternative designs been investigated? Can unnecessary features or some components be eliminated or combined with others? Can the design be made smaller and lighter?
-  Are the dimensional tolerances and surface finish specified necessary? Can they be relaxed?
-  Will the product be difficult to assemble and disassemble for maintenance or recycling? Is the use of fasteners minimized?
-  Does each component have to be manufactured in the plant? Are some parts and subassemblies commercially available as standard products from outside sources?

## II.  Materials

-  Do the materials selected have properties that are significantly above minimum requirements and specifications?
-  Can some materials be replaced by others that are cheaper?
-  Do the materials selected have the proper manufacturing characteristics?
-  Are the materials to be ordered available in standard sizes, dimensions, surface finish, and tolerances?
-  Is material supply reliable? Are there likely to be significant price fluctuations?

## III.  Manufacturing processes

-  Have all alternative manufacturing processes been investigated?
-  Are the methods chosen economical for the type of material, shape to be produced, and desired production rate? Can requirements for tolerances, surface finish, and product quality be met consistently?

- Can the part be formed and shaped to final dimensions rather than made by material-removal processes? Are machining, secondary processes, and finishing operations required? Are they necessary?
- Is tooling available in the plant? Can it be purchased as a standard item?
- Is scrap produced? If so, what is the value of the scrap?
- Are processing parameters optimized? Have all the automation and computer control possibilities been explored for all phases of the manufacturing operation? Can group technology be implemented for parts with similar geometric and manufacturing attributes?
- Are inspection techniques and quality control being implemented properly?

## SUMMARY

- Competitive aspects of production and costs involved are among the most significant considerations in manufacturing. Regardless of how well a product meets design specifications and quality standards, it must also meet economic criteria in order to be competitive in the domestic and global marketplaces. (Section 15.1)
- Selecting an optimal material from among various candidates is one competitive aspect of manufacturing, and depends on the following criteria:

  1. Desired mechanical, physical, and chemical properties
  2. Shapes of commercially available materials
  3. Reliability of supply
  4. Costs of materials and processing (Section 15.2)

- A competitive strategy to contain material costs is to reduce the amount of material required to manufacture a component. (Section 15.3)
- Substitution of materials in existing products can result in significant product improvement and cost reduction, as well as implementation of new or improved processing techniques, better control of processing parameters, and plant automation. (Section 15.4)
- Selection of the most appropriate manufacturing process from among alternatives is perhaps the core competitive decision made by process and manufacturing engineers. The chosen process must successfully meet the requirements for tolerances and surface finish, production rate, and lead time. (Sections 15.5, 15.6)
- A variety of costs have impact on total manufacturing costs and, thus, the ultimate competitiveness of the end product, including:

  1. material costs
  2. tooling costs
  3. fixed costs (such as fuel, taxes, rent, insurance, and capital)
  4. labor costs (Section 15.7)

- Value engineering is a system that evaluates each step in design, materials, processes, and operations to manufacture a product that performs all its intended functions at the lowest possible cost. (Section 15.8)

## BIBLIOGRAPHY

Blaich, R., *Product Design and Corporate Strategy: Managing the Connection for Competitive Advantage.* New York: McGraw-Hill, 1993.

Boothroyd, G., P. Dewhurst, and W. Knight, *Product Design for Manufacture and Assembly.* New York: Marcel Dekker, 1994.

Bralla, J.G. (ed.), *Handbook of Product Design for Manufacturing–Practical Guide for Low-Cost Production.* New York: McGraw-Hill, 1986.

Brown, J., *Value Engineering: A Blueprint.* New York: Industrial Press, 1992.

Corbett, J., M. Donner, J. Meleka, and C. Pym, *Design For Manufacture: Strategies, Principles and Techniques.* Reading, Mass: Addison-Wesley, 1991.

Creese, R.C., et al., *Estimating and Costing for Metal Manufacturing Industries.* New York: Marcel Dekker, 1992.

Dieter, G.E., *Engineering Design,* 2d ed. New York: McGraw-Hill, 1991.

Dixon, J.R., and C. Poli, *Engineering Design and Design for Manufacturing.* Conway, Mass.: Field Stone Publishers, 1995.

Dorf, R.C., and A. Kusiak, (eds.), *Handbook of Design, Manufacturing and Automation.* New York: Wiley, 1995.

Ettlie, J.E., and H.W. Stoll, *Managing the Design-Manufacturing Process.* New York: McGraw-Hill, 1990.

Helander, M., and M. Nagamachi, *Design for Manufacturability: A Systems Approach to Concurrent Engineering and Ergonomics.* Bristol, PA: Taylor & Francis, 1992.

Kmetovicz, R.E., *New Product Development: Design and Analysis.* New York: Wiley, 1992.

Koenig, D.T., *Manufacturing Engineering: Principles for Optimization,* 2d ed. Bristol, PA: Taylor & Francis, 1994.

Koshal, E. (ed.), *Manufacturing Engineer's Reference Book.* 13th ed. Stoneham, Mass.: Butterworth-Heinemann, 1993.

Kusiak, A. (ed.), *Concurrent Engineering: Automation, Tools, and Techniques.* New York: Wiley, 1993.

Lindbeck, J.R., *Product Design and Manufacture.* Englewood Cliffs, NJ: Prentice-Hall, 1995.

Michaels, J.V., *Design to Cost.* New York: Wiley, 1989.

Miller, L., *Concurrent Engineering Design: Integrating the Best Practices for Process Improvement.* Dearborn, Mich.: Society of Manufacturing Engineers, 1993.

Prasad, B., *Concurrent Engineering Fundamentals.* Englewood Cliffs, NJ: Prentice-Hall, 1995.

Pugh, S., *Total Design: Integrated Methods for Successful Product Engineering.* Reading, Mass.: Addison-Wesley, 1991.

Sims, E.R., Jr., *Precision Manufacturing Costing.* New York: Marcel Dekker, 1995.

*Tool and Manufacturing Engineers Handbook.* 4th ed., Vol. 5, *Manufacturing Engineering Management.* Dearborn, Mich.: Society of Manufacturing Engineers, 1988.

*Tool and Manufacturing Engineers Handbook,* 4th ed., Vol. 6, *Design for Manufacturability.* Dearborn, Mich.: Society of Manufacturing Engineers, 1992.

*Tool and Manufacturing Engineers Handbook,* 4th ed., Vol. 7, *Continuous Improvement.* Dearborn, Mich.: Society of Manufacturing Engineers, 1993.

Trucks, H.E., and G. Lewis (eds.), *Designing for Economical Production,* 2d ed. Dearborn, Mich.: Society of Manufacturing Engineers, 1987.

Winchell, W., *Realistic Cost Estimating for Manufacturing,* 2d ed. Dearborn, Mich.: Society of Manufacturing Engineers, 1989.

# QUESTIONS

15.1 List and describe the major considerations involved in selecting materials for products.

15.2 Why is a knowledge of available shapes of materials important? Give several specific examples.

15.3 Describe manufacturing properties of materials. Give three examples demonstrating the importance of this information.

15.4 Why is material substitution an important aspect of manufacturing engineering? Give five examples from your own experience or observations.

15.5 Why has material substitution been particularly critical in the automotive and aerospace industries?

15.6 What factors are involved in the selection of manufacturing processes? Explain why they are important.

15.7 What is meant by process capabilities? Select four different and specific manufacturing processes and describe their capabilities.

15.8 Why is production volume significant in process selection?

15.9 Discuss the disadvantages of long lead times in production.

15.10 What is meant by economic order quantity?

15.11 Describe the various costs involved in manufacturing. Explain how you could reduce each of these costs.

15.12 What is value engineering? What are its benefits?

15.13 What is meant by trade-off? Why is it important in manufacturing?

15.14 Explain why the larger the quantity per package of food products, the lower the cost per unit weight.

15.15 Explain why the value of the scrap produced in a manufacturing process depends on the type of material.

15.16 Comment on the magnitude and range of scrap shown in Table 15.1.

15.17 Explain how the high cost of some of the machinery listed in Table 15.2 can be justified.

15.18 Other than the size of the machine, what factors are involved in the range of prices in each machine category shown in Table 15.2?

15.19 Explain the reasons for the relative positions of the curves shown in Fig. 15.4.

15.20 What factors are involved in the shape of the curve shown in Fig. 15.6?

15.21 It has been stated that, contrary to general public perception, high-quality products do not necessarily cost more than those of poorer quality. Do you think this statement contradicts the concept of return on quality, described in Section 15.7? Explain.

15.22 Is it always desirable to purchase stock that is close to the final dimensions of a part to be manufactured? Explain, and give some examples.

# PROBLEMS

15.23 Make suggestions on how to reduce the dependence of production time on surface finish, as shown in Fig. 15.7.

15.24 What course of action would you take if the supply of a material selected for a product line becomes unreliable?

15.25 Describe the potential problems involved in reducing the quantity of materials in products.

15.26 Discuss your thoughts concerning the replacement of aluminum beverage cans with steel cans.

15.27 Review Fig. 1.2 in Chapter 1 and describe your thoughts concerning the flow chart. Would you want to make any modifications, and if so, what would they be?

15.28 Over the years, numerous consumer products have become obsolete, such as rotary-dial telephones, analog radio tuners, turntables, vacuum tubes, etc. Likewise, new products have entered the market. Make two comprehensive lists of (1) obsolete products and (2) new products. Comment on the reasons for the changes you observe.

15.29 From Problem 15.28, discuss how different manufacturing methods and systems have evolved in order to produce the new products.

15.30 Select three different products and make a survey of the changes in their prices over the past ten years. Discuss the reasons for the changes.

15.31 Discuss the trade-offs involved in selecting one of the two materials for each application listed. Also discuss typical conditions to which these products are subjected in their normal use.

1. Forged vs. cast crankshafts
2. Forged vs. powder-metallurgy connecting rods
3. Plastic vs. sheet-metal light-switch plates
4. Sheet-metal vs. cast hubcaps
5. Wood vs. metal handle for hammers
6. Sheet-metal vs. reinforced-plastic chairs

15.32 Discuss the manufacturing process or processes that are suitable for making the products listed in Problem 15.31. Explain whether they would need additional operations such as coating, plating, heat treating, and finishing. If so, make recommendations and give the reasons for them.

15.33 Figure 2.2 in Chapter 2 shows the shape of a typical tensile-test specimen with round cross-section. Assuming that the starting material (stock) is a round rod and only one specimen is needed, discuss the processes and the machinery by which the specimen can be made, including their relative advantages and limitations. Describe how the process you selected can be changed for economical production as the number of specimens to be made increases.

15.34 Select any one of the following common appliances. State your opinions regarding (1) the materials used in the product and why, and (2) how they are made and assembled, and why those particular methods were used.

1. Toaster
2. Electric iron
3. Automatic coffee maker
4. Electric can opener
5. Tool box
6. Wall clock

15.35 Inspect the components under the hood of your automobile. Identify several parts that have been produced to (1) net-shape and (2) near-net-shape condition. Comment on the design and production aspects of these parts and how the manufacturer achieved the net-shape condition.

15.36 Discuss the factors that influence the choice between the following set of competitive processes:

1. Sand casting vs. die casting of a fractional electric-motor housing
2. Machining vs. forming a large gear
3. Forged vs. powder-metallurgy gear
4. Frying pan made of a casting vs. stamped sheet metal
5. Outdoor furniture made of aluminum tubing vs. cast iron
6. Welded vs. cast machine-tool structures
7. Thread rolled vs. machined bolt for high-strength applications
8. Thermoformed plastic vs. molded thermoset fan blade for household use

15.37 Comment on the differences, if any, between the designs, materials, and processing and assembly methods used for making products such as hand tools and ladders for professional use vs. consumer use.

15.38 The capabilities of some machining processes are shown in Fig. 8.51. Inspect the various shapes produced and suggest alternative processes. Comment on the properties of materials that would influence your suggestions.

15.39 Figure 1.6 in Chapter 1 shows a steel mounting bracket that can be manufactured by either (1) casting or (2) stamping of sheet metal. Assume that the dimensions of the bracket are approximately 100 mm × 100 mm × 25 mm wide (4 in. × 4 in. × 1 in.). Describe the manufacturing sequence and processes involved in making each of these parts. Also comment on the relative costs involved. Can you suggest other methods of manufacturing this part? Explain.

15.40 If the dimensions of the part in Problem 15.39 were (1) ten times larger and (2) ten times smaller, how different would your answer be? Explain.

15.41 In Problem 15.32, what changes could you make in materials or processes to reduce the cost of the product without adversely affecting its functions? Explain your reasons.

15.42 This text began by showing a photograph of a jet engine in Fig. 1.1 in Chapter 1. On the basis of the topics covered in this text, select any three components of this engine and describe the materials and processes that you would use in making them in quantities of, say, 1000. Remember that these parts have to be manufactured at minimum cost while maintaining quality, product integrity, and reliability.

# Index

Ablation, 480, 657
Abrasive
 belts, 604
 grains, 578
 machining, 576
 wear, 162
Abrasive water-jet machining, 626
Abrasive-flow machining, 607
Abrasive-jet machining, 603, 626
Abrasives, 576
 bonded, 578
 coated, 604
Absolute dimensioning, 845
Accelerated strain aging, 112
Acceptance
 quality level, 201
 sampling, 201
Accuracy, 184; see also Tolerances
Acetals, 653
Acoustic emission
 impact, 190
 technique, 190
Acoustic holography, 191
Acoustic impact technique, 190
Acoustic impedance, 419
Acrylic concrete, 562
Acrylics, 654
Acrylonitrile-butadiene-styrene,
 654
Adaptive control, 16, 854
 constraint, 854
 optimization, 855
Addition polymerization, 639
Additives
 extreme pressure, 168
 in ceramics, 743
 in free-machining steels, 493
 in oils, 168
 in polymers, 645
 lubricants, 168

Adhesion theory of friction, 152
Adhesive
 bond (friction), 152
 bonding, 771, 818
 wear, 159
Adhesives, 818
Adsorption, 147, 736
Advanced composites, 668
Age hardening, 273
Aggregation, 178
Agile manufacturing, 21
Aging, 273
Air
 bending, 404, 405
 gages, 184
 Alclad, 172
Alkyds, 656
Alligatoring, 335
Allotropic transformation, 217
Allowable wear land, 486
Allowance, 188
 casting, 281
Alloys, 213, 221
Alloy steels, 123
Alloying, 90
 elements in iron, 219
 mechanical, 711
 of powder metals, 724
Alumina, 736
Alumina-base ceramics, 509
Aluminizing, 176
Aluminum and alloys, 129
Aluminum oxide, 509, 578
 abrasives, 578
 ceramic tools, 509
 coatings, 507
Aminos, 656
Amorphous, 223
 alloys, 140, 268
 polymers, 642

Amplification, 184
Anaerobic adhesive system, 819
Analog sensors, 861
Analog-to-analog converters, 861
Angle
 of acceptance, 334
 gage blocks, 180
 measuring instruments, 180
 milling cutter, 543
 of twist, 49
Anisotropy, 103, 388
 in crystals, 94
 in polymers, 652
 normal, 388, 428
 of thermal expansion, 120, 740
 planar, 388, 429
Annealing, 276
 of glass, 757
Anodic oxidation, 175
Anodizing, 175
Anthropomorphic, 862
Antioch process, 254
Appearance, 12
Apron, 525
Aramids, 655, 662
Arbor cutters, 543
Arc cutting, 625
Arc of contact, 331
Arc strikes, 792
Arc welding processes, 772
 consumable electrode, 772
 nonconsumable electrode, 780
Archard, J.F., 160
Arithmetic
 average, 149
 mean, 194
 mean value, 149
Arm and wrist (robot), 857
Arrays, 862
Arrowhead fracture, 353

Artificial
  aging, 273
  intelligence, 18, 895
  neural network, 18, 897
Aspect ratio, 713
Asperities, 152
Aspiration, 228
Assembly, 9, 864
  automated, 864
  design for, 8, 866
  systems, 864
Assignable variations, 193
ASTM designations, 123
  grain size, 101
Astroloy, 134
Atomic structure (metals), 90
Atomization, 711
Attributes, 883
Attritious wear, 590
Ausforming, 277
Austempering, 277
Austenite, 217
  former, 219
  retained, 271
Austenitic stainless steels, 124
Autoclave, 254
Autocollimator, 181
Autogenous weld, 785
Automachining, 496
Automated
  assembly, 16
  guided vehicle, 17, 856
  inspection, 191
  material handling, 16
  measurement, 185
  storage/retrieval systems, 856
Automatic
  bar machine, 530
  lathe, 526
  pallet changer, 553
  tool changers, 554
Automatically programmed tools, 852
Automation, 15
  hard, 842
  of manufacturing processes, 839
  soft, 844
Availability of materials, 11
Average of averages, 196

Babbitts, 139
Back
  rake angle, 520
  tension, 330

Backlash, 850
Backward
  extrusion, 342
  spinning, 416
Bainite, 270
Bakelite, 637
Balancing, grinding wheels, 602
Ball
  bearing manufacturing, 316, 596
  burnishing, 172
  mill, 711
Ballistic particle manufacturing, 732
Ballizing, 172
Band
  files, 549
  saw, 549
Bar codes, 856
Bar drawing, see Rod drawing
Barrel finishing, 607
Barreling, 45, 297, 318
Basal plane, 90
Base metal (welding), 785
Basic
  hole system, 188
  shaft system, 188
  size, 188
Bat, 747
Batch, 343
  production, 841
Bauschinger effect, 47, 435
Beach marks, 115
Beading, 405
Beads
  draw, 426, 435
  foam molding, 682
Bed, 525
Bed-type machines, 544
Bellows manufacturing, 412
Belt (abrasive), 604
Bench
  grinder, 599
  lathe, 526
Benchmark, 21
Bend
  allowance, 397
  angle, 397
  radius, 397
  tests, 51
  tests, welded joints, 795
Bendability, 397
Bending, 51
  sheet and plate, 396
  tubes and sections, 408

Bernoulli's theorem, 227
Beryllium, 138
  bronze, 133
Bevel
  gear cutting, 552
  protractor, 180
Beveling of punches, 395
Biaxial tension, 75
Bilateral tolerance, 188
Bill of materials, 880
Billet, 337
Binary phase diagram, 214
Binders (for sand), 247, 250
Bioceramics, 743
Biodegradable plastics, 658
Biological
  considerations, 171
  effects (metalworking fluids), 515
Black
  ceramics, 510
  oxide, 175
Blankholder, 424
Blanking, 393
  chemical, 612
  fine, 393
Blasting
  shot, 607
Bleeder cloth, 686
Blending (metal powders), 714
Blends (polymer), 645
Blind riser, 248
Blocker dies, 319
Bloom, 337
Blow molding, 676
Blowholes, 244
Blowing (glass), 755
Blown film, 671
Blue brittleness, 112
Body-centered cubic, 90
Bond fracture, 591
Bonded abrasives, 578
  designations, 581, 582
Bonding,
  adhesive, 771, 818
  of abrasives, 581
  of polymers, 640
Boring, 516, 530
  bar, 530
  machines, 530
  mills, 530
Boron fibers, 662
Boronizing, 274
Borosilicate glass, 750
Boss, 233

Bottle manufacturing, 676, 754
Bottoming,
  in bending, 403
  taps, 535
Boundary
  lubrication, 166
  representation, 876
Branched polymers, 640
Brass, 132
Braze
  metals, 813
  welding, 812, 815
Brazing, 771, 812
  filler metals, 813
  of cutting tools, 503
Breaking down
  ingots, 325
  lubricants, 166
Breaking stress, 29
Bridge die, 356
Bridging, 789
Bridgman, P.W., 69
Brinell hardness, 53
Brinelling, 54
Brittle,
  fracture, 112
  materials testing, 48, 51
Broach, 546
Broaching, 546
Bronze, 132
Buckling, 389
Buckyballs, 757
Buffer, 865
Buffing, 606
Building-block principle, 843
Built-up edge, 462
Bulge test, 437
Bulging (tubes), 411
Bulk modulus, 73
Bulk-deformation processes, 294,
  295
Bulk-molding compound, 684
Bull block, 367
Burden rate, 918
Burning (in grinding), 589
Burnishing, 172, 460
Bur, 549
Burr, 391, 607
Bursting speed, 601
Bus topology, 893
Butt welding, 87

CAD systems, 874
Calcium-deoxidized steels, 494
Calendering, 683

Calibration, 184
Caliper gages, 179
Calipers, 180
Cam plastometer, 46
Camber, 334
Canning, 355
Capacitor-discharge stud welding,
  808
Capacity planning, 880
Capillary action, 513
Capital costs, 917
Capstan, 367
Carbide, 217
Carbides, 501, 737
  coated, 505
Carbon, see also Graphite
  black, 645
  fibers, 662
  in steels, 123
Carbon steels, 123
Carbon-dioxide process, 252
Carbon-fiber reinforced plastic, 662
Carbonitriding, 274
Carbonyls, 711
Carburizing, 274
Carriage, 525
Case hardening, 172, 273, 274, 275
Cast
  irons, 219
  stainless steels, 242
  steels, 242
  structures, 221
  structures in welding, 786
Cast-cobalt alloy tools, 501
Castability, 229, 239
Casting, 211
  alloys, 237
  continuous, 244
  defects, 232
  design, 279
  economics, 282
  expendable mold, 246
  ferrous alloys, 237
  non-ferrous alloys, 243
  of ceramics, 745
  of ingots, 243
  of plastics, 681
  permanent mold, 259
  processes, 246, 247, 251, 283
Cavities, 232; see Porosity
Cellular manufacturing, 17, 887
Cellulosics, 654
Cemented carbides, see Carbides
Cementite, 217
Center cracking, 353, 365

Center drill, 531
Center-burst defect, 353
Center-line average, 149
Center-type grinding, 596
Centerless grinding, 598
Centrifugal casting,
  glass, 755
  metals, 264
Centrifugal spraying (glass), 754
Centrifuge casting, 264
Centrifuging, 264
Ceramic grinding wheel, see
    Vitrified bond
Ceramic-matrix composites, 759
Ceramic-mold casting, 255
Ceramic-shell investment casting,
  257
Ceramics, 734
  applications, 742
  bond, 581
  coatings, 173, 507
  cutting tools, 509
  friction of, 158
  processing, 743, 744
  properties, 739
  structure, 735
  wear of, 164
Cermets, 510, 738
Chain-growth polymerization, 639
Chain-reaction polymerization, 639
Chain-type code, 885
Chamfer (cutting tools), 504
Chance variations, 193
Channel furnace, 235
Chaplets, 249
Charpy test, 60
Chatter, 558
  grinding, 600
  machining, 558
  marks, 600
Cheeks, 248
Chemical
  blanking, 612
  cleaning, 178
  machining, 608
  milling, 608
  properties, 10
  reaction priming, 176
  stability (tools), 497
  synthesis processes, 761
  tempering, 756
  vapor deposition, 174, 506
  vapor infiltration, 761
  wear, 162
Chemically reactive adhesive, 818

Chevron cracking, 353
Chill
    tempering, 756
    zone, 222
Chills, 233
Chip
    breakers, 464, 503
    compression ratio, 458
    curl, 464
    depth of cut, 536
    flow angle, 466
Chip formation
    cutting, 456
    grinding, 583
Chipping (cutting tools), 488
Chips (metal), 459
Chisel-edge angle, 531
Chisel-point drill, 531
Choking, 228
Chromium
    in stainless steels, 124
    plating, 175
Chuck, 525
Chuckers, 526
Chucking machines, 526
Chvorinov's rule, 230
Circular
    cutters, 542
    interpolation, 850
    saws, 549
    tracing, 182
Cladding, 172, 343
Clamps, 863
Classification and coding systems, 883
Clay, 736
Cleaning
    fluids, 178
    of castings, 278
    processes, 177
Clearance, 188, 389
    angle, 458
Cleavage plane, 112
Climb milling, 536, 538
Closed-die forging, 311
Closed-loop control, 847
Cluster mill, 338
Coalescence of voids, 108
Coarse pearlite, 269
Coat-hanger die, 671
Coated abrasives, 604
Coated sheet metal, 389
Coated tools, 505
    multiphase, 507
Coatings, 171

ceramic, 173, 507
    conversion, 173, 367
    for cutting tools, 505
    for electrodes, 779
    for rod and wire drawing, 367
    for sheet metal, 389
Cobalt (in tungsten carbide), 501, 502
Cobalt-base superalloys, 134
Coding systems, 884
Coefficient of adhesion, 153
Coefficient of friction, 153
    in cutting, 468
    in metalworking processes, 155
    in ring compression, 156
    measurement, 156
Cogging, 314
Coining, 313, 403, 728
Co-injection molding, 674
Cold cracking (welds), 791
Cold
    extrusion, 350
    forming, see individual
        processes
    forming of plastics, 682
    isostatic pressing, 720
    roll forming, 408
    saws, 549
    shut, 316
    welding, 799
    working, 102, 106
Cold-box mold, 247
Cold-chamber process, 261
Cold-pressed ceramics, 509
Cold-runner molds, 673
Cold-setting processes, 248
Cold-work steels, 126
Collapsible taps, 535
Collapsibility (cores), 249
Collapsing, 389
Collet, 525
Colloidal, 169
Colorants, 645
Coloring, 176
Columbium, 137; see Niobium
Column-and-knee milling
        machine, 543
Columnar
    dendrites, 221
    grains, 221
Combination square, 180
Communications
    network, 893
    standard, 904
Comminution, 711, 743

Compacts, 511
Compacted-graphite iron, 242
Compaction of metal powders, 715
Compacts (diamond), 511
Comparative length measurement, 180
Competitive aspects of manufacturing, 905
Compilers, 852
Compliant end effector, 862
Composite materials, 659, 758
    ceramic-matrix, 759
    metal-matrix, 759
    processing, 760
Composite molds, 252
Compound
    dies, 395
    rest, 525
Compounded oils, 168
Compressibility
    metal powders, 716
Compression, 45
    molding, 679
Compressive strength,
    various materials, 739
Computer
    numerical control, 16, 526, 846
    simulation, 7, 880
    vision, 862
Computer-aided
    design, 7, 17, 874
    engineering, 7, 17, 874
    manufacturing, 7, 877
    part programming, 851
    process planning, 17, 878
Computer-controlled lathe, 526
Computer-integrated manufacturing, 15, 871
Concurrent engineering, 5
Condensation polymerization, 639
Conditioning of rolls, 337
Constitutional diagrams, 214
Constructive solid geometry, 876
Consumable electrode, 772
Consumer's risk, 201
Contact area, 152
Contact molding, 686
Continuity, 227
Continuous
    casting, 243
    chip, 460
    fibers, 663
    path system, 849

Continuous (*cont.*)
  products, 4
  systems of assembly, 866
Contour roll forming, 408
Contouring lathe, 526
Contouring system, 849,
Contraction (solidification), 231
Control
  adaptive, 854
  charts, 19, 196
  limits, 196
  numerical, 16
  systems (robots), 858
Controllers, 844, 858
Conventional
  milling, 536, 538
  spinning, 413
Conversion coating, 170, 176,
    367
Conveyance card, 892
Coolants, 512
Cooling, see Cutting fluids; indi-
    vidual processes
Cooling rate (castings), 223
Coordinate measuring machine,
    182
Cope, 250
Cope-and-drag investment casting,
    255
Copolymers, 641
Copper and alloys, 131
Copy milling machines, 545
Core
  blowers, 249
  boxes, 249
  drill, 531
  furnace, 235
  prints, 249
Cored dendrites, 224
Cores, 248
Correction for necking, 68
Corrosion, 121
Corrosive wear, 163
Corundum, 736
Cost
  accounting, 916
  justification, 916
  reduction, 11, 13, 20, 916, 919
Costs, see also Economics of man-
    ufacture and individual
    processes
  machinery, 917
Counterblow hammer, 323
Counterboring, 531
Countersinking, 531

Counter-gravity low-pressure
    process, 258
Covalent bond, 639, 735
Crack velocity, 113
Cracks, 147; see Defects and indi-
    vidual processes
  in weld, 790
Crank press, 323
Crankshaft
  drill, 531
  manufacturing, 320
Crater wear, 487
Craters, 147
Crazing, 652
Creep, 59, 651
  rupture, 60
Creep-feed grinding, 599
Crevice corrosion, 122, 825
Crimping, 826
Critical shear stress, 92
Cross-feeding, 594
Cross-linked polymers, 641, 653
Cross-rolling, 315
Cross-tension test (spot welds),
    796
Crown, 334
Crucible furnace, 236
Crush
  dressing, 592
  forming, 592
Crushing, 743
Crystal
  growing, 267
  pulling, 267
Crystalline structure, 90
Crystallinity (polymers), 642
Crystallites, 642
Crystallographic anisotropy, 103,
    388
Crystals, 90
Cubic boron nitride, 510, 578,
    737
Cubic interpolation, 850
Cumulative distribution, 713
Cup (casting), 226
Cup-and-cone fracture, 109
Cupola, 236
Cupping tests, 436
Curing (thermosets), 653
Curl (chip), 464
Cutoff (roughness), 150
Cutters, see individual processes
Cutting, see also Machining
  forces, 467, 521
  ratio, 458

speeds, 522, 533, 544
  temperature, 478
Cutting fluids, 512, 522
  applications, 515
  effects of, 515
Cutting off, 516
  abrasive, 599
  saws, 548
  turning, 516
Cutting processes, 516
Cutting tools
  geometry, 519
  materials, 496, 498, 499
  reconditioning, 512
  tool life curves, 483
  wear, 481
Cutting-edge angle, 520
Cyaniding, 274
Cylindrical grinding, 596
Czochralski process, 267

Dacron, 655
Damping, 559
  in cast iron, 559
  in machine tools, 559
Data acquisition system, 874
Database, 874
Datum, 188
Dead-metal zone, 297, 344
Deburring, 607
Decarburization, 275
Decision-tree coding, 885
Dedicated machines, 841, 919
Deep
  drawability, 428
  drawing, 424
Deep-hole drilling, 531
Defects, 94, 97, 113, 315, 353,
    1102; see also individual
    processes
Defects (casting), 232
Deflection (rolls), 334
Deflocculant, 744
Deformation
  plastic, 28
  processing, 25
  rate, 40, 41
  single crystals, 92
  work, 76
Deformation-zone geometry, 308
Degradation (polymers), 121
Degree of
  crystallinity, 642
  polymerization, 639
Delamination, 665

Delayed tack adhesives, 818
Delivery reel, 330
Delrin, 654
Delta ferrite, 216
Demand scheduling, 891
Deming, W.E., 202
Dendrite multiplication, 226
Dendrites, 221
Density, 117
  various materials, 118, 660, 739
Dent resistance, 441
Deoxidation, 234
Deposition
  chemical vapor, 174, 506
  physical vapor, 174, 506
Depth-of-cut line, 489
Derivative CAPP system, 879
Design attributes, 883
Design considerations, see individual processes
Design for
  assembly, 8, 867
  disassembly, 8, 867
  the environment, 22
  manufacture, 8
  recycling, 22
  service, 8, 867
Design of experiment, 19
Design process, 5
Designations, see individual materials and processes
Desk-top manufacturing, 692
Destructive testing, 191, 795
  welded joints, 795
Detonation spraying, 173
Deviation-type measuring instruments, 180
Devitrification, 752
Dial indicator, 180
Diamond, 758
  abrasives, 578
  coatings, 177, 508
  cutting tools, 511
  film, 177
  pyramid hardness, see Vickers hardness
  turning, 529
  wire saws, 549
Die
  cutting, 393
  distortion, 319
  failures, 371
  geometry, see individual processes

life, see individual processes
manufacturing methods, 369
materials, 126, 127, 128
pressure, see individual processes
quenching, 272
steels, 126, 127, 128
Die casting, 261
  cold chamber, 261
  hot chamber, 261
Dielectric fluid, 616
Dielectrics, 120
Dies, 369; see also individual processes
Diffraction gratings, 179
Diffuse necking, 386
Diffusion
  adhesives, 818
  bonding, 810
  brazing, 815
  coating, 174
  crater wear, 487
  sintering, 724
Digital sensors, 861
Dilatation, 72
Dimensional tolerances, 13, 178, 186
Dimensioning, 845
Dimples (ductile fracture), 108
Dimpling, 406
Dip brazing, 814
Direct
  emulsion, 169
  engineering, 7
  extrusion, 342
  labor costs, 918
  numerical control, 846
Directional solidification, 266
Directionality, 149, 388; see also Anisotropy
Disassembly, 8, 867
Discontinuities,
  castings, 232
  welds, 788
Discontinuous
  chips, 464
  fibers, 663
Discrete products, 4
Disk test, 48
Dislocations, 97
  density, 98
Dispersion, 178, 194
Distortion, see also Warping
  dies, 319
  welds, 904

Distortion-energy criterion, 67
Distributed numerical control, 846
Distribution (statistical), 194
Dividers, 180
Doctor-blade process, 745
Domain, 896
Dopants, 121, 267
Double barreling, 297
Double-action press, 436
Down milling, 536
Downtime, 918
Draft (rolling), 331, 334
Draft angle,
  castings, 280
  forging, 320
Drag, 250
Drain casting, 745
Draw
  beads, 426, 435
  bench, 367
  cut, 546
Drawability, see Deep drawability
Drawing,
  deep, 424
  glass, 753
  rod and wire, 354
  strip, 364
  tubes, 364
Drawing out, 314; see also Cogging
Dressing, 591
Drift, 184
  pin, 824
Drilling, 516, 531
  machines, 533
Drills, 531
  geometry, 532
  life, 533
Drop hammer, 323
Dross, 228
Dry
  drawing, 367
  pressing (ceramics), 746
Drying (ceramics), 747
Dual-phase steels, 124
Ductile fracture, 108
Ductile iron, 241
Ductile-regime
  cutting, 529
  grinding, 594
Ductility, 29
Dummy
  bar, 244
  block, 343

Duplex structure steels, 125
Duplicating
lathe, 526
milling machine, 545
Durometer, 55
Dynamic
loading, 60
stiffness, 559

Earing, 431
Eccentric press, 321
Ecological considerations, 171
Economic
order quantity, 915
production quantity (plastics),
691
Economics of manufacture, 32; see
also Costs and individual
processes
Eddy-current inspection method,
191
Edge dislocation, 97
Edging, 315
Effective
strain, 73
stress, 73
Efficiency (in deformation processing), 77
Ejectors, 164, 261, 262, 673, 679,
680
Elastic
deformation, 92
limit, see Proportional limit
modulus, see Modulus of elasticity
recovery, 29
Elastomers, 658
processing, 683
Electric-arc furnace, 235
Electrical conductivity (various
materials), 120
Electrical resistivity, 120
Electrical-discharge
forming, 420
grinding, 619
machining, 616
wire cutting, 620; see Wire
EDM
Electrically conducting
adhesives, 818, 823
polymers, 657
Electrochemical
grinding, 615
honing, 616
machining, 612

Electrochemical-discharge grind-
ing, 620
Electrodischarge machining, 616
Electrodes
EDM, 617
welding, 778
Electroforming, 175
Electrogas welding, 777
Electrohydraulic forming, 420
Electroless plating, 175
Electrolytic
deposition, 711
grinding, see Electrochemical
grinding
Electron-beam
machining, 624
welding, 784
Electronic gage, 180
Electroplating, 175
Electropolishing, 606
Electroslag welding, 778
Elements, in steel, 69, 171; see
also individual materials
Elongation, 29, 385
various materials, 28
Embossing, 435
Embrittlement, 102
hydrogen, 116
Emery, 736
Emulsification, 178
Emulsions, 168, 515
Enamel, 176
Enameling, 176
Encapsulation, 682
End
cutting edge angle, 467, 518
effector, 858
grain, 317
milling, 542
relief angle, 467, 518
End-feed grinding, 599
End-of-arm tooling, 858
Endurance limit, see Fatigue limit
Energy limited, 323
Energy requirements, see Specific
energy
Engine lathe, 524
Engineered materials, 10, 659
Engineering
metrology, 178
strain, 25
strain rate, 40, 45
stress, 26
Environmental
considerations, 21

effects, 515
stress cracking, 652,
Environmentally conscious design
and manufacturing, 21
Environments (expert systems), 896
Epoxies, 656
Epoxy-based adhesive systems,
818
Equiaxed grains, 221
Equilibrium diagram, 214
Equipment, see individual
processes
Equivalent stress, see Effective
stress
Ergonomics, 22
Erichsen test, 436
Erosion, 163
in forging dies, 164
Etchant, 608
Eutectoid
cementite, 218
ferrite, 218
reaction, 218
Evaporative
adhesives, 818
pattern casting, 253
Execution functions, 873
Expandable pattern casting, 253
Expanded polystyrene process, 253
Expanding, see Bulging
Expendable molds, 246
Experimental design, 203
Expert systems, 9, 18, 895
Explosion welding, 809
Explosive
forming, 418
hardening, 172
External
grinding, see Cylindrical
grinding
spinning, 416
Extreme-pressure additives, 168
Extruder, 669
Extrusion, 342
cold, 350
ceramics, 745
constant, 349
defects, 344, 353
hot, 355
hydrostatic, 351
impact, 351
plastics, 669
powders, 722
ratio, 344
Extrusion blow molding, 676

Face-centered cubic, 90
Face milling, 538
Facing, 516
Factory (future), 18, 899
Failure, 107
Families of part, 883
Fasteners, 823
Fastening, 771, 823
Fatigue, 58
    fracture, 58, 115
    limit, 59
    static, 740
    thermal, 120, 163
    wear, 163
Fatigue strength
    effect of manufacturing
        process, 115, 116
    improvement, 115
Fatigue wear, 163
Faying surfaces, 804
Feature, 188
Feed, see individual processes
Feed
    force, 521
    marks, 492, 538
    rod, 525
Feedback, 847
    devices, 858
Feldspar, 736
Ferrite, 216
    former, 219
Ferritic
    gray iron, 239
    stainless steels, 125
Ferrimagnetism, 121
Ferromagnetism, 121
Ferrous alloys, 122, 237
Ferrule, 808
Fiber pullout, 665
Fiberglass, 751
Fibering (mechanical), 103, 110,
    388
Fibers (reinforcing), 660, 662
Fibrous, 108
Filament winding, 687
Filing, 549
Filled gold, 811
Filler
    metals, 813
    wire, 781
Fillers
    in polymers, 645
Film extrusion, 671
Films
    blown, 671

diamond, 177
Fine
    blanking, 393
    ceramics, 736
    pearlite, 269
Finish, see Surface finish
Finishing cuts, 517
Finishing operations, 604
    casting, 278
    ceramic, 748
    gear, 552
    glass, 757
    powder metallurgy, 728
Finite-element method, 307
Fire polishing, 753, 757
Firing, 748
Fir-tree cracking, 353
Fishmouthing, see Alligatoring
Fishtailing, 353
Fit, 188
Fixed costs, 917
Fixed gages, 183
Fixed-position automation, 842
Fixed-sequence robot, 859
Fixtures, 863
    flexible, 863
Flame
    hardening, 274
    retardants, 646
    wire spraying, 173
Flammability (polymers), 645
Flanging, 406
Flank wear, 482
Flaring, 406
Flash butt welding, 807
Flash, in
    casting, 280
    compression molding, 679
    forging, 320
    friction welding, 802
Flash (butt) welding, 807
Flashless forging, 312
Flask, 248
Flaskless molding, 249
Flat rolling, 325
Flatness, 181
Flattening
    by bending, 407
    by roller leveling, 337
    of rolls, 334
Flaws, see Defects
Flexible
    assembly, 866
    automation, 844
    fixturing, 864

manufacturing cell, 888
manufacturing system, 17,
    866, 889
Flexure, see Bending
Flint, 736
Float glass, 753
Floating-zone method, 267
Flood cooling, 515
Floor-to-floor time, 918
Flow
    lines, 844
    rules, 71
    stress, 33, 39, 76, 299
    turning, 414
Fluid flow (casting), 226, 281
Fluid-forming process, 412
Fluid-to-fluid extrusion, 352
Fluidity, 228
Fluorocarbons, 654
Flute, 531
Flux, 234, 813, 817
Flux-cored arc welding, 776
Fly cutting, 543
Flyer plate, 809
Foam
    casting, 682
    molding, 682
Foil, 325
Folding, 389
Folds, 147
Forced vibration, 558
Forces, see individual processes
Forgeability, 318
Forging, 295
    closed-die, 311
    dies, 319
    impression-die, 309
    isothermal, 312
    metal powders, 722
    open-die, 296
    orbital, 312
    precision, 312
    temperature, 321
Form
    block, 408
    cutting, 550
    grinding, 597
    milling, 542
    tools, 516
Formability, see individual
    processes
    sheet, 436
Forming, 365; see also individual
    processes
Forming-limit diagram, 437

Forming tap, 341
Forward
    extrusion, 342
    slip, 326
    spinning, 416
Foundries, 236
Four-high mill, 338
Four-point bending, 52
Four-slide machine, 405
Fracture, 107; see also Cracking
        and Defects
    abrasive bonds, 591
    abrasive grains, 591
    brittle, 112
    ductile, 108
    modes, 113
    stress, 29
    surfaces, 108, 112, 113, 115
Framework systems, 896
Free
    bending, see Air bending
    space, 651
    volume, 651
Free-form fabrication, 692
Free-machining steels, 493
Freezing range, 222
Frequency distribution, 194, 712
Fretting corrosion, 163
Friability, 577, 591
Friction, 151
    angle, 468
    factor, 154
    hill, 329
    in ceramics, 158
    in metals, 151
    in metalworking, see individual
        processes
    in polymers, 158
    measurement, 156
Friction sawing, 549
Friction welding, 801
    inertia, 802
    linear, 802
Front tension, 330
Fukui test, 436
Full
    annealing, 276
    indicator movement, 182
Full-mold process, 253
Fuller, B., 757
Fullerenes, 757
Fullering, 315
Functional
    dimension, 188
    layout, 883

Furnace brazing, 814
Furnaces (melting), 235
Fused
    aluminum oxide, 578
    deposition modeling, 694
    silica, 750
Fusion,
    weld joints, 785
    welding, 770
Fuzzy logic, 898

Gage
    blocks, 183
    length, 26
    maker's rule, 185
    numbers (rolling), 338
Gages, 183
Gain, 150
Gain scheduling, 854
Galvanic corrosion, 122
Galvanized steel, 389
Galvanizing, 138, 176
Gantry robot, 858
Gap lathe, 526
Garnet, 577
Gas
    metal arc welding, 775
    tungsten arc welding, 781
Gases
    in casting, 233
    shielding, see individual
        processes
Gate, 248
Gating system, 226
Gaussian, 194
Gear manufacturing,
    broaching, 550
    finishing, 552
    form cutting, 550
    generating, 550
    rolling, 340
    shapers, 550
    teeth measurement, 182
Generalized Hooke's law, 65
Generative CAPP system, 879
Gentle grinding, 590
Geometric
    effects (lubrication), 166
    modeling, 875
    softening, 38
    tolerancing, 188
Giant molecules, 639
Glass, 749
    ceramics, 752
    fibers, 662

former, 750
    lubrication, 170
    point, 643
    temperature, 643
    wool, 754
Glass-transition temperature, 643
    various polymers, 644
Glasses
    as bonding agent, 581
    processing, 753
    properties, 751
    strengthening, 756
    types, 750
Glassy behavior, 642
Glazing, 176
    of grinding wheels, 591
Global competitiveness, 20
Globular transfer, 775
Go gage, 183
Goodyear, C., 658
Gold, 139
Grade (bonded abrasives), 582
Grain, 99
    abrasive, 578
    depth of cut, 584
    flow patterns, 317
    fracture, 591
    growth, 106
    size, 100, 101, 388
Grain boundary, 97, 99
    embrittlement, 102
    sliding, 102
Grain structure
    casting, 222, 224, 225
    rolling, 325
Grand average, 196
Granite-epoxy composite, 562
Graphite, 169, 757; see also
        Carbon,
    fibers, 662
    lubricant, 169
Graphitization, 220
Gravity segregation, 225
Gray (cast) iron, 239
Greases, 169
Green
    ceramics, 745
    compact, 715
    density, 716
    design, 22
    molding sand, 247
    strength, 723
Grindability, 593
Grinding
    electrical-discharge, 619

Grinding (*cont.*)
  electrochemical, 615
  fluids, 601
  mechanics, 583
  operations, 594
  ratio, 593
  specific energy , 586
Grinding wheels, 578; see also
      Bonded abrasives,
  dressing, 591
  glazing, 591
  loading, 591
  safety, 601
  selection, 593
  wear, 590
Grit
  blasting, 607
  number (abrasive), 578
Group
  layout, 883
  machine cell, 887
  technology, 17, 881
Gun drilling, 531
Gutter, 320

Hacksaws, 549
Hafnium nitride, 509
Hall-Petch equation, 100
Hammers (forging), 323
Hand lay-up, 686
Hard
  acting wheel, 593
  automation, 842
  chromium plating, 175
  facing, 173
  turning, 529, 602
Hardenability, 271
Hardeners, 235
Hardening
  case, 172, 273, 274, 275
  explosive, 172
  spark, 173
Hardness, 52
  bonded abrasives, 593
  hot, 497
  oxide vs. base metal, 147
  scales, 53
  tests, 53
  various materials, 578, 739
  vs. strength, 56
Hard-wired controls, 846
Hastelloy, 133, 134
Hazards, see individual processes
  metal powder, 715
Headers, 313

Heading, 313
Headstock, 525
Heat, 79
Heat checking, 163, 262, 271
Heat transfer, 229
  casting, 282
Heat treatment, 269
  design considerations, 278
  ferrous alloys, 269
  nonferrous alloys, 272
Heat-affected zone, 147, 787
Heat-resistant alloys, 134; see also
      Superalloys
Helix angle, 531
Hemming, 407
Heterogeneous nucleation, 225
Heuristic, 896
Hexagonal close-packed, 90
Hierarchical coding, 885
High-carbon steel, 123
High-efficiency machining range,
      564
High-energy
  beam machining, 621
  rate forming, 418
High-frequency resistance welding,
      806
High-removal rate machining, 527
High-speed
  automated assembly, 865
  machining, 528
  milling, 542
  steels, 126, 500
High-strength low-alloy steels, 123
High-temperature
  alloys, 134, 243; see
      Superalloys
  polymers, 657
High-velocity oxyfuel gas spray-
      ing, 173
History of
  automation, 15
  manufacturing, 2
Hob, 550
Hold-down ring, 424; see
      Blankholder
Hole preparation (fastening), 823
Hole-basis system, 188
Holemaking processes
  drilling, 531
  punching, 389
  see also Nontraditional
      Processes
Hollow end mill, 542
Holographic interferometry, 190

Holography, 190
Homogeneous deformation, 296
Homogenization, 276
Homologous temperature, 106
Homopolymers, 641
Honeycomb structures (manufac-
      turing), 423
Honing, 604
Hooke's law, 27, 65
Hooker process, 352
Horizontal upsetter, see Headers
Hot
  cracking (welds), 791
  dipping, 176
  extrusion, 355
  hardness, 57, 497
  isostatic pressing, 721, 747
  machining, 496
  shortness, 102, 349
  spots, 280
  tearing, 224, 791
  working, 106
Hot pressing (ceramics), 747
Hot-chamber process, 261
Hot-die forging, 312
Hot-melt adhesives, 818
Hot-plate welding, 828
Hot-runner mold, 673
Hot-twist test, 318
Hot-work steels, 126
Hubbing, 314
Human factors engineering, 22
Hybrid, 661
  code, 885
Hydraulic press, 321, 323
Hydrodynamic
  fluid film, 165
  machining, 625
Hydroform process, 412
Hydrogen embrittlement, 116
Hydroplastic forming, 745
Hydroplasticity, 736
Hydrospinning, 414
Hydrostatic
  extrusion, 343, 351
  tensile stress, 353
Hydrostatic pressure, 43
  effects of, 43, 319, 398
Hygroscopy, 653
Hysteresis, 659
  magnetic, 140
  rubbers, 659

Ideal tensile strength, 94
If-then rules, 896

Image recognition, 863
Impact, 60
    extrusion, 343, 351
    molding, 250
    toughness, 61
    wear, 163
Imperfections (crystals), 96, 97
Impregnation, 728
Impression-die forging, 309
Impurity atom, 97
In good statistical control, 198
Incipient melting, 79
Inclination angle, 466
Inclusions, 97, 109, 147, 228,
        232, 399
Incoloy, 134
Incomplete
    fusion, 789
    penetration, 789
Inconel, 133, 134
Incremental
    deformation, 315
    dimensioning, 845
Indenters, 53
Indexable inserts, 502
Indexing
    assembly, 865
    inserts, 502
Indirect
    emulsion, 169
    extrusion, 342
    labor costs, 918
Induction
    brazing, 814
    furnace, 235
    hardening, 274
    welding, 828
Industrial
    diamond, 758
    engineers, 23
    robots, 16, 856
    Revolution, 2
Inertia friction welding, 802
Infeed grinding, 599
Inference, 896
    engine, 896
Infiltration, 728
Infrared brazing, 814
Ingot casting, 243
Inhomogeneity factor, 360
Inhomogeneous deformation, 297,
        359
Initial Graphics Exchange
        Specification, 875
Injection blow molding, 677

Injection molding
    ceramics, 747
    foam, 676
    polymers, 672
    powders, 722
    reaction, 675
Inlet angle (lubrication), 166
In-line indexing, 865
Inoculant, 225
Inorganic adhesives, 818
In-process
    annealing, 276
    inspection, 191
    measurement, 185
Insert molding, 263, 674
Inserts (cutting tool), 502
Inspection, 188
    automated, 191
    castings, 278
Instability, 36, 51, 75, 385
Insulators, 120
Integral transfer devices, 856
Integrated communications sys-
        tems, 873
Intelligent robot, 859
Interactive computer graphics, 874
Interchangeable manufacturing, 193
Interference, 188
    fit, 188
Interferometry, 181
Intergranular, 113, 148
    corrosion, 122
    fracture, 113
Intermediate
    annealing, 276
    shape, 319
Intermediates (glass), 750
Intermetallic compounds, 213
Internal
    centerless grinding, 599
    grinding, 597
    spinning, 416
International tolerance grade, 188
Interpolation, 850
Interstitial
    atom, 97
    solid solutions, 213
Invar, 120, 133
Inventory,
    control, 880
    zero, 891
Inverse
    lever rule, 216
    segregation, 225
Inverted extrusion, 342

Investment casting, 255
Ion
    implantation, 174, 509
    plating, 174
Ionic bond, 736
Iron, 216
    effects of various elements, 219
Iron-base superalloys, 134
Iron-carbon system, 216
Iron-iron carbide phase diagram,
        216, 218
Ironing, 427
Islands of automation, 894
ISO 9000 standard, 203
ISO/OSI reference model, 895
Isostatic pressing, 720, 747
Isothermal
    forging, 312
    transformation diagram, 269
Izod test, 60

Jacket, 355
Jet engine, 4
Jiggering, 747
Jigs, 864
Job shops, 841
Joggling, 405
Joining
    plastics, 771
    processes, 770, 771, 772
Jolting, 249
Jominy test, 271
Junction growth, 153
Junctions, 152
Juran, J.M., 202
Just-in-time production, 17, 891

KK-3 system, 886
Kanban, 892
Kaolinite, 736
Kelvin model, 649
Kerf, 548, 622, 624, 625
Kevlar, 655, 662
Key-seat cutter, 543
Keyhole technique, 782
Kickout, see Ejectors
Killed steel, 243
Knee, 544
Knockout, see Ejectors
Knoop hardness, 55
Knowledge engineering, 895
Knowledge-based system, 895
Knuckle-joint press, 323
Knurling, 516
Kodel, 655

Labor
  costs, 918
  intensive, 841
Lamellae (pearlite), 218
Lamellar tears, 791
Laminar flow, 228
Laminate strengthening, 757
Laminated
  glass, 757
  object manufacturing, 732
Lancing, 393
Land
  drawing die, 367
  forging die, 320
  negative, 504
Lapping, 605
Laps, 148, 316; see also Defects
Laser beams, 181
Laser types, 622
Laser-assisted forming, 444
Laser-beam
  machining, 621
  torch, 622
  welding, 443, 783
Latex, 659
Lathes, 524
Lattice structure, 90
Lay, 149
Layered structure, 757
Layout
  functional, 883
  group, 883
  machines, 183
Lay-up, 686
Leaching, 122
Lead, 138
  glass, 750
Lead
  angle, 538
  screw, 525
  time, 915
Lead-alkali glass, 750
Lead-free solders, 816
Leaded steels, 493
Lean production (manufacturing),
  20
Leathery behavior, 642
Lee, E.H., 473
Lehr, 753
Leveling rolls, 337
Lever rule, 215
Levitation melting, 236
Lévy-Mises equations, 71
Lexan, 655
Liability (product), 22

Life cycle, 6
Light section microscope, 184
Limit dimensions, 188
Limiting
  dome height test, 441
  drawing ratio, 428
Limits, 188
Line representation, 875
Line-graduated instruments, 179
Linear
  array, 862
  defects, 97
  elastic behavior, 27
  friction welding, 802
  interpolation, 850
  polymers, 640
  variable differential trans-
    former, 180
Linearity, 185
Lip-relief angle, 531
Liquid penetrants, 189
Liquid-metal embrittlement, 102
Liquid-phase
  processing, 760
  sintering, 724
Liquid-surface acoustical hologra-
    phy, 191
Liquidus, 213
Lithium aluminum silicate, 738
Load limited, 321
Loading (grinding wheels), 591
Local area network, 893
Localized necking, 385
Lock seams, 826
Logarithmic strain, see True
    strain
Long fibers, 663
Long-chain molecules, 638
Lost-foam process, 253
Lost-pattern casting, 253
Lost-wax process, 255
Lot size, 193, 841
Low-carbon steel, 123
Low-expansion alloys, 120
Low-melting metals, 138
Low-pressure casting, 260
Low-stress grinding, 590
Lower control limit, 196
Lubricants, 322, 512; see also
    Metalworking fluids and
    individual processes
  selection, 170
Lubrication, 165
  boundary, 166
  geometric effects, 166

  mixed, 166
  regimes, 165
  thick film, 165
  thin film, 166
Lucite, 654
Lueder's bands, 337, 387

Machinability, 493
  metals, 493
  ratings, 493
  steels, 493
  various materials, 239, 495
Machine control systems, 16
Machine tools, see individual
    processes
  structures, 560
Machine vision, 863, 897
Machining, 455; see also individ-
    ual processes
  allowance, 281
  center, 552
  economics, 562
Machinist's rule, 179
Macromolecules, 636
Macrosegregation, 224
Magazine (tool), 554
Magnesium and alloys, 131
Magnetic properties, 121
Magnetic-field assisted polishing,
    607
Magnetic-float polishing, 606
Magnetic-particle inspection, 189
Magnetic-pulse forming, 420
Magnetostriction, 121
Magnification, 185
Main turret, 526
Major strain, 439
Malleable iron, 242
Manganese sulfide inclusions, 493
Manipulators, 856, 857
Mannesmann process, 342
Manual
  assembly, 865
  part programming, 851
Manufacturing, 1
  attributes, 883
  automation protocol, 894
  cell, 887
  communications networks, 893
  costs, 14, 916; see also Costs
  design for, 8
  engineers, 23
  history, 2
  methods, 8
  process selection, 12

Manufacturing (*cont.*)
properties, 10
resource planning, 880
systems, 871
Maraging, 273
Margin (drill), 531
Marking, 623
Marquenching, 277
Martempering, 277
Martensite, 270
Martensitic
gray iron, 241
stainless steels, 125
Maskant, 609
Masking, 609
Masonry drills, 815
Mass production, 841
Master alloys, 235
Masticated, 683
Match-plate pattern, 248
Material
costs, 11
handling and movement, 855
properties, 10
selection, 9, 922
Material removal rates, see individual processes
Material-removal processes, 455
Material-requirements planning, 880
Materials, see specific materials
Materials substitution, 909
Matrix
materials, 663
Matrix array, 862
Mats, 663
Maximum
material condition, 188
reduction, see individual processes
roughness height, 149
Maximum-shear-stress criterion, 66
Maxwell model, 649
Measuring
instruments, 178
machines, 182
Mechanical
alloying, 711
behavior, 25
fastening, 771, 823
fibering, 103, 110, 388
plating, 172
press, 321
properties, 10; see also specific materials

Mechanization, 839
Medium batch, 919
Medium-carbon steel, 123
Melamine, 656
Melt spinning, 268
Melting
furnaces, 235
practice, 234
Melting point, 118
various materials, 118, 644
Mer, 639
Merchant, M.E., 472
Mesh, 308
Metal
bonds (grinding wheels), 582
flow pattern, 344
inert-gas welding, 776
injection molding, 722
powders, 710
Metal-matrix composites, 759
Metallic
coatings, 169
glasses, 140, 268
Metallizing, 173
Metallurgical burn, 589
Metallurgical transformation, 148
Metalworking, see individual processes
Metalworking fluids, 168, 322; see also individual processes
Metastable, 220
Method of
attributes, 194
variables, 194
Metrology, 178
MICLASS, 886
Microabrasive blasting, 607
Microhardness, 55
Micrometer, 179
depth gage, 179
Microporosity, 233
Microscopes, 184
Microsegregation, 224
Microstructure, see Structure and also specific materials
casting, 282
Microwelds, 153
Mild
steel, 123
wear, 160
Mil Standard rule, 185
Mill (rolling), 338
Milling, 535
cutters, 535
machines, 543

Mineral oils, 168
Minimum
bend radius, 397
cost, see individual processes.
dimensions, various processes, 913
various metals, 397
Minor strain, 439
Miscible blends, 645
Mist cooling, 515
Mixed
dimensioning, 845
lubrication, 166
Mobile robot, 858
Mode size, 713
Modeling, 872, 875
casting, 281
sheet forming, 443
Modified austempering, 277
Modifiers (glass), 750
Modular, 843
Module, 553
Modulus of
elasticity, 27
resilience, 27
rigidity, 49
rupture, 52
various materials, 28, 637, 660, 739
Mohr's circles, 66, 67, 74
Mohs hardness, 55
Moisture
in ceramics, 748
effect on polymers, 653
effect on sand, 247
Molding, see individual processes
machines, 249
reinforced plastics, 686
Molds, see individual processes
Molecular weight (polymers), 639
distribution, 639
Molybdenum and alloys, 137
Molybdenum disulfide, 169
Monel, 133
Monocode, 885
Monomer, 639
Move card, 892
Mounted wheels, 579
Mulling machines, 246
Mullite, 737
MultiClass coding, 886
Multilayer blow molding, 677
Multiphase coatings, 507
Multi-component injection molding, 674

Mushy zone, 223
Mylar, 655

Nanofabrication, 13
Nanophase ceramics, 738
Nanotechnology, 13
Natural
  adhesives, 818
  aging, 273
  language processing, 897
  rubber, 659
  strain, see True strain
Near-net shape, 14
Necking, 29
  correction for, 37, 68
  strain, 37
Negative springback, 401
Nesting, 394
Net-shape, 14
  forging, 312
  forming, 707
Network polymers, 641
Neural networks, 897
Neutral point, 325, 329
Newtonian, 42, 649
Nibbling, 394
Nichrome, 133
Nickel and alloys, 133
Nickel-base superalloys, 134
Nimonic, 134
Niobium (columbium), 137
Nitrides, 737
Nitriding, 274
No-bake mold, 248
No-slip point, 325, 329
No-wear EDM, 617
Noble metals, 139
Nodular iron, 241
Noise, 203
Nominal
  size, 186, 188
  strain, 25
  stress, 26
Nonconsumable electrode, 772
Nondestructive testing, 189
  welded joints, 796
Nonferrous metals and alloys,
    128, 243
Nonhomogeneous chips, 464
Nonproductive labor, 918
Nontraditional machining
    processes, 611
Nontransferred arc, 782
Nonsynchronous system (assem-
    bly), 866

Nonuniform deformation, 297
Normal
  anisotropy, 388, 428
  anisotropy (various metals), 429
  distribution curve, 194
  rake angle, 466
  segregation, 225
Normalizing, 276
Nose radius, 520
No-slip point, 325
Not go gage, 183
Notch sensitivity, 61
Notching, 393
  cutting tools, 489
Nucleating agent, 225
Nucleation, 225
Nugget (weld), 804
Numerical control, 845
  programming for, 851
  robot, 859
Nylons, 654

Oblique cutting, 466
Offset, 28
Oils, 168
Olsen test, 436
On-line inspection, 191
One-dimensional defects, 97
One-piece pattern, 248
Open riser, 248
Open-die forging, 296
Open-loop control, 847
Operation sheet, 879
  operational considerations, 13
Opitz system, 885
Optical
  comparator, 182
  fabrication, 694
  flat, 181
  projectors, 182
  Optical properties
  ceramics, 741
  glasses, 752
  polymers, 643
Optimization (adaptive control),
    855; see also individual
    processes
Optimum
  die angle, 347, 362
  cutting speed, 486, 564
  tool life, 564
Orange peel, 106, 388
Orbital
  forging, 312
  welding, 828

Organic
  coatings, 176
  grinding wheels, 582
Organization for manufacture, 23
Orientation (polymers), 652
Orthogonal cutting, 458
Osprey process, 727
Out of control, 198
Out of roundness, 182
Overaging, 273
Overarm, 544
Overbending, 403
Overcut, 617
Overhead, 918
Overlap, 789
Oxalate coatings, 170
Oxidation, 122
  wear, 163
Oxide
  ceramics, 736
  films, 146
Oxides, see specific materials
Oxyfuel gas cutting, 625

Pack rolling, 337
Pallet, 553, 890
Paperless design, 877
Parabolic interpolation, 850
Paraffin, 169
Parison, 677
Part
  families, 883
  programming, 851
Partially
  deoxidized steel, 244
  stabilized zirconia, 737
Particle size distribution, 712
Parting, 393
Parting agent, 248, 250, 321
Parting line
  casting, 280
  forging, 320
Passivation, 122, 124, 147
Patenting, 277, 366
Patternmaker's shrinkage
    allowance, 280
Patterns, 248
Payoff reel, 330
Pearlite, 269
Pearlitic gray iron, 241
Pedestal grinder, 599
Peel test (spot welds), 796
Peeling (adhesives), 820
Peen forming, 422
Peening (shot), 63, 171

Pellets, 670
Penetrants (liquid), 189
Percussion welding, 809
Perforating, 393
Peripheral milling, 536
Permanent deformation, 92
Permanent-mold casting, 259
Permeability (sand), 249
Pewter, 139
Phase
    boundaries, 97
    diagrams, 214
    transformation, 217, 269
Phenolics, 656
Phosphate coating, 170, 176
Phosphor bronze, 133
Photochemical
    blanking, 612
    machining, 732
Photoetching, 612
Physical properties, 10, 117; see
        also specific materials
Physical tempering, 756
Physical vapor deposition, 174, 506
Pick-and-place robot, 859
Pickling, 246, 367
Piece-part production, 841
Piercing, 314, 406
Piezoelectric effect, 121
Pinch rolls, 244
Pipe
    defect, 344, 353
    ingots, 244
Pitting, 122
Pits, 148
Planar anisotropy, 388, 429
    imperfections, 97
Plane
    strain, 47, 70, 299, 326, 364,
        386
    stress, 70, 302
Plane-strain compression test, 46
Planer-type milling machine, 545
Planers, 545
Planing, 545
Planning functions, 873
Plant layout, 855
Plasma, 173, 782
    spraying, 173
Plasma-arc
    cutting, 625
    welding, 782
Plaster-mold casting, 254
Plastic
    anisotropy, 388, 428

deformation, 25, 28, 66, 71,
    92, 102, 148
    forming (ceramics), 745
Plasticizers, 645
Plastics, 636
    for coatings, 169
    friction, 158
    joining, 827
    machining, 495
    processing, 668, 669
    reinforced, 659
    structure, 638
    wear, 164
Plastisols, 678
Plastometer, 46
Plate, 325, 384
Plate glass, 753
Plating
    electroplating, 175
    electroless, 175
    mechanical, 172
Platinum, 139
Playback robot, 859
Plexiglas, 654
Ploughing (plowing),
    in friction, 154
    in grinding, 586
Plug gages, 183
Plugs, 183
    tube bending, 408
    tube bulging, 411
Plunge grinding, 594, 597, 599
Pneumatic gages, 184
Point
    angle, 531
    defects, 97
Point-to-point control, 848
Pointing, 366
Poisson's ratio, 27
    various materials, 28, 637, 739
Polarity (arc welding), 773
Polishing, 606
Polyamides, 654
Polyblends, 645
Polycarbonates, 655
Polycodes, 885
Polycrystalline diamond, 511
Polycrystals, 99
Polyesters, 655, 656
Polyethylenes, 655
Polygonization, 104
Polyimides, 655, 656
Polymer precursor, 761
Polymeric coatings (lubrication),
    169

Polymerization, 639
    degree of, 639
Polymers, 636; see also Plastics
    structure, 638
Polymethylmethacrylate, 654; see
        Acrylics
Polymorphic transformation, see
        Allotropic transformation
Polypropylenes, 655
Polystyrenes, 655
Polysulfones, 655
Polytetrafluoroethylene, see
        Fluorocarbons
Polyurethane, 659
Polyvinyl chloride, 655
Population, 193
Porcelain, 742
    enameling, 176
Porosity
    bonded abrasives, 583
    casting, 233
    ceramics, 739
    ingots, 243
    welds, 788
Portable grinder, 599
Porthole die, 356
Position measurement, 847
Positional tolerancing, 188
Positioning, 848
Postprocess inspection, 191
Postuniform elongation, 385
Potting, 681
Pouring basin, 226, 248
Powder
    metallurgy, 707
    injection molding, 722
    rolling, 722
Powders (metal), 710
Power
    hammer, 323
    requirements, see Specific
        energy
    spinning, 414
Powerhead production units, 843
Precious metals, 139
Precipitates, 272
Precipitation
    hardening, 125, 272
    metal powders, 711
Precision, 185
    casting, 255
    forging, 312
    machining, 13, 529
Precoated sheet steel, 176, 389
Precursor, 662, 761

Preferred orientation, 103
Preforming, see individual processes
Prepregs, 684
Press
 brake, 405
 fit, 827
 working, 384
Presses
 hydraulic, 321, 357
 mechanical, 321
 screw, 323
Pressing
 ceramics, 746
 glass, 755
 powders, 715, 722, 728
Pressure
 casting, 260
 die, see individual processes
 die casting, 261
 pouring, 260
 sensitive adhesives, 818
 sintering, 747
Pressure-multiplying factor, 311
Pressureless compaction, 723
Pressworking, see sheet-metal forming processes
Primary
 bonds, 640
 metals, 235
 shear zone, 460
 working, 294
Primitives of solids, 876
Principal stresses, 66
Probability, 201
Process
 annealing, 276
 capability, 199
 capabilities, 913; see also individual processes
 planning, 878
 selection, 12
Producer's risk, 201
Producibility, 8
Product
 design, 5, 909, 922
 integrity, 19
 liability, 22, 916
 life, 6
 quality, 19, 189
 safety, 22
Product Data Exchange Specification, 875
Production
 card, 892

flow analysis, 884
 methods, 8
 quantity (volume), 915
 rate, 841, 528; see also individual processes
Productive labor, 918
Productivity, 839
Proeutectoid
 cementite, 219
 ferrite, 218
Profile
 measurement, 182
 milling machine, 545
Profilometer, 150
Programmable
 automation, 844
 logic controllers, 845
 Programming
 for NC, 851
 language, 852
Progressive dies, 395
Projection welding, 806
Proof stressing, 793
Properties of materials, see specific materials
Proportional limit, 27
Prototype, 7
Protractor, 180
Pulforming, 688
Pull
 broach, 547
 system, 891
Pulsed arc, 775
Pultrusion, 688
Pulverization, 711
Punch, see individual processes
Punching, 393
Pure
 drawing, 426
 metals (solidification), 221
 shear, 49
Push
 broach, 547
 cut, 546
 system, 891
Pyroceram, 752
Pyrolysis, 662
Pyrophoric, 131

Quality
 assurance, 19, 192
 circle, 202
 control, 193, 202
 engineering, 202
Quantity of

materials, 909
 production, 915
Quartz, 738
Quench cracking, 271
Quenching, 271
 media, 271
 severity, 271

Rack shaper, 551
Radial
 forging, 369
 force, 521
Radiation effects, 44
Radio-frequency sputtering, 174
Radiography, 190
Rake angle, 458, 517
Ram, 343
Rammed graphite molding, 253
Random
 assembly, 864
 sampling, 193
Range (statistics), 195
Rapid
 prototyping, 7, 236, 692, 731
 solidification, 268
Reaction
 bonding, 761
 centering, 761
Reaction-injection molding, 675
Reactive hot melt adhesives, 818
Reagents, 608
Real area of contact, 152
Reamers, 534
Reaming, 534
Reciprocating screw extruder, 672
Reconditioning (cutting tools), 512
Recovery, 104
Recrystallization, 104
 temperature, 104
Recycling, 12, 516
Red hardness, see Hot hardness
Redrawing, 435
Reduction (powder metals), 711
Reduction of area, 30
Redundant work, 77, 347, 359
Refractory metals and alloys, 137
Regenerative chatter, 559
Regulating wheel, 598
Reinforced plastics, 659
 applications, 667
 processing, 683
 properties, 664
Reinforced wheels (grinding), 582

Relative
  costs, 660, 691, 919
  grain force, 586
Relaxation
  stress, 651
  time, 651
Release
  agent, see Parting agent
  cloth, 686
Reliability of supply, 907
Relief angle, 458, 520
René, 134
Repeat accuracy, 185
Repeatability, 850
Replicast CS process, 254
Representative
  strain, 73
  stress, 73
Repressing (powder metal), 722, 728
Residual stresses, 61, 148, 388
  bending, 61
  glasses, 756
  grinding, 590
  reduction of, 63
  rod drawing, 365
  rolling, 335
  welding, 792
Resilience, 27
Resin transfer molding, 687
Resinoid bond, 582
Resins, see Polymers
Resistance brazing, 814
Resistance welding, 803
  high-frequency, 806
  projection, 806
  seam, 805
  spot, 804
Resistive-implant welding, 828
Resistivity (electrical), 120
Resolution, 185, 850
Resulfurized steels, 493
Retained austenite, 271
Retrieval system, 856
Return on investment, 920
Return on quality, 916
Reverse
  drawing, 435
  extrusion, 342
  polarity, 773
Reynolds number, 228
Rheocasting, 226, 266
Rib, 316, 320
Rimmed steel, 244
Ring

gages, 184
rolling, 340
topology, 893
Ring-compression test, 156
Risers, 226, 248, 279
Rivets, 824
Robot assembly, 865
Robots, 16
Rockwell hardness, 54
  superficial hardness, 54
Rod drawing, 357
Roll
  bonding, 799
  compaction, 722
  deflections, 334
  flattening, 334
  forging, 315
  forming, 408
  grinder, 596
Roll-pass design, 339
Roller
  burnishing, 172
  leveling, 337
Rolling
  ceramic (slip), 745
  glass, 753
  metal powders, 722
  metals, 324
Rolling mills, 338
Rolls, 324
Root-mean-square average, 149
Rotary
  bending, 403
  file, 549
  indexing, 865
  swaging, 369
  table grinder, 595
  table milling machine, 545
  transfer, 843
  tube piercing, 342
Rotational molding, 677
Roughing cuts, 517
Roughness, 149; see Surface finish
    and Surface roughness
Roundness, 182
Routing sheet, 878
Roving, 663
Rubber, 658
  bond, 582
  forming, 412
  modified polymers, 645
Rubbers, see Elastomers
Rubbery behavior, 642
Rule of 10, 185, 920
Runnerless mold, 673

Runners, 226, 248
Running-in, 159
Rupture, 60
  modulus, 52

Saddle, 544
Safety, see also individual processes
  in grinding, 601
  in foundries, 237
  in machining magnesium, 495
  in powder metal processing, 715
  in robots, 861
Safety glass, see Laminated glass
Sagging, 756
Sample size, 193
Sampling, 201
Sand
  casting, 246
  molding machines, 249
  molds, 246
  slingers, 249
Sands (casting), 246
Sandwich molding, 674
Saponification, 178
Saran, 655
Sata, T., 474
Sawing, 548
Saws, 548
Scanning
  acoustical holography, 191
  electron microscope, 184
Scarfing, 335
Scheduling, 890
Scleroscope, 55
Scrap, 394, 908
Screw
  dislocation, 97
  extruder, 669
  machines, 530
  press, 323
Screw threads, 529
Seam
  defect, 148, 318, 365
  welding, 805
Seaming, 407, 826
Seamless tube production, 342
Season cracking, 116
Second-phase particles, 214
Secondary
  bonds, 640
  shear zone, 460
  tensile stresses, 318, 353
  working, 294
Sedimentation, 712
Seeded gel abrasives, 578

Segmented
  chips, 464
  dies, 412
Segregation, 224
Seizure, 351
Selection criteria
  materials, 9, 906
  manufacturing processes, 12, 911
Selective
  assembly, 864
  automation, 842
  laser sintering, 731
  leaching, 122
Self-excited vibration, 558
Self-guided vehicle, 857
Self-interstitial, 97
Semi-killed steel, 244
Semicentrifugal casting, 264
Semiconductors, 121
Semipermanent mold, 259
Semisolid metal forming, 266
Semisynthetic solutions, 169
Sendzimir mill, 338
Sensitivity, 185
Sensor fusion, 863
Sensors, 192, 861
Sensory robot, 859
Serrated chips, 464
Service, 8
  design for, 867
  life, 12
Servomotor, 847
Severe wear, 160
Severity of quench, 271
Shaffer, B.W., 473
Shaft-basis system, 188
Shank cutter, 543
Shape factor
  extrusion, 356
  powder metals, 713
Shape rolling, 339
Shape-memory alloys, 139
Shapers, 545
Shaping (grinding wheels), 592
Shaving, 395, 467
Shaw process, 255
Shear
  angle, 458
  die, 355
  factor, 154
  modulus, 49
  on punches, 395
  plane, 458
  spinning, 414

strain, 49, 459
strain rate, 459
stress, 49
stress (critical), 92
strength (adhesives), 819
strength (single crystals), 93
yield stress, 66
Shear-angle relationship, 472
Shearing, 389
Sheet, 325, 384
Sheet extrusion (polymers), 671
Sheet glass, 753
Sheet metal, 384
  formability, 436
  forming, 384; see also individual processes
Sheet (plastic), 671
Sheet-molding compound, 684
Shell
  in casting, 221
  in extrusion, 355
  mills, 543
  mold-casting, 250
Shells (expert system), 896
Shielded metal arc welding, 772
Shielding gas, see individual processes
Shock-resisting steels, 126
Short fibers, 663
Short-circuiting arc, 776
Shot
  blasting, 607
  chamber, 261
  peening, 63, 171
  Shrink
  fit, 827
  flanging, 406
Shrinkage
  allowance, 280
  cavity, 234
  in casting, 231, 280
  in ceramics, 748
  in sintering, 727
Sialon, 511, 738
Side
  cutting edge angle, 467, 518
  rake angle, 520
  relief angle, 458, 467, 518
Silane, 665
Silica, 738
Silicate-bonded sand process, 252
Silicates, 738
Silicon
  carbide, 578, 737
  nitride, 737

nitride-base ceramics, 511
Silicones, 656, 659
Silver, 139
  solder, 815
Simple shear, 49
Simultaneous engineering, see Concurrent engineering
Single-crystal
  blades, 267
  casting, 266
  growing, 267
Single machine cell, 887
Single-action press, 323
Sintered carbides, 501; see Carbides
Sintering
  ceramics, 748
  powder metals, 723
  temperature, 724
Size effect, 117, 587
Sizing, 313, 728
  of fibers, 684
  pass, 366
Skew rolling, 315
Skill required, see individual processes
Skin
  in casting, 221
  pass, 337
  rolling, 387
Skin-dried molds, 247
Skiving, 467
Skull, 355
Slab, 337
  method of analysis, 299, 326
  milling, 536
Slag, 228, 235
  inclusions, 788
Slip
  bands, 94
  casting, 745
  lines, 307, 309
  planes, 92, 98
  system, 94
Slip (ceramics), 745
Slip-line field, 307
Slitting, 394
  saws, 542
Slotters, 546
Slurry
  abrasive, 603
  in plaster molding, 254
  infiltration, 760
Slush
  casting, 230, 260
  molding, 678

Small batch production, 841, 919
Smart
   databases, 906
   end-effector, 862
   sensors, 863
S-N curves, 58
Snagging, 599
Snap gages, 184
Snap-in fasteners, 826
Soaps, 169
Soda-lime glass, 750
Sodium silicate process, 252
Soft automation, 844
Soft forming (ceramics), 745
Soft-acting wheel, 593
Softening (strain), 488
Soils, 178
Soldering, 139, 771
Solders, 816
Solid
   base curing, 694
   lubricants, 169
   model, 875
   solutions, 213
Solid-metal embrittlement, 102
Solid-phase
   forming (plastics), 682
   processing, 760
Solid-state
   bonding, 724
   welding, 771
Solidification, 212
   alloys, 213
   contraction (various metals),
     232
   pattern, 224, 228
   pure metals, 212
   time, 229
   weld metal, 786
Solidus, 213
Solubility (gases), 234
Soluble oils, see Emulsions
Solute, 213
Solution treatment, 273
Solvent, 213
   bonding, 829
   cleaning, 178
   crazing, 652
Solvent stress cracking, 657
Sonotrode, 800
Soot, 645
Spade drill, 531
Spalling, 740
Spark
   hardening, 173
   sintering, 726

Spark-erosion machining, 616
Sparks (grinding), 589
Specific
   gravity, 117
   heat, 119
   heat (various materials), 118
   stiffness, 117
   strength, 117
   volume, 643
Specific energy, 28, 76, 476, 586
   cutting (various metals), 476
   grinding (various metals), 587
Spectra fiber, 663
Spectroscopy, 164
Speed
   cracking, 349, 353
   of response, 185
Speeds, see individual processes
Spheroidite, 269
Spider die, 356
Spin
   forging, 414
   welding, 828
Spinel, 737
Spinnability, 414
Spinning, 413
   conventional, 413
   glass, 755
   shear, 414
   tube, 416
Splat cooling, 140, 268
Splatter, 148
Split pattern, 248
Spot weld (tests), 796
Spray
   deposition, 727
   transfer, 775
Spray-up, 687
Spraying (thermal), 173
Spread, 194
Spreading, 335
Springback, 389, 399
   compensation for, 403
   factor, 400
   negative, 401
Sprue, 226, 248
Sputtering, 174
Square
   die, 355
   turret, 526
Squeeze casting, 265
Stability, 185
Stainless steels, 124
Stamping, see sheet-metal forming
   processes
Stand grinder, 599

Stand-alone machines, 841
Standard
   deviation, 195
   for the Exchange of Product
     Model Data, 875
   hole practice, 188
   shaft practice, 188
   size, 188
Standoff, 418
Stapling, 826
Star topology, 893
Starter bar, 245
Static fatigue, 740
Statistical
   process control, 19, 195
   quality control, 193
Statistics, 193
Steckel rolling, 338
Steel-rule die, 394
Steels, 123
Stellite, 501
Step drill, 531
Step-growth polymerization, 639
Step-reaction polymerization, 639
Stereolithography, 692
Stick welding, 772
Sticking, 153, 304
Stiffness (machine tools), 504,
   558
Stiffness-to-weight ratio, 117
Stitching, 826
Stockless production, 891
Stop-off, 422
Storage, 856
Stored energy, 79, 105
Straddle milling, 542
Straight polarity, 773
Straightening
   rods, 366
   sheet, 337
Straightness, 181
Strain
   effective, 73
   engineering, 25
   nominal, 25
   ratio, 428
   shear, 26, 49
   softening, 435
   true, 31
   volume, 72
Strain aging, 111
   accelerated, 112
Strain hardening, 33, 98
   exponent, 33
   exponent (various materials),
     33

Strain rate, 40
    in metalworking, 41
Strain-rate sensitivity, 40
    exponent, 42
    polymers, 649
    various metals, 43
Strand casting, 244
Strength, see individual materials
    single crystals, 92
Strength coefficient, 33, 42
    various metals, 33, 43
Strength-to-weight ratio, 117
Strengthening (glass), 756
Stress
    cracking, 63, 116
    effective, 73
    engineering, 26
    flow, 33, 39, 76
    relaxation, 60, 62, 651
    relief annealing, 277
    relieving (welds), 793
    shear, 49
    true, 31
    whitening, 652
Stress-corrosion cracking, 63, 116, 122, 388
Stress-relief annealing, 63
Stress-strain curves, 32
    construction, 35
    various metals, 35
    various polymers, 646, 648
Stretch
    bending, 403
    flanging, 406
    forming, 409
Stretcher strains, 337, 387
Striations, 115
Stringers, 399
Strip
    drawing, 364
    rolling, 324
Stripper, 425
Stroke limited, 322
Structural
    defects, 335
    grade steels, 123
    foam molding, 675
Structure, see specific materials
    insensitive, 98
    sensitive, 98
Structure-property relationships (casting), 224
Stud
    arc welding, 808
    welding, 808
Stuffing box, 367

Stylus, 150
Styrofoam, 682
Subcritical annealing, 276
Submerged arc welding, 774
Substitution of materials, 909
Substitutional
    atom, 97
    solid solution, 213
Substrate, 146
Sum dimension, 188
Superabrasives, 578
Superalloys, 133, 134
Superconductivity, 121
Supercooling, 749
Superficial hardness, 54
Superfinishing, 605
Superheat, 229
Superplastic forming, 421, 811
Superplasticity, 30, 42
Surface
    cleaning, 177
    cracking, 315, 349, 353
    damage (welding), 791
    defects, 147, 335; see individual processes
    energy, 96
    fatigue wear, 162
    finish, 13, 490, 627, 914, 915; see also Surface roughness
    grinder, 595
    grinding, 595
    hardening, 274, 275
    integrity, 147, 490
    model, 875
    preparation (adhesive bonding), 820
    profilometer, 150
    rolling, 63, 172
    roughness, 149, 187, 491, 610, 627, 915
    structure, 146
    tension, 228
    texture, 148
    texturing, 173
    treatments, 171
Surfaces, 145
Swaging, 369
Swept volume, 876
Swift test, 436
Swing (lathe), 525
Swing-frame grinder, 599
Synchronous system (assembly), 865
Synthetic
    diamond, 758

organic adhesives, 818
organic polymers, 637
rubbers, 659
solutions, 169

T-slot cutter, 542
Tactile sensing, 862
Taguchi, G., 202
Tailpipe (defect), 353
Tailstock, 525
Tandem
    drawing, 366
    rolling, 338
Tantalum and alloys, 137
Tap, 530, 534
Tap forming, 341
Tapping, 530, 534
Taylor, F.W., 483
Teach pendant, 859
Tear polymers, 641
Tearing
    casting, 224
    deep drawing, 434
    welding, 791
Technical and office protocol, 894
Teemed, 243
Teflon, 163, 654
Telescoping gages, 180
Temper
    designations, 130
    embrittlement, 102, 277
    rolling, 337, 387
Temperature, see also individual processes
    effects, 39
    gradients, 62
    rise, 79, 156, 588
Tempered
    glass, 756
    martensite, 271
Tempering
    in grinding, 589
    of glass, 756
    of metals, 271, 277
Tensile strength, see Ultimate tensile strength
Tension test, 26
Tension-shear test, 795
Tensions in rolling, 330
Terne, 176
Terpolymers, 642
Testing, 188, see also individual methods
    destructive, 191
    nondestructive, 189

Texture, 103
  surface, 148
Theoretical
  shear strength, 93
  tensile strength, 94
Thermal
  barrier, 321
  conductivity, 119
  conductivity (various materials), 118
  cracks, 740
  energy deburring, 608
  expansion, 119
  expansion anisotropy, 120
  fatigue, 120, 163
  inspection, 190
  shock, 120
  spraying, 173
  stresses, 120
  tempering, 756
Thermal properties, see individual materials
Thermally assisted machining, 496
Thermoforming, 678
Thermographic inspection, 190
Thermomechanical processing, 277
Thermoplastics, 646
  applications, 653
Thermosets, 653; see Thermosetting plastics
  applications, 655
Thermosetting plastics, 641
Thick molding compound, 684
Thick-film lubrication, 165
Thin-film lubrication, 166
Thixoforming, 266
Thread
  cutting, 529
  grinding, 597
  measurement, 182
  rolling, 340, 529
Threaded fasteners, 824
Threading, 516, 529
Three-body wear, 162
Three-dimensional printing, 732
Three-high mill, 338
Three-point bending, 52
Through hardening, 273
Through-feed grinding, 594, 598
Thrust force, 467, 521
  cutting, 467, 468, 521
  drilling, 533
Tie line, 216

Time-temperature-transformation diagram, 269
Tin, 139
  plate, 139
Titanium and alloys, 135
Titanium
  aluminide intermetallics, 135
  aluminum nitride, 508
  carbide, 501, 737
  carbonitride, 508
  nitride, 507, 737
Token
  bus, 894
  ring, 894
Tolerances, 178, 187, 280, 610, 914; see also individual processes
Tool
  failure, 481
  geometry, 517
  holders, 543
  materials, 498, 499
  steels, 126, 127, 128
  storage, 553
Tool and die materials, 498, 499
Tool checking station, 554
Tool condition monitoring, 489
Tool post, 525
Tool wear, 481
  measurement, 489
Tool-exchange arm, 554
Tool-life curves, 483
Tool-post grinder, 599
Tooling costs, 916
Toolmaker's microscope, 184, 489
Toolroom lathe, 526
Tools, see individual processes
Torch brazing, 813
Torsion test, 49
Total
  elongation, 29, 387
  indicator reading, 182
  production quantity, 841
  quality control, 202
  quality management, 19, 193
Touch probes, 554
Toughness, 34
Tracer lathe, 526
Tractrix, 435
Transducers, 490, 800, 861
Transfer
  dies, 396
  lines, 553, 843
  machines, 843
  mechanisms, 843, 865

  molding, 680
  systems (assembly), 865
Transferred arc, 782
Transformation
  diagrams, 269
  phase, 217, 269
  toughened zirconia, 737
Transgranular fracture, 113
Transition
  fit, 188
  temperature, 111
Transits, 181
Transverse
  grinding, 594
  rupture strength, 52
Tree (investment casting), 257
Trepanning, 531
Tresca criterion, 66
Triaxial stress, 65
Tribology, 151
Trim loss, 394
Trimming, 319, 395, 679
True
  strain, 31
  strain rate, 40, 46
  stress, 31
  stress-strain curves, 32, 35
True centrifugal casting, 264
Truing, 592
Tube
  bending, 408
  drawing, 364
  production, 342
  sinking, 364
  spinning, 416
  welding, 806
Tumbling, 607
Tundish, 244
Tungsten and alloys, 137
Tungsten carbide, 501, 737
  classification, 505
  production, 729
Tungsten inert-gas welding, 781
Turbine blades, 89, 266
  casting, 266
Turbulence (in casting), 228
Turk's head, 367
Turn broaching, 547
Turning, 516
  centers, 552
Turret lathe, 526
Twinning, 94
Twist
  drill, 531
  test (spot welds), 796

Two-body wear, 162
Two-dimensional imperfections, 97
Two-high mill, 338
Two-phase
    systems, 213
    processing, 760
Tygon, 655

Udimet, 134
Ultimate tensile strength, 29
    ideal, 94
    vs. fatigue, 116
    vs. hardness, 57
    various materials, 28, 637, 660
Ultraprecision machining, 13, 529
Ultrasonic
    inspection, 189
    machining, 602
    soldering, 816
    vibration, 156
    welding, 800
Unconventional machining processes, 611
Undeformed chip
    length, 584
    thickness, 536, 538, 584
Undercutting, 789
Underfilling, 789
Underwater spark forming, 420
Unfused alumina, 578
Uniform
    deformation, 296
    elongation, 29, 385
    strain, 29
Unilateral tolerancing, 188
Unit
    cell, 90
    costs, 283
Universal
    grinder, 597
    machining center, 556
    tool and cutter grinder, 599
Universe, 193
Untended factories, 899
Up
    grinding, 584
    milling, 536
Upper control limit, 196
Upper-bound technique, 307
Upsetters, see Headers
Upsetting, 296
    test, 318
Urea, 656

V dies, 404
V Process, 250
Vacancy, 97
Vacuum
    casting, 258
    evaporation, 174
    forming, 678
    molding, 250
Vacuum-bag molding, 686
Value, 4
    engineering, 921
van der Waals bonds, 640
Vapor
    blanket, 271
    degreasing, 178
Vapor deposition, 173
    chemical, 174
    physical, 174
Variability, 193
Variable blankholder force, 436
Variable-sequence robot, 859
Variant CAPP system, 879
Vents, 248
Vernier calipers, 179
Vertical flaskless molding, 249
Vibration, 558
Vibration welding, 828
Vibratory finishing, 607
Vickers hardness, 54
Vinyls see Polyvinyl chloride
Viscoelastic, 649
    modulus, 649
Viscosity (in casting), 228
    index, 228
Viscous behavior, 642
Visual sensing, 862
Vitreous, see Glasses
Vitrified bond, 581
Voice recognition, 897
Voids, 97; see also Porosity
Voigt model, 649
Volume strain, 72
von Mises criterion, 67
Vulcanization, 641, 658

Warm working, 106
Warping, 62
Waspaloy, 134
Water absorption (polymers), 653
Waterglass, 252
Water-jet machining, 625
Water-soluble fluids, 168
Waterbreak test, 178
Wave soldering, 816
Waviness, 149

    height, 149
    width, 149
Wax patterns, 256
Waxes, 169
Ways, 525
Wear, 158
    abrasive, 162
    adhesive, 159
    attritious, 590
    ceramics, 164
    chemical, 162
    coefficient, 160, 161
    corrosive, 163
    cutting tools, 481
    erosive, 163
    fatigue, 163
    flat, 587
    grinding wheels, 590
    impact, 163
    land (allowable), 486
    measurement, 164
    parts, 159
    plastics, 164
    plates, 159
Web, 316, 320
Weld
    metal, 785
    nugget, 804
    overlay, 173
    profile, 789
    quality, 788
    symbols, 798
Weldability, 793
    various metals, 239, 794
Welded joint, 785
    testing, 794
Welding
    design, 797
    electrodes, 778
    gun, 774
    processes, 770
Welding-chamber method, 356
Wet
    drawing, 367
    forming, 745
    pressing (ceramics), 746
Wheel depth of cut, 584
Whiskers, 117, 512, 663
Whisker-reinforced tool materials, 511
White
    cast iron, 241
    ceramics, 509, 736
    metals, 139
Whitney, E., 193

Wide area network, 893
Wire
    arc, 173
    brushing, 604
    drawing, 357
    EDM, 620
    frame, 875
Work
    envelope, 555, 859
    hardening, 98; see also Strain
        hardening
    in progress, 880
    of deformation, 76
    roll, 338
Workability, see individual
        processes

Workholding devices, see individ-
        ual processes
Working envelope, 859
Workstations, 887
Worldclass manufacturing, 20
Worms, 387
Wrap-around bend test, 795
Wrinkling, 389, 427
Wrought structure, 325, 337

X-rays, 190

Yield criteria, 65
Yield point, 27
Yield stress, 27
    various materials, 28

Yield-point elongation, 337, 387
Young's modulus, see Modulus of
        elasticity

Zero
    inventory, 891
    line, 188
Zinc, 138
    casting alloys, 243, 263
    phosphate coating, 170
Zipper cracks, 336
Zirconia, 737
Zirconium, 138
    nitride, 509
    oxide, 512